T0392542

SILICON AND NANO-SILICON IN ENVIRONMENTAL STRESS MANAGEMENT AND CROP QUALITY IMPROVEMENT

SILICON AND NANO-SILICON IN ENVIRONMENTAL STRESS MANAGEMENT AND CROP QUALITY IMPROVEMENT

Progress and Prospects

Edited by

HASSAN ETESAMI
Soil Science Department, College of Agriculture and Natural Resources, University of Tehran, Tehran, Iran

ABDULLAH H. AL SAEEDI
Department of Environment and Natural Resources, Faculty of Agriculture and Food Science, King Faisal University, Al-Hofuf, Saudi Arabia

HASSAN EL-RAMADY
Soil and Water Department, Faculty of Agriculture, Kafrelsheikh University, Kafr El-Sheikh, Egypt

MASAYUKI FUJITA
Faculty of Agriculture, Kagawa University, Kagawa, Japan

MOHAMMAD PESSARAKLI
University of Arizona, Tucson, AZ, United States

MOHAMMAD ANWAR HOSSAIN
Department of Genetics and Plant Breeding, Bangladesh Agricultural University, Mymensingh, Bangladesh

ELSEVIER

ACADEMIC PRESS
An imprint of Elsevier

Academic Press is an imprint of Elsevier
125 London Wall, London EC2Y 5AS, United Kingdom
525 B Street, Suite 1650, San Diego, CA 92101, United States
50 Hampshire Street, 5th Floor, Cambridge, MA 02139, United States
The Boulevard, Langford Lane, Kidlington, Oxford OX5 1GB, United Kingdom

ISBN: 978-0-323-91225-9

For Information on all Academic Press publications
visit our website at https://www.elsevier.com/books-and-journals

Publisher: Nikki Levy
Acquisitions Editor: Nancy Maragioglio
Editorial Project Manager: Maria Elaine Desamero
Production Project Manager: Sruthi Satheesh
Cover Designer: Greg Harris

Typeset by MPS Limited, Chennai, India

Contents

11. Silicon and nanosilicon mediated heat stress tolerance in plants 153

ABIDA PARVEEN, SAHAR MUMTAZ,
MUHAMMAD HAMZAH SALEEM, IQBAL HUSSAIN,
SHAGUFTA PERVEEN AND SUMAIRA THIND

12. Silicon-mediated cold stress tolerance in plants 161

ROGHIEH HAJIBOLAND

13. Silicon and nano-silicon mediated heavy metal stress tolerance in plants 181

SEYED MAJID MOUSAVI

14. Silicon- and nanosilicon-mediated disease resistance in crop plants 193

KAISAR AHMAD BHAT, ANEESA BATOOL, MADEEHA MANSOOR,
MADHIYA MANZOOR, ZAFFAR BASHIR, MOMINA NAZIR AND
SAJAD MAJEED ZARGAR

Contents

List of contributors

Asim Abbasi Department of Zoology, University of Central Punjab, Punjab Group of College, Bahawalpur, Pakistan

Seyed Abdollah Hosseini Plant Nutrition Department, Abiotic Stress Group, Agro Innovation International, Timac Agro, France

Sobia Afzal Department of Soil Science, University College of Agriculture and Environmental Sciences, The Islamia University of Bahawalpur, Bahawalpur, Pakistan

Adeel Ahmad Institue of Soil and Environmental Science, University of Agriculture Faisalabad, Faisalabad, Pakistan

Iftikhar Ahmad Department of Soil Science, University College of Agriculture and Environmental Sciences, The Islamia University of Bahawalpur, Bahawalpur, Pakistan

Zahoor Ahmad Department of Botany, University of Central Punjab, Punjab Group of College, Bahawalpur, Pakistan

Muhammad Ali Department of Environmental Science, Faculty of Agriculture & Environment, The Islamia University of Bahawalpur, Bahawalpur, Pakistan

Abdullah Alsaeedi Department of Environment and Natural Resources, Faculty of Agriculture and Food Science, King Faisal University, Al Hofuf, Saudi Arabia

Tarek Alshaal Department of Soil and Water, Faculty of Agriculture, Kafrelsheikh University, Kafr El-Sheikh, Egypt; Department of Applied Plant Biology, University of Debrecen, Debrecen, Hungary

Megahed Amer Soils Improvement Department, Soils, Water and Environment Research Institute, Sakha Station, Agricultural Research Center, Kafr El-Sheikh, Egypt

Muhammad Arslan Arshraf Department of Botany, Government College University, Faisalabad, Pakistan

Arkadiusz Artyszak Department of Agronomy, Warsaw University of Life Sciences—SGGW, Warsaw, Poland

Muhammad Ashar Ayub Institute of Soil and Environmental Sciences, University of Agriculture, Faisalabad, Pakistan

Celaleddin Barutçular Department of Filed Crops, Faculty of Agriculture, Cukurova University Adana, Adana, Turkey

Zaffar Bashir Centre of Research for Development and PG Microbiology, University of Kashmir, Srinagar, India

Aneesa Batool Proteomics Laboratory, Division of Plant Biotechnology, Sher-e-Kashmir University of Agricultural Sciences and Technology Kashmir, Srinagar, India; Department of Chemistry, Govt. College for Women, Cluster University Srinagar, Srinagar, India; Department of Chemistry, Baghwant University of Ajmer, Ajmer, India

Kaisar Ahmad Bhat School of Biosciences and Biotechnology, Baba Ghulam Shah Badshah University, Rajouri, India; Proteomics Laboratory, Division of Plant Biotechnology, Sher-e-Kashmir University of Agricultural Sciences and Technology Kashmir, Srinagar, India

Cid Naudi Silva Campos Federal University of Mato Grosso do Sul, Chapadão do Sul, Brazil

Zhong-Liang Chen Key Laboratory of Sugarcane Biotechnology and Genetic Improvement (Guangxi), Ministry of Agriculture and Rural Affairs/Guangxi Key Laboratory of Sugarcane Genetic Improvement/ Sugarcane Research Institute, Guangxi Academy of Agricultural Sciences/Sugarcane Research Center, Chinese Academy of Agricultural Sciences, Nanning, P.R. China; College of Agriculture, Guangxi University, Nanning, P.R. China

Muhammad Dawood Department of Environmental Sciences, Bahauddin Zakariya University, Multan, Pakistan

Renato de Mello Prado School of Agricultural and Veterinary Sciences, São Paulo State University, Jaboticabal, Brazil

Amanda Carolina Prado de Moraes Laboratory of Microbiology and Biomolecules, Department of Morphology and Pathology, Federal University of São Carlos (UFSCar), São Carlos, Brazil; Biotechnology Graduate Program, Federal University of São Carlos (UFSCar), São Carlos, Brazil

Jonas Pereira de Souza Júnior School of Agricultural and Veterinary Sciences, São Paulo State University, Jaboticabal, Brazil

Heba Elbasiony Department of Environmental and Biological Sciences, Home Economy Faculty, Al-Azhar University, Tanta, Egypt

Fathy Elbehery Central Laboratory of Environmental Studies, Kafrelsheikh University, Kafr El-Sheikh, Egypt

Alaa El-Dein Omara Agriculture Microbiology Department, Soil, Water and Environment Research Institute (SWERI), Sakha Agricultural Research Station, Agriculture Research Center, Kafr El-Sheikh, Egypt

Mohamed M. Elgarawani Department of Care Scientific Research Care and Training, Research and training Station, King Faisal University, Al Hofuf, Saudi Arabia

Nevien Elhawat Department of Applied Plant Biology, University of Debrecen, Debrecen, Hungary; Department of Biological and Environmental Sciences, Faculty of Home Economic, Al-Azhar University, Cairo, Egypt

Hassan El-Ramady Soil and Water Department, Faculty of Agriculture, Kafrelsheikh University, Kafr El-Sheikh, Egypt

Tamer Elsakhawy Agriculture Microbiology Department, Soil, Water and Environment Research Institute (SWERI), Sakha Agricultural Research Station, Agriculture Research Center, Kafr El-Sheikh, Egypt

Hugo Fernando Escobar-Sepúlveda Institute of Biological Sciences, The University of Talca, Talca, Chile

Hassan Etesami Soil Science Department, College of Agriculture and Natural Resources, University of Tehran, Karaj, Iran

Saad Farouk Agricultural Botany Department, Faculty of Agriculture, Mansoura University, Mansoura, Egypt

Patrícia Messias Ferreira School of Agricultural and Veterinary Sciences, São Paulo State University, Jaboticabal, Brazil

Fernando Carlos Gómez-Merino Laboratory of Plant Biotechnology, College of Postgraduates in Agricultural Sciences, Veracruz, Mexico

Libia Fernanda Gómez-Trejo Department of Plant Protection, Chapingo Autonomous University, Texcoco, Mexico

Roghieh Hajiboland Department of Plant Sciences, University of Tabriz, Tabriz, Iran

Robert Henry Queensland Alliance for Agriculture and Food Innovation, University of Queensland, Brisbane, QLD, Australia

Mohammad Anwar Hossain Department of Genetics and Plant Breeding, Bangladesh Agricultural University, Mymensingh, Bangladesh

Iqbal Hussain Department of Botany, Government College University, Faisalabad, Pakistan

Muhammad Ammir Iqbal Department of Agronomy, University of Poonch Rawalakot Azad Kashmir, Rawalakot, Pakistan

Muhammad Jafir Department of Entomology, University of Agriculture, Faisalabad, Pakistan

Mallikarjuna Jeer Entomology, ICAR-National Institute of Biotic Stress Management, Raipur, India

Byoung Ryong Jeong Department of Horticulture, Division of Applied Life Science (BK21 Four), Graduate School, Gyeongsang National University, Jinju, Republic of Korea

Danuta Kaczorek Leibniz Centre for Agricultural Landscape Research (ZALF), Müncheberg, Germany; Department of Soil Environment Sciences, Warsaw University of Life Sciences (SGGW), Warsaw, Poland

C.M. Kalleshwaraswamy Department of Agricultural Entomology, College of Agriculture, University of Agricultural and Horticultural Sciences, Shivamogga, India

Muhammad Kamran School of Agriculture, Food and Wine, The University of Adelaide, South Australia, Australia

M. Kannan Department of Nano Science and Technology, Tamil Nadu Agricultural University, Coimbatore, India

Norollah Kheyri Department of Agronomy, Gorgan Branch, Islamic Azad University, Gorgan, Iran

Paulo Teixeira Lacava Laboratory of Microbiology and Biomolecules, Department of Morphology and Pathology, Federal University of São Carlos (UFSCar), São Carlos, Brazil

Yang-Rui Li Key Laboratory of Sugarcane Biotechnology and Genetic Improvement (Guangxi), Ministry of Agriculture and Rural Affairs/Guangxi Key Laboratory of Sugarcane Genetic Improvement/Sugarcane Research Institute, Guangxi Academy of Agricultural Sciences/Sugarcane Research Center, Chinese Academy of Agricultural Sciences, Nanning, P.R. China

Zaffar Malik Department of Soil Science, University College of Agriculture and Environmental Sciences, The Islamia University of Bahawalpur, Bahawalpur, Pakistan

Madeeha Mansoor Proteomics Laboratory, Division of Plant Biotechnology, Sher-e-Kashmir University of Agricultural Sciences and Technology Kashmir, Srinagar, India

Madhiya Manzoor Proteomics Laboratory, Division of Plant Biotechnology, Sher-e-Kashmir University of Agricultural Sciences and Technology Kashmir, Srinagar, India

Piyush Mathur Microbiology Laboratory, Department of Botany, University of North Bengal, Darjeeling, India

Tatiana Minkina Academy of Biology and Biotechnology, Southern Federal University, Rostov-on-Don, Russia

Seyed Majid Mousavi Agricultural Research, Education and Extension Organization (AREEO), Soil and Water Research Institute (SWRI), Department of Soil Fertility and Plant Nutrition, Karaj, Iran

Sahar Mumtaz Department of Botany, Division of Science and Technology, University of Education, Lahore, Pakistan

Momina Nazir Department of Chemistry, Govt. College for Women, Cluster University Srinagar, Srinagar, India

Fatemeh Noori Department of Biotechnology and Plant Breeding, Sari Agricultural Sciences and Natural Resources University, Sari, Iran

Sana Noreen Department of Soil Science, University College of Agriculture and Environmental Sciences, The Islamia University of Bahawalpur, Bahawalpur, Pakistan

Aasma Parveen Department of Soil Science, University College of Agriculture and Environmental Sciences, The Islamia University of Bahawalpur, Bahawalpur, Pakistan

Abida Parveen Department of Botany, Government College University, Faisalabad, Pakistan

Shagufta Perveen Department of Botany, Government College University, Faisalabad, Pakistan

N.B. Prakash Department of Soil Science and Agricultural Chemistry, College of Agriculture, University of Agricultural Sciences, GKVK, Bangalore, India

Daniel Puppe Leibniz Centre for Agricultural Landscape Research (ZALF), Müncheberg, Germany

Vishnu D. Rajput Academy of Biology and Biotechnology, Southern Federal University, Rostov-on-Don, Russia

Rizwan Rasheed Department of Botany, Government College University, Faisalabad, Pakistan

Samiya Rehman Department of Biochemistry, University of Okara, Okara, Pakistan

Muhammad Riaz College of Natural Resources and Environment, South China Agricultural University, Guangzhou, P.R. China

Saima Riaz Department of Botany, Government College University, Faisalabad, Pakistan

Swarnendu Roy Plant Biochemistry Laboratory, Department of Botany, University of North Bengal, Darjeeling, India

Freeha Sabir Department of Soil Science, University College of Agriculture and Environmental Sciences, The Islamia University of Bahawalpur, Bahawalpur, Pakistan

Muhammad Hamzah Saleem College of Plant Science and Technology, Huazhong Agricultural University, Wuhan, P.R. China

Mahima Misti Sarkar Plant Biochemistry Laboratory, Department of Botany, University of North Bengal, Darjeeling, India

Jörg Schaller Leibniz Centre for Agricultural Landscape Research (ZALF), Müncheberg, Germany

Ehsan Shokri Department of Nanotechnology, Agricultural Biotechnology Research Institute of Iran (ABRII), Karaj, Iran

Munna Singh Department of Botany, University of Lucknow, Lucknow, India

Xiu-Peng Song Key Laboratory of Sugarcane Biotechnology and Genetic Improvement (Guangxi), Ministry of Agriculture and Rural Affairs/Guangxi Key Laboratory of Sugarcane Genetic Improvement/Sugarcane Research Institute, Guangxi Academy of Agricultural Sciences/ Sugarcane Research Center, Chinese Academy of Agricultural Sciences, Nanning, P.R. China

Syeda Refat Sultana Department of Filed Crops, Faculty of Agriculture, Cukurova University Adana, Adana, Turkey

Gelza Carliane Marques Teixeira School of Agricultural and Veterinary Sciences, São Paulo State University, Jaboticabal, Brazil

Sumaira Thind Department of Botany, Government College University, Faisalabad, Pakistan

Dan-Dan Tian Institute of Biotechnology, Guangxi Academy of Agricultural Sciences, Nanning, P.R. China

Libia Iris Trejo-Téllez Laboratory of Plant Nutrition, College of Postgraduates in Agricultural Sciences, Texcoco, Mexico

Muhammad Zia ur Rehman Institute of Soil and Environmental Sciences, University of Agriculture, Faisalabad, Pakistan

Krishan K. Verma Key Laboratory of Sugarcane Biotechnology and Genetic Improvement (Guangxi), Ministry of Agriculture and Rural Affairs/Guangxi Key Laboratory of Sugarcane Genetic Improvement/Sugarcane Research Institute, Guangxi Academy of Agricultural Sciences/Sugarcane Research Center, Chinese Academy of Agricultural Sciences, Nanning, P.R. China

Ejaz Ahmad Waraich Department of Agronomy, University of Agriculture Faisalabad, Faisalabad, Pakistan

Danghui Xu State Key Laboratory of Grassland Agro-eco-systems/School of Life Science, Lanzhou University, Lanzhou, P.R. China

Sajad Majeed Zargar Proteomics Laboratory, Division of Plant Biotechnology, Sher-e-Kashmir University of Agricultural Sciences and Technology Kashmir, Srinagar, India

Saman Zulfiqar Department of Botany, The Government Sadiq College Women University, Bahawalpur, Pakistan

About the editors

Dr. Hassan Etesami is a research scientist with 15 years of experience in the field of soil biology and biotechnology. He obtained his PhD degree from the Department of Soil Science, College of Agriculture & Natural Resources, University of Tehran, Iran, where he is currently a member of the faculty as an associate professor. He has also passed a research course as a visiting scholar under the supervision of Prof. Gwyn Beattie at Plant Pathology and Microbiology Department, Iowa State University, Iowa, United States, in 2013. Dr. Etesami has a special interest in developing biofertilizers and biocontrol agents that meet farmers' demands. He has coauthored over 80 publications (research papers, review papers, and book chapters) in various areas including biofertilizers and biocontrol. He is also a reviewer of 115 journals. Dr. Etesami's research areas include stressed agricultural management by silicon, microbial ecology, biofertilizers, soil pollution, integrated management of abiotic (salinity, drought, heavy metals, and nutritional imbalance) and biotic (fungal pathogens) stresses, plant–microbe interactions, environmental microbiology, and bioremediation.

Dr. Abdullah H. Al Saeedi is an associate professor in soil physics and water management, working at the Environment and Natural Resources Department, College of Agriculture and Food Science, King Faisal University, Saudi Arabia. He obtained his PhD from Liverpool John Moores University, UK (1992). Dr. Al-Saeedi has a research interest in soil water relationship, salinity, water management, and improving agriculture practice. He has many publications (research papers and book chapters) in soil physics, salinity, fertilizer requirement, GIS, and using nanosilica in improving agriculture productivity under different abiotic stresses. He reviewed many research paper manuscripts for different journals. He has been awarded Prince Mohammed Bin Fahad Prize for research. During his academic work, he has supervised many master's degree students, worked in national research and academic project.

Dr. Hassan El-Ramady is a professor of plant nutrition and soil fertility, working at the Soil and Water Department, Faculty of Agriculture, Kafrelsheikh University, Egypt. He received his PhD from the Technical University of Braunschweig, Germany (2008). He started his postdoctoral scholarships with ParOwn funded by Egypt to Hungary in 2012, then 2013 and 2014 funded by HSB, Hungary to Debrecen University, again from 2018 to 2019 at Debrecen University. He visited also the United States (2012 and 2014), Austria (2013), Italy (2014), Brazil (2015), and Germany (2014, 2015, 2016, and 2017). His current research program focuses on the biological plant nutrition and its problems including new approaches like nanoparticles under stress. He has over 100 peer-reviewed publications, 30 book chapters, and has edited 5 Arabic books, and he was a lead editor for the book "The Soils of Egypt." He is Editor-in-Chief and associate editor for some journals like *Frontiers in Soil Science*, *Egyptian Journal of Soil Science and Environment*, *Biodiversity*, and *Soil Security*. He is a reviewer for more than 150 journals (https://publons.com/researcher/1671675/hassan-el-ramady/28.08.2021).

Dr. Masayuki Fujita is a professor in the Department of Plant Sciences, Faculty of Agriculture, Kagawa University, Kagawa, Japan. He received his BSc in chemistry from Shizuoka University, Shizuoka, and his M.Agr. and PhD in plant biochemistry from Nagoya University, Nagoya, Japan. His research interests include physiological, biochemical, and molecular biological responses based on secondary metabolism in plants under biotic (pathogenic fungal infection) and abiotic (salinity, drought, extreme temperatures, and heavy metals) stresses, phytoprotectants and biostimulants, phytoalexin, cytochrome P-450, glutathione *S*-transferase, phytochelatin, and redox reaction and antioxidants. He has over 200 peer-reviewed publications and has edited 32 books and special issues of journals.

Dr. Mohammad Pessarakli is a professor in the School of Plant Sciences, College of Agriculture and Life Sciences at the University of Arizona, Tucson, Arizona, United States. His work at the University of Arizona includes research and extension services as well as teaching courses in Turfgrass Science, Management, and Stress Physiology, currently teaching the Plants and Our World course to large classes of over 200 students each semester. He has edited the *Handbook of Plant and Crop Stress* and the *Handbook of Plant and Crop Physiology* (both titles published by, formerly Marcel Dekker, Inc., currently Taylor and Francis Group, CRC Press), and the *Handbook of Photosynthesis*, *Handbook of Turfgrass Management and Physiology*, and the *Handbook of Cucurbits*. He has written 45 book chapters, he is Editor-in-Chief of the *Advances in Plants & Agriculture Research* journal, Editorial Board member of the *Journal of Plant Nutrition* and *Communications in Soil Science and Plant Analysis*, as well as the *Journal of Agricultural Technology*, and a member of the Book Review Committee of the Crop Science Society of America, and Reviewer of the Crop Science, Agronomy, Soil Science Society of America, and *HortScience* journals. He is the author or coauthor of over 200 journal articles in 20 different journals. Dr. Pessarakli is an active member of the Agronomy Society of America, Crop Science Society of America, and Soil Science Society of America, among others. He is an Executive Board member of the American Association of the University Professors, Arizona Chapter. Dr. Pessarakli is a well-known internationally recognized scientist and scholar and an esteemed member (*invited*) of several Who's Who as well as numerous honor societies (i.e., Phi Kappa Phi, Gamma Sigma Delta, Pi Lambda Theta, Alpha Alpha Chapter). He is a Certified Professional Agronomist and Certified Professional Soil Scientist, designated by the American Registry of the Certified Professionals in Agronomy, Crop Science, and Soil Science. Dr. Pessarakli is a United Nations Consultant in Agriculture for underdeveloped countries. He received his BS degree (1977) in Environmental Resources in Agriculture and his MS degree (1978) in Soil Management and Crop Production from Arizona State University, Tempe, and his PhD degree (1981) in Soil and Water Science from the University of Arizona, Tucson. Dr. Pessarakli's environmental stress research work and expertise on plants and crops is internationally recognized.
For more information about Dr. Pessarakli, please visit:
https://cals.arizona.edu/spls/content/mohammad
https://cals.arizona.edu/spls/people/faculty

Dr. Mohammad Anwar Hossain is serving as a professor in the Department of Genetics and Plant Breeding, Bangladesh Agricultural University (BAU), Mymensingh, Bangladesh. He received his BSc in Agriculture and MS in Genetics and Plant Breeding from BAU, Bangladesh. He also received an MS in Agriculture from Kagawa University, Japan, in 2008 and a PhD in Abiotic Stress Physiology and Molecular Biology from Ehime University, Japan in 2011 through Monbukagakusho scholarship. As a JSPS postdoctoral researcher, he has worked on isolating low phosphorus stress-tolerant genes from rice at the University of Tokyo, Japan during the period of 2015–17. His current research program focuses on understanding physiological, biochemical, and molecular mechanisms underlying abiotic stresses in plants and the generation of stress-tolerant and nutrient-efficient plants through breeding and biotechnology. He has over 70 peer-reviewed publications and has edited 13 books, including this one, published by CRC Press, Springer, Elsevier, Wiley, and CABI.

Preface

Crop plants growing under field conditions are constantly exposed to various abiotic and biotic stress factors leading to decreased yield and quality of produce. To achieve sustainable development in agriculture and to increase agricultural production for feeding an increasing global population, it is necessary to use ecologically compatible and environmentally friendly strategies to decrease the adverse effects of stresses on the plant. Silicon is recently recognized as a beneficial plant nutrient element and many growers already include it in their crop fertility programs. It has been widely reported that silicon can promote plant growth and alleviate various stresses as well as increase the quantity and improve the quality of the yield of many plant species. It is probably the only element that can enhance the resistance to multiple stresses. When present in excess, silicon is not noxious to plants and is also free from pollution and noncorrosive. In the last decade nanotechnology has emerged as a prominent tool for enhancing agricultural productivity. The production and applications of nanoparticles have greatly increased in many industries, such as energy production, healthcare, agriculture, and environmental protection. The application of nanoparticles has attracted interest for their potential to alleviate abiotic and biotic stresses in a more rapid, cost-effective, and more sustainable way than conventional treatment technologies. Recently, research related to silicon- and silicon-nanoparticles-mediated abiotic stresses and nutritional improvements in plants has received considerable interest from the scientific community. While significant progress has been made in silicon biochemistry in relation to stress tolerance, an in-depth understanding of the molecular mechanisms associated with the silicon- and nano-silicon-mediated stress tolerance and biofortification in plants is still lacking. Gaining a better knowledge of the regulatory and molecular mechanisms that control silicon uptake, assimilation, and tolerance in plants is, therefore, vital and necessary to develop modern crop varieties that are more resilient to environmental stress.

In this book, *"Silicon and Nano-silicon in Environmental Stress Management and Crop Quality Improvement: Progress and Prospects,"* we present a collection of 25 chapters written by leading experts engaged with silicon- and nano-silicon-mediated environmental stress management and crop quality improvement. This book aims to provide a comprehensive overview of the latest understanding of the physiological, biochemical, and molecular basis of silicon- and nano-silicon-mediated environmental stress tolerance and crop quality improvements in plants. Numerous figures and tables are included in this book to facilitate comprehension of the presented information.

Finally, this book will serve as a unique key source of information and knowledge for graduate and postgraduate students, instructors/educators, and frontline plant scientists around the globe and would be a valuable resource for promoting future research in plant stress tolerance as well as crop quality improvement through biofortification. We believe that the information presented in this book will make a sound contribution to this fascinating area of research and to the agricultural scientific community.

Hassan Etesami[1], Abdullah H. Al Saeedi[2], Hassan El-Ramady[3], Masayuki Fujita[4], Mohammad Pessarakli[5] and Mohammad Anwar Hossain[6]

[1]*Soil Science Department, College of Agriculture and Natural Resources, University of Tehran, Tehran, Iran* [2]*Department of Environment and Natural Resources, Faculty of Agriculture and Food Science, King Faisal University, Al-Hofuf, Saudi Arabia* [3]*Soil and Water Department, Faculty of Agriculture, Kafrelsheikh University, Kafr El-Sheikh, Egypt* [4]*Kagawa University, Kagawa, Japan* [5]*University of Arizona, Tucson, AZ, United States* [6]*Department of Genetics and Plant Breeding, Bangladesh Agricultural University, Mymensingh, Bangladesh*

1

Sources of silicon and nano-silicon in soils and plants

Hassan El-Ramady[1], Krishan K. Verma[2], Vishnu D. Rajput[3], Tatiana Minkina[3], Fathy Elbehery[4], Heba Elbasiony[5], Tamer Elsakhawy[6], Alaa El-Dein Omara[6] and Megahed Amer[7]

[1]Soil and Water Department, Faculty of Agriculture, Kafrelsheikh University, Kafr El-Sheikh, Egypt [2]Key Laboratory of Sugarcane Biotechnology and Genetic Improvement (Guangxi), Ministry of Agriculture and Rural Affairs/Guangxi Key Laboratory of Sugarcane Genetic Improvement/Sugarcane Research Institute, Guangxi Academy of Agricultural Sciences/Sugarcane Research Center, Chinese Academy of Agricultural Sciences, Nanning, P.R. China [3]Academy of Biology and Biotechnology, Southern Federal University, Rostov-on-Don, Russia [4]Central Laboratory of Environmental Studies, Kafrelsheikh University, Kafr El-Sheikh, Egypt [5]Department of Environmental and Biological Sciences, Home Economy Faculty, Al-Azhar University, Tanta, Egypt [6]Agriculture Microbiology Department, Soil, Water and Environment Research Institute (SWERI), Sakha Agricultural Research Station, Agriculture Research Center, Kafr El-Sheikh, Egypt [7]Soils Improvement Department, Soils, Water and Environment Research Institute, Sakha Station, Agricultural Research Center, Kafr El-Sheikh, Egypt

1.1 Introduction

After oxygen, silicon (Si) is the second most abundant element in the earth's crust (28%) and the most abundant element in soils (54%) [1,2]. Silicon is very common in soils and forms a lot of minerals like silicon dioxide (SiO_2), quartz, and alumino silicates [3]. Many fractions of Si could be found in soils as liquid phases, which consist of the dissolved Si forms in soil solution, that is, the complexes of silicic acid-inorganic compounds, the monosilicic acid (H_4SiO_4) and polysilicic acid [4]. Based on many distinguished properties, several applications of Si could be mentioned such as mediating stress tolerance in plants like salinity [5−7], drought [8−10], high-pH stress [11], water deficit stress [12], and metal toxicity [13,14]; as a fertilizer [15,16]; and maintaining soil health [14]. Nano-silicon or SiO_2 nanoparticles (Si-NPs) are considered a very important source of Si, which could be applied to promote plant resistance to many stressful conditions [17]. These Si-NPs have been used in many applications like phytoremediation [18], wastewater treatment [19], nano-pesticides [20,21], food processing units [22], nano-fertilizers [23], industrial applications [24], biomedical issues [25], and biosensors [17] as well as improving crop output during stress conditions [26].

Due to its fascinating roles under stress on plants, Si is well known as a beneficial or quasi-essential element, although it has been classified as a nonessential nutrient because it does not fulfill the criteria of essentiality and has no evidence to be involved in plant metabolism [27,28]. The most important Si form for plant uptake is monosilicic acid (H_4SiO_4) and its solubility in soil and uptake by plant roots depends on clay content, soil organic matter, and Si fractions in soils like Fe/Al oxides/hydroxides [4,29]. Many studies confirmed the role of silicon nano-fertilizers in increasing the biomass and productivity of various plants species [30,31] and others under stresses such as rice under salinity [32], maize under salinity [33], coriander under lead stress [34], rice under

Silicon and Nano-silicon in Environmental Stress Management and Crop Quality Improvement.
DOI: https://doi.org/10.1016/B978-0-323-91225-9.00003-0

fluoride stress [35], and common bean subjected to metal toxicity and sodic soil [26]. These Si-NPs have been applied in agriculture for mitigating different environmental stresses through their utilization as nano-fertilizers, nano-herbicides, and nano-pesticides [17,26]. A great concern recently has been adopted about the production of nano-silica (NS) from different agro-wastes [21,36] as a new concept of nano-management [37].

Therefore this work attempts to update information about the sources of silicon and nano-silicon in soil and agricultural crops. Silicon cycling and its bioavailability in soils, factors affecting this bioavailability to plants, and its uptake by plants particularly under stress will also be handled.

1.2 Sources of silicon and nano-silicon in soils

1.2.1 Silicon in soils and its forms

Silicon is surrounded in the periodic table near as boron (B), carbon (C), nitrogen (N), oxygen (O), phosphorus (P), and sulfur (S) all of which considered as important nutritional elements, as well as aluminum (Al), gallium (Ga), germanium (Ge), and arsenic (As), all considered as nonessential and/or harmful elements. Silicon levels in rocks range from 23% to 47% (basalt to orthoquartzite) [38], depending on soil types. Desilification and fertilization processes are active in certain heavily weathered soils, that is, latosols or latosolic red soils in the tropical regions. Carbonaceous minerals, such as limestones and carbonites, contain trace quantities of Si [38,39]. Silcretes are a form of derived soil that contains a sufficient amount of Si (more than 46%). The percent of Si in the petro-calcic horizon is much smaller than silcretes (8%), and the quantity of Si in minerals present in certain strongly weathered Oxisols is even lower [17,38]. Most of the soils are high in Si, some soils are low, especially the plant-available type of Si [40]. These types of soils include the Oxisols and Ultisols, which are highly weathered, leached, acidic, and low in base saturation [41], and the Histosols, which have a lot of organic matter but very little mineral content [42]. Furthermore, soils with a high proportion of quartz sand and those that have been subjected to long-duration plant productivity have low availability of Si in plants [17,43].

The biogeochemical conditions are the main factors controlling the fractions and availability of Si in soils [44], where Si can increase the availability of phosphorus in Artic soils as nonagricultural or anthropogenic management practices [45]. In soils, Si is separated into three phases: liquid, adsorbed, and solid [39,46,47]. Sauer and Burghardt [47] listed silica as one of the crystalline types of Si in the solid phase fraction. The crystalline of primary and secondary silicates, which are common in mineral soils produced from rocks and sediments, were previously the only crystalline types [43,48,49]. Quartz and disordered silica make up the majority of the silica products (Fig. 1.1). The Si fractions in the solid phase also include amorphous, poorly crystalline, and microcrystalline shapes [39,43]. The soluble and adsorbed phases of Si are identical, except the soluble components are mixed in the soil solution, while the adsorbed phase components are retained on soil particles, Fe and Al oxides/hydroxides. The Si content and abundance in soil are highly dependent on soil-forming processes, and as a result, soil profile. Except for organic soils, mostly mineral soils are composed of sand particles (mostly SiO_2), different types of primary crystalline such as olivine, augite, hornblende, quartz, feldspars-orthoclase, plagioclase, albite, and mica and secondary silicate are clay minerals such as illite, vermiculite, montmorillonite, chlorite, and kaolinite. In most cases, these silicates are very sparingly soluble and bio-geochemically inert. Polymerized silicic acid is only partly water soluble in soil, while monosilicic acid (H_4SiO_4) soluble in water. Inorganic, organic, and organic—inorganic complexes in soil, such as clays, organic matter, and organic—inorganic complex, can be adsorbed with water soluble [40,43,50,51].

1.2.1.1 Silicon in solid forms

The three major types of Si in the solid state are amorphous forms, poorly crystalline and microcrystalline forms, and crystalline forms. The crystalline types of Si, which are primarily used as silicates and silica materials (primary and secondary), account for the majority of Si in the solid phase. Sand and silt particles comprise the main mineral-bearing silicates inherited in soils, while the secondary silicates are found in clay particles formed by pedogenic processes involving phylo-silicates and Al-Fe oxides/hydroxides [4,43]. The Si exists in poorly crystalline and microcrystalline types, that is, short-range ordered silicates and chalcedony and secondary quartz [43]. Short-range ordered silicates in soil horizons are favored when pH H_2O > 5.0 [52], and imogolite is formed by the precipitation of H_4SiO_4 with Al hydroxides [39,53,54]. Amorphous forms include biogenic and lithogenic forms and they are present in soils ranging from 1 to 30 mg g^{-1} on soil mass basis [55]. The biogenic types, which are made up of plant residues and microorganism remains, are referred to as biogenic opal. Plants accumulate Si

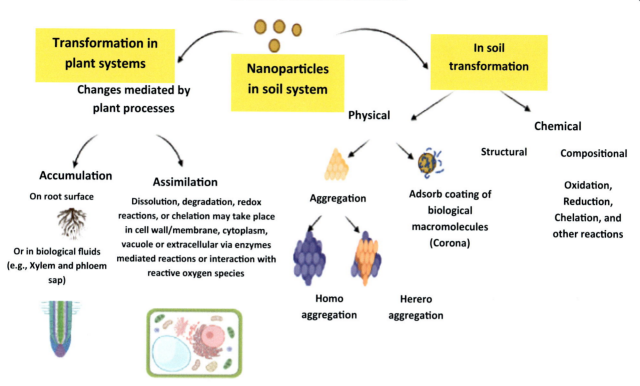

FIGURE 1.1 A brief demonstration for transformation of nanoparticles in soil and plant systems. This transformation includes different pathways in soil and plant systems. This transformation in plant system may include the accumulation on root surfaces or different assimilation methods in plants cells, whereas the transformation in soil system may be happened through the physical or chemical methods.

as silica bodies or phytoliths in their leaves, culms, and stems, while microorganisms contribute as microbial and protozoic Si [43,56]. When soluble Si in the soil is supersaturated, opal A is formed [55]. The solubility of various types of Si in the solid phase has a broad impact on its content in the soil. The solubility of silica-bearing minerals is depending on the density of silica tetrahedral and wide-range crystals [55,57]. Amorphous silica is expected to contribute more than quartz due to its higher solubility (1.8−2 mM silicon). Similar to quartz, amorphous silica dissolution rates increased linearly with saturation but exponentially dependent on the electrolytes [58]. Since quartz is extremely stable and thermodynamically resistant to weathering, its solubility ranged from 0.10 to 0.25 mM silicon [38]. As a result, if quartz is abundant in both residual and transported parent products, it will have a marginal contribution to Si in soil solution. Biogenic-based silica had higher (17-folds) solubility than quartz [59]. Since the release rate from plants litter is unaffected by cellulose hydrolysis and the released silica does not form complexes with organic matter, phytolithic silica is referred to as a pure inorganic lake [60]. The solubility of both crystalline and amorphous silica is approximately constant between pH 2.0 and 8.5, but it rapidly increases at pH 9.0 due to the decrease in H_4SiO_4 concentration in the soil solution [61], allowing crystalline and amorphous silica to dissolve to replenish or buffer the decreased H_4SiO_4 level in soil solution [40]. The plant available forms of Si present in soil ranging from 10 to over $100\,mg\,kg^{-1}$. Less than $20\,mg\,kg^{-1}$ Si are considered as poor and are mostly advised to amendment of Si in soil [51,62].

1.2.1.2 Availability of silicon in soil

Silicon is present in a number of forms in the soil solution, including monomeric (H_4SiO_4, the plant bioavailable form), oligomeric, and polysilicic acid [50]. Some dissolved silicic acid in the soil solution forms complexes with organic and inorganic compounds. Polysilicic acids with a maximum degree of polymerization are classified as polymeric or high-molecular-weight-silica, whereas oligomeric or low-molecular-weight-silica has H_4SiO_4 chains up to 10 Si atoms in length [43]. Different types of oligomeric and polysilicic acids can be found [48]. Plant absorption and nutrition are affected by monosilicic acid, while soil aggregation is affected by polysilicic acid. As per Norton [63], polysilicic acid creates silica bridges between soil particles, which improves soil aggregation and water-holding and buffering capacity, particularly in light-textured soils. After a month of incubation with silicon-rich materials, soils of various textures have increased their water-holding ability [46]. Uncharged H_4SiO_4

is present in typical soils (pH 8 or less) [40]. At pH values greater than 9, H_4SiO_4 dissociates into H^+ + H_3SiO_4 and then $2H^+$ + H_2SiO_4 2 at pH values greater than 11. When the silicic acid content is high and the pH is greater than 9, the formation of stabilized, multiple chains of H_4SiO_4 occurs [64]. The various types of silicon dioxide, silicate minerals, and plant residuum are the main sources of H_4SiO_4 in the soil solution. The amount of H_4SiO_4 produced by various SiO_2 types is determined by their physicochemical properties. The H_4SiO_4 content in the soil influences the amount of SiO_2 in the soil. The minerals that are insoluble and weather resistant, such as feldspar and a variety of silicates complex including circone, garnet, and tourmaline, add a less quantity of silicon to the soil [50,65].

Many factors influence the amount of H_4SiO_4 in the soil solution, which may include soil pH, temperature, size-shape, content of water and organic matter and potential of redox. These factors may influence the solubility of silicon-containing minerals [39,51]. The pH of the soil influences silicon solubility and mobility. The level of H_4SiO_4 in the soil solution depends on adsorption−desorption processes, which are highly dependent on the soil pH [66]. The amount of adsorbed H_4SiO_4 increases in soils with a lot of allophanes, Fe-enriched minerals of crystals, and particularly the more reactive multivalent metal hydroxides. The generation of SiO_2 deposits in the form of crusts is increased during evaporation, transpiration, and freezing processes [67]. Liming and high organic matter content reduce the concentration and mobility of H_4SiO_4 in the soil solution, while acid-producing fertilizer application raises H_4SiO_4 content in soil solution [39,43,68,69].

1.2.1.3 Adsorption of silicon on solid phases

Dissolved silicic acid fractions are absorbed on a variety of solid phases in soils, such as clay particles and Fe and Al hydroxides [70,71]. The adsorption of secondary clay minerals is responsible for a minor loss in the level of Si in the soil solution [43,72]. The Fe and Al hydroxides, on the other hand, have a high adsorption potential and can extract substantial quantities of dissolved silicon from the soil solution [50]. The adsorption of monosilicic acid by oxides influenced the soil pH, redox potential (Eh), and metallic form. From pH 4 to 9, the amount of monosilicic acids adsorbed by oxides increases, and when the oxides of metals in the soil are Al-based rather than Fe, the amount is significantly higher. Ponnamperuma [73] found that increasing the waterlogging period of soil resulted in a loss of Eh, as well as an enhancement in the solubility of Si in soil [74,75]. In general, silicic acid is adsorbed onto secondary Fe-based oxides; a large quantity of silicic acid is adsorbed onto short-range, ordered ferrihydrite than crystalline goethite [39,76]. The group of OH in Fe-oxide surface is replaced with H_4SiO_4, resulting in the formation of a bi-dendatesilicate inner-sphere complex [77−79]. Polysilicic acid is produced when the Fe-oxide surface interacts with the orthosilicic acid in a specific way [50]. Since Fe oxides are abundant in soil, even if their silicon adsorption ability is low efficient than that of Al oxides, the iron oxides can regulate the H_4SiO_4 content in the soluble phase to some extent [40,75,80,81].

1.2.2 Silicon cycle in soil and its bioavailability

In soil, the Si cycle is influenced by solid, liquid, and adsorption rate of silicon. In the soil, the soluble silicon is presented in H_4SiO_4, polymerized and complexed silicic acid, with the form of uncharged H_4SiO_4 being absorbed by the plants and microorganisms [4]. Inside plant tissues or microorganism cell structures, absorbed Si is deposited as polymerized silica. Litter and microorganism remain return these polymerized silica bodies to the top soil, where they eventually join the highly soluble biogenic silica reservoir, which contributes to soil silicon [59,61,82,83]. Silicon is also applied to soils through organic manure and compost applications, and the decomposition of silicon-rich manure will enhance the availability of Si in the soil [40,84]. The biochemistry of Si in the liquid phase is controlled by a number of processes, including (1) the dissolution of silicon containing the minerals of primary and secondary, (2) vegetation and microorganisms absorption of H_4SiO_4 in the soil, (3) Si adsorption on and desorption from different solid forms, and (4) the preservation of stable Si in the soil structure (silica polymer), that is, fertile soil. By atmospheric deposition, wind-blown dust and phytolith particles from savanna fires often add Si to the soil [85−87]. However, as compared to the other silicon inputs to the soil-plant system, the contribution of silicon from the atmosphere to the soil solution is very low [86].

1.2.2.1 Soluble silicon in soils and its bioavailability

Monosilicic acid (H_4SiO_4) is available in soil solution [88]. The silicate solubility depends on H_4SiO_4, which varies thermodynamically from ca. 10^2 to 10^4 M (amorphous to quartz), corresponding to the soil Si 10^3 M [89]. Despite this, the observed content of monosilicic acid (H_4SiO_4) in soil was about 0.1−0.6 mM [40,54,56,88], which is much lower than in saturated monosilicic acid solution and is primarily regulated by pH-dependent processes of adsorption—desorption on sesquioxides [39]. The sufficient quantity of Si can be absorbed by plant roots during the growing season, which is referred to as usable Si in soils, and it is typically used as a measure of the soil's Si-supplying ability. Monosilicic acid is the most common source of Si absorbed and transported by plants. H_4SiO_4 readily polymerizes into polymeric $Si(OH)_4$ in a monosilicic acid-saturated soil solution, where it is in a complex equilibrium with the form of silicates of amorphous and crystalline, exchangeable silicates, and sesquioxides [43]. Usable Si in soils is made up of monosilicic acid in soil solution and fragments of silicate element that can be smoothly converted into monosilicic acids (polymerized silicic acid, exchangeable silicates, and colloidal silicates). Soil Si is mostly found as monosilicic acid (pH 2—9), particularly at physiological values of pH, and transformation of monosilicic to ionic silicates is only possible at pH > 9. The main factors affecting soil Si availability or Si-supplying strength are the types of soil and raw material, historical land-use changes, pH of soil, soil profile, soil Eh, organic matter, ambient air temperature, and corresponding mineral ions [39,50,51,90—92]. The degradation of grassland ecosystem due to intense human disturbance and drying climate can also impact the Si-distribution and bioavailability in soil [93]. Many studies were carried out for better understanding of the bioavailability of silicon in soils (e.g., [94—96]).

1.2.2.2 Soil pH, soil organic matter, and its texture

Monosilicic acid concentration has highly influenced the pH of the soil. The lowest level of monosilicic acid is located at pH 8—9, low or high which the level of monosilicic acid greatly enhances. When the pH of the soil solution drops from 7 to 2, the Si content in the solution can rise dramatically [97]. Various demonstrations have been shown that the availability of Si in soil is strongly linked to soil pH [50,98—102]. The key explanations are that a portion of the carbonate-bound silicates collected by the solution of buffer acetate (pH 4.0) is inaccessible to plants, and acetate buffer approach measured the Si supply potential of these calcareous soil [40,103,104]. Numerous demonstrations have been found that the soil with high or sandy texture is generally low in the availability of Si and thus have minimum Si-uptake strength, while soils with a hard or appropriate Si are in clay texture [57,98,100,101,104]. Minerals of clay soil with the highest specific surface have more efficiency to adsorb, and soil-available silicon level is significantly associated with the clay soil texture [104—106]. According to various studies, soil-available Si concentration potentially associated with physical clay fraction (0.01 mm) but not with smaller clay fraction (0.002 mm) in soil [107,108]. The soil particle size depends on the availability of Si in soil that depends on the acidity of soil condition [98]. Clay content and soil pH were substantially significantly correlated with the available Si content in acid soils (pH below 6.5), while pH, silt, and sand fractions were adversely related to Si available in soils (pH more than 6.5) [65,87]. So far, there have been contradictory demonstrations on the impact of soil organic matter on Si availability. The majority of authors accepted that the available Si in soil is significantly linked with organic matter of soil [101,108], while others conclude that there is no or even a nonsignificant association between available Si in soil and organic matter [101,107,108].

1.2.2.3 Adsorption—desorption balance

The mechanism of Si adsorption on active soil and desorption of soluble Si in the soil solution was found to control the plant-available Si fractions in the soil texture [43,50]. As a result, it was discovered that the characteristics of the soil adsorption complex (the phenomenon of sorption and desorption) have a significant impact on the available Si fraction in plants [109,110]. Soil sorption—desorption properties are largely determined by soil forms and soluble or amorphous Si. It was discovered that the desorption of Si from calcareous silty-loam soil suspension differed between standard and Si applied plants [111]. According to the experimental findings [111,112], the solubility of Si in different soil textures in which equilibrium was approached from under and supersaturation, the SiO_2 (in soil) level was 103.10 M, which was intermediate between quartz and amorphous silica. SiO_2 can be heavily leached out of soil profiles and become depleted due to desilification and fertilization during the weathering and soil-forming phase in highly weathered tropical soil, that is, oxisols. As a result, sesquioxides, not clay minerals of silicate, dominate the residual secondary minerals. H_4SiO_4 solubility in soils is lower than quartz (10^4 M). Soil sorption and desorption characteristics appear to be heavily influenced by soil types, pH, and clay type [43].

1.3 Nano-silicon role in soils

National NPs are common in soil and are highly mobile and chemically reactive. These natural NPs have also a central role in buffering soil systems. They also can serve in limiting the concentration of potential toxic metals and providing a supply of metals for biochemical reactions [113]. However, engineered NPs like NS in agroecosystems still remain largely unknown. NS has several roles in improving the efficiency of agrochemicals, soil health, and crop production [114]. The transformation of these NPs in soil is controlled by several chemical, physical, and biological reactions, which are highly dynamic processes and their fate, behavior, and biological impacts are controlling these NPs in soils [114]. Many drivers can also control the transformation of these NPs (and their processes) in soil such as soil organic matter (through the aggregation), ionic strength (by the aggregation), soil pH (by the aggregation and dissolution), redox potential (Eh; by reduction and oxidation), inorganic ligands (through sulfidation, chlorination, and phosphorylation), and microorganisms, which control all physical and chemical transformation [114]. There is a need for more investigations about the fate and behavior of NS in soils under environmental conditions. This behavior may depend on the properties of silicon NPs and soil characterizations (Table 1.1). The role of soil pH, soil salinity (EC), cation exchange capacity (CEC), soil texture, redox potential (Eh), and other soil properties may control the fate and behavior of nano-silicon in soil.

1.4 Silicon and nano-silicon in plants

The role of silicon in plant nutrition has gained great concerns among several researchers, which confirmed the mitigative role of Si during unfavorable environmental conditions although there is no evidence about the essentiality of Si for higher plants [125]. Si has a distinguished role in mitigating plant nutritional stresses in mineral form or nano-form as reported in the following subsections. The main biological features of Si and nano-Si could also be listed in Table 1.2.

1.4.1 Silicon role and its mechanism in plants

Silicon is a well-known and common mineral constituent of higher plants, but its essentiality still needs evidence. It has distinguished roles for stressful plants including abiotic and biotic stress as reported by several reviews, which confirmed this statement (e.g., [126–133]). Silicon (Si) is a macro-element and acts a vital function in plant cycles. Si is the eighth greatest common element in nature and the second greatest common element in the soil following O_2; however, most of Si is not available to plants [9,29,126,134–137]. Plants can only absorb Si as monosilicic acid (H_4SiO_4), which is naturally present in the soil; however, the concentration varies depending on the soil properties (texture, pH, minerals, organic matter, and other factors). Above a certain concentration, the monosilicic acid accumulation in plant tissues causes SiO_2 precipitation and eventual deposition in cell walls, phytoliths, trichomes, and silica bodies with the exception of strongly growing area of the cell [30,135]. It is reported that different Si transporter genes (e.g., LSi1, LSi2, and LSi6) were stated to help in the transportation of monosilicic in the plant [17]. While Si is not widely recognized as an essential nutrient, it is frequently regarded as a "valuable element" or "structural element" that is helpful for plant growth, physiological/metabolic pathways, cell structure, and mitigation of a broad range of environmental stresses, that is, biotic and abiotic stresses. One of its main functions is to improve the growth and yield of plants growth particularly in stress conditions [10,138,139]. Si encourages plant photosynthesis by preferably exposing leaves to light for achieving plant resistance. However, the macro-element function has been demonstrated to be in response to various biotic and abiotic stressors. Profoundly, some of the utmost important roles of this factor are to increase resistance to pathogens, pests, and diseases, metals toxicity, salinity, and drought stresses. Thus fertilization by Si contributes to improved food safety, increased production with lower inputs cost, and declined adverse effects on health and environmental quality [10,17,30,134,135,140,141].

Suriyaprabha et al. [142] observed that the deficiency of Si leads to imbalance of other nutrients causing poor growth and even plant death. As well, Si plays a potential role in the plant's biochemical functions and intracellular creation of organic compounds. In addition, Rastogi et al. [17] reported that Si has been noticed to strengthen the cell walls of plants and Equisetaceae family' plants cannot survive without Si in nutrient solutions. Thus Si is considered an essential element for this family. Molecular and biochemical investigation reported that Si plays a key role in formulating the mechanical barriers that acts a great role in interrupting the pathogens penetration. In addition, Si can stimulate the expression of defense-associated genes and may have a notable role in managing plant stress signals such as salicylic acid, jasmonic acid, and ethylene [143]. Luyckx et al. [29] reported that plant develops well in the absence of Si; however, in some instances such as triggering higher susceptibility to fungal infection in the silicifier horsetail and rice. However, they reported that Si exerts its protective action by

TABLE 1.1 A survey on some published studies on nano-silicon in soil and its roles.

Details of Si-nanoparticles	Dimension of Si-NPs	Soil properties (country)	The main finding of the study	Reference
Chemical nano-SiO$_2$ fertilizer	80–90 nm	Soil texture (silty clay), pH (7.99), EC (1.1 dS m^{-1}), SOM (9.2 g kg^{-1}) (Iran)	Foliar applied nano-SiO$_2$ recorded the best wheat yield compared to both nano-zinc and nano-boron fertilizers under deficit irrigation	Ahmadian et al. [115]
Chemical nano-SiO$_2$ (at 1.5 mM mg Si L^{-1})	20–35 nm	Soil texture (clay loam), pH (7.84), EC (1.03 dS m^{-1}), SOM (10.3 g kg^{-1}) (Iran)	Foliar nano-SiO$_2$ improved growth, and oil yield of coriander under drought stress	Afshari et al. [116]
Chemical nano-SiO$_2$ (60 mg L^{-1} ~1 mM)	20 nm	Soil N (1.4%), P (1.4%), potash (1.4%), Ca (4%), Mg (0.7%), cellulose (15%), lignin (10%) (India)	Nano-silica could mitigate fluoride stress in rice through enhancing nonenzymatic and enzymatic antioxidants	Banerjee et al. [35]
Chemical chitosan-Si-nano-fertilizer (0.01%–0.16%, w/v)	100 nm	Soil texture (clay), pH (8.2), EC (0.56 dS m^{-1}) (India)	This nano-fertilizer enhanced maize plant growth and yield by inducing antioxidant defense enzymes	Kumaraswamy et al. [31]
Chemical thiol functionalized nano-SiO$_2$ (4%)	20 nm	Soil texture (silt loam), soil pH (7.93), SOM (13.78 g kg^{-1}), Pb, Cd, and Cu in soil (1554, 16.14, and 94.6 mg kg^{-1}, resp.) (China)	This nano-silica remediated polluted soil from heavy metals and improved growth both lettuce and pakchoi	Lian et al. [117]
Biological Si-NPs (2.5 and 5.0 mmol L^{-1})	38.78 nm	EC (7.81 dS m^{-1}), pH (7.30), texture (loamy), Cd, Ni, and Pb were 18.6, 255, and 252 mg kg^{-1} soil, resp. (Egypt)	The role of bio-nano-Si under saline and polluted soil with Pb, Ni, and Cd cultivated with common bean (cv. Bronco)	El-Saadony et al. [26]
Chemical SiO$_2$-NPs (0.75, 1.5, and 2.25 mM)	10–20 nm	Soil texture (sandy loam), pH (7.1), EC (1.2 dS m^{-1}), total Cd (0.22 mg kg^{-1}) (Iran)	Foliar applied SiO$_2$-NPs mitigate Cd stress (20 mg kg^{-1}) by improving antioxidant capacity, and growth of summer savory plants	Memari-Tabrizi et al. [118]
Chemical SiO$_2$-NPs (150, 500, or 2000 mg Si kg^{-1} dry soil)	10 nm	Commercial soil (60% Grainger), Eh (+230 to −230 mV), pH (7.1–7.6), applied As and Cd (5 and 1 kg^{-1}), USA	SiO$_2$-NPs (500 mg kg^{-1}) helped rice seedling shoots to reduce 70% and 50% of As and Cd under water irrigation scheme AWD than CF	Wang et al. [119]
Chemical SiO$_2$-NPs (at 2 mM)	30 nm	Soil texture (clay-loam), pH (7.40), EC (1.7 dS m^{-1}), Si (12.01 mg kg^{-1}) (Iran)	Improving maize production under foliar applied nano-Si in combination with Zn nutrient	Asadpour et al. [120]
Surface-modified nano-silica (3.0%)	18.0 nm	Soil pH (7.61), total Pb, Cd, and As contents (256.3, 3.44, and 114.8 mg kg^{-1}, resp.), China	RNS-SFe could immobilize bioavailability of As, Pb, and Cd, by 85%, 97.1%, and 80.1%, resp., in polluted soil	Cao et al. [121]
Chemical Nano-Si complex with glycine, glutamine, and histidine	10, 30, and 40 nm for Si of three amino acids	Soil texture (silty loam), pH (7.02), EC (0.62 dS m^{-1}) (Iran)	Under drought stress on plant feverfew (irrigation 12 days), foliar at 1.5 or 3.0 mM nano-Si complex was enhanced	Esmaili et al. [122]
Chemical Si-NPs (at 1 and 2 mM)	20 nm	Soil texture (loamy), pH (8.08), EC (1.11 dS m^{-1}) (Iran)	Protection sugar beet plants against water deficit stress by applied nano-Si; improving the antioxidant systems	Namjoyan et al. [123]
Chemical mercapto-functionalized nano-silica (0.2%–0.1%)	20–30 nm	Soil pH (8.12), SOM (19.6 g kg^{-1}), CEC (3.28 cmolc kg^{-1}), and total Cd 3.48 mg kg^{-1} (China)	Nano-silica under Cd stress reduced aggregates stability by 14.8%, increased soil dehydrogenase by 43.4%, wheat grain yield by 33.5%	Wang et al. [124]
Chemical nano-SiO$_2$ (500 mg kg^{-1})	NA	Soil texture (sandy loam), SOM (2.54 g kg^{-1}), pH (7.67), and CEC (7.23 mol kg^{-1}), China	Nano-SiO$_2$ supported the phytoremediation of *Erigeron annuus* L. plants grown in polluted soil by phenanthrene (150 mg kg^{-1}) for 60 days	Zuo et al. [18]

RNS-SFe, Mercapto-propyltrimethoxy silane- and ferrous sulfate-modified nano-silica; *CF*, continuous flooding; *AWD*, alternate wetting and drying; *CEC*, cation exchange capacity; *Eh*, redox potential; *SOM*, soil organic matter; *NA*, not available.

TABLE 1.2   A comparison between biological features of silicon and nano-silicon in plant nutrition [1,2,28–30,126,127].

Comparison item	Silicon (Si)	Silicon nanoparticles (Si-NPs)
The essentiality	Not yet, but a beneficial or quasi-essential metalloid element	Not yet confirmed!
Main uptake form	Monosilicic acid (H_4SiO_4) *via* aquaporin type channel convert into phytoliths as amorphous oxides (SiO_2)	Nano-silica
Translocation from soil solution to roots	Silicic acid [$Si(OH)_4$] uptake by root cells using Si-transporters Lsi1, Lsi2, and Lsi3 (for rice crop);For barley crop (HvLsi1, HvLsi2, and HvLsi6)	Si-NPs less than 20 nm uptake *via* roots symplast or apoplast
Translocation from roots to shoots	Lsi6 from xylem cell to xylem parenchyma cells and to stem by transpiration stream	Nano-silica uptake mainly by apoplast route through an unknown transport
Translocation from shoots to leaves	Si transported and finally polymerized to silica gel ($SiO_2.nH_2O$) in leaves	Nano-silica reaches xylem vessels *via* transporter Lsi6
Main forms	Silica (SiO_2), monosilicic acid (H_4SiO_4), silicon monoxide (SiO)	Silica nanoparticles (SiO_2-NPs) or Si nanoparticles (Si-NPs)
Common mineral	Quartz (SiO_2), kaolinite ($Al_2Si_2O_5(OH)_4$), orthoclase ($KAlSi_3O_8$), plagioclase ($NaAlSi_3O_8$)	Not reported
Mechanism against heavy metal stress	*Meta*-silicic acid (H_2SiO_3) generated from the hydrolysis of a soluble silicate, which retains heavy metals	Nano-silica mediated structural, biochemical, and physiological alterations in plants due to stress
Main roles in plant	- Improve plant resilience under biotic/abiotic stresses - Si-deposition in plant tissues increases abrasiveness; reducing digestibility and palatability for herbivores - Ameliorates to toxicities of chemicals and diseases - Improves physical parameters of plant fibers - Modulates biochemical and physiological processes	- Protect seedling against stress - Stimulate antioxidant system - Mitigate the oxidative stress - Improve seed germination and crop production under stress
Toxicity level	No certain level but 200 ppm may cause phytotoxicity in some nonaccumulators like gerbera and sunflower	There is no global law to determine nanoparticle toxicity
Common level	Mean 0.3%–1.2%, but up to 10% in rice	Nano-silica (2000 mg L^{-1}) did not cause oxidative stress
Deficiency level	May be < 29 mg kg^{-1} in soil causes Si-deficiency	Not reported

forming a physical barrier when it precipitates as SiO_2 and is inserted into biological structures (e.g., the cell wall, see infra). This passive effect, however, is oversimplified and does not explain why plants supplemented with Si are better able to deal with exogenous stresses. There is strong evidence in the literature that such cell wall components cause SiO_2 precipitation.

1.4.2 *Nano-silicon and its role in plants*

Foliar application of nano-Si was recently recommended for improving the plant growth and its productivity in many case studies [26,120,122]. This form of nano-Si has the ability to support cultivated plants under different stresses like drought [122], water stress [123], and soil salinity [26]. The biological synthesis of nano-Si is preferable for foliar application under stress compared to the physical or chemical method of nano-Si preparing. Under saline and contaminated soils with heavy metals (HMs), nano-Si can improve plant growth by promoting the content of carotenoids and chlorophylls, enhancing the rate of transpiration and photosynthesis, membrane stability index, stomatal conductance, free proline, relative water content, and total soluble sugars, and increasing the activities of enzymatic antioxidants such as catalase, peroxidase, ascorbic peroxidase, and superoxide oxide dismutase [26]. It could be concluded the role of applied nano-Si as reported in some studies in Table 1.3.

The use of supersmall-sized fertilizer has many benefits, including being more reactive, reaching the target directly, and only requiring small quantities. With these benefits, the nano-fertilizer is considered to be a game-changing technology for increasing agricultural production in a sustainable and environmentally responsible manner [148]. NPs may exhibit different characteristics compared to their bulk materials as a result of their smaller size, greater surface area:weight ratio, and different features. Likewise, Si-NPs were noticed to show this difference in their properties than their bulk materials [17]. NS has proven a meaningful ability to enhance plant fitness especially in unfavorable environmental variables. Application of NS in soil was displayed an increase in

TABLE 1.3 Summary of some studies on applied nano-silicon on different crops and its role.

Cultivated crop	Experimental conditions	The role of applied nano-Si	Reference
Rice (*Oryza sativa* L.)	Pot experiment using nano-Si ($60 \, mg \, L^{-1}$ for 20 days)	Nano-Si strongly stimulated overall rice growth during fluoride toxicity through enhanced macro- and micro-nutrients uptake during stress	Banerjee et al. [35]
Common bean (*Phaseolus vulgaris* L.)	Field trial using Bio-Si-NPs (2.5 and $5.0 \, mmol \, L^{-1}$)	Applied Si-NPs ($5 \, mmol \, L^{-1}$ was a recommended dose for growth and reduce the content of heavy metals in plants grown on polluted saline soil)	El-Saadony et al. [26]
Maize (*Zea mays* L.)	Field trial, chitosan-silicon nano-fertilizer (0.04%−0.12%, w/v)	Foliar nano-fertilizer promoted total chlorophyll content and leaf area as slow/protective release fertilizer	Kumaraswamy et al. [26]
Lemon balm (*Melissa officinalis* L.)	Pot experiment using Si-NPs (100 and $500 \, mg \, L^{-1}$, 20−30 nm)	Seed priming with Si-NPs along with seedling inoculated *Pseudomonas* sp. increased primary and secondary metabolites in lemon balm plants	Hatami et al. [144]
Lettuce (*Lactuca sativa* L.) and pakchoi (*Brassica chinensis* L.)	Pot experiment using thiol functionalized nano-silica (4% and 20 nm)	This nano-silica could be considered a promising amendment for remediation of heavy metal contaminated soils	Lian et al. [117]
Maize (*Zea mays* L.)	Field experiment using SiO_2-NPs (30 nm and 2 mM)	Foliar applied SiO_2-NPs + Zn ($ZnSO_4$, 0.4%) increased grain yield by 37% compared to the control	Asadpour et al. [120]
Bamboo (*Arundinaria pygmaea* L.)	In vitro trial using 100 mM SiO_2-NPs	SiO_2-NPs with applied 100 and 200 mM of Cu and Mn conferred the optimal rate of growth	Emamverdian et al. [145]
Feverfew (*Tanacetum parthenium* L.)	Greenhouse experiment using nano-Si complex (1.5 and 3.0 mM)	Glycine nano-Si complex was the best compound in reducing the adverse effect of drought stress	Esmaili et al. [122]
Banana (*Musa acuminata*) cv. Grand Nain	In vitro trial, SiO_2-NPs ($50−150 \, mg \, L^{-1}$)	Applied SiO_2-NPs improved the chlorophyll content, induced uptake of K^+, decreased damage in cell wall under salt stress	Mahmoud et al. [146]
Sugar beet (*Beta vulgaris* L.)	Field experiment, applied nano-Si (1, 2 mM and 20 nm)	Under water stress (50%, etc.), nano-Si might protect plants by enhancing antioxidants, and glycine betaine	Namjoyan et al. [123]
Maize (*Zea mays* L.)	Greenhouse trial using nano-SiO_2 ($4 \, mg \, kg^{-1}$ and 5−15 nm)	Nano-SiO_2 did not affect the accumulation of Al, but mitigated its phytotoxicity in maize	de Sousa et al. [147]

the resistance of hawthorn plants to water deficit primarily due to improving the efficiency of photosynthesis and stomatal conductance. Furthermore, seed priming with NS not only reduced the stress of heavy metals toxicity, but also increased the plant development and biomass of wheat plants cultivated in a Cd-stressed condition by activating seed metabolic activities/functions and increasing supply of nutrients to plants [141]. The use of NS either by direct spray or by soil drench has revealed good findings in controlling insects and improving plant features [143]. Nanosilica was also reported to boost germination and also fresh and dry weight in some plants such as a well-known Siexcluder tomato. Furthermore, silica NPs have been shown to shield wheat seedlings from ultraviolet stress by activating the antioxidant protection mechanism. Furthermore, mesoporous NS have been shown to increase the plant development, content of protein, and photosynthesis of lupinin wheat plants, as well as to induce no variations in the antioxidant enzymes activity [135]. El-Naggar et al. [20] reported that Si can be considered a micronutrient that supports plant growth, especially in dry environments, by holding water, binding other nutrients, and increasing the cell strength. Furthermore, as a result of a heavy dose of N fertilizers, Si use allows the plant shoot structure to become more erect, which will enhance chlorophyll content, plant photosynthesis, and yield quality. The efficacy of nano-fertilizers compared to their conventional analogs displayed that the former has better. Thus using NS fertilizer with NPK will improve the absorbability of plants to fertilizers by, thus, it will be more effective compared to conventional chemical fertilizers. Suriyaprabha et al. [142] reported that since amorphous Si particles are commonly considered to be biocompatible, using amorphous NS for food crops is a viable option for overcoming Si deficiency in soil and plants. They added that it is previously stated that studies on the effect of NS in maize showed enhanced growth parameters in addition to improved seed stability. However, the effect of NS on physiological components of maize and the distribution of Si in roots and shoots are critical for determining the precise functional properties of mineral fertilizers. To determine the biotic and abiotic stress resistance pathways adapted through Si fertilization, important regulatory and defensive compounds in maize, such as protein and phenols, must be determined. As a result, the aforementioned NP-mediated abiotic stress amelioration can significantly participate in maize adaptation.

However, the investigations on NS behavior in plants and the interaction mechanisms, its impact, and agricultural applications are still under elementary stage [142]. Also, the knowledge is limited regarding the potential role of NS on stress recovery in plant and the involved cellular mechanisms [141]. Also, recognizing the mode of action of NS on plant growing and development is still lacking [30,140]. The plant roots' uptake of Silicon nanoparticles (SNP) has been suggested to mainly follow the apoplastic pathway, as the silicon carriers are lesser sensitive to SNP. Though the mechanism of uptake is not well known, it is absorbed at a higher rate than other silicates. Mesosporous silica NPs are becoming increasingly important in agricultural sciences [30].

1.5 Conclusion

After oxygen, silicon is the second most abundant nutrient on earth's crust, and it has a number of beneficial effects on plants, supporting crop production under a variety of conditions. Under these stresses, plant cell wall is the key-active part in plant responses, which may establish a signaling cascade toward the cell interior. Therefore Si needs more investigations with a focus on new insights concerning the interaction between Si and plants, which might draw new strategies to improve crop productivity. The advantages of silicon to a broad range of plants have been well established, demonstrating the significance of Si fertilization in sustainable agriculture. To ensure plant productivity, agricultural areas with intensive crop care cultivation systems, particularly those with soils that are naturally low in soluble silicon, are amended with silicon-rich materials. In reality, silicon fertilization is a popular agronomic method in some parts of the world. Although the production and standardization of various approaches for extracting and quantifying various silicon forms in the soil is considered significant progress in silicon research and soil science, their applications in soil fertility and nutrient management have been very limited. A soil interpretation demonstration may be useful to decide whether or not silicon application is necessary, but it does not provide the silicon content needed to increase plant-available silicon to the required level, nor does it indicate the likelihood that the crop in question will react to and gain from silicon fertilization. In near future, soil science-based research demonstrations on silicon are expected to greatly advance current knowledge of silicon in soil and fertilization recommendations for crop production. Further studies are needed concerning the following issues: what is the real role of microorganisms in transformation of nanosilica in soils and its interfaces with different plants? Are there any details about the kinetics of nanosilica and its transformation in different soils? What is the expected impact of NS and its transformation in long term in the agroecosystem? Are there any direct or indirect interconnections between transformation of NS and soil quality and/or climate changes in agroecosystems?

Acknowledgment

V. Rajput and T. Minkina would like to acknowledge funding from the Russian Scientific Foundation, Grant No. 21-77-20089.

References

[1] Kabata-Pendias A. Trace Elements in Soils and Plants. 4th ed. Taylor and Francis Group, LLC; 2011.
[2] Kabata-Pendias A, Szteke B. Trace Elements in Abiotic and Biotic Environments. Taylor & Francis Group, LLC; 2015.
[3] Schaller J, Puppe D. Heat improves silicon availability in mineral soils. Geoderma 2021;386:114909. Available from: http://doi.org/10.1016/j.geoderma.2020.114909.
[4] Schaller J, Puppe D, Kaczorek D, et al. Silicon cycling in soils revisited. Plants 2021;10(2):295. Available from: http://doi.org/10.3390/plants10020295.
[5] Khan A, Khan AL, Muneer S, et al. Silicon and salinity: crosstalk in crop-mediated stress tolerance mechanisms. Front Plant Sci 2019;10:1429. Available from: https://doi.org/10.3389/fpls.2019.01429.
[6] Zhu Y-X, Gong H-J, Yin J-L. Role of silicon in mediating salt tolerance in plants: a review. Plants 2019;8:147. Available from: https://doi.org/10.3390/plants8060147.
[7] Homann J, Berni R, Hausman J-F, et al. A review on the beneficial role of silicon against salinity in non-accumulator crops: tomato as a model. Biomolecules 2020;10:1284. Available from: https://doi.org/10.3390/biom10091284.
[8] Thorne SJ, Hartley SE, Maathuis FJM. Is silicon a panacea for alleviating drought and salt stress in crops? Front Plant Sci 2020;11:1221. Available from: https://doi.org/10.3389/fpls.2020.01221.

[9] Verma KK, Singh P, Song X-P, et al. Mitigating climate change for sugarcane improvement: role of silicon in alleviating abiotic stresses. Sugar Tech 2020;22(5):741—9. Available from: http://doi.org/10.1007/s12355-020-00831-0.

[10] Verma KK, Song XP, Verma CL, et al. Predication of photosynthetic leaf gas exchange of sugarcane (*Saccharum* spp.) leaves in response to leaf positions to foliar spray of potassium salt of active phosphorus under limited water irrigation. ACS Omega 2021;6:2396—409.

[11] Khan A, Kamran M, Imran M, et al. Silicon and salicylic acid confer high-pH stress tolerance in tomato seedlings. Sci Rep 2019;9:19788. Available from: http://doi.org/10.1038/s41598-019-55651-4.

[12] Farahani H, Sajedi NA, Madani H, et al. Effect of foliar-applied silicon on flower yield and essential oil composition of Damask rose (*Rosa damascena* Miller) under water deficit stress. Silicon 2020;. Available from: http://doi.org/10.1007/s12633-020-00762-1.

[13] Li N, Feng A, Liu N, et al. Silicon application improved the yield and nutritional quality while reduced cadmium concentration in rice. Environ Sci Pollut Res Int 2020;27(16):20370—9. Available from: https://doi.org/10.1007/s11356-020-08357-4.

[14] Ma C, Ci K, Zhu J, et al. Impacts of exogenous mineral silicon on cadmium migration and transformation in the soil-rice system and on soil health. Sci Total Environ 2021;759:143501. Available from: http://doi.org/10.1016/j.scitotenv.2020.143501.

[15] Laane H-M. The effects of foliar sprays with different silicon compounds. Plants 2018;7:45. Available from: https://doi.org/10.3390/plants7020045.

[16] Huang C, Wang L, Gong X, et al. Silicon fertilizer and biochar effects on plant and soil PhytOC concentration and soil PhytOC stability and fractionation in subtropical bamboo plantations. Sci Total Environ 2020;715:136846. Available from: http://doi.org/10.1016/j.scitotenv.2020.136846.

[17] Rastogi A, Tripathi DK, Yadav S, et al. Application of silicon nanoparticles in agriculture. 3 Biotech 2019;9:90. Available from: https://doi.org/10.1007/s13205-019-1626-7.

[18] Zuo R, Liu H, Xi Y, et al. Nano-SiO$_2$ combined with a surfactant enhanced phenanthrene phytoremediation by *Erigeron annuus* (L.) Pers. Environ Sci Pollut Res 2020;27:20538—44.

[19] Akhayere E, Essien EA, Kavaz D. Effective and reusable nano-silica synthesized from barley and wheat grass for the removal of nickel from agricultural wastewater. Environ Sci Pollut Res 2019;26:25802—13. Available from: http://doi.org/10.1007/s11356-019-05759-x.

[20] El-Naggar ME, Abdelsalam NR, Fouda MMG, et al. Soil application of nano silica on maize yield and its insecticidal activity against some stored insects after the post-harvest. Nanomaterials 2020;10:739. Available from: https://doi.org/10.3390/nano10040739.

[21] Peerzada JG, Chidambaram R. A statistical approach for biogenic synthesis of nano-silica from different agro-wastes. Silicon 2020;. Available from: http://doi.org/10.1007/s12633-020-00629-5.

[22] Mittal D, Kaur G, Singh P, Yadav K, et al. Nanoparticle-based sustainable agriculture and food science: recent advances and future outlook. Front Nanotechnol 2020;2:579954. Available from: https://doi.org/10.3389/fnano.2020.579954.

[23] Felisberto G, Prado RM, de Oliveira RLL, et al. Are nanosilica, potassium silicate and new soluble sources of silicon effective for silicon foliar application to soybean and rice plants? Silicon 2020;. Available from: http://doi.org/10.1007/s12633-020-00668-y.

[24] AlKhatib A, Maslehuddin M, Al-Dulaijan SU. Development of high-performance concrete using industrial waste materials and nano-silica. J Mater Res Technol 2020;9(3):6696—711. Available from: http://doi.org/10.1016/j.jmrt.2020.04.067.

[25] Selvarajan V, Obuobi S, Ee PLR. Silica nanoparticles—a versatile tool for the treatment of bacterial infections. Front Chem 2020;8:602. Available from: https://doi.org/10.3389/fchem.2020.00602.

[26] El-Saadony MT, Desoky EM, Saad AM, et al. Biological silicon nanoparticles improve *Phaseolus vulgaris* L. yield and minimize its contaminant contents on a heavy metals-contaminated saline soil. J Environ Sci (China) 2021;106:1—14. Available from: http://doi.org/10.1016/j.jes.2021.01.012.

[27] Mandlik R, Thakral V, Raturi G, et al. Significance of silicon uptake, transport, and deposition in plants. J Exp Bot 2020;71(21):6703—18. Available from: http://doi.org/10.1093/jxb/eraa301.

[28] Dhiman P, Rajora N, Bhardwaj S, et al. Fascinating role of silicon to combat salinity stress in plants: an updated overview. Plant Physiol Biochem 2021;162:110—23. Available from: http://doi.org/10.1016/j.plaphy.2021.02.023.

[29] Luyckx M, Hausman J-F, Lutts S, et al. Silicon and plants: current knowledge and technological perspectives. Front Plant Sci 2017;8:411. Available from: https://doi.org/10.3389/fpls.2017.00411.

[30] Mathur P, Roy S. Nanosilica facilitates silica uptake, growth and stress tolerance in plants. Plant Physiol. Biochem 2020;157:114—27. Available from: http://doi.org/10.1016/j.plaphy.2020.10.011.

[31] Kumaraswamy RV, Saharan V, Kumari S, et al. Chitosan-silicon nanofertilizer to enhance plant growth and yield in maize (*Zea mays* L.). Plant Physiol. Biochem 2021;159:53—66. Available from: http://doi.org/10.1016/j.plaphy.2020.11.054.

[32] Abdel-Haliem MEF, Hegazy HS, Hassan NS, et al. Effect of silica ions and nano silica on rice plants under salinity stress. Ecol Eng 2017;99:282—9. Available from: http://doi.org/10.1016/j.ecoleng.2016.11.060.

[33] Naguib DM, Abdalla H. Metabolic status during germination of nano silica primed *Zea mays* seeds under salinity stress. J Crop Sci Biotechnol 2019;22(5):415—23. Available from: https://doi.org/10.1007/s12892-019-0168-0.

[34] Fatemi H, Pour BE, Rizwan M. Isolation and characterization of lead (Pb) resistant microbes and their combined use with silicon nanoparticles improved the growth, photosynthesis and antioxidant capacity of coriander (*Coriandrum sativum* L.) under Pb stress. Environ Pollut 2020;266:114982. Available from: http://doi.org/10.1016/j.envpol.2020.114982.

[35] Banerjee A, Singh A, Sudarshan M, et al. Silicon nanoparticle-pulsing mitigates fluoride stress in rice by finetuning the ionomic and metabolomic balance and refining agronomic traits. Chemosphere 2021;262:127826. Available from: http://doi.org/10.1016/j.chemosphere.2020.127826.

[36] Kauldhar BS, Yadav SK. Turning waste to wealth: a direct process for recovery of nano-silica and lignin from paddy straw agro-waste. J Clean Prod 2018;194:158—66. Available from: http://doi.org/10.1016/j.jclepro.2018.05.136.

[37] El-Ramady H, El-Henawy A, Amer M, et al. Agricultural waste and its nano-management: mini review. Egypt J Soil Sci 2020;60 (4):349—64. Available from: https://doi.org/10.21608/ejss.2020.46807.1397.

[38] Monger HC, Kelly EF. Silica minerals. Soil mineralogy with environmental applications. Madison: Soil Science Society of America; 2002. p. 611—36.

[39] Savvas D, Ntatsi G. Biostimulant activity of silicon in horticulture. Sci Hort 2015;196:66—81. Available from: https://doi.org/10.1016/j.scienta.2015.09.010.

[40] Coskun D, Deshmukh R, Sonah H, et al. The controversies of silicon's role in plant biology. New Phytol 2019;221:67—85. Available from: https://doi.org/10.1111/nph.15343.

[41] Foy CD. Soil chemical factors limiting plant root growth. Adv Soil Sci 1992;19:97—149.

[42] Snyder GH, Jones DB, Gascho GJ. Silicon fertilization of rice on Everglades histosols. Soil Sci Soc Am J 1986;50:1259—63.

[43] Farooq MA, Dietz K-J. Silicon as versatile player in plant and human biology: overlooked and poorly understood. Front Plant Sci 2015;6:994. Available from: https://doi.org/10.3389/fpls.2015.00994.

[44] Wang W, Wei H-Z, Jiang S-Y, et al. Silicon isotope geochemistry: fractionation linked to silicon complexations and its geological applications. Molecules 2019;24:1415. Available from: https://doi.org/10.3390/molecules24071415.

[45] Schaller J, Faucherre S, Joss H, et al. Silicon increases the phosphorus availability of Arctic soils. Sci Rep 2019;9:449. Available from: https://doi.org/10.1038/s41598-018-37104-6.

[46] Matichencov VV, Bocharnikova EA. The relationship between silicon and soil physical and chemical properties. In: Datnoff LE, Snyder GH, Korndörfer GH, editors. Silicon in agriculture. Amsterdam: Elsevier; 2001. p. 209—19.

[47] Sauer D, Burghardt W. The occurrence and distribution of various forms of silica and zeolites in soils developed from wastes of iron production. Catena 2006;65:247—57.

[48] Iler RK. The chemistry of silica. New York: Wiley; 1979. p. 621.

[49] Conley DJ, Sommer M, Meunier JD, et al. Silicon in the terrestrial biogeosphere. In: Ittekot V, Humborg C, Garnier J, editors. Land—ocean nutrient fluxes: silica cycle. SCOPE Series, 66. 2006. p. 13—28.

[50] Frew A, Weston LA, Reynolds OL, et al. The role of silicon in plant biology: a paradigm shift in research approach. Ann Bot 2018;121:1265—73.

[51] Zargar SM, Mahajan R, Bhat JA, et al. Role of silicon in plant stress tolerance: opportunities to achieve a sustainable cropping system. 3 Biotech 2019;9:73. Available from: https://doi.org/10.1007/s13205-019-1613-z.

[52] Wada K. Allophane and imogolite. Minerals in soil environments. Madison: Soil Science Society of America; 1989. p. 1051—87.

[53] Exley C. Silicon in life: a bioinorganic solution to bioorganic essentiality. J Inorg Biochem 1998;69:139—44.

[54] Doucet FJ, Schneider C, Bones SJ, et al. The formation of hydroxyalumino silicates of geochemical and biological significance. Geochim Cosmochim Acta 2001;65:2461—7.

[55] Drees LR, Wilding LP, Smeck NE, et al. Silica in soils: quartz and disorders polymorphs. Minerals in soil environments. Madison: Soil Science Society of America; 1989. p. 914—74.

[56] Aoki Y, Hoshino M, Matsubara T. Silica and testate amoebae in a soil under pine-oak forest. Geoderma 2007;142:29—35.

[57] Epstein E. Silicon. Annu Rev Plant Physiol Plant Mol Biol 1999;50:641—64.

[58] Dove PM, Han N, Wallace AF, et al. Kinetics of amorphous silica dissolution and the paradox of the silica polymorphs. PNAS 2008;105 (29):9903—8. Available from: http://doi.org/10.1073/pnas.0803798105.

[59] Fraysse F, Pokrovsky OS, Schott J, et al. Surface properties, solubility and dissolution kinetics of bamboo Phytoliths. Geochim Cosmochim Acta 2006;70:1939—51.

[60] Fraysse F, Pokrovsky OS, Meunier JD. Experimental study of terrestrial plant litter interaction with aqueous solutions. Geochim Cosmochim Acta 2010;74:70—84.

[61] Dove PM. Kinetic and thermodynamic controls on silica reactivity in weathering environments. Rev. Mineral Geochem 1995;31:235—90.

[62] Liang Y, Nikolic M, Belanger R, et al. Silicon in agriculture: from theory to practice. Springer 2015; Dordrecht.

[63] Norton LD. Micromorphology of silica cementation in soils. In: Ringrose-Voase AJ, Humphreys GS, editors. Soil micromorphology: studies in management and genesis, vol. 22. Elsevier; 1984. p. 811—24.

[64] Knight CTG, Kinrade SD. A primer on the aqueous chemistry of silicon. In: Datnoff LE, Snyder GH, Korndörfer GH, editors. Silicon in agriculture. Amsterdam: Elsevier; 2001. p. 57—84.

[65] Kovda VA. Biogeochemistry of soil cover. Moscow: Nauka Publication; 1985. p. 159—79.

[66] Gerard F, Mayer KU, Hodson MJ, et al. Modelling the biogeochemical cycle of silicon in soils: application to a temperate forest ecosystem. Geochim Cosmochim Acta 2008;72:741—58.

[67] McKeague JA, Cline MG. Silica in soil solutions. I. The form and concentration of dissolved silica in aqueous extracts of some soils. Can J Soil Sci 1963;43:70—82.

[68] Panov NP, Goncharova NA, Rodionova LP. The role of amorphous silicic acid in solonetz soil processes. Vestnik Agr Sci 1982;11:18.

[69] Allmaras RR, Laird DA, Douglas CL, et al. Long-term tillage, residue management and nitrogen fertilizer influences on soluble silica in Haploxerol. Proceedings of the American society of agronomy annual meeting, Madison. 1991. p. 323.

[70] Hansen HCB, Raben-Lange B, Raulund-Rasmussen K, et al. Monosilicate adsorption by ferrihydrite andgoethite at pH 3—6. Soil Sci 1994;158:40—6.

[71] Dietzel M. Interaction of polysilicic and monosilicic acid with mineral surfaces. In: Stober I, Bucher K, editors. Water—rock interaction. Dordrecht: Kluwer; 2002. p. 207—35.

[72] Siever R, Woodford N. Sorption of silica by clay minerals. Geochim Cosmochim Acta 1973;37:1851—80.

[73] Ponnamperuma FN. Dynamic aspects of flooded soils and the nutrition of the rice plant. The mineral nutrition of the rice plant. Baltimore: John Hopkins Press; 1965. p. 295—328.

[74] Jones LHP, Handreck KA. Silica in soils, plants, and animals. Adv Agron 1967;19:107—49.

[75] Delstanche S, Opfergelt S, Cardinal D, et al. Silicon isotopic fractionation during adsorption of aqueous monosilicic acid onto iron oxide. Geochim Cosmochim Acta 2009;73:923—34.

[76] Parfitt RL. Anion adsorption by soils and soil materials. Adv Agron 1978;30:1—50.

[77] Pokrovsky GS, Schott J, Garges F, et al. Iron(III)-silica interactions in aqueous solution: insights from X-ray absorption fine structure spectroscopy. Geochim Cosmochim Acta 2003;67:3559—73.

[78] Hiemstra T, Barnett MO, van Riemsdijk WH. Interaction of silicic acid with goethite. J Colloid Interf Sci 2007;310:8—17.

[79] Schwertmann U, Taylor RM. Iron oxides. Minerals in soil environments. Madison: Soil Science Society of America; 1989. p. 379—438.

[80] Opfergelt S, Bournonville G, Cardinal D, et al. Impact of soil weathering degree on silicon isotopic fractionation during adsorption onto iron oxides in basaltic ash soils, Cameroon. Geochim Cosmochim Acta 2009;73:7226–40.

[81] Farmer V, Delbos E, Miller JD. The role of phytolith formation and dissolution in controlling concentrations of silica in soil solutions and streams. Geoderma 2005;127:71–9.

[82] Saccone L, Conley DJ, Koning E, et al. Assessing the extraction and quantification of amorphous silica in soils of forest and grassland ecosystems. Eur J Soil Sci 2007;58:1446–59.

[83] Song Z, Wang H, Strong PJ, et al. Increase of available silicon by Si-rich manure for sustainable rice production. Agron Sustain Dev 2014;34:813–19. Available from: https://doi.org/10.1007/s13593-013-0202-5.

[84] Kurtz AC, Derry LA, Chadwick OA. Accretion of Asiandust to Hawaiian soils: isotopic, elemental and mineral mass balances. Geochim Cosmochim Acta 1987;65:1971–83.

[85] Street-Perrott FA, Barker P. Biogenic silica: a neglected component of the coupled global continental biogeochemical cycles of carbon and silicon. Earth Surf Proc Land 2008;33:1436–57.

[86] Opfergelt S, Cardinal D, André L, Delvigne C, et al. Variations of 30Si and Ge/Si with weathering and biogenic input in tropical basaltic ash soils under monoculture. Geochim Cosmochim Acta 2010;74:225–40.

[87] Epstein E. The anomaly of silicon in plant biology. Proc Natl Acad Sci USA 1994;91:11–17.

[88] Lindsay WL. Chemical equilibria in soils. NewYork: Wiley Interscience; 1979.

[89] Sumida H. Plant-available silicon in paddy soils. In: Proceedings of the second silicon in agriculture conference, 2002 August 22–26, Tsuruoka, Yamagata, Japan; 2002. p. 43–9.

[90] Husnain, Wakatsuki T, Setyorini D, et al. Silica availability in soils and river water in two watersheds on Java Island, Indonesia. Soil Sci Plant Nutr 2008;54:916–27.

[91] Struyf E, Smis A, Van Damme S, et al. The global biogeochemical silicon cycle. Silicon 2010;1:207–13.

[92] Yang S, Hao Q, Liu H, et al. Impact of grassland degradation on the distribution and bioavailability of soil silicon: implications for the Si cycle in grasslands. Sci Total Environ 2019;657:811–18. Available from: http://doi.org/10.1016/j.scitotenv.2018.12.101.

[93] López-Pérez MC, Pérez-Labrada F, Ramírez-Pérez LJ, et al. Dynamic modeling of silicon bioavailability, uptake, transport, and accumulation: applicability in improving the nutritional quality of tomato. Front Plant Sci 2018;9:647. Available from: https://doi.org/10.3389/fpls.2018.00647.

[94] Sandhya K, Prakash NB. Bioavailability of silicon from different sources and its effect on the yield of rice in acidic, neutral, and alkaline soils of Karnataka, South India. Commun Soil Sci Plant Anal 2019;50(3):295–306. Available from: https://doi.org/10.1080/00103624.2018.1563096.

[95] Caubet M, Cornu S, Saby NPA, et al. Agriculture increases the bioavailability of silicon, a beneficial element for crop, in temperate soils. Sci Rep 2020;10:19999. Available from: http://doi.org/10.1038/s41598-020-77059-1.

[96] Beckwith RS, Reeve R. Studies on soluble silica in soils. I. The sorption of silicic acid by soils and minerals. Aust J Soil Res 1963;1:157–68.

[97] Cai AY, Xue ZZ, Peng JG, et al. Studies of available silica content in the soils of Fujian province and the prerequisite for its variation. J Fujian Acad Agric Sci. 1997;12(4):47–51.

[98] He LY, Wang ZL. Study on relationships between soil particle size or pH and soil available silicon content. Soils 1998;5:243–6.

[99] Zheng L. Available silicon content and its distribution in cultivated soils in Anhui province. Chin J Soil Sci 1998;29:126–8.

[100] Li ZZ, Tao QX, Liu GR, et al. Investigation of available silicon content in cultivated soil in Jiangxi province. Acta Agric Jiangxi 1999;11:1–9.

[101] Qin FJ, Wang F, Lu H, et al. Study on available silicon contents in cultivated land and its influencing factors in Ningbo City. Acta Agric Zhejiangensis 2012;24:263–7.

[102] Ma TS, Feng YJ, Liang YC, et al. Silicon supplying power and silicon fertilizer application in areas along Yangtze river. Soils 1994;26:154–6.

[103] Zhang YL, Li J, Liu MD, et al. A preliminary study of soil silicon status of paddy soils in Liaoning province. Chin J Soil Sci 2003;34:543–7.

[104] Wan XS, He DY, Liao XL. Silicon forms of soils in Hunan province in relation to soil properties. Soils 1993;25:146–51.

[105] Zhang XM, Zhang ZY, Yin KS, et al. Study on the available silicon content and its relationship with soil physico-chemical properties. J Heilongjiang August First Land Reclam Univ 1996;8:42–5.

[106] Dai GL, Duanmu HS, Wang Z, et al. Study on characteristics of available silicon content in Shaanxi province. J Soil Water Conserv 2004;18:51–3.

[107] Shen YZ, Pan WQ, Xu JB, et al. Dynamic changes of soil available silicon over last decade in Yangzhou city and application of silicon fertilizers to rice. Soil Fertil 1994;5:23–6.

[108] Yu QY, Li XL, Zhang YL. Available silicon distribution and affecting factors in paddy soils of Anhui province. J Anhui Agrotech Teachers Coll 1998;12:5–9.

[109] Yu QY, Li XL. Study on adsorption and desorption of silicon in soils. J Anhui Agrotech Teachers Coll 1999;13:1–6.

[110] Yang D, Zhang YL, Liu MD, et al. Characteristics of silicon releasing kinetics in greenhouse soil. Chin J Soil Sci 2012;43:42–6.

[111] Elgawhary SM, Lindsay LW. Solubility of silica in soils. Soil Sci Soc Am Proc 1972;36:439–42.

[112] Ding TP, Ma GR, Shui MX, et al. Silicon isotope study on rice plants from the Zhejiang province, China. Chem Geol 2005; 218: 41–50.

[113] Hartland A, Lead JR, Slaveykova VI, et al. The environmental significance of natural nanoparticles. Nat Educ Knowledge 2013;4(8):7.

[114] Zhang P, Guo Z, Zhang Z, et al. Nanomaterial transformation in the soil–plant system: implications for food safety and application in agriculture. Small 2020;2000705. Available from: http://doi.org/10.1002/smll.202000705.

[115] Ahmadian K, Jalilian J, Pirzad A. Nano-fertilizers improved drought tolerance in wheat under deficit irrigation. Agric Water Manag 2021;244:106544. Available from: http://doi.org/10.1016/j.agwat.2020.106544.

[116] Afshari M, Pazoki A, Sadeghipour O. Foliar-applied silicon and its nanoparticles stimulates physiochemical changes to improve growth, yield and active constituents of coriander (Coriandrum sativum L.) essential oil under different irrigation regimes. Silicon 2021. Available from: http://doi.org/10.21203/rs.3.rs-176146/v1.

[117] Lian M, Wang L, Feng Q, et al. Thiol-functionalized nano-silica for in-situ remediation of Pb, Cd, Cu contaminated soils and improving soil environment. Environ Pollut 2021;280:116879. Available from: https://doi.org/10.1016/j.envpol.2021.116879.

[118] Memari-Tabrizi EF, Yousefpour-Dokhanieh A, Babashpour Asl M. Foliar-applied silicon nanoparticles mitigate cadmium stress through physio-chemical changes to improve growth, antioxidant capacity, and essential oil profile of summer savory (*Satureja hortensis* L.). Plant Physiol Biochem 2021;165:71−9. Available from: https://doi.org/10.1016/j.plaphy.2021.04.040.

[119] Wang X, Jiang J, Dou F, et al. Simultaneous mitigation of arsenic and cadmium accumulation in rice (*Oryza sativa* L.) seedlings by silicon oxide nanoparticles under different water management schemes. Paddy Water Environ 2021. Available from: http://doi.org/10.1007/s10333-021-00855-6.

[120] Asadpour S, Madani H, Mohammadi GN, et al. Improving maize yield with advancing planting time and nano-silicon foliar spray alone or combined with zinc. Silicon 2020;. Available from: http://doi.org/10.1007/s12633-020-00815-5.

[121] Cao P, Qiu K, Zou X, et al. Mercaptopropyl trimethoxysilane- and ferrous sulfate-modified nano-silica for immobilization of lead and cadmium as well as arsenic in heavy metal-contaminated soil. Environ Pollut 2020;266(Pt 3):115152. Available from: https://doi.org/10.1016/j.envpol.2020.115152.

[122] Esmaili S, Tavallali V, Amiri B. Nano-silicon complexes enhance growth, yield, water relations and mineral composition in *Tanacetum parthenium* under water deficit stress. Silicon 2020. Available from: http://doi.org/10.1007/s12633-020-00605-z.

[123] Namjoyan S, Sorooshzadeh A, Rajabi A, et al. Nano-silicon protects sugar beet plants against water deficit stress improving the antioxidant systems and compatible solutes. Acta Physiol Plant 2020;42:157. Available from: http://doi.org/10.1007/s11738-020-03137-6.

[124] Wang Y, Liu Y, Zhan W, et al. Long-term stabilization of Cd in agricultural soil using mercapto-functionalized nano-silica (MPTS/nano-silica): a three-year field study. Ecotoxicol Environ Saf 2020;197:110600. Available from: http://doi.org/10.1016/j.ecoenv.2020.110600.

[125] Ali N, Réthoré E, Yvin J-C, et al. The regulatory role of silicon in mitigating plant nutritional stresses. Plants 2020;9:1779. Available from: https://doi.org/10.3390/plants9121779.

[126] Khan MIR, Ashfaque F, Chhillar H, et al. The intricacy of silicon, plant growth regulators and other signaling molecules for abiotic stress tolerance: an entrancing crosstalk between stress alleviators. Plant Physiol Biochem 2021;162:36−47. Available from: http://doi.org/10.1016/j.plaphy.2021.02.024.

[127] Ranjan A, Sinha R, Bala M, et al. Silicon-mediated abiotic and biotic stress mitigation in plants: underlying mechanisms and potential for stress resilient agriculture. Plant Physiol Biochem 2021;. Available from: http://doi.org/10.1016/j.plaphy.2021.03.044.

[128] Adrees M, Ali S, Rizwan M, et al. Mechanisms of silicon-mediated alleviation of heavy metal toxicity in plants: a review. Ecotoxicol Environ Saf 2015;119:186−97. Available from: http://doi.org/10.1016/j.ecoenv.2015.05.011.

[129] Imtiaz M, Rizwan MS, Mushtaq MA, et al. Silicon occurrence, uptake, transport and mechanisms of heavy metals, minerals and salinity enhanced tolerance in plants with future prospects: a review. J Environ Manage 2016;183:521−9. Available from: http://doi.org/10.1016/j.jenvman.2016.09.009.

[130] Etesami H, Jeong BR. Silicon (Si): review and future prospects on the action mechanisms in alleviating biotic and abiotic stresses in plants. Ecotoxicol Environ Saf 2018;147:881−96. Available from: https://doi.org/10.1016/j.ecoenv.2017.09.063.

[131] Bhat JA, Shivaraj SM, Singh P, et al. Role of silicon in mitigation of heavy metal stresses in crop plants. Plants 2019;8:71. Available from: https://doi.org/10.3390/plants8030071.

[132] Liu B, Soundararajan P, Manivannan A. Mechanisms of silicon-mediated amelioration of salt stress in plants. Plants 2019;8:307. Available from: https://doi.org/10.3390/plants8090307.

[133] Gaur S, Kumar J, Kumar D, et al. Fascinating impact of silicon and silicon transporters in plants: a review. Ecotoxicol Environ Saf 2020;202:110885. Available from: http://doi.org/10.1016/j.ecoenv.2020.110885.

[134] Sahebi M, Hanafi MM, Akmar SNA, et al. Importance of silicon and mechanisms of biosilica formation in plants. Biomed Res Int 2015;16:396010. Available from: http://doi.org/10.1155/2015/396010.

[135] Boroumand N, Behbahani M, Dini G. Combined effects of phosphate solubilizing bacteria and nanosilica on the growth of land cress plant. J Soil Sci Plant Nutr 2020;20:232−43. Available from: http://doi.org/10.1007/s42729-019-00126-8.

[136] Verma KK, Liu XH, Wu KC, et al. The impact of silicon on photosynthetic and biochemical responses of sugarcane under different soil moisture levels. Silicon 2020;12:1355−67.

[137] Verma KK, Song XP, Zeng Y, et al. Characteristics of leaf stomata and their relationship with photosynthesis in *Saccharum officinarum* under drought and silicon application. ACS Omega 2020;5(37):24145−53.

[138] Verma KK, Wu KC, Singh P, et al. The protective role of silicon in sugarcane under water stress: photosynthesis and antioxidant enzymes. Biomed J Sci Tech Res 2019;15(2):002685. Available from: https://doi.org/10.26717/BJSTR.2019.15.002685.

[139] Verma KK, Singh RK, Song QQ, et al. Silicon alleviates drought stress of sugarcane plants by improving antioxidant responses. Biomed J Sci Tech Res 2019;002957. Available from: https://doi.org/10.26717/BJSTR.2019.17.002957.

[140] Qados AMA, Moftah AE. Influence of silicon and nano-silicon on germination, growth and yield of faba bean (*Vicia faba* L.) under salt stress conditions. J Exp Agric Int 2015;509−24. Available from: https://doi.org/10.9734/AJEA/2015/14109.

[141] Ghorbanpour M, Mohammadi H, Kariman K. Nanosilicon-based recovery of barley (*Hordeum vulgare*) plants subjected to drought stress. Environ Sci Nano 2020;7(2):443−61.

[142] Suriyaprabha R, Karunakaran G, Yuvakkumar R, et al. Growth and physiological responses of maize (*Zea mays* L.) to porous silica nanoparticles in soil. J Nanopart Res 2012;14(12):1−14. Available from: https://doi.org/10.1007/s11051-012-1294-6.

[143] Elsharkawy MM, Mousa KM. Induction of systemic resistance against Papaya ring spot virus (PRSV) and its vector Myzuspersicae by *Penicillium simplicissimum* GP17-2 and silica (SiO_2) nanopowder. Int J Pest Manag 2015;61(4):353−8. Available from: https://doi.org/10.1080/09670874.2015.1070930.

[144] Hatami M, Khanizadeh P, Bovand F, et al. Silicon nanoparticle-mediated seed priming and *Pseudomonas* spp. inoculation augment growth, physiology and antioxidant metabolic status in *Melissa officinalis* L. plants. Ind Crops Prod 2021;162:113238. Available from: https://doi.org/10.1016/j.indcrop.2021.113238.

[145] Emamverdian A, Ding Y, Mokhberdoran F, et al. Determination of heavy metal tolerance threshold in a bamboo species (*Arundinaria pygmaea*) as treated with silicon dioxide nanoparticles. Glob Ecol Conserv 2020;24:e01306. Available from: https://doi.org/10.1016/j.gecco.2020.e01306.

[146] Mahmoud LM, Dutt M, Shalan AM, et al. Silicon nanoparticles mitigate oxidative stress of *in vitro*-derived banana (*Musa acuminata* 'Grand Nain') under simulated water deficit or salinity stress. S. Afr. J. Bot. 2020;132:155−63. Available from: https://doi.org/10.1016/j.sajb.2020.04.027.

[147] de Sousa A, Saleh AM, Habeeb TH, et al. Silicon dioxide nanoparticles ameliorate the phytotoxic hazards of aluminum in maize grown on acidic soil. Sci Total Environ 2019;693:133636. Available from: https://doi.org/10.1016/j.scitotenv.2019.133636.

[148] Suciaty T, Purnomo D, Sakya AT. The effect of nano-silica fertilizer concentration and rice hull ash doses on soybean (*Glycine max* (L.) Merrill) growth and yield. IOP Conf Ser Earth Environ Sci 2018;129(1):012009. Available from: https://doi.org/10.1088/1755-1315/129/1/012009.

2

Silicon and nano-silicon: New frontiers of biostimulants for plant growth and stress amelioration

Mahima Misti Sarkar[1], Piyush Mathur[2] and Swarnendu Roy[1]

[1]Plant Biochemistry Laboratory, Department of Botany, University of North Bengal, Darjeeling, India
[2]Microbiology Laboratory, Department of Botany, University of North Bengal, Darjeeling, India

2.1 Introduction

Silicon (Si)—the second most abundant element in the earth's crust only after oxygen—is mainly composed of silicates [1]. Though Si has not been considered an essential element for plant growth and development, but from several research works it is evident that Si is beneficial to plants, especially when the plants are challenged by various environmental stresses [2]. Approximately 28% of the total soil weight is made up of Si, which mainly consists of silicon dioxide (SiO_2), silicate minerals, and aluminosilicates. But none of these can be directly absorbed by plants; Si can be only up taken in the form of monosilicic acid [3]. The uptake of Si by root and its transportation throughout the shoot are regulated by some transporters known as Lsi1, Lsi2, Lsi3, and Lsi6 [4]. Apart from the active transport of Si by these transporters, passive transport is also found to be involved in the transportation of Si in some plants like rice, barley, wheat, maize, pumpkin, cucumber, and soybean [5]. Silicon in plants gets deposited in almost all the parts viz. beneath the cell wall, as phytoliths and Si bodies in the leaves, and trichomes except the actively growing regions [6]. External application of Si has also been shown to impart positive physiological, biochemical, and molecular changes in plants, thereby providing tolerance against a number of abiotic and biotic stresses [7–9].

Application of nanotechnology in the field of agriculture as fertilizers, herbicides, pesticides, and formulations has been known to improve the quality of plants and plant growth and biomass [10]. In this context, nano-silicon is gradually emerging as one of the most potent nanoparticles that are able to stimulate plant growth and boost their ability to adapt under various abiotic and biotic stresses [11]. Nano-silicon also enhances the uptake of Si as well as other beneficial minerals into the plants in an energy-independent manner [2,6,12]. Moreover, the structural flexibility of silica nanoparticles in the form of different shapes and sizes, porosity, etc. enables it to be used as a potential delivery agent of biomolecules and agrochemicals [13–15]. The presence of silanol groups on the surface facilitates the functionalization with a virtually wide range of biomolecules that adds to the beneficial attributes of nano-silicon [16].

The term "biostimulants" has been attributed to any natural or synthetic substances, which can modify the physiological and biochemical processes of plants in a positive way by influencing the hormonal status, nutrient efficiency, and metabolic processes [17]. Considering the immense potential of Si and nano-silicon, they are being projected as new age biostimulants that can not only promote plant growth but also help in mitigating environmental stresses. Silicon and nano-silicon accumulation underneath the cell wall provides mechanical support and resistance to pathogen attack [18]. The ability of Si in enhancing resistance against environmental stresses has been attributed to its ability to regulate the biosynthesis and signaling of phytohormones [19]. Silicon and nano-silicon can induce systemic acquired resistance by increasing the biosynthesis of salicylic acid (SA), a defense hormone thereby enhancing disease resistance [18]. Under osmotic stress, Si has also been known to promote the accumulation of abscisic acid (ABA) that has a direct effect on the upregulation of stress-responsive genes related to stomatal closure, increased water-use efficiency, and decreased rate of transpiration [20].

The present chapter will therefore elaborate the role of Si and nano-silicon as a potent biostimulant for supporting plant growth and development along with their role in alleviation of various abiotic and biotic stresses. This chapter will also focus on the functional attributes of these biostimulants and will provide valuable insights into the role of different phytohormone and molecular responses governing their efficacy in plant growth and stress resilience.

2.2 Prospect of silicon and nano-silicon as biostimulants

Biostimulants may be defined as any nonnutrient substances, which may be organic or inorganic, or any formulated product that can promote plant growth, development, nutrition efficiency, stress tolerance, and crop quality. In addition, they are expected to confer novel properties providing sustainable and cost-effective solutions to improve agricultural practices and crop productivity [21−23]. According to Du Jardin [21], there are many different categories of biostimulants, such as soil organic matter obtained from plant and animal decomposition and microbial residues (humic and fulvic acids); protein hydrolysates and other N-containing compounds (peptides and amino acids obtained from agro-industrial by-products); seaweed extracts and botanicals (purified seaweed and plant extract); chitosan and other biopolymers (biopolymers of chitin, chitosan, laminarin); beneficial fungi (mycorrhiza); plant growth-promoting rhizobacteria (PGPR) and inorganic compounds or chemical elements except the macronutrients and micronutrients.

Among these biostimulants, humic and fulvic acids are known to positively modulate plant growth by increasing micronutrient availability to plant by forming soluble or intact complexes with micronutrients facilitating the easy uptake by plants [24,25]. On the other hand, binding of chitosan with specific cell receptors results in accumulation of hydrogen peroxide and Ca^{2+} in the cell, which are expected to cause several physiological changes and play a key role in the regulation of plant developmental and stress-responsive pathways [26,27]. PGPRs (*Rhizobium*, *Bradyrhizobium*, *Azospirillum*, *Azotobacter*, *Pseudomonas*, and *Bacillus*) are also known to be associated with plant roots and often form a protective biofilm layer on the root surface, facilitating increased water retention and also aid in enhancing tolerance to ionic and osmotic stresses [28].

Apart from the known elements (macronutrients and micronutrients) that play a vital role in stimulating plant growth and development, several others are known for their emerging roles in plant growth, development, yield, as well as adaptation to various challenging environments. These include silicon (Si), aluminum (Al), cerium (Ce), cobalt (Co), sodium (Na), iodine (I), lanthanum (La), selenium (Se), titanium (Ti), and vanadium (V) [29]. Most of these elements are biphasic in their dose response to plants, which means that they are beneficial to plants when applied in a lower dose but can impart considerable toxicity at relatively higher concentrations [29]. Among these elements, Si is one of the most nontoxic elements and also known to mitigate various environmental stresses by modulating various metabolic processes and gene expressions along with phytohormone signaling [19]. Silicon is known to enhance water and nutrient uptake along with its efficacy in cell wall strengthening by increasing cellulose and lignin content, thereby supporting plant growth and resistance against environmental stresses [30−32]. Therefore Si can be considered an efficient biostimulant and its nano-form added for the several beneficial attributes for plant growth and development (Fig. 2.1). The application of nano-silicon has also been observed to modify the composition and nutritional quality of crop plants as well as provide enhanced tolerance toward various stresses [33]. Nano-silicon in most of the cases was observed to be more effective against various abiotic and biotic stresses than the bulk Si and it is known to be easily absorbed by the plants due to its smaller size [34]. Similarly, nano-silicon is also involved in promoting nutrient uptake and nutrient efficacy for better growth and productivity [3]. Most probably Si is the only element that can alleviate the detrimental effects of multiple environmental stresses [35]. On the other hand, nano-silicon (mesoporous type) can be used for targeted and specific delivery of pesticides, herbicides, and fertilizers. In this context, nano-silicon acts as an efficient carrier, thereby reducing the wastage of agrochemicals by ensuring targeted delivery into the plant system [10].

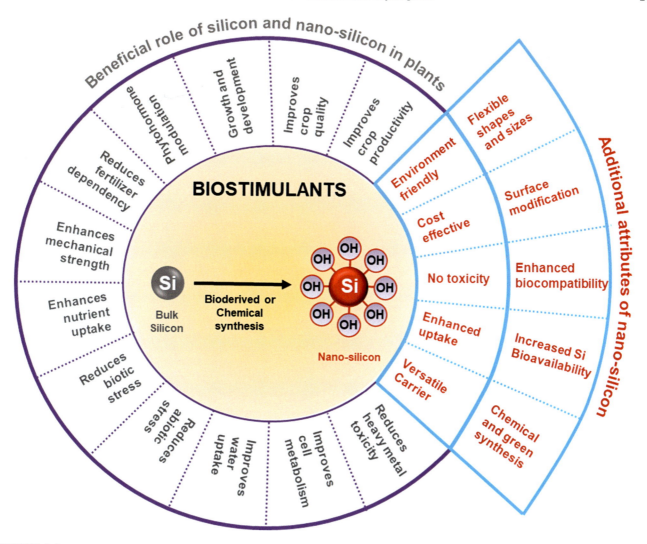

FIGURE 2.1 Attributes of silicon and nano-silicon that establish them as emerging biostimulants for crop growth and development. The additional features of nano-silicon that makes it a valuable proponent for the development of new age biostimulants are also being presented.

2.3 Silicon: an underestimated element for plant growth

There are six macronutrients viz. nitrogen (N), phosphorus (P), potassium (K), sulfur (S), calcium (Ca), and magnesium (Mg), and seven micronutrients viz. iron (Fe), manganese (Mn), zinc (Zn), boron (B), copper (Cu), molybdenum (Mo), and chloride (C1), which have been considered essential for plant growth, development, and yield [36]. Macronutrients are present in a high concentration in the soil and micronutrients are equally essential but present in a lower concentration. Macronutrients and micronutrients are essential for plants because in the absence of these elements, a plant is unable to complete its life cycle properly, also they are directly involved in plant metabolism and their functions cannot be replaced with any other nutrients [37].

Silicon is an element that bears the physical and chemical properties between metals and nonmetals. Silicon is considered to be a quasi-essential element that improves crop yield by promoting some physiological processes and also helps the plant to adapt to different environmental stresses [2,38]. It has been known that Si content in plants may vary from 0.1% to 10% [39]. Silicon is taken up and distributed throughout the plant body by some transporters known as Lsi1, Lsi2, Lsi3, and Lsi6. Both Lsi1 (influx transporter) and Lsi2 (efflux transporter) are present in the root epidermis and help in Si uptake from the soil [40]. Lsi6 is responsible for xylem loading and Lsi2, Lsi3, and Lsi6 are together involved in intravascular silicon transport in the nodal regions [41]. The role of Si in promoting plant growth and stress alleviation has been discussed further in the following sections and also depicted in Fig. 2.2.

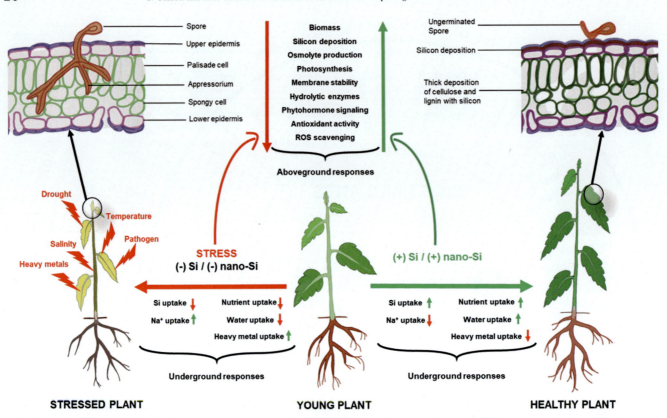

FIGURE 2.2   Effect of silicon and nano-silicon application on plants. The diagrammatic representation shows the impact of different types of stresses on a plant (shown on the left), in the form of poor growth and development, chlorosis, pathogen spread, decreased membrane integrity, ROS production, etc. But silicon/nano-silicon treatment effectively negates these negative effects as shown in the plant (on the right) and induces several positive attributes to defend the plant against stress. These include silica deposition, reduced uptake of toxic ions, and ROS scavenging. The overall underground and aboveground responses of a plant in the presence and absence of silicon and/or nano-silicon are also being presented.

2.3.1 Silicon in plant growth and development

Positive effects of Si on plant growth and development via modulation of physiological and biochemical processes have been demonstrated previously by Abdel-Haliem et al. [12]. They have reported that the application of Si to rice plants resulted in an increased accumulation of osmolytes (proline and soluble carbohydrate), a decrease in oxidative stress indicators (malondialdehyde, plasma membrane permeability, and hydrogen peroxide), and an increase in antioxidative enzyme activities, all of which indicated health status of the plants. Similarly, Hegazy et al. [42] have also reported that exogenous application of Si can improve photosynthetic pigments, proline content, antioxidant enzyme activity along with the decrease in malondialdehyde, and H_2O_2 content of rice plants. Germination parameters viz. germination percentage, germination rate, and mean germination time and growth parameters viz. plant height, leaf number, and leaf area were also found to be increased in *Vicia faba*, treated with Si [43]. Silicon was also known to improve the hypocotyl length, stem diameter, and the flowering rate of *V. faba* [44]. The beneficial effects of Si on plant growth and development can be realized by their ability to form a protective layer by Si deposition, interaction with heavy metal ions, and alteration of plants metabolism under stress [45]. For example, *Low silicon 1 (BdLsi1−1)* mutants of *Brachypodium distachyon* plants were reported to have lower Si uptake and availability as a result of which architectural changes in the cell wall were observed [46]. This study therefore indicated the role of Si in cell wall biosynthesis and development. Moreover, Si plays an important role in the regulation of transpiration rate and hydraulic conductance of roots, thereby improving the efficiency of water usage in plants. Silicon also helps to enhance water uptake by inducing osmotic adjustment, improving the activity of aquaporins, and increasing the root-by-shoot ratio [31].

2.3.2 Role of silicon in stress alleviation

2.3.2.1 Abiotic stress

Application of Si has also been known to alleviate a number of abiotic biotic stresses viz. salinity, drought, temperature, UV, nutrient imbalance, metal toxicity, and many more [47] (Table 2.1). Drought and salinity stress are the major factors that exert pervasive and detrimental effects upon agricultural productivity. Water uptake is significantly reduced due to drought and salinity stress but Si has been known to improve water status and water-use efficiency [48]. In this connection, Si application has been known to increase root and whole plant hydraulic conductance, decrease stomatal conductance and transpiration, and thereby improve leaf water content of sorghum plants under osmotic stress [49,50]. Silicon is also known to alleviate salinity and drought-induced oxidative stress by stimulating the antioxidative defense system of plants. As a result, activities of enzymes like catalase (CAT), peroxidase (POD), superoxide dismutase (SOD), ascorbate peroxidase (APX), dehydroascorbate reductase, guaiacol peroxidase, and glutathione reductase (GR) and nonenzymatic antioxidants like ascorbic acid (AsA), glutathione, phenolic content, carotenoid, and nonprotein amino acids increased, which indicated toward the improvement in stress tolerance [51,52]. Also, osmotic stresses lead to nutrient imbalance in plants by reducing the uptake and transport of nutrients into the plants, which are essential for plant growth and development [53,54]. In this connection, Si can also help in improving the uptake, transport, and distribution of minerals throughout the plant [55,56]. Moreover, Si can mitigate salinity and drought stress by modifying the amount of osmolytes like proline, glycine betaine, total soluble sugars, carbohydrates, and polyols, which help in reducing osmotic shock induced by stress [9]. Si is also capable of reducing the ion toxicity induced by excess salt. Application of Si assists the plants in lowering the concentration of Na^+ in cytosol (through activation of plasma membrane SOS1 transporter and NHX1 in tonoplast) and increasing the uptake of K^+ (through K^+/H^+ symporters—HAK1) [55,56].

Silicon reduces the damages incurred by water-logging in plants as enhanced uptake of Si facilitated the accumulation of silica bodies in vacuoles. At the same time, Si uptake ensures optimum light absorption by keeping the leaf blades erect, thereby increasing the photosynthetic rate [40]. Moreover, increased deposition of Si in the roots reduces the uptake of toxic metals by reducing apoplast flow [40]. Si also helps in amelioration of heavy metal stresses induced by Zn, Al, Mn, and Cd in plants by stimulating the activity of enzymatic and nonenzymatic antioxidants [57]. Si alleviates the combined deleterious effects of drought and UV-B radiation in plants by upregulating the antioxidative defense system, reduction of UV-B absorption, and activating photolyase that helps in repairing DNA damage [58].

2.3.2.2 Biotic stress

Apart from abiotic stresses, biotic stresses are also known to be alleviated by the application of silicon. Biotic stresses of crops include several diseases like damping off, leaf spots, leaf blights, galls, root rots, powdery mildews, wilts, and rusts caused by different species of fungus, bacteria, oomycetes, nematodes, and viruses [59,60] (Table 2.1). Silicon provides resistance against biotic stress by two mechanisms — first by enhancing the deposition of Si underneath the cell wall that provides a physical barrier and second by imparting mechanical resistance to inhibit invasion of pathogen deep inside the plant tissues [39]. Application of Si has been known to reduce anthracnose disease of tomato and sweet paper by increasing the cuticle thickness and firmness of fruits [61,62]. Accumulation of absorbed Si in the epidermal layer of the mango plants also provided a physical barrier, thereby preventing the entry of *Pseudomonas syringae* pv. *Syringae* [63]. Silicon application in sorghum plants also resulted in the prevention of anthracnose severity caused by *Colletotrichum sublineolum* [64]. Foliar spray of potassium silicate on coffee leaf resulted in reduction of leaf rust caused by *Hemileia vastatrix*. Deposition and polymerization of absorbed Si on the upper surface of the leaf provided a physiological barrier against the pathogen [65]. Rice plants treated with Si accumulate electron-dense Si along with cellulose microfibrils in their cell walls, which prevented the colonization of *Pyricularia oryzae* [66]. Silicon treatment kept *Ganoderma boninense* infection below threshold levels by preventing the pathogen from entering and spreading throughout the plant and thus reducing the severity of basal stem rot disease [67].

In addition to providing physical barrier to pathogens, Si also provides resistance against pathogen attack by inducing the production of natural defensive compounds including phenolics, lignins, and phytoalexins [39,68,69]. Silicon application induces defense in cucumber plants by accumulating phenolic compounds in response to the infection by *Podosphaera xanthii* [70]. Silicon treatment increased the concentration of antifungal methylated forms of trans-aconitate (a phytoalexin) in wheat plants and thus provided resistance against powdery mildew disease [71]. Silicon also stimulated the deposition of dopamine, lignin, phenolics, and flavonoids in

TABLE 2.1 List of studies depicting the stress alleviating potential of Si against several abiotic and biotic factors during the last 5 years.

Chemical form of silicon application	Type of abiotic stress/pathogen	Plant/host	Alleviating effects in plants	Reference
Abiotic stresses				
Silicon	Salinity	*Glycine max* cv. Daewon (Soybean)	Improved chlorophyll content; reduced APX and GR activity; upregulation of NO and antioxidant genes; downregulation of S-nitrosothiol; and reduced adverse effect of salinity stress	Chung et al. [75]
Sodium silicate (Na_2SiO_3)	Salinity + cadmium	*Phoenix dactylifera* (Palm)	Decreased metal uptake and enhanced micronutrient uptake; decreased lipid peroxidation rate, POX and CAT activity; significantly increased APX and SOD; decreased ABA, SA, and JA; and reduced Cd toxicity resulted in improved biomass	Khan et al. [76]
Sodium silicate (Na_2SiO_3) + *Rhizophagus clarus* (AMF)	Drought	*Fragaria × annasa* var. Paros (Strawberry)	Increased plant biomass, photosynthetic rate, antioxidant enzyme (CAT, SOD, and POD) activity; improved water and nutritional status; decreased H_2O_2 content; and alleviated water stress with silicon and AMF synergistically	Moradtalab et al. [77]
Sodium metasilicate (Na_2SiO_3)	Drought	*Zea mays* cv. Pearl and cv. Malka (Maize)	Enhanced shoot and root length and biomass; improved photosynthetic pigments; increased SOD, POD, and CAT activities; reduced MDA and H_2O_2 content; and mitigated drought stress leads to maize growth recovery	Parveen et al. [78]
Sodium metasilicate (Na_2SiO_3)	Salinity	*Puccinellia distans*	Improved plant dry weight and water status; increased soluble sugar and amino acids; decreased Na^+ by secretion from leaf; enhanced H^+-ATPase activity; decreased proline and electrolyte leakage; and better performance of plants under stressed conditions	Soleimannejad et al. [79]
Potassium silicate (K_2SiO_3)	Heavy metal (Cd and Zn)	*Oryza sativa* (Rice)	Increased root and shoot dry weight; decreased Cd and Zn concentration in plants, especially in grains; increased chlorophyll content and SOD enzyme activity; decreased plasma membrane permeability, MDA content, CAT and POD enzyme activity; and significantly reduced heavy metal stress	Huang et al. [80]
Metasilicic acid (H_2SiO_3)	Salinity	*Zea mays* (Maize)	Increased shoot and root dry matter; declined Na^+ and enhanced K^+; improved SOD, APX, and CAT activities; improved soluble protein and chlorophyll content; and alleviate salinity-induced osmotic and oxidative stress	Khan et al. [81]
Sodium metasilicate (Na_2SiO_3)	Drought (PEG-6000)	*Lens culinaris* (Lentil)	Improved germination traits; decreased osmolytes (proline, glycine betaine, and sugar), ROS (H_2O_2 and superoxide), and MDH content; increased hydrolytic enzymes (α-amylase, β-amylase, and α-glucosidase); increased APX, POX, SOD, and CAT enzyme activity; and increased silicon content	Biju et al. [82]
Potassium silicate (K_2SiO_3)	Drought	*Mangifera indica* (Mango)	Increased growth and physiological parameters; improved IAA, GA, and CK phytohormones level; reduced ABA content and antioxidant activity; and reduced drought stress through reactive oxygen species scavenging	Helaly et al. [83]

(Continued)

TABLE 2.1 (*Continued*)

Chemical form of silicon application	Type of abiotic stress/ pathogen	Plant/host	Alleviating effects in plants	Reference
Potassium, sodium and calcium silicates (K_2SiO_3, Na_2SiO_3, and $CaSiO_3$)	High temperature	*Fragaria × annasa* (Strawberry)	Retained expression of photosynthetic proteins (PsaA and PsbA); abundantly increased SOD, APX, and CAT; and reduced negative effect of high-temperature stress	Muneer et al. [84]
Potassium silicate (K_2SiO_3)	Salinity + drought (PEG-6000)	*Glycyrrhiza uralensis*	Increased CAT and APX activity; increased glutathione and proline content; decreased membrane permeability and MDA; and reduced adverse effect of salinity and drought stress	Zhang et al. [85]
Biotic stresses				
Potassium silicate (K_2SiO_3)	*Alternaria solani* (Early blight)	*Lycopersicon esculentum* cv. Shalimar-2	Increased defense gene expression level [JA (*PR3*, *LOXD*, and *JERF3*) and SA (*PR1*, *PR2*) marker genes]; increased antioxidant enzyme (SOD, CAT, APX, GR, and POD) activities; and improved defense against pathogen	Gulzar et al. [86]
Soluble silicic acid (SSA) and diatomaceous earth (DE)	*Sesamia inferens* (Pink stem borer PSB)	*Triticum aestivum* cv. GW 273 (Wheat)	Increased photosynthetic rate; decreased transpiration rate; enhanced stomatal conductance; maximized water-use efficiency; decreased intercellular CO_2 concentration; significantly increased yield parameters; and suppressed negative effects of the PSB through silicon accumulation in the stem	Jeer et al. [87]
Sodium silicate (Na_2SiO_3)	*Golovinomyces cichoracearum* (Powdery mildew)	*Arabidopsis thaliana*	Low leaf silicon content induced SA-dependent resistance; high leaf silicon content induced PAD4-dependent but most importantly EDS1and SA-independent resistance; and distinct and multilayered defense to make the plant disease resistance through assimilated silicon	Wang et al. [88]
Potassium silicate (K_2SiO_3)	*Colletotrichum sublineolum* (Anthracnose)	*Sorghum bicolor* (Great millet)	Significantly reduced disease severity ratings due to silicon-induced resistance against anthracnose	De Lima et al. [89]
Sodium metasilicate (Na_2SiO_3)	*Sclerotinia sclerotiorum* (Postharvest rot disease)	*Daucus carota* cv. Chantenay (Carrot)	Rot of carrot reduced significantly due to high inhibition of mycelial growth (92.2%) with inhibition of sclerotia formation (76.3%) and increased POD, PPO, and PAL activity	Elsherbiny and Taher [90]
Silicon derived from *Momordica charantia*	*Pseudoperonospora* sp. (Downy mildew)	*Momordica charantia* L. (Bitter gourd)	Increased number of leaves and flower and fruit; increased phenol and chlorophyll content; significantly increased POD and PPO activity; and increased disease resistance through increased silicon accumulation in the plant	Ratnayake et al. [91]
Potassium silicate (K_2SiO_3)	*Ralstonia solanacearum* (Bacterial wilt)	*Solanum lycopersicum* (Tomato)	Increased silicon content helped in increased root dry weight, leaf photosynthesis, and reduced citric acid content; and suppressed bacterial wilt in HYT tomato variety	Xue-ying et al. [92]
Potassium silicate (K_2SiO_3)	*Calonectria ilicicola* (Black root rot) and *Phytophthora cinnamomic* (Phytophthora root rot)	*Persea americana* (Avocado)	Potassium silicate treatment did not show any improved visual health; potting mix with MM and MD resulted in visible decline in pathogenic infection; and accumulation of silicon in the leaves and fruit peels induced resistance against the pathogens	Dann and Le [93]

banana roots infected by *Fusarium oxysporum* f. sp. *Cubense* [72]. Silicon is also known to induce resistance against pathogens by increased production of defense enzymes like chitinases, β-1,3-glucanases, polyphenol oxidases (PPO), phenylalanine ammonia lyase (PAL), POD, APX, SOD, CAT, GR, lipoxygenase, and glucanase [9]. At the same time, herbivore attack can also be defended by exogenous Si application. For example, crops primed with Si (calcium silicate) increased the mortality rate of whitefly nymphs, which can otherwise reduce crop production [73]. The attack of green leafhopper, plant hopper, and stem maggots on rice plants was also reported to be inhibited by the solicitation of silicon [74]. Caterpillars were also found to be inhibited by silicate [74].

2.4 Emerging role of nano-silicon

Nanomaterials are rapidly gaining importance in various fields including agricultural sector, where they have already been explored for their promising role in imparting better growth, development, and yield [94–96]. Nanoparticles have been used satisfactorily to increase crop yield by reducing the losses incurred due to several environmental stresses viz. salinity, drought, high light intensity, heat, heavy metals, pathogens, and herbivores [97]. The biological role of the nanoparticles depends upon their physiochemical properties, applied concentrations, and application methods [97]. In this context, the application of nano-silicon that appears to be beneficial than bulk silica owing to its superior ability to promote growth, development, and alleviation of environmental stresses to increase crop yield [98–100]. The presence of silanol group on the nano-silicon surface makes it highly versatile with respect to its functionalization with other molecules in order to improve their targeted delivery. Also, the surface chemistry helps nano-silicon to easily interact with the phospholipid bilayer and facilitates endocytosis [101]. Furthermore, the innumerous possibilities of modifying the surface of nano-silicon make it a suitable delivery agent for essential biomolecules viz. proteins, nucleotides, and other compounds in plants that can precisely confer beneficial attributes to the plants [10].

Nano-silicon has been believed to possess greater potential than all the other nanoparticles known with respect to its applications in the agricultural sector. When nano-silicon, nano-zinc, nano-titanium, and nano-aluminum oxides were used against *Lipaphis pseudobrassicae* (major pest of *Brassica* sp.), it was observed that both nano-silicon and nano-aluminum could kill the pest up to a significant level [102]. But nano-aluminum was observed to inhibit the growth of carrot, cabbage, cucumber, soybean, and corn, which is not desirable for crop production. On the contrary, nano-silicon does not exhibit such toxicity against plant growth and development [103]. In another study, the application of silver nanoparticles was shown to increase seed germination of *Thymus kotschyanus* at lower concentration; however, higher concentration had a negative effect. Nano-silicon, on the other hand, effectively increased the germination traits even in higher concentrations, which clearly indicated toward lesser toxicity of nano-silicon compared to other nanomaterials [104]. These studies clearly pointed out the advantages of using nano-silicon over the other nanoparticles for improving plant growth and development which has also been depicted in Fig. 2.2.

2.4.1 Nano-silicon in plant growth and development

From several recent studies, it has been clear that nano-silicon improves the morphological and physiological attributes leading to improved growth and yield [105,106]. Both foliar and soil application of nano-silicon resulted in improvement of the vegetative traits of marigold plants. Soil treatment with 200 mg/L nano-silicon displayed better plant height, root length, and number of branches and foliar treatment with the same concentration displayed a significant increase in the floral traits viz. number of flowers, flower's diameter, and fresh and dry masses of flower in comparison to the untreated plants [107]. Similarly, rice plants irrigated every third and sixth days with nano-silicon showed significant increment in dry matter production, leaf area index, number of panicles per square meter, number of filled grains per panicle, grain yield, grain weight, and chlorophyll content [108]. In a different study, 8 g/L concentration of nano-silicon was found to increase the rate of seed germination, seed germination index, mean germination time, seed vigor index, and seedling fresh weight and dry weight of *Lycopersicum esculentum* [109]. Chlorophyll content, mean plant height, root length, and the number of lateral roots were also increased in *Larix olgensis* plants, treated with nano-silicon [110]. Khan and Ansary [111] emphasized that lower concentration (25 and 50 μg/mL) of nano-silicon was sufficient to promote seed germination, vigor index, and biomass in *Lens culinaris*. Similarly, 5, 10, and 15 kg ha^{-1} concentration of nano-silicon resulted in enhanced proteins, chlorophyll, and phenol contents of maize plant but 20 kg ha^{-1} showed no significant increments [112]. This implies that nano-silicon applications in lower concentrations are effective in sustaining plant growth and productivity. Hypocotyl length and flowering of *V. faba* plants were also reported to be increased in response to nano-silicon treatment [44]. Nano-silicon also increased germination and growth of *Cucumis sativus*, where final germination percentage, germination speed, coefficient of velocity of germination, vigor index, germination index and relative water content of germinant, and root and shoot length and dry weight of seedling were observed to be increased when compared to the control plants [113].

2.4.2 Role of nano-silicon in stress alleviation

2.4.2.1 Abiotic stresses

Abiotic stresses in crop plants incur major losses to yield due to reduced yield and production. Nano-silicon application to the stressed plants has been reported to confer improvement in the performance by improving the physiological and biochemical status of the plant [11] (Table 2.2). Seed priming and seedling treatment with nano-silicon improved growth and resulted in increased germination percentage, germination speed, vigor index, and also the root and shoot length of seedlings of saline-stressed common bean plants [114]. Similar observations were recorded in the case of saline-stressed lentil plants when treated with nano-silicon along with the improvement in seedling weight and cotyledon reserve mobilization [115]. Nano-silicon in combination with exopolysaccharide (EPS)-producing bacterium viz. (*Citrobacter freundii*) and the extracted EPS resulted in increased shoot and root fresh and dry weight and decreased proline content and antioxidant activity of salinity-stressed tomato seedlings, thereby reducing the negative impacts of stress [116]. In saline-stressed basil (*Ocimum basilicum*), treatment of nano-silicon resulted in increased shoot fresh and dry weight and chlorophyll content along with the increase in proline content, which signifies the induction of tolerance in plants [117]. Most importantly, the application of nano-silicon was observed to increase shoot growth, chlorophyll content, along with increased K + uptake and reduction in Na + content, malondialdehyde content, and electrolyte leakage in saline-stressed banana plants in comparison to the control plants [118]. Similarly, nano-silicon also implied a positive role in maintaining physiological and biochemical processes of plants under drought-stressed conditions [119]. The beneficial effect of nano-silicon was also observed when drought-stressed *Prunus mahaleb* plants were pretreated with nano-silicon. Nano-silicon resulted in the restoration of nutritional status (content of N, P, and K) of stressed plants. At the same time, severe effects of stress upon photosynthesis, stomatal conductance, and transpiration rate were significantly reduced by the application of nano-silicon [120]. Furthermore, the synthesis of controlled release fertilizer containing nano-silicon in the core was employed to release nutrients in a control manner along with its enhanced potential to withhold a substantial amount of water in the soil, which helped the plants to cope up with the impacts of drought stress [121].

Apart from above stresses, nano-silicon application has also been shown to alleviate the toxicity of heavy metals in plants. For instance, nano-silicon application reduced Cd concentration in the plant body especially in the grains [122,123]. Nano-silicon also decreased reactive oxygen species (ROS) in the Cd-stressed plant by increasing the activities of antioxidant enzymes [122]. Increased expression of silicon transporter (Lsi1) and PAL was observed when Cd-stressed wheat plants were introduced with nano-silicon. Though the mechanism of increased PAL expression is not clear but the application of nano-silicon increased lignification of xylem and induced the opening of metaxylem, which clearly indicated toward an increased efficiency for nutrient translocation [124]. Nano-silicon reduced both Cd toxicity and drought stress of wheat plants by decreasing the concentration of Cd in plants and simultaneously decreasing the production of hydrogen peroxide, malondialdehyde, and electrolyte leakage and increasing SOD and peroxidase activities [125]. Nano-silicon was also reported to reduce arsenate (As) and chromium (Cr) toxicity of maize cultivars and *Pisum sativum* plants, respectively, by reducing the accumulation of As/Cr and oxidative stress and simultaneously upregulating beneficial nutrients and antioxidant activity [98,126].

2.4.2.2 Biotic stress

Nano-silicon has been known to play a significant role in disease management owing to the presence of antimicrobial and antifungal properties [127] (Table 2.2). Several studies have demonstrated the resistance mechanism of nano-silicon against various disease-causing biotic agents. For example, the severity of root rot disease of *Panax ginseng* (Korean ginseng) caused by the fungus *Ilyonectria mors-panacis* was reduced by 50% upon application of nano-silicon [128]. Nano-silicon reduced the disease severity by regulating and suppressing the expression of membrane-bound sugar efflux transporter known as SWEET (Sugars Will Eventually be Exported

TABLE 2.2 List of studies depicting the stress alleviating potential of nano-silicon against several abiotic and biotic factors during the last 5 years.

Type/size of nano-silicon	Type of abiotic stress/ pathogen	Plant/host	Alleviating effects in plants	Reference
Abiotic stresses				
Nano-silicon + *Bacillus* sp., *Azospirillum lipoferum* and *Azospirillum brasilense*	Drought	*Triticum aestivum* (Wheat)	Nano-silicon and PGPR individually had the potential of pronounced drought alleviation of wheat seedlings; combined application showed increased fresh and dry weight, chlorophyll a, b content; increased gas exchange attributes, nutrient uptake, osmolyte production; increased SOD, POD, and CAT activity; and reduced drought tolerance	Akhtar et al. [134]
Nano-silicon	Salinity	*Mangifera indica* (Mango)	Increased stem diameter, new leaf area, chlorophyll, and carotenoid content; increased mineral (N, P, and K) content, Ca and Mg content; decreased Na content; increased soluble carbohydrate, and phenol content; decreased proline, and DPPH IC_{50} (better antioxidant activity); and significantly reduced salinity symptoms	El-Dengawy et al. [135]
Nano-silicon (20−30 nm)	Flooding induced hypoxia	*Muscadinia Rotundifoli* (Muscadine grape)	Increased enzymatic (SOD, POD, and CAT), nonenzymatic (AsA, glutathione) and antioxidant activities and osmolyte (proline, GB) content; and increased mineral (N, P, K, and Zn) content. Hypoxia stress reduced through better antioxidant defense, osmo-protection and micronutrient accumulation	Iqbal et al. [136]
Nano-silicon	Drought	*Triticum aestivum* cv. HI 1544 (Wheat)	Increased seed germination, seedling growth and vigor, water uptake, and amylase activity; retained suitable balance between ROS production and antioxidant enzyme (POD, CAT, and SOD) activities; improved photosynthetic machinery with increased number of active reaction centers, high trapping, absorbance and electron transport rates; and increased biomass production with reduced drought stress	Rai-Kalal et al. [137]
Nano-silicon (20−35 nm)	Cadmium	*Triticum aestivum* cv. Ujala-2016 (Wheat)	Improved photosynthetic pigment, flavonoid, protein, phenolics, free amino acid, proline, and sugar content; increased APX, CAT, POD, and SOD enzymes activities; reduced endogenous H_2O_2 and MDA level; and better plant growth with reduced Cd-induced stress	Thind et al. [138]
Potassium nano-silicon	Drought	*Zea mays* (Maize)	Increased N, K, and Si in the seed and N, K, Cu, and Si in the shoot; and reduced negative effects of drought by providing resistance	Aqaei et al. [139]
Nano-silicon (20 nm)	Heavy metal (Cu, Mn, and Cd)	*Arundinaria pygmaea* (Bamboo)	Increased plant antioxidant defense enzyme (SOD and CAT) activities and chlorophyll content under Cu and Mn toxicity; thus reduced Cu and Mn toxicity through low accumulation of heavy metals and higher accumulation of silicon	Emamverdian et al. [140]
Nano-silicon synthesized from sodium silicate (Na_2SO_3) (20−35 nm)	Water deficit	*Beta vulgaris* (Sugar beet)	Increased chlorophyll content, net photosynthesis, flavonols (quercetin, rutin), glycine betaine content, and antioxidant enzyme (SOD, CAT, and guaiacol peroxidase) activities; reduced H_2O_2 and MDA; and reduced drought severity	Namjoyan et al. [141]

(Continued)

TABLE 2.2 (*Continued*)

Type/size of nano-silicon	Type of abiotic stress/pathogen	Plant/host	Alleviating effects in plants	Reference
Nano-silicon (10−20 nm)	Salinity	*Fragaria × anansa* cv. Camarosa (Strawberry)	Increased root shoot fresh and dry weight, chlorophyll, and carotenoid content; reduced proline level; maintained epicuticular wax layer; and reduced and limited adverse effect of stress	Avestan et al. [142]
Nano-silicon (20−30 nm)	Drought	*Hordeum vulgare* (Barley)	125 mg/L nano-silicon resulted in maximum shoot biomass; increased chlorophyll and carotenoid content with both 125 and 250 mg/L treatment during stress; increased osmolytes (sugar, protein, flavonoid, and phenol) and antioxidant enzyme activities; and decreased cellular injury (H_2O_2, MDA, and electrolyte leakage) during poststress recovery	Ghorbanpour et al. [143]
Nano-silicon synthesized from sodium silicate (Na_2SO_3) (60−100 nm)	Cadmium	*Oryza sativa* cv. Xiangzaoxian 45 (Rice)	Increased K, Mg, and Fe translocation from the uppermost nodes to rachis; and decreased Cd accumulation	Chen et al. [31]
Nano-silicon (20−30 nm)	Salinity	*Glycine max* cv. M7 (Soybean)	Increased antioxidant enzyme (CAT, SOD, POX, and APX) activities; decreased lipid peroxidation and ROS generation; improved root shoot dry weight, K^+ content, phenolic components, ascorbic acid, and α-tocopherol content; and reduced stress effects	Farhangi-Abriz and Torabian [144]
Nano-silicon	Salinity	*Cucumis sativus* (Cucumber)	Significantly increased plants height, number of leaves, fresh and dry weight of leaves, number of fruits/plants, fruit quality and total yield; reduced salinity stress through increased uptake and content of nitrogen and phosphorus; and decreased uptake and content of Na^+	Yassen et al. [145]

Biotic stresses

Nano-silicon (5−15 nm)	*Phelipanche aegyptiaca* (Broomrape infection)	*Pisum sativum* (Pea)	Improved photosynthesis; increased sugar consumption and metabolism; increased phenylalanine and lignin content, proline and sucrose content; decreased membrane lipid peroxidation and ROS production; and significantly ameliorated negative impacts of disease infection through higher silicon accumulation in pea root and shoot	Shabbaj et al. [146]
Nano-silicon (less than 50 nm)	*Sitophilus oryzae, Rhizopertha dominica, Tribolium castaneum,* and *Orizaephilus surinamenisis*	*Zea mays* (Maize)	Improved plant height, chlorophyll content, total yield and its components, and protein (%); increased peat mortality rate up to 100%; and nano-silicon acted as a fertilizer to improve growth characters and pesticide to defend pest attack	El-Naggar et al. [147]
Nano-silicon synthesized from sodium metasilicate (Na_2SO_3) (30−60 nm)	*Botrytis fabae* (Chocolate spot disease)	*Vicia faba* (Broad bean)	Significantly improved chloroplast, cell wall thickness; maintained cell organelles shape; increased protein, proline and total phenol contents, POD and PPO activity; and reduced disease severity and disease infection	Hasan et al. [148]
Nano-silicon (5 − 15 nm)	*Meloidogyne incognita, Pectobacterium betavasculorum,* and *Rhizoctonia solani* disease complex	*Beta vulgaris* (Beetroot)	Enhanced shoot and root dry weight, chlorophyll content, chlorophyll fluorescence characters, and defense enzyme (SOD, CAT, PPO, and PAL) activities; and reduced root galling, nematode multiplication and disease indices	Khan and Siddiqui [149]
Nano-silicon synthesized from rice straw (73.6 nm) and rice husk (133.7 nm)	*Magnaporthe oryzae* (Blast disease) and *Cochliobolus miyabeanus* (Brown spot disease)	*Oryza sativa* (Rice)	Improved length of panicle, number of filled grains per panicle, grain yield and harvest index; decreased number of unfilled spikelets per panicle; and suppressed disease infection	Gabr et al. [150]

Transporters), as a result of which sugar content reduced in the apoplast [128]. Nano-silicon treatment also suppressed the wilt disease of watermelon (*Citrullus lanatus*) caused by the pathogen *Fusarium oxysporum* f. sp. *niveum*. After getting treatment of engineered nano-silicon, plants significantly increased some of the defense genes like *PR1* (pathogenesis-related protein 1), *PPO* (558 polyphenol oxidase), and *PAO* (polyamine oxidase); and some general plant metabolism genes like *HMA1* (heavy metal APTase 1) and *CDS1* (phosphatidate cytidylyltransferase 1) [129]. In a different study, nano-silicon application and *Rhizobium cicero* inoculation were known to individually improve growth, photosynthetic content, and proline status in *Meloidogyne incognita*, *Rhizoctonia solani*, and *Fusarium solani* infected chickpea plants. However, the combined application of nano-silicon and *R. cicero* further improved the efficacy by reducing black root rot indices, wet root rot, galling, and population of *M. incognita* [130]. Nano-silicon also reduced the disease severity in maize plants by increasing resistance against *Aspergillus* and *Fusarium*, which was mainly attributed to the increasing phenol content and defense enzyme activity [106]. Mesoporous nano-silicon and chitosan-coated mesoporous nano-silicon suppressed disease severity caused by *Fusarium* in watermelon. In this connection, expression of several genes viz. *CSD1* (copper/zinc SOD 1), *MDH* (malate dehydrogenase), *PAO1* (polyamine oxidase 1), *PPO*, *PR1*, and *RAN1* (heavy metal-exporting ATPase) played an important part in reducing the disease burden and thereby increasing the plant yield [131].

Nano-silicon has proven efficacy not only to provide defense against pathogenic attack but also can be used as a pesticide and herbicide. Nano-silicon alleviated the negative effect of tomato plants caused by a parasitic weed *Orobanche* and showed increased growth, photosynthetic rate, higher root lignification, and enzymatic and nonenzymatic activity [132]. The growth of *Alternaria solani* was also shown to be inhibited by the application of mesoporous nano-silicon application. The exact mechanism of fungal infection was not clear but there could be some mechanisms like prevention of direct contact with pathogen DNA, breakdown of fungal cell wall, and induced cell lysis of fungus [133]).

2.5 Crosstalk with phytohormones for the elicitation of enhanced tolerance

Phytohormones are involved in various metabolic processes and help the plants to adapt to multifaceted environmental conditions via different mechanisms [151]. Several studies have demonstrated that Si mitigates abiotic and biotic stresses in plants by regulating the phytohormone levels [152]. The interaction of Si with some phytohormones like ABA, gibberellic acid (GA), auxin (AUX/IAA), cytokinin (CK), jasmonic acid (JA), ethylene (ET), and SA has been studied under stressed conditions [151]. ABA is known to control the water status of plants during stress conditions by regulating stomatal conductance [153], and JA is the key regulator of stress tolerance which synergistically or antagonistically interacts with other hormones like SA [154]. Under salinity stress, genes like *zeaxanthin epoxidase* (*ZEP*) and *9-cis-epoxycarotenoid dioxygenase* genes (*NCED1*, *NCED3*, and *NCED4*) that are involved in ABA biosynthesis are upregulated in the plant body. Moreover, when NaCl-stressed rice plants were treated with silicon, expression of *ZEP* and *NCED 1, 3, 4* was found to be further increased, thereby increasing the ABA concentration [155]. Also, JA content was observed to be decreased upon the application of silicon, but no significant change was observed in SA content. Thus it was concluded that Si provided resistance against salinity stress by upregulating the ABA biosynthesis genes while reducing the accumulation of JA [155]. On the contrary, Si application in saline-stressed cucumber germinant resulted in the inhibition of ABA biosynthesis genes (*NCED1* and *NCED2*) and gibberellin catabolism gene (*GA2ox*). As a result of this, Si-treated germinant maintained a low level of ABA and a higher level of GA and α-amylase, which helped them in breaking seed dormancy [156]. Similarly, Si treatment resulted in decreased accumulation of ABA, JA, and SA content in heavy metal (Cu and Cd) stressed rice plants [157]. Generally, ABA is known to increase during stress to reduce stomatal conductance, which affects photosynthesis and chlorophyll biosynthesis. Reduction of ABA in Si-treated Cu- and Cd-stressed plants indicated toward the amelioration of stress induced by the heavy metals. Similarly, reduction in SA and JA content also pointed toward the positive effect of Si by reducing ROS accumulation and metabolism of α-linolenic acid accounting for lower level of JA accumulation [157]. In saline-stressed rice plants treated with Si, accumulation of JA increased along with the decrease in ABA content. Low ABA and high JA were responsible for improved expression of Si uptake genes (*Lsi1* and *Lsi2*) and induction of antioxidant responses, thereby protecting the plants from the negative effects of salinity [12]. GA is directly involved in various mechanisms of seed germination and vegetative growth. Endogenous GA level was observed to be increased in saline- and water deficit-stressed cucumber plants when treated with Si and thus played an important role in increasing the germination rate [158]. Both GA and IAA are involved in plant growth and development, especially IAA accumulation in higher amount in the growing region of plants influences their responses to salt stress

[159,160]. Therefore increased accumulation of IAA and GA in salt-stressed *Glycyrrhiza uralensis* seedlings after Si treatment clearly indicated the stress alleviating potential of Si [161]. Detached leaf pretreated with Si was also observed to postpone senescence by increasing the biosynthesis of CK in the case of both *Arabidopsis* and sorghum plants. This was also proposed as a general mechanism of Si to ameliorate stress through increased CK biosynthesis [162]. Application of Si in rice plants was observed to increase the production of GA, JA, and SA along with macronutrients and protein expression, which indicated the existing relationship of Si-induced phytohormone signaling with abiotic stress alleviation [163]. This interaction of Si with the major phytohormones that play an important role in modulating plant growth and stress responses have also been depicted in Fig. 2.3.

Phytohormones like SA, JA, and ET are also involved in the defense responses of plant under biotic stresses (Fig. 2.3). SA is involved in inducing defense responses against biotrophic pathogens while JA and ET are involved in defense against necrotrophic pathogens [2]. Application of Si resulted in increased biosynthesis of SA, JA, and ET in the leaves of *Erysiphe cichoracearum* exposed *Arabidopsis* plants, as a result of which the severity of powdery mildew disease showed a significant decline [164]. Silicon was also observed to increase JA-mediated antiherbivore defense against the infestation of rice plants by caterpillar — *Cnaphalocrocis medinalis*, which indicated toward the potential of silicon in inducing phytohormone-mediated defense responses [165]. Similarly, when a model grass *B. distachyon* infested with *Helicoverpa armigera* was introduced with silicon, there was a slight and transient increase in JA content. In this case, Si application was known to provide resistance against the pest by increasing biomechanical defense in the form of increased development of macrohairs and simultaneously JA-induced defense [166]. Silicon provided resistance against *Cochliobolus miyabeanus*, which causes brown spot disease of rice. Silicon prevented the fungus from hijacking the host ET biosynthesis pathway and also prevented the action/production of fungal ET. Thus Si effectively regulated the action of the fungus in suppressing the innate immune system of rice plants [167]. Silicon was also reported to activate JA and ET biosynthesis pathways in tomato plants to establish a defense mechanism against *Ralstonia solanacearum*. Overexpression of ET biosynthesis genes viz. *JERF3* (Jasmonate and ethylene-responsive factor 3), *TSRF1* (tomato stress-responsive factor), and *ACCO* (1-aminocyclopropane-1-carboxylate oxidase) and degradation of JA negative regulator (*JAZ1*) by ubiquitin-protein ligase due to silicon treatment indicated that ET and JA are involved in Si-induced resistance in tomato plants [168]. Transgenic Arabidopsis plants carrying a wheat Si transporter gene (*TaLsi1*), when treated with silicon, also resulted in a higher accumulation of SA that helped to defend powdery mildew disease caused by *Golovinomyces cichoracearum*, by inducing the expression of pathogenesis-related genes viz. *PR1*, *PR2*, and *PR5* [169].

The potential of Si in the modulation of phytohormone biosynthesis and signaling for the alleviation of abiotic and biotic stresses has been well studied. However, very few works have been performed to unravel the

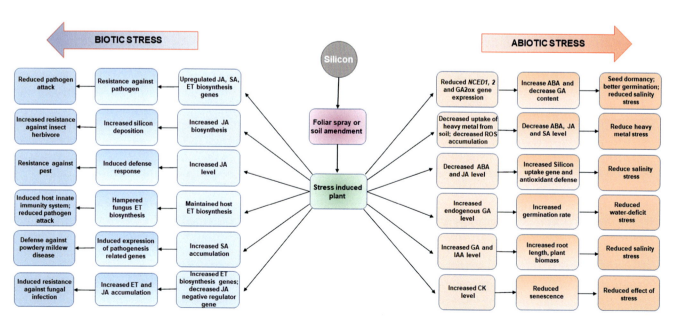

FIGURE 2.3 Interaction of silicon with major phytohormones that has been known to play an important role in eliciting plant responses under abiotic and biotic stresses for providing enhanced level of tolerance.

mechanism of nano-silicon—mediated phytohormone signaling till now. Therefore there is an immediate need to undertake further research in this direction to understand the interaction of nano-silicon with phytohormones under stressed conditions, which will increase our understanding of the alleviating effects of nano-silicon. However, like Si, nano-silicon application in salt-stressed rice plants was also known to reduce the accumulation of ABA when compared to the control plants but no difference was found in the accumulation of JA [12]. On the other hand, IAA was reported to be increased in Bt-transgenic and nontransgenic cotton plants when treated with nano-silicon, thereby establishing its role in promoting growth and development of plants in stressed conditions; however, no significant change in ABA content could be observed [170].

2.6 Molecular mechanism of the alleviation of stress by silicon and nano-silicon

Silicon-mediated abiotic and biotic stress alleviation involves the modulation of plant responses at the molecular level. Though there have been a number of studies depicting the mechanism of silicon in modulating the regulation of genes under stressed conditions, very few studies have focused on the potential of nano-silicon in this regard [171]. One of the first studies describing the silicon-dependent stress responses involving the analysis of gene expression was performed by Watanabe et al. [172] in rice using the microarray approach. Results showed that the application of silicon upregulated the expression of a zinc finger protein homolog, which are the major transcription factors involved in the regulation of stress-responsive genes. Apart from the upregulation of zinc finger protein homolog, there was a downregulation in the expression of chlorophyll a/b binding protein, Xa21 gene family member, metallothionein-like protein, and carbonic anhydrase homolog [172]. Pretreatment with silicon is also involved in increasing the expression level of transcription factors like dehydration responsive element-binding protein (DREB2A) and NAC5 protein, which leads to the increased expression of the OsRDCP1 gene (ring domain containing) and some drought-responsive genes such as OsCMO encoding rice choline monooxygenase and OsRAB16b encoding for dehydrin [173]. This was found to be associated with the production of proline and glycine betaine and simultaneously drought tolerance efficiency increased in rice plants [173]. In a different study, the expression level of photosynthesis-related genes like Os08g02630 (PsbY—a polyprotein of photosystem II), Os05g48630 (PsaH—a subunit of photosystem I), Os07g37030 (PetC—associated with the cytochrome's biological processes in photosynthesis), Os03g57120 (PetH—encoded ferredoxin NADP + reductase), Os09g26810, and Os04g38410 were also found to increase after the application of silicon in the highly Zn-stressed rice plants [174]. As a result of this, the photosynthetic parameters improved under stress due to the improvement in chloroplast structure with better organized thylakoid and increased starch granule size and number [174]. Cu stress can increase the expression level of two transporter genes viz. *copper transporter 1 (COPT1)* and *heavy metal ATPase subunit 5 (HMA5)* in the *Arabidopsis* thereby imparting a toxic effect on the plants; however, application of silicon was shown to suppress the expression level of these genes [175]. In this connection, silicon was mainly involved in limiting the entry of Cu into the cell and also channelizing the excess heavy metals into the vacuoles. Supplementation of silicon on salt-stressed *Sorghum bicolor* plant increased the expression level of *S-adenosyl-Met decarboxylase* and *ornithine decarboxylase* genes, which are involved in polyamines biosynthesis and decreased the expression level of *1-aminocyclopropane-1-1-carboxylic 6 acid synthase* genes involved in ethylene biosynthesis [176]. The increased level of polyamines that are involved in ROS scavenging in coordination with a decreased level of ethylene in turn facilitated the alleviation of salinity stress [176]. Silicon is also observed to ameliorate the salinity stress of maize by allocating more Na + in the leaves than in the root apex and cortex. In this connection, silicon increased the expression of *ZmSOS1* (Salt-overly sensitive 1) and *ZmSOS2* (Salt-overly sensitive 2) along with *ZmHKT1* (High-affinity potassium transporter 1) in the root apex and cortex and the root stele, which were involved in Na + exclusion from the root apex and cortex and also the loading of Na^+ into the xylem [177]. Enhanced sequestration of Na^+ into the vacuoles of leaves also helped in the protection of the photosynthetic apparatus [177]. Apart from these, silicon treatment was shown to improve the relative water content of heat-stressed wheat seedlings by increasing the expression of two aquaporin genes viz. *plasma membrane intrinsic protein 1 (TaPIP1)* and *nodulin 26-like intrinsic protein (TaNIP2)*; however, nano-silicon treatment did not show any such effect on the gene expression levels [178]. On the other hand, nano-silicon was reported to upregulate the expression of *abscisic acid responsive element-binding protein (AREB), abscisic acid and environmental stress-inducible protein (TAS14), 9-cis-epoxycarotenoid dioxygenase (NCED3),* and *cysteine-rich receptor-like protein kinase 42-like (CRK1)* and downregulate the expression of *respiratory burst oxidase (RBOH1), cytosolic ascorbate peroxidase 2 (APX2), mitogen-activated protein kinase 2, ethylene response factor (ERF5), MAPK3* and *dwarf and delayed flowering 2* when applied to salinity-stressed tomato plants [179]. In this connection, the expression of the ABA biosynthesis and signaling genes was primarily involved in the amelioration of the negative effects of stress.

In addition to the abiotic stresses, silicon and nano-silicon impart positive changes at the genetic level to combat the stresses imposed by biotic agents. In this regard, it has been known that the application of silicon on powdery mildew infested *Arabidopsis* plants resulted in the upregulation of several defense-related genes including R genes, genes involved in signal transduction, stress-related transcription factors, stress hormones (SA, JA, ethylene), and enzymes involved in scavenging of ROS [164]. Similarly, application of silicon affected the regulation of stress-responsive genes in wheat plants infected with *Blumeria graminis f.* sp. *tritici* [180]. Also, the presupplementation of silicon to *Alternaria solani* infected tomato plants provided resistance against the pathogen by increasing the expression of some defense-related genes like *pathogenesis-related genes (PR1, PR2, PR3), tomato lipoxygenase D (LOXD),* and *jasmonate and ethylene-responsive factor 3 (JERF3)* [86]. Apart from the defense-related genes, silicon has been shown to affect the expression of three housekeeping genes viz. *Phosphoglycerate kinase,* α-tubulin *(TUB),* and actin *(ACT)* in tomato plants infected by *Ralstonia solanacearum* [181]. Nano-silicon, on the other hand, along with culture filtrate of the plant growth-promoting fungus was observed to activate some defense-related genes like peroxidases (POX), PAL, and *PR1* in order to increase the defense mechanism against papaya ring spot virus and its vector *Myzus persicae* [182].

2.7 Conclusions, current status, and future perspectives

Silicon has now been realized as one of the important beneficial elements due to its multifarious role in plant growth, development, and stress tolerance. Silicon amendments to plants improve their functional attributes both under stressed and nonstressed conditions. Silica forms a mechanical barrier in response to pathogen, insects, and herbivores. With the advancement in nanosciences and their intervention in nanobiotechnology, nano-fertilizers have also gained vast attention of scientists all over the world. Among the different nanoparticles, nano-silicon or silica nanoparticles have shown greater potential in agricultural sector. *In lieu* of this, both Si and nano-silicon are slowly and gradually emerging as promising biostimulants for their prospects in agronomy and agricultural sciences. Although a number of studies have shown positive effects of both Si and nano-silicon, further studies are required to unleash their potential. Therefore more attention should be given toward their practical application in field condition. Concomitantly, future studies should also emphasize the potential of these biostimulants in regulating the plant responses at the molecular level. Though very few studies have reported the toxicity of nano-silicon in plants, these should also be taken into consideration while designing silica-based nanomaterials for the future.

Conflict of interest

The authors declare that they have no competing interests as such.

Acknowledgments

The first author acknowledges the support received from University Grant Commission for providing fellowship [UGC-JRF F. No. 16-6 (DEC.2018)/2019(NET/CSIR)]. All the authors thank the University of North Bengal for providing necessary infrastructure and support for writing the book chapter.

References

[1] Epstein E. The anomaly of silicon in plant biology. Proc Natl Acad Sci USA 1994;91:11−17.
[2] Luyckx M, Hausman JF, Lutts S, et al. Silicon and plants: current knowledge and echnological perspectives. Front Plant Sci 2017;8:411.
[3] Laane HM. The effects of foliar sprays with different silicon compounds. Plants 2018;7:45.
[4] Nawaz MA, Zakharenko AM, Zemchenko IV, et al. Phytolith formation in plants: from soil to cell. Plants 2019;8:249.
[5] Ma JF, Yamaji N. A cooperative system of silicon transport in plants. Trends Plant Sci 2015;20:435−42.
[6] Ma JF, Yamaji N. Silicon uptake and accumulation in higher plants. Trends Plant Sci 2006;11:392−7.
[7] Artyszak A. Effect of silicon fertilization on crop yield quantity and quality—a literature review in Europe. Plants 2018;7:54.
[8] Eneji AE, Inanaga S, Muranaka S, et al. Growth and nutrient use in four grasses under drought stress as mediated by silicon fertilizers. J Plant Nutr 2008;31:355−65.
[9] Etesami H, Jeong BR. Silicon (Si): review and future prospects on the action mechanisms in alleviating biotic and abiotic stresses in plants. Ecotoxicol Environ Saf 2018;147:881−96.

[10] Rastogi A, Tripathi DK, Yadav S, et al. Application of silicon nanoparticles in agriculture. 3 Biotech 2019;9:90.

[11] Mathur P, Roy S. Nanosilica facilitates silica uptake, growth and stress tolerance in plants. Plant Physiol Biochem 2020;157:114—27.

[12] Abdel-Haliem MEF, Hegazy HS, Hassan NS, et al. Effect of silica ions and nano silica on rice plants under salinity stress. Ecol Eng 2017;99:282—9.

[13] Campbell JL, Arora J, Cowell SF, et al. Quasi-cubic magnetite/silica core—shell nanoparticles as enhanced MRI contrast agents for cancer imaging. PLoS One 2011;6:e21857.

[14] Jang HR, Oh HJ, Kim JH, et al. Synthesis of mesoporous spherical silica via spray pyrolysis: pore size control and evaluation of performance in paclitaxel pre purification. Microporous Mesoporous Mater 2013;165:219—27.

[15] Shi YT, Cheng HY, Geng Y, et al. The size-controllable synthesis of nanometer-sized mesoporous silica in extremely dilute surfactant solution. Mater Chem Phys 2010;120:193—8.

[16] Croissant JG, Zhang D, Alsaiari S, et al. Protein-gold clusters-capped mesoporous silica nanoparticles for high drug loading, autonomous gemcitabine/doxorubicin co-delivery, and in-vivo tumor imaging. J Control Release 2016;229:183—91.

[17] Yakhin OI, Lubyanov AA, Yakhin IA. Biostimulants in agrotechnologies: problems, solutions, outlook. Agrochem Her 2016;1:15—21.

[18] El-Shetehy M, Moradi A, Maceroni M, et al. Silica nanoparticles enhance disease resistance in *Arabidopsis* plants. Nat Nanotechnol 2020;16:344—53.

[19] Savvas D, Ntatsi G. Biostimulant activity of silicon in horticulture. Sci Hortic 2015;196:66—81.

[20] Lee SK, Sohn EY, Hamayun M, et al. Effect of silicon on growth and salinity stress of soybean plant grown under hydroponic system. Agrofor Syst 2010;80:333—40.

[21] Du Jardin P. Plant biostimulants: definition, concept, main categories and regulation. Sci Hortic 2015;196:3—14.

[22] Yakhin OI, Lubyanov AA, Yakhin IA, et al. Biostimulants in plant science: a global perspective. Front Plant Sci 2017;7:2049.

[23] Nephali L, Piater LA, Dubery IA, et al. Biostimulants for plant growth and mitigation of abiotic stresses: a metabolomics perspective. Metabolites 2020;10:505.

[24] Chen Y, Magen H, Clapp CE. Mechanisms of plant growth stimulation by humic substances: the role of organo-iron complexes. Soil Sci Plant Nutr 2004;50:1089—95.

[25] Nardi S, Pizzeghello D, Muscolo A, et al. Physiological effects of humic substances on higher plants. Soil Biol Biochem 2002;34:1527—36.

[26] Ferri M, Franceschetti M, Naldrett MJ, et al. Effects of chitosan on the protein profile of grape cell culture subcellular fractions. Electrophoresis 2014;35:1685—92.

[27] Povero G, Loreti E, Pucciariello C, et al. Transcript profiling of chitosan-treated *Arabidopsis* seedlings. J Plant Res 2011;124:619—29.

[28] Van Oosten MJ, Pepe O, De Pascale S, et al. The role of biostimulants and bioeffectors as alleviators of abiotic stress in crop plants. Chem Biol Technol Agric 2017;4:5.

[29] Gómez-Merino FC, Trejo-Téllez LI. The role of beneficial elements in triggering adaptive responses to environmental stressors and improving plant performance. In: Vats S, editor. Biotic and abiotic stress tolerance in plants. Singapore: Springer; 2018. p. 137—72.

[30] Ramírez-Olvera SM, Trejo-Téllez LI, Pérez-Sato JA, et al. Silicon stimulates initial growth and chlorophyll a/b ratio in rice seedlings, and alters the concentrations of Ca, B, and Zn in plant tissues. J Plant Nutr 2018;42:1928—40.

[31] Chen D, Wang S, Yin L, et al. How does silicon mediate plant water uptake and loss under water deficiency? Front Plant Sci 2018;9:281.

[32] Yamamoto T, Nakamura A, Iwai H, et al. Effect of silicon deficiency on secondary cell wall synthesis in rice leaf. J Plant Res 2012;125:771—9.

[33] Juárez-Maldonado A, Ortega-Ortíz H, Morales-Díaz AB, et al. Nanoparticles and nanomaterials as plant biostimulants. Int J Mol Sci 2019;20:162.

[34] Wang S, Wang F, Gao G. Foliar application with nano-silicon alleviates Cd toxicity in rice seedlings. Environ Sci Pollut Res 2015;22:2837—45.

[35] Ma JF. Role of silicon in enhancing the resistance of plant to biotic and abiotic stress. Soil Sci Plant Nutr 2004;50:11—18.

[36] White PJ, Brown PH. Plant nutrition for sustainable development and global health. Ann Bot 2010;105:1073—80.

[37] Kirkby E. Introduction, definition and classification of nutrients. Marschner's mineral nutrition of higher plants. Elsevier; 2012. p. 3—5.

[38] Korndorfer GH, Lepsch I. Effect of silicon on plant growth and crop yield. In: Datnoff LE, Snyder GH, Korndörfer GH, editors. Studies in plant science, vol. 8. Elsevier; 2001. p. 133—47.

[39] Zargar SM, Mahajan R, Bhat JA, et al. Role of silicon in plant stress tolerance: opportunities to achieve a sustainable cropping system. 3 Biotech 2019;9:73.

[40] Ma JF, Yamaji N, Mitani-Ueno N. Transport of silicon from roots to panicles in plants. Proc Jpn Acad Ser 2011;87:377—85.

[41] Yamaji N, Sakurai G, Mitani-Ueno N, et al. Orchestration of three transporters and distinct vascular structures in node for intervascular transfer of silicon in rice. Proc Natl Acad Sci USA 2015;112:11401—6.

[42] Hegazy HS, Hassan NSH, Abdel-Haliem MEF, et al. Biochemical response of rice plant to biotic and abiotic stress under silica ions and nanoparticles application. Egypt J Bot 2015;55:79—103.

[43] Abdul Qados AMS, Moftah AE. Influence of silicon and nano-silicon on germination, growth and yield of faba bean (*Vicia faba* L.) under salt stress conditions. Am J Exp Agric 2015;5:509—24.

[44] Roohizadeh G, Majd A, Arbabian S. The effect of sodium silicate and silica nanoparticles on seed germination and growth in the *Vicia faba* L. Trop Plant Res 2015;2:85—9.

[45] Tubaña BS, Heckman JR. Silicon in soils and plants. In: Rodrigues FA, Datnoff L, editors. Silicon and plant diseases. Cham: Springer; 2015. p. 7—51.

[46] Głazowska S, Baldwin L, Mravec J, et al. The impact of silicon on cell wall composition and enzymatic saccharification of *Brachypodium distachyon*. Biotechnol Biofuels 2018;11:171.

[47] Brahma R, Ahmed R, Choudhury M. Silicon nutrition for alleviation of abiotic stress in plants: a review. J Pharmacogn Phytochem 2020;9:1374—81.

[48] Coskun D, Britto DT, Huynh WQ, et al. The role of silicon in higher plants under salinity and drought stress. Front Plant Sci 2016;7:1072.

[49] Liu P, Yin L, Deng X, et al. Aquaporin- mediated increase in root hydraulic conductance is involved in silicon-induced improved root water uptake under osmotic stress in *Sorghum bicolor* L. J Exp Bot 2014;65:4747−56.

[50] Liu P, Yin L, Wang S. Enhanced root hydraulic conductance by aquaporin regulation accounts for silicon alleviated salt-induced osmotic stress in *Sorghum bicolor* L. Environ Exp Bot 2015;111:42−51.

[51] Debona D, Rodrigues FA, Datnoff LE. Silicon's role in abiotic and biotic plant stresses. Annu Rev Phytopathol 2017;55:85−107.

[52] Kim YH, Khan AL, Waqas M, et al. Silicon regulates antioxidant activities of crop plants under abiotic-induced oxidative stress: a review. Front Plant Sci 2017;8:510.

[53] Gupta B, Huang B. Mechanism of salinity tolerance in plants: physiological, biochemical, and molecular characterization. Int J Genom 2014;2014:701596.

[54] Farooq M, Wahid A, Kobayashi N, et al. Plant drought stress: effects, mechanisms and management. Agron Sustain Dev 2009;29:185−212.

[55] Rizwan M, Ali S, Ibrahim M, et al. Mechanisms of silicon-mediated alleviation of drought and salt stress in plants: a review. Environ Sci Pollut Res 2015;22:15416−31.

[56] Zhu Y, Gong H. Beneficial effects of silicon on salt and drought tolerance in plants. Agron Sustain Dev 2014;34:455−72.

[57] Farooq MA, Dietz KJ. Silicon as versatile player in plant and human biology: overlooked and poorly understood. Front Plant Sci 2015;6:994.

[58] Tripathi DK, Singh VP, Gangwar S, et al. Role of silicon in enrichment of plant nutrients and protection from biotic and abiotic stresses. In: Ahmad P, Wani M, Azooz M, et al., editors. Improvement of crops in the era of climatic changes. New York: Springer; 2014. p. 39−56.

[59] Fortunato AA, Debona D, Bernardeli A, et al. Defence-related enzymes in soybean resistance to target spot. Phytopathology 2015;163:731−42.

[60] Rodrigues FA, Dallagnol LJ, Duarte HSS, et al. Silicon control of foliar diseases in monocots and dicots. In: Rodrigues FA, Datnoff L, editors. Silicon and plant diseases. Cham: Springer; 2015. p. 67−108.

[61] Somapala K, Weerahewa D, Thrikawala S. Silicon rich rice hull amended soil enhances anthracnose resistance in tomato. Proc Food Sci 2016;6:190−3.

[62] Jayawardana HARK, Weerahewa HLD, Saparamadu MDJS. Effect of root or foliar application of soluble silicon on plant growth, fruit quality and anthracnose development of capsicum. Trop Agric Res 2014;26:74−81.

[63] Gutierrez-Barranquero JA, Arrebola E, Bonilla N, et al. Environmentally friendly treatment alternatives to Bordeaux mixture for controlling bacterial apical necrosis (BAN) of mango. Plant Pathol 2012;61:665−76.

[64] Resende RS, Rodrigues F, Costa RV, et al. Silicon and fungicide effects on anthracnose in moderately resistant and susceptible sorghum lines. J Phytopathol 2013;161:11−17.

[65] Carre-Missio V, Rodrigues FA, Schur DA, et al. Effect of foliar-applied potassium silicate on coffee leaf infection by *Hemileia vastatrix*. Ann Appl Biol 2014;164:396−403.

[66] Kim SG, Kim KW, Park EW, et al. Silicon-induced cell wall fortification of rice leaves: a possible cellular mechanism of enhanced host resistance to blast. Phytopathology 2002;92:1095−103.

[67] Najihah NI, Hanafi MM, Idris AS, et al. Silicon treatment in oil palms confers resistance to basal stem rot disease caused by *Ganoderma boninense*. Crop Prot 2015;67:151−9.

[68] Mcginnity P. Silicon and its role in crop production [PhD thesis]; 2015. Available from https://www.planttuff.com/wp-content/uploads/2015/12/silicon-agriculture-iiterature-rvw-1.pdf

[69] Pozza EA, Pozza AAA, Dos Santos Botelho DM. Silicon in plant disease control. Rev Ceres Vicosa 2015;62:323−31.

[70] Menzies JG, Ehret DL, Glass ADM, et al. Effects of soluble silicon on the parasitic fitness of *Sphaerotheca fuliginea* on *Cucumis sativus*. Phytopathology 1991;81:84−8.

[71] Rémus-Borel W, Menzies J, Belanger R. Aconitate and methyl aconitate are modulated by silicon in powdery mildew-infected wheat plants. J Plant Physiol 2009;166:1413−22.

[72] Fortunato AA, Silva WLD, Rodrigues FA. Phenylpropanoid pathway is potentiated by silicon in the roots of banana plants during the infection process of *Fusarium oxysporum* f. sp. *cubense*. Phytopathology 2014;104:597−603.

[73] Correa RSB, Moraes JC, Auad AM, et al. Silicon and acibenzolar-S-methyl as resistance inducers in cucumber, against the whitefly *Bemisia tabaci* (Gennadius) (Hemiptera: Aleyrodidae) biotype B. Neotrop Entomol 2005;34:429−33.

[74] Malhotra C, Kapoor R, Ganjewala D. Alleviation of abiotic and biotic stresses in plants by silicon supplementation. Sci Agric 2016;.

[75] Chung YS, Kim KS, Hamayun M, et al. Silicon confers soybean resistance to salinity stress through regulation of reactive oxygen and reactive nitrogen species. Front Plant Sci 2020;10:1725.

[76] Khan A, Bilal S, Khan AL, et al. Silicon-mediated alleviation of combined salinity and cadmium stress in date palm (*Phoenix dactylifera* L.) by regulating physio-hormonal alteration. Ecotoxicol Environ Saf 2020;188:109885.

[77] Moradtalab N, Hajiboland R, Aliasgharzad N, et al. Silicon and the Association with an arbuscular-mycorrhizal fungus (*Rhizophagus clarus*) mitigate the adverse effects of drought stress on strawberry. Agronomy 2019;9:41.

[78] Parveen A, Liu W, Hussain S, et al. Silicon priming regulates morpho-physiological growth and oxidative metabolism in maize under drought stress. Plants 2019;8:431.

[79] Soleimannejad Z, Abdolzadeh A, Sadeghipour HR. Beneficial effects of silicon application in alleviating salinity stress in halophytic *Puccinellia distans* plants. Silicon 2019;11:1001−10.

[80] Huang F, Wen XH, Cai YX, et al. Silicon-mediated enhancement of heavy metal tolerance in rice at different growth stages. Int J Environ Res Public Health 2018;15:2193.

[81] Khan WUD, Aziz T, Maqsood MA, et al. Silicon nutrition mitigates salinity stress in maize by modulating ion accumulation, photosynthesis, and antioxidants. Photosynthetica 2018;56:1047−57.

[82] Biju S, Fuentes S, Gupta D. Silicon improves seed germination and alleviates drought stress in lentil crops by regulating osmolytes, hydrolytic enzymes and antioxidant defense system. Plant Physiol Biochem 2017;119:250−64.

[83] Helaly MN, El-Hoseiny H, El-Sheery NI, et al. Regulation and physiological role of silicon in alleviating drought stress of mango. Plant Physiol Biochem 2017;118:31—44.

[84] Muneer S, Park YG, Kim S, et al. Foliar or subirrigation silicon supply mitigates high temperature stress in strawberry by maintaining photosynthetic and stress-responsive proteins. J Plant Growth Regul 2017;36:836—45.

[85] Zhang W, Xie Z, Wang L, et al. Silicon alleviates salt and drought stress of *Glycyrrhiza uralensis* seedling by altering antioxidant metabolism and osmotic adjustment. J Plant Res 2017;130:611—24.

[86] Gulzar N, Ali S, Shah MA, et al. Silicon supplementation improves early blight resistance in *Lycopersicon esculentum* Mill. by modulating the expression of defense-related genes and antioxidant enzymes. 3 Biotech 2021;11:232.

[87] Jeer M, Yele Y, Sharma KC, et al. Exogenous application of different silicon sources and potassium reduces pink stem borer damage and improves photosynthesis, yield and related parameters in wheat. Silicon 2021;13:901—10.

[88] Wang L, Dong M, Zhang Q, et al. Silicon modulates multi-layered defense against powdery mildew in *Arabidopsis*. Phytopathol Res 2020;2:7.

[89] De Lima DT, Sampaio MV, Albuquerque CJB, et al. Silicon accumulation and its effect on agricultural traits and anthracnose incidence in lignocellulosic sorghum. Pesq Agropec Trop 2019;49:1—8.

[90] Elsherbiny EA, Taher MA. Silicon induces resistance to postharvest rot of carrot caused by *Sclerotinia sclerotiorum* and the possible of defense mechanisms. Postharvest Biol Technol 2018;140:11—17.

[91] Ratnayake RMRNK, Ganehenege MYU, Ariyarathne HM, et al. Soil application of rice husk as a natural silicon source to enhance some chemical defense responses against foliar fungal pathogens and growth performance of bitter gourd (*Momordica charantia* L.). Ceylon J Sci 2018;47:49—55.

[92] Xue-ying FAN, Wei-peng LIN, Rui LIU, et al. Physiological response and phenolic metabolism in tomato (*Solanum lycopersicum*) mediated by silicon under *Ralstonia solanacearum* infection. J Integr Agric 2018;17:2160—71.

[93] Dann EK, Le DP. Effects of silicon amendment on soilborne and fruit diseases of avocado. Plants 2017;6:51.

[94] Gogos A, Knauer K, Bucheli TD. Nanomaterials in plant protection and fertilization: current state, foreseen applications, and research priorities. J Agric Food Chem 2012;60:97—9792.

[95] Raliya R, Saharan V, Dimkpa C, et al. Nanofertilizer for precision and sustainable agriculture: current state and future perspectives. J Agric Food Chem 2018;66:6487—503.

[96] Kah M, Tufenkji N, White JC. Nano-enabled strategies to enhance crop nutrition and protection. Nat Nanotechnol 2019;14:532—40.

[97] Zhao L, Lu L, Wang A, et al. Nanobiotechnology in agriculture: use of nanomaterials to promote plant growth and stress tolerance. J Agric Food Chem 2020;68:1935—47.

[98] Tripathi DK, Singh VP, Prasad SM, et al. Silicon nanoparticles (SiNp) alleviate chromium (VI) phytotoxicity in *Pisum sativum* (L.) seedlings. Plant Physiol Biochem 2015;96:189—98.

[99] Tripathi DK, Singh S, Singh VP, et al. Silicon nanoparticles more effectively alleviated UV-B stress than silicon in wheat (*Triticum aestivum*) seedlings. Plant Physiol Biochem 2017;110:70—81.

[100] Cui J, Liu T, Li F, et al. Silica nanoparticles alleviate cadmium toxicity in rice cells: mechanisms and size effects. Environ Pollut 2017;228:363—9.

[101] Karimi M, Mirshekari H, Aliakbari M, et al. Smart mesoporous silica nanoparticles for controlled-release drug delivery. Nanotechnol Rev 2016;5:195—207.

[102] Debnath N, Das S, Brahmacharya RL, et al. Entomotoxicity assay of silica, zinc oxide, titanium dioxide, aluminium oxide nanoparticles on *Lipaphis pseudobrassicae*. AIP Conf Proc 2010;1276:307.

[103] Yang L, Watts DJ. Particle surface characteristics may play an important role in phytotoxicity of alumina nanoparticles. Toxicol Lett 2005;158:122—32.

[104] Khalaki MA, Ghorbani A, Moameri M. Effects of silica and silver nanoparticles on seed germination traits of *Thymus kotschyanus* in laboratory conditions. J Range Sci 2016;6:221—31.

[105] Strout G, Russell SD, Pulsifer DP, et al. Silica nanoparticles aid in structural leaf coloration in the Malaysian tropical rainforest understorey herb *Mapania caudata*. Ann Bot 2013;112:1141—8.

[106] Suriyaprabha R, Karunakaran G, Yuvakkumar R, et al. Foliar application of silica nanoparticles on the phytochemical responses of maize (*Zea mays* L.) and its toxicological behavior. Synth React Inorg Met Org Chem 2014;44:1128—31.

[107] Attia EA, Elhawat N. Combined foliar and soil application of silica nanoparticles enhances the growth, flowering period and flower characteristics of marigold (*Tagetes erecta* L.). Sci Hortic 2021;282:110015.

[108] Elshayb OM, Nada AM, Ibrahim HM, et al. Application of silica nanoparticles for improving growth, yield, and enzymatic antioxidant for the hybrid rice ehr1growing under water regime conditions. Materials 2021;14:1150.

[109] Siddiqui MH, Al-Whaibi MH. Role of nano-SiO_2 in germination of tomato (*Lycopersicum esculentum* seeds Mill.). Saud J Biol Sci 2014;21:13—17.

[110] Bao-shan L, Chun-hui L, Li-jun F, et al. Effect of TMS (nanostructured silicon dioxide) on growth of *Changbai larch* seedlings. J For Res 2004;15:138—40.

[111] Khan Z, Ansari MYK. Impact of engineered Si nanoparticles on seed germination, vigour index and genotoxicity assessment via DNA damage of root tip cells in *Lens culinaris*. Plant Biochem Physiol 2018;6:3.

[112] Suriyaprabha R, Karunakaran G, Yuvakkumar R, et al. Growth and physiological responses of maize (*Zea mays* L.) to porous silica nanoparticles in soil. J Nanopart Res 2012;14:1294.

[113] Alsaeedi AH, Elgarawany MM, El-Ramady H, et al. Application of silica nanoparticles induces seed germination and growth of cucumber (*Cucumis sativus*). Environ Arid Land Agric Sci 2019;28:57—68.

[114] Alsaeedi AH, El-Ramady H, Alshaal T, et al. Engineered silica nanoparticles alleviate the detrimental effects of Na^+ stress on germination and growth of common bean (*Phaseolus vulgaris*). Environ Sci Pollut Res 2017;24:21917—28.

[115] Janmohammadi M, Sabaghnia N, Ahadnezhad A. Impact of silicon dioxide nanoparticles on seedling early growth of lentil (*Lens culinaris* Medik.) genotypes with various origins. Agric For 2015;61:19—33.

[116] Isfahani FM, Tahmourespour A, Hoodaji M, et al. Influence of exopolysaccharide-producing bacteria and SiO_2 nanoparticles on proline content and antioxidant enzyme activities of tomato seedlings (*Solanum lycopersicum* L.) under salinity stress. Pol J Environ Stud 2019;28:153—63.

[117] Kalteh M, Alipour ZT, Ashraf S, et al. Effect of silica nanoparticles on basil (*Ocimum basilicum*) under salinity stress. J Chem Health Risks 2014;4:49–55.

[118] Mahmoud LM, Dutt M, Shalan AM, et al. Silicon nanoparticles mitigate oxidative stress of in vitro-derived banana (*Musa acuminata* 'Grand Nain') under simulated water deficit or salinity stress. S Afr J Bot 2020;132:155–63.

[119] Ashkavand P, Tabari M, Zarafshar M, et al. Effect of SiO_2 nanoparticles on drought resistance in hawthorn seedlings. For Res Pap 2015;76:350–9.

[120] Ashkavand P, Zarafshar M, Tabari M, et al. Application of SiO_2 nanoparticles as pretreatment alleviates the impact of drought on the physiological performance of *Prunus mahaleb* (Rosaceae). Bol Soc Argent Bot 2018;53:207–19.

[121] Mushtaq A, Jamil N, Rizwan S, et al. Engineered silica nanoparticles and silica nanoparticles containing controlled release fertilizer for drought and saline areas. IOP Conf Ser: Mater Sci Eng 2018;414:012029.

[122] Hussain A, Rizwan M, Ali Q, Ali S. Seed priming with silicon nanoparticles improved the biomass and yield while reduced the oxidative stress and cadmium concentration in wheat grains. Environ Sci Pollut Res 2019;26:1–10.

[123] Gao M, Zhou J, Liu H, et al. Foliar spraying with silicon and selenium reduces cadmium uptake and mitigates cadmium toxicity in rice. Sci Total Environ 2018;631:1100–8.

[124] Asgari F, Majd A, Jonoubi P, et al. Effects of silicon nanoparticles on molecular, chemical, structural and ultrastructural characteristics of oat (*Avena sativa* L.). Plant Physiol Biochem 2018;127:152–60.

[125] Khan ZS, Rizwan M, Hafeez M, et al. Effects of silicon nanoparticles on growth and physiology of wheat in cadmium contaminated soil under different soil moisture levels. Environ Sci Pollut Res 2020;27:4958–68.

[126] Tripathi DK, Singh S, Singh VP, et al. Silicon nanoparticles more efficiently alleviate arsenate toxicity than silicon in maize cultiver and hybrid differing in arsenate tolerance. Front Environ Sci 2016;4:46.

[127] Rajwade JM, Chikte RG, Paknikar KM. Nanomaterials: new weapons in a crusade against phytopathogens. Appl Microbiol Biotechnol 2020;104:1437–61.

[128] Abbai R, Ahn JC, Mohanan P, et al. Silica nanoparticles suppress the root rot of Panax ginseg from *Ilyonectria mors-panacis* infection by reducing sugar efflux into apoplast. In: Proceedings of The Plant Resources Society of Korea Conference, Chungcheongbuk-do: The Plant Resources Society of Korea; 2018. p. 59.

[129] Kang H, Elmer W, Shen Y, et al. Silica nanoparticle dissolution rate controls the suppression of Fusarium wilt of watermelon (*Citrullus lanatus*). Environ Sci Technol 2021;.

[130] Siddiqui ZA, Aziz S. Management of *Rhizoctonia solani*, *Fusarium solani* and Meloidogyne incognita by silicon dioxide nanoparticles and *Rhizobium ciceri* alone and in combination on chickpea. J Biosens Renew Sci 2020;1:35–42.

[131] Buchman JT, Elmer W, Ma C, et al. Chitosan-coated mesoporous silica nanoparticle treatment of *Citrullus lanatus* (watermelon): enhanced fungal disease suppression and modulated expression of stress-related genes. ACS Sustain Chem Eng 2019;7:19649–59.

[132] Madany M, Saleh A, Habeeb T, et al. Silicon dioxide nanoparticles alleviate the threats of broomrape infection in tomato by inducing cell wall fortification and modulating ROS homeostasis. Environ Sci Nano 2020;7.

[133] Derbalah A, Shenashen M, Hamza A, et al. Antifungal activity of fabricated mesoporous silica nanoparticles against early blight of tomato. Egypt J Basic Appl Sci 2018;5:145–50.

[134] Akhtar N, Ilyas N, Mashwani ZUR, et al. Synergistic effects of plant growth promoting rhizobacteria and silicon dioxide nano-particles for amelioration of drought stress in wheat. Plant Physiol Biochem 2021;166:160–76.

[135] El-Dengawy EFA, EL-Abbasy UK, El-Gobba MH. Influence of nano-silicon treatment on growth behavior of 'sukkary' and 'gahrawy' mango root-stocks under salinity stress. J Plant Production, Mansoura Univ 2021;12:49–61.

[136] Iqbal Z, Sarkhosh A, Balal RM, et al. Silicon alleviate hypoxia stress by improving enzymatic and non-enzymatic antioxidants and regulating nutrient uptake in muscadine grape (*Muscadinia rotundifolia* Michx.). Front Plant Sci 2021;11:618873.

[137] Rai-Kalal P, Tomar RS, Jajoo A. Seed nanopriming by silicon oxide improves drought stress alleviation potential in wheat plants. Funct Plant Biol 2021;48.

[138] Thind S, Hussain I, Rasheed R, et al. Alleviation of cadmium stress by silicon nanoparticles during different phenological stages of ujala wheat variety. Arab J Geosci 2021;14:1028.

[139] Aqaei P, Weisany W, Diyanat M, et al. Response of maize (*Zea mays* L.) to potassium nano-silica application under drought stress. J Plant Nutr 2020;43:1205–16.

[140] Emamverdian A, Ding Y, Mokhberdoran F, et al. Determination of heavy metal tolerance threshold in a bamboo species (*Arundinaria pygmaea*) as treated with silicon dioxide nanoparticles. Glob Ecol Conserv 2020;24:e01306.

[141] Namjoyan S, Sorooshzadeh A, Rajabi A, et al. Nano-silicon protects sugar beet plants against water deficit stress by improving the antioxidant systems and compatible solutes. Acta Physiol Plant 2020;42:157.

[142] Avestan A, Ghasemnezhad M, Esfahani M, et al. Application of nano-silicon dioxide improves salt stress tolerance in strawberry plants. Agronomy 2019;9:246.

[143] Ghorbanpour M, Mohammadi H, Kariman K. Nanosilicon-based recovery of barley (*Hordeum vulgare*) plants subjected to drought stress. Environ Sci Nano 2019;7.

[144] Farhangi-Abriz S, Torabian S. Nano-silicon alters antioxidant activities of soybean seedlings under salt toxicity. Protoplasma 2018;255:953–62.

[145] Yassen A, Abdallah E, Gaballah M, et al. Role of silicon dioxide nano fertilizer in mitigating salt stress on growth, yield and chemical composition of cucumber (*Cucumis sativus* L). Int J Agric Res 2017;12:130–5.

[146] Shabbaj II, Madany MMY, Tammar A, et al. Silicon dioxide nanoparticles orchestrate carbon and nitrogen metabolism in pea seedlings to cope with broomrape infection. Environ Sci Nano 2021;8.

[147] El-Naggar ME, Abdelsalam NR, Fouda MMG, et al. Soil application of nano silica on maize yield and its insecticidal activity against some stored insects after the post-harvest. Nanomaterials 2020;10:739.

[148] Hasan KA, Soliman H, Baka Z, et al. Efficacy of nano-silicon in the control of chocolate spot disease of *Vicia faba* L. caused by *Botrytis fabae*. Egypt J Basic Appl Sci 2020;7:53–66.

[149] Khan MR, Siddiqui ZA. Use of silicon dioxide nanoparticles for the management of *Meloidogyne incognita*, *Pectobacterium betavasculorum* and *Rhizoctonia solani* disease complex of beetroot (*Beta vulgaris* L.). Sci Hortic (Amsterdam) 2020;265:109211.

[150] Gabr WE, Hassan AA, Hashem IM, et al. Effect of biogenic silica nanoparticles on blast and brown spot diseases of rice and yield component. J Plant Production, Mansoura Univ 2017;8:869–76.

[151] Fahad S, Hussain S, Matloob A, et al. Phytohormones and plant responses to salinity stress: a review. J Plant Growth Regul 2015;750:391–404.

[152] Kim YH, Khan AL, Lee IJ. Silicon: a duo synergy for regulating crop growth and hormonal signaling under abiotic stress conditions. Crit Rev Biotechnol 2016;36:1099–109.

[153] Lee SC, Luan S. ABA signal transduction at the crossroad of biotic and abiotic stress responses. Plant Cell Environ 2012;35:53–60.

[154] Kim YH, Khan AL, Hamayun M, et al. Influence of short-term silicon application on endogenous physiohormonal levels of *Oryza sativa* L. under wounding stress. Biol Trace Elem Res 2011;144:1175–85.

[155] Kim YH, Khan AL, Waqas M, et al. Silicon application to rice root zone influenced the phytohormonal and antioxidant responses under salinity stress. J Plant Growth Regul 2014;33:137–49.

[156] Gou T, Chen X, Han R, et al. Silicon can improve seed germination and ameliorate oxidative damage of bud seedlings in cucumber under salt stress. Acta Physiol Plant 2020;42:12.

[157] Kim YH, Khan AL, Kim DH, et al. Silicon mitigates heavy metal stress by regulating P-type heavy metal ATPases, *Oryza sativa* low silicon genes, and endogenous phytohormones. BMC Plant Biol 2014;14:13.

[158] Hamayun M, Sohn EY, Khan SA, et al. Silicon alleviates the adverse effects of salinity and drought stress on growth and endogenous plant growth hormones of soybean (*Glycine max* L.). Pak J Bot 2010;42:1713–22.

[159] Iqbal N, Umar S, Khan NA, et al. A new perspective of phytohormones in salinity tolerance: regulation of proline metabolism. Environ Exp Bot 2014;100:34–42.

[160] Ryu H, Cho YG. Plant hormones in salt stress tolerance. J Plant Biol 2015;58:147–55.

[161] Lang DY, Fei PX, Cao GY, et al. Silicon promotes seedling growth and alters endogenous IAA, GA3 and ABA concentrations in *Glycyrrhiza uralensis* under 100 mM NaCl stress. J Hortic Sci Biotechnol 2019;94:87–93.

[162] Markovich O, Steiner E, Kouřil S, et al. Silicon promotes cytokinin biosynthesis and delays senescence in *Arabidopsis* and Sorghum. Plant Cell Environ 2017;40:1189–96.

[163] Jang SW, Kim Y, Khan AL, et al. Exogenous short-term silicon application regulates macro-nutrients, endogenous phytohormones, and protein expression in *Oryza sativa* L. BMC Plant Biol 2018;18:4.

[164] Fauteux F, Chain F, Belzile F, et al. The protective role of silicon in the *Arabidopsis*-powdery mildew pathosystem. Proc Natl Acad Sci USA 2006;103:17554–9.

[165] Ye M, Song Y, Long J, et al. Priming of jasmonate-mediated antiherbivore defense responses in rice by silicon. Proc Natl Acad Sci 2013;110:3631–9.

[166] Hall CR, Waterman JM, Vandegeer RK, et al. The role of silicon in antiherbivore phytohormonal signalling. Front Plant Sci 2019;10:1132.

[167] Bockhaven JV, Spichal L, Novak O, et al. Silicon induces resistance to the brown spot fungus *Cochliobolus miyabeanus* by preventing the pathogen from hijacking the rice ethylene pathway. New Phytol 2015;206:761–73.

[168] Ghareeb H, Bozsó Z, Ott PG, et al. Transcriptome of silicon-induced resistance against *Ralstonia solanacearum* in the silicon non-accumulator tomato implicates priming effect. Physiol Mol Plant Pathol 2011;75:83–9.

[169] Vivancos J, Labbé C, Menzies JG, et al. Silicon-mediated resistance of *Arabidopsis* against powdery mildew involves mechanisms other than the salicylic acid (SA)-dependent defence pathway. Mol Plant Pathol 2015;16:572–82.

[170] Le VN, Rui Y, Gui X, et al. Uptake, transport, distribution and bio-effects of SiO2 nanoparticles in Bt-transgenic cotton. J Nanobiotechnol 2014;12:50.

[171] Manivannan A, Ahn YK. Silicon regulates potential genes involved in major physiological processes in plants to combat stress. Front Plant Sci 2017;8:1346.

[172] Watanabe S, Shimoi E, Ohkama N, et al. Identification of several rice genes regulated by Si nutrition. Soil Sci Plant Nutr 2004;50:1273–6.

[173] Khattab HI, Emam MA, Emam MM, et al. Effect of selenium and silicon on transcription factors *NAC5* and *DREB2A* involved in drought-responsive gene expression in rice. Biol Plant 2014;58:265–73.

[174] Song A, Li P, Fan F, et al. The effect of silicon on photosynthesis and expression of its relevant genes in rice (*Oryza sativa* L.) under high-zinc stress. PLoS One 2014;9:e113782.

[175] Li J, Leisner SM, Frantz J. Alleviation of copper toxicity in *Arabidopsis thaliana* by silicon addition to hydroponic solutions. J Am Soc Hortic Sci 2008;133:670–7.

[176] Yin L, Wang S, Tanaka K, et al. Silicon-mediated changes in polyamines participate in silicon-induced salt tolerance in *Sorghum bicolor* L. Plant Cell Environ 2016;39:245–58.

[177] Bosnic P, Bosnic D, Jasnic J, et al. Silicon mediates sodium transport and partitioning in maize under moderate salt stress. Environ Exp Bot 2018;155:681–7.

[178] Younis AA, Khattab H, Emam MM. Impacts of silicon and silicon nanoparticles on leaf ultrastructure and *TaPIP1* and *TaNIP2* gene expressions in heat stressed wheat seedlings. Biol Plant 2020;64:343–52.

[179] Almutairi ZM. Effect of nano-silicon application on the expression of salt tolerance genes in germinating tomato (*Solanum lycopersicum* L.) seedlings under salt stress. POJ 2016;9:106–14.

[180] Chain F, Côté-Beaulieu C, Belzile F, et al. A comprehensive transcriptomic analysis of the effect of silicon on wheat plants under control and pathogen stress conditions. Mol Plant Microbe Interact 2009;22:1323–30.

[181] Ghareeb H, Bozsó Z, Ott PG, et al. Silicon and *Ralstonia solanacearum* modulate expression stability of housekeeping genes in tomato. Physiol Mol Plant Pathol 2011;75:176–9.

[182] Elsharkawy MM, Mousa KM. Induction of systemic resistance against Papaya ringspot virus (PRSV) and its vector *Myzus persicae* by *Penicillium simplicissimum* GP17-2 and silica (SiO2) nanopowder. Int J Pest Manag 2015;61.

Silicon uptake, acquisition, and accumulation in plants

Seyed Abdollah Hosseini

Plant Nutrition Department, Abiotic Stress Group, Agro Innovation International, Timac Agro, France

3.1 Introduction

3.1.1 Si in soil and plant

After oxygen, Si is the most abundant element in the Earth's crust and in the soil [1]. The average total Si content in the soil can be reached up to 30% and is highly dependent on soil type and the soil weathering [1,2]. The bioavailability of Si in the soil depends on its biogeochemical cycling, its mineral form, and its solubility [3]. In the field, the pool of available Si tends to decrease, since the Si accumulated in crops is usually exported and not recycled in soil [4], which raises the question of Si fertilization. Si is present in the soil in various forms of which most types are in the form of insoluble crystalline aluminosilicates and are only partially soluble, hence are not directly available for plants [5]. Monosilicic acid ($Si(OH)_4$) is the main water-soluble form of Si [1], although it can be converted to $H_3SiO_4^{2-}$ at pH above 9 and to $H_2SiO_4^{2-}$ above pH 11 [4]. Thus available Si in soil includes monosilicic acid in soil solution (0.1−0.6 mM) and parts of silicate components that can be easily converted into $Si(OH)_4$ such as polymerized silicic acid, exchangeable silicates, and part of colloidal silicates. Si is usually less prevalent in the soil solution at high pH (8−9) because of the high adsorption of $H_3SiO_4^{2-}$ to soil colloids through interactions with iron and aluminum oxides [1]. The concentration of available Si is dominated by soil parent materials and factors that could affect Si adsorption-desorption process in the soil such as pH, water status, temperature, and accompanied ions [6]. However, the effect of pH on Si availability is more complex and is highly dependent on other parameters, such as soil texture, temperature, organic matter, and accompanying ions [3].

Silicon has been found almost in all terrestrial plants in which its concentration largely varies among different plant species ranging from 0.1% to 10% of dry weight [1,7,8]. For example, the accumulation of Si in Bryophyta, Lycopsida, and Equisetopsida is substantially higher compared to Filicopsida in Pteridophyta, Gymnospermae, and most Angiospermae which accumulate less Si [9]. *Meta*-analysis performed by Hodson et al. for different species also showed these differences [10]. There are genotypical differences in shoot Si concentration and variation within species which is usually much lower than the variation among species. For instance, compared to chickpea (*Cicer arietinum*) which contains 3.0 mg Si g^{-1}, rice (*Oryza sativa*) which is from the Poaceae family, contains 39 mg Si g^{-1}. Plant species are categorized as accumulators ($>1\%$ Si), intermediate accumulators (0.5%−1% Si), and nonaccumulators or excluders ($<0.5\%$ Si) based on their ability to accumulate Si [11−14].

3.2 Silicon uptake, acquisition, and accumulation in higher plants

Silicon is present in the soil as an inert element [15]. Plant roots take up Si in the form of silicic acid $Si(OH)_4$ [7]. Monosilicic acid in the form of undissociated molecule, which is the sole molecular species likely to cross

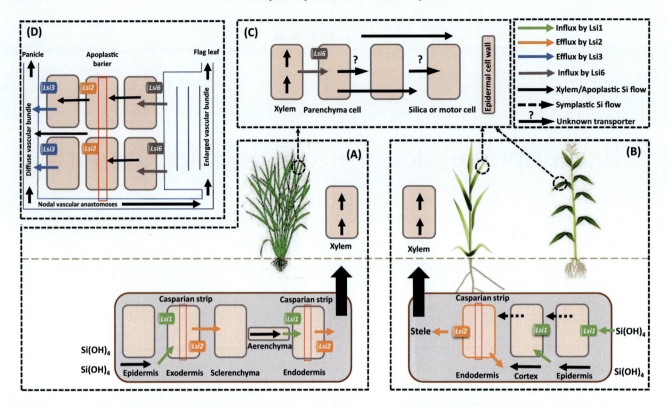

FIGURE 3.1 A schematic model representing the Si uptake and transport in rice, maize, and barley plants. (A) Si uptake in roots of rice. (B) Si uptake in roots of maize and barely. (C) Possible Si unloading and deposition pathway in leaf. (D) Possible Si transport and distribution in rice node. *Source: Modified after J.F. Ma, N. Yamaji, A cooperative system of silicon transport in plants, Trends Plant Sci 20 (2015) 435−442. https:// doi.org/10.1016/j.tplants.2015.04.007; N. Yamaji, G. Sakurai, N. Mitani-Ueno, J.F. Ma, Orchestration of three transporters and distinct vascular structures in node for intervascular transfer of silicon in rice, Proc Natl Acad Sci.112 (2015) 11401−11406. https://doi.org/10.1073/pnas.1508987112; G.C. Yan, M. Nikolic, M.J. Ye, Z.X. Xiao, Y.C. Liang, Silicon acquisition and accumulation in plant and its significance for agriculture, J Integr Agric 17 (2018) 2138−2150. https://doi.org/10.1016/S2095-3119(18)62037-4.*

plant root plasma membrane at physiological pH ranges [16]. Wide variation for Si accumulation has been observed among plant species [15]. This has been attributed to the specific ability of roots to take up Si [1].

Three different types of Si uptake mechanisms have been proposed based on the uptake rates relative to water. The active uptake which is usually even faster than water uptake and pose substantial depletion of Si in the uptake solution. There is also a passive uptake like water uptake where Si levels remain unchanged. The third possible way for Si uptake is the rejective uptake which is slower than water uptake [1]. Indeed, both active and passive mechanisms of Si uptake cooperatively function and mediate Si uptake in high accumulator and intermediate Si plants [17]. To benefit from Si, plants must transport it from soil and many transporters are shown to be involved in the uptake, translocation, and distribution of Si within the crop plants [18]. The discovery of Si transporters by Ma et al. was a great endeavor to reveal molecular mechanisms involved in silicon uptake [7,19]. In this context, a model of Si uptake, transport, and distribution system in rice, maize, and barley has been provided based on the cooperation of influx and efflux transport proteins (Fig. 3.1).

3.2.1 Si uptake by root system

A pioneering study performed by a Japanese group identified the first silicon transporter *OsLsi1* using an induced mutagenesis approach using a rice mutant genotype lacking silicon uptake (*lsi1*) [19]. *Lsi1* is involved in Si influx from external solution into root cells (Fig. 3.1A and B). The expression of this gene was shown to be decreased by Si [19]. In the root, the *Lsi1* expression is lower in the apical compared to basal root regions indicating a higher Si uptake within the mature root than root tips [20]. Also, *Lsi1* is mainly expressed in main root and lateral roots and physiological evidence demonstrated the fact that root hairs did not contribute to Si uptake [21]. They are influx silicon transporter belonging to the nodulin 26-like intrinsic protein (NIP), a subfamily of plant aquaporin. All aquaporins have a conserved hourglass structure with six alpha helix transmembrane domains

(H1−H6) which are joined by five loops (LA−LE). Two of those loops, LB and LE, contain a conserved motif: the Asn-Pro-Ala motifs (NPA) domain. These domains give rise to one of the two constricts. The second constriction is formed by the ar/R selectivity filter consisting of four amino acids. These amino acids are from the helices H2, H5, and loop LE. These conserved NPA domains and selectivity filters are responsible for the specificity of the channel. The *Lsi1*, which transports uncharged silicic acid in particular, belongs to the NIP III group comprising a unique selectivity filter Gly (G), Ser (S), Gly (G), and Arg (R) [7]. This gene is localized on chromosome 2 and comprises five exons and four introns. The cDNA of this gene was 1409 bp and the deduced protein comprised 298 amino acids [19].

The same Japanese group identified the second Si transporter *Lsi2*, which was cloned by using a novel rice mutant (*lsi2*) that is defective in Si uptake [22]. *Lsi2* is an efflux silicon transporter that mediates actively the load of Si out of the cell against the concentration gradient (Fig. 3.1A and B). As it is an active efflux transporter, *Lsi2* is predicted to have 11 transmembrane domains, belonging to the putative anion-channel transporter family where its homology and mechanism are different from the *Lsi1* [9,20]. The transport activity of *Lsi2* is declined by low pH while its efflux activity is decreased by low temperature [12]. *Lsi2* can transport silicic acid against the concentration gradient [9]. *Lsi2* has the same expression pattern and cellular localization like *Lsi1*, but it can be expressed to a lower extent in the root tips with a possibility of higher accumulation in the mature root zones [12]. *Lsi2* is also localized to the plasma membranes of both exodermal and endodermal root cells but at the proximal side [7].

Both *Lsi1* and *Lsi2* have also been identified in barley (*HvLsi1* and *HvLsi2*) [23,24], maize (*ZmLsi1* and *ZmLsi2*) [24,25], wheat (*TaLsi1*) [26], pumpkin (*CmLsi1*) [7], soybean (*GmNIP2−1* and *GmNIP2−2*) [27], cucumber (*CSiT1*, *CSiT2*, and *CsLsi1*) [28,29], and horsetail (*EaNIP3−1*, *EaNIP3−3*, *EaNIP3−4*, *EaLsi2−1*, and *EaLsi2−2*) [30,31].

Both influx and efflux transporters in barley and maize (*HvLsi1/ZmLsi1* and *HvLsi2/ZmLsi2*) are localized in different cells. *HvLsi1/ZmLsi1* is polarly localized to epidermal, hypodermal, and cortical cells at the distal side, while *HvLsi2/ZmLsi2* is localized exclusively on the endodermis without polarities [23,32]. These differences of localizations result in two differential routes of Si transport from the external solution to the xylem vessels in barley/maize compared to rice [7,18].

In rice, Si is taken up from the external solution by *OsLsi1* at the distal side and then is released into the apoplast of aerenchyma which is mediated by *OsLsi2* at the proximal side of exodermal cells. Si is then transported into the stele by both *OsLsi1* and *OsLsi2* through the endodermal cell layer (Fig. 3.1A). While, in barley and maize roots, Si can be taken up from external solution by *HvLsi1/ZmLsi1* localized at the distal side of epidermal and cortical cells and then transported symplastically to the endodermis and then released by Si efflux transporter (*HvLsi2/ZmLsi2*) followed by transport to the stele apoplastically (Fig. 3.1B). The different responses of these plants to Si uptake are described by Ma and Yamaji [18]. They described that rice contains two Casparian strips and both *OsLsi1* and *OsLsi2* cooperate for Si uptake and transport between symplastic and apoplastic pathways, while barley and maize comprises of only one Casparian strip and Si is first taken up from external solution by *HvLsi1* or *ZmLsi1* and then released to the stele by *HvLsi2* or *ZmLsi2*.

3.2.2 Si transport in vascular tissue

Nearly, 90% of Si taken up by roots is translocated to shoots through xylem and via transpiration stream [33]. Besides, higher concentration of Si in xylem (20 mM) and its deposition in the cell wall of xylem vessels, Si must be transported out of xylem prior to its deposition in the shoot [33,34]. In this context, an influx Si transporter *Lsi6* which is responsible for loading of silicic acid from the xylem into xylem parenchyma cells has been identified by Yamaji et al. [14]. *Lsi6* is localized in the adaxial side of xylem parenchyma cells in the leaf sheaths and leaf blades [14] and influencing the Si distribution and deposition in shoots (Fig. 3.1C). The knockout of *Lsi6* does not affect the root uptake of Si, while it causes an increase of Si deposition in the silicified epidermal cells of leaf blades and sheaths and thus mediates increased excretion of Si in the guttation fluid [21]. *Lsi6* has been identified in rice (*OsLsi6*), barley (*HsLsi6*), and maize (*ZmLsi6*) and is expressed in both roots and shoots of these plants [14,24,32].

In rice, distribution of Si in the node and its accumulation in the husk is of high importance for its proper productivity. Indeed, the intervascular transfer of Si between enlarged and diffuse vascular bundles in the node of rice is needed for the delivery of Si to the panicle [21] (Fig. 3.1D). Notably, *Lsi6* is also highly expressed in the first node below the panicles where it is mainly localized at the proximal side of xylem transfer cells located at the outer boundary region of the enlarged large vascular bundles [14]. The knockout of *Lsi6* resulted in decreased

Si accumulation in the panicles, but increased Si accumulation in the flag leaf [14]. Hence *Lsi6* seems to contribute as Si transporter involved in transfer of silicic acid from the enlarged large vascular bundles coming from the roots to the xylem transfer cells. Besides *Lsi6*, Yamaji et al. showed that both *Lsi2* and *Lsi3* are located at the node and involved in the intervascular transfer in rice [35]. *Lsi2* is polarly localized to the bundle sheath cell next to enlarged vascular bundle and xylem transfer cell, while *Lsi3* is in the parenchyma tissues between enlarged vascular bundle and diffuse vascular bundle [25].

3.3 Si accumulation and deposition in different parts of plant

The accumulation of Si in different zones of plant depends on the plant species and have been largely investigated [36]. Roots endodermis was shown to be the main and persistent site of silicification. The accumulation of Si is relatively lower compared to other parts of plant [37]. Lux et al. have found that the Si content of the endodermal cell wall in the basal part of the root was 3.48% after 8 h of silica application to the apical half of the root which was comparable with the Si content in roots completely immersed in + Si solution for 4 h [37]. Moore et al. also showed the same localization of Si in endodermal cell walls with no differences between the proximal and distal sides [38]. The higher concentration was shown either in the inner tangential wall (ITW) and in the endodermis radial walls. This deposition was shown not only in rice root but also in the roots of other crops [15]. However, sorghum deposits Si in its root in a different way either with the ITW of endodermis or via phytolith [39,40].

Si accumulation in shoots depends mainly on its uptake from external solution and its transport via Si transporters and later Si release into xylem. In rice, silicon accumulation in xylem sap can reach upto 6.0 mM in first 30 min of exposure to only 0.5 mM silicon in the growth medium. Furthermore, the silicon concentration increased to 18 mM in the next 8.5 h suggesting that the xylem loading is a rapid process that is accomplished against a concentration gradient in rice [13,22]. In aerial tissue, silicon is unloaded from xylem and transported to peripheral tissues in inflorescence and leaf [15]. Once Si is transported to the shoots, silicic acid is polymerized to amorphous silica in higher concentration via the transpiration stream [13]. Amorphous silica can be deposited in higher concentration in the cell wall of different organs like stem, leave, and hulls [19,41]. This silicon deposited in the cell walls is thought to strengthen the plant and protect the plants against different biotic and abiotic stresses (Fig. 3.1C).

3.4 Conclusion and future perspective

There is no doubt that Si chemistry is complex due to certain soil influencing factors such as pH, texture, and organic matter which need to be further explored experimentally. This is basically due to alkalinization of soil pH because of the type of silicates used in different experiments and the fact that even the freely available form of Si monosilicic acid might be soluble up to a certain concentration. Hence further research on the impact of Si on soil properties need to be considered. In addition, the discovery of Si transporter provides the opportunities to further investigate the Si uptake and accumulation in different Si-accumulating and nonaccumulating species and to profoundly decipher the mechanism(s) by which Si positively influences on growth and development of crop plants particularly under different biotic and abiotic stresses.

References

[1] Liang Y, Nikolic M, Bélanger R, Gong H, Song A. Silicon in agriculture: from theory to practice. Netherlands: Springer; 2015. Available from: https://doi.org/10.1007/978-94-017-9978-2.

[2] Cheong YWY, Halais P. Needs of sugar cane for silicon when growing in highly weathered latosols. Exp Agric 1970;6:99–106. Available from: https://doi.org/10.1017/S0014479700000144.

[3] Imtiaz M, Rizwan MS, Mushtaq MA, Ashraf M, Shahzad SM, Yousaf B, et al. Silicon occurrence, uptake, transport and mechanisms of heavy metals, minerals and salinity enhanced tolerance in plants with future prospects: a review. J Environ Manage 2016;183:521–9. Available from: https://doi.org/10.1016/j.jenvman.2016.09.009.

[4] Coskun D, Deshmukh R, Sonah H, Menzies JG, Reynolds O, Ma JF, et al. The controversies of silicon's role in plant biology. New Phytol 2019;221:67–85. Available from: https://doi.org/10.1111/nph.15343.

[5] Richmond KE, Sussman M. Got silicon? The non-essential beneficial plant nutrient. Curr Opin Plant Biol 2003;6:268–72. Available from: https://doi.org/10.1016/S1369-5266(03)00041-4.

[6] Yan GC, Nikolic M, Ye MJ, Xiao ZX, Liang YC. Silicon acquisition and accumulation in plant and its significance for agriculture. J Integr Agric 2018;17:2138−50. Available from: https://doi.org/10.1016/S2095-3119(18)62037-4.

[7] Ma JF, Yamaji N, Mitani-Ueno N. Transport of silicon from roots to panicles in plants. Proc Jpn Acad Ser B 2011;87:377−85. Available from: https://doi.org/10.2183/pjab.87.377.

[8] Epstein E. SILICON. Annu Rev Plant Physiol Plant Mol Biol 1999;50:641−64. Available from: https://doi.org/10.1146/annurev.arplant.50.1.641.

[9] Yamaji N, Ma JF. Further characterization of a rice silicon efflux transporter, Lsi2. Soil Sci Plant Nutr 2011;57:259−64. Available from: https://doi.org/10.1080/00380768.2011.565480.

[10] Hodson MJ, White PJ, Mead A, Broadley MR. Phylogenetic variation in the silicon composition of plants. Ann Bot 2005;96:1027−46. Available from: https://doi.org/10.1093/aob/mci255.

[11] Ma JF, Miyake Y, Takahashi E. Chapter 2 Silicon as a beneficial element for crop plants. In: Datnoff LE, Snyder GH, Korndörfer GH, editors. Silicon in Agriculture, Studies in Plant Science. Amsterdam: Elsevier; 2001. p. 17−39. Available from: https://doi.org/10.1016/S0928-3420(01)80006-9.

[12] Ma JF, Yamaji N, Tamai K, Mitani N. Genotypic difference in silicon uptake and expression of silicon transporter genes in rice. Plant Physiol 2007;145:919−24. Available from: https://doi.org/10.1104/pp.107.107599.

[13] Mitani N, Ma JF. Uptake system of silicon in different plant species. J Exp Bot 2005;56:1255−61. Available from: https://doi.org/10.1093/jxb/eri121.

[14] Yamaji N, Mitatni N, Ma JF. A transporter regulating silicon distribution in rice shoots. Plant Cell 2008;20:1381−9. Available from: https://doi.org/10.1105/tpc.108.059311.

[15] Mandlik R, Thakral V, Raturi G, Shinde S, Nikolić M, Tripathi DK, et al. Significance of silicon uptake, transport, and deposition in plants. J Exp Bot 2020;71:6703−18. Available from: https://doi.org/10.1093/jxb/eraa301.

[16] Raven JA. Cycling silicon—the role of accumulation in plants. New Phytol 2003;158:419−21.

[17] Liang Y, Sun W, Zhu YG, Christie P. Mechanisms of silicon-mediated alleviation of abiotic stresses in higher plants: a review. Environ Pollut 2007;147:422−8. Available from: https://doi.org/10.1016/j.envpol.2006.06.008.

[18] Ma JF, Yamaji N. A cooperative system of silicon transport in plants. Trends Plant Sci 2015;20:435−42. Available from: https://doi.org/10.1016/j.tplants.2015.04.007.

[19] Ma JF, Tamai K, Yamaji N, Mitani N, Konishi S, Katsuhara M, et al. A silicon transporter in rice. Nature 2006;440:688−91. Available from: https://doi.org/10.1038/nature04590.

[20] Yamaji N, Ma JF. Spatial distribution and temporal variation of the rice silicon transporter Lsi1. Plant Physiol 2007;143:1306−13. Available from: https://doi.org/10.1104/pp.106.093005.

[21] Ma JF, Goto S, Tamai K, Ichii M. Role of root hairs and lateral roots in silicon uptake by rice. Plant Physiol 2001;127:1773−80. Available from: https://doi.org/10.1104/pp.010271.

[22] Ma JF, Tamai K, Ichii M, Wu GF. A rice mutant defective in Si uptake. Plant Physiol 2002;130:2111−17. Available from: https://doi.org/10.1104/pp.010348.

[23] Chiba Y, Mitani N, Yamaji N, Ma JF. HvLsi1 is a silicon influx transporter in barley. Plant J 2009;57:810−18. Available from: https://doi.org/10.1111/j.1365-313X.2008.03728.x.

[24] Mitani N, Yamaji N, Ma JF. Identification of maize silicon influx transporters. Plant Cell Physiol 2009;50:5−12. Available from: https://doi.org/10.1093/pcp/pcn110.

[25] Mitani N, Chiba Y, Yamaji N, Ma JF. Identification and characterization of maize and barley Lsi2-like silicon efflux transporters reveals a distinct silicon uptake system from that in rice. Plant Cell 2009;21:2133−42. Available from: https://doi.org/10.1105/tpc.109.067884.

[26] Montpetit J, Vivancos J, Mitani-Ueno N, Yamaji N, Rémus-Borel W, Belzile F, et al. Cloning, functional characterization and heterologous expression of TaLsi1, a wheat silicon transporter gene. Plant Mol Biol 2012;79:35−46. Available from: https://doi.org/10.1007/s11103-012-9892-3.

[27] Deshmukh RK, Vivancos J. Identification and functional characterization of silicon transporters in soybean using comparative genomics of major intrinsic proteins in Arabidopsis and rice. Plant Mol Biol 2013;83:303−15. Available from: https://doi.org/10.1007/s11103-013-0087-3.

[28] Wang H-S, Yu C, Fan P-P, Bao B-F, Li T, Zhu Z-J. Identification of two cucumber putative silicon transporter genes in Cucumis sativus. J Plant Growth Regul 2015;34:332−8. Available from: https://doi.org/10.1007/s00344-014-9466-5.

[29] Sun H, Guo J, Duan Y, Zhang T, Huo H, Gong H. Isolation and functional characterization of CsLsi1, a silicon transporter gene in Cucumis sativus. Physiol Plant 2017;159:201−14.

[30] Grégoire C, Rémus-Borel W, Vivancos J, Labbé C, Belzile F, Bélanger RR. Discovery of a multigene family of aquaporin silicon transporters in the primitive plant Equisetum arvense. Plant J 2012;72:320−30. Available from: https://doi.org/10.1111/j.1365-313X.2012.05082.x.

[31] Vivancos J, Deshmukh R, Grégoire C, Rémus-Borel W, Belzile F, Bélanger RR. Identification and characterization of silicon efflux transporters in horsetail (Equisetum arvense). J Plant Physiol 2016;200:82−9. Available from: https://doi.org/10.1016/j.jplph.2016.06.011.

[32] Yamaji N, Chiba Y, Mitani-Ueno N, Ma JF. Functional characterization of a silicon transporter gene implicated in silicon distribution in barley. Plant Physiol 2012;160:1491−7. Available from: https://doi.org/10.1104/pp.112.204578.

[33] Yoshimoto N, Takahashi H, Smith FW, Yamaya T, Saito K. Two distinct high-affinity sulfate transporters with different inducibilities mediate uptake of sulfate in Arabidopsis roots. Plant J 2002;29:465−73.

[34] Maksimović JD, Bogdanović J, Maksimović V, Nikolic M. Silicon modulates the metabolism and utilization of phenolic compounds in cucumber (Cucumis sativus L.) grown at excess manganese. J Plant Nutr Soil Sci 2007;170:739−44. Available from: https://doi.org/10.1002/jpln.200700101.

[35] Yamaji N, Sakurai G, Mitani-Ueno N, Ma JF. Orchestration of three transporters and distinct vascular structures in node for intervascular transfer of silicon in rice. Proc Natl Acad Sci. 2015;112:11401−6. Available from: https://doi.org/10.1073/pnas.1508987112.

[36] Parry DW, Soni SL. Electron-probe microanalysis of silicon in the roots of Oryza sativa L. Ann Bot 1972;36:781−3. Available from: https://doi.org/10.1093/oxfordjournals.aob.a084633.

[37] Lux A, Luxová M, Abe J, Tanimoto E, Hattori T, Inanaga S. The dynamics of silicon deposition in the sorghum root endodermis. New Phytol 2003;158:437−41. Available from: https://doi.org/10.1046/j.1469-8137.2003.00764.x.

[38] Moore KL, Schröder M, Wu Z, Martin BGH, Hawes CR, McGrath SP, et al. High-resolution secondary ion mass spectrometry reveals the contrasting subcellular distribution of arsenic and silicon in rice roots. Plant Physiol 2011;156:913−24. Available from: https://doi.org/10.1104/pp.111.173088.

[39] Parry DW, Kelso M. The distribution of silicon deposits in the roots of *Molinia caerulea* (L.) Moench. and *Sorghum bicolor* (L.) Moench. Ann Bot 1975;39:995−1001. Available from: https://doi.org/10.1093/oxfordjournals.aob.a085043.

[40] Metcalfe CR. Anatomy of the monocotyledons. 1. Gramineae. Oxford: Clarendon Press; 1960.

[41] Prychid CJ, Rudall PJ, Gregory M. Systematics and biology of silica bodies in monocotyledons. Bot Rev 2003;69:377−440. Available from: https://doi.org/10.1663/0006-8101(2004)069[0377:SABOSB]2.0.CO;2.

4

Biological function of silicon in a grassland ecosystem

Danghui Xu[1], Mohammad Anwar Hossain[2] and Robert Henry[3]

[1]State Key Laboratory of Grassland Agro-ecosystems/School of Life Science, Lanzhou University, Lanzhou, P.R. China
[2]Department of Genetics and Plant Breeding, Bangladesh Agricultural University, Mymensingh, Bangladesh
[3]Queensland Alliance for Agriculture and Food Innovation, University of Queensland, Brisbane, QLD, Australia

4.1 Introduction

Silicon (Si) is one of the most abundant elements in the soil and the second most abundant element in the earth's crust [1] and a "quasi-essential" element for plants. Plants take up Si in the form of silicic acid, $Si(OH)_4$ [2,3] with concentrations in the soil solution from 0.1 to 0.6 mmol L^{-1}[1,4]. The Si concentrations in plants range from 1 to 100 g kg^{-1} by weight [5], and this is the largest range in concentration found among the mineral elements and greater than that for some micronutrients [6]. These differences may be related to the difference in Si uptake among difference species. There are three modes of Si uptake, exclusive, passive and active uptake, which correspond to the plant species of low, medium and high Si accumulation [4,7]. It is well known that silicon is not only a beneficial element for plants growth [8,9], but also plays an important role in improving resistance against biotic stress (e.g., pest and pathogen) and alleviating abiotic stresses (drought, salt, temperature, strong wind and heavy rain) [10−12]. Si is therefore highly beneficial for growth of individual plants and productivity of plant communities [13,14].

Alpine meadow is one of the most important types of grassland ecosystems on the Qinghai−Tibet plateau (QTP), covering approximately 40% of the QTP's territory [15−17]. Alpine meadow plays an important role in many ecosystem functions such as carbon storage, biodiversity conservation, climate control, nutrients cycling, and provides important ecosystems services for millions of inhabitants living both upstream and downstream of the QTP [18,19]. However, the QTP has been experiencing an increase in fertilizer production and N deposition [20,21], which has enhanced the availability of N in soils and has greatly influenced the ecological processes in alpine ecosystem [22]. In the future, this issue may be much more evident with global climate change, and cause serious impacts on ecosystem structure and function, making it important to examine the effects of N fertilization with Si and without Si on ecosystem functions of the QTP's alpine meadow [23−25].

Phosphorus (P) fertilization has been widely employed to restore degraded grasslands and improve annual aboveground net primary productivity (ANPP) in the QTP's alpine meadow [26−28]. Some studies have reported that P fertilizer application had little impact on plant community composition [29−32], while others have shown that P addition decreased plant species diversity and richness [33−35] and altered the performance of different functional groups [25,27]. Despite this, to date, the influence of P enrichment on alpine ecosystem plant community composition and productivity and the performance of different functional groups remains unclear.

Si is a key element for metabolic pathways and turnover of carbon (C), N and P in many terrestrial ecosystem [36,37], so the interest in the impacts of Si on N and P concentrations in plants and productivity is increasing [38–40]. Most studies have reported that there was no significant correlation between Si and N concentrations [41]. However, our previous study reported that Si application promoted N levels in nonleguminous plants in alpine meadows [40], despite the fact that the N content in wetland plants showed negative correlations with the content of Si [36,37]. Similarly, the application of Si in a P-deficient habitat also increased P concentrations in plants [38,39,42] but suppressed P levels in an environment of excessive P [43,44].

Some studies found that Si facilitated grasses in positively altering the utilization of nutrients (N and P) and particularly alleviating P deficiency in grasslands improving both the growth of vegetation and litter decomposition in soil [41,45]. Furthermore, many studies have observed that Si applications enhanced the net photosynthetic rate of plants [40,46–49] and could improve plant biomass production [39,45,50]. Examining the responses of plant community composition and productivity and performance of different functional groups to N and P amendments with or without Si addition can not only contribute to deepening the insights into the interactions among N, P, and Si but also provide essential guidelines for grassland management. However, the possible role of silicon in promoting plant N and P uptake, enhancing net photosynthetic rate of plants and improving plant biomass production in alpine grassland has not been resolved to date. In this chapter, we will give an overview of the recent progress made in alpine grassland about N or P addition with or without Si in influencing plant communities, functional groups and individual plants. We also address the interaction of N and Si and P and Si in alpine grassland.

4.2 Silicon distribution in meadow plants

Grasslands are among the most significant ecosystem types in the world, with a high net primary productivity and Si content [51–53]. They play a significant role in terrestrial Si cycle and global carbon sequestration [54,55]. At the plant community level, Schaller et al. [37] found that Si content varied from 1.96 to 24.9 g kg^{-1} in Central European temperate grasslands. Ji et al. [56] recently reported that Si content varied from 0.12 to 36.8 g kg^{-1} in typical steppe and from 0.18 to 57.7 g kg^{-1} in meadow steppe in eastern Inner Mongolia and northern Hebei, China. The range of variation in Si content in European temperate grasslands is narrower than that in the Chinese steppe. This may be due to the difference in species and environmental conditions. Ji et al. [56] also reported that the average Si content in plant above-ground parts is 2.15 ± 0.92 g kg^{-1} in meadow steppe and is 3.92 ± 0.10 g kg^{-1} in typical steppe. In addition, the Si content was negatively correlated with the mean annual precipitation, but positively correlated with the mean annual temperature. So, not only mean annual precipitation and mean annual temperature influence Si content and absorption of plant above-ground parts, but also soil pH, soil Si availability and other soil conditions directly and indirectly influence the Si content and absorption. This suggests that future studies should continue to examine the effects of other factors on grassland Si distribution.

The Si content of plants was found to vary considerably among the different functional groups and Si content was increased with species richness and functional group richness. But Schaller et al. [37] reported that species richness had a slightly negative effect on Si concentrations, while functional group richness did not affect Si concentrations and Si concentrations depended on functional group identity. The presence of grasses had positive effects on Si stocks for grasses are known as Si accumulators [6] and legume presence slightly decreased Si stocks as legumes are calcium accumulators [57–59]. In contrast, Si concentrations were higher if grasses were present than if grasses were absent. Si concentrations were lower aboveground parts in communities with legumes. Grasses presence increased the Si stocks by 140%. Both the presence of specific plant functional groups and species diversity altered Si stocks, whereas Si concentration was affected mostly by the presence of specific plant functional groups [37,60]. Species diversity had a negative effect on plant Si accumulation [37]. The relationship between plant Si accumulation and the structure of the plant community and plant functional groups indicated that plant Si accumulation may in turn affect ecosystem processes such as plant litter decomposition and nutrient cycling in grasslands.

The Si content of plants varies considerably between different species. Based on the mechanisms of Si uptake, plant species can be divided into three types, Si accumulator, Si intermediate and Si excluder [7]. Ji et al. [56] investigated Si content in the aboveground parts of 108 plants from typical steppe and 76 plants from meadow steppe, and found that Si content of plants varied considerably from 0.18 ± 0.01 g kg^{-1} (*Aquilegia yabeana*) to 57.7 ± 0.22 g kg^{-1} (*Equisetum pratense*). The average Si content in Equisetopsida was 57.7 ± 0.22 g kg^{-1}, in Monocotyledoneae was 9.29 ± 2.1 g kg^{-1} and in Dicotyledoneae was 3.69 ± 1.6 g kg^{-1}, respectively (Fig. 4.1).

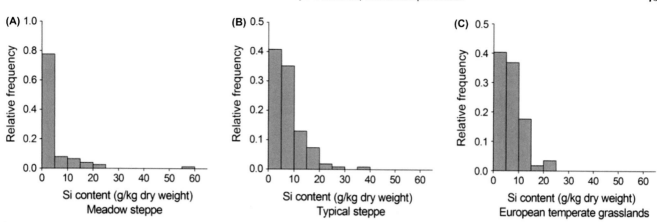

FIGURE 4.1 Relative frequency of Si content of (A) 76 species in meadow steppe, (B) 108 species in typical steppe and (C) 57 species in European temperate grasslands. The data for European temperate grasslands are from Schaller et al. [37] and for meadow steppe and typical steppe northern China are from Ji et al. [56].

This indicated that Si content decreased from the Equisetopsida to Dicotyledoneae. Some studies found that ferns accumulate more Si than angiosperms [5,61]. This may be because many ferns use a meshwork of SiO_2 fibres in the cell walls to reinforce mechanical support of the frond weight as ferns lack stable mechanical tissues for support [61]. Additionally, most monocot species are Si accumulators and their Si content was higher than that of eudicot species (Graminales are the higher Si accumulators) [1,56,62]. Other studies also reported that high Si content and absorption was not a general feature of monocot species [5,63]. For example, Ji et al. [56] reported that two monocot species of Liliales and Asparagales have lower Si content than that of eudicot species. So, plant phylogeny may be the main factor to influences the Si content of individual species.

4.3 Silicon in relation to plant community structure in alpine meadow

Many studies have demonstrated that N fertilization strongly contributes to the effective restoration of degraded grasslands and the increase of the cover of grasses and aboveground biomass as N is the main limitation factor for plant growth [37,49,64–66]. Excessive N supply to grasses often decreased plant species diversity and richness and caused changes in plant community structure and species loss in grassland ecosystems [25,32,67,68]. Moreover, N addition altered plant functional group richness and biomass due to different plant nutrient use efficiencies [69,70].

A study found that the application of Si only to grassland ecosystem did not affect the species richness of the community. Silicon in combination with N resulted in higher species richness than N addition only. Compared to N addition only with 70, 140 and 210 kg·ha^{-1}, 14.36 Si kg·ha^{-1} application with 70, 140 and 210 kg·ha^{-1} N increased species richness 2.75, 3.75 and 1.58, respectively (Fig. 4.2). A highly significant interaction between Si and N was found in species richness, suggesting that Si incorporation with N may protect species diversity in plant community [25,70]. Some research suggested that Si availability during plant growth may strongly impact species richness and the relationship between legumes and grasses in mixed grasslands [37,38,60,70]. Several mechanisms have been used to account for the increase in species richness under elevated N with Si rather than N addition only. Neu et al. [39] and Klotzbücher et al. [71] showed that Si application enhances Si accumulation in plant biomass and regulates plant carbon accumulation and lignin biosynthesis, which in turn regulates plant growth rate and plant height and determines plant species richness and species diversity of the community [72].

The Si fertilization of grassland can enhance plant aboveground biomass of whole plant community. Compared to control, Si fertilization with 7.18, 14.36 and 21.54 kg ha^{-1}increased plant aboveground biomass 60.1, 87.7 and 112.5 g m^{-2} respectively. Additional, Si in combination with N resulted in higher aboveground biomass than N addition only. Compared to N addition only with 70, 140 and 210 kg ha^{-1}, 14.36 Si kg·ha^{-1} application with 0, 140 and 210 kg ha^{-1} N increased aboveground biomass 56.1, 43.2 and 67.8 g m^{-2}, respectively (Fig. 4.3) [25]. A highly significant interaction between Si and N was found in the aboveground biomass [25,40]. Some studies found that Si application to crops enhanced plant biomass

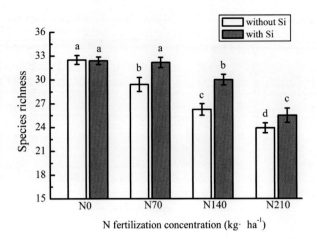

FIGURE 4.2   Mean species richness of the community after N addition only, Si addition only, and N addition with Si in alpine meadow, n = 6. Different letters above bars indicate significant difference between different treatments. *Source: From Xu DH, Fang XW, Zhang RY, et al. Influences of nitrogen, phosphorus and silicon addition on plant productivity and species richness in an alpine meadow. AoB Plants, 2015; 7: plv125.*

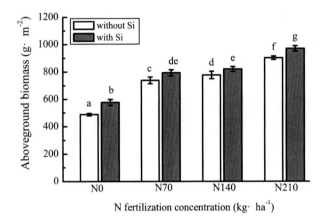

FIGURE 4.3   Mean aboveground biomass of the community after N addition only, Si addition only, and N addition with Si in alpine meadow, *n* = 6. Different letters above bars indicate significant difference between different treatments. *Source: From Xu DH, Fang XW, Zhang RY, et al. Influences of nitrogen, phosphorus and silicon addition on plant productivity and species richness in an alpine meadow. AoB Plants, 2015; 7: plv125.*

accumulation, including aboveground and belowground biomass and crop yield, especially under stressful conditions [45,73,74]. Thus Si accumulation in plants can exert a feed forward effect on carbon capture in plant biomass.

For the four functional groups, species richness differed in their response to Si and N fertilization [25,70]. Silicon or N fertilization only significantly increased species richness of grasses. N fertilization only significantly decreased species richness of legumes and forbs. Nitrogen and Si addition together led to significantly higher species richness of grasses than N or Si addition alone. There were no significant differences of species richness of sedges among the different treatments. The positive response of grasses could have been primarily ascribed to the enhancement of the dominant species owing to their ability to quickly explore available resources relative to other species [60]. In contrast, the species of sedges and forbs are slower-growing or shorter, so they had the disadvantage of having to compete for light more than the grasses. The competitive advantage of legume species is proposed to decline with increased N availability because of their inherent N-fixing characteristics [39].

Silicon fertilization alone had no effect on aboveground biomass of grasses and sedges. Aboveground biomass of grasses and sedges was significantly higher in Si fertilization with N than with the same amount of N only. Increasing soil bioavailable Si can enhance the competitiveness of Si-accumulating grasses while maintaining or improving productivity of the grasslands in arid and semiarid regions [75]. Silicon in combination with N resulted in higher aboveground biomass of forbs and legumes than N fertilization only. Two-way ANOVA revealed that there was a significant interaction effects between Si and N in aboveground biomass of forbs, and between different years in all functional groups [25,70]. Some studies found that the potential of Si in mediating plant carbon assimilation provided a reference for potential manipulation of long-term carbon sequestration via biomass carbon accumulation in Si-accumulator dominated grasslands [48]. Si-accumulator plants can deposit silica in plant tissues, which can strengthen cell walls, improve leaf erectness or light capture and enhance plant growth [6,76]. This provides additional evidence for Si regulation of plant growth and aboveground biomass accumulation in alpine meadows and under environmental stresses [38−40,47−49].

4.4 Silicon in relation to plant carbon, nitrogen and phosphorus concentration

Plant carbon concentrations and stocks differ in their response to the Si fertilization in different ecosystem. Our study found that Si fertilization had no effect on plant carbon content both in individual species and whole plant communities [77−79]. Si application enhanced plant Si accumulation. Studies on the tradeoff between biomass carbon reduction and Si accumulation indicated that an increase in Si content of plant by 1% on average caused a reduction of biomass carbon by less than 1% [39,71]. Liu et al. [48] found that the carbon content of plants was negatively correlated with the plant Si content under salt stress condition. They suggested the potential of Si in mediating plant salinity and C assimilation, which provides a reference for potential manipulation of long-term C sequestration and biomass C accumulation in Si-accumulator dominated grasslands. The differences in these findings may be attributed to (1) different soil texture and (2) different Si fertilizer addition processes. Our study focused on Si-addition studies on the Qinghai-Tibetan Plateau, whereas Liu et al. [48] focused on the influence of Si availability on the biomass carbon and phytolith-occluded carbon in grassland under high-salinity conditions at northeastern Inner Mongolia. Moreover, the Si was added directly to soil in our study, rather than being Si differences in the soil.

In our previous studies, we selected five common species that were present in all plots (Si fertilization, N fertilization and N plus Si fertilization) from each of four functional groups: *Kobresia capillifolia* for the sedges, *Elymus nutans* for the grasses, *Oxytropis kansuensis* for the legumes, *Anemone rivularis* and *Potentilla fragarioides* for the forbs. Compared to without Si, Si incorporation significantly increased leaf N concentration by 12.4%, 17.9%, 10.4% and 11.9% in species of *K. capillifolia*, *E. nutans*, *A. rivularis* and *P. fragarioides*, respectively under N addition of 210 kg ha^{-1} (Fig. 4.4). Two-way ANOVA revealed a significant interaction effect between N and Si on leaf N concentration in species of *K. capillifolia*, *E. nutans*, *A. rivularis* and *P. fragarioides*. In summary, our data suggested that the improvement of the five species can be achieved through a small Si supply in N application [40]. This finding agrees with a previous report examining the role of Si in growth and leaf N concentrations of *Phragmites australis* [38,77−79]. Neu et al. [39] found that Si supply influenced nitrogen use efficiency, by increasing crops grain yield and plant aboveground biomass production. In grasses, current research addressed the beneficial role of Si for nitrogen concentration, nutrient cycling and plant biomass production [38,40,70,80]. Schaller et al. [38] found that grasses show lower N uptake under low Si conditions, potentially further hampering their ability to compete initially in a low Si, low available N environment. Johnson et al. [81] reported that although N addition did not affect the N concentration in legumes, silicon addition can increase root nodulation and synthesis of essential amino acids in a legume. Similar results have been reported in other studies. Si has been shown to increase root nodulation and nitrogen fixation in legumes [71,82−84], but the mechanisms for this have yet to be demonstrated. The simplest explanation is that increased root growth increased potential invasion sites for rhizobial bacteria [84] and increased nodule density [81,82].

The application of Si in an alpine meadow experiment indicates that phosphorus (P) concentrations in grass and legume functional groups were higher with Si than without Si fertilization (Fig. 4.5). Si fertilization only or Si fertilization with N can increase plant P concentration and soil net N nitrification, ammonification, and mineralization [25,40,70,85]. This finding agrees with many studies which reported that Si application to crops increased plant P uptake, especially under P deficient soil conditions [28,38,42,86,87]. However, others studies found that high Si supply can decrease plant P uptake at both medium and high P levels [28,43,88]. So, Li et al. [45] predicted that the application of Si to plants can suppress P uptake into the plant shoots under luxurious and excessive P levels but improve plant P absorption under P deficient soil conditions. The mechanisms by which Si suppresses the uptake of excessive P under medium and high P levels and improvement of P uptake under P deficient soil conditions are as follows: Si accumulation in rice shoots can decrease plant P uptake by down-regulating inorganic P transporter genes in the roots under medium and high P levels [88]. Under P deficient soil conditions, Si fertilization can enhance plant P uptake by increasing organic acids exudation from roots to activate inorganic P in the rhizosphere and up-regulating inorganic P transporter genes in the roots [42]. However, the quantity of P required varies largely from gramineous plants such as *K. capillifolia* and *E. nutans* to

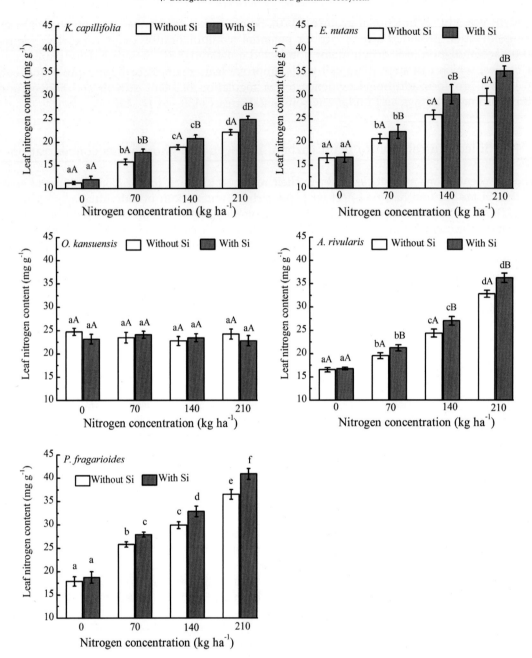

FIGURE 4.4 Leaf N concentration in five species under different nitrogen incorporated with silicon (grey) and without silicon (white). Vertical bars indicate standard deviation of the mean ($n = 6$). Different small letters indicate significant differences among nitrogen rates and different capital letters indicate significant differences between with silicon and without silicon addition ($P < 0.05$). *Source: From Xu DH, Gao XG, Gao TP, et al. Interactive effects of nitrogen and silicon addition on growth of five common plant species and structure of plant community in alpine meadow. Catena, 2018; 169: 80–89; https://doi.org/10.1016/j.catena.2018.05.017.*

dicotyledonous plants such as *A. rivularis* and *P. fragarioides* [89]. The beneficial effects of Si on modulation of P uptake suggest that the fertilization with Si in alpine meadow strongly impacts the plant P concentration and biogeochemical cycle of P [45].

4.5 Silicon in relation to plant physiological aspects in presence of N-fertilization

Silicon fertilization alone increased net photosynthetic rate (Pn) of *K. capillifolia, E. nutans, O. kansuensis, A. rivularis* and *P. fragarioides* and increased instantaneous water-use efficiency (WUEi) of *K. capillifolia* and *E. nutans,*

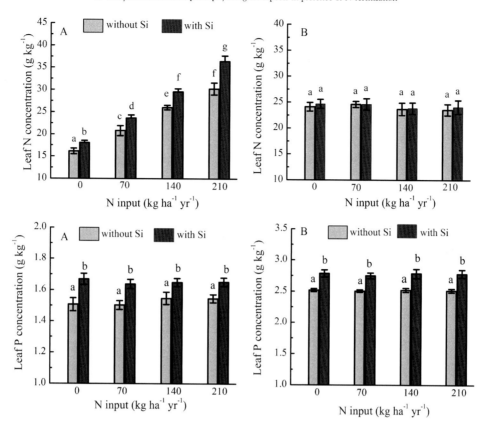

FIGURE 4.5 P concentration of grass (A) and legume (B) plant functional groups under fertilization by N with Si (N + Si) and without Si (N-only).

TABLE 4.1 Plant net photosynthetic rate (Pn), stomatal conductance (gs), transpiration rate (Tr), and instantaneous water-use efficiency (WUEi) in the five common species under nitrogen with incorporation of silicon or without silicon addition. Mean values ± SD, $n = 6$.

Species	Levels of N (kg ha^{-1})	Pn (μmolCO$_2$ m^{-2} s^{-1})		gs (mol m^{-2} s^{-1})		Tr (mmol H$_2$O m^{-2} s^{-1})		WUEi (μmol CO$_2$ mmol H$_2$O)	
		Without Si	**With Si**	**Without Si**	**With Si**	**Without Si**	**With Si**	**Without Si**	**With Si**
Kobresia capillifolia	0	42 ± 1.45aA	48 ± 1.58aB	0.36 ± 0.02aA	0.43 ± 0.02aB	6.9 ± 0.24aA	6.1 ± 0.26aB	6.1 ± 0.09aA	8.0 ± 0.32aB
	70	56 ± 0.88bA	63 ± 0.87bB	0.44 ± 0.01bA	0.45 ± 0.01aA	8.2 ± 0.42bA	7.7 ± 0.31bB	6.8 ± 0.45bA	8.3 ± 0.33aB
	140	64 ± 0.85cA	69 ± 1.73cB	0.46 ± 0.01bA	0.46 ± 0.01aA	9.2 ± 0.28cA	8.0 ± 0.44cB	6.9 ± 0.39bA	9.3 ± 0.25bB
	210	67 ± 1.51dA	74 ± 1.76dA	0.51 ± 0.01cA	0.57 ± 0.01bB	9.3 ± 0.31cA	8.3 ± 0.36cB	7.3 ± 0.47cA	9.5 ± 0.14bB
Euphorbia nutans	0	34 ± 0.88aA	43 ± 1.23aB	0.39 ± 0.01aA	0.41 ± 0.01aA	6.3 ± 0.32aA	5.6 ± 0.11aB	5.4 ± 0.29aA	7.7 ± 0.17aB
	70	43 ± 1.15bA	57 ± 0.73bB	0.42 ± 0.01bA	0.55 ± 0.02bB	7.2 ± 0.28bA	7.2 ± 0.36bA	5.9 ± 0.21aA	7.9 ± 0.21bB
	140	51 ± 1.58cA	57 ± 1.53bB	0.47 ± 0.01cA	0.54 ± 0.02bB	8.2 ± 0.18cA	7.5 ± 0.25bB	6.2 ± 0.17bA	7.6 ± 0.13aB
	210	54 ± 1.15dA	63 ± 1.11cB	0.50 ± 0.01dA	0.56 ± 0.01cB	8.3 ± 0.15cA	8.1 ± 0.26cA	6.5 ± 0.21bA	7.7 ± 0.11aB
Oxytropis kansuensis	0	14 ± 0.41dA	15 ± 0.47bA	0.32 ± 0.01cA	0.43 ± 0.01dB	4.3 ± 0.27dA	4.8 ± 0.17bB	3.1 ± 0.14aA	3.2 ± 0.21aA
	70	9.6 ± 0.58cA	13 ± 0.34bB	0.35 ± 0.01dA	0.37 ± 0.01cA	3.2 ± 0.12cA	3.5 ± 0.26aA	3.1 ± 0.13aA	3.2 ± 0.16aA
	140	8.7 ± 0.76bA	10 ± 0.24aB	0.29 ± 0.01bA	0.31 ± 0.01aB	2.9 ± 0.26bA	3.5 ± 0.25aB	3.1 ± 0.26aA	3.0 ± 0.15aA
	210	6.9 ± 0.61aA	9.8 ± 0.64aB	0.22 ± 0.01aA	0.32 ± 0.01aB	2.4 ± 0.16aA	3.4 ± 0.22aB	2.9 ± 0.16aA	3.0 ± 0.22aA
Anemone rivularis	0	21 ± 0.67cA	25 ± 0.71dB	0.31 ± 0.01cA	0.35 ± 0.01cB	6.1 ± 0.27aB	6.7 ± 0.29dA	3.5 ± 0.17cA	3.6 ± 0.29cB
	70	17 ± 0.63bA	21 ± 0.67cB	0.27 ± 0.01bA	0.31 ± 0.01bB	6.7 ± 0.33bB	6.4 ± 0.17cA	2.6 ± 0.23bA	3.3 ± 0.22bB

(Continued)

TABLE 4.1 *(Continued)*

Species	Levels of N (kg ha^{-1})	Pn (μmolCO$_2$ m^{-2} s^{-1})		gs (mol m^{-2} s^{-1})		Tr (mmol H$_2$O m^{-2} s^{-1})		WUEi (μmol CO$_2$ mmol H$_2$O)	
		Without Si	With Si	Without Si	With Si	Without Si	With Si	Without Si	With Si
	140	16 ± 0.85bA	19 ± 0.91bB	0.28 ± 0.01bA	0.30 ± 0.01bB	6.5 ± 0.25aB	5.8 ± 0.31bA	2.4 ± 0.15aA	3.2 ± 0.11aB
	210	14 ± 0.77aA	16 ± 0.74aB	0.23 ± 0.01aA	0.28 ± 0.01aB	6.1 ± 0.22aB	5.0 ± 0.25aA	2.3 ± 0.17aA	3.2 ± 0.13aB
Potentilla fragarioides	0	13 ± 0.74cA	14 ± 0.65cB	0.27 ± 0.01cA	0.33 ± 0.01dB	3.6 ± 0.26bA	3.7 ± 0.11bA	3.5 ± 0.14cA	3.8 ± 0.26bB
	70	9.9 ± 0.39bA	12 ± 0.54cB	0.22 ± 0.01bA	0.24 ± 0.01cB	3.9 ± 0.16cB	3.2 ± 0.15aA	2.5 ± 0.16bA	3.0 ± 0.16aB
	140	8.7 ± 0.37aA	10 ± 0.29bB	0.19 ± 0.01aA	0.22 ± 0.01bB	3.9 ± 0.14cB	3.6 ± 0.25bA	2.3 ± 0.17aA	2.8 ± 0.22aB
	210	8.8 ± 0.61aA	9.9 ± 0.37bB	0.18 ± 0.01aA	0.19 ± 0.01aA	3.2 ± 0.15aA	3.4 ± 0.26aA	2.1 ± 0.17aA	2.9 ± 0.15aB

Different small letters within a column indicate significant differences across nitrogen rates and different capital letters within a line indicate significant differences between with silicon or without silicon addition according to LSD test ($P < 0.05$) with one-way ANOVA.
From Xu DH, Gao XG, Gao TP, et al. Interactive effects of nitrogen and silicon addition on growth of five common plant species and structure of plant community in alpine meadow. Catena, 2018; 169: 80−89; Doi: 10.1016/j.catena.2018.05.017.

but did not affect that of *O. kansuensis, A. rivularis* and *P. fragarioides* (Table 4.1). Compared to without Si, N incorporation with Si significantly increased Pn by 10.4, 16.7, 42.0, 14.3 and 12.5% in *K. capillifolia, E. nutans, O. kansuensis, A. rivularis* and *P. fragarioides*, respectively under N addition of 210 kg ha^{-1}[40]. The observation that Si application increases plant net photosynthesis rate and water use efficiency has been widely confirmed [76,90,91]. With exogenous Si application, the expression of some key genes related to photosynthesis (e.g., photosynthetic pigment) is increased even under unstressed conditions [42,92]. The feed-forward stimulation of Si on leaf gas exchange characteristics maybe fundamentally associated with enhanced mesophyll conductance and photosynthesis pigments, which leads to higher chloroplastic CO$_2$ concentrations and an increased rate of photosynthetic [76]. Taken together, Si fertilization in alpine meadow imposes a positive effect on plant leaf gas exchange characteristics.

4.6 Conclusions and perspective

Based on Si distribution in meadow plants and the multiple regulation of Si on plant community structure, plant C, N and P accumulation and nutrients adsorption and plant gas exchange characteristics, this chapter highlighted the impacts of Si fertilization alone, or with N fertilization, on plant community structure, aboveground biomass, plant N and P concentration and gas exchange characteristics of individual plant species and plant functional groups. Nitrogen fertilization decreased the community species richness; Si fertilization with of N alleviated the loss of species richness of the whole community and the forbs group. Our findings highlight the importance of Si in improving aboveground biomass and alleviating N fertilization-induced biodiversity loss in grassland. Additionally, Si fertilization in alpine meadow usually increased plant N and P accumulation, plant-available nutrients, plant leaf net photosynthetic rate and instantaneous water-use efficiency. Si accumulation in plants has multiple impacts on plant community structure, growth and biogeochemical cycles of N and P nutrients in alpine meadow. However, there are some areas needing to be further investigated:

1. Biogeochemical cycle of Si and the amount of Si accumulation in soil, plants community, plant functional groups and individual under Si fertilization with different levels of N fertilization should be investigated to make Si management possible in alpine meadow.
2. Tradeoff analyses for reduction of nonstructural carbohydrate concentration such as starch and sucrose versus increase in the concentration of structural carbohydrate such as cellulose and lignin in response to Si accumulation need to be further investigated to ascertain the comprehensive impacts of Si on aboveground biomass accumulation and plant community structure.
3. The effect of Si on lignin synthesis in plant roots should be elucidated to evaluate the contribution of the increased root biomass to soil organic carbon turnover.
4. The mechanisms of Si-mediated regulation of plant nutrient uptake such as N and P across different plant functional groups and individual, and the effects on their biogeochemical processes.

Acknowledgements

The authors (Danghui Xu) would like to acknowledge funding from the Natural Science Foundation of China (32171611), the Second Tibetan Plateau Scientific Expedition and Research Program (2019QZKK0301) and the National Key R & D Program of China (2016YFC0501906).

References

[1] Epstein E. Silicon. Annu Rev Plant Physiol Plant Mol Biol 1999;50:641−64. Available from: https://doi.org/10.1146/annurev.arplant.50.1.641.

[2] Ma JF, Yamaji N. Silicon uptake and accumulation in higher plants. Trends Plant Sci 2006;11:392−7. Available from: https://doi.org/10.1016/j.tplants.2006.06.007.

[3] Siddiqui H, Ahmed KBM, Sami F, et al. Silicon nanoparticles and plants: current knowledge and future perspectives. Cham: Springer; 2020.

[4] Wu JW, Shi Y, Zhu YX, et al. Mechanisms of enhanced heavy metal tolerance in plants by silicon: a review. Pedosphere 2013;23(6):815−25. Available from: https://doi.org/10.1016/S1002-0160(13)60073-9.

[5] Hodson MJ, White PJ, Mead A, et al. Phylogenetic variation in the silicon composition of plants. Ann Bot 2005;96:1027−46. Available from: https://doi.org/10.1093/aob/mci255.

[6] Epstein E. The anomaly of silicon in plant biology. P Natl Acad Sci 1994;91:11−17. Available from: https://doi.org/10.1073/pnas.91.1.11.

[7] Takahashi E, Ma J, Miyake Y. The possibility of silicon as an essential element for higher plants. Comm Agric Food Chem 1990;2:99−102.

[8] Fauteux F, Chain F, Belzile F, et al. The protective role of silicon in the Arabidophytogenics−powdery mildew pathosystem. P Natl Acad Sci 2006;103:17554−9. Available from: https://doi.org/10.1073/pnas.0606330103.

[9] Ma JF, Tamai K, Yamaji N, et al. A silicon transporter in rice. Nature 2006;440:688−91. Available from: https://doi.org/10.1038/nature04590.

[10] Imtiaz M, Rizwan MS, Mushtaq MA, et al. Silicon occurrence, uptake, transport and mechanisms of heavy metals, minerals and salinity enhanced tolerance in plants with future prospects: a review. J Environ Manage 2016;183:521−9. Available from: https://doi.org/10.1016/j.jenvman.2016.09.009.

[11] Kim YH, Khan AL, Waqas M, et al. Silicon regulates antioxidant activities of crop plants under abiotic-induced oxidative stress: a review. Front Plant Sci 2017;6(8):510. Available from: https://doi.org/10.3389/fpls.2017.00510.

[12] Emamverdian A, Ding Y, Mokhberdoran F, et al. Determination of heavy metal tolerance threshold in a bamboo species (Arundinaria pygmaea) as treated with silicon dioxide nanoparticles. Glob Ecol Conserv 2020;24:e01306. Available from: https://doi.org/10.1016/j.gecco.2020.e01306.

[13] Van Bockhaven J, De Vleesschauwer D, Höfte M. Towards establishing broad-spectrum disease resistance in plants: silicon leads the way. J Exp Bot 2013;64:1281−93. Available from: https://doi.org/10.1093/jxb/ers329.

[14] Hao Q, Yang S, Song Z, et al. Silicon affects plant stoichiometry and accumulation of C, N, and P in grasslands. Front Plant Sci 2020;11:1304. Available from: https://doi.org/10.3389/fpls.2020.01304.

[15] Zhao Z, Dong S, Jiang X, et al. Effects of warming and nitrogen deposition on CH_4, CO_2 and N_2O emissions in alpine grassland ecosystems of the Qinghai-Tibetan Plateau. Sci Total Environ 2017;592:565−72. Available from: https://doi.org/10.1016/j.scitotenv.2017.03.082.

[16] Zhang H, Yao Z, Wang K, et al. Annual N_2O emissions from conventionally grazed typical alpine grass meadows in the eastern Qinghai-Tibetan Plateau. Sci Total Environ 2018;625:885−99. Available from: https://doi.org/10.1016/j.scitotenv.2017.12.216.

[17] Dong S, Shang Z, Gao J, et al. Enhancing sustainability of grassland ecosystems through ecological restoration and grazing management in an era of climate change on Qinghai-Tibetan Plateau. Agric Ecosyst Environ 2020;287:106684. Available from: https://doi.org/10.1016/j.agee.2019.106684.

[18] Dong Q, Ma Y, Zhao X. Study on management technology for "black soil type" degraded artificial grassland in Yangtze and Yellow River Headwater Region. Pratacult Sci 2007;24(8):9−14.

[19] Tang L, Dong S, Sherman R, et al. Changes in vegetation composition and plant diversity with rangeland degradation in the alpine region of Qinghai-Tibet Plateau. Rangel J 2015;37:107e115. Available from: https://doi.org/10.1071/RJ14077.

[20] Galloway JN, Dentener FJ, Capone DG, et al. Nitrogen cycles: past, present, and future. Biogeochemistry 2004;70:153−226. Available from: https://doi.org/10.1007/s10533-004-0370-0.

[21] Shen H, Dong S, Li S, et al. Effects of simulated N deposition on photosynthesis and productivity of key plants from different functional groups of alpine meadow on Qinghai-Tibetan plateau. Environ Pollut 2019;251:731−7. Available from: https://doi.org/10.1016/j.envpol.2019.05.045.

[22] Chen X, Wang G, Huang K, et al. The effect of nitrogen deposition rather than warming on carbon flux in alpine meadows depends on precipitation variations. Ecol Eng 2017;107:183−91. Available from: https://doi.org/10.1016/j.ecoleng.2017.07.018.

[23] Zhang X, Han X. Nitrogen deposition alters soil chemical properties and bacterial communities in the Inner Mongolia grassland. J Environ Sci 2012;24:1483−91.

[24] Sala OE, Chaplin FS, Armesto JJ, et al. Global biodiversity scenarios for the year 2100. Science 2000;287:1770−4. Available from: https://doi.org/10.1126/science.287.5459.1770.

[25] Xu DH, Fang XW, Zhang RY, et al. Influences of nitrogen, phosphorus and silicon addition on plant productivity and species richness in an alpine meadow. AoB Plants 2015;7.

[26] Stöcklin J, Schweizer K, Körner C. Effects of elevated CO_2 and phosphorus addition on productivity and community composition of intact monoliths from calcareous grassland. Oecologia 1998;116:50−6. Available from: https://doi.org/10.1007/s004420050562.

[27] Ren F, Song W, Chen L, et al. Phosphorus does not alleviate the negative effect of nitrogen enrichment on legume performance in an alpine grassland. J Plant Ecol 2017;10(5):822−30. Available from: https://doi.org/10.1093/jpe/rtw089.

[28] Zhao Y, Yang B, Li M, et al. Community composition, structure and productivity in response to nitrogen and phosphorus additions in a temperate meadow. Sci Total Environ 2019;654:863–71. Available from: https://doi.org/10.1016/j.scitotenv.2018.11.155.

[29] Goldberg DE, Miller TE. Effects of different resource additions of species diversity in an annual plant community. Ecology 1990;71:213–25. Available from: https://doi.org/10.2307/1940261.

[30] Avolio ML, Koerner SE, La Pierre KJ, et al. Changes in plant community composition, not diversity, during a decade of nitrogen and phosphorus additions drive above-ground productivity in a tallgrass prairie. J Ecol 2014;102:1649–60. Available from: https://doi.org/10.1111/1365-2745.12312.

[31] Sundqvist MK, Liu Z, Giesler R, et al. Plant and microbial responses to nitrogen and phosphorus addition across an elevational gradient in subarctic tundra. Ecology 2014;95:1819–35. Available from: https://doi.org/10.1890/13-0869.1.

[32] Soons MB, Hefting MM, Dorland E, et al. Nitrogen effects on plant species richness in herbaceous communities are more wide-spread and stronger than those of phosphorus. Biol Conserv 2017;212:390–7. Available from: https://doi.org/10.1016/j.biocon.2016.12.006.

[33] Carpenter AT, Moore JC, Redente EF, et al. Plant community dynamics in a semi-arid ecosystem in relation to nutrient addition following a major disturbance. Plant Soil 1990;126:91–9. Available from: https://doi.org/10.1007/BF00041373.

[34] Kirkham F, Mountford J, Wilkins R. The effects of nitrogen, potassium and phosphorus addition on the vegetation of a Somerset peat moor under cutting management. J Appl Ecol 1996;33:1013–29. Available from: https://doi.org/10.2307/2404682.

[35] Ceulemans T, Merckx R, Hens M, et al. Plant species loss from European semi-natural grasslands following nutrient enrichment–is it nitrogen or is it phosphorus? Glob Ecol Biogeogr 2013;22:73–82. Available from: https://doi.org/10.1111/j.1466-8238.2012.00771.x.

[36] Schoelynck J, Bal K, Backx H, et al. Silica uptake in aquatic and wetland macrophytes: A strategic choice between silica, lignin and cellulose? New Phytol 2010;186:385–91. Available from: https://doi.org/10.1111/j.1469-8137.2009.03176.x.

[37] Schaller J, Roscher H, Hillebrand H, et al. Plant diversity and functional groups affect Si and Ca pools in above ground biomass of grassland systems. Oecologia 2016;182:277–86. Available from: https://doi.org/10.1007/s00442-016-3647-9.

[38] Schaller J, Brackhage C, Gessner MO, et al. Silicon supply modifies C:N:P stoichiometry and growth of *Phragmites australis*. Plant Biol 2012;14:392–6. Available from: https://doi.org/10.1111/j.1438-8677.2011.00537.x.

[39] Neu S, Schaller J, Dudel EG. Silicon availability modifies nutrient use efficiency and content, C:N:P stoichiometry, and productivity of winter wheat (*Triticum aestivum* L.). Sci Rep 2017;7:40829. Available from: https://doi.org/10.1038/srep40829.

[40] Xu DH, Gao XG, Gao TP, et al. Interactive effects of nitrogen and silicon addition on growth of five common plant species and structure of plant community in alpine meadow. Catena 2018;169:80–9. Available from: https://doi.org/10.1016/j.catena.2018.05.017.

[41] Song Z, Liu C, Müller K, et al. Silicon regulation of soil organic carbon stabilization and its potential to mitigate climate change. Earth-Sci Rev 2018;185:463–75. Available from: https://doi.org/10.1016/j.earscirev.2018.06.020.

[42] Kostic L, Nikolic N, Bosnic D, et al. Silicon increases phosphorus (P) uptake by wheat under low P acid soil conditions. Plant Soil 2017;419:447–55. Available from: https://doi.org/10.1007/s11104-017-3364-0.

[43] Ma JF, Takahashi E. Effect of silicon on the growth and phosphorus uptake of rice. Plant Soil 1990;126:115–19. Available from: https://doi.org/10.1007/BF00041376.

[44] Hu AY, Che J, Shao JF, et al. Silicon accumulated in the shoots results in down-regulation of phosphorus transporter gene expression and decrease of phosphorus uptake in rice. Plant Soil 2018;423:317–25. Available from: https://doi.org/10.1007/s11104-017-3512-6.

[45] Li ZC, Song ZL, Yang XM, et al. Impacts of silicon on biogeochemical cycles of carbon and nutrients in croplands. J Integr Agr 2018;17:2182–95 CNKI:SUN:ZGNX.0.2018-10-006.

[46] Eneji E, Inanaga S, Muranaka S, et al. Effect of calcium silicate on growth and dry matter yield of *Chloris gayana* and *Sorghum sudanense* under two soil water regimes. Grass Forage Sci 2005;60:393–8. Available from: https://doi.org/10.1111/j.1365-2494.2005.00491.x.

[47] Xu DH, Gao TP, Li QX, et al. Research advances on biological function of silicon and its application in grassland ecosystem. Acta Ecol Sin 2020;40(22):8347–53. Available from: https://doi.org/10.5846/stxb201910142141.

[48] Liu L, Song Z, Yu C, et al. Silicon effects on biomass carbon and phytolith-occluded carbon in grasslands under high-salinity conditions. Front Plant Sci 2020;11:657. Available from: https://doi.org/10.3389/fpls.2020.00657.

[49] Xu DH, Li QX, Zhang RY. Effects of silicon and nitrogen fertilization on the growth and net photosynthetic rate of Thermopsis lanceolata in an alpine meadow. Pratacultural Sci 2020;37(9):1681–7. Available from: https://doi.org/10.11829/j.issn.1001-0629.2019-0239.

[50] Liang YC, Nikolic M, Bélanger R, et al. Silicon biogeochemistry and bioavailability in soil. In: Silicon in agriculture: from theory to practice. 2015. Springer, Netherlands. Doi: 10.1007/978-94-017-9978-2_3

[51] Blecker SW, McCulley RL, Chadwick OA, et al. Biologic cycling of silica across a grassland bioclimosequence. Global Biogeochem Cycl 2006;20:1–11. Available from: https://doi.org/10.1029/2006GB002690.

[52] Carnelli AL, Madella M, Theurillat JP. Biogenic silica production in selected alpine plant species and plant communities. Ann Bot 2001;87:425–34.

[53] Scurlock JMO, Hall DO. The global carbon sink: a grassland perspective. Glob Chang Biol 1998;4:229–33. Available from: https://doi.org/10.1046/j.1365-2486.1998.00151.x.

[54] Piao S, Fang J, Zhou L, et al. Changes in biomass carbon stocks in China's grasslands between 1982 and 1999. Global Biogeochem Cycl 2007;21:GB2002.

[55] Song Z, Liu H, Si Y, et al. The production of phytoliths in China's grasslands: Implications to biogeochemical sequestration of atmospheric CO_2. Glob Chang Biol 2012;18:3647–53. Available from: https://doi.org/10.1111/gcb.12017.

[56] Ji Z, Yang X, Song Z, et al. Silicon distribution in meadow steppe and typical steppe of northern China and its implications for phytolith carbon sequestration. Grass Forage Sci 2018;73:482–92. Available from: https://doi.org/10.1111/gfs.12316.

[57] Bauer P, Elbaum R, Weiss IM. Calcium and silicon mineralization in land plants: transport, structure and function. Plant Sci 2011;180:746–56. Available from: https://doi.org/10.1016/j.plantsci.2011.01.019.

[58] Broadley M, Brown P, Cakmak I, et al. Beneficial elements. In: Marschner P, (ed.), Marschner's mineral nutrition of higher plants. 3rd (ed.) 2012. Science Press.

[59] Larcher W. Physiological plant ecology: ecophysiology and stress physiology of functional groups. Berlin: Springer; 2003.

[60] Schaller J, Hodson MJ, Struyf E. Is relative Si/Ca availability crucial to the performance of grassland ecosystems? Ecosphere 2017;8(3): e01726. Available from: https://doi.org/10.1002/ecs2.1726.

[61] Zhang S, Zhang J, Slik JW, et al. Leaf element concentrations of terrestrial plants across China are influenced by taxonomy and the environment. Global Ecol Biogeogr 2012;21:809−18. Available from: https://doi.org/10.1111/j.1466-8238.2011.00729.x.

[62] Richmond KE, Sussman M. Got silicon? The non-essential beneficial plant nutrient. Curr Opin Plant Biol 2003;6:268−72. Available from: https://doi.org/10.1016/S1369-5266(03)00041-4.

[63] Yang X, Song Z, Liu H, et al. Plant silicon content in forests of north China and its implications for phytolith carbon sequestration. Ecol. Res. 2015;30:347−55. Available from: https://doi.org/10.1007/s11284-014-1228-0.

[64] LeBauer DS, Treseder KK. Nitrogen limitation of net primary productivity in terrestrial ecosystems is globally distributed. Ecology 2008;89:371−9. Available from: https://doi.org/10.1890/06-2057.1.

[65] Bai Y, Wu J, Clark CM, et al. Tradeoffs and thresholds in the effects of nitrogen addition on biodiversity and ecosystem functioning: evidence from inner Mongolia Grasslands. Glob Chang Biol 2010;16:358−72. Available from: https://doi.org/10.1111/j.1365-2486.2009.01950.x.

[66] Song L, Bao X, Liu X, et al. Nitrogen enrichment enhances the dominance of grasses over forbs in a temperate steppe ecosystem. Biogeosciences 2011;8:2341−50. Available from: https://doi.org/10.5194/bgd-8-5057-2011.

[67] Stevens CJ, Dise NB, Mountford JO, et al. Impact of nitrogen deposition on the species richness of grasslands. Science 2004;303:1876−9. Available from: https://doi.org/10.1126/science.1094678.

[68] Clark CM, Tilman D. Loss of plant species after chronic low-level nitrogen deposition to prairie grasslands. Nature 2008;451 (7179):712−15. Available from: https://doi.org/10.1038/nature06503.

[69] Yang H, Li Y, Wu M, et al. Plant community responses to nitrogen addition and increased precipitation: the importance of water availability and species traits. Glob Chang Biol 2011;17:2936−44. Available from: https://doi.org/10.1111/j.1365-2486.2011.02423.x.

[70] Xu DH, Gao TP, Fang XW, et al. Silicon addition improves plant productivity and soil nutrient availability without changing the grass: legume ratio response to nitrogen fertilization. Sci Rep 2020;10:10295. Available from: https://doi.org/10.1038/s41598-020-67333-7.

[71] Klotzbücher T, Klotzbücher A, Kaiser K, et al. Variable silicon accumulation in plants affects terrestrial carbon cycling by controlling lignin synthesis. Glob Chang Biol 2018;24:183−9. Available from: https://doi.org/10.1111/gcb.13845.

[72] Vojtech E, Turnbull LA, Hector A. Differences in light interception in grass monocultures predict short-term competitive outcomes under productive conditions. PLoS One 2007;2:e499. Available from: https://doi.org/10.1371/journal.pone.0000499.

[73] Tuna AL, Kaya C, Higgs D, et al. Silicon improves salinity tolerance in wheat plants. Environ Exp Bot 2008;62:10−16. Available from: https://doi.org/10.1016/j.envexpbot.2007.06.006.

[74] Chen W, Yao XQ, Cai KZ, et al. Silicon alleviates drought stress of rice plants by improving plant water status, photosynthesis and mineral nutrient absorption. Biol Trace Elem Res 2011;142:67−76. Available from: https://doi.org/10.1007/s12011-010-8742-x.

[75] Etesami H, Jeong BR. Silicon (Si): review and future prospects on the action mechanisms in alleviating biotic and abiotic stresses in plants. Ecotox Environ Saf 2018;147:881−96. Available from: https://doi.org/10.1016/j.ecoenv.2017.09.063.

[76] Detmann KC, Araujo WL, Martins SCV, et al. Silicon nutrition increases grain yield, which, in turn, exerts a feed-forward stimulation of photosynthetic rates via enhanced mesophyll conductance and alters primary metabolism in rice. New Phytol 2012;196:752−62. Available from: https://doi.org/10.1111/j.1469-8137.2012.04299.x.

[77] Bin ZJ, Wang JJ, Zhang WP, et al. Effects of N addition on ecological stoichiometric characteristics in six dominant plant species of alpine meadow on the Qinghai-Xizang Plateau, China. Chin J Plant Ecol 2014;38(3):231−7. Available from: https://doi.org/10.3724/SP. J.1258.2014.00020.

[78] Bin ZJ, Zhang RY, Zhang RY, et al. Effects of nitrogen, phosphorus and silicon addition on leaf carbon, nitrogen, and phosphorus concentration of *Elymus nutans* of alpine meadow on Qinghai-Tibetan Plateau, China. Acta Ecol Sin 2015;35(14):4699−706. Available from: https://doi.org/10.5846/stxb201311142729.

[79] Si XL, Wang WY, Gao XG, et al. Effects of nitrogen and silicon application on leaf nitrogen content and net photosynthetic rate of *Elymus nutans* in alpine meadow. Chin J Plant Ecol 2016;40:1238−44. Available from: https://doi.org/10.17521/cjpe.2015.0398.

[80] Schoelynck J, Müller F, Vandevenne F, et al. Silicon−vegetation interaction in multiple ecosystems: a review. J Veg Sci 2014;25:301−13. Available from: https://doi.org/10.1111/jvs.12055.

[81] Johnson SN, Hartley SE, Ryalls JMW, et al. Silicon-induced root nodulation and synthesis of essential amino acids in a legume is associated with higher herbivore abundance. Funct Ecol 2017;31:1903−9. Available from: https://doi.org/10.1111/1365-2435.12893.

[82] Nelwamondo A, Dakora FD. Silicon promotes nodule formation and nodule function in symbiotic cowpea (*Vigna unguiculata*). New Phytol 1999;142:463−7. Available from: https://doi.org/10.1046/j.1469-8137.1999.00409.x.

[83] Dakora FD, Nelwamondo A. Silicon nutrition promotes root growth and tissue mechanical strength in symbiotic cowpea. Funct Plant Biol 2003;30:947−53.

[84] Mali M, Aery NC. Silicon effects on nodule growth, dry-matter production, and mineral nutrition of cowpea (*Vigna unguiculata*). J Plant Nutr Soil Sci 2008;171:835−40. Available from: https://doi.org/10.1002/jpln.200700362.

[85] Mou J, Bin ZJ, Li QX, et al. Effects of nitrogen and silicon addition on soil nitrogen mineralization in alpine meadows of Qinghai-Xizang Plateau. Chin J Plant Ecol 2019;43:77−84. Available from: https://doi.org/10.17521/cjpe.2018.0218.

[86] Shi XJ, Mao ZY, Shi XH. The effect of combined application of silicon, zinc and magnesium on the nutrition of rice. J Southwest Agric Univ 1996;18:440−3.

[87] Marxen A, Klotzbücher T, Jahn R, et al. Interaction between silicon cycling and straw decomposition in a silicon deficient rice production system. Plant Soil 2016;398:153−63. Available from: https://doi.org/10.1007/s11104-015-2645-8.

[88] Zhang JL, Zhu CH, Dou P, et al. Effect of phosphorus and silicon application on the uptake and utilization of nitrogen, phosphorus and potassium by maize seedlings. Chin J Eco-Agric 2017;25:677−88.

[89] Hua KK, Zhang WJ, Guo ZB, et al. Evaluating crop response and environmental impact of the accumulation of phosphorus due to long-term manuring of vertisol soil in northern China. Agric Ecosyst Environ 2016;219:101−10. Available from: https://doi.org/10.1016/j.agee.2015.12.008.

[90] Gong HJ, Randall DP, Flowers TJ. Silicon deposition in the root reduces sodium uptake in rice seedlings by reducing bypass flow. Plant Cell Environ 2006;111:1−9. Available from: https://doi.org/10.1111/j.1365-3040.2006.01572.x.

[91] Ouzounidou G, Giannakoula A, Ilias I, et al. Alleviation of drought and salinity stresses on growth, physiology, biochemistry and quality of two *Cucumis sativus* L. cultivars by Si application. Braz J Bot 2016;39:531−9. Available from: https://doi.org/10.1007/s40415-016-0274-y.

[92] Ashfaque F, Inam A, Iqbal S, et al. Response of silicon on metal accumulation, photosynthetic inhibition and oxidative stress in chromium-induced mustard (*Brassica juncea* L.). S Afr J Bot 2017;111:153−60. Available from: https://doi.org/10.1016/j.sajb.2017.03.002.

5

Use of silicon and nano-silicon in agro-biotechnologies

Amanda Carolina Prado de Moraes[1,2] *and Paulo Teixeira Lacava*[1]

[1]Laboratory of Microbiology and Biomolecules, Department of Morphology and Pathology, Federal University of São Carlos (UFSCar), São Carlos, Brazil [2]Biotechnology Graduate Program, Federal University of São Carlos (UFSCar), São Carlos, Brazil

5.1 Introduction

Agricultural biotechnology has significantly advanced to meet global food supply and climatic-environmental changes during the past few decades [1,2]. In this scenario, new biotechnological strategies for plant adaptation and tolerance to abiotic and biotic stresses in a more sustainable agricultural production are constantly investigated [2].

The nutrient bioavailability of conventional fertilizers is drastically lost when reaching plant tissues due to chemical leaching, hydrolysis by soil moisture, evaporation, and degradation by photolytic and microbial processes [3]. Thus, the excessive amount of fertilizer that remains in the soil causes groundwater pollution, soil deterioration, and biodiversity loss [4,5]. Likewise, the use of chemical pesticides comes with risks to the environment and human health [6]. Supplementation with silicon (Si) has been a potentially sustainable option for solving such problems. It is naturally abundant in soils and plays a beneficial role in the growth and production of a broad range of plant species. Si has been exploited especially to improve plant resistance to environmental and pathogen stresses effectively [7].

Nanotechnology is one of the novel strategies employed in agro-systems for fertilization and plant disease management, demonstrating significant potential [8]. The materials at the nanoscale have specific physical, optical, mechanical, and chemical characteristics compared to their bulk form [9]. Therefore, studies on the application of nano-silicon in plants have shown various advantages, including improved nutrient uptake, resistance to abiotic and biotic stresses, and slow release of molecules in plant tissues. Furthermore, nano-silicon has the potential to enhance the efficiency of microbial biofertilizers. In this regard, the present chapter addresses recent findings on the use of silicon and nano-silicon for crop improvement.

5.2 Silicon for plant health

All plants grown in soil contain some silicon (Si) in their tissues as it is the second most abundant element after oxygen in the soil [10]. Plants require different amounts of silicon, according to their species. The highest Si accumulators are the monocotyledons, such as bamboo (*Bambuseae*), barley (*Hordeum vulgare*), rice (*Oryza sativa*), sorghum (*Sorghum bicolor*), sugarcane (*Saccharum officinarum*), and wheat (*Triticum aestivum*) [11–13]. They uptake different forms of silicon, being the monosilicic acid the most common form absorbed by plants [14]. The uptake process of monosilicic acid into the roots is energy-dependent and facilitated by the *Lsi1* and *LSi2* silicon transporters [10,15].

Even though Si is not an essential element for plants, it plays a critical role in the plant's resistance to biotic and abiotic stresses [16]. The deposition of Si in the form of amorphous silica (SiO_2) in the cell wall acts as a physical barrier since it enhances plant tissues' mechanical strength and rigidity, hindering penetration by fungi and chewing by insects [10]. Furthermore, the larger and thicker leaves prevent water loss during transpiration and increase plant resistance to strong wind and rain [17,18].

Under environmental stresses, such as drought, salinity, heat, metal toxicity, and nutrient imbalance, Si may perform at many levels, either in plant functioning and in the soil, through regulation of nutrient uptake (NPK), alleviation of metal toxicity, and increase in antioxidant enzymes activity [10,18]. Si facilitates the homeostasis of phytohormones and root cell wall extension, thus increasing the root surface area and, consequently, nutrient uptake [19–22].

A new type of silicon fertilizer, the organosilicon fertilizer (OSiF), was reported to reduce metal accumulation in rice (*O. sativa* L.), maize (*Zea mays* L.), and wheat (*T. aestivum* L.) [23–25]. The OSiF enhanced leaf gas exchange characteristics and chlorophyll content and decreased oxidative damage and health risk index of Cd and Pb [24]. Si application has recently been found to reduce significantly stem borer infestation, blast disease, and false smut disease in different rice genotypes [26]. Helaly et al. [27] reported that Si positively influenced the growth of mango (*Mangifera indica*) under drought stress by improving the concentration of auxins (IAA), gibberellins (GA), and cytokinins (CK). In the last few years, the studies focusing on plant responses at transcriptomics and proteomics levels have been elucidative in the role of Si. The activation or deactivation of specific genes responsible for phytohormonal and antioxidant responses under adverse conditions is crucial for plant resilience [28]. For example, Si was able to shift the transcriptome of salt-stressed cucumber (*Cucumis sativus*) back to that in normal conditions [29]. A proteomic study revealed that Si acts at the protein level to induce tolerance to Cd toxicity in *Arabidopsis thaliana* plants [30]. Another proteomic investigation demonstrated that proteins related to stress responses, photosynthesis, and signal transduction were differently expressed under Si application in Carnation (*Dianthus caryophyllus* L.) plants under hyperhydricity stress [31].

Given the potential of Si in agricultural systems, modern technologies have been explored to improve its performance and explore its potential for other applications. In this way, Si at the nanoscale level, the nano-silicon, has gained much attention due to its improved features and several functions to benefit plants.

5.3 Nano-silicon

Nanotechnology has been transforming modern agriculture over the last decade by providing efficient tools to optimize and reduce waste in the food chain. Nanoscale materials (<100 nm) have a larger surface-area-to-volume ratio than bulk materials, which give them a higher solubility, reactivity, and adherence [32,33]. They may be the solution for most conventional farming obstacles concerning levels of inputs, systems efficacy, waste, and economic loss. Yet, it provides a wide range of new possibilities to be explored. Plant responses to nanoparticles rely on several factors, including size, shape, method of application, and chemical/physical properties of the nanomaterial applied [34].

Nano-silicon (NSi) is known for its broad benefits in modern agriculture [35]. The improved properties of NSi over bulk material provide greater efficiency for several applications in crop plants, including direct and indirect impacts on plant growth and development [36]. Silica (SiO_2) nanoparticles (SiNPs) have been proved to enhance crop yield and protection in different ways due to their unique and adjustable physicochemical features, such as large surface area, biocompatibility, nontoxicity, and chemical stability [37–39]. Amorphous silica NPs, which are nonporous and composed of spherical particles [40], have been reported to enhance plant resistance and growth under abiotic and biotic stress by improving plant cell wall rigidity, balancing nutrient uptake, inducing the synthesis of phytohormones and enzymes, and expression of genes related to pathogen defense [41–45]. On the other hand, mesoporous silica NPs, which have tunable nanoscale pores, are considered excellent carriers for the controlled delivery of active ingredients to plants, thus providing a more specific action with minimum product loss and soil damage [46,47].

Several chemical methods synthesize silica NPs, being the most common the reverse microemulsion, flame synthesis, and sol-gel methods. However, chemical approaches can be costly, ecologically unsafe, and depend on many controlled conditions [48]. Consequently, the green synthesis of SiNPs has recently emerged as a profitable eco-friendly strategy, especially for agricultural waste management, since it is generally a complex and expensive public service [49]. Amorphous SiNPs can be biosynthesized from diverse waste sources, including bamboo, rice husk, sugar beet bagasse, maize stalk, cassava periderm, among others [49–52]. Another green practice in producing SiNPs is using living organisms. The fungus *Fusarium oxysporum*, for example, is capable of leaching out hollow SiNPs when exposed to sand by a process involving specific proteins present in the fungal biomass [53]. Owing to the high amounts of enzymes, proteins, and stabilizing molecules excreted by the fungi, they are considered an excellent biological factory for commercial green synthesis of NPs [54]. Likewise, the *Saccharomyces cerevisiae* yeast in the presence of sodium silicate as a precursor has been recently reported to synthesize 25 nm SiNPs in a cost-effective and contaminant-free manner [48]. Hence, the synthesis of NPs from agro-waste materials and microorganisms would be a promising low-cost and sustainable approach combining innovative use of plant parts that would be discarded, prospection of molecules produced by microbes, and reduced environmental pollution.

5.3.1 Nano-silicon as nanoregulators, nanopesticides, and nanofertilizers

The agricultural biotechnology sector has focused on strategies that merge high crop yields and low adverse impacts. As a result, nano-silicon has been explored to reduce the use of chemicals in agriculture. Studies show that SiNPs are multifunctional materials due to their role as nanoregulators (increase plant resistance to abiotic stress), nanopesticides (enhance plant resistance to biotic stress), and nanofertilizers (support plant health by providing essential nutrients or by other mechanisms) [55].

When absorbed, SiNPs are deposited on the cell walls of roots and leaves, providing enough mechanical strength to prevent infections by fungi, bacteria, and nematodes. Furthermore, the nano-silicon layer at the epidermal cell wall contributes to the structural leaf coloration and considerably decreases cuticular transpiration, alleviating drought stress [15,36,56,57]. Alsaeedi et al. [43] reported that SiNPs significantly improved the leaf area, chlorophyll content, and height of cucumber plants under water stress. The plants also showed higher amounts of N, K, and Si. Similarly, Aqaei et al. [57] found that potassium silica NPs increased N, K, Cu, and Si in maize shoots, increasing plants' resistance against drought stress. The same authors suggested that Si increases the concentration of some elements by stimulating their binding in plant tissues and aiding their translocation to the shoot [57].

Biotic and abiotic stresses, such as pests, drought, salinity, and heavy metal, trigger the accumulation of reactive oxygen species (ROS) resulted from the plant defense system [58]. Although it participates in plant signaling for adaptation to adverse conditions and regulates plant growth [59], it provokes cellular damage and inhibits plant development when excessively accumulated [60]. Nano-silicon alleviates the toxic effects of ROS in plants by inducing the activity of antioxidant enzymes, expression of critical genes associated with oxidative stress mitigation, and hormone metabolism [61]. Elshayb et al. [62] developed a field experiment with foliar application of green SiNPs in rice plants submitted to water stress. They found that SiNPs increased the levels of proline and antioxidant enzymes, which are molecules involved in controlling plant cell water loss and mitigation of oxidative damage, respectively, thus providing plant tolerance to drought. As a result, the growth and yield parameters like leaf area index, dry matter, grain weight, and chlorophyll content were higher in water-stressed plants treated with SiNPs [62].

Industrial waste and sewage disposal have caused the excessive accumulation of heavy metals in soils, negatively influencing plant growth, physiology, metabolism, and senescence [63]. Fatemi et al. [64] reported that foliar application of SiNPs eased the toxic impacts of lead (Pb) on coriander plants (*Coriandrum sativum* L.) by adjusting activities of the antioxidant enzymes peroxidase (POD), catalase (CAT), and superoxide dismutase (SOD), and probably by other mechanisms, such as the stimulation of root exudates that chelate with metals and consequently reduce their concentration in the plant [65] and lessen cell wall porosity [66]. SiNPs also showed efficiency in reducing cadmium (Cd) concentrations in roots, shoots, and grains of wheat, plus improved dry biomass of shoots, roots, spikes, grains, and chlorophyll concentration. The reduction of Cd levels and enhanced plant growth traits are explained by the increase in SOD and POD activities and the preservation of cell membrane integrity, both induced by SiNPs, leading to mitigation oxidative stress [67]. Other recent studies reported the nano-silicon influencing higher activities of antioxidant enzymes, and thus alleviating phytotoxicity of chromium (Cr) in pea (*Pisum sativum*) seedlings [68] and arsenic (As) stress in maize (*Z. mays*) [69].

Salinity stress negatively impacts plant growth due to the salt accumulation in the apoplast that results in cellular dehydration. Consequently, the chloroplast levels and the subsequent photosynthetic activity decrease [70]. Mahmoud et al. [71] verified that the lowest concentration of SiNPs applied (200 mg L^{-1}) was sufficient to improve chlorophyll content in banana (*Musa acuminate* "Grand Nain") plants under salinity conditions. Furthermore, it reduced Na^+ levels and increased K^+ content in the banana leaves, resulting in the improved K^+/Na^+ ratio and diminished cell wall damage. Decreased Na^+ levels and increased K^+ contents in the banana leaves can have happened because silicon reduces the membrane permeability to Na^+ in the leaf cells [71,72]. In addition, Si arranges Na^+ and Cl^- ions dispersion over the root system, providing plant resistance to salt toxicity [73]. The oxidative stress induced by the presence of excessive NaCl promotes lipid peroxidation along with the increase in electron leakage (EL) and the decrease in the membrane stability index (MSI) [74]. Hence, the cell membrane damage caused by salinity stress can be predicted by values of EL and MSI [42]. Nano-silicon proved to increase MSI significantly and decrease EL in salt-stressed Faba bean (*Vicia faba* L.) plants due to activation of antioxidant enzymes that reduced the ROS [74]. In the same study, nano-silicon enhanced the plant sugar accumulation, which plays an essential role in the plant response to salt stress. This organic solute diminishes osmotic potential in the cytoplasm and helps maintain water homeostasis among cellular compartments [74].

Plant diseases are one of the main factors limiting crop productivity [75]. Besides mechanical reinforcement of plant cell walls [76], the defense mechanism against pathogens mediated by silica involves enhanced accumulation of phenolic compounds and enzymes such as chitinases, peroxidase, and polyphenol oxidases (PPOs) [77]. Nevertheless, it is still unclear whether SiNPs can induce resistance in plants, how their performance differs from bulk Si, and the molecular pathways they may influence [44]. A recent study using the model plant *A. thaliana* against a bacterial pathogen proved that SiNPs could induce systemic acquired resistance involving the defense hormone salicylic acid. In addition, they promoted the slow release of $Si(OH)_4$ on the leaves and did not cause phytotoxicity, being safer, and more effective than direct application of $Si(OH)_4$ [44]. Likewise, Suriyaprabha et al. [78] found that 20−40 nm nanosilica promoted higher fungal resistance than bulk silica in maize plants. The plants treated with nanosilica presented higher expression of phenolic compounds and fewer disease lesions [78]. Mesoporous silica NPs were reported to inhibit the pathogenic fungus *Alternaria solani* along with the increase in fresh and dry weight of tomato (*Lycopersicon esculentum* L. H. Karst.) plants [79]. The positively charged surface of these mesoporous SiNPs interacts with the protein thiol groups on the fungal cell surface, causing cell lysis [80] and subsequent shutdown of proteins or mutations in the pathogen DNA, stopping its replication [81,82]. It has been demonstrated that SiNPs promote xylem cell wall lignification and cell wall thickness [83]. The thickening of the plant cell wall protects plants from insect attacks as it is a physical barrier to insect feeding [84]. Vani and Brindhaa [85] showed that amorphous nanosilica can be a potent nanocide because it caused 100% mortality of the stored grain pest *Corcyra cephalonica*.

The application of SiNPs might be a satisfactory alternative to mineral N and P fertilization in crop production. Studies have shown that Si promotes nutrient acquisition and biomass production at different Si levels depending on the plant species. For example, in winter wheat plants (*T. aestivum* L.), biomass production and N concentration improved at low to medium silica supply levels. P accumulation was enhanced, and carbon decreased as the silica level increased [86]. In common reed (*Phragmites australis*), the P status and biomass production increased with moderate silica concentrations, while greater silica levels induced the opposite effect on both [87]. Yassen et al. [88] found that different concentrations of SiNPs (from 15 to 120 mg L^{-1}) increased the growth and yield of cucumber plants (*C. sativus*), 60 mg L^{-1} being the optimal dose for biomass improvement. One of the most critical factors influencing plant biomass is nutrient availability; therefore plant biomass enhances as its nutrient acquisition improves. In addition, the fixation of SiNPs in the leaf cells might affect the structure of chloroplasts and increase the plant's ability to absorb light, thus positively impacting the plant biomass [64]. Likewise, nano-silicon has been reported to influence either positively or negatively plant height, micronutrients, and enzymes, determined by the concentration of Si. Le et al. [89] observed a significant decrease in Bt-transgenic and nontransgenic cotton plant height as the SiNPs concentration increased. In contrast, the Fe amount was greater in plants treated with the higher SiNPs concentration (2000 mg L^{-1}). Lemongrass [*Cymbopogon flexuosus* (Steud.) Wats] plants treated with SiNPs at a much lower concentration (150 mg L^{-1}) had enhanced activities of CAT, POD, and SOD, which are necessary enzymes in the plant antioxidant defense system, and nitrate reductase enzyme, involved in nitrogen metabolism. These effects consequently led to enhanced lemongrass plant growth and yield [90]. These findings indicate that the concentration of nanoparticles is an essential factor to be considered when applying to plants, as high NP concentrations can be disadvantageous for some parameters. Furthermore, each plant species demands its own optimal NP levels to obtain benefits. Besides concentration, the plant responses to nanoparticles depend on various components, including the nanoparticle shape, size, chemical properties, physical properties, and application method [34].

SiNPs are generally applied to the plants through different modes. When applied to the soil, the nanoparticles interact with root exudates and bind to membrane transporters or carrier proteins, then are taken up via plant roots, move through the symplastic and apoplastic pathways, and are transported to aerial parts via xylem sap [89,91,92]. Among the factors influencing the success of the soil application method, there are soil texture, pH, and salt content. Another technique is foliar spraying, which consists of applying a solution containing the nanoparticles directly to plant leaves, absorbed by stomata or leaf epidermal cells [92]. Foliar feeding of nanomaterials is considered more effective and economical over soil application as the latter triggers nanoparticle agglomeration in soil, causing material loss [32]. Nevertheless, the uptake, translocation, and accumulation of nano-silicon rely on many other variables, including plant species, plant age, environmental conditions, size, shape, stability, and chemical composition of the nanoparticle [93].

5.3.2 Nano-silicon as delivery systems

The use of nanostructured materials for delivery systems has become a viable way to obtain sustainable gains in agriculture. In this way, mesoporous silica nanoparticles (MSNs) can successfully act as target-specific or controlled-release carriers for agrochemicals, DNA, proteins, and other molecules into plants due to their biocompatibility and biodegradability [46,94–96]. The main goals of these smart delivery systems are to enhance nutrient use efficiency, protect molecules from premature deterioration in a biological environment, and reduce loss and costs in crop production while preventing environmental degradation [97,98].

The MSNs are composed of mesopores of various sizes organized in a hexagonal array. These pores host the molecules and release them in a controlled and directed manner [98]. For this, the exterior and interior MSN pore surfaces are covalently modified for acting as gating devices that regulate the release of hosted molecules according to internal (e.g., enzymatic activity) or external (e.g., pH, temperature, and light) stimuli [98–100]. Therefore, the release of molecules happens when these factors trigger the removal of the gating device (cap) from the pores [98]. Sun et al. [101] entrapped the phytohormone abscisic acid (ABA) in MSN mesopores using intracellular glutathione (GSH) as a stimulus for ABA release from the pores. The application of MSNs loaded with ABA in *A. thaliana* seedlings reduced stomatal leaf aperture, inhibited water loss, and prolonged the expression of the *AtGALK2* gene, which is responsible for inducing ABA. Consequently, the seedlings were more resistant under drought stress [101].

Approaches based on MSNs have been advantageous in crop genetic engineering. The delivery of foreign DNA into *A. thaliana* roots was successfully executed using MSNs as carriers without mechanical force. It was confirmed by the gene expression detected in the epidermal layer and the inner cortical and endodermal root tissues [96]. This modern approach is advantageous over the conventional gene-gun method as it does not cause physical damage to the plant tissues. Furthermore, a novel transient gene expression technique in tomato (*Solanum lycopersicum*) was introduced by Hajiahmadi et al. [102]. In this study, MSNs as nanocarriers of pDNA enhanced plant resistance against the pest *Tuta absoluta* and presented biocompatibility, reduction in time, and energy consumption as advantages [102].

The release of active ingredients in a controlled way allows slow absorption, thus avoiding an overdose in the plant, reducing the threat to nontarget organisms, and attenuating waste [103]. MSNs carrying the pesticide spirotetramat improved its deposition, uptake, and translocation performance in cucumber (*C. sativus*) plants. Moreover, the residues and metabolites of the pesticide presented a low risk to the edible parts of the plants [6]. Bravo-Cadena [104] encapsulated natural antimicrobial essential oils into MSNs to address bacterial phytopathogens and found that this nano-biocide prevented the quick volatilization and degradation of the essential oils and maintained their slow-release and prolonged effect. Abdelrahman et al. [105] reported the better efficiency of prochloraz fungicide delivered by MSNs than the conventional one in rice (*O. sativa* L.) plants. The MSNs (70.89 nm) promoted a better uptake and translocation, longer duration, and greater antifungal activity against *Magnaporthe oryzae*, the rice blast disease [105].

Nano-formulations of nutrients enhance solubility and support the distribution of insoluble nutrients in the soil, slow down soil absorption and fixation rate, thus increasing their bioavailability. In contrast, a reduced micronutrient bioavailability is found when applying conventional fertilizers because of their low solubility and larger particle sizes [106]. Therefore, nano-formulations might boost the efficiency of fertilizer at the same time that spares the fertilizer resource. Moreover, it raises the uptake of nutrients by the plant and consequently improves crop production [3]. Sun et al. [107] detected a significant increase in the growth, plant biomass, photosynthetic activity, total protein content, and germination of wheat (*T. aestivum*) and lupin (*Lupinus angustifolius*) treated with MSNs. These authors also reported that even the highest concentration of the nanoparticles inoculated (2000 mg L^{-1}) did not cause oxidative stress and membrane damage, proved by no changes in the levels of H_2O_2, electrolyte leakage, and lipid peroxidation. Wanyika et al. [46] found that urea-loaded MSNs presented a slow release profile in water and soil. The nano-fertilizer had a release period five times slower than pure urea.

Similarly, Kioni et al. [108] tested MSNs as carriers for the controlled release of urea, used in this study as a model fertilizer. It was demonstrated a slow and sustained release of urea, proving the potential of MSNs as fertilizer nanocarriers [108].

5.3.3 Nano-silicon associated with plant growth-promoting bacteria

The plant microbiota, which involves all microorganisms, and the plant microbiome, comprising all microbial genomes, present in the rhizosphere, phyllosphere, and endosphere, have an essential role in the plant's growth and health [109–111]. This microbiota includes beneficial, neutral, or pathogenic microorganisms. Beneficial bacteria, the so-called plant growth-promoting bacteria (PGPB), can promote plant growth through direct or indirect mechanisms [112]. Some direct benefits are the uptake of essential nutrients, like nitrogen, phosphorous, iron, and the synthesis of phytohormones [113,114]. The indirect benefits include the production of antagonistic substances and induced resistance against phytopathogens [113]. Biofertilizers containing PGPB, such as nitrogen fixers, phosphorous solubilizers, potassium solubilizers, and biocontrol agents, are widely employed in sustainable agriculture [115,116]. However, their practical application may not have the expected result in the plant as there are some obstacles regarding their stability in the soil, microbial interaction, field applications, and delivery systems [117]. In this way, silica NPs can stabilize and improve the performance of PGPB in the environment, thus overcoming the limitations of biofertilizers and biopesticides [118].

Recently, studies have proved that SiNPs stimulate the growth, viability, and beneficial properties of PGPB, helping their performance in the plant. Karunakaran et al. [119] found that 50 nm SiNPs increased the viability and population of PGPB in the soil and maintained an optimal pH for the bacteria. Consequently, they increased the nitrogen, phosphorus, and potassium (NPK) content in maize seeds, achieving 100% seed germination [119]. Likewise, Rangaraj et al. [120] attested that nanosilica increased the bacterial population, total biomass, and soil nutrient contents in an experiment with maize. Both studies reported that SiNPs promoted better plant response than other sources of silicon. Another study with maize plants showed that cotreatment of plant growth-promoting rhizobacteria (PGPR) with SiNPs promoted improvement in soil and leaf health, and plant development, resulting in enhanced N, P, and K uptake, yield-related traits, and productivity of maize plants under water deficit and soil salinity [121]. Nanosilica associated with PGPB were also reported to enhance germination parameters and seedling weight in tomato (*Lycopersicum esculentum*) [122] and promote the growth of land cress (*Barbarea verna*), and increase N and P concentrations in the soil [123]. Moreover, in an *in-vitro* experiment, SiNPs increased cytokinin synthesis in *Bacillus subtilis* by 85% [124].

Some PGPB can support plants against pathogen attacks by producing antibiotics, antioxidant enzymes, decreasing nutrient availability for pathogens, promoting induced systemic resistance in the plant, among other mechanisms [125–127]. Djaya et al. [128] developed a biocontrol delivery system (BDS) containing antagonistic PGPR + graphite + amorphous silica NPs against *Ralstonia solanacearum* that causes bacterial wilt on potatoes. As a result, the SiNPs in the formulation maintained the PGPR population. The largest inhibition zones were seen in the BDS treatments, demonstrating that this delivery mechanism enhanced the antagonistic activity. Furthermore, the authors verified that SiNPs formed agglomerated micron-sized particles, which allowed the antagonistic bacteria to be encapsulated or absorbed on the surface of the particles [128]. Hence, nanoparticles can improve the bacteria already selected for biocontrol. Nevertheless, it is difficult to predict whether NPs also stimulate harmful microbes; therefore massive investigation on target beneficial bacteria, NPs, pathogens, and plants is essential to assure the safety and effectiveness of nanobiofertilizers/nanobiopesticides [129]. Pour et al. [130] developed an encapsulation method based on alginate-silica NPs and carbon nanotubes to deliver the PGPB *Pseudomonas fluorescens* and *Bacillus subtilis* into UCB-1 pistachio explants. The encapsulated bacteria increased the micropropagation of pistachio, root length, bud length, and plant biomass. Accordingly, the study revealed that PGPB, together with SiNPs, can increase the proliferation rate in the tissue culture of pistachio, thus potentially boosting its commercial reproduction [130].

The hydration properties of silica NP surfaces facilitate the attraction between the NP and bacterium, improving bacterial acid resistance [131,132]. The stimulating effect of mineral NPs on bacterial growth may be due to the enhanced oxygen mass transfer and ion exchange processes in the media [133] and the possible change in the size and shape of cells induced by the NPs, resulting in higher bacterial growth [134].

In summary, there is consistent evidence of beneficial interaction among SiNPs, PGPB, and plants, indicating the potential of these nanoparticles to improve the action of biofertilizers and biopesticides in several crop species.

5.4 Conclusions and perspectives

Silicon improves plant resistance to environmental stress and pathogen attacks through diverse mechanisms. SiNPs have enhanced characteristics compared to bulk silicon, allowing them to penetrate and interact better within plant tissues, increasing nutrient uptake and promoting plant growth under stress, including drought, salinity, metal toxicity, insect pests, and microbial pathogens. It involves cell wall thickening, modulation of antioxidant enzymes, increased photosynthesis, activation or deactivation of genes, etc.

Mesoporous silica has proven to efficiently deliver DNA, enzymes, fertilizers, and pesticides with controlled release into the plant, acting in targeted tissues and releasing the molecules according to intracellular or external stimuli. This system is advantageous because the molecule is discharged inside the plant only at the desirable moment (e.g., stress condition), thus being more effective and preventing product loss. In addition, SiNPs can improve PGPB activity, resulting in enhanced efficiency of microbial biofertilizers in the field. However, the interaction among nanoparticles, plants, environments, and living organisms must be investigated to determine the ideal concentrations and sizes of NPs to achieve successful results and avoid toxic effects.

The use of nano-silicon can transform conventional agriculture and launch better crop yields to meet global food demands while reducing adverse environmental consequences.

Acknowledgments

The authors would like to acknowledge funding from the Coordenação de Aperfeiçoamento de Pessoal de Nível Superior (CAPES), grant no. 88882.426494/2019-01.

References

[1] Altman A, Hasegawa PM. Introduction to plant biotechnology 2011: basic aspects and agricultural implications. Plant Biotechnol Agric 2012. Available from: https://doi.org/10.1016/B978-0-12-381466-1.00050-X.

[2] Moshelion M, Altman A. Current challenges and future perspectives of plant and agricultural biotechnology. Trends Biotechnol 2015;33:337–42. Available from: https://doi.org/10.1016/j.tibtech.2015.03.001.

[3] Kalra T, Tomar PC, Arora K. Micronutrient encapsulation using nanotechnology: nanofertilizers. Plant Arch 2020;20:1748–53.

[4] Tilman D, Cassman KG, Matson PA, Naylor R, Polasky S. Agricultural sustainability and intensive production practices. Nature 2002;418:671–7. Available from: https://doi.org/10.1038/nature01014.

[5] Mozumder P, Berrens RP. Inorganic fertilizer use and biodiversity risk: an empirical investigation. Ecol Econ 2007;62:538–43. Available from: https://doi.org/10.1016/j.ecolecon.2006.07.016.

[6] Zhao P, Yuan W, Xu C, Li F, Cao L, Huang Q. Enhancement of spirotetramat transfer in cucumber plant using mesoporous silica nanoparticles as carriers. J Agric Food Chem 2018;66:11592–600. Available from: https://doi.org/10.1021/acs.jafc.8b04415.

[7] Wang M, Gao L, Dong S, Sun Y, Shen Q, Guo S. Role of silicon on plant–pathogen interactions. Front Plant Sci 2017;8:701. Available from: https://doi.org/10.3389/fpls.2017.00701.

[8] Adisa IO, Pullagurala VLR, Peralta-Videa JR, Dimkpa CO, Elmer WH, Gardea-Torresdey JL, et al. Recent advances in nano-enabled fertilizers and pesticides: a critical review of mechanisms of action. Environ Sci: Nano 2019;6:2002–30. Available from: https://doi.org/10.1039/c9en00265k.

[9] Guo LJ. Recent progress in nanoimprint technology and its applications. J Phys D Appl Phys 2004;37:R123. Available from: https://doi.org/10.1088/0022-3727/37/11/R01.

[10] Ma JF, Yamaji N. A cooperative system of silicon transport in plants. Trends Plant Sci 2015;20:435–42. Available from: https://doi.org/10.1016/j.tplants.2015.04.007.

[11] Ma JF, Miyake Y, Takahashi E. Chapter 2 silicon as a beneficial element for crop plants. Stud Plant Sci 2001;8:17–39. Available from: https://doi.org/10.1016/S0928-3420(01)80006-9.

[12] Ma JF, Takahashi E. Soil, fertilizer, and plant silicon research in Japan. Elsevier; 2002. Available from: http://doi.org/10.1016/b978-0-444-51166-9.x5000-3.

[13] Liang Y, Sun W, Zhu YG, Christie P. Mechanisms of silicon-mediated alleviation of abiotic stresses in higher plants: a review. Environ Pollut 2007;147:422–8. Available from: https://doi.org/10.1016/j.envpol.2006.06.008.

[14] Richmond KE, Sussman M. Got silicon? The non-essential beneficial plant nutrient. Curr Opin Plant Biol 2003;6:268–72. Available from: https://doi.org/10.1016/S1369-5266(03)00041-4.

[15] Asgari F, Majd A, Jonoubi P, Najafi F. Effects of silicon nanoparticles on molecular, chemical, structural and ultrastructural characteristics of oat (*Avena sativa* L.). Plant Physiol Biochem 2018;127:152–60. Available from: https://doi.org/10.1016/j.plaphy.2018.03.021.

[16] Ma JF, Yamaji N. Silicon uptake and accumulation in higher plants. Trends Plant Sci 2006;11:392–7. Available from: https://doi.org/10.1016/j.tplants.2006.06.007.

[17] Hattori T, Inanaga S, Tanimoto E, Lux A, Luxová M, Sugimoto Y. Silicon-induced changes in viscoelastic properties of sorghum root cell walls. Plant Cell Physiol 2003;44:743–9. Available from: https://doi.org/10.1093/pcp/pcg090.

[18] Guntzer F, Keller C, Meunier JD. Benefits of plant silicon for crops: a review. Agron Sustain Dev 2012;32:201–13. Available from: https://doi.org/10.1007/s13593-011-0039-8.

[19] Vaculík M, Lux A, Luxová M, Tanimoto E, Lichtscheidl I. Silicon mitigates cadmium inhibitory effects in young maize plants. Environ Exp Bot 2009;67:52–8. Available from: https://doi.org/10.1016/j.envexpbot.2009.06.012.

[20] Hameed A, Sheikh MA, Jamil A, Basra SMA. Seed priming with sodium silicate enhances seed germination and seedling growth in wheat (*Triticum aestivum* L.) under water deficit stress induced by polyethylene glycol. Pak J Life Soc Sci 2013;11:19–24.

[21] Yin L, Wang S, Li J, Tanaka K, Oka M. Application of silicon improves salt tolerance through ameliorating osmotic and ionic stresses in the seedling of *Sorghum bicolor*. Acta Physiol Plant 2013;35:3099–107. Available from: https://doi.org/10.1007/s11738-013-1343-5.

[22] Etesami H, Jeong BR. Silicon (Si): review and future prospects on the action mechanisms in alleviating biotic and abiotic stresses in plants. Ecotoxicol Environ Saf 2018;147:881–96. Available from: https://doi.org/10.1016/j.ecoenv.2017.09.063.

[23] Huang H, Yang F, Song F, et al. Effect of organic silicon modified fertilizers on absorption of nutrients and cadmium in maize. Soil Fertil Sci China 2019;4:156–63.

[24] Huang H, Rizwan M, Li M, Song F, Zhou S, He X, et al. Comparative efficacy of organic and inorganic silicon fertilizers on antioxidant response, Cd/Pb accumulation and health risk assessment in wheat (*Triticum aestivum* L.). Environ Pollut 2019;255:113146. Available from: https://doi.org/10.1016/j.envpol.2019.113146.

[25] Yang FW, Huang HL, Song FR, et al. The effect and mechanism of rice cadmium remediation by silicone modified compound fertilizers. J Nucl Agric Sci 2019.

[26] Sarma RS, Shankhdhar D, Shankhdhar SC. Beneficial effects of silicon fertilizers on disease and Insect-Pest management in rice genotypes (*Oryza sativa* L.). J Pharmacog Phytochem 2019;8:358–62.

[27] Helaly MN, El-Hoseiny H, El-Sheery NI, Rastogi A, Kalaji HM. Regulation and physiological role of silicon in alleviating drought stress of mango. Plant Physiol Biochem 2017;118:31–44. Available from: https://doi.org/10.1016/j.plaphy.2017.05.021.

[28] Dhiman P, Rajora N, Bhardwaj S, Sudhakaran SS, Kumar A, Raturi G, et al. Fascinating role of silicon to combat salinity stress in plants: an updated overview. Plant Physiol Biochem 2021;162. Available from: https://doi.org/10.1016/j.plaphy.2021.02.023.

[29] Zhu Y, Yin J, Liang Y, Liu J, Jia J, Huo H, et al. Transcriptomic dynamics provide an insight into the mechanism for silicon-mediated alleviation of salt stress in cucumber plants. Ecotoxicol Environ Saf 2019;174:245–54. Available from: https://doi.org/10.1016/j.ecoenv.2019.02.075.

[30] Carneiro JMT, Chacón-Madrid K, Galazzi RM, Campos BK, Arruda SCC, Azevedo RA, et al. Evaluation of silicon influence on the mitigation of cadmium-stress in the development of *Arabidopsis thaliana* through total metal content, proteomic and enzymatic approaches. J Trace Elem Med Biol 2017;44:50–8. Available from: https://doi.org/10.1016/j.jtemb.2017.05.010.

[31] Muneer S, Wei H, Park YG, Jeong HK, Jeong BR. Proteomic analysis reveals the dynamic role of silicon in alleviation of hyperhydricity in carnation grown *in vitro*. Int J Mol Sci 2018;19:50. Available from: https://doi.org/10.3390/ijms19010050.

[32] Ram P, Kumar R, Rawat A, Singh VP, Pandey P. Nanomaterials for efficient plant nutrition. Int J Chem Stud 2018;6:867–71.

[33] Monica RC, Cremonini R. Nanoparticles and higher plants. Caryologia 2009;62:161–5. Available from: https://doi.org/10.1080/00087114.2004.10589681.

[34] Rastogi A, Zivcak M, Sytar O, Kalaji HM, He X, Mbarki S, et al. Impact of metal and metal oxide nanoparticles on plant: a critical review. Front Chem 2017;5:70. Available from: https://doi.org/10.3389/fchem.2017.00078.

[35] Torney F, Trewyn BG, Lin VSY, Wang K. Mesoporous silica nanoparticles deliver DNA and chemicals into plants. Nat Nanotechnol 2007;2:295–300. Available from: https://doi.org/10.1038/nnano.2007.108.

[36] Rastogi A, Tripathi DK, Yadav S, Chauhan DK, Živčák M, Ghorbanpour M, et al. Application of silicon nanoparticles in agriculture. 3 Biotech 2019;9:1–11. Available from: https://doi.org/10.1007/s13205-019-1626-7.

[37] Chen Z, Li Z, Lin Y, Yin M, Ren J, Qu X. Bioresponsive hyaluronic acid-capped mesoporous silica nanoparticles for targeted drug delivery. Chem Eur J 2013;19:1778–83. Available from: https://doi.org/10.1002/chem.201202038.

[38] Rasouli S, Davaran S, Rasouli F, Mahkam M, Salehi R. Positively charged functionalized silica nanoparticles as nontoxic carriers for triggered anticancer drug release. Des Monomers Polym 2014;17:227–37. Available from: https://doi.org/10.1080/15685551.2013.840475.

[39] Dinker MK, Kulkarni PS. Recent advances in silica-based materials for the removal of hexavalent chromium: a review. J Chem Eng Data 2015;60:2521–40. Available from: https://doi.org/10.1021/acs.jced.5b00292.

[40] Plumeré N, Ruff A, Speiser B, Feldmann V, Mayer HA. Stöber silica particles as basis for redox modifications: particle shape, size, polydispersity, and porosity. J Colloid Interface Sci 2012;368:208–19. Available from: https://doi.org/10.1016/j.jcis.2011.10.070.

[41] Elsharkawy MM, Mousa KM. Induction of systemic resistance against Papaya ring spot virus (PRSV) and its vector *Myzus persicae* by *Penicillium simplicissimum* GP17-2 and silica (SiO_2) nanopowder. Int J Pest Manag 2015;61:353–8. Available from: https://doi.org/10.1080/09670874.2015.1070930.

[42] Alsaeedi AH, El-Ramady H, Alshaal T, El-Garawani M, Elhawat N, Almohsen M. Engineered silica nanoparticles alleviate the detrimental effects of Na$^+$ stress on germination and growth of common bean (*Phaseolus vulgaris*). Environ Sci Pollut Res 2017;24:21917–28. Available from: https://doi.org/10.1007/s11356-017-9847-y.

[43] Alsaeedi A, El-Ramady H, Alshaal T, El-Garawany M, Elhawat N, Al-Otaibi A. Silica nanoparticles boost growth and productivity of cucumber under water deficit and salinity stresses by balancing nutrients uptake. Plant Physiol Biochem 2019;139:1–10. Available from: https://doi.org/10.1016/j.plaphy.2019.03.008.

[44] El-Shetehy M, Moradi A, Maceroni M, Reinhardt D, Petri-Fink A, Rothen-Rutishauser B, et al. Silica nanoparticles enhance disease resistance in *Arabidopsis* plants. Nat Nanotechnol 2021;16:344–53. Available from: https://doi.org/10.1038/s41565-020-00812-0.

[45] Cui J, Li Y, Jin Q, Li F. Silica nanoparticles inhibit arsenic uptake into rice suspension cells: via improving pectin synthesis and the mechanical force of the cell wall. Environ Sci Nano 2020;7:162–71. Available from: https://doi.org/10.1039/c9en01035a.

[46] Wanyika H, Gatebe E, Kioni P, Tang Z, Gao Y. Mesoporous silica nanoparticles carrier for urea: potential applications in agrochemical delivery systems. J Nanosci Nanotechnol 2012;12:2221–8. Available from: https://doi.org/10.1166/jnn.2012.5801.

[47] Cao L, Zhang H, Cao C, Zhang J, Li F, Huang Q. Quaternized chitosan-capped mesoporous silica nanoparticles as nanocarriers for controlled pesticide release. Nanomaterials 2016;6:126. Available from: https://doi.org/10.3390/nano6070126.

[48] Zamani H, Jafari A, Mousavi SM, Darezereshki E. Biosynthesis of silica nanoparticle using *Saccharomyces cervisiae* and its application on enhanced oil recovery. J Pet Sci Eng 2020;190:107002. Available from: https://doi.org/10.1016/j.petrol.2020.107002.

[49] Snehal S, Sc MA, Lohani P, Correspondence S, Snehal M, Sc A. Silica nanoparticles: its green synthesis and importance in agriculture. J Pharmacogn Phytochem 2018;7:3383–93.

[50] Mor S, Manchanda CK, Kansal SK, Ravindra K. Nanosilica extraction from processed agricultural residue using green technology. J Cleaner Prod 2017;143:1284–90. Available from: https://doi.org/10.1016/j.jclepro.2016.11.142.

[51] Adebisi JA, Agunsoye JO, Bello SA, Haris M, Ramakokovhu MM, Daramola MO, et al. Green production of silica nanoparticles from maize stalk. Part Sci Technol 2020;38:667–75. Available from: https://doi.org/10.1080/02726351.2019.1578845.

[52] Adebisi JA, Agunsoye JO, Ahmed II, Bello SA, Haris M, Ramakokovhu MM, et al. Production of silicon nanoparticles from selected agricultural wastes. Mater Today Proc 2021;38:669–74. Available from: https://doi.org/10.1016/j.matpr.2020.03.658.

[53] Bansal V, Sanyal A, Rautaray D, Ahmad A, Sastry M. Bioleaching of sand by the fungus *Fusarium oxysporum* as a means of producing extracellular silica nanoparticles. Adv Mater 2005;17:889–92. Available from: https://doi.org/10.1002/adma.200401176.

[54] Chhipa H. Mycosynthesis of nanoparticles for smart agricultural practice: a green and eco-friendly approach. Green synthesis, characterization and applications of nanoparticles. 2019. p. 87–109. Available from: http://doi.org/10.1016/b978-0-08-102579-6.00005-8.

[55] Zhao L, Lu L, Wang A, Zhang H, Huang M, Wu H, et al. Nano-biotechnology in agriculture: use of nanomaterials to promote plant growth and stress tolerance. J Agric Food Chem 2020;68:1935–47. Available from: https://doi.org/10.1021/acs.jafc.9b06615.

[56] Strout G, Russell SD, Pulsifer DP, Erten S, Lakhtakia A, Lee DW. Silica nanoparticles aid in structural leaf coloration in the Malaysian tropical rainforest understorey herb *Mapania caudata*. Ann Bot (Lond) 2013;112:1141–8. Available from: https://doi.org/10.1093/aob/mct172.

[57] Aqaei P, Weisany W, Diyanat M, Razmi J, Struik PC. Response of maize (*Zea mays* L.) to potassium nano-silica application under drought stress. J Plant Nutr 2020;43:1205–16. Available from: https://doi.org/10.1080/01904167.2020.1727508.

[58] Pan X, Geng Y, Zhang W, Li B, Chen J. The influence of abiotic stress and phenotypic plasticity on the distribution of invasive *Alternanthera philoxeroides* along a riparian zone. Acta Oecol 2006;30:333–41. Available from: https://doi.org/10.1016/j.actao.2006.03.003.

[59] Qi J, Wang J, Gong Z, Zhou JM. Apoplastic ROS signaling in plant immunity. Curr Opin Plant Biol 2017;38:92–100. Available from: https://doi.org/10.1016/j.pbi.2017.04.022.

[60] Nxele X, Klein A, Ndimba BK. Drought and salinity stress alters ROS accumulation, water retention, and osmolyte content in sorghum plants. South Afr J Bot 2017;108:261–6. Available from: https://doi.org/10.1016/j.sajb.2016.11.003.

[61] Mostofa MG, Rahman MM, Ansary MMU, Keya SS, Abdelrahman M, Miah MG, et al. Silicon in mitigation of abiotic stress-induced oxidative damage in plants. Crit Rev Biotechnol 2021;1–17. Available from: https://doi.org/10.1080/07388551.2021.1892582.

[62] Elshayb OM, Nada AM, Ibrahim HM, Amin HE, Atta AM. Application of silica nanoparticles for improving growth, yield, and enzymatic antioxidant for the hybrid rice ehr1 growing under water regime conditions. Materials 2021;14:1150. Available from: https://doi.org/10.3390/ma14051150.

[63] Ghori NH, Ghori T, Hayat MQ, Imadi SR, Gul A, Altay V, et al. Heavy metal stress and responses in plants. Int J Environ Sci Technol 2019;16:1807–28. Available from: https://doi.org/10.1007/s13762-019-02215-8.

[64] Fatemi H, Esmaiel Pour B, Rizwan M. Foliar application of silicon nanoparticles affected the growth, vitamin C, flavonoid, and antioxidant enzyme activities of coriander (*Coriandrum sativum* L.) plants grown in lead (Pb)-spiked soil. Environ Sci Pollut Res 2021;28:1417–25. Available from: https://doi.org/10.1007/s11356-020-10549-x.

[65] Kidd PS, Llugany M, Poschenrieder C, Gunsé B, Barceló J. The role of root exudates in aluminium resistance and silicon-induced amelioration of aluminium toxicity in three varieties of maize (*Zea mays* L.). J Exp Bot 2001;52:1339–52. Available from: https://doi.org/10.1093/jxb/52.359.1339.

[66] da Cunha KPV, do Nascimento CWA. Silicon effects on metal tolerance and structural changes in maize (*Zea mays* L.) grown on a cadmium and zinc enriched soil. Water Air Soil Pollut 2009;197:323–30. Available from: https://doi.org/10.1007/s11270-008-9814-9.

[67] Ali S, Rizwan M, Hussain A, Zia ur Rehman M, Ali B, Yousaf B, et al. Silicon nanoparticles enhanced the growth and reduced the cadmium accumulation in grains of wheat (*Triticum aestivum* L.). Plant Physiol Biochem 2019;140:1–8. Available from: https://doi.org/10.1016/j.plaphy.2019.04.041.

[68] Tripathi DK, Singh VP, Prasad SM, Chauhan DK, Dubey NK. Silicon nanoparticles (SiNp) alleviate chromium (VI) phytotoxicity in *Pisum sativum* (L.) seedlings. Plant Physiol Biochem 2015;96:189–98. Available from: https://doi.org/10.1016/j.plaphy.2015.07.026.

[69] Tripathi DK, Singh S, Singh VP, Prasad SM, Chauhan DK, Dubey NK. Silicon nanoparticles more efficiently alleviate arsenate toxicity than silicon in maize cultivar and hybrid differing in arsenate tolerance. Front Environ Sci 2016;4:46. Available from: https://doi.org/10.3389/fenvs.2016.00046.

[70] Munns R, Tester M. Mechanisms of salinity tolerance. Annu Rev Plant Biol 2008;59:651–81. Available from: https://doi.org/10.1146/annurev.arplant.59.032607.092911.

[71] Mahmoud LM, Dutt M, Shalan AM, El-Kady ME, El-Boray MS, Shabana YM, et al. Silicon nanoparticles mitigate oxidative stress of in vitro-derived banana (*Musa acuminata* 'Grand Nain') under simulated water deficit or salinity stress. South Afr J Bot 2020;132:155–63. Available from: https://doi.org/10.1016/j.sajb.2020.04.027.

[72] Guerriero G, Hausman JF, Legay S. Silicon and the plant extracellular matrix. Front Plant Sci 2016;7:463. Available from: https://doi.org/10.3389/fpls.2016.00463.

[73] Farhangi-Abriz S, Torabian S. Nano-silicon alters antioxidant activities of soybean seedlings under salt toxicity. Protoplasma 2018;255:953–62. Available from: https://doi.org/10.1007/s00709-017-1202-0.

[74] Qados A. Mechanism of nanosilicon-mediated alleviation of salinity stress in faba bean (*Vicia faba* L.) plants. Am J Exp Agric 2015;7:78–95. Available from: https://doi.org/10.9734/ajea/2015/15110.

[75] Khiyami MA, Almoammar H, Awad YM, Alghuthaymi MA, Abd-Elsalam KA. Plant pathogen nanodiagnostic techniques: forthcoming changes? Biotechnol Biotechnol Equip 2014;28:775–85. Available from: https://doi.org/10.1080/13102818.2014.960739.

[76] Kim SG, Kim KW, Park EW, Choi D. Silicon-induced cell wall fortification of rice leaves: a possible cellular mechanism of enhanced host resistance to blast. Phytopathology 2002;92:1095–103. Available from: https://doi.org/10.1094/PHYTO.2002.92.10.1095.

[77] Cherif M, Asselin A, Belanger RR. Defense responses induced by soluble silicon in cucumber roots infected by *Pythium* spp. Phytopathology 1994;84:236–42. Available from: https://doi.org/10.1094/Phyto-84-236.

[78] Suriyaprabha R, Karunakaran G, Kavitha K, Yuvakkumar R, Rajendran V, Kannan N. Application of silica nanoparticles in maize to enhance fungal resistance. IET Nanobiotechnol 2014;8:133−7. Available from: https://doi.org/10.1049/iet-nbt.2013.0004.

[79] Derbalah A, Shenashen M, Hamza A, Mohamed A, el Safty S. Antifungal activity of fabricated mesoporous silica nanoparticles against early blight of tomato. Egypt J Basic Appl Sci 2018;5:145−50. Available from: https://doi.org/10.1016/j.ejbas.2018.05.002.

[80] Zhang H, Chen G. Potent antibacterial activities of Ag/TiO$_2$ nanocomposite powders synthesized by a one-pot sol-gel method. Environ Sci Technol 2009;43:2905−10. Available from: https://doi.org/10.1021/es803450f.

[81] Petica A, Gavriliu S, Lungu M, Buruntea N, Panzaru C. Colloidal silver solutions with antimicrobial properties. Mater Sci Eng B Solid State Mater Adv Technol 2008;152:22−7. Available from: https://doi.org/10.1016/j.mseb.2008.06.021.

[82] Salem HF, Eid KAM, Sharaf MA. Formulation and evaluation of silver nanoparticles as antibacterial and antifungal agents with a minimal cytotoxic effect. Int J Drug Deliv 2011;1:293. Available from: https://doi.org/10.5138/ijdd.v3i2.224.

[83] Nazaralian S, Majd A, Irian S, Najafi F, Ghahremaninejad F, Landberg T, et al. Comparison of silicon nanoparticles and silicate treatments in fenugreek. Plant Physiol Biochem 2017;115:25−33. Available from: https://doi.org/10.1016/j.plaphy.2017.03.009.

[84] Reynolds OL, Keeping MG, Meyer JH. Silicon-augmented resistance of plants to herbivorous insects: a review. Ann Appl Biol 2009;155:171−86. Available from: https://doi.org/10.1111/j.1744-7348.2009.00348.x.

[85] Vani C, Brindhaa U. Silica nanoparticles as nanocides against *Corcyra cephalonica* (S.), the stored grain pest. Int J Pharma Bio Sci 2013;4.

[86] Neu S, Schaller J, Dudel EG. Silicon availability modifies nutrient use efficiency and content, C:N:P stoichiometry, and productivity of winter wheat (*Triticum aestivum* L.). Sci Rep 2017;7:1−8. Available from: https://doi.org/10.1038/srep40829.

[87] Schaller J, Brackhage C, Gessner MO, Bäuker E, Gert Dudel E. Silicon supply modifies C:N:P stoichiometry and growth of *Phragmites australis*. Plant Biol 2012;14:392−6. Available from: https://doi.org/10.1111/j.1438-8677.2011.00537.x.

[88] Yassen A, Abdallah E, Gaballah M, Zaghloul S. Role of silicon dioxide nano fertilizer in mitigating salt stress on growth, yield and chemical composition of cucumber (*Cucumis sativus* L.). Int J Agric Res 2017;12:130−5. Available from: https://doi.org/10.3923/ijar.2017.130.135.

[89] Le VN, Rui Y, Gui X, Li X, Liu S, Han Y. Uptake, transport, distribution and bio-effects of SiO$_2$ nanoparticles in Bt-transgenic cotton. J Nanobiotechnol 2014;12:1−15. Available from: https://doi.org/10.1186/s12951-014-0050-8.

[90] Mukarram M, Khan MMA, Corpas FJ. Silicon nanoparticles elicit an increase in lemongrass (*Cymbopogon flexuosus* (Steud.) Wats) agronomic parameters with a higher essential oil yield. J Hazard Mater 2021;412:125254. Available from: https://doi.org/10.1016/j.jhazmat.2021.125254.

[91] Sun D, Hussain HI, Yi Z, Siegele R, Cresswell T, Kong L, et al. Uptake and cellular distribution, in four plant species, of fluorescently labeled mesoporous silica nanoparticles. Plant Cell Rep 2014;33:1389−402. Available from: https://doi.org/10.1007/s00299-014-1624-5.

[92] Roychoudhury A. Silicon-nanoparticles in crop improvement and agriculture. Int J Rec Adv Biotechnol Nanotechnol 2020;3.

[93] Rico CM, Majumdar S, Duarte-Gardea M, Peralta-Videa JR, Gardea-Torresdey JL. Interaction of nanoparticles with edible plants and their possible implications in the food chain. J Agric Food Chem 2011;59:3485−98. Available from: https://doi.org/10.1021/jf104517j.

[94] Trewyn BG, Giri S, Slowing II, Lin VSY. Mesoporous silica nanoparticle based controlled release, drug delivery, and biosensor systems. Chem Commun 2007;3236−45. Available from: https://doi.org/10.1039/b701744h.

[95] Nair R, Varghese SH, Nair BG, Maekawa T, Yoshida Y, Kumar DS. Nanoparticulate material delivery to plants. Plant Sci 2010;179:154−63. Available from: https://doi.org/10.1016/j.plantsci.2010.04.012.

[96] Chang FP, Kuang LY, Huang CA, Jane WN, Hung Y, Hsing YIC, et al. A simple plant gene delivery system using mesoporous silica nanoparticles as carriers. J Mater Chem B 2013;1:5279−87. Available from: https://doi.org/10.1039/c3tb20529k.

[97] Chinnamuthu C, Boopathi P. Nanotechnology and agroecosystem. Madras Agric J 2009;96:17−31.

[98] Deodhar Gv, Adams ML, Trewyn BG. Controlled release and intracellular protein delivery from mesoporous silica nanoparticles. Biotechnol J 2017;12:1600408. Available from: https://doi.org/10.1002/biot.201600408.

[99] Slowing II, Vivero-Escoto JL, Wu CW, Lin VSY. Mesoporous silica nanoparticles as controlled release drug delivery and gene transfection carriers. Adv Drug Deliv Rev 2008;60:1278−88. Available from: https://doi.org/10.1016/j.addr.2008.03.012.

[100] Slowing II, Vivero-Escoto JL, Trewyn BG, Lin VSY. Mesoporous silica nanoparticles: structural design and applications. J Mater Chem 2010;20:7924−37. Available from: https://doi.org/10.1039/c0jm00554a.

[101] Sun D, Hussain HI, Yi Z, Rookes JE, Kong L, Cahill DM. Delivery of abscisic acid to plants using glutathione responsive mesoporous silica nanoparticles. J Nanosci Nanotechnol 2017;18:1615−25. Available from: https://doi.org/10.1166/jnn.2018.14262.

[102] Hajiahmadi Z, Shirzadian-Khorramabad R, Kazemzad M, Sohani MM. Enhancement of tomato resistance to *Tuta absoluta* using a new efficient mesoporous silica nanoparticle-mediated plant transient gene expression approach. Sci Hortic (Amsterdam) 2019;243:367−75. Available from: https://doi.org/10.1016/j.scienta.2018.08.040.

[103] Panpatte DG, Jhala YG, Shelat HN, Vyas RV. Nanoparticles: the next generation technology for sustainable agriculture. Microbial inoculants in sustainable agricultural productivity: vol. 2: functional applications. 2016. p. 289−300. Available from: http://doi.org/10.1007/978-81-322-2644-4_18.

[104] Bravo-Cadena M. Application of mesoporous silica nanoparticles for biocide delivery to plants to prevent pre-harvest losses [thesis]. University of Oxford; 2018.

[105] Abdelrahman TM, Qin X, Li D, Senosy IA, Mmby M, Wan H, et al. Pectinase-responsive carriers based on mesoporous silica nanoparticles for improving the translocation and fungicidal activity of prochloraz in rice plants. Chem Eng J 2021;404:126440. Available from: https://doi.org/10.1016/j.cej.2020.126440.

[106] Cui HX, Sun CJ, Liu Q, Jiang J, Gu W. Applications of nanotechnology in agrochemical formulation. perspectives, challenges and strategies. International Conference on Nanoagriculture, São Pedro, Brazil. 2010. p. 28−33.

[107] Sun D, Hussain HI, Yi Z, Rookes JE, Kong L, Cahill DM. Mesoporous silica nanoparticles enhance seedling growth and photosynthesis in wheat and lupin. Chemosphere 2016;152:81−91. Available from: https://doi.org/10.1016/j.chemosphere.2016.02.096.

[108] Kioni PN, Wanyika H, Gatebe E, Tang Z, Gao Y. Controlled release of fertiliser using mesoporous silica nanoparticles. In: Scientific conference proceedings. Dedan Kimathi University of Technology; 2017. Availble from: < http://repository.dkut.ac.ke:8080/xmlui/handle/123456789/1020.html > [accessed 06.09.21].

[109] Hardoim PR, van Overbeek LS, Berg G, Pirttilä AM, Compant S, Campisano A, et al. The hidden world within plants: Ecological and evolutionary considerations for defining functioning of microbial endophytes. Microbiol Mol Biol Rev 2015;79:293−320. Available from: https://doi.org/10.1128/mmbr.00050-14.

[110] Brader G, Compant S, Vescio K, Mitter B, Trognitz F, Ma LJ, et al. Ecology and genomic insights into plant-pathogenic and plant-nonpathogenic endophytes. Annu Rev Phytopathol 2017;55:61−83. Available from: https://doi.org/10.1146/annurev-phyto-080516-035641.

[111] Lemanceau P, Blouin M, Muller D, Moënne-Loccoz Y. Let the core microbiota be functional. Trends Plant Sci 2017;22:583−95. Available from: https://doi.org/10.1016/j.tplants.2017.04.008.

[112] Compant S, Samad A, Faist H, Sessitsch A. A review on the plant microbiome: ecology, functions, and emerging trends in microbial application. J Adv Res 2019;19:29−37. Available from: https://doi.org/10.1016/j.jare.2019.03.004.

[113] Glick BR. Plant growth-promoting bacteria: mechanisms and applications. Scientifica 2012;2012. Available from: https://doi.org/10.6064/2012/963401.

[114] Gond SK, Torres MS, Bergen MS, Helsel Z, White JF. Induction of salt tolerance and up-regulation of aquaporin genes in tropical corn by rhizobacterium Pantoea agglomerans. Lett Appl Microbiol 2015;60:392−9. Available from: https://doi.org/10.1111/lam.12385.

[115] Mohammadi K, Sohrabi Y. Bacterial biofertilizers for sustainable crop production: a review. J Agric Biol Sci 2012;7:307−16.

[116] Sahu PK, Brahmaprakash GP. Formulations of biofertilizers—approaches and advances. Microbial inoculants in sustainable agricultural productivity: vol. 2: functional applications. 2016. p. 179−98. Available from: http://doi.org/10.1007/978-81-322-2644-4_12.

[117] Koul O. Nano-biopesticides today and future perspectives. Academic Press; 2019. Available from: http://doi.org/10.1016/C2017-0-03028-8.

[118] Gouda S, Kerry RG, Das G, Paramithiotis S, Shin HS, Patra JK. Revitalization of plant growth promoting rhizobacteria for sustainable development in agriculture. Microbiol Res 2018;206:131−40. Available from: https://doi.org/10.1016/j.micres.2017.08.016.

[119] Karunakaran G, Suriyaprabha R, Manivasakan P, Yuvakkumar R, Rajendran V, Prabu P, et al. Effect of nanosilica and silicon sources on plant growth promoting rhizobacteria, soil nutrients and maize seed germination. IET Nanobiotechnol 2013;7:70−7. Available from: https://doi.org/10.1049/iet-nbt.2012.0048.

[120] Rangaraj S, Gopalu K, Rathinam Y, Periasamy P, Venkatachalam R, Narayanasamy K. Effect of silica nanoparticles on microbial biomass and silica availability in maize rhizosphere. Biotechnol Appl Biochem 2014;61:668−75. Available from: https://doi.org/10.1002/bab.1191.

[121] Hafez EM, Osman HS, Gowayed SM, Okasha SA, Omara AED, Sami R, et al. Minimizing the adverse impacts of water deficit and soil salinity on maize growth and productivity in response to the application of plant growth-promoting rhizobacteria and silica nanoparticles. Agronomy 2021;11:676. Available from: https://doi.org/10.3390/agronomy11040676.

[122] Siddiqui MH, Al-Whaibi MH. Role of nano-SiO$_2$ in germination of tomato (Lycopersicum esculentum seeds Mill.). Saudi J Biol Sci 2014;21:13−17. Available from: https://doi.org/10.1016/j.sjbs.2013.04.005.

[123] Boroumand N, Behbahani M, Dini G. Combined effects of phosphate solubilizing bacteria and nanosilica on the growth of land cress plant. J Soil Sci Plant Nutr 2020;20:232−43. Available from: https://doi.org/10.1007/s42729-019-00126-8.

[124] Kurdish I, Roy A, Chobotarov A, Herasimenko I, Plotnikov V, Gylchuk V, et al. Free-flowing complex bacterial preparation for crop and efficiency of its use in agroecosystems. J Microbiol Biotechnol Food Sci 2018;2020:527−31.

[125] Santoyo G, del Orozco-Mosqueda MC, Govindappa M. Mechanisms of biocontrol and plant growth-promoting activity in soil bacterial species of Bacillus and Pseudomonas: a review. Biocontrol Sci Technol 2012;22:855−72. Available from: https://doi.org/10.1080/09583157.2012.694413.

[126] Li H, Ding X, Wang C, Ke H, Wu Z, Wang Y, et al. Control of tomato yellow leaf curl virus disease by Enterobacter asburiae BQ9 as a result of priming plant resistance in tomatoes. Turk J Biol 2016;40:150−9. Available from: https://doi.org/10.3906/biy-1502-12.

[127] Srivastava S, Bist V, Srivastava S, Singh PC, Trivedi PK, Asif MH, et al. Unraveling aspects of Bacillus amyloliquefaciens mediated enhanced production of rice under biotic stress of Rhizoctonia solani. Front Plant Sci 2016;7:587. Available from: https://doi.org/10.3389/fpls.2016.00587.

[128] Djaya L, Hersanti, Istifadah N, Hartati S, Joni IM. In vitro study of plant growth promoting rhizobacteria (PGPR) and endophytic bacteria antagonistic to Ralstonia solanacearum formulated with graphite and silica nano particles as a biocontrol delivery system (BDS). Biocatal Agric Biotechnol 2019;19:101153. Available from: https://doi.org/10.1016/j.bcab.2019.101153.

[129] de Moraes ACP, Ribeiro L, da S, de Camargo ER, Lacava PT. The potential of nanomaterials associated with plant growth-promoting bacteria in agriculture. 3 Biotech 2021;11:1−17. Available from: https://doi.org/10.1007/s13205-021-02870-0.

[130] Pour MM, Saberi-Riseh R, Mohammadinejad R, Hosseini A. Nano-encapsulation of plant growth-promoting rhizobacteria and their metabolites using alginate-silica nanoparticles and carbon nanotube improves UCB1 pistachio micropropagation. J Microbiol Biotechnol 2019;29:1096−103. Available from: https://doi.org/10.4014/jmb.1903.03022.

[131] Gordienko AS, Kurdish IK. Surface electrical properties of Bacillus subtilis cells and the effect of interaction with silicon dioxide particles. Biophysics (Oxf) 2007;52:217−19. Available from: https://doi.org/10.1134/S0006350907020121.

[132] Hirota R, Hata Y, Ikeda T, Ishida T, Kuroda A. The silicon layer supports acid resistance of Bacillus cereus spores. J Bacteriol 2010;192:111−16. Available from: https://doi.org/10.1128/JB.00954-09.

[133] Kurdish IK. Interaction of microorganisms with nanomaterials as a basis for creation of high-efficiency biotechnological preparations. Nanobiotechnology in bioformulations. Springer; 2019. p. 259−87. Available from: http://doi.org/10.1007/978-3-030-17061-5_11.

[134] Phenrat T, Long TC, Lowry GV, Veronesi B. Partial oxidation ("aging") and surface modification decrease the toxicity of nanosized zerovalent iron. Environ Sci Technol 2009;43:195−200. Available from: https://doi.org/10.1021/es801955n.

6

The genetics of silicon accumulation in plants

Libia Iris Trejo-Téllez[1], Libia Fernanda Gómez-Trejo[2],
Hugo Fernando Escobar-Sepúlveda[3] and Fernando Carlos Gómez-Merino[4]

[1]Laboratory of Plant Nutrition, College of Postgraduates in Agricultural Sciences, Texcoco, Mexico
[2]Department of Plant Protection, Chapingo Autonomous University, Texcoco, Mexico [3]Institute of Biological Sciences, The University of Talca, Talca, Chile [4]Laboratory of Plant Biotechnology, College of Postgraduates in Agricultural Sciences, Córdoba, Mexico

6.1 Introduction

Silicon (Si) has largely been regarded as a *quasi*-essential element for plants and recently categorized within the group of beneficial elements and as an inorganic biostimulant with paramount implications for modern sustainable agriculture [1,2]. With a relative abundance of approximately 282,000 ppm, accounting for approximately 28.2% of the earth's crust content, Si is the second most abundant element on the planet, just after oxygen. Importantly, more than 90% of the earth's crust is composed of silicate minerals (e.g., quartz, feldspar, mica, amphibole, pyroxene, olivine, and diverse clay minerals), aluminosilicates (e.g., andalusite, kyanite, and sillimanite with the composition Al_2SiO_5), and silicon dioxide or silica (SiO_2), all of them being biologically inert, and very few organisms use these forms directly [3].

Silicon is widely present in practically all three major branches of life, including Archaea [4,5], Eubacteria [6,7], and Eukaryotes [8—10]. However, the molecular and biochemical mechanisms that control silicon metabolism in living organisms remain elusive and represent a pressing need for scientists involved in biological and environmental studies to better understand them.

In plants, Si is found in virtually every taxonomic unit, though its concentration varies greatly among species, ranging from 0.1% to 10% Si on a dry weight basis, thus displaying an uneven distribution among different plant taxa [11,12]. Takahashi et al. [13] and Epstein [14] developed the first schematic representations of such a distribution. Ma et al. [15], Ma [16], and Liang et al. [12] then proposed further analysis and novel approaches adding new data. In Fig. 6.1, we present an updated approach of such distribution based on current data and findings. Among Bryophyta, liverworts, mosses, and hornwort clades present substantial Si accumulation [17]. In Pteridophyta, the divisions Lycopsida, Equisetopsida, and Spheropsida and to a lesser degree Pteropsida (i.e., Filicopsida) are also Si accumulators. Among Spermatophyta, a lower capacity to absorb and accumulate Si is observed in both Gymnospermae and Angiospermae. Among Angiospermae, the families Cyperaceae, Poaceae, and Balsaminaceae display enhanced capacities to accumulate Si, whereas Cucurbitaceae, Urticaceae, and Commelinaceae display intermediate concentrations of Si in their tissues [12,16,18,19].

Plant roots can absorb Si exclusively in the form of silicic acid [$Si(OH)_4$ or H_4SiO_4], also known as monosilicic acid or ororthosilicic acid, which is scarce in the soil solution (0.1—0.6 mM in most agricultural soils) [20]. The concentration of this Si form available for plants is significantly affected by soil pH and the amounts of clay, minerals, organic matter, and iron/aluminum oxides/hydroxides present in the soil [21]. Furthermore, plants exhibit differential capabilities to absorb and transfer Si, so they have been categorized into three groups: high-Si, intermediate-Si, or non-Si accumulators (Si excluders), respectively [16,22]. High-Si accumulator plants exhibit Si contents higher than 1.0%, whereas non-accumulator species have less than 0.5%, and intermediate-Si

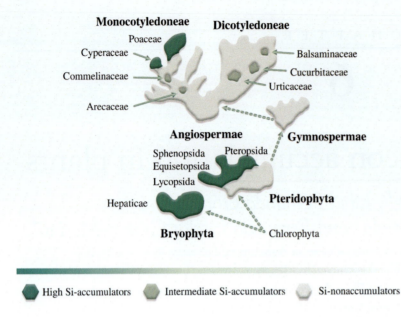

accumulators display Si concentrations between 0.5% and 1.0% on a dry weight basis [12,15,16,23–25]. This ability of plants to accumulate Si is associated with different mechanisms by which this element is absorbed: active, passive, and rejective [26,27]. Regarding the active mechanism, the magnitude of Si uptake is usually larger than that predicted based on the mass flow and is attributed to the density of silicon channels in the roots and shoots that facilitate the absorption process. Furthermore, diffusion can also take place during Si absorption. Consequently, in some plant species, Si concentrations in the xylem are frequently higher than those found in the soil solution, implying that Si uptake is a metabolically driven process [15,28]. For the passive mechanism, it is predicted to be completely driven by mass flow. In the rejective mechanism, the buildup of the concentration of silicic acid in the soil solution typically results from the low concentrations of Si that are absorbed by plants [21]. Hence, plants that rely primarily on active, passive, or rejective mechanisms have been classified as high-Si, intermediate-Si, or non-Si accumulators, respectively [15,22]. Such capability to absorb and transport Si is facilitated by an efficient activity of influx and efflux Lsi channels [29], which must be controlled coordinately to ensure that all steps involved in Si metabolism function in a concerted manner. Though the dual action of both influx and efflux channels is a prerequisite to accumulate Si within plant tissues, the presence of the former proteins (i.e., influx channels) is a *sinequanon* condition for Si uptake in plants [30].

6.2 Genetic and molecular basis of Si uptake and movement of Si within plant cells

The molecular cloning and characterization of the first silicon transporter (Low silicon 1 or Lsi1) in higher plants [31] opened new avenues to explore the silicon uptake system in plants and to develop strategies for sustainable use efficiency of this beneficial element. Furthermore, significant progress in novel sequencing technologies and advances in omics sciences are allowing a deeper understanding of Si uptake and transport in plants. Such tremendous strides are enabling a more precise classification of plants according to their capacities to absorb, transport, and accumulate Si, and thus determine whether they are Si competent or not in order to establish efficient Si application programs.

The silicon channel Lsi1 is a nodulin 26-like intrinsic protein (NIP), which integrates a group of highly conserved multifunctional major intrinsic proteins (MIPs) belonging to the plant-specific aquaporin channel protein superfamily [32,33]. This ancient protein superfamily is present in all living organisms [34], with tens of thousands of homologs in archaeal, bacterial, and eukaryotic organisms, thus encompassing the three domains of life [35].

As members of the NIP family of aquaporins, plant Lsi1 proteins harbor a MIP domain and six transmembrane (TM) domains, as well as a conserved aromatic/arginine (ar/R) selective filter and two Asn-Pro-Ala (NPA) motifs. The ar/R and the NPA motifs are critical factors determining the selectivity of Si by the Lsi1 channels [36,37]. Furthermore, high- and intermediate-Si accumulator species contain a Gly/Ala-Ser-Gly-Arg [(G/A)SGR] motif, which is absent in most plants categorized as low-Si accumulators or Si excluders. In addition to the

(G/A)SGR, another Ser-Thr-Ala-Arg (STAR) motif may be present in high and intermediate accumulators [38,39]. Importantly, a specific space of 108 amino acids between the two NPA motifs is also essential for Si permeability in plant cells [36]. These plasma membrane proteins enable the uptake and movement of silicic acid from the soil solution through the rhizosphere and the epidermal cells [40]. Just recently, 80 putative Lsi1 proteins were identified through an in silico exploration of sequenced Viridiplantae genomes. In streptophytes, these channel-type Lsi1 proteins display two NPA motifs, a GSGR or STAR selectivity filter, and 108 amino acids between two NPA motifs, which are absent in chlorophytes, while streptophytes evolved two different types of Lsi1 channels with different selectivity filters [41].

Once Si has entered the plant cell, the efflux Lsi channels facilitate Si movement further across the endodermis, epidermis, and hypodermis in seminal and crown roots and also in the cortex in lateral roots [42,43]. Interestingly, 133 putative efflux Lsi channels were identified in the aforementioned genome-wide approach and both Lsi1 and Lsi2 evolved two types of gene structures each; however, Lsi2s are ancient and were also found in chlorophytes [41]. Lsi2 proteins exhibit a simpler domain organization as compared to Lsi1; they harbor 9−11 conserved TM domain and belong to an anion transporter family characterized by the presence of a Citrate-$Mg^{2+}:H^+(CitM) − Citrate-Ca^{2+}:H^+(CitH)$ Symporter (CitMHS) domain, involved in ion transport, and nine conserved TM domains [44]. Though Lsi2 channels lack homology with Lsi1 ones, the cooperative system of both channel-type Lsi1 and Lsi2 enables plants to translocate Si to the aerial parts of the plants where it is deposited as silica phytoliths, microscopic structures of amorphous hydrated silica ($SiO_2 \bullet nH_2O$) [45].

Since the discovery of Lsi1 in rice [31], an increasing number of genes and their corresponding protein products have been identified, cloned, and characterized in maize [43], barley [46,47], pumpkin [48], wheat [49], horsetail [39], soybean [50], wild rice species [51], sorghum [52], tobacco [53], and tomato [54]. Furthermore, genome-wide analyses have revealed the presence and transcriptional activity of *Lsi1* genes in potato [55], strawberry [56], *indica* rice cultivars [57], flax [58], *Arachis* [59], poinsettia [60], pepper [44], and ryegrass [61]. Nawaz et al. [41] reported 80 putative influx Lsi channels in the genome of 80 plant species explored. Whether the corresponding protein products of such genes are enzymatically active and whether they are bonafide influx Lsi channels remain open questions.

Efflux Lsi channels have also been identified and characterized in diverse plant species, including rice [62], maize and barley [42], pumpkin [63], horsetail [64], and tomato [54]. Moreover, in silico analyses have predicted the existence and the activity of additional *Lsi2* genes in tomato, potato, *Arabidopsis*, maize, rice, and sorghum [44]. When surveying the genome of 80 plant species, Nawaz et al. [41] found 133 putative *Lsi2* genes and protein products.

In addition to Lsi1 and Lsi2, the rice proteome contains the Lsi3 and Lsi6 channels. In coordination with Lsi2, Lsi3 and Lsi6 are implicated in the intervascular transfer of Si, which is required to have an adequate distribution of this element, resulting in a unique hyperaccumulation of Si in rice husk [65,66]. Similar to Lsi2, Lsi3 is an active efflux Si channel enzymatically stimulated by a proton gradient [62,66−68]. Lsi3 orthologs have been found in tomatoes [69] and finger millet [70].

In rice, the Lsi6 channel is a member of the NIP subgroup of aquaporins, implicated in the distribution of Si in the shoots. It is transcriptionally expressed in the leaf sheath and blades, as well as in the root tips, and the protein product is responsible for the transport of Si out of the xylem and subsequently affects the distribution of Si in the leaf [71]. Homologs of Lsi6 have been identified and characterized in maize [43], barley [47], and soybean [50].

Just recently, Kumar et al. [72] identified the sorghum Siliplant1 (Slp1), a novel basic protein with seven repeat units rich in proline, lysine, and glutamic acid, transcriptionally active in immature leaf and inflorescence, and found implicated in Si precipitation in vitro at a biologically relevant silicic acid concentration. Interestingly, 225 Slp1 homologs were present in almost all streptophytes analyzed, regardless of their Si accumulation capacity [41]. Taking into consideration the findings and models described elsewhere [2,9,37,50,66,72−79], we developed Fig. 6.2, which depicts the current picture of Si uptake, translocation, and accumulation in plant cells.

Since the list of sequenced plant genomes is continuously increasing, a more in-depth understanding of the mechanisms behind Si absorption and transport within plant cells is becoming accessible, and more genes and proteins involved in Si uptake and movement are being unraveled. It is worth mentioning that, despite the divergence in the capacities to absorb and transport silicic acid among plant species, Lsi1 channels represent the most determining factor for such processes [30].

Just recently, Exley et al. [80] reexamined the evidence of the biochemical and molecular mechanisms of Si movement in plants and have challenged the current paradigm involving exclusive influx and efflux channels for Si uptake and transport. They argue that the use of molecular dynamics simulations will enable the unequivocal identification of channels involved in silicon transport in plants and thereby pave the way to better evidence-based approaches.

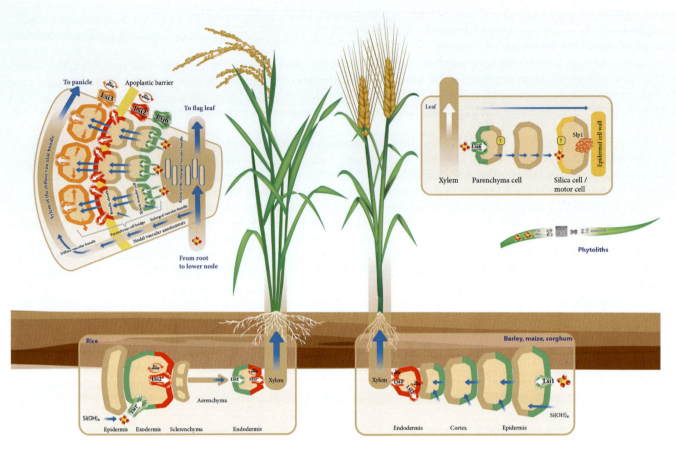

FIGURE 6.2 Representative model of Si uptake by roots from the soil solution and its translocation to shoots, leaves, and seeds. The silicic acid molecule [$Si(OH)_4$], which consists of a silicon atom surrounded by four hydroxyl groups, displays a tetrahedral molecular structure. From root epidermis cells, $Si(OH)_4$ is transported through exodermis cells by the influx Lsi1 and the efflux Lsi2 channels. In the aerenchyma of rice roots, $Si(OH)_4$ moves through the apoplast until it reaches the endodermis where the Lsi1 and Lsi2 channels load the molecule into the stele. An undefined transporter loads $Si(OH)_4$ into the xylem. Via the xylem, $Si(OH)_4$ arrives in the shoots, where the Lsi6 channel unloads it into the xylem parenchyma cells. In the node, $Si(OH)_4$ is transferred to the parenchyma cell bridge and exported and reloaded to the xylem of diffuse vascular bundles, which is mediated by Lsi3. An undefined protein transports $Si(OH)_4$ in the leaf cells where it is precipitated by the action of Slp1.

6.3 Distribution of Lsi channels and Silp1 proteins in plants

The influx Lsi1 and Lsi6 channels belong to the aquaporin superfamily. Aquaporins are present in all living organisms and have been classified into six major groups in Eukaryotes [81]. Interestingly, the number of aquaporins in plant genomes is higher than that found in mammalian genomes [50,82], which may have resulted from the higher degree of compartmentalization at subcellular and tissue levels for the fine regulation of water transport and from genome duplications observed in plants [50,83]. Accordingly, plant species exhibiting genome duplications such as soybean [84,85] and *Brassica rapa* [86] contain nearly 60 aquaporins each, almost twice the number found in *Arabidopsis* (35) or rice (33) [50,87,88]. We explored the genomes and proteomes of different crop and model plant species in order to identify putative Lsi influx and efflux channels and siliplant1 protein homologs involved in Si uptake and transport. The protein sequences were obtained by using Pfam (pfam.xfam. org) taking into consideration the protein sequences exhibiting the highest homology to those involved in Si uptake and transport previously reported from pepper (*Capsicum annuum*) [44] (Table 6.1).

Though environmental factors such as growth conditions, soil pH, and Si availability in the soil solution may affect Si accumulation in plant tissues, the genetic background and molecular mechanisms attributed to the existence of bonafide influx and efflux Lsi channels seem to be the most critical factors for plants to absorb and transport Si. When analyzing plant capacities to absorb and accumulate Si, interspecific variations are much more common as compared to intraspecific variations. However, rice cultivars belonging to the *japonica* subspecies will usually exhibit higher Si concentrations than cultivars from the *indica* subspecies [89—91]. Several efforts have

TABLE 6.1 List of Lsi influx and efflux channels and siliplant1 proteins from different plant species retrieved from a genome-wide exploration.

Plant species	Lsi influx channels	Lsi efflux channels	Siliplant1
Arabidopsis (Arabidopsis thaliana)	55	3	32
Turnip (Brassica rapa)	60	3	51
Pepper (Capsicum annuum)	64	2	31
Melon (Cucumis melo)	34	2	23
Cucumber (Cucumis sativus)	35	2	23
Soja (Glycine soja)	107	13	51
Soybean (Glycine max)	112	13	43
Barley (Hordeum vulgare)	80	45	44
Wild apple (Malus baccata)	50	1	33
Apple (Malus domestica)	52	4	32
Rice (Oryza sativa)	91	17	90
Peach (Prunus persica)	33	2	25
Almond (Prunus dulcis)	34	1	23
Tomato (Solanum lycopersicum)	44	3	27
Potato (Solanum tuberosum)	64	7	38
Sorghum (Sorghum bicolor)	42	7	30
Wheat (Triticum aestivum)	136	19	102
Grape (Vitis vinifera)	36	4	18
Maize (Zea mays)	123	14	28

Data were retrieved from Pfam (pfam.xfam.org). This table presents the list of sequences from the highest homology protein family to CaLsi1 (for Lsi influx [44]), CaLsi2 (for Lsi efflux [44]) and SbSlp1 (for Siliplant1 [72]) with E-value 3.0e-52 (MIP family; PF00230), 8.0e-08 (CitMHS family; PF03600), and 8.0e-08 (Pollen_Ole_e_I family; PF01190), respectively.

been made to identify genetic loci governing these differential Si accumulation patterns by using QTL mapping approaches, which have covered almost every rice chromosome [e.g., Refs. [92–94]]. Nevertheless, none of the QTL determined is collocated within the *Lsi* genes, suggesting that such genome-wide inferences do not explain the genetic variation for Si uptake observed in vivo [30,31,67].

The pivotal role of Si lies in its ability to stimulate plant growth and metabolism, while helping to alleviate deleterious effects of abiotic and biotic stresses in plants. Importantly, silicic acid, the only form of Si that plant cells can absorb, has impacted life establishment and maintenance on Earth, affecting the biochemical evolution of life by excluding aluminum from biota and providing adventitious benefits through biological silicification [76].

In the 21st century, great challenges related to population growth rates, the food supply, mounting pressures on natural resources, and climate change threaten the sustainability of food systems at large [95]. Therefore the future of food security and nutrition, rural wealth, efficiency of food systems, and sustainability and resilience of rural livelihoods, agricultural systems, and their natural resource base depend on measures and decisions taken and implemented at present. Si supplementation might be one of the most promising strategies to produce more resilient crops in sustainable agriculture in the context of challenging environments. Importantly, Si has been found to stimulate photosynthesis in a hormetic dose–response manner [96–98], thus contributing to ensuring food security and sustaining life on the planet [99].

6.4 Conclusion

Silicon (Si), the most studied beneficial element categorized as a potent inorganic biostimulant, has the capacity to stimulate plant growth and metabolism under both nonstressful conditions [98,100,101], and stressful conditions [101−105]. However, more evident benefits of Si applications are frequently observed under stressful conditions.

Si is absorbed and distributed within the plant through a cooperative system of influx (Lsi1 and Lsi6) and efflux (Lsi2 and Lsi3) channels, in coordination with siliplant 1 (Slp1), a silica mineralizer protein that actively participates in silica deposition in vivo. Lsi1 and Lsi6 belong to the NIP family of aquaporins, whereas Lsi2 and Lsi3 are members of an anion transporter family. Siliplant 1 belongs to a family of proline-rich proteins, transcriptionally active in young leaves and inflorescences, where it participates in the processes of Si deposition as phytoliths during biosilicification. The structure of NIP aquaporins in chlorophytes differs from those found in streptophytes, the latter exhibiting a more complex motif organization. In general, Lsi1 and Lsi6 are more abundant than Lsi2 and Lsi3. Interestingly, more than 220 Slp1 homologs have been identified in almost all streptophytes, regardless of their Si accumulation capacity. The biological and environmental importance of Slp1 is such that it may help improve a plant's ability to withstand different stresses. Currently, Slp1 is gaining importance as a key factor for determining biosilicification in plants; moreover, Slp1 and the influx and efflux Lsi channels represent efficient molecular mechanisms to cope with different environmental challenges resulting from global climate change.

References

[1] Gómez-Merino FC, Trejo-Téllez LI. The role of beneficial elements in triggering adaptive responses to environmental stressors and improving plant performance. In: Vats S, editor. Biotic and abiotic stress tolerance in plants. Singapore: Springer; 2018. p. 137−72.

[2] Katz O, Puppe D, Kaczorek D, Prakash NB, Schaller J. Silicon in the soil−plant continuum: intricate feedback mechanisms within ecosystems. Plants 2021;10:652. Available from: https://doi.org/10.3390/plants10040652.

[3] Laane HM. The effects of foliar sprays with different silicon compounds. Plants 2018;7:45. Available from: https://doi.org/10.3390/plants7020045.

[4] Orange F, Westall F, Disnar JR, Prieur D, Bienvenu N, Le Romancer M, et al. Experimental silicification of the extremophilic Archaea *Pyrococcus abyssi* and *Methanocaldococcus jannaschii*: applications in the search for evidence of life in early Earth and extraterrestrial rocks. Geobiology 2009;7(4):403−18. Available from: https://doi.org/10.1111/j.1472-4669.2009.00212.x.

[5] Campbell KA, Lynne BY, Handley KM, Jordan S, Farmer JD, Guido DM, et al. Tracing biosignature preservation of geothermally silicified microbial textures into the geological record. Astrobiology 2015;15(10):858−82. Available from: https://doi.org/10.1089/ast.2015.1307.

[6] Bist V, Niranjan A, Ranjan M, Lehri A, Seem K, Srivastava S. Silicon-solubilizing media and its implication for characterization of bacteria to mitigate biotic stress. Front Plant Sci 2020;11:28. Available from: https://doi.org/10.3389/fpls.2020.00028.

[7] Pastore G, Kernchen S, Spohn M. Microbial solubilization of silicon and phosphorus from bedrock in relation to abundance of phosphorus-solubilizing bacteria in temperate forest soils. Soil Biol Biochem 2020;151:108050. Available from: https://doi.org/10.1016/j.soilbio.2020.108050.

[8] Hrast M, Obreza A. The role of silicon compounds in living organisms. Farm Vest 2010;61(1):37−41.

[9] Farooq MA, Dietz KJ. Silicon as a versatile player in plant and human biology: overlooked and poorly understood. Front Plant Sci 2015;6:994. Available from: https://doi.org/10.3389/fpls.2015.00994.

[10] Petkowski JJ, Bains W, Seager S. On the potential of silicon as a building block for life. Life 2020;10(6):84. Available from: https://doi.org/10.3390/life10060084.

[11] Richmond KE, Sussman M. Got silicon? The non-essential beneficial plant nutrient. Curr Opin Plant Biol 2003;6:268−72. Available from: https://doi.org/10.1016/s1369-5266(03)00041-4.

[12] Liang Y, Nikolic M, Bélanger R, Gong H, Song A. Silicon in agriculture. Dordrecht: Springer; 2015. p. 235.

[13] Takahashi E, Tanaka H, Miyake Y. Distribution of silicon accumulating plants in the plant kingdom. Jpn J Soil Sci Plant Nutr 1981;52:511−15.

[14] Epstein E. Silicon. Ann Rev Plant Physiol Plant Mol Biol 1999;50:641−64. Available from: https://doi.org/10.1146/annurev.arplant.50.1.641.

[15] Ma JF, Miyake Y, Takahashi E. Silicon as a beneficial element for crop plants. In: Datnoff LE, Snyder GH, Korndörfer GH, editors. Studies in plant science. Silicon in agriculture. Amsterdam: Elsevier; 2001. p. 17−39.

[16] Ma JF. Functions of silicon in higher plants. In: Müller WEG, editor. Silicon biomineralization. Progress in molecular and subcellular biology, vol. 33. Berlin: Springer; 2003. p. 27−47.

[17] Thummel RV, Brightly WH, Strömberg CAE. Evolution of phytolith deposition in modern bryophytes, and implications for the fossil record and influence on silica cycle in early land plant evolution. New Phytol 2019;221(4):2273−85. Available from: https://doi.org/10.1111/nph.15559.

[18] Frantz JM, Pitchay DDS, Locke JC, Horst LE, Krause CR. Silicon is deposited in the leaves of New Guinea impatients. Plant Health Prog 2005;6(1). Available from: https://doi.org/10.1094/PHP-2005-0217-01-RS.

[19] Yan G, Nikolić M, Ye M, Xiao Z, Liang Y. Silicon acquisition and accumulation in plant and its significance for agriculture. J Integr Agric 2018;17:2138−50. Available from: https://doi.org/10.1016/S2095-3119(18)62037-4.

[20] Knight CTG, Kinrade SD. A primer on the aqueous chemistry of silicon. In: Datnoff LE, Snyder GH, Korndörfer GH, editors. Studies in plant science, vol. 8. Amsterdam: Elsevier; 2001. p. 57−84.

[21] Tubaña BS, Heckman JR. Silicon in soils and plants. In: Rodrigues FA, Datnoff LE, editors. Silicon and plant diseases. Cham: Springer; 2015. p. 7−51.

[22] Takahashi E, Ma JF, Miyake Y. The possibility of silicon as an essential element for higher plants. Comm Agric Food Chem 1990;2:99–122.

[23] Ma JF, Takahashi E. Soil, fertilizer, and plant silicon research in Japan. Amsterdam: Elsevier; 2002. p. 294.

[24] Hodson MJ, White PJ, Mead A, Broadley MR. Phylogenetic variation in the silicon composition of plants. Ann Bot 2005;96:1027–46. Available from: https://doi.org/10.1093/aob/mci255.

[25] Smis A, Ancin-Murguzur FJ, Struyf E, Soininen EM, Herranz-Jusdado JG, Meire P, et al. Determination of plant silicon content with near-infrared reflectance spectroscopy. Front Plant Sci 2014;5:496. Available from: https://doi.org/10.3389/fpls.2014.00496.

[26] Liang YC, Hua HX, Zhu YG, Zhang J, Cheng CM, Romheld V. Importance of plant species and external silicon concentration to active silicon uptake and transport. New Phytol 2006;172:63–72. Available from: https://doi.org/10.1111/j.1469-8137.2006.01797.x.

[27] Cornelis JT, Delvaux B, Georg RB, Lucas Y, Ranger J, Opfergelt S. Tracing the origin of dissolved silicon transferred from various soil-plant systems towards rivers: a review. Biogeoscience 2011;8:89–112. Available from: https://doi.org/10.5194/bg-8-89-2011.

[28] Imtiaz M, Rizwan MS, Mushtaq MA, Ashraf M, Shahzad SM, Yousaf B, et al. Silicon occurrence, uptake, transport and mechanisms of heavy metals, minerals and salinity enhanced tolerance in plants with future prospects: a review. J Environ Manag 2016;183(3):521–9. Available from: https://doi.org/10.1016/j.jenvman.2016.09.009.

[29] Marron AO, Ratcliffe S, Wheeler GL, Goldstein RE, King N, Not F, et al. The evolution of silicon transport in eukaryotes. Mol Biol Evol 2016;33(12):3226–48. Available from: https://doi.org/10.1093/molbev/msw209.

[30] Deshmukh R, Bélanger R. Molecular evolution of aquaporins and silicon influx in plants. Funct Ecol 2016;30:1277–85. Available from: https://doi.org/10.1111/1365-2435.12570.

[31] Ma JF, Tamai K, Yamaji N, Mitani N, Konishi S, Katsuhara M, et al. A silicon transporter in rice. Nature 2006;440:688–91. Available from: https://doi.org/10.1038/nature04590.

[32] Liu Q, Wang H, Zhang Z, Wu J, Feng Y, Zhu Z. Divergence in function and expression of the NOD26-like intrinsic proteins in plants. BMC Genomics 2009;10:313. Available from: https://doi.org/10.1186/1471-2164-10-313.

[33] Pommerrenig B, Diehn TA, Bienert GP. Metalloid-porins: essentiality of Nodulin 26-like intrinsic proteins in metalloid transport. Plant Sci 2015;238:212–27. Available from: https://doi.org/10.1016/j.plantsci.2015.06.002.

[34] Gomes D, Agasse A, Thiébaud P, Delrot S, Gerós H, Chaumont F. Aquaporins are multifunctional water and solute transporters highly divergent in living organisms. Biochim Biophys Acta 2009;1788(6):1213–28. Available from: https://doi.org/10.1016/j.bbamem.2009.03.009.

[35] Finn RN, Cerdà J. Evolution and functional diversity of aquaporins. Biol Bull 2015;229(1):6–23. Available from: https://doi.org/10.1086/BBLv229n1p6.

[36] Deshmukh RK, Vivancos J, Ramakrishnan G, Guérin V, Carpentier G, Sona H, et al. A precise spacing between NPA domains of aquaporins is essential for silicon permeability in plants. Plant J 2015;83(3):489–500. Available from: https://doi.org/10.1111/tpj.12904.

[37] Vatansever R, Ozyigit II, Filiz E, Gozukara N. Genome-wide exploration of silicon (Si) transporter genes, Lsi1 and Lsi2 in plants; insights into Si-accumulation status/capacity of plants. Biometals 2017;30(2):185–200. Available from: https://doi.org/10.1007/s10534-017-9992-2.

[38] Mitani-Ueno N, Yamaji N, Zhao FJ, Ma JF. The aromatic/arginine selectivity filter of NIP aquaporins plays a critical role in substrate selectivity for silicon, boron, and arsenic. J Exp Bot 2011;62:4391–8. Available from: https://doi.org/10.1093/jxb/err158.

[39] Grégoire C, Remus-Borel W, Vivancos J, Labbe C, Belzile F, Bélanger RR. Discovery of a multigene family of aquaporin silicon transporters in the primitive plant Equisetumarvense. Plant J 2012;72:320–30. Available from: https://doi.org/10.1111/j.1365-313X.2012.05082.x.

[40] Tripathi P, Sangita S, Abdul LK, Yong-Suk C, Yoonha K. Silicon effects on the root system of diverse crop species using root phenotyping technology. Plants 2021;10(5):885. Available from: https://doi.org/10.3390/plants10050885.

[41] Nawaz MA, Azeem F, Zakharenko AM, Lin X, Atif RM, Baloch FS, et al. In-silico exploration of channel type and efflux silicon transporters and silicification proteins in 80 sequenced viridiplantae genomes. Plants 2020;9(11):1612. Available from: https://doi.org/10.3390/plants9111612.

[42] Mitani N, Chiba Y, Yamaji N, Ma JF. Identification and characterization of maize and barley Lsi2-like silicon efflux transporters reveal a distinct silicon uptake system from that in rice. Plant Cell 2009;21:2133–42. Available from: https://doi.org/10.1105/tpc.109.067884.

[43] Mitani N, Yamaji N, Ma JF. Identification of maize silicon influx transporters. Plant Cell Physiol 2009;50(1):5–12. Available from: https://doi.org/10.1093/pcp/pcn110.

[44] Gómez-Merino FC, Trejo-Téllez LI, García-Jiménez A, Escobar-Sepúlveda HF, Ramírez-Olvera SM. Silicon flow from root to shoot in pepper: a comprehensive in silico analysis reveals a potential linkage between gene expression and hormone signaling that stimulates plant growth and metabolism. Peer J 2020;8:e10053. Available from: https://doi.org/10.7717/peerj.10053.

[45] Kumar S, Soukup M, Elbaum R. Silicification in grasses: variation between different cell types. Front Plant Sci 2017;8:438. Available from: https://doi.org/10.3389/fpls.2017.00438.

[46] Chiba Y, Mitani N, Yamaji N, Ma JF. HvLsi1 is a silicon influx transporter in barley. Plant J 2009;57(5):810–18. Available from: https://doi.org/10.1111/j.1365-313X.2008.03728.x.

[47] Yamaji N, Chiba Y, Mitani-Ueno N, Feng Ma J. Functional characterization of a silicon transporter gene implicated in silicon distribution in barley. Plant Physiol 2012;160(3):1491–7. Available from: https://doi.org/10.1104/pp.112.204578.

[48] Mitani N, Yamaji N, Ago Y, Iwasaki K, Ma JF. Isolation and functional characterization of an influx silicon transporter in two pumpkin cultivars contrasting in silicon accumulation. Plant J 2011;66:231–40. Available from: https://doi.org/10.1111/j.1365-313X.2011.04483.x.

[49] Montpetit J, Vivancos J, Mitani-Ueno N, Yamaji N, Rémus-Borel W, Belzile F, et al. Cloning, functional characterization and heterologous expression of TaLsi1, a wheat silicon transporter gene. Plant Mol Biol 2012;79:35–46.

[50] Deshmukh RK, Vivancos J, Guérin V, Sonah H, Labbé C, Belzile F, et al. Identification and functional characterization of silicon transporters in soybean using comparative genomics of major intrinsic proteins in Arabidopsis and rice. Plant Mol Biol 2013;83:303–15. Available from: https://doi.org/10.1007/s11103-013-0087-3.

[51] Mitani-Ueno N, Ogai H, Yamaji N, Ma JF. Physiological and molecular characterization of Si uptake in wild rice species. Physiol Plant 2014;151(3):200–7. Available from: https://doi.org/10.1111/ppl.12125.

[52] Markovich O, Kumar S, Cohen D, Addadi S, Fridman E, Elbaum R. Silicification in leaves of sorghum mutant with low silicon accumulation. Silicon 2015;11:2385—91. Available from: https://doi.org/10.1007/s12633-015-9348-x.

[53] Coskun D, Deshmukh R, Sonah H, Shivaraj SM, Frenette-Cotton R, Tremblay L, et al. Si permeability of a deficient Lsi1 aquaporin in tobacco can be enhanced through a conserved residue substitution. Plant Direct 2019;3(8):e00163. Available from: https://doi.org/10.1002/pld3.163.

[54] Sun H, Duan Y, Mitani-Ueno N, Che J, Jia J, Liu J, et al. Tomato roots have a functional silicon influx transporter but not a functional silicon efflux transporter. Plant Cell Environ 2020;43(3):732—44. Available from: https://doi.org/10.1111/pce.13679.

[55] Vulavala VK, Elbaum R, Yermiyahu U, Fogelman E, Kumar A, Ginzberg I. Silicon fertilization of potato: expression of putative transporters and tuber skin quality. Planta 2016;243(1):217—29. Available from: https://doi.org/10.1007/s00425-015-2401-6.

[56] Ouellette S, Goyette MH, Labbé C, Laur J, Gaudreau L, Gosselin A, et al. Silicon transporters and effects of silicon amendments in strawberry under high tunnel and field conditions. Front Plant Sci 2017;8:949. Available from: https://doi.org/10.3389/fpls.2017.00949.

[57] Sahebi M, Hanafi MM, Rafii MY, Azizi P, Abiri R, Kalhori N, et al. Screening and expression of a silicon transporter gene (Lsi1) in wild-type indica rice cultivars. Biomed Res Int 2017;2017:9064129. Available from: https://doi.org/10.1155/2017/9064129.

[58] Shivaraj SM, Deshmukh RK, Rai R, Bélanger R, Agrawal PK, Dash PK. Genome-wide identification, characterization, and expression profile of aquaporin gene family in flax (Linum usitatissimum). Sci Rep 2017;7:46137. Available from: https://doi.org/10.1038/srep46137.

[59] Shivaraj SM, Deshmukh R, Sonah H, Bélanger RR. Identification and characterization of aquaporin genes in Arachis duranensis and Arachis ipaensis genomes, the diploid progenitors of peanut. BMC Genomics 2019;20(1):222. Available from: https://doi.org/10.1186/s12864-019-5606-4.

[60] Hu J, Li Y, Jeong BR. Putative silicon transporters and effect of temperature stresses and silicon supplementation on their expressions and tissue silicon content in poinsettia. Plants 2020;9(5):569. Available from: https://doi.org/10.3390/plants9050569.

[61] Pontigo S, Larama G, Parra-Almuna L, Nunes-Nesi A, Mora ML, Cartes P. Physiological and molecular insights involved in silicon uptake and transport in ryegrass. Plant Physiol Biochem 2021;163:308—16. Available from: https://doi.org/10.1016/j.plaphy.2021.04.013.

[62] Ma JF, Yamaji N, Mitani N, Tamai K, Konishi S, Fujiwara T, et al. An efflux transporter of silicon in rice. Nature 2007;448:209—12. Available from: https://doi.org/10.1038/nature05964.

[63] Mitani-Ueno N, Yamaji N, Ma JF. Silicon efflux transporters isolated from two pumpkin cultivars contrasting in Si uptake. Plant Signal Behav 2011;6(7):991—4. Available from: https://doi.org/10.4161/psb.6.7.15462.

[64] Vivancos J, Deshmukh R, Grégoire C, Rémus-Borel W, Belzile F, Bélanger RR. Identification and characterization of silicon efflux transporters in horsetail (Equisetum arvense). J Plant Physiol 2016;200:82—9. Available from: https://doi.org/10.1016/j.jplph.2016.06.011.

[65] Ma JF, Yamaji N. A cooperated system of silicon transport in plants. Trends Plant Sci 2015;20:435—42. Available from: https://doi.org/10.1016/j.tplants.2015.04.007.

[66] Yamaji N, Sakurai G, Mitani-Ueno N, Ma J. Orchestration of three transporters and distinct vascular structures in node for intervascular transfer of silicon in rice. Proc Natl Acad Sci USA 2015;112:11401—6. Available from: https://doi.org/10.1073/pnas.1508987112.

[67] Ma JF, Yamaji N, Tamai K, Mitani N. Genotypic difference in silicon uptake and expression of silicon transporter genes in rice. Plant Physiol 2007;145(3):919—24. Available from: https://doi.org/10.1104/pp.107.107599.

[68] Yamaji N, Ma JF. Further characterization of a rice Si efflux transporter, Lsi2. Soil Sci Plant Nutr 2011;57:259—564. Available from: https://doi.org/10.1080/00380768.2011.565480.

[69] Muneer S, Jeong BR. Proteomic analysis of salt-stress responsive proteins in roots of tomato (Lycopersicon esculentum L.) plants towards silicon efficiency. Plant Growth Reg 2015;77(2):133—46. Available from: https://doi.org/10.1007/s10725-015-0045-y.

[70] Jadhao KR, Bansal A, Rout GR. Silicon amendment induces synergistic plant defense mechanism against pink stem borer (Sesamia inferens Walker.) in finger millet (Eleusine coracana Gaertn.). Sci Rep 2020;10:4229. Available from: https://doi.org/10.1038/s41598-020-61182-0.

[71] Yamaji N, Mitatni N, Ma JF. A transporter regulating silicon distribution in rice shoots. Plant Cell 2008;20(5):1381—9. Available from: https://doi.org/10.1105/tpc.108.059311.

[72] Kumar S, Adiram-Filiba N, Blum S, Sanchez-Lopez JA, Tzfadia O, Omid A, et al. Siliplant1 protein precipitates silica in sorghum silica cells. J Exp Bot 2020;71(21):6830—43. Available from: https://doi.org/10.1093/jxb/eraa258.

[73] Ma JF, Yamaji N. Silicon uptake and accumulation in higher plants. Trends Plant Sci 2006;11(8):392—7. Available from: https://doi.org/10.1016/j.tplants.2006.06.007.

[74] Van Bockhaven J, De Vleesschauwer D, Höfte M. Towards establishing broad-spectrum disease resistance in plants: silicon leads the way. J Exp Bot 2013;64(5):1281—93. Available from: https://doi.org/10.1093/jxb/ers329.

[75] Seal P, Das P, Biswas AK. Versatile potentiality of silicon in mitigation of biotic and abiotic stresses in plants: a review. Am J Plant Sci 2018;9:1433—54. Available from: https://doi.org/10.4236/ajps.2018.97105.

[76] Exley C, Guerriero G, Lopez X. Silicic acid: the omniscient molecule. Sci Total Environ 2019;665:432—7. Available from: https://doi.org/10.1016/j.scitotenv.2019.02.197.

[77] Wang Y, Xiao X, Xu Y, Chen B. Environmental effects of silicon within biochar (sichar) and carbon—silicon coupling mechanisms: a critical review. Environ Sci Technol 2019;53(23):13570—82. Available from: https://doi.org/10.1021/acs.est.9b03607.

[78] Gaur S, Kumar J, Kumar D, Chauhan D, Prasad S, Srivastava P. Fascinating impact of silicon and silicon transporters in plants: a review. Ecotoxicol Environ Saf 2020;202:110885. Available from: https://doi.org/10.1016/j.ecoenv.2020.110885.

[79] Coskun D, Deshmukh R, Shivaraj SM, Isenring P, Bélanger RR. Lsi2: a black box in plant silicon transport. Plant Soil 2021;. Available from: https://doi.org/10.1007/s11104-021-05061-1.

[80] Exley C, Guerriero G, Lopez X. How is silicic acid transported in plants? Silicon 2020;12:2641—5. Available from: https://doi.org/10.1007/s12633-019-00360-w.

[81] Zardoya R, Villalba S. A phylogenetic framework for the aquaporin family in eukaryotes. J Mol Evol 2001;52(5):391—404. Available from: https://doi.org/10.1007/s002390010169.

[82] King LS, Kozono D, Agre P. From structure to disease: the evolving tale of aquaporin biology. Nat Rev Mol Cell Biol 2004;5:687—98. Available from: https://doi.org/10.1038/nrm1469.

[83] Wudick MM, Luu DT, Maurel C. A look inside: localization patterns and functions of intracellular plant aquaporins. New Phytol 2009;184:289−302. Available from: https://doi.org/10.1111/j.1469-8137.2009.02985.x.

[84] Xu C, Nadon BD, Kim KD, Jackson SA. Genetic and epigenetic divergence of duplicate genes in two legume species. Plant Cell Environ 2018;41(9):2033−44. Available from: https://doi.org/10.1111/pce.13127.

[85] Nadon B.D. The evolution of gene and genome duplication in soybean [PhD dissertation]. Athens, Georgia: The University of Georgia. < http://getd.libs.uga.edu/pdfs/nadon_brian_d_201905_phd.pdfView > 2019.

[86] Cheng F, Wu J, Wang X. Genome triplication drove the diversification of *Brassica* plants. Hortic Res 2014;1:14024. Available from: https://doi.org/10.1038/hortres.2014.24.

[87] Quigley F, Rosenberg JM, Shachar-Hill Y, Bohnert HJ. From genome to function: the Arabidopsis aquaporins. Genome Biol 2002;3(1). Available from: https://doi.org/10.1186/gb-2001-3-1-research0001 research0001.1−research0001.17.

[88] Sakurai J, Ishikawa F, Yamaguchi T, Uemura M, Maeshima M. Identification of 33 rice aquaporin genes and analysis of their expression and function. Plant Cell Physiol 2005;46(9):1568−77. Available from: https://doi.org/10.1093/pcp/pci172.

[89] Dai WM, Zhang KQ, Duan BW, Zheng KL, Zhuang JY, Cai R. Genetic dissection of silicon content in different organs of rice. Crop Sci 2005;45:1345−52. Available from: https://doi.org/10.2135/cropsci2004.0505.

[90] Biradar H, Bhargavi M, Sasalwad R, Parama R, Hittalmani S. Identification of QTL associated with silicon and zinc content in rice (*Oryza sativa* L.) and their role in blast disease resistance. Ind J Gen Plant Breed 2007;67:105−9.

[91] Wu JR, Dai WM, Zhang KQ, Duan BW, Zheng KL, Cai R, et al. Mapping of QTLs for silicon contents in the stem and flag leaf recombinant inbred rice population. Sci Agric Sin 2007;40(1):13−18.

[92] Wu QS, Wan XY, Su N, Cheng ZJ, Wang JK, Lei CL, et al. Genetic dissection of silicon uptake ability in rice (*Oryza sativa* L.). Plant Sci 2006;171(4):441−8. Available from: https://doi.org/10.1016/j.plantsci.2006.05.001.

[93] Wu J, Fan F, Du J, Fan Y, Zhuang J. Dissection of QTLs for hull silicon content on the short arm of rice chromosome 6. Rice Sci 2010;17:99−104. Available from: https://doi.org/10.1016/S1672-6308(08)60111-0.

[94] Talukdar P, Douglas A, Price AH, Norton GJ. Biallelic and genome-wide association mapping of germanium tolerant loci in rice (*Oryza sativa* L.). PLoS ONE 2015;10(9):e0137577. Available from: https://doi.org/10.1371/journal.pone.0137577.

[95] FAO (Food and Agriculture Organization of the United Nations). The future of food and agriculture. Trends and challenges. Rome, Italy. <http://www.fao.org/3/i6583e/i6583e.pdf>; 2017.

[96] Cooke J, Leishman MR. Consistent alleviation of abiotic stress with silicon addition: a meta-analysis. Funct Ecol 2016;30(8):1340−57. Available from: https://doi.org/10.1111/1365-2435.12713.

[97] Li Z, Song Z, Yan Z, Hao Q, Song A, Liu L, et al. Silicon enhancement of estimated plant biomass carbon accumulation under abiotic and biotic stresses. A meta-analysis. Agron Sustain Dev 2018;38:1−19. Available from: https://doi.org/10.1007/s13593-018-0496-4.

[98] Xu J, Zhang J, Lv Y, Xu K, Lu S, Liu X, et al. Effect of soil mercury pollution on ginger (*Zingiber officinale* Roscoe): growth, product quality, health risks and silicon mitigation. Ecotoxicol Environ Saf 2020;195:110472. Available from: https://doi.org/10.1016/j.ecoenv.2020.110472.

[99] Agathokleous E. The rise and fall of photosynthesis: hormetic dose-response in plants. J For Res 2021;32:889−98. Available from: https://doi.org/10.1007/s11676-020-01252-1.

[100] Trejo-Téllez LI, García-Jiménez A, Escobar-Sepúlveda HF, Ramírez-Olvera SM, Bello-Bello JJ, Gómez-Merino FC. Silicon induces hormetic dose-response effects on growth and concentrations of chlorophylls, amino acids and sugars in pepper plants during the early developmental stage. Peer J 2020;9(8):e9224. Available from: https://doi.org/10.7717/peerj.9224.

[101] Ramírez-Olvera SM, Trejo-Téllez LI, Gómez-Merino FC, Ruíz-Posadas LM, Alcántar-González EG, Saucedo-Veloz C. Silicon stimulates plant growth and metabolism in rice plants under conventional and osmotic stress conditions. Plants 2021;10(4):777. Available from: https://doi.org/10.3390/plants10040777.

[102] Cassol JC, Sponchiado D, Dornelles SHB, Tabaldi LA, Barreto EPM, Pivetta M, et al. Silicon as an attenuator of drought stress in plants of *Oryza sativa* L. treated with diethotale. Braz J Biol 2021;81(4):1061−72. Available from: https://doi.org/10.1590/1519-6984.235052.

[103] Khan I, Awan SA, Rizwan M, Ali S, Hassan MJ, Brestic M, et al. Effects of silicon on heavy metal uptake at the soil-plant interphase: a review. Ecotoxicol Environ Saf 2021;222:112510. Available from: https://doi.org/10.1016/j.ecoenv.2021.112510.

[104] Semenova NA, Smirnov AA, Grishin AA, Pishchalnikov RY, Chesalin DD, Gudkov SV, et al. The effect of plant growth compensation by adding silicon-containing fertilizer under light stress conditions. Plants 2021;10(7):1287. Available from: https://doi.org/10.3390/plants10071287.

[105] Song XP, Verma KK, Tian DD, Zhang XQ, Liang YJ, Huang X, et al. Exploration of silicon functions to integrate with biotic stress tolerance and crop improvement. Biol Res 2021;54(1):19. Available from: https://doi.org/10.1186/s40659-021-00344-4.

7

Silicon-mediated modulations of genes and secondary metabolites in plants

Saad Farouk

Agricultural Botany Department, Faculty of Agriculture, Mansoura University, Mansoura, Egypt

7.1 Introduction

Plant growth is modulated by different environmental stresses and extents of variations among these factors that either benefit or harm normal plant development [1]. These stresses adversely restrict plant growth and development of major economic plants leading to a considerable decline in productivity (about 70%), hence an enormous risk to worldwide food accompanying rising temperatures, excessive pollution, and sporadic or intermittent drought [2].

Plants respond to various stresses by exhibiting typical morpho-biochemical characteristics and molecular responses [3–6]. The beginning of anxiety and resistance reactions is mediated via signaling pathways that elicit the assimilation of secondary metabolite (SM) intermediates, which occupy a diversity of functions in response to altering the environment on plants [7].

SMs have been introduced by Kossel [8] that is frequently restricted to small fragments and products of metabolism include organic substances occurring in the whole biota whereas generally disseminated in plants, which are not frankly implicated in the regulation of plant development and some biochemical processes like protein assimilation, photosynthesis and photoassimilate transport, and reproduction [9]. SMs are usually occupied in crop protection under stressful conditions [10]. Besides their significance in biotic stress tolerance, SMs are in addition occupied in nullifying abiotic factors, that is, temperature, heavy metals, drought, and salinity [11,12]. SMs like other protective molecules were reported to defend critical biomolecules from stress-provoked oxidative burst [13]. Certainly, crops have a dissimilar display of protection systems to permit them for coping with stressful situations.

Recent evidence has revealed that the cellular stress reaction connected with SMs assimilation occupies widespread cross-talk and signaling processes between pathways in plant cells, which possibly will occupy the contribution of molecules [14]. Conversely, the biochemical motivation of the signal transduction scheme implicated is hitherto uncertain. Many of the researchers applied metabolomics and transcriptomic technologies in examining and understanding stress-associated genes implicated in SMs assimilation [15,16]. Metabolomics and genomic research data with utilizes of the assimilation pathways enzymes technologies have eased considerable processes occupied in SMs accumulation [15].

Numerous researches have a statement that Si-mediated regulation of diverse crop stress tolerance indirectly persuades the SMs production in plants [17,18]. Silicon is known as the second most occurrence nutrients in the earth's crust [19]. Furthermore, Si application was realistically established to recover the growth of numerous crops, and positive impacts of Si are recorded during environmental stresses [20]. Studies have intensively reported alleviation of different plant stresses by Si such as drought stress [4], salt stress [3], high/low temperature [21], plant diseases and pest attack [22], and heavy metals [19]. The Si-mediated stress tolerance is believed to be caused by two chief routes: (1) SiO_2 deposition-induced physical and mechanical protection and (2) biochemical response-elicited metabolic modification [19]. However, recent research studies have recommended that

Silicon and Nano-silicon in Environmental Stress Management and Crop Quality Improvement.
DOI: **https://doi.org/10.1016/B978-0-323-91225-9.00014-5**

prospective Si-research should be supposed to be centered on precise features of the Si—plant interactions to develop agricultural strategies designed at enhancing plant productivity [23].

Recently, nanoparticles (NPs) have paid awareness to the distinctive properties they possess for agricultural purposes [24]. The prime purpose of NPs, in plant productivity, is the decline of the exploit of agrochemicals and raises of the production via pest and nutrients management, in addition to enhancing nutritional and nutraceutical significances [24,25]. Among the different NPs, the nanoparticles of silicon (Si NPs) can be utilized in NP type for agricultural rationales that can be believed a novel resource of this nutrient [26]. Generally, Si NPs can boost photosynthetic processes and plant biomass magnitude and reduce evaporation and transpiration; in addition, it motivates antioxidant enzyme activities and decreases the plant's susceptibility to plant pathogens and then increase plant productivity [27,28]. Furthermore, several types of research have recorded the impact of Si NPs on ion uptake [26]; however, less data are accessible in the literature regarding the effects of Si NPs on SMs metabolism and assimilation.

By keeping the above in view, we try to describe the role and mechanism involved in Si-mediated regulation of diverse environmental stresses ultimately through influencing SMs assimilation in plants.

7.2 Overview and assortment of plant secondary metabolites

Plants create an unrestricted and diverse variety of organic substances; the enormous mainstream of which do not play in growth directly. These compounds frequently referred to SMs are differentially dispersed in the plant kingdom [29]. About 100,000 SMs are documented in the plant kingdom restricted to definite groups. Plant SMs are classified into four major categories as indicated in Fig. 7.1 [30].

Phenolic compounds, a chief group of SMs that are everywhere in plants, are essential for the defense system against parasites and pests [5]. Phenolic compounds typically occur in plants' insoluble or bound structures but they can additionally be classified into subgroups along with their chemical structures [31]. They are extremely structurally varied, containing simple molecules (gallic and caffeic acid) and polyphenols (flavonoids and polymers).

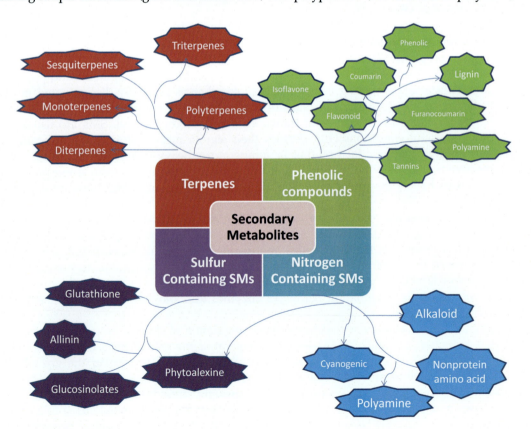

FIGURE 7.1   Diversity of plant secondary metabolites (SMs).

Terpenes or terpenoids (50,000 dissimilar structures) are the main group of SMs formed from the derivatives of glycolytic or acetyl CoA intermediates and are discovered in the whole biota [32]. Terpenes are the vital construction blocks of diverse complex phytohormones and pigments, and they are chiefly accountable for their fragrances and biochemical impacts. Terpenes also mediate plant stress reactions, as these are toxic to mammals and insects. Some plants possess saponins as terpene glycosides and essential oils as volatile terpenes.

Sulfur-containing SMs (about 200 compounds) are comparatively negligible fraction of plant SMs. The group encompasses the familiar glucosinolates and their breakdown products [33]. Sulfur-containing SMs have directly or obliquely associated with the plant defense system [34].

Nitrogen-containing SMs (about 12,000 compounds) are distributed in several plant cultivars and induce stress tolerance against salinity, drought, and herbivores [35]. Alkaloids are small molecular weight substances and their biosynthesis is connected with the accessibility of a few amino acids notably tryptophan, tyrosin, and lysine [31].

7.3 Stress and protection reactions in relation to the secondary metabolites production

Stress reaction in plants comprises a range of molecular, cellular cross-talk, and signaling responses instigated throughout the discovery of explicit or collective stressful impact that may accelerate the SMs induction [36]. Although earlier times SMs were considered not as useful, it is now a well-established fact that the absence or lack of plant potential to SMs synthesis resulted in impeded growth and survivability, and even death under stresses [37]. Plants' immune systems have developed several stress discovery mechanisms, which include transmembrane recognition, polymorphic nucleotide-binding site and leucine-rich repeat domains (NB-LRR) protein production by most R-genes, and the creation of SMs to cope with the stressful conditions, and consequently become changed to endure the circumstance [38]. This could be attained through an effect on plant biochemical pathways, generated by signal transduction procedure to accommodate the motivation concerning adjustments in SMs that facilitate regulation of cell osmotic pressure, avoid cell components oxidation, pathogenic growth, and frighten herbivores [14] through physio-biochemical connected pathways regulation. Generation of the stress may motivate expression or despotism of the stress-genes network through accurate regulation, which may cause the assembly of efficient cellular molecules to accommodate the stress influence [39] that may be in the outline of assimilation of osmoprotectants, detoxification enzymes, transporters, chaperones, and proteases that serve as the earliest line of plant defense [40].

Plants' terpenoid metabolites were revealed to play an important role in adaptation to unfavorable environments [41]. Terpenoids are inducibly relisted in response to herbivore attacks in plants and not single occupation directly as protective phytoalexins for deterring detrimental invaders but also indirectly attract natural enemies [42,43]. It is evident that plant—insect, plant—herbivore, or plant—injury interaction mostly requires terpenes; however, they may have finely described functions in plant growth regulation [34]. The influences of water deficit on SMs in conifers have also been deliberate; total terpenes in the seedlings of *Pinus sylvestris* and *P. abies* were established to be about 32%—39% and 35%—45% elevated in water stress-affected plants than in control plants [44]. Likewise, accretion of terpenes in *Salvia officinalis* raises in water deficit situation whereas accumulation of dry mass is declined [45]. Water deficit encourages oxidative anxiety that consecutively improves flavonoid assembly as antioxidant solutes. Under abiotic stress conditions, terpenes such as tocopherol and carotenoids (zeaxanthin, neoxanthin, and lutein) act as antioxidants and may directly scavenge reactive oxygen species (ROS) in response to photoinhibition [46]. Lately, it is demonstrated that terpenoid phytoalexins additionally defend root injures under drought or salinity stress. Carnosic acid, a diterpene protects Labiatae cultivars from drought-evoked oxidative stress in combination with that of other low-molecular-weight antioxidants (α-tocopherol and ascorbate) in chloroplasts [47]. In addition, abscisic acid (ABA) is a terpene-derived phytohormone that plays a critical role in foremost plant metabolic processes occupied in stress response [48]. In stress conditions, plant ABA concentration rises significantly, exciting stress-tolerance impacts that assist plants to adapt and endure under stress factors [49]. Terpenoids recognized in *Commelina benghalensis* have been a statement to nullify poisonous radicals like 2, 2-diphenyl-1-picrylhydrazyl [50]. These observations propose that terpenes defend plants from stress injures because of their antioxidant properties or direct quenching of oxidants.

Phenolics are one of the majority ever-present groups of SMs that are assimilating in plants and acquire metabolic functions [51,52]. Phenolics are frequently produced and accumulated in the subepidermal layers of plant tissues, depicting stress and pathogen attack [53]. Within an ordinary environment, crops possess little concentration of phenolics; however, their buildup is increased noticeably under stressful conditions [54,55]. Phenolics are identified to achieve diverse vital occupations in plants, that is, production and tolerance of plants alongside different stresses [56]. Phenolic compounds such as flavonoids and hydroxycinnamates have redox properties [57], which permit them to perform as antioxidants and thus detoxify singlet oxygen and they play a key function in stabilizing lipid peroxidation [58]. Isoflavones, flavonols, flavones, anthocyanins, and coumarins represent the major phenolic, which mediate plant growth and resistance response alongside pests [59]. In addition, several phenolics like phenolic acid, flavonoids, and tannins were demonstrated to give protection against pest attacks by acting as antimicrobial and insecticidal mediators [60]. Definite flavonoids display the aptitude to offer heavy metal stress protection by chelation of transition metals (e.g., Fe, Cu, Ni, and Zn). As Fe and Cu produce hydroxyl radical by Fenton's reaction, the chelation of these metals in the soil may be an efficient outline of resistance against the impacts of high metals toxicity. Cadmium-stressed *Brassica juncea* plants accumulated an elevated quantity of rutin polyphenol than control plants to avoid oxidative injures [61]. For example, the levels of phenolics in *Hypericum brasilience* and *Pisum sativum* rise once they are exposed to water deficit [62]. Phenolic compounds normally build up within water stress owing to modifications in the phenylpropanoid pathway: most of the key genes of this pathway are inspired by water stress. For instance, water deficit provokes the activation of the phenylalanine ammonia lyase (PAL) gene in lettuce plants and the expression of numerous genes linked to flavonoid assembly in *Scutellaria baicalensis* [63].

A native immune system is developed in plants under pathogen assault. In the basal immunity system, infected cells recognize pathogens throughout microbe-associated molecular models, which are apparent by pattern detection receptors in host cells. Moreover, host cells, in addition, distinguish pathogen assault through effectors-triggered immunity in reply to effectors owing to pathogen toxins. Plants perceive signals from such effectors and motivate numerous metabolic processes to generate SMs. The level of SMs is considerably reduced through stress revival [64]. High temperatures enhance the biosynthesis of alkaloids; for instance, the alkaloid hydroxycamptothecin is occupied in heat-shock tolerance; hydroxycamptothecin buildup was accounted to boost sixfold in *Camptotheca acuminata* seedlings within incubation at 40°C for 2 h [65]. A variety of cultivars of *Lupinus angustifolius* exposed to high temperatures in the field also drastically provokes alkaloid accretion [66]. Nitrogen-containing SMs stand for the vital division of SMs serving key functions in plant protection particularly for the lessening of oxidative injury. Additionally, alkaloids are nitrogenous organic SMs that have been revealed to have antimicrobial activity through inhibiting enzyme activity or throughout a detergent-like mode of action alongside Gram-negative bacteria, leading to the disturbance of their outer membranes [67]. To acclimatize in the water deficit stress, *Senecio jacobaea*, *S. aquaticus*, and their hybrids amplified the accretion of pyrrolizidine alkaloids [68]. In plants, nitrogenous defensive substances, that is, cyanogenic glycosides and glucosinolates, liberate poison substances to defend plants alongside herbivorous [69]. Hydrogen cyanide is the best example liberated by cyanogenic glycosides, which deters insect and herbivore feeding, thereby mediating resistance against harmful biotic components. There are S-containing phytoalexins in response to plant pathogen and mechanical breaks build up in the infection spots. Phytoalexins buildup limits the distribution of pathogens by accelerating cell death known as the hypersensitive response in various plants [31]. Additionally, defensins, thionins, and lectins S-rich SMs demonstrated an extensive array of inhibition of microbial pathogen [70]. Glucosinolates are S-rich SMs, extensively synthesized in oilseed crops of Brassicales (*Brassica oleracea*). These products have noxious properties, which restrain growth (antibiosis) and proceed as feeding restrictions (antixenosis) alongside a variety of insects from leaf chewing lepidopteran larvae to phloem-feeding aphids [71]. Many publications have reviewed that environmental stresses affect gene expression as well as plant SMs assimilation [72]; for example, increased PAL activity is one of the core stress acclamatory reactions. The endogenous levels and storage of SM are also influenced by genetic factors. Several factors affect the expression of genes correlated to the SMs assembly.

7.4 Silicon modulation of secondary metabolism within stress condition

SMs control plant development by interacting with existing ecological circumstances. They achieve imperative occupations like antimicrobial and phytoalexins [39], defense from UV radiations, and extra plant stresses [34]. Although, various functions of SMs in plant protection systems have been broadly demonstrated in plants; however, regulation of plant stress tolerance by Si indirectly via influencing SMs is largely unknown.

Several investigations have demonstrated the effect of Si on terpene assimilation under a stressful environment. Silicon application lessens salt stress in *Rosa hybrida* by regulating terpene metabolism [73]. Al-Garni et al. [74] revealed that Si application boosts concentrations of photosynthetic pigments to preserve the standard photosynthetic process and suggested as a mechanism of salinity tolerance in *Coriandrum sativum*; these findings are similar as earlier reported in mung bean [75]. Hussain et al. [26] stated that Si application averts lessening of carotenoids resulting from high-temperature stress in barley and suggested that Si fertilization is an effective strategy to enhance heat stress tolerance. Si supply was also revealed to drastically raise photosynthetic pigment concentration in water deficit conditions to maintain proper photosynthetic activity in barley [76]. Kim et al. [77] revealed that Si addition significantly increased expression of ABA synthesis genes as well ABA accretion under salinity in rice. Si treatment improved assimilation of momilactones, resulting in growth inhibition of *Magnaporthe grisea* and, therefore, resistance to rice blast disease [78].

Among the SMs, Si-mediated regulation of phenolics is more extensively studied [17,79]. Numerous researches have established Si-mediated mitigation of environmental stresses by altering plant SMs production [80]. Si uptake increased phenolics assimilation with antioxidant and/or structural function within stressful situations [81]. Numerous investigations discovered that the positive impacts of Si on plant disease resistance are ascribed to the generation of phenolic molecules [82]. Accordingly, Shetty et al. [60] revealed that Si amendment motivated accretion of fungi toxic phenolic substances such as chlorogenic acid and rutin in *R. hybrida* in response to powdery mildew pathogen. They also established supplementation of chlorogenic acid and rutin improved resistance against powdery mildew in rose plants [60]. Equally, Si enhances resistance against leaf bacterial blight induced by *Xanthomonas oryzae pv. oryzae* (Xoo) by motivating leaf buildup of phenolics and lignin, in addition, to activate PPO and PAL in rice [81]. Silicon enhanced resistance against *Phytophthora cinnamomi* and *Cryphonectria parasitica* in chestnut plants by the rising assembly of phenolic compounds in plant tissues [79]. Yang et al. [83] statement that the Si amendment drastically boosts phenols and lignins and improves protection against brown plant-hopper infestation in rice. Vega-Mas et al. [18] suggested that Si alleviates Al-toxicity in barley possibly by accelerating phenol and lignin accretion at the root level. Silicon calms Al-induced oxidative burst via rising phenols content and modulating the enzymatic antioxidant activities [84]. Reports suggested that Si improves Al tolerance via rising assembly of phenolics with Al-chelating aptitude [85]. For example, Adrees et al. [86] observed that phenolic compounds alleviate Al-toxicity in maize through Al-chelation by flavonoid phenolics and reduce Al uptake in plants. Heavy metal-chelating drastically contributes to the detoxification of heavy metals, and it is produced through the chelation of heavy metal with flavonoid phenolics or further organic acids [87]. Parveen et al. [17] reported that Si lightens depressing impacts of water deficit via affecting phenolics accretion in maize. Si-treated barley plants showed better growth due to increased accumulation of total soluble proteins, anthocyanins, flavonoids, and phenolics as well as enzymatic antioxidant activities within elevated temperature anxiety [26]. Hence, the above outcomes confirm the function of Si in regulating stress tolerance through its influence on phenolic synthesis and accumulation.

Water deficit accelerated the accumulation of several SMs, that is, rutin, quercetin, and betulinic acid in *H. brasiliense*, and that of artemisinin in *Artemisia* [88]. Similarly, exposure of St. John's wort plants to drought resulted in a noteworthy decline in photosynthesis with attendant amplification in SMs level including pseudohypericin, hypericin, and hyperforin [89]. Equally, a raise in total flavonoids was recorded in drought-affected *Glechoma longituba* plants [90]. Several reports are available about the prophylactic effects of Si by enhancing accretion of phytoalexin exhibiting antifungal activity. Studies have revealed that Si-mediated disease resistance is connected with improved enzymatic activities occupied in SMs assimilation like peroxidase, polyphenol oxidase, and PAL [91]. These are the main enzymes occupied in secondary metabolism like phenylpropanoid pathway, leading to the assembly of phenolic compounds in plants [92]. In the synthesis of SMs, PAL is a formerly dedicated enzyme in the phenylpropanoid pathway controlling the diversion of carbon fluxes from primary metabolism to synthesize phenolics [93]. Likewise, supplied Si increased resistance to powdery mildew disease in cucumber by improving buildup of new identified antifungal phytoalexin viz. flavonol aglycone rhamnetin (3,5,3',4'-tetrahydroxy-7-O-methoxyflavone) in infected leaves [94]. This antifungal specific flavonol was identified to be assimilation through phenylpropanoid pathway [95]. Commonly, Si induced lignin—carbohydrate complexes in epidermal cells as well as momilactone phytoalexin assembly in rice [78]. Silicon-mediated antifungal activity in plants was ascribed to the improved accretion of phenolics and flavonoids principally rutin and chlorogenic acid resulting in reduced conidial germination and appressorium formation of the pathogen [60]. Flavonoids are important in the absorption of UV radiation and thus may protect plant tissues [96]. Silicon accumulation and the recently found reverse content of Si and phenols suggest a protective function of Si and phenols under UV-B stress [96,97]. A higher Lsi1 expression level is important for increased tolerance to UV-B irradiation in rice [98]. The overexpression of Lsi1 genes leads to upregulation of PAL, 4-coumarate-CoA ligase (4CL), and photolyase (PL) and may regulate other genes responsible for resistance to UV-B. The upregulation of PAL and 4CL results in the enlarged level of UV-B absorptive substances such as flavonoids and total phenolics, and the upregulation of PL plays a role in DNA reparation [98]. Overexpression of Lsi1 also causes strengthened photosynthesis and detoxification-related pathways that contribute to the enhanced defense against UV-B irradiation in rice [98].

Direct confirmation of Si-mediated regulation of plant stress tolerance through nitrogen-containing SMs was less, only little researches have given straight proof. In this concern, Khan et al. [99] revealed that Si addition raises alkaloids concentration as a strategy to mitigate ROS accumulation under salinity. Although, a prospective function of polyamines in plant stress tolerance was well demonstrated; however, Yin et al. [100] provide evidence about the molecular insights into Si-dependent modulation of polyamines in *Sorghum bicolor*. Silicon improved upregulation of gene encoding *S*-adenosyl-methionine decarboxylase, the prime enzyme regulating the synthesis of Put, Spm, and Spd [100]. Furthermore, Yin et al. [100] recommended that Si-mediated salinity tolerance was connected to polyamine alkaloids and ethylene assimilation in sorghum. This fact is clarified as ethylene synthesis shares a common precursor with the polyamines viz. Spd and Spm [101], and therefore a competitive environment exists between polyamines and ethylene synthesis [101], and Si could have favored polyamines assimilation in place of ethylene synthesis [100]. Hence, Si application equilibrates the metabolism of polyamines and ethylene to alleviate environmental stress [100]. So, Si-mediated polyamine biosynthesis gene regulation does not merely support stress mitigation but also improves essential pathways in cells upon stress and increases plant development [100]. Furthermore, upon infection of pathogens to plant such as bacteria and viruses, the plant builds up huge quantities of polyamines that restrict the growth and reproduction of plant pathogens [102]. Plant cells invaded by pathogens evoke assimilation of polyamine and polyamines oxidase activity, which raises H_2O_2 concentration [103]. The ADC-gene overexpression from trifoliate orange enlarged ulcer disease resistance considerably in citrus [104]. Wang [105] found that greater insect resistance was associated with a higher Put level in Chinese cabbage. Interestingly, glucosinolates and benzoxazinoids appear to control the SMs buildup [106]. In *Arabidopsis*, mutants imperfect in the CYP83B1 enzyme requisite for indole glucosinolate assimilation confirm lesser production of the phenylpropanoid sinapoylmalate [106]. The phenylpropanoid phenotype is released in mutants that no longer generate the substrate of CYP83B1, indole-3-acetaldoxime [106], suggesting that it may be the aldoxime excess accumulation before the lack of downstream glucosinolates that represses sinapoylmalate. The latest investigation reveals that clusters of Kelch Domain F-Box (KFB) genes that are occupied in PAL inactivation [107] are unregulated in indole glucosinolate mutants in a MED5-dependent manner, while PAL activity is repressed [108]. PAL activity and sinapoylmalate buildup are (somewhat) rescued in glucosinolate-deficient KBF mutants [108]. Although, the above findings revealed the role of Si in modulating SMs and related genes under different stress factors. However, the detailed role and mechanism involved in Si-mediated secondary metabolism regulation under various plant stresses as well as different plant species are largely unknown. Hence, efforts are desirable to explore the detailed function of Si in plant stress regulation via secondary metabolism in plants.

7.5 Silicon-mediated expression of transcription factors and some associated secondary metabolite responsive genes

Protection response system occupies assembly of a range of chemical, structural, and protein-based physio-molecular response(s) against invading unfamiliar substances or organisms into plants' systems via the diversity of responses established in different plants [109]. The defense response system becomes activated when intra- or extracellular signal is received by receptors in the cell plasma membrane, connecting their binding accompanied by signal transduction cascade beginning which may give rise to de novo synthesis or activation of transcription factors (TFs) accountable for regulating SMs assimilation genes expression [110]. Consequently, higher assimilation of SMs could be a marker of elevated expression of genes and metabolic pathway for its biosynthesis in cells, although translocation of a bioactive compound from the site of assimilation to storage site occupies noteworthy function with several SMs [111]. For example, participation of membrane transportation system via ATP-binding cassette transport has been implicated in the SMs assimilation in several medicinal plants [112].

The stress stimuli are supposed by plants and activate a cascade of signal transduction pathways that upregulate gene expression and therefore accelerate metabolic modifications, which attained plant stress tolerance [113]. A great number of other TFs are contributed to the abiotic stress responses, which are responsible for stress tolerance. It was recommended that the mechanism that participates in the interaction of plants to stress is extremely complex. Several genes are overexpressed under abiotic stresses and so induce direct defense of the plant cells whereas others are occupied in different signaling pathways [114]. Additionally, several TFs, that can modify diverse stress-inducible genes, contribute to the control of plants under stress conditions [115]. The products of these stress-inducible genes are supposed to be occupied in stress tolerance [116]. The overexpression of stress-inducible genes may be motivated directly by the stress circumstances or endorsed by the secondary stresses [113]. Similarly, the excess expression of several TFs can manage an extensive array of signaling pathways that are occupied in stress tolerance achievement [117]. Additionally, abiotic stresses convince the assimilation of several imperative metabolic proteins counting those required for the creation of osmolytes and regulatory proteins occupied in signal transduction pathways like kinases and/or the transcription factors [118]. The TFs, that is, CBF1/DREB1B, CBF2/DREB1C, and CBF3/DREB1A, are occupied in the transcriptional responses to abiotic stresses [119]. It was accounted that the extra-expression of CBF1/DREB1B, CBF2/DREB1C, and CBF3/DREB1A transcription factors in *Arabidopsis* improved the resistance to abiotic stresses [120]. It was observed that the transcription factor OsDREB2A was faintly expressed in control plants and was motivated by dehydration [121]. The treatment with Si upregulated the expression of OsDREB2A in two drought-stressed rice species relative to the untreated and drought-stressed ones [121]. Such outcomes might be recognized as the improvement of phosphorylation of the DREB2A protein to their energetic forms in Si-pretreated shoots. To control the expression of enzyme genes, TFs can integrate interior and exterior indications and so supervise the buildup of SMs. Regulation of genes connected with the SMs assembly is under the influence of several TFs at different ranks [122]. The recognition of TFs and research on their regulatory system in the SMs assembly pathways have amplified in the latest decades. Accordingly, we present the TFs connected with the regulation of the SMs pathway in different plants.

The WRKY TFs family has been extensively deliberated in plants within stress environment. The inducible expression pattern of WRKY genes supports their contribution in the modulation of defense-related SMs assimilation. In tobacco plants, WRKY3 and WRKY6 are connected with the assimilation of volatile terpenes [123]. In *Artemisia annua*, WRKY1 regulates artemisinin assimilation via binding to the promoter of the sesquiterpene assimilation gene [124]. In *V. vinifera*, VviWRKY24 improves the expression of the VviSTS29 gene, which is accountable for resveratrol assimilation that sequentially confers antimicrobial resistance [125]. Additionally, SsWRKY18, SsWRKY40, and SsMYC2 are accountable for the regulation of abietane-type diterpene production in *Salvia sclarea*; these compounds have antibacterial and antifungal properties [126]. Additionally, ZmWRKY79 raises the buildup of phytoalexins in maize under stress conditions [127].

Among the diverse TFs, myeloblastosis (MYB) are occupied in the assimilation of SMs and contribute to different plant metabolic pathways, that is, growth, reproduction, and stress responses. They can be classified into four subclasses, that is, R1, R2, R3, and R4, conditional on DNA-binding domain repeats. The R2R3 family of MYB TFs is drastically connected with the regulation of diverse SMs pathways in the diverse plant. Accordingly, AtMYB113, AtMYB114, AtMYB75, and AtMYB90 in *Arabidopsis thaliana* are potentially occupied in the regulation of anthocyanin deposits through modification of the phenylpropanoid pathway [128]. Furthermore, MYB TFs may be occupied in the assembly of glutaminase (GLs), flavonoids, hydroxycinnamic acid amides (HCAAs), and proanthocyanins. The latest research proved that CsMYB2/CsMYB26 from *Camellia sinensis* plants connects to the promoter district of the genes CsF30H and CsLAR and so improves the production of flavonoids that are accountable for tolerance against blister blight disease induced by *Exobasidium vexans* [129]. At the molecular altitude, the expression intensities of 3057 differentially expressed genes (DEGs) in the wheat genome were changed following Si application [130]. These genes are principally occupied in secondary metabolism pathways. Gene ontology (GO) enhancement analysis demonstrated that in response to Si, the majority of the upregulated genes were occupied in nicotianamine metabolic and biosynthetic processes. The GO terms connected to membrane, phenylpropanoid, and suberin assembly processes were downregulated. These outcomes are parallel to what has been recorded in rice and tomato, where the expression levels of the mainstream of genes altered under Si application [131]. Of the 3057 DEGs origin, 191 genes encoded 25 diverse types of TFs that adjust plant development in response to Si application. These results propose that Si application treatment influences several metabolic pathways via changing the expression of genes that encode TFs. Accordingly, 411 R2R3-MYB TFs were well known in wheat, accounting for roughly 0.54% of all annotated wheat genes, which is further than that of rice (0.3934%) [132] and *Brachypodium distachyon* (0.39%) [133], but less than that of *Arabidopsis* (0.60%) [132]. The MYB cluster of TFs proceeds as regulators to regulators of plant development [90]. Accordingly, wheat R2R3-MYB protein TaPL1 and TaMYB1 occupied as constructive regulators of anthocyanin assimilation [134]; in rice, R2R3-MYB gene OsMYB103L influences leaf rolling and mechanical potency [135]; and in *Arabidopsis*, MYB21, MYB24, MYB33, MYB57, MYB65, and MYB103 are regulators of numerous pathways that influence anther development [136]. In addition, R2R3-MYB genes are occupied in response to different stresses. The overexpression of wheat R2R3-MYB genes TaMYB30-B, TaMYB33, TaMYB56-B, and TaSIM confers tolerance to salinity and drought in transgenic *Arabidopsis* [90].

The bHLH TFs are prospective regulators of stress response strategies and typically cooperate with MYB proteins to produce complexes that improve the expression of precise genes. As imperative modulators of stress responses, the bHLH TFs control the assimilation of SMs, that is, anthocyanin, alkaloids, glucosinolate, diterpenoid phytoalexin, and saponins [137]. In rice plants, DPF has been recognized as a bHLH TF that normalizes the assimilation of diterpenoid phytoalexin by diterpenoid phytoalexin-related genes activation [138]. Accordingly, NbbHLH1, NbbHLH2, and NbbHLH3 were connected with nicotine assimilation by virus-induced gene silencing; of these TFs, NbbHLH1 and NbbHLH2 create optimistic regulation by binding to G-box elements in the putrescine N-methyltransferase promoter, whereas NbbHLH3 is a negative regulator [139].

The bZIP TFs are dimeric, transcriptional enhancer proteins, with a preserved leucine zipper and completely charged DNA binding place. They are normally occupied in the modulation of plant metabolic pathways. Members of the bZIP TF family are connected with stress responses, mostly with tolerance to ROS and osmotic imbalance. RsmA is a bZIP-like protein that represses the assimilation of SMs in a mutant line; on the contrary, in an overexpressor line, RsamA is an active contributor in the reinstatement of SMs [140]. Additionally, the light-responsive bZIP protein, HY5, is implicated in the assimilation of anthocyanin within a stress environment, for example, MdHY5 in apple plants either cooperates with MdMYB10 or proceeds separately as an anthocyanin regulator. Phytoalexins, which are blast pathogen-resistant elements in rice, are also associated with bZIP proteins. For example, OsTGAP1 is a bZIP protein in rice that binds to the promoter site of the genes OsKSL4 and OsCPS4 and improves phytoalexins assimilation. Moreover, OsTGAP1 moderates the expression of other genes connected to terpenoid assimilation by modulating the MEP pathway [141].

AP2/ERF TFs are largely occupied in mediating plant stress responses through the regulation of plant SMs [142]. ORCA and ORCA2 are AP2/ERF proteins that bind to the promoter sites of genes responsible for the biosynthesis of terpenoid indole alkaloids [143]. The jasmonic acid-responsive elicitation of nicotine assimilation in tobacco occupies the exploit of AP2/ERF and bHLH proteins, suggesting that interaction of TFs is capable of contributing to SMs assimilation [144]. Furthermore, the AP2/ERF protein has been shown to attach with CrPRX1 and plays a vital function in vinblastine assembly. Yet, a lack of extra studies has led to predictions that presently anonymous AP2/ERF TFs may be accountable for the regulation of bisindole alkaloids. Additionally, the Ap2/ERF GBERF1 controls the expression of PAL, C4H, C3H, CCR, HCT, CoMT, and F5H genes to improve the biosynthesis of lignin and defend against Verticilium dahliae [145].

NAC TFs comprise huge groups of plant-specific proteins, which organize different metabolic pathways in plant development and stress responses [146]. There are 140 supposed NAC or NAC-like genes in different plant genomes. Twenty of these genes are recognized as stress-responsive genes, counting OsNAC5 that is induced by stress factors [147]. Overexpression of OsNAC5 increased stress tolerance and accretion of osmolytes, detoxification, and redox homeostasis without causing growth defects [148]. The overexpression of OsNACs induces stress resistance through upregulation of the expression of stress receptive rice genes like LEA3 genes that encode LEA protein [148]. Correspondingly, PtrNAC72 regulates the expression of the arginine decarboxylase (ADC) gene that is accountable for the assimilation of Put and can mediate ROS homeostasis in Poncirus trifoliate [149]. Additionally, PaNAC03 is a NAC TF, which depressingly regulates several genes in the flavonoid biosynthesis, that is, CHS, F30 H, and LAR3, and so improves plant tolerance against Heterobasidion annosum. Expression of the transcription factors OsNAC5 was distinguished in water-stressed rice plants untreated or treated with Si [121]. Remarkably, the rank of expression was much elevated in Si-treated and water-stressed rice plants.

There are some researchers who advocate that Si NPs are impeding with diverse signaling processes and able to change plant SMs metabolism, the precise mechanisms through which this modulation could happen is not implicit. Accordingly, Si NPs application (25 mM) boosts parthenolide in Feverfew plants by 378.61 μg mg^{-1} after 24 h, which established the Si NPs as elicitors are most efficient [150]. Additionally, Hegazi et al. [151] reported that foliar application with 5 ppm Si NPs improves α-tocopherol assembly in vitro of Argan plant. Fatemi et al. [28] proved that under normal or contamination with lead, application of Si NPs commonly increased flavonoids and ascorbic acid accumulation of coriander plants relative to untreated plants. The results of some reports proved that the preliminary retorts of plants to NPs might comprise superior altitudes of ROS, cytoplasmic Ca^{2+}, and upregulation of mitogen-activated protein kinase (MAPK) cascades close to additional abiotic stresses. MAPK phosphorylation and activation of downstream TFs commonly cause the transcriptional reprogramming of plants' secondary metabolism [152].

7.6 Conclusion and perspective

Environmental stresses are a great risk to global food safety, and the occurrence of stressed events is intensified by global climate change. However, from the past few decades' function of Si in mitigation of both biotic and abiotic stress has been convincingly established in various plant species. Furthermore, plant SMs themselves represent an imperative constituent of the plant defense system against various stresses. Recently, Si was accounted to reinforce plant stress tolerance by modulating SMs production. Modulation of stress tolerance by Si through secondary metabolism was reported in both biotic and abiotic stress; however, efforts were focused mostly on biotic stress especially pathogen—plant interactions. Furthermore, the exhaustive molecular mechanism of Si-mediated regulation of plant stress tolerance via secondary metabolism is rarely known in most plant species, and most focus was made on biotic stress especially plant pathogens.

References

[1] Charrier G, Ngao J, Saudreau M, Améglio T. Effects of environmental factors and management practices on microclimate, winter physiology, and frost resistance in trees. Front Plant Sci 2015;6:259.

[2] Mahalingam R. Combined stresses in plants. Switzerland: Springer International Publishing; 2015.

[3] Farouk S, Elhindi KM, Alotaibi MA. Silicon supplementation mitigates salinity stress on *Ocimum basilicum* L. via improving water balance, ion homeostasis, and antioxidant defense system. Ecotoxicol Environ Saf 2020;206:111396. Available from: https://doi.org/10.1016/j.ecoenv.2020.111396.

[4] Farouk S, Al-Huqail AA. Sodium nitroprusside application regulates antioxidant capacity, improves phytopharmaceutical production and essential oil yield of marjoram herb under drought. Ind Crops Prod 2020;158:113034. Available from: https://doi.org/10.1016/j.indcrop.2020.113034.

[5] Farouk S, Almutairi AB, Alharbi YO, Al-Bassam WI. Acaricidal efficacy of Jasmine and lavender essential or mustard fixed oils against two-spotted spider mite and their impact on growth and yield of eggplants. Biology MDPI 2021;10(5):410. Available from: https://doi.org/10.3390/biology10050410.

[6] Farouk S, Al-Ghamdi AAM. Sodium nitroprusside application enhances drought tolerance in marjoram herb by promoting chlorophyll biosynthesis, sustaining ion homeostasis, and enhancing osmotic adjustment capacity. Arab J Geosci 2021;14:430. Available from: https://doi.org/10.1007/s12517-021-06846-5.

[7] Khan MIR, Jahan B, Alajmi MF, Rehman MT, Khan NA. Exogenously-sourced ethylene modulates defense mechanisms and promotes tolerance to zinc stress in mustard (*Brassica juncea* L.). Plants 2019;8:540.

[8] Kossel A. Ueber Schleim und schleimbildende Stoffe1. DMW Deut Med Wochenschr 1891;17:1297—9.

[9] Chrysargyris A, Papakyriakou E, Petropoulos SA, Tzortzakis N. The combined and single effect of salinity and copper stress on growth and quality of *Mentha spicata* plants. J Hazard Mater 2019;368:584—93.

[10] Hartmann T. From waste products to ecochemicals: fifty years research of plant secondary metabolism. Phytochemistry 2007;68:2831—46.

[11] Kurepin LV, Ivanov AG, Zaman M, Pharis RP, Hurry V, Hüner NPA. Interaction of glycine betaine and plant hormones: protection of the photosynthetic apparatus during abiotic stress. In: Hou HJM, Najafpour MM, Moore GF, Allakhverdiev SI, editors. Photosynthesis: structures, mechanisms, and applications. Cham: Springer International Publishing; 2017. p. 185—202. Available from: http://doi.org/10.1007/978-3-319-48873-8_9.

[12] Zandalinas SI, Mittler R, Balfagón D, Arbona V, Gómez-Cadenas A. Plant adaptations to the combination of drought and high temperatures. Physiol Plant 2017. Available from: http://doi.org/10.1111/ppl.12540.

[13] Ahmad R, Hussain S, Anjum MA, Khalid MF, Saqib M, Zakir I, et al. Oxidative stress and antioxidant defense mechanisms in plants under salt stress. In: Hasanuzzaman S, Hakeem KR, Nahar K, Alharby HF, editors. Plant abiotic stress tolerance. Cham: Springer; 2019. p. 191—205.

[14] Rejeb IB, Pastor V, Mauch-Mani B. Plant responses to simultaneous biotic and abiotic stress: molecular mechanisms. Plants 2014;3 (4):458—75. Available from: https://doi.org/10.3390/plants3040458.

[15] Rai M, Rai A, Kawano N, et al. De novo RNA sequencing and expression analysis of *Aconitum carmichaeliito* analyze key genes involved in the biosynthesis of diterpene alkaloids. Molecules 2017;22:2155. Available from: https://doi.org/10.3390/molecules22122155.

[16] Isah T, Umar S, Mujib A, et al. Secondary metabolism of pharmaceuticals in the plant in vitro cultures: strategies, approaches, and limitations to achieving higher yield. Plant Cell Tiss Organ Cult 2018;132(2):239—65. Available from: https://doi.org/10.1007/s11240-017-1332-2.

[17] Parveen A, Liu W, Hussain S, Asghar J, Perveen S, Xiong Y. Silicon priming regulates morpho-physiological growth and oxidative metabolism in maize under drought stress. Plants (Basel), 8. 2019. p. 431.

[18] Vega-Mas I, Rossi MT, Gupta KJ, González-Murua C, Ratcliffe RG, Estavillo JM, et al. Tomato roots exhibit in vivo glutamate dehydrogenase aminating capacity in response to excess ammonium supply. J Plant Physiol 2019;239:83—91.

[19] Bhat JA, Shivaraj SM, Singh P, Navadagi DB, Tripathi DK, Dash PK, et al. Role of silicon in mitigation of heavy metal stresses in crop plants. Plants (Basel) 2019;8:71.

[20] Guntzer F, Keller C, Meunier JD. Benefits of plant silicon for crops: a review. Agron Sustain Dev 2012;32:201—13. Available from: https://doi.org/10.1007/s13593-011-0039-8.

[21] Muneer S, Park YG, Kim S, Jeong BR. Foliar or sub-irrigation silicon supply mitigates high-temperature stress in strawberry by maintaining photosynthetic and stress-responsive proteins. J Plant Growth Regul 2017;36:836—45.

[22] Reynolds OL, Padula MP, Zeng R, Gurr GF. Silicon: potential to promote direct and indirect effects on plant defense against arthropod pests in agriculture. Front Plant Sci 2016;7:744.

[23] Luyckx M, Hausman JF, Lutts S, Guerriero G. Silicon and plants: current knowledge and technological perspectives. Front Plant Sci 2017;8:411.

[24] Rivero-Montejo SdJ, Vargas-Hernandez M, Torres Pacheco I. Nanoparticles as novel elicitors to improve bioactive compounds in plants. Agriculture 2021;11:134. Available from: https://doi.org/10.3390/agriculture11020134.

[25] Prasad R, Bhattacharyya A, Nguyen QD. Nanotechnology in sustainable agriculture: recent developments, challenges, and perspectives. Front Microbiol 2017;8:1014.

[26] Hussain A, Rizwan M, Ali Q, Ali S. Seed priming with silicon nanoparticles improved the biomass and yield while reduced the oxidative stress and cadmium concentration in wheat grains. Environ Sci Pollut Res 2019;26:7579—88.

[27] Wang S, Wang F, Gao S. Foliar application with nano-silicon alleviates Cd toxicity in rice seedlings. Environ Sci Pollut Res 2015;22:2837—45.

[28] Fatemi H, Pour BE, Rizwan M. Foliar application of silicon nanoparticles affected the growth, vitamin C, flavonoid, and antioxidant enzyme activities of coriander (Coriandrum sativum L.) plants grown in lead (Pb)-spiked soil. Environ Sci Pollut Res 2021;28:1417—25. Available from: https://doi.org/10.1007/s11356-020-10549-x.

[29] Jain C, Khatana S, Vijayvergia R. Bioactivity of secondary metabolites of various plants: a review. Int J Pharm Sci Res 2019;10:494—504. Available from: https://doi.org/10.13040/IJPSR.0975-8232.10.494-04.

[30] Ashraf MA, Iqbal M, Rasheed R, Hussain I, Riaz M, Arif MS. Environmental stress and secondary metabolites in plants: an overview. Plant metabolites and regulation under environmental stress. Academic Press; 2018. Available from: http://doi.org/10.1016/B978-0-12-812689-9.00008-X.

[31] Taiz L, Zeiger E. Plant physiology. 5th ed. Sunderland, MA: Sinauer Associates; 2015.

[32] Pichersky E, Raguso RA. Why do plants produce so many terpenoid compounds? New Phytol 2018;220:692—702.

[33] Venditti A, Bianco A. Sulfur-containing secondary metabolites as neuroprotective agents. Curr Med Chem 2020;27:4421—36.

[34] Mazid M, Khan T, Mohammad F. Role of secondary metabolites in defense mechanisms of plants. Biol Med 2011;3:232—49.

[35] Rivero J, Álvarez D, Flors V, Azcón-Aguilar C, Pozo MJ. Root metabolic plasticity underlies functional diversity in mycorrhiza-enhanced stress tolerance in tomato. New Phytol 2018;220:1322—36.

[36] Rojas CM, Senthil-Kumar M, Tzin V, Mysore KS. Regulation of primary plant metabolism during plant-pathogen interactions and its contribution to plant defense. Front Plant Sci 2014;5:17.

[37] Li YL, Zhu RX, Li G, Wang NN, Liu CY, Zhao ZT, et al. Secondary metabolites from the endolichenic fungus Ophiosphaerella korrae. RSC Adv 2019;9:4140—9.

[38] Jones JDG, Dangl JL. The plant immune system. Nature 2006;444:323—9. Available from: https://doi.org/10.1038/nature0528.

[39] Clements T, Ndlovu T, Khan W. Broad-spectrum antimicrobial activity of secondary metabolites produced by Serratia marcescens strains. Microbiol Res 2019;229:126329.

[40] Fraire-Velázquez S, Balderas-Hernández VE. Abiotic stress in plants and metabolic responses. In: Vahdati K, Leslie C, editors. Abiotic stress-plant responses and applications in agriculture. New York: InTech Open Science; 2013. p. 25—48. Available from: https://doi.org/10.5772/54859.

[41] Zhang X, van Doan C, Arce CCM, Hu L, Gruenig S, Parisod C, et al. Plant defense resistance in natural enemies of a specialist insect herbivore. Proc Natl Acad Sci USA 2019;116:23174—81.

[42] Taniguchi S, Miyoshi S, Tamaoki D, Yamada S, Tanaka K, Uji Y, et al. Isolation of jasmonate-induced sesquiterpene synthase of rice: product of which has an antifungal activity against Magnaporthe oryzae. J Plant Physiol 2014;171:625—32.

[43] Huber M, Epping J, Schulze Gronover C, Fricke J, Aziz Z, Brillatz T, et al. A latex metabolite benefits plant fitness under root herbivore attack. PLoS Biol 2016;14:e1002332.

[44] Turtola S, Manninen AM, Rikala R, Kainulainen P. Drought stress alters the concentration of wood terpenoids in Scots pine and Norway spruce seedlings. J Chem Ecol 2003;29:1981—95.

[45] Delano-Frier JP, Aviles-Arnaut H, Casarrubias-Castillo K, et al. Transcriptomic analysis of grain amaranth (Amaranthus hypochondriacus) using pyrosequencing: comparison with A. tuberculatus, expression profling in stems and in response to biotic and abiotic stress. BMC Genomics 2011;12:363. Available from: https://doi.org/10.1186/1471-2164-12-363.

[46] Dall'Osto L, Fiore A, Cazzaniga S, Giuliano G, Bassi R. Different roles of α-and β-branch xanthophylls in photosystem assembly and photoprotection. J Biol Chem 2007;282:35056—68. Available from: https://doi.org/10.1074/jbc.M704729200.

[47] Munné-Bosch S, Alegre L. Drought induced changes in the redox state of α-tocopherol, ascorbate, and the diterpene carnosic acid in chloroplasts of Labiatae species differing in carnosic acid contents. Plant Physiol 2003;131:1816—25. Available from: https://doi.org/10.1104/pp.102.019265.

[48] Sah SK, Reddy KR, Li J. Abscisic acid and abiotic stress tolerance in crop plants. Front Plant Sci 2016;7:571.

[49] Ng LM, Melcher K, Teh BT, Xu HE. Abscisic acid perception and signaling: structural mechanisms and applications. Acta Pharmacol Sin 2014;35:567—84.

[50] Khatun A, Rahman M, Rahman MS, Hossain MK, Rashid MA. Terpenoids and phytosteroids isolated from Commelina benghalensis Linn. with antioxidant activity. J Basic Clin Physiol Pharmacol 2019;31:20180218.

[51] Thakur AV, Ambwani S, Ambwani TK, Ahmad AH, Rawat DS. Evaluation of phytochemicals in the leaf extract of Clitoria ternatea Willd. through GC-MS analysis. Trop Plant Res 2018;5:200—6. Available from: https://doi.org/10.22271/tpr.2018.v5.i2.025.

[52] Isah T. Stress and defense responses in plant secondary metabolites production. Isah Biol Res 2019;52:39. Available from: https://doi.org/10.1186/s40659-019-0246-3.

[53] Clé C, Hill LM, Niggeweg R, Martin CR, Guisez Y, Prinsen E, et al. Modulation of chlorogenic acid biosynthesis in Solanum lycopersicum; consequences for phenolic accumulation and UV-tolerance. Phytochemistry 2008;69:2149—56. Available from: https://doi.org/10.1016/j.phytochem.2008.04.024.

[54] Ahanger MA, Alyemeni MN, Wijaya L, Alamri SA, Alam P, Ashraf M, et al. Potential of exogenously sourced kinetin in protecting Solanum lycopersicum from NaCl-induced oxidative stress through up-regulation of the antioxidant system, ascorbate-glutathione cycle and glyoxalase system. PLoS One 2018;13:e0202175.

[55] Ahanger M.A., Gul F., Ahmad P., Akram N.A. Environmental stresses and metabolomics—deciphering the role of stress responsive metabolites. In: Ahmad P, Ahanger MA, Singh VP, Tripathi DK, Alam P, Alyemeni MN, editors. Plant metabolites and regulation under environmental stress. Academic Press, 53-67; 2018b.

[56] Giménez MJ, Valverde JM, Valero D, Guillén F, Martínez-Romero D, Serrano M, et al. Quality and antioxidant properties on sweet cherries as affected by preharvest salicylic and acetylsalicylic acids treatments. Food Chem 2014;160:226−32. Available from: http://doi.org/10.1016/j.foodchem.2014.03.107.

[57] Vallverdú-Queralt A, Regueiro J, Martínez-Huélamo M, Rinaldi Alvarenga JF, Leal LN, Lamuela-Raventos RM. A comprehensive study on the phenolic profile of widely used culinary herbs and spices: rosemary, thyme, oregano, cinnamon, cumin and bay. Food Chem 2014;154:299−307. Available from: http://doi.org/10.1016/j.foodchem.2013.12.106.

[58] Huang X, Bie Z. Cinnamic acid-inhibited ribulose-1,5-bisphosphate carboxylase activity is mediated through decreased spermine and changes in the ratio of polyamines in cowpea. J Plant Physiol 2010;167(1):47−53. Available from: https://doi.org/10.1016/j.jplph.2009.07.002.

[59] Stringlis IA, De Jonge R, Pieterse CM. The age of coumarins in plant−microbe interactions. Plant Cell Physiol 2019;60:1405−14019. Available from: https://doi.org/10.1093/pcp/pcz076.

[60] Shetty R, Fretté X, Jensen B, Shetty NP, Jensen JD, Jørgensen HJL, et al. Silicon-induced changes in antifungal phenolic acids, flavonoids, and key phenylpropanoid pathway genes during the interaction between miniature roses and the biotrophic pathogen *Podosphaera pannosa*. Plant Physiol 2011;157:2194−205.

[61] Kapoor D. Antioxidative defense responses and activation of phenolic compounds in *Brassica juncea* plants exposed to cadmium stress. Int J Green Pharm 2016;10:228−34. Available from: https://doi.org/10.22377/IJGP.V10I04.760.

[62] Dawid C, Hille K. Functional metabolomics—a useful tool to characterize stress-induced metabolome alterations opening new avenues towards tailoring food crop quality. Agronomy 2018;8(8):138. Available from: https://doi.org/10.3390/agronomy8080138.

[63] Yuan Y, Liu Y, Wu C, Chen S, Wang Z, Yang Z, et al. Water deficit affected flavonoid accumulation by regulating hormone metabolism in *Scutellaria baicalensis* Georgi roots. PLoS One 2012;7:e42946.

[64] Wojakowska A, Muth D, Narożna D, Mądrzak C, Stobiecki M, Kachlicki P. Changes of phenolic secondary metabolite profiles in the reaction of narrow leaf lupin (*Lupinus angustifolius*) plants to infections with *Colletotrichum lupinifungus* or treatment with its toxin. Metabolomics 2013;9(3):575−89. Available from: http://doi.org/10.1007/s11306-012-0475-8.

[65] Karwasara VS, Dixit VK. Culture medium optimization for camptothecin production in cell suspension cultures of *Nothapodytes nimmoniana* (J. Grah.) Mabberley. Plant Biotechnol Rep 2013;7(3):357−69. Available from: https://doi.org/10.1007/s11816-012-0270-z.

[66] Jansen JJ, Allwood JW, Marsden-Edwards E, van der Putten WH, Goodacre R, van Dam NM. Metabolomic analysis of the interaction between plants and herbivores. Metabolomics 2009;5:150−61.

[67] Alhanout K, Malesinki S, Vidal N, Peyrot V, Rolain JM, Brunel JM. New insights into the antibacterial mechanism of action of squalamine. J Antimicrob Chemother 2010;65:1688−93. Available from: https://doi.org/10.1093/jac/dkq213.

[68] Kirk H, Vrieling K, Van Der Meijden E, Klinkhamer PG. Species by environment interactions affect pyrrolizidine alkaloid expression in *Senecio jacobaea*, *Senecio aquaticus*, and their hybrids. J Chem Ecol 2010;36:378−87. Available from: https://doi.org/10.1007/s10886-010-9772-8.

[69] Vetter J. Plant cyanogenic glycosides. Toxicon 2000;38:11−36.

[70] Pagare S, Bhatia M, Tripathi N, Pagare S, Bansal Y. Secondary metabolites of plants and their role: overview. Curr Trends Biotechnol Pharm 2015;9:293−304.

[71] Santolamazza-Carbone S, Sotelo T, Velasco P, Cartea ME. Antibiotic properties of the glucosinolates of *Brassica oleracea* var. acephala similarly affect generalist and specialist larvae of two lepidopteran pests. J Pest Sci 2016;89:195−206. Available from: https://doi.org/10.1007/s10340-015-0658-y.

[72] Sharma M, Koul A, Sharma D, Kaul S, Swamy MK, Dhar MK. Metabolic engineering strategies for enhancing the production of bioactive compounds from medicinal plants. Natural bio-active compounds. Singapore: Springer; 2019. p. 287−316.

[73] Soundararajan P, Manivannan A, Cho YS, Jeong BR. Exogenous supplementation of silicon improved the recovery of hyperhydric shoots in *Dianthus caryophyllus* L. by stabilizing the physiology and protein expression. Front Plant Sci 2017;8. Available from: https://doi.org/10.3389/fpls.2017.00738.

[74] Al-Garni SMS, Khan MMA, Bahieldin A. Plant growth-promoting bacteria and silicon fertilizer enhance plant growth and salinity tolerance in *Coriandrum sativum*. J Plant Interact 2019;14:386−96.

[75] Mahmood S, Daur I, Al-Solaimani SG, Ahmad S, Madkour MH, Yasir M, et al. Plant growth promoting rhizobacteria and silicon synergistically enhance salinity tolerance of mung bean. Front Plant Sci 2016;7:876 -876.

[76] Ghorbanpour M, Mohammadi H, Kariman K. Nanosilicon-based recovery of barley (*Hordeum vulgare*) plants subjected to drought stress. Environ Sci Nano 2020;7:443−61.

[77] Kim YH, Khan AL, Waqas M, Shim JK, Kim DH. Silicon application to rice root zone influenced the phytohormonal and antioxidant responses under salinity stress. J Plant Growth Regul 2014;33:137−49.

[78] Rodrigues FÁ, McNally DJ, Datnoff LE, Jones JB, Labbé C, Benhamou N. Silicon enhances the accumulation of diterpenoid phytoalexins in rice: a potential mechanism for blast resistance. Phytopathology 2004;94:177−83. Available from: https://doi.org/10.1094/PHYTO.2004.94.2.177.

[79] Carneiro-Carvalho A, Vilela A, Ferreira-Cardoso J, Marques T, Anjos R, GomesLaranjo J, et al. Productivity, chemical composition and sensory quality of "Martaínha" chestnut variety treated with silicon. CyTA J Food 2019;17:316−23.

[80] Fleck AT, Nye T, Repenning C, Stahl F, Zahn M, Schenk MK. Silicon enhances suberization and lignification in roots of rice (*Oryza sativa*). J Exp Bot 2011;62:2001−11. Available from: https://doi.org/10.1093/jxb/erq392.

[81] Song A, Xue G, Cui P, Fan F, Liu H, Yin C. The role of silicon in enhancing resistance to bacterial blight of hydroponic-and soil-cultured rice. Sci Rep 2016;6:24640. Available from: https://doi.org/10.1038/srep24640.

[82] Sakr N. The role of silicon (Si) in increasing plant resistance against fungal diseases. Hell Plant Prot J 2016;9:1−15.

[83] Yang L, Han Y, Li P, Li F, Ali S, Hou M. Silicon amendment is involved in the induction of plant defense responses to a phloem feeder. Sci Rep 2017;7:4232.

[84] Pontigo S, Godoy K, Jiménez H, Gutiérrez-Moraga A, Mora MdlL, Cartes P. Silicon-mediated alleviation of aluminum toxicity by modulation of Al/Si uptake and antioxidant performance in ryegrass plants. Front Plant Sci 2017;8:642.

[85] Shahnaz G, Shekoofeh E, Kourosh D, Moohamadbagher B. Interactive effects of silicon and aluminum on the malondialdehyde (MDA), proline, protein and phenolic compounds in *Borago officinalis* L. J Med Plant Res 2011;5:5818−27.

[86] Adrees M, Ali S, Rizwan M, Rehman MZ, Ibrahim M, Abbas F, et al. Mechanisms of silicon-mediated alleviation of heavy metal toxicity in plants: a review. Ecotoxicol Environ Saf 2015;119:186−97.

[87] Wu L, Wang H, Lan H, Liu H, Qu J. Adsorption of Cu(II)−EDTA chelates on triammonium-functionalized mesoporous silica from aqueous solution. Sep Purif Technol 2013;117:118−23.

[88] Verma N, Shukla S. Impact of various factors responsible for fluctuation in plant secondary metabolites. J Appl Res Med Aromatic Plants 2015;2(4):105−13. Available from: http://doi.org/10.1016/j.jarmap.2015.09.002.

[89] Zobayed SMA, Afreen F, Kozai T. Phytochemical and physiological changes in the leaves of St. John's wort plants under a water stress condition. Environ Exp Bot 2007;59(2):109−16. Available from: http://doi.org/10.1016/j.envexpbot.2005.10.002.

[90] Zhang L, Wang Q, Guo Q, Chang Q, Zhu Z, Liu L, et al. Growth, physiological characteristics and total flavonoid content of *Glechoma longitubain* response to water stress. J Med Plants Res 2012;6(6):1015−24.

[91] Hayasaka T, Fujii H, Ishiguro K. The role of silicon in preventing appressorial penetration by the rice blast fungus. Phytopathology 2008;98:1038−44.

[92] Kurabachew H, Wydra K. Induction of systemic resistance and defense-related enzymes after elicitation of resistance by rhizobacteria and silicon application against *Ralstonia solanacearum* in tomato (*Solanum lycopersicum*). Crop Prot 2014;57:1−7.

[93] Rahman MA, Alam I, Kim YG. Screening for salt responsive proteins in two contrasting alfalfa cultivars using a comparative proteome approach. Plant Physiol Biochem 2015;89(1):112−22.

[94] Fawe A, Abou-Zaid M, Menzies JG, Belanger RR. Silicon-mediated accumulation of flavonoid phytoalexins in cucumber. Phytopathology 1998;88:396−401.

[95] Currie HA, Perry CC. Silica in plants: biological, biochemical and chemical studies. Ann Bot (Lond.) 2007;100:1383−9.

[96] Schaller J, Brackhage C, Bäucker E, Dudel EG. UV screening of grasses by plant silica layer? J Biosci 2013;38:413−16.

[97] Schaller J, Brackhage C, Dudel E. Silicon availability changes structural carbon ratio and phenol content of grasses. Environ Exp Bot 2012;77:283−7.

[98] Fang X, Yang CQ, Wei YK, Ma QX, Yang L, Chen XY. Genomics grand for diversified plant secondary metabolites. Plant Div Res 2011;33(1):53−64.

[99] Khan A, Khan AL, Muneer S, Kim Y-H, Al-Rawahi A, Al-Harrasi A. Silicon and salinity: crosstalk in crop-mediated stress tolerance mechanisms. Front Plant Sci 2019;10:1429.

[100] Yin L, Wang S, Tanaka K, Fujihara S, Itai A, Den X, et al. Silicon-mediated changes in polyamines participate in silicon-induced salt tolerance in *Sorghum bicolor* L. Plant Cell Environ 2016;39:245−58. Available from: https://doi.org/10.1111/pce.12521.

[101] Gill SS, Tuteja N. Reactive oxygen species and antioxidant machinery in abiotic stress tolerance in crop plants. Plant Physiol Biochem 2010;48:909−30. Available from: https://doi.org/10.1016/j.plaphy.2010.08.016.

[102] Chen D, Shao Q, Yin L, Younis A, Zheng B. Polyamine function in plants: metabolism, regulation on development, and roles in abiotic stress responses. Front Plant Sci 2019;9:1945.

[103] Yordanova R, Alexieva V, Popova L. Influence of root oxygen deficiency on photosynthesis and antioxidant status in barley plants. Russ J Plant Physiol 2003;50:163−7.

[104] Wang J. Changes in polyamine contents in Citrus and its closely related species under abiotic stresses and isolation, characterization of two polyamine biosynthetic genes. Huazhong Agric Univ 2009;1−8.

[105] Wang X. Studies on the evaluation methods and the mechanism of resistance of Chinese cabbage (*Brassica campestris* L.) to Diamondback Moth (*Plutella xylostella*). Chin Acad Agric Sci 2007;24−31.

[106] Kim JI, Dolan WL, Anderson NA, Chapple C. Indole glucosinolate biosynthesis limits phenylpropanoid accumulation in *Arabidopsis thaliana*. Plant Cell 2015;27:1529−46.

[107] Zhang X, Gou M, Liu CJ. *Arabidopsis* Kelch repeat F-box proteins regulate phenylpropanoid biosynthesis via controlling the turnover of phenylalanine ammonia-lyase. Plant Cell 2013;25:4994−5010.

[108] Kim JI, Zhang X, Pascuzzi PE, Liu CJ, Chapple C. Glucosinolate and phenylpropanoid biosynthesis are linked by proteasome-dependent degradation of PAL. New Phytol 2020;225:154−68.

[109] Goyal S, Lambert C, Cluzet S, Mérillon JM, Ramawat KG. Secondary metabolites and plant defense. In: Mérillon JM, Ramawat KG, editors. Plant defense: biological control. Netherlands: Springer; 2012. p. 109−38. Available from: https://doi.org/10.1007/978-94-007-1933-0_5.

[110] Zhao J, Davis LC, Verpoorte R. Elicitor signal transduction leading to production of secondary metabolites. Biotechnol Adv 2005;23:283−333.

[111] Yang L, Wen K-S, Ruan X, Zhao Y-X, Wei F, Wang Q. Response of plant secondary metabolites to environmental factors. Molecules 2018;23:762.

[112] Yazaki K. ABC transporters involved in the transport of plant secondary metabolites. FEBS Lett 2006;580(4):1183−91. Available from: https://doi.org/10.1016/j.febslet.2005.12.009.

[113] Agarwal P, Agarwal PK, Nair S, Sopory SK, Reddy MK. Stress inducible DREB2A transcription factor from *Pennisetum glaucum* is a phosphoprotein and its phosphorylation negatively regulates its DNA binding activity. Mol Genet Genom 2007;277:189−98.

[114] Blumwald E., Grover A., Good A.G. Breeding for abiotic stress resistance: challenges and opportunities. In: New directions for adiverse planet. Proceeding of the 4th international Crop Science Congress, 26 Sep−1 Oct 2004, Brisbane, Australia. CDROM <http://www.cropscience.org.au>.

[115] Bartels D, Sunkar R. Drought and salt tolerance in plants. Crit Rev Plant Sci 2005;24:23−58.

[116] Kavar T, Maras M, Kidric M, Sustar-Vozlic J, Meglic V. Identification of genes involved in the response of leaves of *Phaseolus vulgaris* to drought stress. Mol Breed 2007;21:159−72.

[117] Umezawa T, Fujita M, Fujita Y, Yamaguchi-Shinozaki K, Shinozaki K. Engineering drought tolerance in plants: discovering and tailoring genes to unlock the future. Curr Opin Biotechnol 2006;17:113−22. Available from: https://doi.org/10.1016/j.copbio.2006.02.002.

[118] Chaves MM, Oliveira MM. Mechanisms underlying plant resilience to water deficits: prospects for water-saving agriculture. J Exp Bot 2004;55:2365–84. Available from: https://doi.org/10.1093/jxb/erh269.

[119] Thomashow MF. Molecular basis of plant cold acclimation: insights gained from studying the CBF cold response pathway. Plant Physiol 2010;154:571–7.

[120] Gilmour SJ, Fowler SG, Thomashow MF. *Arabidopsis* transcriptional activators CBF1, CBF2, and CBF3 have matching functional activities. Plant Mol Biol 2004;54(5):767–81.

[121] Khattab HI, Emam MA, Emam MM, Helal NM, Mohamed MR. Effect of selenium and silicon on transcription factors NAC5 and DREB2A involved in drought-responsive gene expression in rice. Biol Plant 2014;58:265–73. Available from: https://doi.org/10.1007/s10535-014-0391-z.

[122] Yang CQ, Fang X, Wu XM, Mao YB, Wang LJ, Chen XY. Transcriptional regulation of plant secondary metabolism. FJ Integr Plant Biol 2012;54:703–12.

[123] Skibbe M, Qu N, Galis I, Baldwin IT. Induced plant defenses in the natural environment: *Nicotiana attenuata* WRKY3 and WRKY6 coordinate responses to herbivory. Plant Cell 2008;20:1984–2000.

[124] Ma CH, Yang L, Hu SY. Silicon supplying ability of soil and advances of silicon fertilizer research. Hubei Agri Sci 2009;4:987–9. Available from: https://doi.org/10.3969/j.issn.0439-8114.2009.04.066 (in Chinese with English abstract).

[125] Vannozzi A, Wong DCJ, Höll J, Hmmam I, Matus JT, Bogs J, et al. Combinatorial regulation of stilbene synthase genes by WRKY and MYB transcription factors in grapevine (*Vitis vinifera* L.). Plant Cell Physiol 2018;59:1043–59.

[126] Alfieri M, Vaccaro MC, Cappetta E, Ambrosone A, De Tommasi N, Leone A. Coactivation of MEP-biosynthetic genes and accumulation of abietane diterpenes in *Salvia sclarea* by heterologous expression of WRKY and MYC2 transcription factors. Sci Rep 2018;8:11009.

[127] Fu F, Zhang W, Li YY, Wang HL. Establishment of the model system between phytochemicals and gene expression profiles in Macroscleeid cells of *Medicago truncatula*. Sci Rep 2017;7:2580.

[128] Gonzalez A, Zhao M, Leavitt JM, Lloyd AM. Regulation of the anthocyanin biosynthetic pathway by the TTG1/bHLH/Myb transcriptional complex in *Arabidopsis* seedlings. Plant J 2008;53:814–27.

[129] Nisha SN, Prabu G, Mandal AKA. Biochemical and molecular studies on the resistance mechanisms in tea [*Camellia sinensis* (L.) O. Kuntze] against blister blight disease. Physiol Mol Biol Plants 2018;24:867–80.

[130] Hao L, Shi S, Guo H, Zhang J, Li P, Feng Y. Transcriptome analysis reveals differentially expressed MYB transcription factors associated with silicon response in wheat. Sci Rep 2021;11:4330. Available from: https://doi.org/10.1038/s41598-021-83912-8.

[131] Yoo YH, et al. Genome-wide transcriptome analysis of rice seedlings afer seed dressing with *Paenibacillus yonginensis* DCY84(T) and silicon. Int J Mol Sci 2019. Available from: https://doi.org/10.3390/ijms20235883.

[132] Katiyar A, et al. Genome-wide classification and expression analysis of MYB transcription factor families in rice and Arabidopsis. BMC Genomics 2012;13:544.

[133] Chen S, Niu X, Guan Y, Li H. Genome-wide analysis and expression profile of the MYB genes in *Brachypodium distachyon*. Plant Cell Physiol 2017;58:1777–88.

[134] Zhang L, et al. The wheat MYB transcription factor TaMYB18 regulates leaf rolling in rice. Biochem Biophys Res Commun 2016;481:77–83.

[135] Yang C, et al. OsMYB103L, an R2R3-MYB transcription factor, influences leaf rolling and mechanical strength in rice (*Oryza sativa* L.). BMC Plant Biol 2014;14:1–15.

[136] Zhu J, et al. AtMYB103is a crucial regulator of several pathways affecting Arabidopsis anther development. Sci China Life Sci 2010;53:1112–22.

[137] Dubos C, Stracke R, Grotewold E, Weisshaar B, Martin C, Lepiniec L. MYB transcription factors in *Arabidopsis*. Trends Plant Sci 2010;15:573–81.

[138] Yamamura C, Mizutani E, Okada K, Nakagawa H, Fukushima S, Tanaka A, et al. Diterpenoid phytoalexin factor, a bHLH transcription factor, plays a central role in the biosynthesis of diterpenoid phytoalexins in rice. Plant J 2015;84:1100–13.

[139] Todd AT, Liu E, Polvi SL, Pammett RT, Page JE. A functional genomics screen identifies diverse transcription factors that regulate alkaloid biosynthesis in *Nicotiana benthamiana*. Plant J 2010;62:589–600.

[140] Shaaban MI, Bok JW, Lauer C, Keller NP. Suppressor mutagenesis identifies a velvet complex remediator of *Aspergillus nidulans* secondary metabolism. Eukaryot Cell 2010;9:1816–24.

[141] Yoshida Y, Miyamoto K, Yamane H, Nishizawa Y, Minami E, Nojiri H, et al. OsTGAP1 is responsible for JA-inducible diterpenoid phytoalexin biosynthesis in rice roots with biological impacts on allelopathic interaction. Physiol Plant 2017;161:532–44.

[142] Wasternack C, Song S. Jasmonates: biosynthesis, metabolism, and signaling by proteins activating and repressing transcription. J Exp Bot 2017;68:1303–21.

[143] De Sutter V, Vanderhaeghen R, Tilleman S, Lammertyn F, Vanhoutte I, Karimi M, et al. Exploration of jasmonate signaling via automated and standardized transient expression assays in tobacco cells. Plant J 2005;44:1065–76.

[144] Shoji T, Hashimoto T. Stress-induced expression of NICOTINE2-locus genes and their homologs encoding ethylene response factor transcription factors in tobacco. Phytochemistry 2015;113:41–9.

[145] Guo W, Jin L, Miao Y, He X, Hu Q, Guo K, et al. An ethylene response-related factor, GbERF1-like, from *Gossypium barbadense* improves resistance to *Verticillium dahliae* via activating lignin synthesis. Plant Mol Biol 2016;91:305–18.

[146] Tran LSP, Nishiyama R, Yamaguchi-Shinozaki K, Shinozaki K. Potential utilization of NAC transcription factors to enhance abiotic stress tolerance in plants by biotechnological approach. GM Crops 2010;1:32–9. Available from: https://doi.org/10.4161/gmcr.1.1.10569.

[147] Fang Y, You J, Xie K, Xie W, Xiong L. Systematic sequence analysis, and identification of tissue-specific or stress-responsive genes of NAC transcription factor family in rice. Mol Genet Genom 2008;280:547–63.

[148] Takasaki H, Maruyama K, Kidokoro S, Ito Y, Fujita Y, Shinozaki K. The abiotic stress-responsive NAC-type transcription factor OsNAC5 regulates stress-inducible genes and stress tolerance in rice. Mol Genet Genom 2010;284:173–83. Available from: https://doi.org/10.1007/s00438-010-0557-0.

[149] Wu G, Johnson SK, Bornman JF, Bennett SJ, Clarke MW, Singh V, et al. Growth temperature and genotype both play important roles in sorghum grain phenolic composition. Sci Rep 2016;6:21835. Available from: http://doi.org/10.1038/srep21835.

[150] Khajavi M, Rahaie M, Ebrahimi A. The effect of TiO_2 and SiO_2 nanoparticles and salinity stress on expression of genes involved in parthenolide biosynthesis in Feverfew (*Tanacetum parthenium* L.). Caryologia Int J Cytol Cytosyst Cytogenet 2019;72:3−14.

[151] Hegazi GA, Ibrahim WM, Hendawy MH, Salem HM, Ghareb HE. Improving α-tocopherol accumulation in *Argania spinosa* suspension cultures by precursor and nanoparticles feeding. Plant Arch 2020;20:2431−7.

[152] Phukan UJ, Jeena GS, Shukla RK. WRKY transcription factors: molecular regulation and stress responses in plants. Front Plant Sci 2016;7:760. Available from: https://doi.org/10.3389/fpls.2016.00760.

8

Silicon improves salinity tolerance in crop plants: Insights into photosynthesis, defense system, and production of phytohormones

Freeha Sabir[1], Sana Noreen[1], Zaffar Malik[1], Muhammad Kamran[2], Muhammad Riaz[3], Muhammad Dawood[4], Aasma Parveen[1], Sobia Afzal[1], Iftikhar Ahmad[1] and Muhammad Ali[5]

[1]Department of Soil Science, University College of Agriculture and Environmental Sciences, The Islamia University of Bahawalpur, Bahawalpur, Pakistan [2]School of Agriculture, Food and Wine, The University of Adelaide, South Australia, Australia [3]College of Natural Resources and Environment, South China Agricultural University, Guangzhou, P.R. China [4]Department of Environmental Sciences, Bahauddin Zakariya University, Multan, Pakistan [5]Department of Environmental Science, Faculty of Agriculture & Environment, The Islamia University of Bahawalpur, Bahawalpur, Pakistan

8.1 Introduction

The extent of agricultural land that is affected by high salt stress is increasing worldwide and as a major abiotic stress. Salinity causes a restriction in crop productivity due to its adverse effects on plant morphology and physiobiochemistry. About 20% of the world's total area is damaged by soil salinity which causes the loss of 10 million hectares of arable land annually [1]. The salt-affected land had restricted food production and has contributed to world poverty, especially in underdeveloped countries [2]. High-salinity level imparts uncompromising constraints and disturbs plant metabolism by triggering ion imbalance reflecting in obstructed transport of essential ions and solutes [3]. Salt stress induces cytosolic Na^+ toxicity, which results in the activation of depolarization activated K^+ outward rectifying channels and reactive oxygen species (ROS) activated nonselective cation channels to increase K^+-efflux.

Developing crops to grow successfully in saline lands has been a concern for a long time. Plant growth regulators are extensively used to regulate plant growth and to enhance stress tolerance. Therefore exploring potential growth regulators and their mechanisms is highly important for improving salt tolerance in crops. To improve salt tolerance, many efforts have been made to mitigate the salt-induced phytotoxicity by various exogenous substances like salicylic acid (SA), polyamine, nitric oxide, and glycine betaine. Silicon (Si) is an important constituent in soil and effectively alleviate various abiotic stress like salinity, temperature, and heavy metal stress on plants [4,5]. Silicon is a major component of plant cell walls and acts as a resistant component parallel to lignin in some plant species [6]. Plants can access Si in the form of silicic acid through the roots and transport its upper parts via the xylem [7]. Plants can absorb Si in the form of silicic acid through the roots and transport its upper parts via the xylem [8]. The exogenous Si application could induce antioxidative responses and regulate ion balance in crop plants against abiotic stresses [9,10]. Singh and Roychoudhury also discussed that the Si (Na_2SiO_3) had a significant role in the augmentation of various antioxidant, osmolyte defense, and methyl glyoxal detoxification system that helps in reducing oxidative damages caused by fluoride in rice and barley plants [11].

Silicon has also been observed to generate various downstream genes to produce resistance against both abiotic and biotic stresses in crop plants [12]. Silicon, being a beneficial metalloid, is also involved in upregulating and metabolism of other macro and microelements [13]. In addition, Si and selenium (Se) are the two perfectly related trace elements that generate tolerance in higher plants subjected to abiotic stress [14]. Several crop species, that is, barley (*Hordeum vulgare* L.) [15], wheat (*Triticum aestivum* L.) [16], grapevine rootstock (*Vitis vinifera* L.) [17], cucumber (*Cucumis sativus* L.) [18], and tomato (*Solanum lycopersicum* L.) [19] have shown significant result against salinity by the application of Si [20]. Many studies reported provision of silicon under salt stress improved photosynthesis, stimulated the vegetative growth and dry matter content. In addition to that, the accumulation of silica in the leaf biomass narrowed the transpiration that could reduce sodium uptake [21].

This chapter focuses on the recent findings about the Si as a prominent constituent in mediated salt stress tolerance for plant growth and development. Special emphasis will be given to their role in photosynthesis, antioxidant defense system, and increment in constituent of various phytohormones.

8.2 Salinity-induced injuries in plants

Soil secondary salinization is the major environmental constrain limiting the production of agronomic crops and causing considerable crop losses. The extent of saline land is predicted to worsen in many areas of the world due to brackish irrigation water, faulty drainage of irrigated lands, and global warming. Plants are affected by salt stress at their growth stages due to two main reasons: (1) osmotic stress or water scarcity (drought); owing to excessive salts manifestation in soil solution. Excess salt concentration in the rhizosphere hampers the plant's capability to uptake water via roots to above-ground parts, which consequently lowers growth rate and (2) salt-specific or ion-excess impact of salinity; in which the high levels of salt move into the plants through a transpiration stream and cause stunted growth [22].

8.2.1 Osmotic injury in plants

Salt stress increases the uptake of sodium ion (Na^+) and chloride ion (Cl^-) from the soil, which results in the suppression of transport of some essential nutrients such as nitrogen, phosphorus, potassium, and calcium [23]. Moreover, high salt accumulation in the root zone disturbs the nutritional and osmotic balance which results in physiological salt stress that restricts the water uptake [24]. Halophytic plants with high resistance against sodium toxicity also face retarded growth due to osmotic stress [25]. Under osmotic stress, plant roots cannot uptake water, and high energy is needed to sustain the osmotic balance via the accumulation of compatible solutes [26].

It is also reported that salinity-induced osmotic stress adversely affects the ion transport, enzymatic activity of solutes, cell structure, uptake of nutrients, shoot and root growth, and plant morphology [27]. Plant seedlings in salt-stressed soils have less osmotic potential, altering the imbibition of seed with efficient water [28]. Salt-induced toxicity hinders the enzymes activity and nucleic acid metabolism [29], disturbs protein metabolism [30], alters hormonal balance [31], and impedes seed reserves utilization [32].

8.2.2 Specific ion toxicity

Ionic toxicity disturbs the cell biochemical reactions and thus leads to nutrient imbalance and poor plant growth. Under a saline environment, the presence of high Na^+ and Cl^- competitively restrict the transport activity and uptake of K^+ and NO_3^-. Chemically, Na^+ and Cl^- are similar to some of the essential nutrients such as K^+ and NO_3^-. Hence in the cell cytosol, Na^+ ion substitutes K^+, however, K^+ is essential for enzymatic activities and played an indirect role in synthesizing protein contents in plants. Moreover, Na^+ accumulation in apoplast disrupts the Ca^{+2} interactions with membrane phospholipids or cell wall proteins [33,34]. Na^+ ion toxicity causes impaired fertility, suspension in flowering with some fractional or ample grain loss which subsequently gives rise to poor-quality panicle formation in rice [35,36], restricts the photosynthetic activity in major agronomic crops [37,38], and minimizes phosphorus (P), potassium (K), and calcium (Ca) concentrations [39]. Photosystem

II is initially affected by salt stress due to chlorophyll contents reduction and affects stomatal conductivity [25]. High Na^+/K^+ ratio causes stomata closure, disrupts photosynthetic tissue, and decreases the substomatal CO_2 concentration. Due to the limited availability of CO_2 for carboxylation, plants are triggered toward senescence [40]. Ion toxicity also causes overproduction of ROS which cause damage to membrane proteins and lipids and disrupt normal plant growth.

8.2.2.1 Overproduction of reactive oxygen species

Overproduction of ROS occurred due to excess molecular oxygen that behaves like an electron acceptor. ROS production involved the singlet oxygen, hydroxyl radical, superoxide radicals, and hydrogen peroxide in plants that are powerful oxidizing compounds [41,42]. ROS formation occurs in plants both in radical or nonradical forms under normal physiological circumstances in cytoplasm, mitochondria, and peroxisomes [43,44]. Under salinity stress, overproduction of ROS irregulates the electron transport chain (etc.) in cytoplasm and mitochondria by altering or completely distorting the reproductive and regulatory processes by restricting microspore genesis, endorsing cell death, extending stamen filaments, and accelerating the senescence of fully fertilized embryos and may abort the ovule [45,46]. Several antioxidative enzymes, such as catalase (CAT), superoxide dismutase (SOD), glutathione peroxidase (GPX), peroxidase (POD), ascorbate peroxidase (APX), and glutathione reductase (GR), and nonenzymatic compounds, such as ascorbic acid, nonprotein amino acids, glutathione, and phenolic compounds are produced in plants to maintain the level of ROS in cells [47–49]. Higher ROS production due to salinity stress consequently limits the availability of P and K^+ with the constitution of Na^+ and Cl^-, which negatively affects plants' growth and productivity and may also be triggered toward plant death [50,51].

8.2.2.2 Hormonal imbalance

Phytohormones have the ability to alleviate the various abiotic stresses via improving growth, development, nutrient balance, and source/sink transition. These are the signaling molecular compounds produced endogenously and are critical in restraining physiological responses that ultimately lead to optimal plant growth under a saline environment. Various studies reported different levels of production of several phytohormones in different plant species in response to salt stress which relates to their biosynthesis due to changes in gene expression and responses they regulate [52].

To overcome the hormonal imbalance, plants limit the ionic influx by minimizing the unidirectional uptake of ions or increasing the efflux of ions [53]. Detoxification involves the phenomena of vacuolar sequestration of Na^+ and Cl^- via membrane transporters, lowering the water potential [54]. Unidirectional transport of Na^+ from the soil in excess amount is mainly mediated by various systems, which involve the high-affinity potassium transporter (HKT)-type carriers and nonselective ion channels [55]. H^+ antiporters likewise, salt overly sensitive 1 (SOS1) usually involve in extrusion of Na^+ from cytoplasm, while, similar antiporter but from NHX (Na^+/H^+ antiporters) subfamily carried out the vacuolar sequestration of Na^+. Salt translocation from soil to above-ground part, such as shoot, is limited by HKTs, which also restrict the excess Na^+ loading from xylem. Main feature of the SOS pathway is SOS1 activity, which raises up immediately after phosphorylation by kinase SOS2. SOS2 remained attached with SOS3- a typical Ca^{2+} binding protein [56]. Na^+ uptake is reduced via cGMP (cyclic guanosine monophosphate) which limits the functioning of the main uptake pathway, that is, nonselective ion channels. All these functions described before are directly related to ABA regulatory role [57].

8.3 Regulatory role of Si to mitigate salt stress

Salinity stress obstructed the growth and productivity of crop plants by disturbing osmotic and ionic concentration, influencing the rate of photosynthesis, producing excessive oxidants and radicals, counteracting essential metabolic pathways, interrupting the endogenous phytohormonal functions, enfeebled the antioxidant defense ability of the crops, and manipulating the pattern of gene expression [4]. It has been reported that silicon application may enhance the plant tolerance to abiotic stresses by adjusting plant physiology, biochemistry, and antioxidant capacity along with regulation of phytohormones [58]. To sustain their life cycle under a saline environment, the crop plants had adopted retrospective mechanistic cascades of physiobiochemical and molecular pathways to lessen the menace of salinity stress; however, persistent exposure can devastate the defense system that leads to cell death and the collapse of essential apparatus [59]. Furthermore, a high concentration of Na^+ and Cl^- proved toxic to plants as they can create an ionic imbalance in the cell which leads to membrane disability and cell injury.

To complete a life cycle, various abiotic and sometimes multifaceted stresses have been exposed to plants. In return, these stresses can generate various ROS, like singlet oxygen, superoxide, hydrogen peroxide, or hydroxyl radicals in plant cells [60,61]. Previous literature [62,63] explained that in plants serious oxidative damages to the protein, DNA, and lipids of the cell component have been generated due to these ROS. Consequently ROS enhancement levels decline the growth and yield of plants [64,65]. To cope with salt stress, status of the plant ROS scavenging is the priority defense mechanism [66]. Certain studies portray that with the addition of exogenous Si, the ability of ROS scavenging can be improved by the regulation of antioxidant defense activity [63,67]. To mitigate the salinity stress in plants, several organic and inorganic substances can be used and Si application is one of them. There are ample evidences on how Si plays a promising role in plant growth promotion, nutrient balance, and upregulating the rate of photosynthesis that eventually helps plants resist salinity [68,69]. Silicon exerts beneficial effects on crops through the deposition of Si in the form of SiO_2 nH_2O in stems and leaves of the plants [70].

Table 8.1 shows several crop plants exhibiting different Si applications under different salt levels.

8.3.1 Silicon-induced salt tolerance and photosynthesis restoration

Under saline conditions, reduction in photosynthesis mainly occurs as plants close their stomata to reduce water loss, which leads to a lessening in leaf transpiration rates and dropping of internal CO_2 concentration in the leaves of plants [10,64]. Silicon accelerates the growth and development of several plant species by performing several physiological functions and altering many plant characteristics including germination, growth of seedlings, flowering, and grain yield. Silicon application improves photosynthesis by inducing changes in stomatal conductance under salinity stress [75,89,90] explained that photosynthetic functioning under a saline soil status highly depends on the adequate leaf water potential and its maintenance of sufficient turgor pressure. To lessen the salinity menace, the application of Si prominently alleviates the salinity stress to upgrade the plant water status. In their work, Abbas et al. reported that Si application had increased the stomatal conductance, number, and size of stomata in okra plants subjected to salinity stress [91]. In another study, Altuntas et al. exhibited that the exogenous application of Si considerably upregulated the stomatal conductance, leaf area, and leaf water status in salinity-induced plants [64]. These improvements were directly associated with an increase in gas exchange in leaves[10]. The efficient photosynthetic activity and the transpiration rate were increased in plants under salinity stress due to the concealed Si barrier created on the outer layer of plants [64]. The physical barrier of silica gel produced by applying exogenous Si on the outer layer of leaves, roots, and vascular tissues of stems reduces the evapotranspiration and electrolyte leakage from the leaf membranes [92]. This fact is further delineated by Al-aghabary et al. in their research that Si application in a stressed plant can uphold cells by upregulating the permeability of their plasma membranes, which in turn improves the activity of the antioxidative enzymes into the cells [93].

The growth of salt-stressed plants has prominently increased with the inclusion of the Si solution in the growth medium. Si boosts up the rate of photosynthesis, which was directly associated with leaf ultrastructure, chlorophyll contents, and ribulose biphosphate carboxylase activity. However, the biochemical and molecular effects of Si are still under examination [94–96]. Literature showed that the incorporation of Si solution to stressed plants produces several compatible solutes or osmolytes, such as glycine betaine, total free amino acids, proline, total soluble sugars, and antioxidant compounds (such as phenolics), which had remarkably enhanced the osmotic adjustment capability and antioxidant activity in salt-stressed plants [91,97]. In their study, explained that sugars might act as osmoprotectants that have been significantly found in the cucumber leaves and roots in the salt-stressed medium.

Moreover, several enzymes activation subjected to silicon application also take part in sugar synthesis and in the degradation of sugars into starch [98]. Proteomic study of [99] in pepper and [100] in tomato subjected to salty status of soil identified the fact that the levels of rubisco (ribulose-1-bisphosphate carboxylase-oxygenase) and other proteins were declined in the light-harvesting complex. They further set forth that the introduction of Si solutes mitigated these declines. Stress markers in their research established that rubisco was an ally of salt tolerance species [101]. Considering this into the investigation, the study of [64] concludes that Si improves plant growth in the salt-stressed condition by enhancing the photosynthesis, stomatal conductance, germination, leaf water status, and cell membrane stability and growth all of which in turn leads to higher biomass production of a plant [102] Also supported this speculation that silicon application could promptly alleviate the deteriorating effects of salinity on the quantum efficiency of photosystem II PSII.

TABLE 8.1 Exogenous application of Si on various crop plants under different salinity stress.

NaCl stress level	Crop grown	Source of Si	Si doze	Si-mediated tolerance mechanism	References
1.8, 4, and 6 dSm^{-1}	Geranium plant (Pelargonium X hortorum)	$K_2SiO_3 \cdot nH_2O$	1 mM	Growth parameters and oil yield.	[71]
200 mM	Wheat (Triticum aestivum L.)	$K_2SiO_3 \cdot nH_2O$	4 mM	Elevated antioxidant and osmoprotectants.	[72]
200 mM	Talh trees (Vachellia seyal L.)	$K_2SiO_3 \cdot nH_2O$	2 mM	Improved osmolytes and antioxidant system.	[73]
0, 25, 50, and 75 mM	Purple coneflower (Echinacea angustifolia L.)	$K_2SiO_3 \cdot nH_2O$	0, 0.75, 1.5, and 2.25 mM	Significantly enhanced physiological parameters and shoot antioxidant ability.	[74]
150 mM	French bean (Phaseolus vulgaris L.)	$CaSiO_3$	5, 10, and 15 g Kg^{-1}	Enhanced growth, yield, biochemical parameters, and enzyme activity.	[54]
0, 25, and 50 mM	Cherry tomatoes (Solanum lycopersicum L.)	Silicate	0, 1, and 2 mM	Enhanced water content and protect photosynthesis activity.	[75]
125 mM	Wheat (Triticum aestivum L.)	Na_2SiO_3	1 mM	Improved antioxidant specially glutathione.	[76]
21, 42, and 63 dSm^{-1}	Licorice (Glycyrrhiza glabra L.)	K_2SiO_3	0, 1.4, 2.8, and 4.2 dSm^{-1}	Enhanced total sugar and N metabolism-related enzyme.	[77]
100 mM	Rice (Oryza sativa L.)	Salicylic acid (SA)	0.5 and 1.0 mM	Enhanced antioxidant enzyme and S-nitrosothiol (SNO) activity.	[78]
100 mM	Rice (Oryza sativa L.)	$Na_2SiO_3 5H_2O$	0.5, 1.0, or 2.0 mM	Modulating phytohormonal and enzymatic antioxidants responses.	[58]
0 and 200 mmol L^{-1}	Weeping alkali grass (Puccinellia distans L.)	Na_2SiO_3	0 and 1.5 mmol L^{-1}	Greater soluble sugars and amino acids, increase osmotic adjustment, enhanced the activity of H$^+$-ATPase in roots and shoots, lower proline and reduced electrolyte leakage.	[79]
0, 50, 100, and 150 mM	Sunflower (Helianthus annuus L.)	Na_2SiO_3	0.0, 1.0, 1.5, and 2.0 mM	Silicon positively modulates nitrogen metabolism and antioxidant enzyme activities.	[80]
75 mM NaCl	Cucumber (Cucumis sativus L.)	$Na_2SiO_3 \cdot 9H_2O$	0.3 mM silicon	Enhanced polyamine accumulation and decreased free putrescine concentrations, but increased spermidine and spermine concentrations in both leaves androots.	[81]
0, 50, and 60 mM	Syngenta 8441 Maize genotype and EV 1089	H_2SiO_3	0 and 2 mM	Changes in ion accumulation, improvement in photosynthetic rate, enzyme activities, and phenolic compounds.	[82]
50 mM NaCl	Rock Fire (rose) (Lagonosticta sanguinodorsalis L.)	K_2SiO_3	0 or 1.8 mM	Si enhanced water uptake, aids in the mitigation of osmotic imbalance and equilibrate the ionic balance between intra- and extracellular regions.	[83]
0.46, 4, 8, and 12 dSm^{-1}	Wheat (Triticum aestivum L.)	Na_2SiO_3	8, 50, 100, and 150 mg kg^{-1}	Wheat growth and antioxidant enzymes.	[84]

(Continued)

TABLE 8.1 (*Continued*)

NaCl stress level	Crop grown	Source of Si	Si doze	Si-mediated tolerance mechanism	References
2.0 and 6.0 dSm^{-1}	Okra (*Abelmoschus esculentus* L.)	Si(OH)$_4$	150 mg·L^{-1}	Si maintained the nutritional adjustment in salinized okra, enhancement the availability of beneficial nutrient element, reduce membrane permeability.	[85]
0.5% and 1.0%	Golden berry (*Physalis peruviana* L.)	SiO$_2$.xH$_2$0	0, 0.5, and 1.0 g L^{-1}	It efficiently mitigates the injury to the photosynthetic pigments, no. of stomata and leaf blade thickness.	[86]
200 mM	Red Thorn (*Vachellia reficiens* L.)	K$_2$SiO$_3$	2 mM	Significantly improved enzymatic and nonenzymatic antioxidant defense systems, enhanced production of proline, and glycine betaine.	[73]
2.74, 5.96, 10.74, and 13.38 dSm^{-1}	Wheat (*Triticum aestivum* L.)	K$_2$SiO$_3$	0, 2.1, 4.2, 6.3, and 8.4 mg	Upregulation of micro and macronutrients.	[87]
0 and 120 mM	Starflower (*Helianthus annuus* L.)	Na$_2$SiO$_3$	0, 0.5, 1, 1.5, 2, and 2.5 mM	Increase production of proline, water status, and photosynthesis and antioxidant machinery.	[67]
120 mM	Wheat (*Triticum aestivum* L.)	Na$_2$SiO$_3$	0, 10, 20, 30, 40, or 50 mM	Increase germination percentage, plant water status, and mitigate-specific ionic effect	[88]

8.3.2 Si and enhancement of phytohormones

Silicon, being an essential nutrient element shows complex association with certain phytohormones such as cytokinins (CKs), auxins (AXs), and brassinosteroids (BRs) [103]. Two major plant hormones are AXs and CKs that potentially enhance root and shoot growth, delay senescence, and upregulated cell division and expansion. In addition, Si enhances indole-3-acetic acid (IAA) by mediating PIN2 and PIN3 protein that help in AX transportation [104]. Moreover, Hosseini et al., in their study, also found out that Si upregulated CKs content in plants subjected to salt stress as it declines stress-induced senescence [105]. When the plant is subjected to magnesium deficiency, Si escalates plant growth and development by triggering CK synthesis that exhibits the plants' morphological attributes [106]. Hormones like CKs and AXs, produced in the root meristem are involved in signaling [107,108]. Several phytohormones signaling molecules are critical in modulating plant growth in the presence of Si. The response of plants generally involves fluctuations in the levels of several phytohormones that relate to the changes in gene expression helps in their biosynthesis [109]. The significance of silicon cannot be disregarded owing to its valuable effects on plant growth status and development, as Si upgraded plants growth modifies the levels of endogenous growth hormones, such as gibberellins (GAs), jasmonic acid (JA), SA, and abscisic acid (ABA). These growth hormones play a very affirmative role in the growth and development of plants [110]. The study of [111] showed its significant results that the application of Si might induce stress resistance by modulating the phytohormone homeostasis (Fig. 8.1).

It is indicated in the literature that the endogenous bioactive gibberellins 1 and gibberellins 4 (GA1 and GA4) contents were upregulated with the application of basic (100 mg L^{-1}) and double (200 mg L^{-1}) doses of Si helped in building the plant growth status and development under salinity stress, though the basic Si addition proved

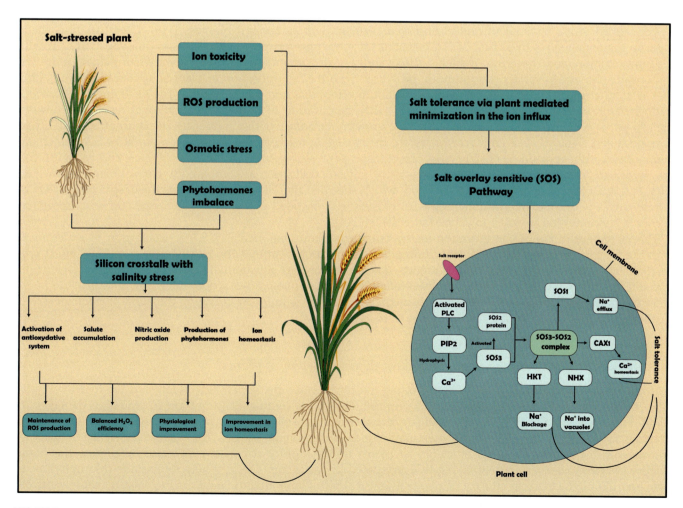

FIGURE 8.1   Silicon cross-talk with salt stress and plant-mediated salt overlay sensitive (SOS) pathway responsible to maintain ion homeostasis (Na$^+$, K$^+$, and Ca^{2+}) in crop plant grown under salinity stress.

more effective than double Si addition [112]. Moreover, the contents of endogenous JA are significantly enhanced with Si nutrition, and this can further reinforce the role of Si as a proficient element for the improvement of plants subjected to salinity [58]. Phytohormone like SA portrays different effects with the addition of Si.

Under salt stress, Si addition in plants showed a significant upregulation of SA contents in plants. ABA, in general, is known to be a major stress hormone that proved important during cellular signaling [113]. Application of Si enhances ABA synthesis, nutrient uptake, and proline biosynthesis. Moreover, [114] showed a positive result that Si supplementation increases ABA and antioxidant enzyme level in maize under salinity. Likewise, BR also has an ameliorative effect on plant growth status subjected to different abiotic stresses, like salinity stress [115], metal stress [74], oxidative stress [82], and low-temperature stress [116]. It has been reported in the study of [117] that the soil plant analysis development value of a plant has been significantly increased with the combined treatment of Si and BR. Hence the Si addition in salinity-stressed plants upgraded the levels of several phytohormones that help in plant growth status and development.

8.3.2.1 Silicon-mediated hormonal response under saline conditions

The plant regulates the most important adaptive strategies to mitigate salinity stress by perceiving the stress signals due to environmental changes and/or others and transferring them to different plant parts. This action is done by the plant to induce stress-related responses. The chain from receiving stress to alleviating stress is complex, and it has long been exposed that phytohormones are the strategic player in this process. ABA, JA, and SA, besides their regulatory role in the development and growth of plant parts, also govern stress responses [118]. In general, ABA is stated as a "stress hormone," which is known to conserve plant water by stomata closure under salt stress by persuading the expression of stress-related genes [107]. In this regard, Si has proved its criticality in mitigating salt stress and its role in stimulating phytohormones. Lang et al. revealed that Si amplified plants under salinity stress have significantly improved plant growth status and biomass [119]. Literature also sheds light on the endogenous induction of GA essential for numerous plant life phases, including vegetative growth, seed germination, and flowering. Evidence discovered the downregulation of GA1 level when plants are grown under salt stress; however, its concentration recovered with the Si application to plant subjected to salinity [120]. IAA is another plant hormone accountable for plant development and apical growth. The declining IAA concentration was recovered when augmented with Si [121]. Application of Si-mediated hormonal response to lessen the salt stress is restricted to the increment of the growth hormones. It also influences the concentration of stress hormones [58]. Showed their endeavor to resolve the cross-talk between the hormones brought by Si supplementation. Zeaxanthin epoxidase (ZEP) in ABA biosynthesis precursor encoded by the OsZEP gene and its concentration increased in salinity-stressed plants subjected to silicon application as compared to control plants. However, after 24 h, the concentration of OsZEP declined, which specified that Si could ameliorate the destructive effects of salinity stress with time. Interestingly, the ABA biosynthesis encoded genes were downregulated with upregulation of Si concentration from 0.5 to 2.0 mM after 24 h [120].

The increment in the Si concentration in the salt stress plant also increases SA concentration in antagonism with the JA concentration [121,122]. Concluding the above discussion that the Si arbitrated hormonal effect will be challenging due to many conflicting reports. Therefore detailed substantiations are needed to answer many opposite questions linking the molecular association of hormonal responses synchronized by Si to provide resistance in plants. So far, Si could ameliorate the negative effects of salinity by changing the plant physiological and biochemical reactions, including the hormonal response.

8.3.3 Role of Si in strengthening antioxidant defense system of plants

In plants, the overproduction of ROS, including hydrogen peroxide, superoxide anion, and the hydroxyl radicals, predominantly in mitochondria and chloroplast, showed a significant indication of oxidative damages in plants subjected to salinity [123]. To mitigate the damage caused by ROS, plants engaged with antioxidant enzymes to safeguard the nucleic acid, membrane lipids, and proteins [77,124,125] suggested that the Si application magnifies the photochemical efficiency of plants under salinity stress. It could also ameliorate the plant defense systems and assist in detoxifying ROS promoted by salt stress. This can help to upgrade chlorophyll contents and the phytochemical efficacy of photosystem II (PSII). Plants that are subjected to salinity, show reduced chlorophyll contents, while its concentration is slightly upgraded in the presence of Si [10]. These findings suggest that the silicate levels partially compensate the negative effect of salinity, which helps the plants resist salinity by raising SOD and CAT activities, chlorophyll, and photochemical efficiency of PSII. This can overall strengthen the antioxidant defense ability of crop plants with Si application under salinity [126]. Salinity stress inhibits the carbon uptake and weakens the photosynthetic machinery, which impairs the transport chain of the electron. It produces photon energy channelization, which further causes the generation of ROS. Production of ROS, that is, (OH^-) hydroxyl anion, (O_2^-) super oxidase, and (H_2O_2) hydrogen peroxide is an irreversible process. ROS can affect metabolic responses and plant developmental processes by oxidizing DNA, proteins, and lipids [127]. Antioxidative enzymes system like CAT, APX, POX, SOD, and GR play a crucial role in mitigating stress by dropping ROS and reducing lipid peroxidation. In addition, CAT exists in the peroxisome, whereas APX is present in peroxisome, chloroplast, cytosol, and mitochondria. The application of Si plays a major role in upregulating the CAT, SOD, and APX for the scavenging of ROS-induced oxidative stress [128]. However, to mitigate the low phosphorous-induced ROS accretion, exogenous Si application is applied, which helps in improving enzymatic antioxidant activity, that is, CAT, SOD, and POX. Besides, the application of Si act as the signaling molecule that downregulates the ROS, magnifies metal carriers, reduces electrolyte leakage, and inflates membrane stability [129]. Recent literature of [2] evidently proved that Si stimulates carotenoid accumulation, an antioxidant that scavenges ROS content and impedes oxidative stress.

8.4 Conclusion and future prospects

Agricultural land is largely affected by salinity stress either due to natural phenomena or by poor agricultural practices. Silicon is known to reduce the accumulation of Na + in roots/shoots and its application significantly upregulated the plant biomass, enhanced water status, increased K^+ retention in plant tissues and the ability of antioxidant enzymes under salt stress. Exogenous Si application inhibits the salinity stress and performed a multifunctional role in the plant. This chapter highlighted that Si's application might trigger salt resistance and upregulation of chlorophyll contents and photosynthetic activity of plants. In addition, exogenous Si application could also mediate plant growth under a saline medium by modulating antioxidant enzymes activities, including SOD, CAT, POD, APX, GR, and the fusion of certain phytohormones that proved beneficial for the upregulation of plant physiology and the final yield. Besides the maintenance of significant improvement in the uptake of mineral elements, Si application retard excess salt accumulation that could trigger the decline of the important mineral elements, enzyme activities, chlorophyll contents, and photosynthesis rate. Further studies are required for future investigation to develop the Si responsive crop cultivars and maintain Si's significance as a critical element of crop growth status.

References

[1] Hamayun M, Sohn EY, Khan SA, Shinwari ZK, Khan AL, Lee IJ. Silicon alleviates the adverse effects of salinity and drought stress on growth and endogenous plant growth hormones of soybean (*Glycine max* L.). Pak J Bot 2010;42(3):1713–22 Jun 1.

[2] Khan WU, Aziz T, Maqsood MA, Farooq M, Abdullah Y, Ramzani PM, et al. Silicon nutrition mitigates salinity stress in maize by modulating ion accumulation, photosynthesis, and antioxidants. Photosynthetica 2018;56(4):1047–57 Dec.

[3] Naveed M, Sajid H, Mustafa A, Niamat B, Ahmad Z, Yaseen M, et al. Alleviation of salinity-induced oxidative stress, improvement in growth, physiology and mineral nutrition of canola (*Brassica napus* L.) through calcium-fortified composted animal manure. Sustainability. 2020;12(3):846. Available from: https://doi.org/10.3390/su12030846.

[4] Liu B, Soundararajan P, Manivannan A. Mechanisms of silicon-mediated amelioration of salt stress in plants. Plants. 2019;8(9):307 Sep.

[5] Riaz M, Kamran M, Rizwan M, Ali S, Parveen A, et al. Cadmium uptake and translocation: synergetic roles of selenium and silicon in Cd detoxification for the production of low Cd crops: A critical review. Chemosphere. 2021;21:129690. Available from: https://doi.org/10.1016/j.chemosphere.2021.129690.

[6] Hasanuzzaman M, Nahar K, Fujita M. Silicon and selenium: two vital trace elements that confer abiotic stress tolerance to plants. Emerging technologies and management of crop stress tolerance. San Diego, CA: Academic Press; 2014. p. 377–422.

[7] Roychoudhury A. Silicon-nanoparticles in crop improvement and agriculture. Int J Recent Adv Biotechnol Nanotechnol 2020;3(1).

[8] Cooke J, Leishman MR. Consistent alleviation of abiotic stress with silicon addition: a meta-analysis. Funct Ecol 2016;30(8):1340–57 Aug.

[9] Emamverdian A, Ding Y, Xie Y, Sangari S. Silicon mechanisms to ameliorate heavy metal stress in plants. BioMed Res Int 2018;2018 Apr 22.

[10] Ali M, Sobia A, Aasma P, Muhammad K, Muhammad RJ, Ghulam HA, et al. Silicon mediated improvement in the growth and ion homeostasis by decreasing Na^+ uptake in maize (*Zea mays* L.) cultivars exposed to salinity stress. Plant Physiology and Biochemistry 2021;158:208–18. Available from: https://doi.org/10.1016/j.plaphy.2020.10.040.

[11] Singh A, Roychoudhury A. Silicon-regulated antioxidant and osmolyte defense and methylglyoxal detoxification functions co-ordinately in attenuating fluoride toxicity and conferring protection to rice seedlings. Plant Physiol Biochem 2020;154:758–69 Sep 1.

[12] Manivannan A, Ahn YK. Silicon regulates potential genes involved in major physiological processes in plants to combat stress. Front Plant Sci 2017;8:1346 Aug 3.

[13] Tripathi DK, Singh S, Singh VP, Prasad SM, Chauhan DK, Dubey NK. Silicon nanoparticles more efficiently alleviate arsenate toxicity than silicon in maize cultiver and hybrid differing in arsenate tolerance. Front Environ Sci 2016;4:46 Jul 7.

[14] Ahmad P, Azooz MM, Prasad MN. Ecophysiology and responses of plants under salt stress. New York: Springer Science & Business Media; 2012. Nov 9.

[15] He Q, Li P, Zhang W, Bi Y. Cytoplasmic glucose-6-phosphate dehydrogenase plays an important role in the silicon-enhanced alkaline tolerance in highland barley. Funct Plant Biol 2020;48(2):119−30 Aug 11.

[16] Tuna AL, Kaya C, Higgs D, Murillo-Amador B, Aydemir S, Girgin AR. Silicon improves salinity tolerance in wheat plants. Environ Exp Bot 2008;62(1):10−16 Jan 1.

[17] Liu P, Yin L, Wang S, Zhang M, Deng X, Zhang S, et al. Enhanced root hydraulic conductance by aquaporin regulation accounts for silicon alleviated salt-induced osmotic stress in *Sorghum bicolor* L. Environ Exp Bot 2015;111:42−51 Mar 1.

[18] Zhu Z, Wei G, Li J, Qian Q, Yu J. Silicon alleviates salt stress and increases antioxidant enzymes activity in leaves of salt-stressed cucumber (*Cucumis sativus* L.). Plant Sci 2004;167(3):527−33 Sep 1.

[19] Gunes A, Inal A, Bagci EG, Pilbeam DJ. Silicon-mediated changes of some physiological and enzymatic parameters symptomatic for oxidative stress in spinach and tomato grown in sodic-B toxic soil. Plant Soil 2007;290(1):103−14 Jan.

[20] Kabir AH, Hossain MM, Khatun MA, Mandal A, Haider SA. Role of silicon counteracting cadmium toxicity in alfalfa (*Medicago sativa* L.). Front Plant Sci 2016;7:1117 Jul 27.

[21] Puppe D, Sommer M. Experiments, uptake mechanisms, and functioning of silicon foliar fertilization—a review focusing on maize, rice, and wheat. Adv Agron 2018;152:1−4 Jan 1.

[22] Parihar P, Singh S, Singh R, Singh VP, Prasad SM. Effect of salinity stress on plants and its tolerance strategies: a review. Environ Sci Pollut Res 2015;22(6):4056−75 Mar.

[23] Safdar H, Amin A, Shafiq Y, Ali A, Yasin R, Shoukat A, et al. A review: Impact of salinity on plant growth. Nat Sci 2019;17(1):34−40.

[24] Ali M. Melatonin-Induced Salinity Tolerance by Ameliorating Osmotic and Oxidative Stress in the Seedlings of Two Tomato (*Solanum lycopersicum* L.) Cultivars. Journal of Plant Growth Regulation 2021;. Available from: https://doi.org/10.1007/s00344-020-10273-3.

[25] Najar R, Aydi S, Sassi-Aydi S, Zarai A, Abdelly C. Effect of salt stress on photosynthesis and chlorophyll fluorescence in *Medicago truncatula*. Plant Biosyst 2019;153(1):88−97 Jan 2.

[26] Acosta-Motos JR, Ortuño MF, Bernal-Vicente A, Diaz-Vivancos P, Sanchez-Blanco MJ, Hernandez JA. Plant responses to salt stress: adaptive mechanisms. Agronomy. 2017;7(1):18 Mar.

[27] Zulfiqar H, Shahbaz M, Ahsan M, Nafees M, Nadeem H, Akram M, et al. Strigolactone (GR24) induced salinity tolerance in sunflower (*Helianthus annuus* L.) by ameliorating morpho-physiological and biochemical attributes under in vitro conditions. Journal of Plant Growth Regulation. 2021;40(5):2079−91. Available from: https://doi.org/10.1007/s00344-020-10256-4.

[28] Khan MA, Khān MA, Weber DJ, editors. Ecophysiology of high salinity tolerant plants. Springer Science & Business Media; 2006.

[29] Gomes-Filho E, Lima CR, Costa JH, da Silva AC, Lima MD, de Lacerda CF, et al. Cowpea ribonuclease: properties and effect of NaCl-salinity on its activation during seed germination and seedling establishment. Plant Cell Rep 2008;27(1):147−57 Jan.

[30] Dantas BF, Ribeiro LD, Aragão CA. Germination, initial growth and cotyledon protein content of bean cultivars under salinity stress. Rev Bras Sementes. 2007;29(2):106−10 Aug.

[31] Khan MA, Rizvi Y. Effect of salinity, temperature, and growth regulators on the germination and early seedling growth of *Atriplex griffithii* var. stocksii. Can J Bot 1994;72(4):475−9 Apr 1.

[32] Othman Y, Al-Karaki G, Al-Tawaha AR, Al-Horani A. Variation in germination and ion uptake in barley genotypes under salinity conditions. World J Agric Sci 2006;2(1):11−15.

[33] Thorne SJ, Hartley SE, Maathuis FJ. Is silicon a panacea for alleviating drought and salt stress in crops? Front Plant Sci 2020;11:1221 Aug 18.

[34] Kamran M, Parveen A, Ahmar S, Malik Z, Hussain S, Chattha MS, et al. An overview of hazardous impacts of soil salinity in crops, tolerance mechanisms, and amelioration through selenium supplementation. International journal of molecular sciences. 2020;21(1):148. Available from: https://doi.org/10.3390/ijms21010148.

[35] Abdullah ZK, Khan MA, Flowers TJ. Causes of sterility in seed set of rice under salinity stress. J Agron Crop Sci 2001;187(1):25−32 Jul 13.

[36] Rao PS, Mishra B, Gupta SR, et al. Reproductive stage tolerance to salinity and alkalinity stresses in rice genotypes. Plant Breed 2008;127:256−61.

[37] Chaves MM, Flexas J, Pinheiro C. Photosynthesis under drought and salt stress: regulation mechanisms from whole plant to cell. Ann bot 2009;103:551−60.

[38] Moradi F, Ismail AM. Responses of photosynthesis, chlorophyll fluorescence and ROS-scavenging systems to salt stress during seedling and reproductive stages in rice. Ann Bot 2007;99:1161−73.

[39] Fageria NK, Stone LF, dos Santos AB. Breeding for salinity tolerance. Plant breeding for abiotic stress tolerance. Heidelberg: Springer-Verlag Berlin; 2012. p. 103−22.

[40] Pessarakli M. Handbook of plant and crop stress. New York: Dekker; 1994.

[41] Ahanger MA, Agarwal R. Salinity stress induced alterations in antioxidant metabolism and nitrogen assimilation in wheat (*Triticum aestivum* L) as influenced by potassium supplementation. Plant Physiol Biochem 2017;115:449−60.

[42] Foyer CH. Reactive oxygen species, oxidative signaling and the regulation of photosynthesis. Environ Exp Bot 2018;154:134−42.

[43] Winterbourn CC. Reactive oxygen species in biological systems. Vitamin E: chemistry and nutritional benefits. RSC Publishing; 2019. p. 98−117.

[44] Saini P, Gani M, Kaur JJ, et al. Reactive oxygen species (ROS): a way to stress survival in plants. Abiotic stress-mediated sensing and signaling in plants: an omics perspective. Singapore: Springer; 2018. p. 127−53.

[45] Suo J, Zhao Q, David L, et al. Salinity response in chloroplasts: insights from gene characterization. Int J Mol Sci 2017;18:1011.

[46] Numan M, Bashir S, Khan Y, et al. Plant growth promoting bacteria as an alternative strategy for salt tolerance in plants: a review. Microbiol Res 2018;209:21−32.

[47] Caverzan A, Casassola A, Brammer SP. Reactive oxygen species and antioxidant enzymes involved in plant tolerance to stress. Embrapa TrigoCapítulo Em Livro Científico (ALICE) 2016;463−80.

[48] Ahanger MA, Alyemeni MN, Wijaya L, et al. Potential of exogenously sourced kinetin in protecting Solanum lycopersicum from NaCl-induced oxidative stress through up-regulation of the antioxidant system, ascorbate-glutathione cycle and glyoxalase system. PLoS One 2018;13:e0202175.

[49] Ur Rahman S, Xuebin Q, Kamran M, Yasin G, Cheng H, et al. Silicon elevated cadmium tolerance in wheat (*Triticum aestivum* L.) by endorsing nutrients uptake and antioxidative defense mechanisms in the leaves. Plant Physiology and Biochemistry. 2021; 31. Available from: https://doi.org/10.1016/j.plaphy.2021.05.038.

[50] Munns R. Comparative physiology of salt and water stress. Plant Cell Environ 2002;25:239−50.

[51] Ahanger MA, Tomar NS, Tittal M, Argal S, Agarwal R. Plant growth under water/salt stress: ROS production; antioxidants and significance of added potassium under such conditions. Physiol Mol Biol Plants 2017;2017(23):731−44.

[52] Fahad S, Hussain S, Matloob A, et al. Phytohormones and plant responses to salinity stress: a review. Plant Growth Regul 2015;75:391−404.

[53] Madhaiyan M, Poonguzhali S, Sa T. Characterization of 1-aminocyclopropane-1-carboxylate (ACC) deaminase containing Methylobacterium oryzae and interactions with auxins and ACC regulation of ethylene in canola (*Brassica campestris*). Planta 2007;226:867−76.

[54] Kumar SB. Salinity stress, its physiological response and mitigating effects of microbial bio inoculants and organic compounds. J Pharmacogn Phytochem 2020;9:1397 -1303.

[55] Isayenkov SV, Maathuis FJ. Plant salinity stress: many unanswered questions remain. Front Plant Sci 2019;10:80.

[56] Liu J, Zhu JK. An Arabidopsis mutant that requires increased calcium for potassium nutrition and salt tolerance. Proc Natl Acad Sci U S A 1997;94(26):14960−4.

[57] Chinnusamy V, Schumaker K, Zhu JK. Molecular genetic perspectives on cross-talk and specificity in abiotic stress signalling in plants. J Exp Bot 2004;55:225−36.

[58] Kim YH, Khan AL, Waqas M, Jeong HJ, Kim DH, Shin JS, et al. Regulation of jasmonic acid biosynthesis by silicon application during physical injury to *Oryza sativa* L. J Plant Res 2014;127(4):525−32 Jul.

[59] Khan A, Khan AL, Muneer S, Kim YH, Al-Rawahi A, Al-Harrasi A. Silicon and salinity: crosstalk in crop-mediated stress tolerance mechanisms. Front Plant Sc. 2019;10:1429 Nov 7.

[60] Sharma P, Jha AÁ, Dubey RS, PessarakliM RO. Oxidative damage and antioxidative defense mechanism in plants under stressful conditions. J Bot 2012;2012:1−26.

[61] Das B, Dadhich P, Pal P, Srivas PK, Bankoti K, Dhara S. Carbon nanodots from date molasses: new nanolights for the in vitro scavenging of reactive oxygen species. J Mater Chem B. 2014;2(39):6839−47.

[62] Lobo V, Patil A, Phatak A, Chandra N. Free radicals, antioxidants and functional foods: impact on human health. Pharmacogn Rev 2010;4(8):118 Jul.

[63] Tripathi DK, Singh S, Singh VP, Prasad SM, Dubey NK, Chauhan DK. Silicon nanoparticles more effectively alleviated UV-B stress than silicon in wheat (*Triticum aestivum*) seedlings. Plant Physiol Biochem 2017;110:70−81 Jan 1.

[64] Altuntas O, Dasgan HY, Akhoundnejad Y. Silicon-induced salinity tolerance improves photosynthesis, leaf water status, membrane stability, and growth in pepper (*Capsicum annuum* L.). HortScience 2018;53(12):1820−6 Dec 1.

[65] Liang X, Wang H, Hu Y, Mao L, Sun L, Dong T, et al. Silicon does not mitigate cell death in cultured tobacco BY-2 cells subjected to salinity without ethylene emission. Plant Cell Rep 2015;34(2):331−43 Feb.

[66] Baxter A, Mittler R, Suzuki N. ROS as key players in plant stress signalling. J Exp Bot 2014;65(5):1229−40 Mar 1.

[67] Torabi F, Majd A, Enteshari S. The effect of silicon on alleviation of salt stress in borage (*Borago officinalis* L.). Soil Sci Plant Nut. 2015; 61(5):788−98 Sep 3.

[68] Liang Y, Sun W, Zhu YG, Christie P. Mechanisms of silicon-mediated alleviation of abiotic stresses in higher plants: a review. Environ Pollut 2007;147(2):422−8 May 1.

[69] Ma JF, Yamaji N. Functions and transport of silicon in plants. Cell Mol Life Sci 2008;65(19):3049−57 Oct.

[70] Ma JF, Miyake Y, Takahashi E. Silicon as a beneficial element for crop plants. Stud Plant Sci 2001;8:17−39 Jan 1.

[71] Hassanvand F, Nejad AR, Fanourakis D. Morphological and physiological components mediating the silicon-induced enhancement of geranium essential oil yield under saline conditions. Ind Crops Prod 2019;134:19−25 Aug 1.

[72] Alzahrani Y, Kuşvuran A, Alharby HF, Kuşvuran S, Rady MM. The defensive role of silicon in wheat against stress conditions induced by drought, salinity or cadmium. Ecotoxicol Environ Saf 2018;154:187−96 Jun 15.

[73] Al-Huqail AA, Alqarawi AA, Hashem A, Malik JA, Abd_Allah EF. Silicon supplementation modulates antioxidant system and osmolyte accumulation to balance salt stress in *Acacia gerrardii* Benth. Saudi J Biol Sci 2019;26(7):1856−64 Nov 1.

[74] Fariduddin Q, Yusuf M, Chalkoo S, Hayat S, Ahmad A. 28-homobrassinolide improves growth and photosynthesis in *Cucumis sativus* L. through an enhanced antioxidant system in the presence of chilling stress. Photosynthetica 2011;49(1):55−64.

[75] Haghighi M, Pessarakli M. Influence of silicon and nano-silicon on salinity tolerance of cherry tomatoes (*Solanum lycopersicum* L.) at early growth stage. Sci Hortic 2013;161:111−17 Sep 24.

[76] Saqib M, Zörb C, Schubert S. Silicon-mediated improvement in the salt resistance of wheat (*Triticum aestivum*) results from increased sodium exclusion and resistance to oxidative stress. Funct Plant Biol 2008;35(7):633−9 Aug 21.

[77] Cui J, Zhang E, Zhang X, Wang Q. Silicon alleviates salinity stress in licorice (*Glycyrrhiza uralensis*) by regulating carbon and nitrogen metabolism. Sci Rep 2021;11(1):1−2 Jan 13.

[78] Kim Y, Mun BG, Khan AL, Waqas M, Kim HH, Shahzad R, et al. Regulation of reactive oxygen and nitrogen species by salicylic acid in rice plants under saline stress conditions. PLoS One 2018;13(3):e0192650 Mar 20.

[79] Soleimannejad Z, Abdolzadeh A, Sadeghipour HR. Beneficial effects of silicon application in alleviating salinity stress in halophytic puccinellia distans plants. Silicon. 2019;11(2):1001−10 Apr.

[80] Conceição SS, Oliveira Neto CF, Marques EC, Barbosa AV, Galvão JR, Oliveira TB, et al. Silicon modulates the activity of antioxidant enzymes and nitrogen compounds in sunflower plants under salt stress. Arch Agron Soil Sci 2019;65(9):1237−47 Jul 29.

[81] Yin J, Jia J, Lian Z, Hu Y, Guo J, Huo H, et al. Silicon enhances the salt tolerance of cucumber through increasing polyamine accumulation and decreasing oxidative damage. Ecotoxicol Environ Saf 2019;169:8−17 Mar 1.

[82] Khan TA, Fariduddin Q, Yusuf M. Lycopersicon esculentum under low temperature stress: an approach toward enhanced antioxidants and yield. Environ Sci Pollut Res 2015;22(18):14178−88.

[83] Soundararajan P, Manivannan A, Ko CH, Jeong BR. Silicon enhanced redox homeostasis and protein expression to mitigate the salinity stress in *Rosa hybrida* 'Rock Fire'. J Plant Growth Regul 2018;37(1):16−34 Mar.

[84]   Saleh J, Najafi N, Oustan S. Effects of silicon application on wheat growth and some physiological characteristics under different levels and sources of salinity. Commun Soil Sci Plant Anal 2017;48(10):1114–22 May 31.

[85]   Abbas T, Sattar A, Ijaz M, Aatif M, Khalid S, Sher A. Exogenous silicon application alleviates salt stress in okra. Hortic Environ Biotechnol 2017;58(4):342–9 Aug.

[86]   Rezende RA, Rodrigues FA, Soares JD, Silveira HR, Pasqual M, Dias GD. Salt stress and exogenous silicon influence physiological and anatomical features of in vitro-grown cape gooseberry. Ciência Rural. 2017;48 Apr 18.

[87]   Ibrahim MA, Merwad AM, Elnaka EA, Burras CL, Follett L. Application of silicon ameliorated salinity stress and improved wheat yield. J Soil Sci Environ Manage 2016;7(7):81–91.

[88]   Azeem M, Iqbal N, Kausar S, Javed MT, Akram MS, Sajid MA. Efficacy of silicon priming and fertigation to modulate seedling's vigor and ion homeostasis of wheat (*Triticum aestivum* L.) under saline environment. Environ Sci Pollut Res 2015;22(18):14367–71.

[89]   Parveen NU, Ashraf MU. Role of silicon in mitigating the adverse effects of salt stress on growth and photosynthetic attributes of two maize (*Zea mays* L.) cultivars grown hydroponically. Pak J Bot 2010;42(3):1675–84 Jun 1.

[90]   Xu HL, Gauthier L, Gosselin A. Photosynthetic responses of greenhouse tomato plants to high solution electrical conductivity and low soil water content. J Hortic Sci 1994;69(5):821–32 Jan 1.

[91]   Abbas T, Balal RM, Shahid MA, Pervez MA, Ayyub CM, Aqueel MA, et al. Silicon-induced alleviation of NaCl toxicity in okra (*Abelmoschus esculentus*) is associated with enhanced photosynthesis, osmoprotectants and antioxidant metabolism. Acta Physiologiae Plantarum 2015;37(2):6 Feb 1.

[92]   Ma JF, Yamaji N. Silicon uptake and accumulation in higher plants. Trends Plant Sci 2006;11(8):392–7 Aug 1.

[93]   Al-aghabary K, Zhu Z, Shi Q. Influence of silicon supply on chlorophyll content, chlorophyll fluorescence, and antioxidative enzyme activities in tomato plants under salt stress. J Plant Nutr 2005;27(12):2101–15 Jan 2.

[94]   Nahar K, Hasanuzzaman M, Alam M, Fujita M. Glutathione-induced drought stress tolerance in mung bean: coordinated roles of the antioxidant defence and methylglyoxal detoxification systems. AoB Plants 2015;7 Jan 1.

[95]   Nahar K, Hasanuzzaman M, Alam MM, Rahman A, Mahmud JA, Suzuki T, et al. Insights into spermine-induced combined high temperature and drought tolerance in mung bean: osmoregulation and roles of antioxidant and glyoxalase system. Protoplasma 2017;254 (1):445–60 Jan 1.

[96]   Riaz M, Kamran M, Rizwan M, Ali S, Wang X. Foliar application of silica sol alleviates boron toxicity in rice (*Oryza sativa*) seedlings. Journal of Hazardous Materials. 2022;423:127175. Available from: https://doi.org/10.1016/j.jhazmat.2021.127175.

[97]   Zhu Y, Guo J, Feng R, Jia J, Han W, Gong H. The regulatory role of silicon on carbohydrate metabolism in *Cucumis sativus* L. under salt stress. Plant Soil 2016;406(1):231–49 Sep.

[98]   Manivannan A, Soundararajan P, Muneer S, Ko CH, Jeong BR. Silicon mitigates salinity stress by regulating the physiology, antioxidant enzyme activities, and protein expression in *Capsicum annuum* 'Bugwang'. BioMed Res Int 2016;2016 Oct.

[99]   Muneer S, Park YG, Manivannan A, Soundararajan P, Jeong BR. Physiological and proteomic analysis in chloroplasts of *Solanum lycopersicum* L. under silicon efficiency and salinity stress. Int J Mol Sci 2014;15(12):21803–24 Dec.

[100] Sadder MT, Alsadon A, Wahb-Allah M. Transcriptomic analysis of tomato lines reveals putative stress-specific biomarkers. Turk J Agric For 2014;38(5):700–15 Sep 3.

[101] Mateos-Naranjo E, Gallé A, Florez-Sarasa I, Perdomo JA, Galmés J, Ribas-Carbó M, et al. Assessment of the role of silicon in the Cu-tolerance of the C4 grass *Spartina densiflora*. J Plant Physiol 2015;178:74–83 Apr 15.

[102] Kim YH, Khan AL, Lee IJ. Silicon: a duo synergy for regulating crop growth and hormonal signaling under abiotic stress conditions. Crit Rev Biotechnol 2016;36(6):1099–109 Nov 1.

[103] Moradtalab N, Weinmann M, Walker F, Höglinger B, Ludewig U, Neumann G. Silicon improves chilling tolerance during early growth of maize by effects on micronutrient homeostasis and hormonal balances. Front Plant Sci 2018;9:420 Apr 26.

[104] Souri Z, Khanna K, Karimi N, Ahmad P. Silicon and plants: current knowledge and future prospects. J Plant Growth Regul 2021;40 (3):906–25.

[105] Hosseini SA, Naseri Rad S, Ali N, Yvin JC. The ameliorative effect of silicon on maize plants grown in Mg-deficient conditions. Int J Mol Sci 2019;20(4):969.

[106] Tripathi DK, Vishwakarma K, Singh VP, Prakash V, Sharma S, Muneer S, et al. Silicon crosstalk with reactive oxygen species, phytohormones and other signaling molecules. J Hazard Mater 2020;124820 Dec 13.

[107] Fahad S, Hussain S, Bano A, Saud S, Hassan S, Shan D, et al. Potential role of phytohormones and plant growth-promoting rhizobacteria in abiotic stresses: consequences for changing environment. Environ Sci Pollut Res 2015;22(7):4907–21.

[108] Peleg Z, Blumwald E. Hormone balance and abiotic stress tolerance in crop plants. Curr Opin Plant Biol 2011;14(3):290–5 Jun 1.

[109] Iqbal N, Umar S, Khan NA, Khan MI. A new perspective of phytohormones in salinity tolerance: regulation of proline metabolism. Environ Exp Bot 2014;100:34–42 Apr 1.

[110] Etesami H, Jeong BR. Silicon (Si): review and future prospects on the action mechanisms in alleviating biotic and abiotic stresses in plants. Ecotoxicol Environ Saf 2018;147:881–96 Jan 1.

[111] Kim YH, Khan AL, Shinwari ZK, Kim DH, Waqas MU, Kamran MU, et al. Silicon treatment to rice (*Oryza sativa* L. cv.'Gopumbyeo') plants during different growth periods and its effects on growth and grain yield. Pak J Bot 2012;44(3):891–7 Jun 1.

[112] Merhij IE, Al-Timmen WM, Jasim AH. The effect of silicon, tillage and the interaction between them on some antioxidants and phytohormones during drought stress of maize (*Zea mays* L.) plants. Plant Arch. 2019;19:67–74.

[113] Parveen A, Ahmar S, Kamran M, Malik Z, Ali A, Riaz M, et al. Abscisic acid signaling reduced transpiration flow, regulated Na$^+$ ion homeostasis and antioxidant enzyme activities to induce salinity tolerance in wheat (*Triticum aestivum* L.) seedlings. Environmental Technology & Innovation. 2021;24:101808. Available from: https://doi.org/10.1016/j.eti.2021.101808.

[114] Mir BA, Khan TA, Fariduddin Q. 24-epibrassinolide and spermidine modulate photosynthesis and antioxidant systems in *Vigna radiata* under salt and zinc stress. Int J Adv Res 2015;3(5):592–608.

[115] Yusuf M, Fariduddin Q, Hayat S, Hasan SA, Ahmad A. Protective response of 28-homobrassinolide in cultivars of *Triticum aestivum* with different levels of nickel. Arch Environ Contam Toxicol 2011;60(1):68–76.

[116] Hussain M, Khan TA, Yusuf M, Fariduddin Q. Silicon-mediated role of 24-epibrassinolide in wheat under high-temperature stress. Environ Sci Pollut Res 2019;26(17):17163−72.

[117] Lee CW, Mahendra S, Zodrow K, Li D, Tsai YC, Braam J, et al. Developmental phytotoxicity of metal oxide nanoparticles to *Arabidopsis thaliana*. Environ Toxicol Chem 2010;29(3):669−75.

[118] Dhiman P, Rajora N, Bhardwaj S, Sudhakaran SS, Kumar A, Raturi G, et al. Fascinating role of silicon to combat salinity stress in plants: an updated overview. Plant Physiol Biochem 2021;162:110−23.

[119] Lang DY, Fei PX, Cao GY, Jia XX, Li YT, Zhang XH. Silicon promotes seedling growth and alters endogenous IAA, GA3 and ABA concentrations in *Glycyrrhiza uralensis* under 100 mM NaCl stress. J Hortic Sci Biotechnol 2019;94(1):87−93 Jan 2.

[120] Abdel-Haliem ME, Hegazy HS, Hassan NS, Naguib DM. Effect of silica ions and nano silica on rice plants under salinity stress. Ecol Eng 2017;99:282−9 Feb 1.

[121] Devinar G, Llanes A, Masciarelli O, Luna V. Different relative humidity conditions combined with chloride and sulfate salinity treatments modify abscisic acid and salicylic acid levels in the halophyte *Prosopis strombulifera*. Plant Growth Regul 2013;70(3):247−56.

[122] Mauad M, Crusciol CA, Nascente AS, Grassi H, Lima GP. Effects of silicon and drought stress on biochemical characteristics of leaves of upland rice cultivars. Revista Ciência Agronômica 2016;47:532−9.

[123] Prochazkova D, Wilhelmova N. Leaf senescence and activities of the antioxidant enzymes. Biol Plant 2007;51(3):401−6 Sep 1.

[124] Munns R, Tester M. Mechanisms of salinity tolerance. Annu Rev Plant Biol 2008;59:651−81 Jun 2.

[125] Kamran M, Danish M, Saleem MH, Malik Z, Parveen A, Abbasi GH, et al. Application of abscisic acid and 6-benzylaminopurine modulated morpho-physiological and antioxidative defense responses of tomato (*Solanum lycopersicum* L.) by minimizing cobalt uptake. Chemosphere. 2021;263:128169. Available from: https://doi.org/10.1016/j.chemosphere.2020.128169.

[126] Zhang Y, Liang Y, Zhao X, Jin X, Hou L, Shi Y, et al. Silicon compensates phosphorus deficit-induced growth inhibition by improving photosynthetic capacity, antioxidant potential, and nutrient homeostasis in tomato. Agronomy. 2019;9(11):733.

[127] Frazão JJ, de Mello Prado R, de Souza Júnior JP, Rossatto DR. Silicon changes C: N: P stoichiometry of sugarcane and its consequences for photosynthesis, biomass partitioning and plant growth. Sci Rep 2020;10(1):1 Jul 27.

[128] Fatemi H, Pour BE, Rizwan M. Isolation and characterization of lead (Pb) resistant microbes and their combined use with silicon nanoparticles improved the growth, photosynthesis and antioxidant capacity of coriander (*Coriandrum sativum* L.) under Pb stress. Environ Pollut 2020;266:114982 Nov 1.

[129] Khorasaninejad S, Hemmati K. Effects of silicon on some phytochemical traits of purple coneflower (*Echinacea purpurea* L.) under salinity. Sci Hortic (Amsterdam) 2020;264:108954 Apr 5.

9

Nanosilicon-mediated salt stress tolerance in plants

Muhammad Jafir[1], Muhammad Ashar Ayub[2] and Muhammad Zia ur Rehman[3]

[1]Department of Entomology, University of Agriculture, Faisalabad, Pakistan [2]Institute of Soil and Environmental Sciences, University of Agriculture, Faisalabad, Pakistan [3]Institute of Soil and Environmental Sciences, University of Agriculture, Faisalabad, Pakistan

9.1 Introduction

Salinity is the major abiotic stress imposing factor responsible for limiting crop production and has become an alarming issue worldwide [1,2]. Currently, over 20% of the global agriculture is affected by soil salinity and is affected by salinity. This problem is supposed to exacerbate further globally because of the inappropriate nutritional inputs and irrigation practices [3,4]. Plant's defense response toward salinity stress can be enhanced via application of beneficial elements such as silicon (Si) helping in cellular boast and physiological strengthening. Si can be considered a broad-spectrum beneficial element that can ameliorate the adverse effect of stress in plants as well as boast their nutritional homeostasis [5–7]. After oxygen, Si is known as second plentiful element on the earth. Despite nonessential elements for the plant, it has been evaluated as a plant growth-promoting element, especially under various abiotic and biotic stresses [5,8–10].

Recently, nanotechnology has revolutionized agriculture by achieving ecofriendly and cost-effective goals in crop production and crop protection [11–13]. Silicon nanoparticles (SiNPs) have been introduced to improve the quantity as well as quality traits of the plants and crop production performance [14]. Reports indicate that an appropriate quantity of nanosilica significantly enhances seed germination [15], development of plant and production [16] and helped in mitigation of abiotic stresses [17]. Currently, SiNPs have attracted the attention of scientists because of their ability to neutralize the adverse effects of salinity on plant growth. Various publications reported that SiNPs improve the seed germination as well as the morphology of plants under saline soil [18,19].

Under salinity stress, plants tend to cope with adverse effects via modulation in gene expression and adoption of salinity escape and tolerance mechanisms ([20–23]). Sever salinity effects plants Na:K balance due to higher Na + contents in soil solution making K homeostasis an issue [24,25] and boasting plant defense can help in withstanding this imbalance. First plant cellular response toward abiotic stress in induction of oxidative stress which can be devastating for plant cellular health and SiNPs can be helpful in this regard as they can ameliorate oxidative stress as well as can help in osmotic adjustments in plants [26,27]. This chapter is an effort to review the role of NPs especially of Si in the mitigation of salinity stress in plants and exploration of their potential in enhancement of crop growth and production.

9.2 Effect of salt stress on plants

Soil salinity has become an alarming issue worldwide and has severe morphological, physiological, and biochemical effects on plants (Fig. 9.1).

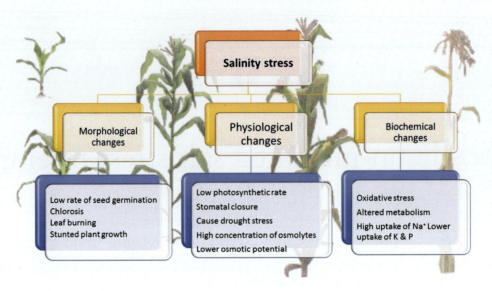

FIGURE 9.1   Physiological and biochemical response of plants under salinity.

Salinity/salt stress affects the growth and development of plants which is the main constraint in sustainable saline agriculture. Salinity causes certain changes in the plant's physiology and metabolism and its type, exposure concentration and time can have different effects on crop health [28]. The first response of plants under salinity stress is lowering of water potential hampering plant's capability to uptake water triggering major homeostasis crisis in the plant tissues. At cellular level, if observed, we can monitor a major effect on both the root and shoot cells [29–31]. To counter this effect, plants tend to accumulate salts in their roots thus managing cellular water potential difference [32,33]. Salinity induced osmotic stress can also hinder plant's ability of carbon assimilation which can be due to rapid drop in xylem pressure [34] which can be experienced even after very short period of salt exposure, effecting shoot metabolism [35]. Another effect is stomatal conductance deregulation via disturbance in osmotic endurance of the guard cell ion channels [36,37]. Salinization impedes transpiration by reducing the stomatal conductance which causes the poor gaseous exchange, photosynthetic rate chlorophyll contents [38–40].

The second effect of salinity on plant physiology is an ionic imbalance (accompanied by ion toxicity of specific ion stress) and Na and Cl toxicity are prominent in this regard. As salinity is hypertonic stress, maintanace of water balance is an issue for the plant as well as Na and Cl specific ion toxicity, which can be lethal for plant growth [41]. The Na^+ can be very toxic for plants as not only it disrupts ionic homeostasis but is also involved in enzymatic deregulation [42] as Na interferes with K homeostasis, which is an integral part of many cellular enzymes [43]. Besides this, Na toxicity can interfere with biomolecules metabolism, cytoskeleton functioning, and protein biosynthesis in cytosol which is toxic for plants [44,45]. Ion, chloride (Cl) toxicity is also evident under severe saline conditions [46]. The ionic homeostasis can be vital response of plant to escape salinity stress as explained in Fig. 9.2.

Accumulation of salts in the body of plants can cause biochemical disturbance via specific ion toxicity, membrane instability, free radical deposition, and other metabolic disorders [47,48]. Biochemical parameters of plants are disturbed by the less adsorption of water because of high accumulation of toxic ions such as sodium and chloride ions. This ionic toxicity shrinks the cell volume, which enhances the oxidative stress that reduces the plant viability and efficiency [49]. Upon interaction with plant roots, excessive salinity initiates the plant's cellular defense/signaling response in terms of overproduction of reactive oxygen species (ROS) (resulting in higher cellular O_2^-, OH^-, and H_2O_2 contents) and ionic imbalance in the cell [28,41,50]. Excessive salts influence the activity of photosystem II (PSII) more than PSI by disintegrating the PSII reaction center and oxygen evolution complex (OEC), decreasing the quinine receptors activity and disrupting the electron transport chain [38,51,52]. Wang et al. [53] reported that salinization reduces the total chlorophyll contents and disintegrates the chloroplast by minimizing the grana that results in decreased photosynthetic rate. Soori et al. reported the decreased mesophyll conductance by reducing the carboxylation in different pomegranate genotypes under salinity. Moreover, it reduces the use of NADPH in C3 cycle, reduction in rubisco and chlorophyll contents utilization [54,55].

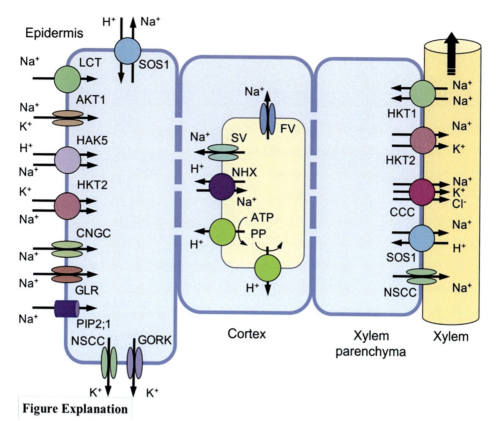

Figure Explanation

The major pathways for Na⁺ uptake in the root epidermis are glutamate receptor-like (GLRs) channels or cyclic nucleotide-gated (CNGCs) non-selective cation channels and HKT2 high-affinity K⁺ transporters. Other possible pathways for Na⁺ uptake may involve AKT1 Shaker-type K⁺ channels, HAK5 high-affinity K⁺ transporters, the low-affinity cation transporter LCT1, and PIP2;1 aquaporins. The uptake of Na⁺ is counterbalanced by active Na⁺ extrusion via SOS1 Na⁺/H⁺ exchangers. Vacuolar Na⁺ sequestration is conferred by tonoplast-based Na⁺/H⁺ exchangers from the NHX family fueled by either H⁺-ATPase or H⁺-PPase pumps. Another component of vacuolar Na⁺ sequestration is efficient control over tonoplast slow- (SV) and fast- (FV) activating ion channels that may allow Na⁺ to leak back to the cytosol. Passive Na⁺ loading into the xylem is mediated by non-selective cation channels (NSCCs), and its active loading requires operation of cotransporters such as SOS1, CCC (cation-chloride cotransporters), and HKT2 (K⁺/Na⁺ symporter). Na⁺ withdrawal from the xylem is achieved by HKT1 high-affinity K⁺ transporters. Salinity-induced K⁺ loss from the root epidermis is mediated by NSCCs and depolarization-activated outward-rectifying GORK K⁺ channels.

FIGURE 9.2 Na homeostasis response in plants. *Source: Reprinted from Zhao C, et al. Mechanisms of plant responses and adaptation to soil salinity. The Innovation 2020;1(1):100017 with permission and license to reuse.*

To overcome the biochemical disturbances produced by salinity stress, plants also stimulate the defense mechanism biochemically. Plants also produce compounds that protect the oxygen-evolving complex of PII stability, and maintenance of protein and membrane integrity during salt stress [56]. Furthermore, mannitol, sorbitol, glycerol, and GB are classified as compatible solutes to scavenge the ROS [57]. Mannitol in higher plants is the major compatible solutes that are involved in tolerance to osmotic stress under salinity [58]. Mannitol is synthesized by sucrose and raffinose, and its higher concentration protects the plant against stress by scavenging ROS [59,60]. Under abiotic stress, mannitol, inositol accumulates polyols in genetically modified plants that are resistant to stresses are used as biochemical markers [61,62].

Under salinity stress, these compatible solutes maintain the structure of macromolecules in the cytosol, which alter because of these charged ions accumulation [63]. Compatible solutes are named so due to their consistency with the metabolism of plant cells, and which causes the change in cell water content by lowering the water potential [64]. These compatible solutes are capable of replacing water on proteins structure due to their hydrophilic nature without causing any change in their structure and function [65,66]. These compounds maintain the enzyme activity by protecting them from toxic damage of high ionic concentration activities [67,68]. These compounds are involved in osmoregulation under salt stress.

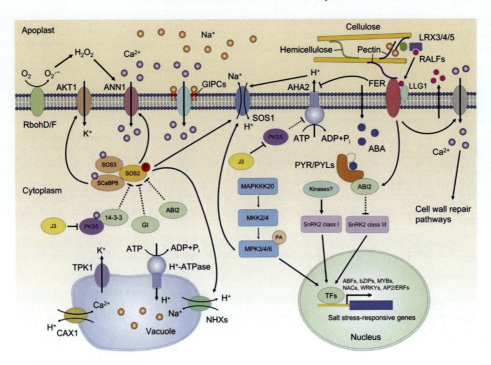

Figure Explanation

The SOS signaling pathway, consisting of SOS3/SOS3-like calcium-binding protein 8 (SCaBP8), SOS2, and SOS1, is important for sensing salt-induced Ca²⁺ signals and in the regulation of ion homeostasis by extruding excessive Na⁺ out of cells. 14-3-3, GI, and ABI2 negatively regulate the kinase activity of SOS2. Ca²⁺-mediated binding of PKS5 with 14-3-3 releases the inhibition on SOS2. GIPCs act as putative salt stress sensors that directly bind to Na⁺ and trigger Ca²⁺ influx via an unknown Ca²⁺ channel. GIPCs-mediated Ca²⁺ influx is required for the activation of the SOS signaling pathway. RbohD/F are involved in the production of ROS at the plasma membrane, and ROS can activate the ANN1-mediated Ca²⁺ signaling pathway. AKT1, which is regulated by SCaBP8, mediates the influx of K⁺ to the cytosol under salt stress. MAP kinase cascades, including MAPKKK20, MKK2, MKK4, MPK3, MPK4, and MPK6, are involved in the relay of salt stress signals. Salt stress-induced accumulation of ABA activates subclass III SNF1-related protein kinase 2s (SnRK2s) via the PYR/PYLs-PP2Cs-mediated regulatory module. Subclass I SnRK2s are activated via an ABA-independent pathway under osmotic stress. Activated MPKs and SnRK2s transduce signals to downstream transcription factors, including ABFs, zips, MYBs, NACs, WRKYs, and AP2/ERFs, in the nucleus to induce the expression of stress-responsive genes. In the apoplast, cell wall-localized leucine-rich repeat extensins LRX3, LRX4, and LRX5, together with secreted peptides RALF22/23 and receptor-like kinase FER, function as a module to sense salt stress-induced cell wall changes. FER, RALFs, and LLG1 form a complex at the plasma membrane to trigger Ca²⁺ signaling and consequently activate the cell wall repair pathway. FER also inhibits the activity of AHA2 to regulate apoplastic pH. In the vacuole, NHXs, CAX1, TPK1, and H⁺-ATPase are involved in the regulation of ion homeostasis under high salinity. The dashed lines indicate that the negative regulatory roles are released under salt stress.

FIGURE 9.3 Cellular signaling in plant cell under salinity stress. *Source: Reprinted from Zhao C, et al. Mechanisms of plant responses and adaptation to soil salinity. The Innovation 2020;1(1):100017, with permission and license to reuse.*

By affecting the physiological and biochemical morphological parameters of plants under salt stress, plants experience a variety of morphological changes. The toxicity of salts caused may lead to depressions in plants. Plants grown in salt-affected soils showed stunted growth, leaf tip burn, even at high levels of salinity, leaves of plant with drying damages accumulation of the salt leading to the toxic effects, that is, accelerated leaf senescence and/or necrosis [69–73]. Upon salinity interaction, the plant's salt stress signaling plays a vital role and can help plant escape salinity stress efficiently (Fig. 9.3) [152].

9.3 Silicon: a beneficial nutrient in saline agriculture

A metalloid is an element that has transitional physical as well as chemical properties between metals and nonmetals. Among various metalloids, Si is the most common on the earth's crust and is the second most abundant element after oxygen. Natural Si exists as silicates either in the form of crystalline or amorphous [74,75]. In soil, Si is present as silicic acid ($Si(OH)_2$) with a concentration in the range of 0.1−0.6 mM [76]. It is a vital element for plant nutrition as well in the nutrient cycle, especially for the grassland ecosystems [77,78]. Si fertilizer in a natural

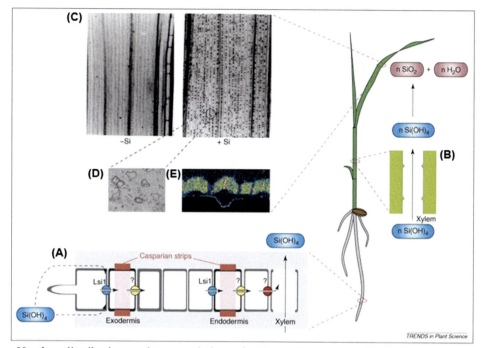

Uptake, distribution and accumulation of silicon (Si) in rice. Si is taken up via transporters in the form of silicic acid (a) and then translocated to the shoot in the same form (b). In the shoot, Si is polymerized into silica and deposited in the bulliform cells (silica body) (c,d) and under the cuticle (e). (c) Silicon detected by soft X-ray and (e) by SEM.

FIGURE 9.4 Uptake and accumulation mechanism of Silicon in rice plant. *Source: Reprinted with permission from Ma JF, Yamaji N. Silicon uptake and accumulation in higher plants. Trends Plant Sci 2006;11(8):392—97.*

ecosystem improves plant growth and nitrogen use proficiency [79—81] and reduces the biodiversity loss induced by the deprived addition of nitrogen [82]. Si availability to plants regulates the nutrient's dynamics and affects the plant growth and abundance in the grassland ecosystem [83,84]. Therefore Si plays a significant role in plant abundance and biomass, but the research area is little explored in this regard [85]. In addition, the recent research shows that Si proved effective in reducing abiotic stresses in plants [81,86]. The role of Si in saline agriculture is well documented and studied phenomenon as it helps in Na detoxification, osmotic adjustment, cell wall strengthening, and upregulation of physiological and biochemical responses of plants [87]. When Si is applied in the soil, its absorption in plant takes place via transporters (in form of silicic acid) from roots, then it is translocated to shoot where it can be polymerized into silica as well as it can be deposited in the cuticle [88] (Fig. 9.4).

9.4 Nanosilica: types, sources, synthesis, and uptake mechanism

9.4.1 Types of nanosilica

Soil contains various portions of Si out of which nano-Si is an important portion and can be of immense importance due to higher surface area and beneficial role in plants [89]. Mostly the SiNPs exist as silicon dioxide (SiO_2), quartz, and silica sand that are inert and relatively hard minerals chemically derived from the SiO_4 subunits arranged in the tetrahedral symmetry. In contrast, nanosilica is amorphous with 5—100 nm particle size [90], and has a comparatively large surface area, thermal properties, and permeability in plants [91]. Nanosilica has certain benefits on bulk silica, like comparatively high surface-to-volume ratio, distinctive electrical and thermal characteristics, and higher influx in plant cells [91]. The nanoSi as well as other nano materials have unique role in agriculture and industry thus widely being studied [92—94]. Based on the architectural constitution, nanosilica can be divided into different types. Most synthesized nanosilica can be distinguished into monodisperse spherical and spongy type (microporous and mesoporous) hollow, colloidal, and etched types [91,95,96]. Among these, mesoporous nanosilica has gained attention in the field of agriculture because of its functionalized selectivity with distinct groups that can direly increase its solubility along with absorption. Moreover, functionalization verifies the inertness to act as a transporter of essential nutrients and psychostimulants [97,98].

9.4.2 Nanosilica, sources, and synthesis

Nanoparticles have certain unique properties compared with bulk because of their small size, variety in shape, and high surface area-to-volume ratio [99]. Si exhibits certain physiological and chemical properties that make it superior in action to bulk materials [100]. Natural sources include rice husk, bamboo, sugar cane, reed plant, and waste materials from which Si is obtained are waste of solar power industry and waste of industries [101]. There are several methods through which synthesis of nanosilica takes place such as laser-assisted electrochemical etching, plasma-enhanced chemical vapor deposition, pulse laser deposition, and combustion analysis [102,103]. The synthesis of nanosilica using organic waste makes it economical and sustainable. The most commonly used organic wastes are rice straw and rice husk [104,105]. Different acids can be used to precipitate the rice husk for the synthesis of mesoporous nanosilica with the size 5−30 nm having 3−9 nm of pore diameter. Jelly-like materials obtained after acidification are aged and dried into nanosilica powder [106]. Gu et al. [104] prepared the nanosilica from rice husk through pyrolysis and calcination methods. Pyrolysis of acidified rice husk was done with the help of CO_2 and N_2 followed by oxidation bypassing O_2 gas. Finally, the decayed rice husk is calcined at 610°C to get a nanosilica powder. Peres et al. [107] reported that the nanosilica could also be synthesized using microwaves. Rice husk is initially leached with hydrochloric acid and heated microwave oven, then cleaned, followed by calcination at 700°C to obtain nanosilica.

Similarly, Hassan et al. [108] adopted the sol−gel technique for the synthesis of nanosilica using rice husk. In this procedure, raw material is initially burned in a muffle oven to eliminate carbon contamination, followed by acidification with hydrochloric acid, and then washed to remove the other contamination. The residues of the rice husk were washed with sodium hydroxide having strong acidification strength NH_4OH was added to raise the pH to 8.7 and then centrifuged to get a nanosilica powder. Nanosilica production from rice straw is also common and carried out through delignification of sliced straw with some alkaline solution followed by the neutralization with H_2SO_4 [105]. Various other substrates, including agricultural wastes such as sugarcane waste, sugarbeet brasse, barley husk, teff straw, corn cob husks, bamboo leaves, etc., as a raw material for the synthesis of nanosilica has also been documented [109,110].

The chemical production of nanosilica from the inorganic source has also been documented. Torkashv and Turkish Bagheri-Mohagheghi synthesize the nanosilica by Si-enriched bentonite [111]. The clay is heated at 682°C for 1 h and silica leached with hydrochloric acid. Afterward, silica-enriched clay is treated with sodium hydroxide followed by acidic ethanol precipitation that yielded nanosilica powder. Jafari et al. [112] recycled the silica fumes for the production of colloidal nanosilica. Sodium hydroxide is used to treat the silica fumes to get sodium silicate, which is treated with H_2SO_4, and then NaOH is added to increase the pH up to 10.5, followed by ultrasonication to get nanosilica. Yan et al. [113] used the fly ash for its synthesis, initially calcination at 800°C to get a ride from organic matter. Afterward, the calcinated ash is treated with sodium hydroxide and filtered to get silica. Cetyltrimethylammonium bromide (CTAB) is added to the silica followed by continuous stirring to obtain nanosilica. Tetraethyl orthosilicate (TEOS) is the most commonly used starter for the synthesis of nanosilica. Pham et al. [114] followed the microemulsion sol−gel method to synthesize nanosilica with ethyl orthosilicate. The TEOS emulsion is prepared by treating with sodium lauryl sulfate and carbonic acid. The homogenized microemulsion is then reacted with hydrochloric acid or NH_4OH by maintaining the pH 4−10 and stirrer at 80°C to get nanosilica. The Stober method is most commonly used to synthesize Si-based nanomaterials, which has been continuously improved. In fact, in this technique, TEOS is hydrolyzed and then reacted with a blend of NH_3 and absolute alcohol with constant mixing at the boiling point to condense. These particles are centrifuged and dried to obtain a nanosilica powder. Among all types of SiNPs, amorphous nanoSi is very important and well-studied. An investigation by Mourhly et al. [115] show preparatory method for nano-Si using silica rich rocks under alkaline conditions and low temperature (Fig. 9.5).

9.4.3 Absorption pathways of nanosilica

The role of Si in crop production, as well as protection, has been well studied. Different modes have been documented through which mesoporous silica nanoparticles can be applied about 20 nm in size [116].

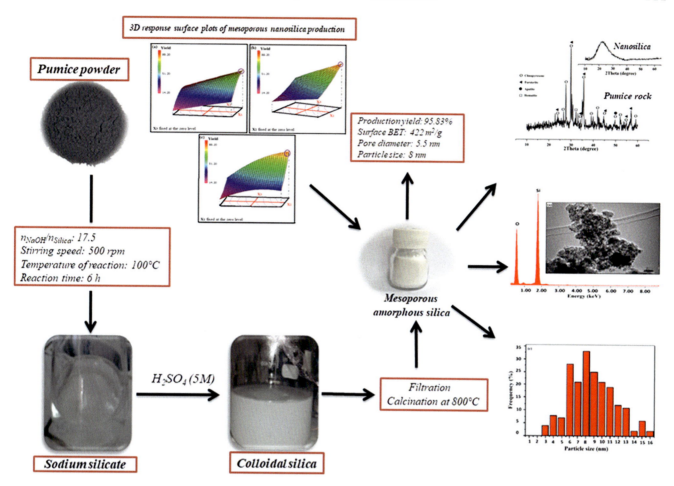

FIGURE 9.5 Schematic representation of preparation and characterization of nanosilica. *Source: Graphical abstract of research paper published by Mourhly A, et al., Highly efficient production of mesoporous nano-silica from unconventional resource: process optimization using a central composite design. Microchem J 2019;145:139–145, with permission and license to reuse.*

Commonly, Si is absorbed by the plants in the form of monosilica through roots and carried out to all parts of the plants via xylem tissues and deposited in the cell wall in the form of silica [117,118]. Schaller et al. investigated that translocation of nanosilica in plants follow the same routes as that of other silicates their chemistry of uptake is still unknown [119]. Moreover, scanning electron microscopy of the rice leaves confirmed the uptake of nanosilica in the form of amorphous silica bodies deposited in the flag leaves [120]. Mainly, Si is translocated through two pathways: symplastic and apoplastic routes. The symplastic pathway is controlled by the influx of LSi1 and efflux of LSi2 transporter genes [121]. Both the genes exist in the same cell. In the xylem vessel, the loading of Si is carried out by an unknown transporter known as LSi6 to the leaf tissues of the plants [122]. However, there are few studies on the absorption mechanism of Si by plants.

9.5 Chemistry of nano-Si in salt-contaminated soil

9.5.1 The fate of SiNPs in soil

The behavior of metal-based nanoparticles discharged by the environment helps in evaluating their role in the agriculture industry by determining their mobility and ability to reduce potential risks. Metal nanoparticles change their properties by reacting with physical and chemical compounds either by aggregation and precipitation into soil matrix or by dispersing in soil. In the agriculture sector, metal-based nanoparticle's environmental behavior is controlled by properties or by characteristics of the soil. Quantum effects are dependent on the size-assigned nanoparticles with unique physical and chemical properties [123]. Moreover, certain features such as chemical compound morphology, surface coating, and functionality used in manufacturing metal-based nanoparticles increased their effectiveness. Due to the complex nature of metal-based nanoparticles, their environmental behavior is still a challenge. Due to their pH 5, all particles were positively charged, which showed alpha-hetero values near 1, but at pH 8, all the particles were negatively charged; their alpha-hetero values were affected by solution conditions [124]. At a high concentration of sodium chloride and calcium chloride, these alpha hetero values were between < 0.001 to 1 Mesoporous silica nanoparticles are effective carriers for urea, boron, and nitrogen-based fertilizers because of their specific pore size (2−10 nm) [125,126]. Zinc, copper, and titanium oxide nanoparticles were reported to be increased due to the presence of SiO_2, which in turn decreases the effectiveness of soil layers [127].

9.5.2 Transportation assimilation and intertissue dynamics of nano-Si in plants

Mathur and Roy [128] investigated that the absorption of nanosilica is the same as that of other silicates. Nanosilica migrates from roots to the leaf blade in different condensed states. The plant roots absorb silicic acid from the soil solution, which is transported by LSi1 and then released to the exosomes through LSi2. Afterward, it enters into the endodermis of the roots via LSi1 and is circulated to the central portion of the roots through LSi2. Ultimately, silicic acid penetrates the xylem through an unknown transporter and is transported to the shoots via a transpirational pull. Silicic acid is transported into the leaves via LSi6 and localized in the xylem parenchyma cells of leaf blades and leaf sheaths. Si is converted from liquid form to the solid amorphous silica in the canopy (shoots and leaves) and mainly accumulated in the cell walls of various tissues such as epidermal cells [129−131] (Table 9.1).

9.6 Nano-Si-mediated tolerance in plants under salinity stress

9.6.1 Physiological modulation

Salinity can cause physiological stress, leading to excessive production of ROS, disrupt metabolic activity, and cause damage to the endomembrane system and plasmalemma [132]. Plants have developed two ROS removal systems (one is enzymatic removal and the other is nonenzymatic removal). Lipid peroxidation is a significant damaging process in plants caused by the ROS production as a result of stress. The concentration of its end-product, malondialdehyde, is frequently used to determine its severity [133]. The application of nanosilica in the salt-affected soil, significantly enhanced the growth and physiology of plants. Furthermore, nanosilica improves the water use efficiency of plants under salinity stress. It was reported that plants treated with nanosilica possessed high chlorophyll contents, photosynthetic pigments, plant height, large leaf area, and maximum nitrogen contents. Further reported that Si enhanced the dry mass of cucumber, common bean, maize, wheat, and soybean [134−138]. Si treatment to various plant species under saline circumstances was reported to reduce lipid peroxidation [139,140]. The favorable effect of Si on modulating antioxidant responses is ascribed to the reduction of lipid peroxidation and its application to salt-treated plants. It was discovered that it protects membrane integrity and reduces plasmalemma permeability to electrolytes under salt stress [141].

9.6.2 Biochemical effects

The induction of root endodermal silicification, antioxidant activity, and cellular water balance are aided by the exogenous application of nanoparticles [142]. With substantial modifications in the structure organelles, and activation of the defense system and scavenging of particular ions, nanosilica treatment has increased plant growth, yield, and plant water status [143]. Moreover, nanosilica is a plant growth inducer, increasing roots endodermal silicification, improving the antioxidant activity under salinity [144]. Compared with the untreated plants under abiotic stress, single-bonded SiO_2 nanoparticles treatment can enhance the chlorophyll contents, K^+ absorbance, reduce the Na^+ level, and minimize the cell wall damage [17]. As a result, raising K^+ concentration and reducing Na^+ will improve the seed germination, growth, and development by activating several enzymes in the cytosol, which are involved in ROS cleaning and so preserving cytosolic K^+ levels [69,145], promote cell turgor, and reduce cell stiffness, preventing stomatal closure and therefore increasing CO_2 assimilation, increasing water absorption, and thereby maintaining nutritive position in the cells. With the increasing level of K^+ inside the cell than Na^+, excellent homeostasis is retained [146]. Nanosilica was also discovered to assist in

TABLE 9.1 A list of studies carried out summarizing the application of nanosilica against salinity stress tolerance in various plants.

Crop	Application rate	Application method	Plants response	Reference
Common bean	100, 200 and 300 mg L^{-1}	Seed treatment	Improved growth, germination rate, and germination process. Alleviated the detrimental impacts of salinity on common bean.	[134]
Cucumber	100, 200 and 300 mg L^{-1}	Through irrigation	Improved the germination rate, germination index, fresh mass, K$^+$/Na$^+$ ratio, shoot dry mass, root dry mass, and plant height.	[135]
Maize and faba bean	300 mg L^{-1}	Foliar spray	Improved grain and straw yield, increased chlorophyll contents, nitrogen uptake, and nitrogen use efficiency.	[153]
Strawberry	50 mg L^{-1}	As suspension as nutrient solution	Maintained the epicuticular was structure, carotenoid contents, chlorophyll contents, and fewer proline contents.	[142]
Maize	10 mg mL^{-1}	Seed treatment	Seed treatment projected a higher germination rate and seedling vigor index via increased antioxidant enzyme activity, suppressing ROS and lipid peroxidase production. Also, increase the gibberellin content.	[130]
Maize	5, 10 and 15 kh ha^{-1}	Soil treatment	Enhanced phenolic activity conferring a protective physical barrier as well as induced disease resistance.	[154]
Land cress plant	0.05 and 0.07 mg kg^{-1} soil	Soil application	Increased root and shoots dry weight. Enhanced the soil nitrogen and phosphorous content and vegetative growth of the plants.	[131]
Wheat	80 mg kg^{-1}, 600 mg L^{-1}	Soil application and foliar spray	Increased thousand grains weight, increased activity of peroxidase in the cell wall, improved physiological features of plant, increased N, P, K absorption, and antioxidant activity.	[155]
Sweet pepper	1 and 2 cm^3 L^{-1}	Through irrigation	Significantly enhances the growth and yield parameters.	[156]
Tomato	01 and 02 mM	Nutrient solution	Improved water use efficiency, photosynthetic rate, and mesophyll conductance.	[157]
Field pumpkin	6 mg L^{-1}	Seed treatment	Increased antioxidant enzyme activity, seed germination rate, stomatal conductance, water use efficiency, and photosynthesis.	[144]
Faba bean	1, 2, and 3 mM	Soil application	Enhanced seed germination seed quality, relative water contents, and plant growth and yield parameters.	[18]
Faba bean	1, 2, and 3 mM	Soil application	increased ascorbate peroxidase (APX), catalase (CAT), peroxidase (POD), and antioxidant activity in plant leaves.	[158]
Maize	1 mg/1.8 g	Soil application	Increased root elongation	[15]
Lentil	1 mM	Seed treatment	Improved seed germination time, seed germination index, seed vigor index, fresh seedling weight, and dry weight.	[159]
Tomato	8 g L^{-1}	Seed dressing	Increased germination rate, germination time, seed germination index, seed vigor index, fresh seedling weight, and dry weight	[160]
Basil	various SI sources	Foliar spray	Significant increase in dry weight, fresh weight, chlorophyll contents, and proline.	[19]
Soybean	0.5, 1, and 2 mM	Foliar application	Alleviated the salinity by increasing K$^+$, antioxidant activity, nonenzymatic compounds, and decrease in Na$^+$ uptake, lipid peroxidation, and reactive oxygen species (ROS) production.	[27]
Maize	10 g kg^{-1} soil	Soil application	Significantly increased uptake of NPK with nSiO$_2$. The combination of nSiO$_2$ and NPK improved the photosynthetic activity, yield, and productivity of maize plants. Improved plant physiology.	[161]
Banana	150 mg L^{-1}	Soil application	Enhanced shoot growth and chlorophyll content, improved photosynthesis, maintain K$^+$ and Na$^+$ balance decreases cell wall damage.	[17]
Wheat	10 μM	Nutrient solution	Enhanced antioxidants to protect against UV-B-generated oxidative stress.	[162]
Rice	150 g L^{-1}	Soil application	Improved the free amino acid contents, proline contents, and total carbohydrate contents under salinity. Malondialdehyde contents and water was controlled by silica ions.	[141]

removing free radicals, which inhibit the seed germination after penetrating the seed. Resultantly, activation of superoxide dismutase (SOD) and catalase (CAT) enzymes occurs by triggering some oxidation—reduction reactions followed by the production of superoxide radicals which play a significant role in removing free radicals found in the seed [147]. Furthermore, nanosilica can help break the seed dormancy by increasing gibberellin secretion while suppressing abscisic acid synthesis [16].

9.6.3 Gene expression

Phytohormones play a key role in plant stress adaption by mediating several adaptive responses [148]. In response to water stress, abscisic acid is known to accumulate. As a result, abscisic acid may influence the regulation of LSi1 and LSi2. LSi1 and LSi2 both include abscisic acid-responsive motifs in their promoter regions [86]. According to [149], the comparative manifestation of the two genes (Lsi1 and Lsi2) was downregulated in plants exposed to salt, which is linked to high abscisic acid production [86]. Ma [121] discovered that a high level of expression of both Lsi1 and Lsi2 is required to increase Si absorption and stress tolerance. Devoto and Turner [150] investigated that jasmonic acids are involved in the upregulation of both the genes (Lsi1 and Lsi2) under salinity because both the jasmonic acid and both Lsi1 and Lsi2 gene are overexpressed by the application of nanosilica under salinity. As a result, plants were more sensitive to ozone stress due to a lack of antioxidant gene activation following a mutation in the jasmonic acid biosynthesis gene, opr3 [151].

9.7 Conclusion

Salinity is the major abiotic stress imposing factor responsible for decreasing crop production. Over 20% of agriculture is affected by soil salinity worldwide, and this is supposed to exacerbate further at the global level due to inappropriate nutritional inputs and irrigation practices. Under the saline condition, plants experience osmotic stress that causes certain physiological changes followed by membrane nutrient disruption, inhibiting the capability to remove ROS, reduction in photosynthetic pigments, and the stomatal opening is reduced. Oxidative stress induced by ROS causes damage to cellular components such as proteins, lipids, and genetic material, which ultimately damage cellular structures. To overcome the salinity stress, Si has been introduced as an essential element for plant nutrition as well in the nutrient cycle, especially for grassland ecosystems. Si-based fertilizer in a natural ecosystem improves plant growth and nitrogen use proficiency and reduces the biodiversity loss induced by the addition of nitrogen. Recently, nanosilica has attracted the attention of scientists because of its ability to neutralize the counter effect of salinity on plant growth. Nanosilica is an amorphous solid with particle size ranging 5—100 nm. Compared with bulk silica, it has a higher surface-to-volume ratio, unique electrical and thermal properties, and higher air permeability in plant cells. Although, it is classified as a nonessential element for higher plants. The positive effect of this metalloid on plant growth and reproduction has been investigated in different crops. Among the various types of nanosilica, mesoporous nanosilica has gained attention in the field of agriculture because of its functionalized selectivity with multiple groups that can direly increase its solubility and absorption rate.

Commonly, Si is absorbed by the plants in the form of monosilicic acid via xylene and distributed to all parts of the plant, and then deposited in the cell wall in the form of silica. Mesoporous nanosilica are effective carriers for urea, boron, and nitrogen-based fertilizers due to their special pore size (2—10 nm). Compared to the untreated plants under abiotic stress, single-bonded SiO_2 nanoparticles treatment can enhance chlorophyll content, promote K^+ uptake, modify Na^+ levels, and reduce cell wall damage. As a result, raising K^+ content and decreasing Na^+ will improve germination and growth dynamic by activating several enzymes in the cytosol, which are involved in ROS detoxification and so preserving cytosolic K^+ levels, promote cell turgor and reduce cell stiffness, preventing stomatal closure and therefore increasing CO_2 assimilation, increasing water uptake, and thereby maintaining nutritional status in cells; when K^+ is higher inside the cell than Na^+, excellent homeostasis is maintained. Nanosilica is discovered to assist in the removal of free radicals, which inhibit germination. After infiltrating inside the seed, nanosilica activates the CAT and SOD enzymes by triggering some oxidation—reduction reactions, resulting in the production of superoxide ion radicals, which play a key role in removing free radicals present within the seed. Additionally, nanosilica promotes gibberellin secretion and abscisic acid suppression, which can help break the seed dormancy.

9.8 Future direction

Identifying and characterizing the regulatory factors and regulatory programs of Si in plant salt tolerance will help improve the salt tolerance of crops. Many factors, however, have not been well investigated, and certain areas demand further investigation. Based on the currently documented findings, we propose that future research should focus on the following issues.

1. As omics technologies have advanced, research on the molecular processes nanosilica-based mitigation of salinity stress has expanded to the transcriptomic and proteomic levels.
2. A recent study was first to indicate that nanosilica directly affects the expression of abscisic acid and jasmonic acid along with various transporters proteins in an experimental setting. Though, the precise regulatory mechanisms of nanosilica on cell signaling pathways and the potential interaction of nanosilica with other salt stress sensors are unknown.
3. The detailed mechanisms and fat of nanosilica in the saline soil are not well-studied.
4. More research on the regulation mechanisms of nanosilica in salt-induced osmotic stress is needed. Si transporters are also members of the aquaporin family's NIP subfamily. Cloning and functional studies of Si transporters and AQPs in various species can help researchers understand how water metabolism is regulated better.
5. Research has shown that under stress situations, Si can enhance root length and the creation of suberin structures in the endodermis and exodermis of the roots, as well as govern the allocation of ions to various areas of the root. However, the molecular processes through which nanosilica influences root structure and ionic absorption or dissemination in the root are unknown and need to be researched further.
6. Under salinity stress, the deposition of carbohydrates can play a crucial role in energy storage, osmoregulation, and signaling pathways. As a result, more research is needed in the regulatory functions and mechanisms of carbohydrates during Si application in stressful situations. To summarize, it is a dire need to improve the baseline study of the molecular pathways through which nanosilica overcome salinity stress and lay a theoretical foundation for the use of nanosilica in crop production.

References

[1] Zhu Y-X, et al. Identification of cucumber circular RNAs responsive to salt stress. BMC Plant Biol 2019;19(1):1−18.
[2] Ayub MA, et al. Plant life under changing environment. London: Academic Press; 2020.
[3] Zhu Y, et al. Silicon confers cucumber resistance to salinity stress through regulation of proline and cytokinins. Plant Physiol Biochem 2020;156:209−20.
[4] Farooqi ZUR, et al. Chapter 10: Threats to arable land of the world: current and future perspectives on land use. In:. Examining international land use policies. Pennsylvania: IGI Global; 2021.
[5] Wu J, et al. Distinct physiological responses of tomato and cucumber plants in silicon-mediated alleviation of cadmium stress. Front Plant Sci 2015;6:453.
[6] Zargar SM, et al. Role of silicon in plant stress tolerance: opportunities to achieve a sustainable cropping system. 3 Biotech 2019;9(3):73.
[7] Khan A, et al. Silicon and salinity: crosstalk in crop-mediated stress tolerance mechanisms. Front Plant Sci 2019;10:1429.
[8] Etesami H, Jeong BR. Silicon (Si): review and future prospects on the action mechanisms in alleviating biotic and abiotic stresses in plants. Ecotoxicol Environ Saf 2018;147:881−96.
[9] Almutairi ZM. Effect of nano-silicon application on the expression of salt tolerance genes in germinating tomato (*Solanum lycopersicum* L.) seedlings under salt stress. Plant Omics 2016;9(1):106−14.
[10] ur Rehman MZ, et al. Split application of silicon in cadmium (Cd) spiked alkaline soil plays a vital role in decreasing Cd accumulation in rice (*Oryza sativa* L.) grains. Chemosphere 2019;226:454−62.
[11] Benelli G. Mode of action of nanoparticles against insects. Environ Sci Pollut Res Int 2018;25(13):12329−41.
[12] Jafir M, et al. Characterization of Ocimum basilicum synthesized silver nanoparticles and its relative toxicity to some insecticides against tobacco cutworm, *Spodoptera litura* Feb. (Lepidoptera; Noctuidae). Ecotoxicol Environ Saf 2021;218:112278.
[13] Sohail MI, et al. Comprehensive analytical chemistry. Amsterdam: Elsevier; 2019 [chapter 1].
[14] Rastogi A, et al. Application of silicon nanoparticles in agriculture. 3 Biotech 2019;9(3):1−11.
[15] Karunakaran G, et al. Effect of nanosilica and silicon sources on plant growth promoting rhizobacteria, soil nutrients and maize seed germination. IET Nanobiotechnol 2013;7(3):70−7.
[16] Yuvakkumar R, et al. Influence of nanosilica powder on the growth of maize crop (*Zea mays* L.). Int J Green Nanotechnol 2011;3(3):180−90.
[17] Mahmoud LM, et al. Silicon nanoparticles mitigate oxidative stress of in vitro-derived banana (Musa acuminata 'Grand Nain') under simulated water deficit or salinity stress. South African J Bot 2020;132:155−63.
[18] Qados AMA, Moftah AE. Influence of silicon and nano-silicon on germination, growth and yield of faba bean (*Vicia faba* L.) under salt stress conditions. J Exp Agr Int 2015;5(6):509−24.
[19] Kalteh M, et al. Effect of silica nanoparticles on basil (*Ocimum basilicum*) under salinity stress. J Chem Health Risks 2014;4(3):49−55.

[20] Orellana S, et al. The transcription factor SlAREB1 confers drought, salt stress tolerance and regulates biotic and abiotic stress-related genes in tomato. Plant Cell Environ 2010;33(12):2191–208.

[21] Pan Y, et al. An ethylene response factor (ERF5) promoting adaptation to drought and salt tolerance in tomato. Plant Cell Rep 2012;31(2):349–60.

[22] Bastías A, et al. The transcription factor AREB1 regulates primary metabolic pathways in tomato fruits. J Exp Bot 2014;65(9):2351–63.

[23] Hong Z, et al. Removal of feedback inhibition of Δ1-pyrroline-5-carboxylate synthetase results in increased proline accumulation and protection of plants from osmotic stress. Plant Physiol 2000;122(4):1129–36.

[24] Flam-Shepherd R, et al. Membrane fluxes, bypass flows, and sodium stress in rice: the influence of silicon. J Exp Bot 2018;69(7):1679–92.

[25] Bosnic P, et al. Silicon mediates sodium transport and partitioning in maize under moderate salt stress. Environ Exp Bot 2018;155:681–7.

[26] Zou L-P, et al. Molecular cloning, expression and mapping analysis of a novel cytosolic ascorbate peroxidase gene from tomato: full length research paper. DNA Seq 2005;16(6):456–61.

[27] Farhangi-Abriz S, Torabian S. Nano-silicon alters antioxidant activities of soybean seedlings under salt toxicity. Protoplasma 2018;255(3):953–62.

[28] Arif Y, et al. Salinity induced physiological and biochemical changes in plants: an omic approach towards salt stress tolerance. Plant Physiol Biochem 2020;156:64–77.

[29] Munns R, et al. Leaf water status controls day-time but not daily rates of leaf expansion in salt-treated barley. Funct Plant Biol 2000;27(10):949–57.

[30] Fricke W, et al. Rapid and tissue-specific changes in ABA and in growth rate in response to salinity in barley leaves. J Exp Bot 2004;55(399):1115–23.

[31] İbrahimova U, et al. Progress in understanding salt stress response in plants using biotechnological tools. J Biotechnol 2021;329:180–91.

[32] Shabala SN, Lew RR. Turgor regulation in osmotically stressed Arabidopsis epidermal root cells. Direct support for the role of inorganic ion uptake as revealed by concurrent flux and cell turgor measurements. Plant Physiol 2002;129(1):290–9.

[33] Byrt CS, et al. Root cell wall solutions for crop plants in saline soils. Plant Sci 2018;269:47–55.

[34] Wegner LH, et al. Sequential depolarization of root cortical and stelar cells induced by an acute salt shock—implications for Na + and K + transport into xylem vessels. Plant Cell Environ 2011;34(5):859–69.

[35] Shabala L, et al. Cell-type-specific H + -ATPase activity in root tissues enables K + retention and mediates acclimation of barley (*Hordeum vulgare*) to salinity stress. Plant Physiol 2016;172(4):2445–58.

[36] Cosgrove DJ, Hedrich R. Stretch-activated chloride, potassium, and calcium channels coexisting in plasma membranes of guard cells of *Vicia faba* L. Planta 1991;186(1):143–53.

[37] Furuichi T, Tatsumi H, Sokabe M. Mechano-sensitive channels regulate the stomatal aperture in *Vicia faba*. Biochem Biophys Res Commun 2008;366(3):758–62.

[38] Betzen BM, et al. Effects of increasing salinity on photosynthesis and plant water potential in Kansas salt marsh species. Trans Kansas Acad Sci 2019;122(1–2):49–58.

[39] Methenni K, et al. Salicylic acid and calcium pretreatments alleviate the toxic effect of salinity in the Oueslati olive variety. Sci Hortic 2018;233:349–58.

[40] Soori N, et al. Effect of salinity stress on some physiological characteristics and photosynthetic parameters of several Iranian commercial pomegranate genotypes. J Plant Process Funct 2019;8:155–70.

[41] James RA, et al. Major genes for Na + exclusion, Nax1 and Nax2 (wheat HKT1; 4 and HKT1; 5), decrease Na + accumulation in bread wheat leaves under saline and waterlogged conditions. J Exp Bot 2011;62(8):2939–47.

[42] Cheeseman JM. The integration of activity in saline environments: problems and perspectives. Funct Plant Bio 2013;40(9):759–74.

[43] Wu H, et al. It is not all about sodium: revealing tissue specificity and signalling roles of potassium in plant responses to salt stress. Plant Soil 2018;431(1):1–17.

[44] Shabala S, Cuin TA. Potassium transport and plant salt tolerance. Physiol Plant 2008;133(4):651–69.

[45] Spitzer J, Poolman B. The role of biomacromolecular crowding, ionic strength, and physicochemical gradients in the complexities of life's emergence. Microbiol Mol Bio Rev 2009;73(2):371–88.

[46] Bazihizina N, et al. Friend or foe? Chloride patterning in halophytes. Trends Plant Sci 2019;24(2):142–51.

[47] Hanin M, et al. New insights on plant salt tolerance mechanisms and their potential use for breeding. Front Plant Sci 2016;7:1787.

[48] Atzori G, et al. Effects of increased seawater salinity irrigation on growth and quality of the edible halophyte *Mesembryanthemum crystallinum* L. under field conditions. Agri Water Manag 2017;187:37–46.

[49] Carillo P, et al. Morpho-anatomical, physiological and biochemical adaptive responses to saline water of *Bougainvillea spectabilis* Willd. trained to different canopy shapes. Agri Water Manag 2019;212:12–22.

[50] Cambridge M, et al. Effects of high salinity from desalination brine on growth, photosynthesis, water relations and osmolyte concentrations of seagrass *Posidonia australis*. Mar Pollut Bull 2017;115(1–2):252–60.

[51] Gao E-B, Huang Y, Ning D. Metabolic genes within cyanophage genomes: implications for diversity and evolution. Genes 2016;7(10):80.

[52] Çiçek N, et al. Salt stress effects on the photosynthetic electron transport chain in two chickpea lines differing in their salt stress tolerance. Photosynth Res 2018;136(3):291–301.

[53] Wang P, et al. Low salinity promotes the growth of broccoli sprouts by regulating hormonal homeostasis and photosynthesis. Hortic Environ Biotechnol 2019;60(1):19–30.

[54] Li H, et al. Exogenous melatonin confers salt stress tolerance to watermelon by improving photosynthesis and redox homeostasis. Front Plant Sci 2017;8:295.

[55] Xu H, Lu Y, Tong S. Effects of arbuscular mycorrhizal fungi on photosynthesis and chlorophyll fluorescence of maize seedlings under salt stress. Em J Food Agri 2018;30(3):199–204.

[56] Papageorgiou GC, Murata N. The unusually strong stabilizing effects of glycine betaine on the structure and function of the oxygen-evolving photosystem II complex. Photosynth Res 1995;44(3):243–52.

[57] Bose J, Rodrigo-Moreno A, Shabala S. ROS homeostasis in halophytes in the context of salinity stress tolerance. J Exp Bot 2014;65(5):1241–57.

[58] Stoop JM, Williamson JD, Pharr DM. Mannitol metabolism in plants: a method for coping with stress. Trends Plant Sci 1996;1(5):139−44.

[59] Moran JF, et al. Drought induces oxidative stress in pea plants. Planta 1994;194(3):346−52.

[60] Managbanag JR, Torzilli AP. An analysis of trehalose, glycerol, and mannitol accumulation during heat and salt stress in a salt marsh isolate of *Aureobasidium pullulans*. Mycologia 2002;94(3):384−91.

[61] Ashraf M, Harris P. Potential biochemical indicators of salinity tolerance in plants. Plant Sci 2004;166(1):3−16.

[62] Sandhu D, et al. Variable salinity responses of 12 alfalfa genotypes and comparative expression analyses of salt-response genes. Sci Rep 2017;7(1):1−18.

[63] Stępień P, Kłbus G. Water relations and photosynthesis in *Cucumis sativus* L. leaves under salt stress. Biol Plant 2006;50(4):610−16.

[64] Farouk S. Osmotic adjustment in wheat flag leaf in relation to flag leaf area and grain yield per plant. J Stress Physiol Biochem 2011;7(2):117−38.

[65] Galinski EA. Compatible solutes of halophilic eubacteria: molecular principles, water-solute interaction, stress protection. Experientia 1993;49(6):487−96.

[66] Hare PD, Cress WA, Van Staden J. Dissecting the roles of osmolyte accumulation during stress. Plant Cell Environ 1998;21(6):535−53.

[67] Lutts S, Lefèvre I. How can we take advantage of halophyte properties to cope with heavy metal toxicity in salt-affected areas? Ann Bot (Lond.) 2015;115(3):509−28.

[68] Khan N, Bano A, Zandi P. Effects of exogenously applied plant growth regulators in combination with PGPR on the physiology and root growth of chickpea (*Cicer arietinum*) and their role in drought tolerance. J Plant Int 2018;13(1):239−47.

[69] Siddiqui MN, et al. Impact of salt-induced toxicity on growth and yield-potential of local wheat cultivars: oxidative stress and ion toxicity are among the major determinants of salt-tolerant capacity. Chemosphere 2017;187:385−94.

[70] Shahid MA, et al. Salt stress effects on some morphological and physiological characteristics of okra (*Abelmoschus esculentus* L.). Soil Environ 2011;30(1):66−73.

[71] Bhattarai S, et al. Morphological, physiological, and genetic responses to salt stress in alfalfa: a review. Agronomy 2020;10(4):577.

[72] Dogan M. Effect of salt stress on in vitro organogenesis from nodal explant of *Limnophila aromatica* (Lamk.) Merr. and *Bacopa monnieri* (L.) Wettst. and their physio-morphological and biochemical responses. Physiol Mol Bio Plants 2020;26(4):803.

[73] Kumar A, et al. Effect of salt stress on seed germination, morphology, biochemical parameters, genomic template stability, and bioactive constituents of *Andrographis paniculata* Nees. Acta Physiol Plant 2021;43(4):1−14.

[74] Sommer M, et al. Silicon pools and fluxes in soils and landscapes—a review. J Plant Nutr Soil Sci 2006;169(3):310−29.

[75] Frew A, et al. The role of silicon in plant biology: a paradigm shift in research approach. Ann Bot (Lond.) 2018;121(7):1265−73.

[76] Luyckx M, et al. Silicon and plants: current knowledge and technological perspectives. Front Plant Sci 2017;8:411.

[77] Schaller J, et al. Silica decouples fungal growth and litter decomposition without changing responses to climate warming and N enrichment. Ecology 2014;95(11):3181−9.

[78] Marxen A, et al. Interaction between silicon cycling and straw decomposition in a silicon deficient rice production system. Plant Soil 2016;398(1−2):153−63.

[79] Bruning B, Rozema J. Symbiotic nitrogen fixation in legumes: perspectives for saline agriculture. Environ Exp Bot 2013;92:134−43.

[80] Dhamala NR, et al. N transfer in three-species grass-clover mixtures with chicory, ribwort plantain or caraway. Plant Soil 2017;413 (1−2):217−30.

[81] Hoffmann J, et al. A review on the beneficial role of silicon against salinity in non-accumulator crops: tomato as a model. Biomolecules 2020;10(9):1284.

[82] Xu D, et al. Influences of nitrogen, phosphorus and silicon addition on plant productivity and species richness in an alpine meadow. AoB Plants 2015;7:plv125.

[83] Schaller J, et al. Silicon supply modifies C:N:P stoichiometry and growth of *Phragmites australis*. Plant Biol 2012;14(2):392−6.

[84] Schaller J, et al. Plant diversity and functional groups affect Si and Ca pools in aboveground biomass of grassland systems. Oecologia 2016;182(1):277−86.

[85] Song Z, Müller K, Wang H. Biogeochemical silicon cycle and carbon sequestration in agricultural ecosystems. Earth-Sci Rev 2014;139:268−78.

[86] Ma J, Yamaji N. Functions and transport of silicon in plants. Cell Mol Life Sci 2008;65(19):3049−57.

[87] Dhiman P, et al. Fascinating role of silicon to combat salinity stress in plants: an updated overview. Plant Physiol Biochem 2021;162:110−23.

[88] Ma JF, Yamaji N. Silicon uptake and accumulation in higher plants. Trends Plant Sci 2006;11(8):392−7.

[89] Yang X, et al. Distinguishing the sources of silica nanoparticles by dual isotopic fingerprinting and machine learning. Nat Commun 2019;10(1):1−9.

[90] Imai S, et al. Size and surface modification of amorphous silica particles determine their effects on the activity of human CYP3A4 in vitro. Nanoscale Res Lett 2014;9(1):1−7.

[91] Jeelani PG, et al. Multifaceted application of silica nanoparticles. A review. Silicon 2020;12(6):1337−54.

[92] Sohail MI, et al. Sufficiency and toxicity limits of metallic oxide nanoparticles in the biosphere. Nanomaterials: synthesis, characterization, hazards and safety. Amsterdam: Elsevier; 2021. p. 145−221.

[93] Barik T, Sahu B, Swain V. Nanosilica—from medicine to pest control. Parasitol Res 2008;103(2):253−8.

[94] Lazaro A, et al. The properties of amorphous nano-silica synthesized by the dissolution of olivine. Chem Eng J 2012;211:112−21.

[95] Rosenberg DJ, et al. Synthesis of microporous silica nanoparticles to study water phase transitions by vibrational spectroscopy. Nanoscale Adv 2019;1(12):4878−87.

[96] Potapov V, Fediuk R, Gorev D. Obtaining sols, gels and mesoporous nanopowders of hydrothermal nanosilica. J Sol-Gel Sci Technol 2020;94(3):1−14.

[97] Kokina I, et al. Target transportation of auxin on mesoporous Au/SiO$_2$ nanoparticles as a method for somaclonal variation increasing in flax (*L. usitatissimum* L.). J Nanomat 2017;7143269.

[98] Narayan R, et al. Mesoporous silica nanoparticles: a comprehensive review on synthesis and recent advances. Pharmaceutics 2018;10 (3):118.

[99] Roduner E. Size matters: why nanomaterials are different. Chem Soc Rev 2006;35(7):583—92.

[100] O'Farrell N, Houlton A, Horrocks BR. Silicon nanoparticles: applications in cell biology and medicine. Int J Nanomed 2006;1(4):451.

[101] Rehman WU, et al. When silicon materials meet natural sources: opportunities and challenges for low-cost lithium storage. Small 2021;17(9):1904508.

[102] Li Y, et al. Expression of jasmonic ethylene responsive factor gene in transgenic poplar tree leads to increased salt tolerance. Tree Physiol 2009;29(2):273—9.

[103] Eisenhawer B, et al. Growth of doped silicon nanowires by pulsed laser deposition and their analysis by electron beam induced current imaging. Nanotechnology 2011;22(7):075706.

[104] Gu S, et al. A novel two-staged thermal synthesis method of generating nanosilica from rice husk via pre-pyrolysis combined with calcination. Indus Crops Prod 2015;65:1—6.

[105] Bhattacharya M, Mandal MK. Synthesis of rice straw extracted nano-silica-composite membrane for CO_2 separation. J Clean Prod 2018;186:241—52.

[106] Liou T-H, Yang C-C. Synthesis and surface characteristics of nanosilica produced from alkali-extracted rice husk ash. Mat Sci Eng B 2011;176(7):521—9.

[107] Peres EC, et al. Microwave synthesis of silica nanoparticles and its application for methylene blue adsorption. J Environ Chem Eng 2018;6(1):649—59.

[108] Hassan A, et al. Synthesis and characterization of high surface area nanosilica from rice husk ash by surfactant-free sol—gel method. J Sol-Gel Sci Technol 2014;69(3):465—72.

[109] Rovani S, et al. Highly pure silica nanoparticles with high adsorption capacity obtained from sugarcane waste ash. ACS Omega 2018;3 (3):2618—27.

[110] Pieła A, et al. Biogenic synthesis of silica nanoparticles from corn cobs husks. Dependence of the productivity on the method of raw material processing. Bioorg Chem 2020;99:103773.

[111] Torkashvand H, Bagheri-Mohagheghi M. Purification, synthesis and structural, optical characterizations of silicon (Si) nano-particles from bentonite mineral: the effect of magnesium-thermic chemical reduction. Silicon 2021;13:1367—79.

[112] Jafari V, Allahverdi A, Vafaei M. Ultrasound-assisted synthesis of colloidal nanosilica from silica fume: effect of sonication time on the properties of product. Adv Powder Technol 2014;25(5):1571—7.

[113] Yan F, et al. A green and facile synthesis of ordered mesoporous nanosilica using coal fly ash. ACS Sustain Chem Eng 2016;4 (9):4654—61.

[114] Pham TD, Vu CM, Choi HJ. Enhanced fracture toughness and mechanical properties of epoxy resin with rice husk-based nano-silica. Polym Sci Ser A 2017;59(3):437—44.

[115] Mourhly A, et al. Highly efficient production of mesoporous nano-silica from unconventional resource: process optimization using a central composite design. Microchem J 2019;145:139—45.

[116] Schoelynck J, et al. Silica uptake in aquatic and wetland macrophytes: a strategic choice between silica, lignin and cellulose? New Phytolog 2010;186(2):385—91.

[117] Sahebi M, et al. Importance of silicon and mechanisms of biosilica formation in plants. BioMed Res Int 2015;2015.

[118] Nawaz MA, et al. Phytolith formation in plants: from soil to cell. Plants 2019;8(8):249.

[119] Schaller J, et al. Silica uptake from nanoparticles and silica condensation state in different tissues of *Phragmites australis*. Sci Total Environ 2013;442:6—9.

[120] Alvarez RdCF, et al. Effects of soluble silicate and nanosilica application on rice nutrition in an Oxisol. Pedosphere 2018;28(4):597—606.

[121] Ma JF. MIPs and their role in the exchange of metalloids. New York: Springer; 2010. p. 99—109.

[122] Yamaji N, et al. Orchestration of three transporters and distinct vascular structures in node for intervascular transfer of silicon in rice. Proc Natl Acad Sci U S A 2015;112(36):11401—6.

[123] Schmid G. Nanoparticles: from theory to application. New York: John Wiley & Sons; 2011.

[124] Lorenzo J, Montaña ÁM. The molecular shape and the field similarities as criteria to interpret SAR studies for fragment-based design of platinum (IV) anticancer agents. Correlation of physicochemical properties with cytotoxicity. J Mol Graphics Model 2016;69:39—60.

[125] Torney F, et al. Mesoporous silica nanoparticles deliver DNA and chemicals into plants. Nat Nanotechnol 2007;2(5):295—300.

[126] Wanyika H, et al. Mesoporous silica nanoparticles carrier for urea: potential applications in agrochemical delivery systems. J Nanosci Nanotechnol 2012;12(3):2221—8.

[127] Oliveira EMd, et al. Effects of the silica nanoparticles ($NPSiO_2$) on the stabilization and transport of hazardous nanoparticle suspensions into landfill soil columns. REM-Int Eng J 2017;70(3):317—23.

[128] Mathur P, Roy S. Nanosilica facilitates silica uptake, growth and stress tolerance in plants. Plant Physiol Biochem 2020;157.

[129] Raiesi Ardali T, et al. Silicon and silica nanoparticles: uptake and transport mechanism in plants and their effects on plant yield. J Biosaf 2021;13(4):77—94.

[130] Naguib DM, Abdalla H. Metabolic status during germination of nano silica primed *Zea mays* seeds under salinity stress. J Crop Sci Biotechnol 2019;22(5):415—23.

[131] Boroumand N, Behbahani M, Dini G. Combined effects of phosphate solubilizing bacteria and nanosilica on the growth of land cress plant. J Soil Sci Plant Nut 2020;20(1):232—43.

[132] Ahmad R, et al. Oxidative stress and antioxidant defense mechanisms in plants under salt stress. Plant abiotic stress tolerance. Cham: Springer; 2019. p. 191—205.

[133] Gill SS, Tuteja N. Reactive oxygen species and antioxidant machinery in abiotic stress tolerance in crop plants. Plant Physiol Biochem 2010;48(12):909—30.

[134] Alsaeedi AH, et al. Engineered silica nanoparticles alleviate the detrimental effects of Na^+ stress on germination and growth of common bean (*Phaseolus vulgaris*). Environ Sci Poll Res 2017;24(27):21917—28.

[135] Alsaeedi A, et al. Exogenous nanosilica improves germination and growth of cucumber by maintaining K^+/Na^+ ratio under elevated Na^+ stress. Plant Physiol Biochem 2018;125:164—71.

[136] Hamayun M, et al. Silicon alleviates the adverse effects of salinity and drought stress on growth and endogenous plant growth hormones of soybean (*Glycine max* L.). Pak J Bot 2010;42(3):1713−22.

[137] Kaya MD, et al. Seed treatments to overcome salt and drought stress during germination in sunflower (*Helianthus annuus* L.). European J Agron 2006;24(4):291−5.

[138] Gong S, et al. A gene expression atlas of the central nervous system based on bacterial artificial chromosomes. Nature 2003;425 (6961):917−25.

[139] Moussa HR. Influence of exogenous application of silicon on physiological response of salt-stressed maize (*Zea mays* L.). Int J Agric Biol 2006;8(3):293−7.

[140] Soylemezoglu G, et al. Effect of silicon on antioxidant and stomatal response of two grapevine (*Vitis vinifera* L.) rootstocks grown in boron toxic, saline and boron toxic-saline soil. Sci Hortic 2009;123(2):240−6.

[141] Abdel-Haliem ME, et al. Effect of silica ions and nano silica on rice plants under salinity stress. Ecol Eng 2017;99:282−9.

[142] Avestan S, et al. Application of nano-silicon dioxide improves salt stress tolerance in strawberry plants. Agronomy 2019;9(5):246.

[143] Parveen N, Ashraf M. Role of silicon in mitigating the adverse effects of salt stress on growth and photosynthetic attributes of two maize (*Zea mays* L.) cultivars grown hydroponically. Pak J Bot 2010;42(3):1675−84.

[144] Siddiqui MH, et al. Nano-silicon dioxide mitigates the adverse effects of salt stress on *Cucurbita pepo* L. Environ Toxicol Chem 2014;33 (11):2429−37.

[145] Kader MA, Lindberg S. Cytosolic calcium and pH signaling in plants under salinity stress. Plant Signal Behav 2010;5(3):233−8.

[146] Parida AK, Das AB. Salt tolerance and salinity effects on plants: a review. Ecotoxicol Environ Saf 2005;60(3):324−49.

[147] Gengmao Z, et al. The role of silicon in physiology of the medicinal plant (*Lonicera japonica* L.) under salt stress. Sci Rep 2015;5(1):1−11.

[148] Eyidogan F, et al. Signal transduction of phytohormones under abiotic stresses. Phytohormones and abiotic stress tolerance in plants. Berlin, Heidelberg: Springer; 2012. p. 1−48.

[149] Ranjan A, et al. Silicon-mediated abiotic and biotic stress mitigation in plants: underlying mechanisms and potential for stress resilient agriculture. Plant Physiol Biochem 2021;163:15−25.

[150] Devoto A, Turner JG. Regulation of jasmonate-mediated plant responses in Arabidopsis. Ann Bot (Lond.) 2003;92(3):329−37.

[151] Sasaki-Sekimoto Y, et al. Coordinated activation of metabolic pathways for antioxidants and defence compounds by jasmonates and their roles in stress tolerance in Arabidopsis. Plant J 2005;44(4):653−68.

[152] Zhao C, et al. Mechanisms of plant responses and adaptation to soil salinity. The Innovation 2020;1(1):100017.

[153] Amer M, El-Emary FA. Impact of foliar with nano-silica in mitigation of salt stress on some soil properties, crop-water productivity and anatomical structure of maize and faba bean. Environ Biodivers Soil Sec 2018;2(2018):25−38.

[154] Rangaraj S, et al. Augmented biocontrol action of silica nanoparticles and *Pseudomonas fluorescens* bioformulant in maize (*Zea mays* L.). RSC Adv 2014;4(17):8461−5.

[155] Ayman M, et al. Influence of nano-silica on wheat plants grown in salt-affected soil. J Prod Dev 2020;25(3):20279−96.

[156] Tantawy A, et al. Nano silicon application improves salinity tolerance of sweet pepper plants. Int J ChemTech Res 2015;8(10):11−17.

[157] Haghighi M, Pessarakli M. Influence of silicon and nano-silicon on salinity tolerance of cherry tomatoes (*Solanum lycopersicum* L.) at early growth stage. Sci Hortic 2013;161:111−17.

[158] Qados AMA. Mechanism of nanosilicon-mediated alleviation of salinity stress in faba bean (*Vicia faba* L.) plants. Am J Exp Agric 2015;7 (2):78−95.

[159] Sabaghnia N, Janmohammadi M. Effect of nano-silicon particles application on salinity tolerance in early growth of some lentil genotypes. Ann Univ Mariae Curie-Sklodowska, C Biol 2015;69(2):2014.

[160] Siddiqui MH, Al-Whaibi MH. Role of nano-SiO₂ in germination of tomato (*Lycopersicum esculentum* seeds Mill.). Saudi J Biol Sci 2014;21 (1):13−17.

[161] El-Naggar ME, et al. Soil application of nano silica on maize yield and its insecticidal activity against some stored insects after the postharvest. Nanomaterial 2020;10(4):739.

[162] Tripathi DK, et al. Silicon nanoparticles more effectively alleviated UV-B stress than silicon in wheat (*Triticum aestivum*) seedlings. Plant Physiol Biochem 2017;110:70−81.

10

Silicon- and nanosilicon-mediated drought and waterlogging stress tolerance in plants

Abdullah Alsaeedi[1], Mohamed M. Elgarawani[2], Tarek Alshaal[3,4] and Nevien Elhawat[4,5]

[1]Department of Environment and Natural Resources, Faculty of Agriculture and Food Science, King Faisal University, Al Hofuf, Saudi Arabia [2]Department of Care Scientific Research Care and Training, Research and training Station, King Faisal University, Al Hofuf, Saudi Arabia [3]Department of Soil and Water, Faculty of Agriculture, Kafrelsheikh University, Kafr El-Sheikh, Egypt [4]Department of Applied Plant Biology, University of Debrecen, Debrecen, Hungary [5]Department of Biological and Environmental Sciences, Faculty of Home Economic, Al-Azhar University, Cairo, Egypt

10.1 Introduction

Drought and waterlogging possibly occur for the plant in one season and harmfully affect the plant's metabolism. One-third of the world's arable land has experienced yield reduction due to cyclic or unpredictable drought [1]. Drought is a meteorological term and is commonly defined as a period without significant rainfall or a water supply deficiency. Moreover, it could be also defined as a shortage of water availability, including precipitation, irrigation, and soil moisture storage. During the alleviation of a season or the plant growth stage, the nonuniform moisture distribution has been termed as drought [2,3]. Timing, intensity, and duration determine the magnitude of the plant's drought effect [4].

Over the last few decades, the number of waterlogging incidents on cropland has increased worldwide, mainly due to more intense and unpredictable rainfalls due to the dramatic climate change [1,5]. About 10% of the world's arable land is affected by waterlogging caused by heavy rainfall, excess water, and poor soil drainage [6–8].

Generally, environmental stress dominates the causes of crop losses worldwide, reducing average yield dramatically by more than 50% for all major crops [9]. Drought stress is often interrelated, either individually or in combination; it causes severe morphological, physiological, biochemical, and molecular changes that adversely affect plant growth, productivity, and ultimate yield. Drought and waterlogging share the same mechanism of limited gas exchange in plants and all consequences and damage caused by them. Drought could be more complicated than waterlogging due to the potential of combined effects of drought, salinity, and heat stresses.

In the earth's crust, silicon (Si) is the second most plentiful element after oxygen. Silicon is a major inorganic component in higher plants, and a large amount of evidence demonstrating the value of Si in improving crop yields [10]. However, in higher plants, the Si essentiality is not yet proved since there is no obvious evidence on its function in plant metabolism [11]. Despite Si, to date, is not essential to plant growth, it has several benefits to plants such as enhancing growth, yield, crop quality, N_2 fixation, photosynthesis, and water retention through diminishing transpiration, particularly under abiotic and biotic stresses, that is, drought, high temperature, UV radiation, nutrient deficiency, salinity, metal toxicity, and pathogen and fungus attack [12]. Enhancement in the physiological processes, plant growth, and enlargement under several stresses with Si supplementation is well documented. The exogenous application of Si enhances plant tolerance to drought stress by different

Silicon and Nano-silicon in Environmental Stress Management and Crop Quality Improvement.
DOI: https://doi.org/10.1016/B978-0-323-91225-9.00005-4

mechanisms. Silicon application can decrease drought stress by increasing plant uptake of inorganic nutrients and shifting gas exchange features in plants [13]. Soil application of Si increased the uptake of macronutrients (i.e., P, K, Ca, and Mg) and micronutrients (i.e., Fe, Cu, and Mn) in crop under water-deficit stress [14].

10.2 Drought and waterlogging stress definition and forms

In agriculture, drought is known for representing a negative imbalance between plant transpiration, soil evaporation against water, and soil moisture supply, resulting in low soil potential that hinders the water flow into the root's system or causes reverse flow, commonly known as "dehydration." Drought is characterized by the reduction of water content, turgor, total water potential, wilting, closure of stomata, and a decrease in cell enlargement and growth [15]. The drought (soil or atmospheric water deficits) alongside the conjunction of high temperature and radiation has been recorded on a global basis [16,17]. There are several reasons leading to water deficit in the plant such as low rainfall, soil moisture, salinity, high or low temperature, dry wind, and high intensity of light that in some conditions provide enough water in the soil, but the plants are not capable of uptake it [18,19]. The most comprehensive definition of drought or water stress is defined as a deficiency of water supply to crops for meeting the evapotranspiration rate at the right timing regardless of the source of water, causing stress or highly unpredictable fluctuation imposed on average plant growth, metabolism, and productivity [20,21].

Waterlogging occurs when rainfall or irrigation water deposits on the soil surface or subsoil for a prolonged period; it can also happen when the amount of added water through rainfall or irrigation is more than soil percolation capacity within 1 or 2 days [22]. In waterlogging status, all soil pores are saturated with water with rare pores for oxygen; this makes waterlogging stress always combined with oxygen deficiency stress. These phenomena termed in the physiological study with "hypoxia," the complete blockage of oxygen termed "anoxia" [23—25].

10.3 Ecological grouping of plant according to drought and waterlogging stress tolerance

Plants are classified into five ecological groups according to their drought or waterlogging tolerance as follows: hydrophytes, hygrophytes, mesophytes, xerophytes, and halophytes. The last definition of the term "hydrophyte" introduced by the Federal Manual for Identifying and Delineating Jurisdictional Wetlands is "any macrophyte that grows in water or on a substrate that is at least periodically deficient in oxygen as a result of excessive water content; plants typically found in wetlands and other aquatic habitats" [26]. A hygrophyte is a plant growing above ground that is acclimatized to ambient air saturated with moisture. The main habitats of these plants are wet and dark forests, very shaded swamps, and very humid meadows. Hygrophytes are the least tolerant group to drought among all types of terrestrial plants [27].

Mesophytes are plant species with a moderate affinity toward water supply. Most of the known plant species belong to mesophytes that require suitable growth conditions. Their root system and leaves are well-developed. They possess big leaves covered with a cuticle layer; they also have high growth rates. They are distinguished with unprotected stomata, well-developed vascular tissues, and herbaceous or woody stem. Consequently, they cannot grow on waterlogged soils or in water; also, they are unable to survive in dry lands. Grasses and herbs are the simplest mesophytes species, while the richer populations include herbs and bushes. The richest mesophytic communities comprise of trees in tropical forests [27].

A xerophyte is a plant that thrives in environments with low soil moisture availability. These plants flourish in deserts or where the rainfall is naturally low; however, some xerophytes can also exist in wet habitats. These plants have unique adaptive features to stand desiccation such as reduced transpiration, high root osmotic pressure, deep roots, short vegetative period, low vegetation, and intracellular CO_2 cycle (https://www.brainkart.com/article/Adaptations-of-plants_39974/).

10.4 Response of plant physiology, biochemistry, and molecular biology of drought and waterlogging stress tolerance in plants

Drought is a physiological form of water deficit where soil water available to the plant is insufficient, which harmfully affects the plant's metabolism. However, plants possess multiple morphological (i.e., reduced leaf area, reduced shoot length, leaf molding, wax content, efficient rooting system, and stability in yield and number of branches), physiological (i.e., transpiration, water use efficiency, stomatal activity, and osmotic adjustment), and biochemical responses (i.e., accumulation of proline, polyamine, trehalose, increase of nitrate reductase activity, and storage of carbohydrate at cellular and organism levels) under drought stress, making it a more complex phenomenon to decipher [28–30]. Out of various plant responses to water lack, enhanced ABA accumulation is one of the key mechanisms of adaptation to water stress [31,32]. The ABA, as a plant growth regulator, plays an important role in the response and tolerance against dehydration conditions. It is well documented that dehydration prompts the synthesis of ABA, which consequently triggers the expression of several genes, that is, rd22 [33], RD29A, RD29B, KIN2, RAB18 [34], and PYL8 [35].

10.4.1 Physiological response to drought stress

Waterlessness causes stress in plant growth with a 30%–50% decrease in yield because little humidity happens as a result of the high evapotranspiration, temperature, and high concentration of sunlight [36]. High temperatures caused by drought stress increase respiration, photosynthesis, and enzyme activity in the plant.

Waterlessness causes stress in plant growth with a 30%–50% decrease in yield because little humidity happens as a result of the high evapotranspiration, temperature, and high concentration of sunlight. High temperatures caused by drought stress increase respiration, photosynthesis, and enzyme activity in the plant [37].

The prolonged drought stress conditions lead to the accumulation of salts and ions in the upper layer of the soil around the root system, causing osmotic stress and ion toxicity. The first reaction to stress is a biophysical response. Actually, in increasing drought stress, cell wall withered and loose, then a decrease in cell volume, pressure, and the relative development of the cell occurs; consequently, growth is decreased [38].

10.4.2 Molecular response to drought stress

In drought environments, reduced water potential and increased cell content of abscisic acid (ABA) regulate the metabolism of cells. Increased substances such as proline, glycine, and betaine can be one of the major molecular responses to drought stress [39]. Accretion of solutes in cells under stress conditions, in order to sustain cell volume against the loss of water, is called the osmotic adaptation [40]. Drought stress prompts the generation of the free radicals that cause lipid peroxidation and membrane deterioration in plants [41]. Drought stress leads to an imbalance between antioxidant defenses and the amount of reactive oxygen species (ROS), resulting in oxidative stress. The ROS are required for intracellular signals but at high concentration can cause destruction at various levels of the organization including chloroplasts [42]. The ROS have the capacity to initiate lipid peroxidation and destroy proteins, lipids, and nucleic acids [43]. Mechanism of retardation of lipid peroxidation consists of free radical rummaging enzymes such as catalase, peroxidase, and superoxide dismutase [9]. A number of enzymatic and nonenzymatic antioxidants are present in chloroplasts that serve to stop ROS accumulation [4]. Under water stress condition, the development of ROS increased and the antioxidant system protects the cell by controlling the intracellular ROS concentration. One of the expected consequences of water stress-induced cellular build-up of ROS is an increase in lipid peroxidation. The peroxidation of lipids in the cell membrane is one of the most damaging cellular responses spotted in response to water stress [41]. The quantity of lipid peroxidation has also long been reflected as one of the factors, which show the sternness of stress described by a plant [44].

10.4.3 Waterlogging

In low-lying rainy areas, waterlogging is a serious problem that threatens crop growth and productivity. The main cause of plant injuries under waterlogging is oxygen deficiency (hypoxia) that influences nutrients and water uptake, so the plants wither even when their roots are enclosed by extra water. Deficiency of oxygen converts the energy metabolism from the aerobic pathway to the anaerobic one. Plants, adapted to water-logged conditions, have different mechanisms, that is, aerenchyma formation, increased availability of soluble sugars, greater activity of glycolytic pathway and fermentation enzymes, and involvement of antioxidant defense mechanism to cope with the posthypoxia/anoxia oxidative stress. Gaseous plant hormone, ethylene, plays a vital role in enhancing plant response to hypoxia. It has been reported that ethylene induces the gene expression of enzymes associated with aerenchyma formation, glycolysis, and fermentation pathway. Moreover, nonsymbiotic hemoglobin genes and nitric oxide have also been suggested as an alternative way for fermentation for keeping lower redox possible (low $NADH/NAD^+$ ratio) and thereby playing an important role in anaerobic stress tolerance [45].

10.4.4 Physiological response to waterlogging

Plants grown under waterlogged conditions suffer from the low activity of ribulose 1,5-bisphosphate carboxylase-oxygenase (Rubisco) and photosynthesis of mesophyll cells. Irfan et al. [46] reported an increase in Rubisco activity in bitter melon exposed to flooding condition in the first day compared to control followed by a decrease to lower levels. The level of Rubisco carbamylation follows a direct relationship with the level of Rubisco activation [47], which indeed is controlled by internal carbon dioxide (C_i) and light intensity [48]. Increasing C_i, in general, reduces the activation of Rubisco [49]. Elevating C_i first causes an initial increase in Rubisco activity, while the extended increase in C_i leads to a decline in Rubisco activity [48]. The production of ribulose 1,5-bisphosphate (RuBP) decreased under the elevated C_i, whereas an increase in 3-phosphoglycerate (PGA) level was reported in different plants [48,50]. The active site in Rubisco may be inhibited by the excessive consumption of RuBP. This maintains equilibrium between the consumption and generation of RuBP and the controlling of Rubisco activity [49]. It has been proposed that high CO_2 concentration results in the overproduction of PGA causing acidification of the stroma that consequently reduces the Rubisco carbamylation [50]. The photosynthesis rate is strongly correlated to the activity of the Rubisco enzyme since Rubisco is the cornerstone in inserting CO_2 onto RuBP yielding PGA that is later converted to carbohydrates [50]. The downregulation of Rubisco protein and/or lowered activity of the existing enzyme might reduce the activity of Rubisco under waterlogged conditions. Moreover, under flooded conditions, the transportation of photosynthates via the phloem is blocked lowering the demand for sucrose [51]. This leads to the accumulation of starch in chloroplasts [48,52]. It is suggested that the accumulation of starch may reduce the exchange rate of CO_2 in plants under waterlogged conditions. These observed physiological changes demonstrate that the stomatal and metabolic factors are the major controllers for the decline in the CO_2 exchange rate under waterlogging stress.

10.4.5 Biochemical changes under waterlogging

Organisms utilize fermentation pathways, that is, alcoholic and lactate fermentation, as a mechanism to oxidize the glycolytic substrate and synthesize adenosine triphosphate (ATP), maintaining the viability of the cells under deficient-oxygen supply conditions. The increased rate of glycolysis increases the supply of these substrates that feed these pathways. Under aerobic conditions, ATP synthesis occurs through the oxidation of carbon compounds using oxygen as a final electron acceptor. On the contrary, oxidation of substrates is restricted by the lack of oxygen under hypoxia conditions. So far, two pathways are proposed under anaerobic conditions: (1) fermentation pathway and (2) the hemoglobin/nitric oxide cycle (Hb-NO cycle) [45].

10.4.6 Molecular response to waterlogging

Due to the limited dissolving of oxygen gas in freshwater, plants grown under waterlogged conditions suffer from a hypoxic state. Higher plants are aerobic organisms. The response of plants to external hypoxia has been intensively studied for decades. Under anaerobic conditions, plants utilize glycolysis to produce ATP and ethanol through the fermentation passage to produce NAD^+ that is essential for supporting the Embden-Meyerhof-Parnas pathway [53]. Plants tolerate waterlogged conditions not only through metabolic changes but also via different morphological adaptations such as aerenchyma formation [54]. The comprehensive studying of the key genes and processes controlling the waterlogging tolerance of plants is important to uncover the possible resistance mechanisms and ultimately transferring them into crops [55]. Proteomics analyses of several plants exposed to hypoxia conditions have recognized many anaerobically induced polypeptides (ANPs) that are substantial for waterlogging tolerance [56,57]. Also, several studies reported that the ANPs are mainly utilized in glycolysis and fermentation passageways [58]. Sachs et al. [59] in their study on maize under hypoxia conditions identified several anaerobic peptides, that is, glucose-6-phosphate isomerase, enolase, aldolase, glyceraldehyde-3-phosphate dehydrogenase, alcohol dehydrogenase, and sucrose synthase that have been selectively induced by hypoxia. Microarray measurements have been done to study the response of different plants to hypoxia, that is, cotton [56], *Arabidopsis thaliana* [60], maize [53] and other plants [61]. These investigations showed an increase in the expression of different proteins, that is, Zinc finger-like protein, SKP1/ASK1-like protein, and 20S proteasome subunit α-3, upon exposure to hypoxia after 2 h [58].

10.5 Effect of drought and waterlogging stress on plant and yield components

The drought effect cycle can be shortened as drought stress sinks the water potential of the growing plants leading to dehydration, decreased stomatal conductance and closure of stomatal apertures, reduction in CO_2, decline in photosynthesis metabolism [62,63], altered chlorophyll biosynthesis, photo-inhibition of photosystem II (PSII) [64,65], changes in membrane-bound ATP-ase enzyme complex, in addition to a decline in the activity and the concentration of Rubisco enzyme, and magnitudes the formation of ROS as a result of respiration metabolism shortage [65–68].

10.5.1 Morphological and anatomical changes

Plant growth and development are highly impacted by the metabolism disruption caused by drought stress, which triggers many responses that act adversely on germination, cell division, and elongation, of aboveground architecture and root proliferation [63,69–71]. Drought stress reduces the germination percentage and all associated parameters in most of the plants and crops [72]. At water potential of 0.5 MPa (moderate drought stress), germination percentage and rate of soybean (*Glycine max* L.) were dropped by 50%, while no germination was noticed at water potential of 1 MPa (severe drought stress) [73,74]. Likewise, germination percentage of wheat (*Triticum aestivum* L.) reduced by 50% at 0.5 MPa, whereas no germination was seen at 1 MPa; also, seedling vigor and coleoptile length reduced by 87% and 90%, respectively, at 1 MPa [75–77]. In barley (*Hordeum vulgare*), germination percentage and vigor index declined by 25% and 65%, respectively, when drought stress increased to 0.5 MPa [78]. Some studies concluded that corn (*Zea mays* L.) lost 50% of its germination percentage and 70% of seed vigor at water potential of 0.5 MPa [79,80]. Sorghum (*Sorghum bicolor*) showed the same response to drought stress in germination parameters [79–83]. Sunflower (*Helianthus annuus* L.) did not germinate at water potential of 1.2 MPa [84]. Potato (*Solanum tuberosum* L.) showed high sensitivity to drought during the establishment stage [85–88]. Germination percentage of tomato (*Solanum lycopersicum*) declined by up to 70% due the increase in water potential stress to 0.2 MPa [89,90]. The reduction in seed germination under drought stress could be ascribed to changes in the enzymatic and metabolic processes existing in the seeds, such as the development of metabolites triggered by drought stress in the generation of ROS, reduction in water diffusion through the tegument, and the absorption of water by the seeds that induces a hydration deficit [91]. Water-deficit stress severely impairs seed germination by 50% and increase to complete inhibition in some crops like cotton (*Gossypium hirsutum var. latifolium L*) [92], alfalfa (*Medicago sativa*) [93], and other plants [94–96].

Growth parameters including plant height and leaf area are major features reflecting plant flourishing in terms of physiological processing and yield productivity. Plant height and leaf area are led by the frequent cell division, cell elongation, and cell enlargement, which they are limited by water availability for plants. Leaf rolling, to avoid further dehydration, and descent in leaf flag area in cereals; different wheat cultivars showed an extreme leaf rolling and more than 50% reduction in leaf area under water deficit [97,98]. Flag leaf length and plant height also declined to less than 20% under water-deficit stress [98–100]. Furthermore, leaf area and parameters of aboveground part of maize [83], soybean, sorghum [64], [101], rice [102], tomato [103], cucumber [104], onion [105], grapes [106], apple [65,107], and other crops [71] were significantly stricken under water-deficit stress condition.

The reduction in biomass and yield is the ultimate outcome of the dysfunction of plant metabolism including photosynthesis, respiration, enzymes secretion, and ROS generation. Under drought stress, yield penalty could be revealed in quantity or quality of biomass, grains, or fruits. The yield reduction depends on plant growth stage, drought severity and duration, the response of plant after stress removal, and interaction between stress and other factors [83,108,109]. For instance, imposing water deficit at silking, grain-filling, and maturity stages in maize led to reduction in the total dry biomass by 37%, 34%, and 21%, respectively; also, general grain yield declined by 90% compared to well-watered plants [110]. Similar examples of yield depletion were reported by many researchers [65,109,111,112]. The myth of "less water less yield" is not valid in the quality aspects; a wealth of researches have conducted in several species water deficit that typically results in diminished plant growth but enhanced fruit quality [113–115]. Characteristics of fruit organoleptic quality include fruit's external appearance, size, texture, and taste [114,116–118]. A considerable number of studies on the water-deficit effect accentuate highly variable responses of plants/fruits depending on the duration and severity of the deficit, the plant/fruit development stage, the genotype, and the presence of other stress factors [116,119,120]. Water-deficit stress, regulated or genuine, affects adversely fruit production due to the low water availability and the depletion of assimilates translocation to the fruits [121], reduction in fruit size and yield during the season for apple [121–123], peach [124,125], nectarine [126], pear [127,128], grapes vine and berry [129], pomegranate [130], date palm [131], mango [132], banana [133,134], tomato [115,135], olive [136], and other fruits [137]. Similar results were obtained by many researches for vegetables crops like onion [138], chili peppers [139,140], cucumber [104], and potato [141] and for field crops like wheat, whereas number of grain per plant and weight of 1000-grain were terribly reduced [65,142]. Maize showed a devastating decrease in its grain yield and kernel number in ear [83,110,143]. The dehydration during water-deficit stress increases the concentration of soluble sugar, acid content, and solid mass of fruit and grain. Apple fruit showed a significant increase in soluble sugar, whereas a decrease in acidity and amino acids content, number of flowering and fruits for next season under water deficit [114,116,144]. It is well documented that, as apart from plant adaptive mechanism under drought stress, plant activates the translocation and accumulation of total soluble sugar (TSS), including sucrose, glucose, fructose, and sorbitol, which constitutes more than 80% from the total solutes involved in the osmotic adjustment process in the cell [123,145−148]. This increases the soluble solid content (SSC) in fruits with sweetness taste. Also, water-deficit stress governs the distribution of sugar in fruits by enzyme activity; this plays an important role in maintaining the stability of cell turgor, controlling fruit hardness, and forming fruit flavor [149]. As fruit enters the climacteric respiration peak under water-deficit stress, acids (i.e., malic, citric, and ascorbic acid) and soluble carbohydrates from starch hydrolysis are needed to maintain the elevated respiratory rate during ripening; this consequently lowers acidity in fruits [123,150−154]. Water-deficit stress in apple during the early stage of fruit development could increase the SSC and total sugar concentration and lower the acidity [123,146,153,155−157], with improved firmness, red color, and aroma [114,158]. This has been reported in most of orchard fruits like peach [113,159,160], pear [161], and plum [162]. Potato (*S. tuberosum L.*) exhibited a significant increment in TSS in tubers under drought stress; this is attributed to the high level of sucrose, also an increase in crude protein content. Date palm, also, showed a significant increase in TSS under water-deficit stress [131,163]. Tomato (*Lycopersicon esculentum* Mill.) fruit firmness and color index were positively correlated with drought stress; taste and nutritional value parameters, such as TSS, organic acids, and vitamin C (ascorbic acid), were improved in response to water-deficit stress [115,164,165]. An increase in the concentration of the active components, that is, capsaicinoids and vitamin C, in chili pepper was reported [140,166]. Protein content in legume and grain crops responded positively to drought stress; on the other hand, carbohydrate content declined due to the deficiency of starch synthesis [167−169]. Cotton quality parameters, that is, fiber length, strength, fineness, and elongation, were negatively affected by water-deficit stress [170].

10.5.2 Morphological and anatomical changes to waterlogging stress

Waterlogging stress is also known to cause a number of morphological and anatomical changes in plants. For instance, some woody species grown under waterlogging stress possessed hypertrophied lenticels as a common anatomical adaptation to these conditions [171]. Moreover, hypertrophic growth corresponded to the radical cell division and expansion of cells close to the stem base. This is proposed to be linked to the production of ethylene and auxin [172]. Another possible reason for the lenticel formation is the downward movement of the oxygen, in addition to the secondary metabolites of the fermentation pathway, that is, ethanol, CO_2, and methane. The biological function of lenticels is still covered; however, their existence is often due to the waterlogging resistance in plants [173]. The number of the hypertrophied lenticels on the submerged part of the plant is higher; this supports the hypothesis on their role in maintaining the water homeostasis in the plant and denies the theory about its function in facilitating oxygen entry to the root system. The evidence for the proposed role of the hypertrophied lenticels in the conservation of water homeostasis is the effective involvement in the replacement of decaying roots and helping in water uptake and translocate to the shoot [174]. Another morphological adaptation of plants to waterlogging stress is the emergence of adventitious roots that substitute the basal roots [175]. These special roots ensure the continuous uptake of water and nutrients when basal roots become unable to do so [176]. Moreover, under waterlogged conditions and losing the main root its function, plants sacrifice by the main root to save the energy for the well-adapted new root system [177]. Furthermore, the appearance of the adventitious roots is a sign of waterlogging tolerance [178]. The formation of lacunae gas spaces (aerenchyma) in the root cortex is another morphological adaptation to waterlogging stress. The establishment of parenchyma is also classified as an adaptive trait of plats to waterlogging stress [179]. There are two types of aerenchyma formation. The first type is constitutive aerenchyma where it is independent of abiotic stress. It is created during the development of tissues; this type is called shizogeny. It is formed by the highly controlled specific patter of tissues during cell separation. The second type is referred to as Isogeny; it is induced by the abiotic stress where the cortex layer is partially broken down similar to programmed cell death [180].

10.5.3 Effect of drought on nutritional status

The distribution of plant species depends on their hydrophobicity, nutrient absorption, and carbon fixation, as well as how these aspects are organized. Among the different abiotic forms of stress, drought is a major constraining factor regarding crop yields and productivity worldwide [181]. Drought is a temporary severe environmental condition that affects plant growth and development [9,104]. In semiarid regions of the tropics, the occurrence of drought or lack of water in the soil is common, while in temperate and other tropical regions crop plants are subject to monsoon periods of water stress, especially during summer [182]. Various physiological and biochemical processes are altered by dehydration, such as relativism [183], gas exchange, photosynthesis [184] and metabolism of carbohydrates, protein, amino acids, and other organic compounds [9]. A little piece of literature studied the effect of dehydration on nutrient absorption and the consequences of physiological processes. Under drought stress, plants also suffer from limited supply with nutrients due to the decrease in mobility and uptake of nutrients [185]. Except K and Ca, the other macronutrients (N, S, P, and Mg) are incorporated into vital biomolecules such as amino acids, proteins, nucleic acids, phospholipids, and chlorophyll [185]. Uptake of nutrients by plant's root is controlled by supply and demand at the root surface [186]. Although plant's roots are able to absorb nutrients found even at very low concentration in soil solution due to their very large surface area, the bioavailability of these nutrients depends on soil moisture content that allows nutrients to flow from soil matrix to roots [187].Consequently, under water stress condition, roots become unable to uptake many nutrients from the soil because of the low root activity as well as slow mobility of nutrients [188]. Furthermore, drought stress indirectly affects the nutrients bioavailability where the mineralization process depends on microorganisms and enzymes activity, which require adequate soil water supply for their growth. Moreover, water molecules are adhered on the surfaces of soil particles forming hydration shells that influence the physiochemical reactions. Water is necessary for the diffusion of solutes and nutrients within plant tissues; therefore it plays an important role in distribution of metabolites and solutes within the whole plant [189].

Taken up water and nutrients by plant roots are mainly transferred via the xylem to the aboveground parts of plant; however, the flow rate of water throughout the root and xylem heavily depends on the root pressure and transpiration rate [189]. The increase in the transpiration rate increases the uptake and transportation of nutrients through the xylem [189]. Drought stress reduces nutrient uptake due to the reduction in soil moisture, which decreases the diffusion rate of nutrient from the soil matrix to the absorbing root surface and translocation to the leaves [190]. A number of studies have shown a decrease in accumulation of some elements beside other physiological effects under water stress. Drought stress enhances the stomatal closure to reduces water loss via transpiration [191]; thus limited nutrient transport from the root to the shoot occurs alongside imbalance in active transport and membrane permeability, resulting in a reduced absorption power in the roots [38,109,190]. Therefore drought causes low nutrient availability in the soil and lesser nutrient transport in plants [190]. These factors acting together have critical consequences for plant development, affecting different physiological processes.

10.5.4 Effect of waterlogging on nutritional status

Waterlogging not only causes hypoxia in the root zone [192] but also drops down the level of the endogenous nutrients [193]. Also, some pieces of literature reported that hypoxia increases the permeability of Na^+ through the root cell membrane [192]. Moreover, waterlogging leads to an extreme shortage of indispensable nutrients such as N, P, K, Mg, and Ca [194]. Similarly, a substantial decrease in the uptake of N, P, K, and Ca by rapeseed (Brassica napus) exposed to short waterlogged stress was reported Boem et al. [195]. A decrease in the content of P, K, and Mg in wheat (T. aestivum) sprouts exposed to waterlogging stress was reported by Stieger and Feller [196]. Otherwise, endogenous Ca content in wheat plants under waterlogged conditions showed no change. However, a reduction in Ca concentration alongside N, P, K, and Mg was noticed in wheat grown under water-saturated conditions [68]. Likewise, an obvious difference in uptake of Cu, Zn, P, and K by waterlogging-sensitive wheat genotype was reported by Fritioff et al. [197] compared to waterlogging-tolerant genotypes.

10.6 Mechanisms of drought and waterlogging stress in plants

Drought stress induces visual symptoms on plant starts with loss of leaf turgor, drooping, wilting, yellowing, necrosis of the tip tissues, and leaf death [68,198–200]. Symptoms progressing depends on drought severity, plant species, growth stage, and soil physiochemical properties [201]. As soil surrounding root continue drying and soil potential decrease, the root generates two warning message signals to the whole plant, particularly leaf, that is, hydraulic gradient signal and biosynthesis compound signals. Regardless of its type, the signal is the first step in the plant's drought tolerance mechanism, considering that biosynthesized signals dominate this effect and process in advance drought stress [202–205].

Stomatal closure is the primary action of these signals to balance the outflux transpiration with the influx of moisture to prevent further stresses. Drought stress is involved in the disintegration of the vital bioprocesses in the plant, that is, photosynthesis, respiration, carboxylation enzymes, ATP synthesis, and CO_2 assimilation [206–208]. Under mild and moderate drought, the stomatal limitation is the primary influencer in the reduction of photosynthesis due to low CO_2 concentration and low water retention in leaf [107,209–211]. Once the drought stress is relieved, stomata and photosynthesis return to their original functionality [212–214]. During this process, root xylem sap is acidic due to the generation of ABA [215].

Contrarily, prolonged drought stress completely closes stomata resulting in total blocking for CO_2 and O_2; consequently, none stomatal limitation will induce the generation of ROS causing changes in the chloroplast structure and damage to plant membrane systems [216,217]. Also, membrane lipid peroxidation aggravates the production of superoxide free radicals that degrade photosynthetic pigments, thereby destroying the photosynthetic electron transfer system, hindering the transfer of electrons, and destroying the physiological functions of photosynthetic organs, which make it challenging to return the plant to life after relieving drought stress [216,218,219].

Flooding, generally, is classified into two forms depending on water existing intensity, that is, waterlogging and submergence. Waterlogging is the condition in which water exists on the soil surface, and plant roots only are submerged in water; while submergence is the state that the whole plant partially or completely is immersed in water [7,220,221]. Excess water in the rhizosphere reduces the oxygen diffusion to 1/10,000 of the value in empty pores, which is $0.240 \ cm^2 \ s^{-1}$; in other words, the flux of O_2 into soils is approximately 320,000 times less when the soil pores are filled with water than when they are filled with gas [222,223].

Hypoxia is a phenomenon that occurs in the rhizosphere as oxygen in soil pore becomes limited to a point below the optimum level. In plant physiology, the term "hypoxia" is reserved for situations in which the oxygen concentration is a limiting factor, and it is the most common form of waterlogging stress on the plant [224–226]. Anoxia is the extreme form of hypoxia used to qualify the lacking of oxygen in physiological studies, and it is a usual form of stress in soil that experiences long-term flooding or waterlogging. It occurs in plants entirely submerged by water, in deep roots below flood water [58,225,227].

Waterlogging restricts the gases exchange between plant root and the atmosphere; thus oxygen will not be available for plant respiration. Respiration stoppage leads to the devastation of all oxidization and reduction process. The plant responses to oxygen deprivation by shifting from aerobic respiration to the low ATP-yielding fermentation, leading to an "energy crisis" [221,228]. In plants, respiration is the main metabolic process affected by soil waterlogging. In aerobic respiration, the oxidative phosphorylation process generates energy as ATP, lack of oxygen will block Krebs cycle and electron transport system, and decrease NAD^+ level, pyruvate accumulation, and ATP level [229–231]. In the absence of O_2, plants use an alternate pathway to produce ATP, the anaerobic metabolism or fermentative metabolism extracts energy mainly from glycolysis to produce ATP and NAD^+ [229,230,232]. The fermentation process reduces ATP production efficiency as only three ATP molecules compared to 39 ATP molecules through aerobic respiration of hexose molecule that impairs cellular metabolism and function [229,231,233,234].

10.6.1 Signaling and stomatal behavior

Signaling is the immediate and first response of the plant to any biotic and abiotic stress, including drought or waterlogging stress. Sensing tissues—mainly roots in drought or waterlogging stress—generate a multicascaded systemic signal that travels to other plant parts in an attempt to trigger acclimation and defense mechanisms in these parts, even if they did not yet sense the stress [235–240]. Stomatal and nonstomatal limitations working in the integration process, the nonstomatal limitation is exaggerated by the effect of stomatal limitation, make the segregation of effect difficult [241,242].

Under drought stress conditions, soil dehydration or soil water deficit is the most approved driver directly inducing stomatal closure [243–245]. Direct water evaporation from leaf surface may cause closure of stomata from the guard cells without any metabolic involvement (stomatal limitation). This process of stomatal closure is referred to as hydro passive closure. On the other hand, stomatal closure may also metabolically involve processes (nonstomatal) that require ions and metabolites, which is known as hydro active closure. This process is regulated by phytohormones and enzymes, mainly ABA [244]. Stomatal limitation starts at soil moisture of 50%

from available water (field capacity moisture—wilting point moisture), while the nonstomatal start under severe dryness at 30% or less [209,246−249]. The root tip is sensing the high resistance of moisture moving from soil matrices to root tissues due to the dryness effect, and this creates a reverse hydraulic potential gradient causing root dehydration [250]. ABA is synthesized in the dehydrating root tissues from ß-carotene through several enzymatic steps [251−253], and accumulated in root vascular parenchyma, then quickly transported into the apoplastic space of the leaf tissues through xylem by long-distance transport signal. Afterward, it enters the cytoplasm of leaf cells by simple diffusion without specific transporters [254]. Although 25%−30% of the ABA in xylem sap might come from shoots due to recirculation of basipetally transported ABA in the phloem [255], leaf ABA concentration increases in response to soil drying due to an increase in ABA biosynthesis by roots and ABA recirculation from shoots via phloem transport [256], where it induces and maintains stomatal closure [250,257−259]. Root xylem pH dropping was due to the accumulation of the naturally weak acid endogenous ABA [215,259]; on the other hand, xylem sap and leaf apoplast pH increased as water-deficit effect increases [202,215,260−262]. ABA is a weak acid in the well-watered plant; it exists in two forms, that is, anionic (ABA$^-$) and protonated ABAH [218,263,264]. Due to the alkaline condition of the xylem sap, ABA that arrives from the root will remain deprotonated (anionic form) and will not diffuse through the plasma membrane and taken up passively by mesophyll. This leads to reducing the ABA transported into the mesophyll cells and a build-up of ABA in the apoplast, which in turn leads to stomatal closure [202,218,263,265−270]. Activation of ABA receptors in guard cells induces Ca^{2+}-dependent and Ca^{2+}-independent signaling to regulate ion channels activity to mediate the closure of stomata pores [218,271,272]. The rapid activation of nonselective ion channels by ABA triggers membrane depolarization and allows Ca^{2+} influx from the extracellular space. The elevation of cytosolic Ca^{2+} may be due to the release of Ca^{2+} from intracellular stores and/or influx through the plasma membrane and tonoplast [273−275] or by the activation of phospholipase C, which generates inositol 1,4,5-trisphosphate (IP$_3$) and diacylglycerol by hydrolysis of phosphatidylinositol 4,5-bisphosphate (PIP$_2$) leading to release Ca^{2+} from an intracellular store [274−278]. There are two major K^+ channels in the guard cells; ABA inhibits K_{in} channels, while it activates K_{out} channels. ABA-induced membrane depolarization coupled with upregulation of K_{out} channel activity induces net K^+ efflux from guard cells and a parallel loss of Cl^- and malate and/or conversion of malate to starch, which consequently shrinks cell turgor and causes stomatal aperture closure [279−282].

Under waterlogging conditions, oxygen deprivation is the substantial cause beyond plant wilting, even when the root system is surrounded by an excess of water. Many researchers attributed the stomatal closure to ABA level in leaves but emphasizing that the source of this ABA is not the root [283−287]. Low-oxygen status (hypoxia) triggers many cellular responses in roots system such as a decline in the production of ATP, drop the cytoplasmic pH below seven, and impaired ethylene biosynthesis process [233,237,262,288−290]. The effect of hypoxia or anoxia becomes clearly evident within few minutes to few hours [228,233,262,289−291]. Inhibition of respiration and loss of ATP synthesis induce root pores blockage (low hydraulic conductance), which stops ion transport systems that normally create the gradient in water potential across the root endodermis leads to wilting phase. The disturbance in root ATP synthesis, in response to root hydraulic conductivity, showed a decline in *Arabidopsis thaliana* (*Arabidopsis*) *thaliana* [292,293], *T. aestivum* (wheat) [292,293], *Z. mays* (maize) [294], and *S. lycopersicum* (tomato) [295]. The downregulation of hydraulic conductivity by hypoxia was attributed to the inhibition of aquaporins (AQP) activities caused by cellular acidosis [262,293]. The reduction of root hydraulic conductivity may be the reason behind stomatal closure at the beginning of the waterlogging stress [295−297].

10.6.2 Mechanisms of drought resistance

Drought resistance, actually, is the capability of species or cultivars for growth and production under water scarcity conditions. The consequences of the extended drought stress on physiological and morphological traits of plant and ultimately the yield depend on many factors such as the exposure time, growth stage of plant, plant species, and water holding capacity of the soil in the root zone [37]. To reduce water loss, plants tend to close the stomata, decrease sweating, or a combination of all two levels to cut down the amount of transpiration [298]. With increasing water shortages, some plant species can clog pores; this reduces transpiration, especially when the stomata are completely blocked and cuticular resistance is truer. One of the drought tolerances in crop plants is through water conservation and sustaining water absorption. The important feature is that this requires one to have deep roots and branches and a low resistance to flow of water inside the plant [71].

10.6.3 Mechanisms of resistance to waterlogging

The antioxidant system plays a key role in the plant's tolerance to waterlogging stress. The ROS are generated at the transmission when the whole plant or part of it is exposed to hypoxia or returned to an aerobic environment. Overproduction of ROS, that is, superoxide (O_2^-), hydrogen peroxide (H_2O_2), hydroxyl radical (OH^-) [9,48], and NO [299] occurs when plants are exposed to stress. After waterlogging, rhizomes and roots are more susceptible to oxidative shock under these conditions. The ROS in higher concentrations may injury plants by oxidizing the biomolecules such as proteins, lipids, and nucleic acids leading to serious mutations [9]. However, plants possess different mechanisms to scavenge the deleterious ROS such as enzymatic antioxidants, that is, peroxidase, catalase (CAT), superoxide dismutase (SOD), and nonenzymatic antioxidants, that is, glutathione, ascorbate, proline, and tocopherol [300]. The synthesis of the antioxidants is more obvious through rearranging the biochemical setup throughout stress, in which the plant stunts the growth to redirect these molecules for new functions. Therefore in the interval between return to aerobic and reactivation of the electron transport chain, ROS production is preferred. An imbalance between the scavenging and the production of ROS due to a change in the biochemical composition under stress may lead to the leakage of excess ROS. This, indeed, causes severe lipid peroxidation and declines the photosynthesis rate. The constant ROS level in different plant organs is controlled by both ROS-producing passageways and ROS-scavenging techniques [9].

The recent findings of the novel roles of these species (ROS and free radicals) have been attributed to controlling and regulating various biological processes. Nicotinamide adenine dinucleotide phosphate oxidases, amine oxidases, and cell wall-bound peroxidases are recently recognized as new centers of ROS production. The NO is well known as a plant growth regulator on the inter/intracellular level that acts as nonenzymatic antioxidants [301,302]. At higher levels, NO, however, can injury the plants since it acts as a potent oxidant that mainly disrupts proteins [303]. O_2^-, as an important ROS species, targets the proteins rich in Fe-S cluster, heme groups, or disulfide bond and oxidizes them [304]. Thus O^{2-} is destructive to electron transfer in photosynthesis. Rubisco, the main enzyme in photosynthesis, is very susceptible to oxidative stress [305]. O^{2-} enhances the programmed cell death (apoptosis), lipid peroxidation, and subsequently membrane injury [306]. Sequestration of metal ions, for example, Fe^{3+} and Cu^{2+}, by ferritin and Cu-binding proteins is expected to be an important mechanism to avoid the OH^--derived toxicity through the metal-dependent Haber Weiss or Fenton reactions [307].

10.7 Role of silicon and nanosilicon in alleviating the deleterious effect of drought and waterlogging stress

The soil Si content ranges between 50 and 400 g kg^{-1}; however, it heavily depends on the soil type where in clay soil it varies from 200 to 350 g kg^{-1}, while from 450 to 480 g kg^{-1} in sandy soil [104,308]. The major earth-Si compounds are the inactive quartz and/or crystalline silicates that make up the soil basic. The major active Si compounds that are represented physically and chemically in the soil are soluble monosilicate acids, polysilicate acids, and organosilicon complexes [309].

Although having abundant obtainability, silicic acid (H_4SiO_4) is the major phytoavailable form of Si in the soil, and it is typically a limiting factor. Consequently, to increase Si uptake and Si-derived benefits in the plants, the molecular basis of Si-absorb and transport within the tissues of plants is the way of understanding its mechanism. Silicon is deliberated as a multitalented micronutrient because of its multipurpose role in providing numerous benefits for plant growth mostly under stress conditions [96].

The use of Si fertilizers today is very common in many crop production systems around the world. Seven crops out of the most produced crops worldwide in 2012 are classified as Si-accumulators with more than 1.0% (on dry mass basis) of Si in their tissues including wheat (*T. aestivum*), rice (*Oryza sativa*), and sugarcane (*Saccharum officinarum*) [310].

Despite it is abundant in soil; Si is never found in a phytoavailable form and is always found in complexes with other elements forming oxides or silicates [14]. Plants and microorganisms are able to uptake Si in the form of monosilicic acid which is soluble Si form [311] and finally precipitated in plant tissues as an amorphous form [312]. Another additional benefit of Si is that Si controls the chemical and biological characteristics of soil such as the content of P, Al, Fe, and Mn; the mobility of heavy metals; microbial activity; soil organic matter; and construction of polysilicic acids [313]. Polysilicic acid vitally affects soil structure, water retention, adsorption capacity, and stabilization of soil corrosion [313]. Soil solution contains silicic acid at very varied concentration ranged from 0.1 to 0.6 mM. In the aboveground part of plants, a wide range of Si content was cited from 1.0 to

100.0 g kg^{-1} (on dry mass basis). This depends mainly on the uptake and transportation features of Si, plant species, soil Si content, and chemical form of Si [314]. Species of the Gramineae family are characterized with active Si uptake such as wheat [314], rice [315], barley [316], and ryegrass [317]. Though, some Gramineae species such as oats passively absorb Si [318]. Passive uptake of Si has been confirmed in some dicots such as melon, cucumber, strawberry, and soybean [319].

Silicon provides strong suit to the plant by making the plant tissues stronger, rigid, and massive [320]. Soluble Si can improve disease resistance through altering the signaling system in plant under biotic stress where Si interacts with many significant compounds [321]. Moreover, the accumulation of Si in plants tissues induces the synthesis of phenolic compounds and phytoalexins that support the plant resistance against several pathogens [322].

Water availability in soil is one of the main environmental factors that limit the development and yield of plants. Water deficit may result from the scarcity of water in soil (drought) or inability of plant to absorb water (physiological drought). In the physiological drought, there is enough water in the soil solution around plant roots but plants cannot take it up due to some physiological reason such as high salt concentrations (salinity), excess of water (flooding), or low temperatures. All these factors, consequently, cause water-deficit stress and changes in cell water relations as water potential reduced and turgor of plant cells lowered. These changes result in disruption of the most important processes and reduction in growth rates. When plants are exposed to salinity, they suffer additionally from a toxic level of salts in cells beside the osmotic stress [323].

Unfortunately, molecular mechanisms underlying Si uptake in plants are unidentified yet [324]. Investigations into the mechanisms by which plant takes up Si by Emamverdian et al. [325] indicated that Si interacts with polyphenols in xylem cell walls and affected lignin deposition and biosynthesis. The addition of Si to rice grown under polyethylene glycol-induced water deficit reduced the rate of transpiration and enhanced membrane permeability [326]. Exogenous application of Si to sorghum (*So. bicolor* L.) plants exposed to elevated Zn doses in hydroponic culture alleviated Zn toxicity and increased the dry mass and relative water content (RWC). It was suggested that the development of drought tolerance in sorghum achieved by the addition of Si might be related to improvement of water uptake ability [323]. Similarly, Si application enhanced water retention and increased dry mass of wheat plants in pots under drought stress [326]. Also, Si was placed in the cell walls of roots, leaves, stems, and hulls reducing water loss via transpiration [327]. Richmond and Sussman [290] and Ma et al. [324] have reported that this might be a useful result of Si on plant growth during stress conditions, because it is improbable that Si affects the activity of antioxidant enzymes. Applying Si modifies the cell wall construction, which may be responsible for the increase in the cell wall stretching [328].

Plants can uptake Si either by active or passive mechanisms [329]. The quantity of uptake of Si by the active mechanism is usually larger than that expected based on the mass movement and is attributed to the concreteness of Si transporters in shoots and roots that facilitate the absorption process across the membranes of root cells. In rice, the transporters mediate both the radial transport and the xylem loading of Si [330]. Additionally, these transporters were recently identified and were coded by low-Si genes such as the Lsi1 and Lsi2 in roots and the Lsi6 in shoots [315]. Lsi1 may encode a membrane protein similar to the water channel proteins, also known as aquaporin [324]. The amount of Si taken up by the plant via the passive mechanism is likely entirely driven by mass flow. In the rejective mechanism, the build-up of the concentration of H_4SiO_4 in the soil solution typically results from the low concentrations of Si that are absorbed by plants. Ma et al. [331] categorized plant species based on the mechanisms of Si uptake. The plants that rely primarily on active, passive or rejective mechanisms are classified as high-, intermediate- or non-accumulators, respectively. The plants in the high-accumulator category have a Si content in the shoot that ranges from 1.0% to 10% dry weight and are primarily monocotyledons such as bamboo (Bambuseae), barley (*H. vulgare*), rice (*O. sativa*), sorghum (*S. bicolor*), sugarcane (*S. officinarum*), and wheat (*T. aestivum*) [332]. Because of the efficient Si uptake system in the high-accumulators, the amount of Si taken up by plant from the soil is several times higher than the uptake of some of the essential macro- or micronutrients. For example, the uptake of N is the largest among the essential nutrients, but the accumulation of Si may be twice the amount of N in rice. The intermediate-accumulators are mostly dryland Gramineae species with shoot Si content that ranges between 0.5% and 1.5% dry weight. Cucumber and tomato as dicots are classified as low-accumulators since they accumulate Si at the rate of <0.2% shoot dry mass. Mitani and Ma [330] attributed the low accumulation of Si in low-accumulators to a lack of specific transporters to facilitate the radial transport and the xylem loading of Si and suggested that the transport of Si across cells was accomplished via a passive diffusion mechanism. Later, Liang et al. [319] reported that both the active or passive uptake mechanisms of Si that occur in high-accumulator plants also found in the intermediate-accumulator plants (e.g., sunflower and wax gourd).

Absorbed silicic acid is deposited in epidermal layers in leaf cells in the form of hard, polymerized silica gel ($SiO_2 \cdot nH_2O$) that is also termed as phytolith [333]. In wheat, the Si is deposited in all tissues but high concentrations are found in the inner tangential and radial walls of the endodermis [334]. Silica cells are specific cells that contain phytoliths; they found in vascular bundles and in silica bodies in bulliform cells, fusoid cells, or prickle hairs in rice, wheat, and bamboo, respectively [335]. Phytoliths are best recognized as biogenic opal (Si-O-Si bonding) [336]. The deposition of Si in plant tissues occurs when silicic acid concentration is above 2 mol m^{-3} [337] and takes place initially in the epidermis of the shoots as well as the vascular system and the endodermis of roots of some plant species [327]. The precipitated Si in plant cells is immobile and cannot be transferred to newly growing tissues [338]. Transpiration remains a viable option as one of the primary drivers in Si transport and deposition in plants; therefore the duration of plant growth significantly affects the concentration of Si in plant; for example, older leaves contain more Si than younger leaves [339]. Based on earlier studies, the silica gel possibly binds with organic components [336]. Conversely, the studies by Casey et al. [340] and Ma et al. [331] confirmed that only the mono- and the di-silicic acids but not the organosilicic complexes were found in the xylem exudates of rice and wheat.

10.8 Mechanisms of silicon- and nanosilicon-mediated drought and waterlogging stress tolerance in plants

Exposure growing crops to abiotic stress causes a reduction of more than 50% of crop productivity worldwide [341]. Physiological responses such as ion uptake, photosynthesis, translocation, respiration, transpiration rate, stomatal behavior and conductivity, mineral nutrition, seed germination, and water relation are affected by different abiotic stresses such as drought [342], salinity [343], and heavy metal [344]. Enhancement in the physiological processes, plant growth, and enlargement under several stresses with Si supplementation is well documented. The exogenous application of Si enhances plant tolerance to drought stress by different mechanisms. Under drought stress, the upregulation of aquaporin gene (PIP; Plasma membrane Intrinsic Protein) may occur upon the addition of Si that also alleviates the inhibition of ROS-induced aquaporin activity. Silicon supply, under drought stress, adjusts the osmosis through increasing the accumulation of soluble sugars and/or amino acids in the xylem sap, which rises osmotic driving force or by motivating the K^+ translocation to xylem sap by SKOR (Stelar K^+ Outward Rectifer) gene. Silicon supply can improve the root hydraulic conductivity via amending the root growth and increasing root:shoot ratio along with elevating aquaporin activity and osmotic driving power. Higher root hydraulic conductance increases the uptake and transportation of water, which helps to preserve a higher photosynthetic rate and improve plant resistance to water deficit [267,345].

Silicon application can also decrease drought stress by increasing plant uptake of inorganic nutrients and shifting gas exchange features in plants [13]. Under drought stress, exogenous application of Si enhanced the seed germination, biochemical processes, and keeps the plantlets away from the oxidative stress by enhancing antioxidant defense [104,309,346,347]. In the maize plants, soil application of calcium silicate increased seed germination under drought stress [96]. An increase in the yield of cucumber plants under salinity [347] or drought stress [104] was noticed upon the application of Si in the form of silica nanoparticles (SiNPs) that enhanced the K^+/Na^+ ratio. Silicon application increases the photosynthetic rate, leaf and root water retention, osmotic potential, water use efficiency (WUE), even though decreases transpiration rate, and membrane permeability under water-deficit stress in different crop species viz. Kentucky bluegrass (*Poa pratensis* L.) [342], wheat [348,349], maize [350], tomato [351], rice [352], Fennel [353], oil palm [354], cucumber [104], common bean [346], and white lupin [355]. However, in some plants such as pepper (*Capsicum annuum* L.), soybean and rice, Si supply increased the transpiration rate and net photosynthetic rate under water deficiency stress [13] (Table 10.1).

One of the main effects of drought stress on plant growth is the limited uptake of essential nutrients [356]. Yet, soil application of Si increased the uptake of macronutrients (i.e., P, K, Ca, and Mg) and micronutrients (i.e., Fe, Cu, and Mn) in crop under water-deficit stress [14]. An increase in K and total P content in rice straw of Si-treated plants compared to control under drought conditions has been reported [356]. The regulation of gas exchange in plants is attributed to Si addition under drought stress [354]. Silicon addition has been extensively reported to reduce the oxidative damage via inducing the activity of several antioxidant enzyme (i.e., SOD, CAT, peroxidase, and ascorbate peroxidase) under drought stress in tomato [351], chickpea [357], wheat [358], and sunflower [14]. In drought-stressed rice plants, Si application was reported to upregulate the expression of both ring field containing protein OsRDCP1 gene and drought-specific genes, OsCMO coding rice choline monooxygenase and dehydrin OsRAB16b compared to control [359].

TABLE 10.1 The benefits of silicon to crops grown under drought and waterlogging stress as cited by several studies.

Plant	Abiotic stress	Experiment type	Growth condition and rate of Si-compound(s)	Significant findings	References
Sorghum (*Sorghum bicolor* L. var. Moench)	Drought	Pot experiment	Plants were subjected to two watering regimes, i.e., dry treatment (0.03 g g^{-1} soil water content) and wet treatment (0.08 g g^{-1} soil water content). Si was applied at a rate of 0.3 g kg^{-1} sand dune Regosol	• Si alleviated the reduction in plant dry weight • Si lowered the shoot/root ratio • Si maintained the photosynthetic rate and stomatal conductance • Si increased water uptake	[364]
Sorghum (cv. Gadambalia)	Drought	Hydroponic solution	Drought stress was simulated by mixing sorbitol with the culture solution. Si was added to cultural solution at a rate of 0 or 1.67 mM	• Si increased the stomatal conductance • Si increased leaf water potential	[323]
Upland rice (*Oryza sativa* L.)	Drought	Pot experiment	Plants were subjected to three soil moisture levels (i.e., 60%, 70%, or 80% water field capacity; WFC). Si was applied at a rate of 0, 200, 400, or 600 kg ha^{-1}	• Si increased the Si concentration in soil and rice shoots • Si content in the clay soil was higher than that in the sandy soil • Si increased the grain yield and plant height but reduced the number of stalks	[365]
Rice	Drought	Pot experiment under greenhouse conditions	Drought stress was applied at -0.050 MPa or -0.025 MPa soil water potential values. Si was added at a rate of 0 or 350 kg ha^{-1}	• Si decreased the proline content in the vegetative and reproductive phases of upland rice plants • Si increased the activity of peroxidase (POD) enzyme in the plants' reproductive phase	[366]
Two rice lines, i.e., w-14 (drought susceptible) and w-20 (drought resistant)	Drought	Pot experiment	Drought stress was simulated by keeping the soil moisture at 25.1% of saturation % for 15 days. Si was applied at a rate of 1.5 mM (added as potassium metasilicate; K_2SiO_3)	• Si significantly increased the photosynthetic rate (Pr), transpiration rate (Tr), Fv/F0, and Fv/Fm • Si reduced the content of K, Na, Ca, Mg, and Fe	[367]
Wheat (*Triticum aestivum* L.)	Drought	Pot experiment	Four watering regimes were applied, i.e., 100%, 75%, 50%, and 25% WFC. Si was applied at a rate of 2, 4, 6, and 8 mM	• Si increased morphology and physiochemical characters (i.e., total free amino acid, total soluble sugar, total soluble protein, and total proline) • Si increased the activity of catalase (CAT), superoxide dismutase (SOD), ascorbate peroxidase(APX), and POD enzymes	[368]
Wheat (var. Longchun 8139)	Drought	Pot experiment	Drought stress was applied by maintaining 50% of WFC. Si was applied at a rate of 2.11 mmol sodium metasilicate (Na_2SiO_3) kg^{-1} soil	• Si enhanced the water status • Si increased the activity of SOD, CAT and glutathione reductase (GR) enzymes • Si increased the unsaturation of fatty acids of lipids, the contents of photosynthetic pigments, and total thiols • Si decreased the content of hydrogen peroxide (H_2O_2), activity of acid phospholipase (AP) and oxidative stress of proteins • Si increased the net rate of CO_2 assimilation	[369]

(Continued)

TABLE 10.1   (*Continued*)

Plant	Abiotic stress	Experiment type	Growth condition and rate of Si-compound(s)	Significant findings	References
Wheat (var. Verynack)	Drought	Pot experiment	Drought stress was simulated by gypsum block (potential -1.0 MPa). Si was applied in the form of Na_2SiO_3 at a rate of 2 mM kg^{-1} soil	• Si increased the content of proline, glycine betaine, and soluble protein • Si significantly increased the activities of SOD, CAT, APX, and POD enzymes • Si increased the relative water content (RWC) and chlorophyll content	[370]
Wheat (var. Sehar-2006)	Drought	Field experiment	Si was applied at a rate of 12 kg ha^{-1} (added as K_2SiO_3)	• Si increased the K^+ content in shoot and grains • Si increased the biomass and grain yield	[371]
Wheat landraces	Drought	• Pot experiment under glasshouse conditions • Hydroponic culture with 1/2 strength • Hoagland's solution	8% (w/v) PEG-6000 was applied to generate drought stress of -0.12 MPa. Si was applied at a rate of 100 mL 1.5 mM Na_2SiO_3 (for pot experiment) and 0.2, 0.9, or 1.8 mM (for hydroponic culture)	• Si did not affect plant growth under drought stress • The positive effects of Si on plant growth and yield are not universal and depend on genotype-specific responses	[372]
Maize (*Zea mays* L.) cultivars, i.e., BR1010 (sensitive to drought stress) and DKB-390 (tolerant to drought stress)	Drought	Pot experiment	Si was applied at a rate of 0 or 27 g Si per pot (19 dm^{-3}) (added as calcium silicate; Ca_2SiO_4)	• Si increased maize yield by 12% for BR1010 and 14% for DKB-390 • Si increased the net photosynthetic rate, transpiration rate, and stomatal conductance in DKB-390	[373]
Maize (cv. Nongda108)	Drought	Pot experiment	Drought stress was applied as 20% polyethylene glycol (PEG-6000) in the nutrient solutionSi was applied at a rate of 0 or 2 mmol L^{-1} silicic acid (H_4SiO_4)	• Si decreased the transpiration rate and conductance for both adaxial and abaxial leaf surfaces • Si increased Si content in shoots and roots of maize • Si deposition in cell walls of the leaf epidermis was mostly in the form of polymerized SiO_2	[374]
Maize (cv. Nongda108)	Drought	Hydroponic culture (2 L)	Drought stress was applied as 0%, 20%, and 30% PEG-6000 in nutrient solution. Two Si levels were applied (i.e., 0 and 2 mmol L^{-1} as H_4SiO_4)	• Si increased water use efficiency (WUE) by 20% • Si reduced the transpiration rate through stomata by 2. Fivefold lower compared to control plants • Si reduced the water flow rate in xylem vessels by 20% compared to control due to silica deposits	[375]
Maize	Drought	Pot experiment	Soil moisture was kept as 40% water holding capacity (WHC) as drought-induced condition. Si was applied at a rate of 100 mg kg^{-1} soil	• Si increased the dry mass of shoot and root, length of shoot and root, and the total seedling biomass • Si increased the net photosynthetic rate, while decreased the transpiration rate • Si increased the content of the photosynthetic pigments and RWC • Si increased the activity of CAT, SOD, and POD enzymes	[376]
Maize	Drought	Field experiments	Drought stress was applied through irrigating plants at a rate of 80% of crop evapotranspiration. Si was applied at a rate of 0, 2, and 4 mM (added as Na_2SiO_3).	• Si increased photosynthetic efficiency, stomatal conductance, and cell membrane integrity	[377]

(*Continued*)

TABLE 10.1   (*Continued*)

Plant	Abiotic stress	Experiment type	Growth condition and rate of Si-compound(s)	Significant findings	References
Maize (hybrid DH 605)	Drought	Soil column field experiment	Drought stress was applied by keeping soil moisture at 50% of WFC for 7 days. Si was applied at a rate of 0 and 0.06 mg kg^{-1} dry soil (added as $Na_2SiO_3 \cdot 9H_2O$)	• Si significantly increased leaf area, chlorophyll content, photosynthetic rate, and osmolyte content • Si increased the activities of SOD, POD, and CAT enzymes • Si decreased malondialdehyde (MDA) content and superoxide radical accumulation • Si increased the maize grain yield	[378]
Sunflower (*Helianthus annuus* L.)	Drought	Pot experiment under glasshouse conditions	Twelve sunflower cultivars were subjected to drought stress (40% of WFC) for 3 weeks. Si was applied to the soil as Na_2SiO_3 at a rate of 100 mg Si kg^{-1} soil.	• Si increased the uptake of Si, K, S, Mg, Fe, Cu, Mn, Na, Cl, V, Al, Sr, Rb, Ti, Cr, and Ba	[379]
Sunflower	Drought	Pot experiment under glasshouse conditions	Twelve sunflower cultivars were subjected to drought stress (40% of WFC) for 3 weeks. Si was applied to the soil as Na_2SiO_3 at a rate of 100 mg Si kg^{-1} soil	• Si decreased the content of H_2O_2, proline, and MDA as well as stomatal resistance • Si mitigated the membrane damage through increasing leaf relative water content • Si increased the activity of CAT, while it decreased the SOD and APX activity • Si increased the nonenzymatic antioxidant activity	[380]
Pistachio (*Pistacia vera* L. var. Ahmadaghaii)	Drought	Pot experiment under field conditions	Plants were subjected to two watering regimes, i.e., 35% and 75% WFC. Si was applied at a rate of 0.35 g kg^{-1} soil (added as Na_2SiO_3)	• Si significantly increased plant dry weight, RWC, and stomatal conductance • Si improved the maximum quantum yield of PSII • Si maintained the reduction in the net assimilation rate due to drought stress • Si increased the activity of CAT and SOD enzyme and lowered lipid peroxidation in the plant leaves	[381]
Mango (*Mangifera indica* L.)	Drought	Field experiment	Four mango cultivars were subjected to two water regimes where each tree received 20 m^3 (control) and 10 m^3 (drought stress) during the whole season. Si was applied at a rate of 30 g Si (control) and 15 g Si (drought) added as K_2SiO_3	• Si declined the activity of POX, CAT, and SOD enzymes in mango leaves • Si improved the fruit quality	[382]
Zygophyllum (*Zygophyllum xanthoxylum*)	Drought	pot experiment	Plants were exposed to drought stress (30% of WFC). Different Si concentrations (1.5−7.5 mM) were applied as K_2SiO_3	• Si (2.5 mM) increased the activity of SOD, POD, and CAT enzymes • Si reduced the contents of MDA, soluble sugar, and free proline • Si increased the RWC in leaf tissues, leaf area, and chlorophyll a content • Si (2.5 mM) significantly increased Si content in root and shoot of *Z. xanthoxylum* • Si (2.5 mM) significantly lowered the K$^+$ content in root and shoot by 65.2% and 42%, respectively	[383]

(*Continued*)

TABLE 10.1   (*Continued*)

Plant	Abiotic stress	Experiment type	Growth condition and rate of Si-compound(s)	Significant findings	References
Lentil (*Lens culinaris* Medik)	Drought	Petri dishes (9 cm diameter)	PEG-6000 was used to simulate drought stress with final osmotic potential of −0.58 MPa. Si was applied at a rate of 2 mM Si (added as Na_2SiO_3)	• Si increased the Si content in plant tissue, the activity of antioxidant enzymes (APX, POD, CAT, and SOD), and hydrolytic enzymes (α-amylase, β-amylase, and α-glucosidase) • Si reduced the content of osmolytes (proline, glycine betaine, and total soluble sugar) and reactive oxygen species (superoxide anion O_2^- and H_2O_2)	[384]
Peanut (*Arachis hypogaea* L.)	Drought	Hydroponic solution(1 L capacity) containing 750 mL of Hoagland's nutrient medium	Drought stress was applied as 10% and 15% PEG.Si was applied at a rate of 2 mM Si (added as Na_2SiO_3)	• Si leaf chlorophyll content, RWC %, growth and biomass • Si induced the uptake and transport of mineral nutrients • Si reduced membrane lipid peroxidation and H_2O_2 accumulation • Si increased the activity of SOD, CAT, APX, GPX, and GR as well as nonenzymatic antioxidants, i.e., ascorbate (AsA) and glutathione (GSH) • Si increased the content of sugars and sugar alcohols (talose, mannose, fructose, sucrose, cellobiose, trehalose, pinitol, and myo-inositol), amino acids (glutamic acid, serine, histidine, threonine, tyrosine, valine, isoleucine, and leucine) • Si increased the level of indole-3-acetic acid (IAA), gibberellic acid (GA3), jasmonic acid (JA), and zeatin • Si increased the content of polyphenols (chlorogenic acid, caffeic acid, ellagic acid, rosmarinic acid, quercetin, coumarin, naringenin, and kaempferol)	[385]
Rapeseed (*Brassica napus* cv. Binasharisha 3)	Drought	Semihydroponic culture	Drought stress was maintained at two levels, i.e., 10% and 20% of PEG. One mM of silicon dioxide (SiO_2) was applied	• Si decreased the lipid peroxidation and H_2O_2 accumulation • Si increased the activity of AsA and GSH and the antioxidant enzymes, i.e., CAT, APX, glutathione S-transferase (GST), monodehydroascorbate reductase (MDHAR), dehydroascorbate reductase (DHAR), glutathione reductase (GR) • Si increased the content of RWC and proline	[386]

(Continued)

TABLE 10.1 (*Continued*)

Plant	Abiotic stress	Experiment type	Growth condition and rate of Si-compound(s)	Significant findings	References
Rapeseed (cv. Markus)	Drought	Pot experiment	Soil moisture was kept at 30% of WFC for 10 days (drought). Si was applied at a rate of 60 mL Si solution (orthosilicic acid tetraethyl ester) per pot	• Si increased the Si uptake by rapeseed • Si induced the water status • Si decreased the transpiration rate • Si reduced the stomatal conductance • Si increased the content of abscisic acid	[387]
Cowpea (*Vigna unguiculata*) and Mungbean (*Vigna radiata*)	Drought	Pot experiment under field conditions	Four different soil moisture contents were tested, i.e., 100%, 75%, 50%, and 25% WHC. Si was applied at a rate of 20, 40, or 460 ppm (added as magnesium metasilicate; $MgSi_3$)	• Si increased the content of total carbohydrate and protein	[388]
Soybean (*Glycine max* L.) cultivar, Zhonghuang 13	Drought	Hydroponic culture containing full strength Hoagland nutrient solution in a growth chamber	Plants were subjected to drought stress of −0.5 MPa that was simulated with 20% PEG. Si was added to the nutrient solution at a rate of 1.7 mM	• Si increased the RWC by 19% • Si application significantly reduced the membrane damage • Si increased the photosynthesis by 18.3% • Si increased the dry mass of shoot and root • Si decreased the activity of CAT, SOD, and POD as well as the H_2O_2 content	[389]
Tomato (*Solanum lycopersicum*; cv. "Jinpeng 1#")	Drought	Hydroponic culture with Hoagland solution under greenhouse conditions	Drought stress was simulated by 1.0% PEG. Si was applied at a rate of 0.6 mM (added as Na_2SiO_3)	• Si improved hydraulic conductivity in radial direction enhancing water uptake • Si increased the accumulation of proline, soluble sugar, and soluble protein, and the osmotic adjustment ability • Si increased the activity of SOD and CAT, while it reduced the production of O_2^-, H_2O_2, and MDA	[390]
Cucumber (*Cucumis sativus*)	Drought	Field experiment under greenhouse conditions	Three watering regimes were calculated based on crop evapotranspiration (etc.) (i.e., 100%, 85%, and 70% of etc.). Si was applied at a rate of 100, 200, 300, and 400 mg kg^{-1} (added as silica nanoparticles; SiNPs)	• Si enhanced the growth and increased the fruit yield • Si increased the plant height and chlorophyll content • Si increased the content of N, K, and Si in root, stem, and leaf • Si reduced the uptake of Na and its accumulation in root, stem, and leaf	[104]
—	Drought and waterlogging	Pot experiment	Two levels of Si (8 and 200 mg kg^{-1} soil) and two soil moisture regimes (−20 kPa and waterlogged)	• Waterlogging significantly increased Si phytoavailability • Si increased the soil Si content	[391]
Rice	Waterlogging	Field experiment	Silicon ($SiO_2 \geq 55\%$) was applied at 45 kg ha^{-1}, and spray Si was applied at 0.5 kg ha^{-1} solution	• Si enhanced rice growth and development • Si increased the number of leaves and tillers, dry mass of root and shoot, sugar content, and the activity of antioxidant enzymes (SOD, CAT, POD, and GR)	[392]

(*Continued*)

TABLE 10.1 *(Continued)*

Plant	Abiotic stress	Experiment type	Growth condition and rate of Si-compound(s)	Significant findings	References
Rice (cv. Oochikara, WT)	Waterlogging	Pot experiment	Rice plants were placed in a bucket for 7 days; the submerged depth was two-thirds of the average plant height. Si was added before transplantation at a rate of 2 mmol kg^{-1} (added as K_2SiO_3)	• Si enhanced root morphological traits • Si increased plant biomass and Si content in plant tissues • Si had no significant effects on photosynthetic rate, transpiration rate, stomatal conductance, and intercellular carbon dioxide concentration • Si improved the photochemical quenching and the integrity of cell structure • Si declined MDA and increased the peroxidase and CAT activity in leaves	[393]
Barley (*Hordeum vulgare* L.)	Waterlogging	Pot experiment	Flooding conditions were maintained by 1 cm water layer stagnating on the soil surface for 7 days. Si was applied at a rate of 1 g kg^{-1} soil (added as SiO_2)	• Si application increased Si content in barely tissues (shoot and root) and induced the shoot and root growth under optimum water supply • Si reduced the oxidative destruction in the roots and leaves • Si increased the activity of guaiacol peroxidase only under optimum water supply	[394]
Maize (cv Fard 4)	Waterlogging	Pot experiment under greenhouse conditions	Plants were flooded by maintaining the water 1−2 cm above the soil surface. Si was applied at a rate of 1 mM (added as Na_2SiO_3)	• Si significantly increased chlorophyll content and enhanced water retention • Si significantly reduced the content of H_2O_2, MDA, and proline and leaf membrane injury	[395]
Barley (cv. Poldek)	Waterlogging	Pot experiment	Plants were watered with 3.5 mmol of Si solution (added as H_4SiO_4) up to field capacity	• Si diminished the unlimited carbon denitrification (uC-D N_2O) emission • Si increased the potential denitrification (PD) N emission • Si declined the $N_2O:N_2$ ratio • Si induced the N_2 formation	[396]

Cellular water deficiency, primarily, rises from salinity, drought, or low temperature, while secondarily because of high radiation, heavy metals, or other reasons. Therefore it is very important to determine the resistance mechanisms that plants use to tolerate water stress and also to find the ways for increasing this tolerance. For overcoming the negative effect of water stress, addition of Si to the growth medium may have a valuable effect on plants. Results of experiments conducted by Kaya et al. [360] on maize, growing under water stress, showed that Si (added at the rate of 1 and 2 mmol dm^{-3} Na_2SiO_3) significantly increased shoot growth, while the root growth did not change. The higher Si dose was more efficient than the lower one; however, regarding the leaf RWC, both Si concentrations resulted in a similar increase compared to control plants. Improved plant water status (higher RWC index) may result from reduced water loss by transpiration due to deposition of Si (forming silica gel layer) on epidermal cell walls [104]. Both Si treatments lowered proline concentration in maize plants grown under water stress [360]. Similar response was detected in wheat growing under salinity stress [361]. Amino acid, proline, exists broadly in proteins but it may also accumulate in the cytosol in response to environmental stresses, particularly under osmotic stress. Accumulated free proline shares substantially to osmotic adjustment and may protect and stabilize some subcellular components (e.g., proteins and membranes).

Water stress, usually, leads to damaging of mineral nutrition and distractions in ion homeostasis. Under drought stress, maize leaves contained approximately 50% less Ca than control plants, while, in roots, its amount was higher compared to the control [360]. Lowering Ca concentration in plant cells is detrimental because Ca plays an essential role in maintaining the structural and functional integrity of plant membranes and regulation of their permeability and selectivity [362]. The ability of plants to maintain membrane stability is a crucial trait of stress resistance. Some investigations indicate that the addition of Si may increase concentrations of Ca in plant tissues and thus restore membrane integrity in water-stressed plants [360]. Interruption of ion homeostasis may result from reduced K^+ concentrations in water-stressed plants. Potassium plays an important role in processes involving osmotic adjustment and its adequate level in plants may improve water stress tolerance. Under water stress conditions, the presence of Si may result in better supply of K^+ [360]. This beneficial effect may be attributed to the stimulating action of Si on H^+-ATP-ase [363]. In plants growing under salt-stress conditions, Si addition helped in maintaining an adequate supply of essential nutrients and reduced Na uptake and its transport to shoots [361,363]. In experiment with salt-stressed barley, Liang [363] indicated that Si (added at the rate of 1 mmol dm^{-3} K_2SiO_3) decreased Na but increased K concentration in both roots and shoots. Selective uptake of mineral ions is associated with the activity of H^+-ATP-ase [362]. This membrane-located enzyme generates proton motive force that is used for ion transport inside the cell.

10.9 Conclusion and future perspectives

Silicon, albeit is not essential, plays an important role in plant growth and development, particularly under biotic and/or abiotic stress. Silicon accumulation in plants can improve the salt and drought tolerance by regulating both physiological and biochemical processes. Plant Si research has made great strides since the designation of Si as "dispensable." While this element is still generally excluded from most plant growth-media formulations, it is now widely accepted to benefit many plant species, including many agriculturally prominent crops. Although these benefits may be particularly pronounced under stresses such as drought and salinity, growing evidence indicates that Si may also improve growth under relatively benign conditions. Today, the multiple threats faced by our species, including rapid human population growth, changing climate, and increasing salinity, add urgency to the investigation of crop improvement by Si.

First, it has to be not ignored that plants differ among them regarding Si uptake and accumulation from their growth medium and, consequently, differ largely in their response to Si and obtained benefits. The ecological sorting of plants only based on their uptake and accumulation of Si can possibly lead to fatal mistakes where several factors affecting the accumulation of Si in plant tissues such as soil physicochemical properties, form and concentration of Si in soil solution, plant species, and growth stage. However, the recent advances in describing precisely the Si transporters have made it possible to depend on the molecular variations to set plants in groups according to their accumulation of Si. The list of Si-induced benefits to plants, particularly under abiotic stress, recently grows fast whether directly or indirectly. With the growing world population and consequently increased demand for food/feed, crops will always grow under some stressors. Taken together, we propose a unifying model, termed the apoplastic obstruction hypothesis, by which Si can exert its multitude of beneficial effects. Through this model, our aim is to stimulate critical thinking and positive advances toward a better understanding of Si properties. Recent researches have started helping in raising the Si benefits to the level of the beneficial elements. The futuristic studies should focus more on the comprehensive analysis of mechanisms by which Si is taken up, distributed in different plant tissues, alleviating stress conditions. Also, the optimum applied Si concentration and application method have to gain more attention, particularly for different plant species.

Acknowledgment

Tarek Alshaal and Nevien Elhawat would like to thank the Tempus Public Foundation (TPF), Budapest, Hungary, for their cofinancial support and TKP2020-IKA-04 program which provided by the National Research, Development and Innovation Fund of Hungary under the 2020-4.1.1-TKP2020 funding scheme.

References

[1] IPCC Land—Climate Interactions. Climate Change and Land: an IPCC special report on climate change, desertification, land degradation, sustainable land management, food security, and greenhouse gas fluxes in terrestrial ecosystems; 2019.

[2] Rao NKS, Laxman RH, Shivashankara KS. Physiological and morphological responses of horticultural crops to abiotic stresses. In: Rao NKS, Shivashankara KS, Laxman RH, editors. Abiotic stress physiology of horticultural crops. New Delhi: Springer India; 2016, p. 3—17. ISBN 978-81-322-2723-6.

[3]   Hayes MJ, Svoboda MD, Wardlow BD, Anderson MC, Kogan F. Drought monitoring: historical and current perspectives. Remote sensing of drought: innovative monitoring approaches. CRC Press; 2012, p. 1–19. Available from: http://doi.org/10.1201/b11863.

[4]   Manivannan P, Jaleel CA, Somasundaram R, Panneerselvam R. Osmoregulation and antioxidant metabolism in drought-stressed helianthus annuus under triadimefon drenching. C R Biol 2008;331:418–25. Available from: https://doi.org/10.1016/j.crvi.2008.03.003.

[5]   Hirabayashi Y, Mahendran R, Koirala S, Konoshima L, Yamazaki D, Watanabe S, et al. Global flood risk under climate change. Nat Clim Change 2013;3:816–21. Available from: https://doi.org/10.1038/nclimate1911.

[6]   Liang K, Tang K, Fang T, Qiu F. Waterlogging tolerance in maize: genetic and molecular basis. Mol Breed 2020;40. Available from: https://doi.org/10.1007/s11032-020-01190-0.

[7]   Setter TL, Waters I. Review of prospects for germplasm improvement for waterlogging tolerance in wheat, barley and oats. Plant Soil 2003;253.

[8]   Voesenek LACJ, Sasidharan R. Ethylene—and oxygen signalling—drive plant survival during flooding. Plant Biol 2013;15:426–35. Available from: https://doi.org/10.1111/plb.12014.

[9]   Abdelaal KAA, Hafez YM, El-Afry MM, Tantawy DS, Alshaal T. Effect of some osmoregulators on photosynthesis, lipid peroxidation, antioxidative capacity, and productivity of barley (*Hordeum vulgare* L.) under water deficit stress. Environ Sci Pollut Res 2018;25:30199–211. Available from: https://doi.org/10.1007/s11356-018-3023-x.

[10] Keeping MG, Kvedaras OL, Bruton AG. Epidermal silicon in sugarcane: cultivar differences and role in resistance to sugarcane borer *Eldana saccharina*. Environ Exp Bot 2009;66:54–60. Available from: https://doi.org/10.1016/j.envexpbot.2008.12.012.

[11] Ma, J.F.; Miyake, Y.; Takahashi, E. Chapter 2 Silicon as a beneficial element for crop plants. Studies in plant science; Elsevier, 2001; 8, 17–39. ISBN 978-0-444-50262-9.

[12] Bockhaven JV, Spíchal L, Novák O, Strnad M, Asano T, Kikuchi S, et al. Silicon induces resistance to the brown spot fungus *Cochliobolus miyabeanus* by preventing the pathogen from hijacking the rice ethylene pathway. New Phytol 2015;206:761–73. Available from: https://doi.org/10.1111/nph.13270.

[13] Rizwan M, Ali S, Ibrahim M, Farid M, Adrees M, Bharwana SA, et al. Mechanisms of silicon-mediated alleviation of drought and salt stress in plants: a review. Environ Sci Pollut Res Int 2015;22:15416–31. Available from: https://doi.org/10.1007/s11356-015-5305-x.

[14] Gunes A, Pilbeam DJ, Inal A, Coban S. Influence of silicon on sunflower cultivars under drought stress, I: growth, antioxidant mechanisms, and lipid peroxidation. Commun Soil Sci Plant Anal 2008;39:1885–903. Available from: https://doi.org/10.1080/00103620802134651.

[15] Ceccarelli S, Grando S, Baum M. Participatory plant breeding in water-limited environments. Exp Agric 2007;43. Available from: https://doi.org/10.1017/S0014479707005327.

[16] Passioura JB. Review: environmental biology and crop improvement. Funct Plant Biol 2002;29:537. Available from: https://doi.org/10.1071/FP02020.

[17] Turner NC. Imposing and maintaining soil water deficits in drought studies in pots. Plant Soil 2019;439:45–55. Available from: https://doi.org/10.1007/s11104-018-3893-1.

[18] Salehi-Lisar SY, Bakhshayeshan-Agdam H. Drought stress in plants: causes, consequences, and tolerance. In: Hossain MA, Wani SH, Bhattacharjee S, Burritt DJ, Tran L-SP, editors. Drought stress tolerance in plants, vol. 1: physiology and biochemistry. Cham: Springer International Publishing; 2016, p. 1–16. ISBN 978-3-319-28899-4.

[19] Arbona V, Manzi M, Ollas C, Gómez-Cadenas A. Metabolomics as a tool to investigate abiotic stress tolerance in plants. IJMS 2013;14:4885–911. Available from: https://doi.org/10.3390/ijms14034885.

[20] Hu YC, Shao HB, Chu LY, Gang W. Relationship between water use efficiency (WUE) and production of different wheat genotypes at soil water deficit. Colloids Surf B Biointerfaces 2006;53. Available from: https://doi.org/10.1016/j.colsurfb.2006.10.002.

[21] Medina V. Drought physiology of beans: dissection of drought response mechanisms in soybean (*Glycine max*), tepary (*Phaseolus Acutifolius*), lima (*P. Lunatus*) and common bean (*P. vulgaris*) [Ph.D. thesis]. Davis, United States: University of California; 2016.

[22] Hingane AJ, Saxena KB, Patil SB, Sultana R, Srikanth S, Mallikarjuna N, et al. Mechanism of water-logging tolerance in pigeon pea. Indian J Genet Plant Breed 2015;75:208–14. Available from: https://doi.org/10.5958/0975-6906.2015.00032.2.

[23] Kęska K, Szcześniak MW, Makałowska I, Czernicka M. Long-term waterlogging as factor contributing to hypoxia stress tolerance enhancement in cucumber: comparative transcriptome analysis of waterlogging sensitive and tolerant accessions. Genes (Basel) 2021;12:189. Available from: https://doi.org/10.3390/genes12020189.

[24] Araki H, Hossain MA, Takahashi T. Waterlogging and hypoxia have permanent effects on wheat root growth and respiration. J Agron Crop Sci 2012;198:264–75. Available from: https://doi.org/10.1111/j.1439-037X.2012.00510.x.

[25] Barrett-Lennard EG, van Ratingen P, Mathie MH. The developing pattern of damage in wheat (*Triticum aestivum* L.) due to the combined stresses of salinity and hypoxia: experiments under controlled conditions suggest a methodology for plant selection. Aust J Agric Res 1999;50:129–36.

[26] Tiner RW. Wetland indicators: a guide to wetland identification, delineation, classification, and mapping. Boca Raton, FL: Lewis Publishers; 1999. ISBN 978-0-87371-892-9.

[27] Lawrence E. Henderson's dictionary of biological terms. 11th ed. Harlow: Longman Scientific & Technical; 1998. ISBN 978-0-582-22708-8.

[28] Ammar MH, Anwar F, El-Harty EH, Migdadi HM, Abdel-Khalik SM, Al-Faifi SA, et al. Physiological and yield responses of faba bean (*Vicia faba* L.) to drought stress in managed and open field environments. J Agron Crop Sci 2015;201:280–7. Available from: https://doi.org/10.1111/jac.12112.

[29] Conesa MR, de la Rosa JM, Domingo R, Bañon S, Pérez-Pastor A. Changes induced by water stress on water relations, stomatal behaviour and morphology of table grapes (Cv. crimson seedless) grown in pots. Sci Hortic (Amsterdam) 2016;202:9–16. Available from: https://doi.org/10.1016/j.scienta.2016.02.002.

[30] Haworth M, Killi D, Materassi A, Raschi A. Coordination of stomatal physiological behavior and morphology with carbon dioxide determines stomatal control. Am J Bot 2015;102:677–88. Available from: https://doi.org/10.3732/ajb.1400508.

[31] Bano A, Ullah F, Nosheen A. Role of abscisic acid and drought stress on the activities of antioxidant enzymes in wheat. Plant Soil Environ 2012;58(2012):181–5. Available from: https://doi.org/10.17221/210/2011-PSE.

[32] Brodribb TJ, McAdam SAM. Abscisic acid mediates a divergence in the drought response of two conifers. Plant Physiol 2013;162:1370−7. Available from: https://doi.org/10.1104/pp.113.217877.

[33] Abe H, Yamaguchi-Shinozaki K, Urao T, Iwasaki T, Hosokawa D, Shinozaki K. Role of arabidopsis MYC and MYB homologs in drought- and abscisic acid-regulated gene expression. Plant Cell 1997;9:1859−68. Available from: https://doi.org/10.1105/tpc.9.10.1859.

[34] Yao X, Xiong W, Ye T, Wu Y. Overexpression of the aspartic protease ASPG1 gene confers drought avoidance in arabidopsis. J Exp Bot 2012;63:2579−93. Available from: https://doi.org/10.1093/jxb/err433.

[35] Lim CW, Baek W, Han S-W, Lee SC. Arabidopsis PYL8 plays an important role for ABA signaling and drought stress responses. Plant Pathol J 2013;29:471−6. Available from: https://doi.org/10.5423/PPJ.NT.07.2013.0071.

[36] Stagnari F, Perpetuini G, Tofalo R, Campanelli G, Leteo F, Della Vella U, et al. Long-term impact of farm management and crops on soil microorganisms assessed by combined DGGE and PLFA analyses. Front Microbiol 2014;5. Available from: https://doi.org/10.3389/fmicb.2014.00644.

[37] Salehi-Lisar SY, Bakhshayeshan-Agdam H. Drought stress in plants: causes, consequences, and tolerance. In: Hossain M, Wani S, Bhattacharjee S, Burritt D, Tran PLS, editors. Drought stress tolerance in plants, vol. 1: physiology and biochemistry. Springer International Publishing; 2016, p. 1−16. ISBN 9783319288994.

[38] Hu Y, Schmidhalter U. Drought and salinity: a comparison of their effects on mineral nutrition of plants. J Plant Nutr Soil Sci 2005;168:541−9. Available from: https://doi.org/10.1002/jpln.200420516.

[39] Sofy MR, Elhawat N. Tarek alshaal glycine betaine counters salinity stress by maintaining high K^+/Na^+ ratio and antioxidant defense via limiting Na + uptake in common bean (*Phaseolus vulgaris* L.). Ecotoxicol Environ Saf 2020;200:110732. Available from: https://doi.org/10.1016/j.ecoenv.2020.110732.

[40] Heidari Y, Moaveni P. Study of drought stress on ABA accumulation and proline among in different genotypes forage corn. Res J Biol Sci 2009;4:1121−4.

[41] Nair AS, Abraham TK, Jaya DS. Studies on the changes in lipid peroxidation and antioxidants in drought stress induced cowpea (*Vigna unguiculata* L.) Varieties. J Environ Biol 2008;29:689−91.

[42] Smirnoff N. The role of active oxygen in the response of plants to water deficit and desiccation. New Phytol 1993;125:27−58. Available from: https://doi.org/10.1111/j.1469-8137.1993.tb03863.x.

[43] Hendry GAF. Oxygen, free radical processes and seed longevity. Seed Sci Res 1993;3:141−53. Available from: https://doi.org/10.1017/S0960258500001720.

[44] Chowdhury SR, Choudhuri MA. Hydrogen peroxide metabolism as an index of water stress tolerance in jute. Physiol Plant 1985;65:476−80. Available from: https://doi.org/10.1111/j.1399-3054.1985.tb08676.x.

[45] Sairam RK, Kumutha D, Ezhilmathi K, Deshmukh PS, Srivastava GC. Physiology and biochemistry of waterlogging tolerance in plants. Biol Plant 2008;52:401−12. Available from: https://doi.org/10.1007/s10535-008-0084-6.

[46] Irfan M, Hayat S, Hayat Q, Afroz S, Ahmad A. Physiological and biochemical changes in plants under waterlogging. Protoplasma 2010;241:3−17. Available from: https://doi.org/10.1007/s00709-009-0098-8.

[47] Stec B. Structural mechanism of RuBisCO activation by carbamylation of the active site lysine. PNAS 2012;109:18785−90.

[48] Pan T, Wang Y, Wang L, Ding J, Cao Y, Qin G, et al. Increased CO_2 and light intensity regulate growth and leaf gas exchange in tomato. Physiol Plant 2020;168:694−708. Available from: https://doi.org/10.1111/ppl.13015.

[49] Kitaya Y, Okayama T, Murakami K, Takeuchi T. Effects of CO_2 concentration and light intensity on photosynthesis of a rootless submerged plant, *Ceratophyllum demersum* L., used for aquatic food production in bioregenerative life support systems. Adv Space Res 2003;31:1743−9. Available from: https://doi.org/10.1016/s0273-1177(03)00113-3.

[50] Westram A, Lloyd JR, Roessner U, Riesmeier JW, Kossmann J. Increases of 3-phosphoglyceric acid in potato plants through antisense reduction of cytoplasmic phosphoglycerate mutase impairs photosynthesis and growth, but does not increase starch contents. Plant Cell Environ 2002;25:1133−43. Available from: https://doi.org/10.1046/j.1365-3040.2002.00893.x.

[51] Kaiser C, Kilburn MR, Clode PL, Fuchslueger L, Koranda M, Cliff JB, et al. Exploring the transfer of recent plant photosynthates to soil microbes: mycorrhizal pathway vs direct root exudation. New Phytol 2015;205:1537−51. Available from: https://doi.org/10.1111/nph.13138.

[52] Vettermann W. Mechanism of the light-dependent accumulation of starch in chloroplasts of Acetabularia, and its regulation. Protoplasma 1973;76:261−78. Available from: https://doi.org/10.1007/BF01280702.

[53] Bailey-Serres J, Fukao T, Gibbs DJ, Holdsworth MJ, Lee SC, Licausi F, et al. Making sense of low oxygen sensing. Trends Plant Sci 2012;17:129−38. Available from: https://doi.org/10.1016/j.tplants.2011.12.004.

[54] Loreti E, van Veen H, Perata P. Plant responses to flooding stress. Curr Opin Plant Biol 2016;33:64−71. Available from: https://doi.org/10.1016/j.pbi.2016.06.005.

[55] Ren C-G, Kong C-C, Yan K, Zhang H, Luo Y-M, Xie Z-H. Elucidation of the molecular responses to waterlogging in *Sesbania cannabina* roots by transcriptome profiling. Sci Rep 2017;7:9256. Available from: https://doi.org/10.1038/s41598-017-07740-5.

[56] Christianson JA;, Llewellyn DJ;, Dennis ES;, Wilson IW. Global gene expression responses to waterlogging in roots and leaves of cotton (*Gossypium hirsutum* L.). Plant Cell Physiol 2010;51:21−37. Available from: https://doi.org/10.1093/pcp/pcp163.

[57] Ellis MH, Dennis ES, Peacock WJ. Arabidopsis roots and shoots have different mechanisms for hypoxic stress tolerance. Plant Physiol 1999;119:57−64. Available from: https://doi.org/10.1104/pp.119.1.57.

[58] Zhang Z, Zhang D, Zheng Y. Transcriptional and post-transcriptional regulation of gene expression in submerged root cells of maize. Plant Signal Behav 2009;4:132−5. Available from: https://doi.org/10.4161/psb.4.2.7629.

[59] Sachs MM, Freeling M, Okimoto R. The anaerobic proteins of maize. Cell 1980;20:761−7. Available from: https://doi.org/10.1016/0092-8674(80)90322-0.

[60] Klok EJ, Wilson IW, Wilson D, Chapman SC, Ewing RM, Somerville SC, et al. Expression profile analysis of the low-oxygen response in arabidopsis root cultures. Plant Cell 2002;14:2481−94. Available from: https://doi.org/10.1105/tpc.004747.

[61] Sasidharan R, Mustroph A, Boonman A, Akman M, Ammerlaan AMH, Breit T, et al. Root transcript profiling of two Rorippa species reveals gene clusters associated with extreme submergence tolerance. Plant Physiol 2013;163:1277−92. Available from: https://doi.org/10.1104/pp.113.222588.

[62] Athar HUR, Ashraf M. Photosynthesis under drought stress. Handbook of photosynthesis. 2nd ed. 2005. 893−809, doi: 10.1201/9781420027877.

[63] Kapoor D, Bhardwaj S, Landi M, Sharma A, Ramakrishnan M, Sharma A. The impact of drought in plant metabolism: how to exploit tolerance mechanisms to increase crop production. Appl Sci (Switzerland) 2020;10. Available from: https://doi.org/10.3390/app10165692.

[64] Zhang F, Zhu K, Wang YQ, Zhang ZP, Lu F, Yu HQ, et al. Changes in photosynthetic and chlorophyll fluorescence characteristics of sorghum under drought and waterlogging stress. Photosynthetica 2019;57:1156−64. Available from: https://doi.org/10.32615/ps.2019.136.

[65] Hussain S, Hussain S, Qadir T, Khaliq A, Ashraf U, Parveen A, et al. Drought stress in plants: an overview on implications, tolerance mechanisms and agronomic mitigation strategies. Plant Sci Today 2019;6:389−402. Available from: https://doi.org/10.14719/pst.2019.6.4.578.

[66] Lawlor DW, Cornic G. Photosynthetic carbon assimilation and associated metabolism in relation to water deficits in higher plants. Plant Cell Environ 2002;25. Available from: https://doi.org/10.1046/j.0016-8025.2001.00814.x.

[67] Jaleel CA, Manivannan P, Wahid A, Farooq M, Al-Juburi HJ, Somasundaram R, et al. Drought stress in plants: a review on morphological characteristics and pigments composition. Int J Agric Biol 2009;11:100−5.

[68] Sharma M, Kumar P, Kumar T. Drought stress in crop plants: physiology aspects. J Emerg Technol Innov Res 2019;6:1346−66.

[69] Elizamar C, da S, Manoel B, de A, Andre D, de AN, et al. Drought and its consequences to plants—from individual to ecosystem. In: Akinci, S, editor. Responses of organisms to water stress. IntechOpen; 2013. ISBN 978-953-51-5346-7.

[70] Fathi A, Tari DB. Effect of drought stress and its mechanism in plants. International J Life Sci 2016;10:1−6. Available from: https://doi.org/10.3126/ijls.v10i1.14509.

[71] Zargar SM, Gupta N, Nazir M, Mahajan R, Malik FA, Sofi NR, et al. Impact of drought on photosynthesis: molecular perspective. Plant Genet 2017;11:154−9. Available from: https://doi.org/10.1016/j.plgene.2017.04.003.

[72] Harris D, Joshi A, Khan PA, Gothkar P, Sodhi PS. On-farm seed priming in semi-arid agriculture: development and evaluation in maize, rice and chickpea in India using participatory methods. Exp Agric 1999;35:15−29. Available from: https://doi.org/10.1017/S0014479799001027.

[73] Wijewardana C, Raja Reddy K, Jason Krutz L, Gao W, Bellaloui N. Drought stress has transgenerational effects on soybean seed germination and seedling vigor. PLoS One 2019;14:1−20. Available from: https://doi.org/10.1371/journal.pone.0214977.

[74] Basal O, Szabó A, Veres S. PEG-induced drought stress effects on soybean germination parameters. J Plant Nutr 2020;43. Available from: https://doi.org/10.1080/01904167.2020.1750638.

[75] Baque A, Nahar M, Yeasmin M, Quamruzzaman Md, Rahman A, Azad MdJ, et al. Germination behavior of wheat (*Triticum aestivum* L.) as influenced by polyethylene glycol (PEG). Ujar 2016;4:86−91. Available from: https://doi.org/10.13189/ujar.2016.040304.

[76] Worku A, Ayalew D, Tadesse T. Germination and early seedling growth of bread wheat (*Triticum aestivum* L.) as affected by seed priming and coating. J Biol 2016;7.

[77] Ghafoor U, Chaudhry S, Abrar MM, Saeed AM, Quyyam A. Germination of two spring wheat (*Triticum aestivum* L.) cultivers under salt stress condition in pot trial. Int J Sci Eng Res 2014;5:6.

[78] Khafagy MAM. Effect of pre-treatment of barley grain on germination and seedling growth under drought stress. AAS 2017;2:33. Available from: https://doi.org/10.11648/j.aas.20170203.12.

[79] Khodarahmpour Z. Effect of drought stress induced by polyethylene glycol (PEG) on germination indices in corn (*Zea mays* L.) hybrids. Afr J Biotechnol 2011;10:18222−7. Available from: https://doi.org/10.5897/AJB11.2639.

[80] Golbashy M, Ebrahimi M, Khavari Khorasani S, Mostafavi K. Effects of drought stress on germination indices of corn hybrids (*Zea mays* L.). Electron J Plant Breed 2012;3:664−70.

[81] Queiroz MS, Oliveira CES, Steiner F, Zuffo AM, Zoz T, Vendruscolo EP, et al. Drought stresses on seed germination and early growth of maize and sorghum. J Agric Sci 2019;11:310. Available from: https://doi.org/10.5539/jas.v11n2p310.

[82] Radić V, Balalić I, Cvejić S, Jocić S, Marjanović-Jeromela A, Miladinović D. Drought effect on maize seedling development. Ratar Povrt 2018;55. Available from: https://doi.org/10.5937/RatPov1803135R.

[83] Abu Sayeed MdH, Farzana A, Shamim AB, Nilima H, Tahmina A, Shalim Uddin M. Morpho-physiological mechanisms of maize for drought tolerance. In: Hossain, A, editor. Plant stress physiology. IntechOpen; 2021.

[84] Kaya MD, Okçu G, Atak M, Çikili Y, Kolsarici Ö. Seed treatments to overcome salt and drought stress during germination in sunflower (*Helianthus annuus* L.). Eur J Agron 2006;24:291−5. Available from: https://doi.org/10.1016/j.eja.2005.08.001.

[85] Obidiegwu JE, Bryan GJ, Jones HG, Prashar A. Coping with drought: stress and adaptive responses in potato and perspectives for improvement. Front Plant Sci 2015;6:1−23. Available from: https://doi.org/10.3389/fpls.2015.00542.

[86] Muthoni J, Kabira JN. Potato production under drought conditions: identification of adaptive traits. Int J Hortic 2016, Available from: https://doi.org/10.5376/ijh.2016.06.0012.

[87] Muthoni J, Shimelis H. Heat and drought stress and their implications on potato production under dry African tropics. Aust J Crop Sci 2020;14:1405−14. Available from: https://doi.org/10.21475/ajcs.20.14.09.p2402.

[88] Sharma N, Kumar P, Kadian MS, Pandey SK, Singh SV, Luthra SK. Performance of potato (*Solanum tuberosum*) clones under water stress. Indian J Agric Sci 2011;81:825−9.

[89] Mudila H, Ansari MW. Antioxidant metabolism of tomato (*Lycopersicon esculentum* L.) seedlings under polyethylene glycol (PEG) induced drought stress condition. Mater Int 2020. Available from: https://doi.org/10.33263/Materials23.412420.

[90] Ishola Esan V, Ayanniyin Ayanbamiji T, Omoyemi Adeyemo J, Oluwafemi S. Effect of drought on seed germination and early seedling of tomato genotypes using polyethylene glycol 6000. Int J Sci 2018;4:36−43. Available from: https://doi.org/10.18483/ijsci.1533.

[91] Florido M, Bao L, Lara RM, Castro Y, Acosta R. Effect of water stress simulated with PEG 6000 on tomato seed germination (Solanum section lycopersicon). Cult Trop 2018;39:87−92.

[92] Meneses CHSG, Bruno R, de LA, Fernandes PD, Pereira WE, Lima LHG, et al. Germination of cotton cultivar seeds under water stress induced by polyethylene glycol-6000. Sci Agric 2011;68:131−8. Available from: https://doi.org/10.1590/S0103-90162011000200001.

[93] Zeid IM, Shedeed ZA. Response of Alfalfa to putrescine treatment under drought stress. Biol Plant 2006;50:635−40. Available from: https://doi.org/10.1007/s10535-006-0099-9.

[94] Fan HF, Ding L, Du CX, Wu X. Effect of short-term water deficit stress on antioxidative systems in cucumber seedling roots. Bot Studies 2014;55:1−7. Available from: https://doi.org/10.1186/s40529-014-0046-6.

[95] Toscano S, Romano D, Tribulato A, Patanè C. Effects of drought stress on seed germination of ornamental sunflowers. Acta Physiol Plant 2017;39:184. Available from: https://doi.org/10.1007/s11738-017-2484-8.

[96] Zargar SM, Macha MA, Nazir M, Agrawal GK, Rakwal R. Silicon: a multitalented micronutrient in OMICS perspective—an update. Curr Proteomics 2012;9:245−54. Available from: https://doi.org/10.2174/157016412805219152.

[97] Amal B-A, Said M, Abdelaziz B, Mouhammed M, Nasser EN, Keltoum EB. Relationship between leaf rolling and some physiological parameters in durum wheat under water stress. Afr J Agric Res 2020;16:1061−8. Available from: https://doi.org/10.5897/AJAR2020.14939.

[98] Bazzaz MM, AlMahmud A, Khan AS. Effects of water stress on morpho-phenological changes in wheat genotypes. Glob J Sci Front Res 2014;XIV:93−101.

[99] Pour-Aboughadareh A, Mohammadi R, Etminan A, Shooshtari L, Maleki-Tabrizi N, Poczai P. Effects of drought stress on some agronomic and morpho-physiological traits in durum wheat genotypes. Sustainability (Switzerland) 2020;12:1−14. Available from: https://doi.org/10.3390/su12145610.

[100] Amal B-A, Anne-Aliénor V, Hervé S, Abdelaziz B, Said M, Nasser EN, et al. Role of leaf rolling on agronomic performances of durum wheat subjected to water stress. Afr J Agric Res 2020;16:791−810. Available from: https://doi.org/10.5897/ajar2019.14620.

[101] Wijewardana C, Alsajri FA, Irby JT, Krutz LJ, Golden B, Henry WB, et al. Physiological assessment of water deficit in soybean using midday leaf water potential and spectral features. J of Plant Interact 2019;14:533−43. Available from: https://doi.org/10.1080/17429145.2019.1662499.

[102] Zubaer MA, Chowdhury AKMMB, Islam MZ, Hasan MA, Ahmed T. Effects of water stress on growth and yield attributes of aman rice genotypes. Int J Sustain Crop Prod 2007;2:25−30.

[103] Litvin AG, Van Iersel MW, Malladi A. Drought stress reduces stem elongation and alters gibberellin-related gene expression during vegetative growth of tomato. J Am Soc Hortic Sci 2016;141:591−7. Available from: https://doi.org/10.21273/JASHS03913-16.

[104] Alsaeedi A, El-Ramady H, Alshaal T, El-Garawany M, Elhawat N, Al-Otaibi A. Silica nanoparticles boost growth and productivity of cucumber under water deficit and salinity stresses by balancing nutrients uptake. Plant Physiol Biochem 2019;139:1−10. Available from: https://doi.org/10.1016/j.plaphy.2019.03.008.

[105] Ali MH, Haque MA, Hossain MM, Hassan R. Effect of soil moisture stress on growth and yield of onion. Bangladesh J. Crop Sci. 2005;16:65−72. Available from: https://doi.org/10.3923/pjbs.2007.3085.3090.

[106] Jogaiah S. Grapes. Abiotic stress physiology of horticultural crops. New Delhi: Springer India; 2016.

[107] Bhusal N, Han SG, Yoon TM. Impact of drought stress on photosynthetic response, leaf water potential, and stem sap flow in two cultivars of bi-leader apple trees (Malus × Domestica Borkh.). Sci Hortic (Amsterdam) 2019;246. Available from: https://doi.org/10.1016/j.scienta.2018.11.021.

[108] Parveen A, Rai GK, Bagati S, Rai PK, Singh P. Morphological, physiological, biochemical and molecular responses of plants to drought stress. Abiotic stress tolerance mechanisms in plants. CRC Press; 2020, p. 321−39. Available from: http://doi.org/10.1201/9781003163831-9.

[109] Farooq M, Wahid A, Kobayashi N, Fujita D, Basra SMA. Plant drought stress: effects, mechanisms and management. Agron Sustain Dev 2009;29:185−212.

[110] Kamara AY, Menkir A, Badu-Apraku B, Ibikunle O. The influence of drought stress on growth, yield and yield components of selected maize genotypes. J Agric Sci 2003;141:43−50. Available from: https://doi.org/10.1017/S0021859603003423.

[111] Farooq M, Hussain M, Wahid A, Siddique KHM. Drought stress in plants: an overview. Plant responses to drought stress: from morphological to molecular features, 9783642326. Berlin: Springer-Verlag; 2012, p. 1−33. ISBN 9783642326530.

[112] Steduto P, Hsiao TC, Fereres E, Raes D. Crop yield response to water. FAO; 2012.

[113] Mirás-Avalos JM, Alcobendas R, Alarcón JJ, Valsesia P, Génard M, Nicolás E. Assessment of the water stress effects on peach fruit quality and size using a fruit tree model, QualiTree. Agric Water Manag 2013;128:1−12. Available from: https://doi.org/10.1016/j.agwat.2013.06.008.

[114] Lopez G, Hossein Behboudian M, Girona J, Marsal J. Drought in deciduous fruit trees: implications for yield and fruit quality. Plant responses to drought stress. Berlin: Springer; 2012.

[115] Van de Wal B. Measuring and modelling plant-fruit interactions and fruit quality under changing water availability in tomato and grape [Ph.D. dissertation]. Belgium: Ghent University; 2017.

[116] Ripoll J, Urban L, Staudt M, Lopez-Lauri F, Bidel LPR, Bertin N. Water Shortage and quality of fleshy fruits-making the most of the unavoidable. J Exp Bot 2014;65:4097−117. Available from: https://doi.org/10.1093/jxb/eru197.

[117] Crisosto CH, Costa G. Preharvest factors affecting peach quality. The peach: botany, production and uses. CAB eBooks; 2008, p. 536−49.

[118] Shewfelt RL. What is quality? Postharvest Biol Technol 1999;15:197−200.

[119] Bertin N, Guichard S, Leonardi C, Longuenesse JJ, Langlois D, Navez B. Seasonal evolution of the quality of fresh glasshouse tomatoes under Mediterranean conditions, as affected by air vapour pressure deficit and plant fruit load. Ann Bot (Lond.) 2000;85:741−50. Available from: https://doi.org/10.1006/anbo.2000.1123.

[120] Samarah NH. Soybean yield, yield components, seed quality, dehydrin-like proteins, soluble sugars, and mineral nutrients in response to drought stress imposed prior to severe stress [retrospective theses and dissertations]; 2000.

[121] Failla O, Zoccffl G, Treccani C, Cocuccl' S. Growth, development and mineral content of apple fruit in different water status conditions. J Hortic Sci 1992;67:265−71. Available from: https://doi.org/10.1080/00221589.1992.11516247.

[122] Naor A, Naschitz S, Peres M, Gal Y. Responses of apple fruit size to tree water status and crop load. Tree Physiol 2008;28:1255–61. Available from: https://doi.org/10.1093/treephys/28.8.1255.

[123] Wang Y, Liu L, Wang Y, Tao H, Fan J, Zhao Z, et al. Effects of soil water stress on fruit yield, quality and their relationship with sugar metabolism in 'Gala' apple. Sci Hortic (Amsterdam) 2019;258:108753. Available from: https://doi.org/10.1016/j.scienta.2019.108753.

[124] Najla S, Vercambre G, Génard M. Effects of water deficit and variations of fruit microclimate on peach fruit growth and quality. Plant Stress 2011;2005–10.

[125] Lopez G, Mata M, Arbones A, Solans JR, Girona J, Marsal J. Mitigation of effects of extreme drought during stage III of peach fruit development by summer pruning and fruit thinning. Tree Physiol 2006;26:469–77. Available from: https://doi.org/10.1093/treephys/26.4.469.

[126] Naor A, Stern R, Peres M, Greenblat Y, Gal Y, Flaishman MA. Timing and severity of postharvest water stress affect following-year productivity and fruit quality of field-grown "Snow Queen" Nectarine. J Am Soc Hortic Sci 2005;130:806–12. Available from: https://doi.org/10.21273/jashs.130.6.806.

[127] Marsal J, Rapoport HF, Manrique T, Girona J. Pear fruit growth under regulated deficit irrigation in container-grown trees. Sci Hortic (Amsterdam) 2000;85:243–59. Available from: https://doi.org/10.1016/S0304-4238(99)00151-X.

[128] Marsal J, Behboudian MH, Mata M, Basile B, del Campo J, Girona J, et al. Fruit thinning in 'conference' pear grown under deficit irrigation to optimise yield and to improve tree water status. J Hortic Sci Biotechnol 2010;85:125–30.

[129] Gaudin R, Kansou K, Payan JC, Pellegrino A, Gary C. A water stress index based on water balance modelling for discrimination of grapevine quality and yield. J Int Sci Vigne Vin 2014;48:1–9. Available from: https://doi.org/10.20870/oeno-one.2014.48.1.1655.

[130] Galindo A, Calín-Sánchez A, Griñán I, Rodríguez P, Cruz ZN, Girón IF, et al. Water stress at the end of the pomegranate fruit ripening stage produces earlier harvest and improves fruit quality. Sci Hortic (Amsterdam) 2017;226:68–74. Available from: https://doi.org/10.1016/j.scienta.2017.08.029.

[131] Gribaa A, Dardelle F, Lehner A, Rihouey C, Burel C, Ferchichi A, et al. Effect of water deficit on the cell wall of the date palm (*Phoenix dactylifera* "Deglet Nour," Arecales) fruit during development. Plant Cell Environ 2013;36:1056–70. Available from: https://doi.org/10.1111/pce.12042.

[132] Wei J, Liu G, Liu D, Chen Y. Influence of irrigation during the growth stage on yield and quality in mango (*Mangifera Indica* L). PLoS One 2017;12:1–14. Available from: https://doi.org/10.1371/journal.pone.0174498.

[133] Surendar KK, Devi DD, Ravi I, Krishnakumar S, Kumar SR, Velayudham K. Water stress in banana—a review. Bull Environ Pharmacol Life Sci 2013;26:1–18.

[134] Ravi I, Vaganan MM. Abiotic stress tolerance in banana. Abiotic stress physiology of horticultural crops. New Delhi: Springer India; 2016, p. 207–22.

[135] Sadashiva AT, Singh A, Kumar RP, Sowmya V, D'mello DP. Tomato. Abiotic stress physiology of horticultural crops. New Delhi: Springer India; 2016, p. 121–31.

[136] Gucci R, Lodolini E, Rapoport HF. Productivity of olive trees with different water status and crop load. J Hortic Sci Biotechnol 2007;82:648–56. Available from: https://doi.org/10.1080/14620316.2007.11512286.

[137] Rao NKS. Arid zone fruit crops. Abiotic stress physiology of horticultural crops. New Delhi: Springer India; 2016, p. 223–34.

[138] Rao NKS. Onion. Abiotic stress physiology of horticultural crops. New Delhi: Springer India; 2016, p. 133–49.

[139] Madhavi Reddy K, Shivashankara KS, Geetha GA, Pavithra KC. Capsicum (hot pepper and bell pepper). Abiotic stress physiology of horticultural crops. New Delhi: Springer India; 2016, p. 151–66.

[140] Ruiz-Lau N, Medina-Lara F, Minero-García Y, Zamudio-Moreno E, Guzmán-Antonio A, Echevarría-Machado I, et al. Water deficit affects the accumulation of capsaicinoids in fruits of capsicum Chinense Jacq. HortScience 2011;46:487–92. Available from: https://doi.org/10.21273/hortsci.46.3.487.

[141] van Loon CD. The effect of water stress on potato growth, development, and yield. Am Potato J 1981;58:51–69.

[142] Zhao W, Liu L, Shen Q, Yang J, Han X, Tian F, et al. Effects of water stress on photosynthesis, yield, and water use efficiency in winter wheat. Water (Switzerland) 2020;12:1–19. Available from: https://doi.org/10.3390/W12082127.

[143] Ge T, Sui F, Bai L, Tong C, Sun N. Effects of water stress on growth, biomass partitioning, and water-use efficiency in summer maize (*Zea mays* L.) throughout the growth cycle. Acta Physiol Plant 2012;34:1043–53. Available from: https://doi.org/10.1007/s11738-011-0901-y.

[144] Shackel KA. Water relations of woody perennial plant species. J Int Sci Vigne Vin 2007;41:121–9. Available from: https://doi.org/10.20870/oeno-one.2007.41.3.847.

[145] Lo Bianco R, Rieger M, Sung SJS. Effect of drought on sorbitol and sucrose metabolism in sinks and sources of peach. Physiol Plant 2000;108:71–8. Available from: https://doi.org/10.1034/j.1399-3054.2000.108001071.x.

[146] Francaviglia D, Farina V, Avellone G, Lo Bianco R. Fruit yield and quality responses of apple cvars Gala and Fuji to partial rootzone drying under Mediterranean conditions. J Agric Sci 2013;151.

[147] Lakso AN. Morphological and physiological adaptations for maintaining photosynthesis under water stress in apple trees. Effects of stress on photosynthesis. Springer; 1983.

[148] Chaves MM, Maroco JP, Pereira JS. Understanding plant responses to drought—from genes to the whole plant. Funct Plant Biol 2003;30:239–64. Available from: https://doi.org/10.1071/FP02076.

[149] Yamaki S, Ishikawa K. Roles of four sorbitol related enzymes and invertase in the seasonal alteration of sugar metabolism in apple tissue. J Am Soc Hortic Sci 1986;111:134–7.

[150] Nahar K, Ullah S. Fruit quality and osmotic adjustment of four tomato cultivars under drought stress. Asian J Soil Sci Plant Nutr 2017;2:1–8. Available from: https://doi.org/10.9734/ajsspn/2017/36861.

[151] Meigh DF, Jones JD, Hulme AC. The respiration climacteric in the apple. Phytochemistry 1967;6:1507–15. Available from: https://doi.org/10.1016/s0031-9422(00)82943-x.

[152] Ebel RC, Proebsting EL, Patterson ME. Regulated deficit irrigation may alter apple maturity, quality, and storage life. HortScience 2019;28:141–3. Available from: https://doi.org/10.21273/hortsci.28.2.141.

[153] Kowitcharoen L, Wongs-Aree C, Setha S, Komkhuntod R, Kondo S, Srilaong V. Pre-harvest drought stress treatment improves antioxidant activity and sugar accumulation of sugar apple at harvest and during storage. Agric Nat Resour 2018;52:146–54. Available from: https://doi.org/10.1016/j.anres.2018.06.003.

[154] Mossad A, Farina V, Bianco R. Fruit yield and quality of "Valencia" orange trees under long-term partial rootzone drying. Agronomy 2020;10. Available from: https://doi.org/10.3390/agronomy10020164.

[155] Mpelasoka BS, Behboudian MH, Mills TM. Water relations, photosynthesis, growth, yield and fruit size of "Braeburn" apple: responses to deficit irrigation and to crop load. J Hortic Sci Biotechnol 2001;76:150–6. Available from: https://doi.org/10.1080/14620316.2001.11511342.

[156] Mills T, Behboudian M, Clothier B. Water relations, growth, and the composition of "Braeburn" apple fruit under deficit irrigation materials and methods. J Am Soc Hortic Sci 1996;121:286–91.

[157] Talluto G, Farina V, Volpe G, Lo Bianco R. Effects of partial rootzone drying and rootstock vigour on growth and fruit quality of "Pink Lady" apple trees in mediterranean environments. Aust J Agric Res 2008;59:785–94. Available from: https://doi.org/10.1071/AR07458.

[158] Naor A. Irrigation scheduling and evaluation of tree water status in deciduous orchards, 32. Wiley; 2010. ISBN 9780470767986.

[159] Razouk R, Ibijbijen J, Kajji A, Karrou M. Response of peach, plum and almond to water restrictions applied during slowdown periods of fruit growth. Am J Plant Sci 2013;04:561–70. Available from: https://doi.org/10.4236/ajps.2013.43073.

[160] Lopez G, Behboudian MH, Vallverdu X, Mata M, Girona J, Marsal J. Mitigation of severe water stress by fruit thinning in "O'Henry" peach: implications for fruit quality. Sci Hortic, 125. 2010, p. 294–300. Available from: http://doi.org/10.1016/j.scienta.2010.04.003.

[161] Lopez G, Larrigaudière C, Girona J, Behboudian MH, Marsal J. Fruit thinning in "Conference" pear grown under deficit irrigation: implications for fruit quality at harvest and after cold storage. Sci Hortic (Amsterdam) 2011;129:64–70. Available from: https://doi.org/10.1016/j.scienta.2011.03.007.

[162] Intrigliolo DS, Castel JR. Response of plum trees to deficit irrigation under two crop levels: tree growth, yield and fruit quality. Irrig Sci 2010;28:525–34. Available from: https://doi.org/10.1007/s00271-010-0212-x.

[163] Mattar MA, Soliman SS, Al-Obeed RS. Effects of various quantities of three irrigation water types on yield and fruit quality of 'succary' date palm. Agronomy 2021;11:796. Available from: https://doi.org/10.3390/agronomy11040796.

[164] Cui J, Shao G, Lu J, Keabetswe L, Hoogenboom G. Yield, quality and drought sensitivity of tomato to water deficit during different growth stages. Sci Agric 2020;77. Available from: https://doi.org/10.1590/1678-992x-2018-0390.

[165] Li B, Wim V, Shukla MK, Du T. Drip irrigation provides a trade-off between yield and nutritional quality of tomato in the solar greenhouse. Agric Water Manag 2021;249.

[166] Duah SA, Silva C, Nagy Z. Effect of water supply on physiological response and phytonutrient composition of chili peppers. Water 2021;13. Available from: https://doi.org/10.3390/w13091284.

[167] Gao S, Gao J, Zhu X, Song Y, Li Z, Ren G, et al. ABF2, ABF3, and ABF4 promote ABA-mediated chlorophyll degradation and leaf senescence by transcriptional activation of chlorophyll catabolic genes and senescence-associated genes in arabidopsis. Mol Plant 2016;9:1272–85. Available from: https://doi.org/10.1016/j.molp.2016.06.006.

[168] Gharti Chhetri GB, Lales J. Effect of drought on yield and yield components of nine spring wheat (Triticum aestivum) cultivars at reproductive stage under tropical environmental conditions. Belg J Bot 1990;123:19–26.

[169] Silva A, do N, Ramos MLG, Ribeiro WQ, de Alencar ER, da Silva PC, et al. Water stress alters physical and chemical quality in grains of common bean, triticale and wheat. Agric Water Manag 2020;231:106023. Available from: https://doi.org/10.1016/j.agwat.2020.106023.

[170] Karademir C, Karademir E, Ekinci R, Berekatoğlu K. Yield and fiber quality properties of cotton (Gossypium hirsutum L.) under water stress and non-stress conditions. Afr J Biotechnol 2011;10:12575–83. Available from: https://doi.org/10.5897/ajb11.1118.

[171] Yamamoto F, Sakata T, Terazawa K. Physiological, morphological and anatomical responses of Fraxinus mandshurica seedlings to flooding. Tree Physiol 1995;15:713–19. Available from: https://doi.org/10.1093/treephys/15.11.713.

[172] Eckert C, Xu W, Xiong W, Lynch S, Ungerer J, Tao L, et al. Ethylene-forming enzyme and bioethylene production. Biotechnol Biofuels 2014;7:33. Available from: https://doi.org/10.1186/1754-6834-7-33.

[173] Parelle J, Brendel O, Bodénès C, Berveiller D, Dizengremel P, Jolivet Y, et al. Differences in morphological and physiological responses to water-logging between two sympatric oak species (Quercus petraea [Matt.] Liebl., Quercus robur L.). Ann For Sci 2006;63:849–59. Available from: https://doi.org/10.1051/forest:2006068.

[174] Shimamura S, Yamamoto R, Nakamura T, Shimada S, Komatsu S. Stem hypertrophic lenticels and secondary aerenchyma enable oxygen transport to roots of soybean in flooded soil. Ann Bot 2010;106:277–84. Available from: https://doi.org/10.1093/aob/mcq123.

[175] Malik AI, Colmer TD, Lambers H, Schortemeyer M. Changes in physiological and morphological traits of roots and shoots of wheat in response to different depths of waterlogging. Aust J Plant Physiol 2001;28.

[176] Mergemann H, Sauter M. Ethylene induces epidermal cell death at the site of adventitious root emergence in rice. Plant Physiol 2000;124:609–14. Available from: https://doi.org/10.1104/pp.124.2.609.

[177] Lösch R, Busch J. Plant functioning under waterlogged conditions. In: Esser K, Kadereit JW, Lüttge U, Runge M, editors. Progress in botany, 61. Berlin: Springer; 2000, p. 255–68. ISBN 978-3-642-52371-7.

[178] Steffens B, Wang J, Sauter M. Interactions between ethylene, gibberellin and abscisic acid regulate emergence and growth rate of adventitious roots in deepwater rice. Planta 2006;223:604–12. Available from: https://doi.org/10.1007/s00425-005-0111-1.

[179] Videmšek U, Turk B, Vodnik D. Root aerenchyma—formation and function. Acta Agric Slov 2006;87(2):445–53.

[180] Pellinen R, Palva T, Kangasjärvi J. Subcellular localization of ozone-induced hydrogen peroxide production in birch (Betula pendula) leaf cells. Plant J 1999;20:349–56. Available from: https://doi.org/10.1046/j.1365-313X.1999.00613.x.

[181] Valliyodan B, Nguyen HT. Understanding regulatory networks and engineering for enhanced drought tolerance in plants. Curr Opin Plant Biol 2006;9:189–95. Available from: https://doi.org/10.1016/j.pbi.2006.01.019.

[182] El-Ramady H, Abdalla N, Alshaal T, El-Henawy A, Elmahrouk M, Bayoumi Y, et al. Plant nano-nutrition: perspectives and challenges Environmental Chemistry for a Sustainable World In: Gothandam KM, Ranjan S, Dasgupta N, Ramalingam C, Lichtfouse E, editors. Nanotechnology, food security and water treatment. Cham: Springer International Publishing; 2018, p. 129–61. ISBN 978-3-319-70165-3.

[183] Silva EC, Nogueira RJMC, Vale FHA, Araújo FP, de Pimenta MA. Stomatal changes induced by intermittent drought in four umbu tree genotypes. Braz J Plant Physiol 2009;21:33−42.

[184] Pagter M, Bragato C, Brix H. Tolerance and physiological responses of phragmites australis to water deficit. Aquat Bot 2005;81:285−99. Available from: https://doi.org/10.1016/j.aquabot.2005.01.002.

[185] Amtmann A, Blatt MR. Regulation of macronutrient transport. New Phytol 2009;181:35−52. Available from: https://doi.org/10.1111/j.1469-8137.2008.02666.x.

[186] Pugnaire FI, Valladares F. Functional plant ecology. 2nd ed. Boca Raton: CRC Press; 2007. ISBN 978-0-429-12247-7.

[187] Alshaal T, El-Ramady H, Al-Saeedi AH, Shalaby T, Elsakhawy T, Omara AE-D, et al. The rhizosphere and plant nutrition under climate change. In: Naeem M, Ansari AA, Gill SS, editors. Essential plant nutrients. Cham: Springer International Publishing; 2017, p. 275−308. ISBN 978-3-319-58840-7.

[188] El-Ramady H, Alshaal T, Elhawat N, Ghazi A, Elsakhawy T, Omara AE-D, et al. Plant nutrients and their roles under saline soil conditions. In: Hasanuzzaman M, Fujita M, Oku H, Nahar K, Hawrylak-Nowak B, editors. Plant nutrients and abiotic stress tolerance. Singapore: Springer; 2018, p. 297−324. ISBN 978-981-10-9043-1.

[189] Lawlor DW, Mengel K, Kirkby EA. Principles of plant nutrition. Ann Bot (Lond.) 2004;93:479−80. Available from: https://doi.org/10.1093/aob/mch063.

[190] Hu Y, Burucs Z, von Tucher S, Schmidhalter U. Short-term effects of drought and salinity on mineral nutrient distribution along growing leaves of maize seedlings. Environ Exp Bot 2007;60:268−75. Available from: https://doi.org/10.1016/j.envexpbot.2006.11.003.

[191] Silva da EC, Nogueira RJMC, Neto de ADA, Brito de JZ, Cabral EL. Aspectos ecofisiológicos de dez espécies em uma área de caatinga no município de Cabaceiras, Paraíba, Brasil. Iheringia Sér Bot 2004;59:201−6.

[192] Armstrong W, Drew MC. Root growth and metabolism under oxygen deficiency. In: Wasel, Y, et al., editors. Plant roots: the hidden half. New York: Dekker; 2002, p. 729−61.

[193] Ashraf MA, Ahmad MSA, Ashraf M, Al-Qurainy F, Ashraf MY, Ashraf MA, et al. Alleviation of waterlogging stress in upland cotton (Gossypium hirsutum L.) by exogenous application of potassium in soil and as a foliar spray. Crop Pasture Sci. 2011;62:25−38. Available from: https://doi.org/10.1071/CP09225.

[194] Smethurst CF, Garnett T, Shabala S. Nutritional and chlorophyll fluorescence responses of lucerne (Medicago sativa) to waterlogging and subsequent recovery. Plant Soil 2005;270:31−45. Available from: https://doi.org/10.1007/s11104-004-1082-x.

[195] Boem FHG, Lavado RS, Porcelli CA. Note on the effects of winter and spring waterlogging on growth, chemical composition and yield of rapeseed. Field Crops Res 1996;47:175−9. Available from: https://doi.org/10.1016/0378-4290(96)00025-1.

[196] Stieger PA, Feller U. Nutrient accumulation and translocation in maturing wheat plants grown on waterlogged soil. Plant Soil 1994;160:87−95. Available from: https://doi.org/10.1007/BF00150349.

[197] Fritioff Å, Greger M. Uptake and distribution of Zn, Cu, Cd, and Pb in an aquatic plant Potamogeton natans. Chemosphere 2006;63:220−7. Available from: https://doi.org/10.1016/j.chemosphere.2005.08.018.

[198] Bernacchia G, Furini A. Biochemical and molecular responses to water stress in resurrection plants. Physiol Plant 2004;121:175−81.

[199] Akhtar I, Nazir N. Effect of waterlogging and drought stress in plants. Int J Water Resources Environ Sci 2013;2:34−40. Available from: https://doi.org/10.5829/idosi.ijwres.2013.2.2.11125.

[200] Arbona V, Manzi M, de Ollas C, Gómez-Cadenas A. Metabolomics as a tool to investigate abiotic stress tolerance in plants. Int J Mol Sci 2013;14.

[201] Bhargava S, Sawant K. Drought stress adaptation: metabolic adjustment and regulation of gene expression. Plant Breed 2013;132:21−32. Available from: https://doi.org/10.1111/pbr.12004.

[202] Schachtman DP, Goodger JQD. Chemical root to shoot signaling under drought. Trends Plant Sci 2008;13:281−7. Available from: https://doi.org/10.1016/j.tplants.2008.04.003.

[203] Davies WJ, Zhang J. Root signals and the regulation of growth and development of plants in drying soil. Annu Rev Plant Physiol Plant Mol Biol 1991;42:55−76. Available from: https://doi.org/10.1146/annurev.pp.42.060191.000415.

[204] Biswal AK, Misra AN, Misra M. Physiological, biochemical and molecular aspects of water stress responses in plants, and the biotechnological applications. Proc Natl Acad Sci India B Biol Sci 2002;72:115−34.

[205] Gollan T, Passioura JB, Munns R. Soil water status affects the stomatal conductance of fully turgid wheat and sunflower leaves. Austr J Plant Physiol 1986;13:459−64.

[206] Pietrowska-Borek M, Dobrogojski J, Sobieszczuk-Nowicka E, Borek S. New insight into plant signaling: extracellular ATP and uncommon nucleotides. Cells 2020;9:345. Available from: https://doi.org/10.3390/cells9020345.

[207] Time A, Acevedo E. Effects of water deficits on prosopis tamarugo growth, water status and stomata functioning. Plants 2020;10:53. Available from: https://doi.org/10.3390/plants10010053.

[208] Yamane K, Hayakawa K, Kawasaki M, Taniguchi M, Miyake H. Bundle sheath chloroplasts of rice are more sensitive to drought stress than mesophyll chloroplasts. J Plant Physiol 2003;160:1319−27. Available from: https://doi.org/10.1078/0176-1617-01180.

[209] Tombesi S, Nardini A, Frioni T, Soccolini M, Zadra C, Farinelli D, et al. Stomatal closure is induced by hydraulic signals and maintained by aba in drought-stressed grapevine. Sci Rep 2015;5:12449. Available from: https://doi.org/10.1038/srep12449.

[210] Song X, Zhou G, He Q, Zhou H. Stomatal limitations to photosynthesis and their critical water conditions in different growth stages of maize under water stress. Agric Water Manag 2020;241. Available from: https://doi.org/10.1016/j.agwat.2020.106330.

[211] Cai Y, Wang J, Li S, Zhang L, Peng L, Xie W, et al. Photosynthetic response of an alpine plant, rhododendron delavayi franch, to water stress and recovery: the role of mesophyll conductance. Front Plant Sci 2015;6:1089. Available from: https://doi.org/10.3389/fpls.2015.01089.

[212] Flexas J, Barón M, Bota J, Ducruet JM, Gallé A, Galmés J, et al. Photosynthesis limitations during water stress acclimation and recovery in the drought-adapted vitis Hybrid Richter-110 (V. berlandieri × V. rupestris). J Exp Bot 2009;60:2361−77. Available from: https://doi.org/10.1093/jxb/erp069.

[213] Yi X-P, Zhang Y-L, Yao H-S, Luo H-H, Gou L, Chow WS, et al. Rapid recovery of photosynthetic rate following soil water deficit and watering in cotton plants (Gossypium herbaceum L.) is related tothe stability of the photosystems. J Plant Physiol 2016;194:23−34. Available from: https://doi.org/10.1016/j.jplph.2016.01.016.

[214] Campos H, Trejo C, Peña CB, Peña-Valdivia P, García-Nava R, Víctor Conde-Martínez F, et al. Stomatal and non-stomatal limitations of bell pepper (Capsicum annuum L.) plants under water stress and re-watering:delayed restoration of photosynthesis during recovery. Environ Exp Bot 2014;98:56−64. Available from: https://doi.org/10.1016/j.envexpbot.2013.10.015.

[215] Karuppanapandian T, Geilfus CM, Mühling KH, Novák O, Gloser V. Early Changes of the pH of the apoplast are different in leaves, stem and roots of Vicia faba L. under declining water availability. Plant Sci 2017;255:51−8. Available from: https://doi.org/10.1016/j.plantsci.2016.11.010.

[216] Grassi G, Magnani F. Stomatal, mesophyll conductance and biochemical limitations to photosynthesis as affected by drought and leaf ontogeny in ash and oak trees. Plant Cell Environ 2005;28.

[217] Maxwell K, Johnson GN. Chlorophyll fluorescence—a practical guide. J Exp Bot 2000;51:659−68. Available from: https://doi.org/10.1093/jexbot/51.345.659.

[218] Chen K, Li GJ, Bressan RA, Song CP, Zhu JK, Zhao Y. Abscisic acid dynamics, signaling, and functions in plants. J Integ Plant Biol 2020;62:25−54. Available from: https://doi.org/10.1111/jipb.12899.

[219] Maxwell K, Johnson GN. Chlorophyll fluorescence—a practical guide. J Exp Bot 2000;51.

[220] Jia W, Ma M, Chen J, Wu S. Plant morphological, physiological and anatomical adaption to flooding stress and the underlying molecular mechanisms. Int J Mol Sci 2021;22:1−24. Available from: https://doi.org/10.3390/ijms22031088.

[221] Nishiuchi S, Yamauchi T, Takahashi H, Kotula L, Nakazono M. Mechanisms for coping with submergence and waterlogging in rice. Rice 2012;5:2. Available from: https://doi.org/10.1186/1939-8433-5-2.

[222] Drew M, Armstrong W. Root growth and metabolism under oxygen deficiency. In: Kafkafi, U, Waisel, Y, Eshel, A, editors. Plant roots. CRC Press; 2002, p. 729−61. ISBN 978-0-8247-0631-9.

[223] Fairbanks DF, Wilke CR. Diffusion coefficients in multicomponent gas mixtures. Ind Eng Chem 1950;42:471−5. Available from: https://doi.org/10.1021/ie50483a022.

[224] Fukao T, Barrera-Figueroa BE, Juntawong P, Peña-Castro JM. Submergence and waterlogging stress in plants: a review highlighting research opportunities and understudied aspects. Front Plant Sci 2019;10:1−24. Available from: https://doi.org/10.3389/fpls.2019.00340.

[225] Morard P, Silvestre J. Plant injury due to oxygen deficiency in the root environment of soilless culture: a review. Plant Soil 1996;184.

[226] Sasidharan R, Bailey-Serres J, Ashikari M, Atwell BJ, Colmer TD, Fagerstedt K, et al. Community recommendations on terminology and procedures used in flooding and low oxygen stress research. New Phytol 2017;214:1403−7. Available from: https://doi.org/10.1111/nph.14519.

[227] Ahmed F, Rafii MY, Ismail MR, Juraimi AS, Rahim HA, Asfaliza R, et al. Waterlogging tolerance of crops: breeding, mechanism of tolerance, molecular approaches, and future prospects. BioMed Res Int 2013;2013.

[228] Gibbs J, Greenway H. Review: mechanisms of anoxia tolerance in plants. I. Growth, survival and anaerobic catabolism. Funct Plant Biol 2003;30:353. Available from: https://doi.org/10.1071/pp98095_er.

[229] Borella J, Amarante Ldo, Oliveira Ddos SCde, Oliveira ACBde, Braga EJB. Waterlogging-induced changes in fermentative metabolism in roots and nodules of soybean genotypes. Sci Agric (Piracicaba, Braz.) 2014;71:499−508. Available from: https://doi.org/10.1590/0103-9016-2014-0044.

[230] Horchani F, Khayati H, Raymond P, Brouquisse R, Aschi-Smiti S. Contrasted effects of prolonged root hypoxia on tomato root and fruit (Solanum lycopersicum) metabolism. J Agron Crop Sci 2009;195:313−18. Available from: https://doi.org/10.1111/j.1439-037X.2009.00363.x.

[231] Kaur G, Singh G, Motavalli PP, Nelson KA, Orlowski JM, Golden BR. Impacts and management strategies for crop production in waterlogged or flooded soils: a review. Agron J 2020;112:1475−501. Available from: https://doi.org/10.1002/agj2.20093.

[232] Zabalza A, van Dongen JT, Froehlich A, Oliver SN, Faix B, Gupta KJ, et al. Regulation of respiration and fermentation to control the plant internal oxygen concentration. Plant Physiol 2009;149:1087−98. Available from: https://doi.org/10.1104/pp.108.129288.

[233] Cho HY, Loreti E, Shih MC, Perata P. Energy and sugar signaling during hypoxia. New Phytol 2021;229:57−63. Available from: https://doi.org/10.1111/nph.16326.

[234] Geigenberger P. Response of plant metabolism to too little oxygen. Curr Opin Plant Biol 2003;6:247−56. Available from: https://doi.org/10.1016/S1369-5266(03)00038-4.

[235] Pirasteh-Anosheh H, Saed-Moucheshi A, Pakniyat H, Pessarakli M. Stomatal responses to drought stress. Water stress and crop plants: a sustainable approach, vol. 1. Wiley; 2016, p. 24−40.

[236] Fichman Y, Mittler R. Rapid systemic signaling during abiotic and biotic stresses: is the ROS wave master of all trades? Plant J 2020;102:887−96. Available from: https://doi.org/10.1111/tpj.14685.

[237] Dat JF, Capelli N, Folzer H, Bourgeade P, Badot P-M. Sensing and signalling during plant flooding. Plant Physiol Biochem 2004;42:273−82. Available from: https://doi.org/10.1016/j.plaphy.2004.02.003.

[238] Devireddy AR, Arbogast J, Mittler R. Coordinated and rapid whole-plant systemic stomatal responses. New Phytol 2020;225:21−5. Available from: https://doi.org/10.1111/nph.16143.

[239] Kollist H, Zandalinas SI, Sengupta S, Nuhkat M, Kangasjärvi J, Mittler R. Rapid responses to abiotic stress: priming the landscape for the signal transduction network. Trends Plant Sci 2019;24:25−37. Available from: https://doi.org/10.1016/j.tplants.2018.10.003.

[240] Zandalinas SI, Sengupta S, Burks D, Azad RK, Mittler R. Identification and characterization of a core set of ROS wave-associated transcripts involved in the systemic acquired acclimation response of arabidopsis to excess light. Plant J 2019;98:126−41. Available from: https://doi.org/10.1111/tpj.14205.

[241] Osakabe Y, Osakabe K, Shinozaki K, Tran LSP. Response of plants to water stress. Front Plant Sci 2014;5:1−8. Available from: https://doi.org/10.3389/fpls.2014.00086.

[242] Perez-Martin A, Michelazzo C, Torres-Ruiz JM, Flexas J, Fernández JE, Sebastiani L, et al. Regulation of photosynthesis and stomatal and mesophyll conductance under water stress and recovery in olive trees: correlation with gene expression of carbonic anhydrase and aquaporins. J Exp Bot 2014;65. Available from: https://doi.org/10.1093/jxb/eru160.

[243] Lisar SYS, Motafakkerazad R, Hossain, Mosharraf M, Rahman IMM. Water stress in plants: causes, effects and responses. Water Stress 2012, Available from: https://doi.org/10.5772/39363.

[244] Mahajan S, Tuteja N. Cold, salinity and drought stresses: an overview. Arch Biochem Biophys 2005;444:139−58. Available from: https://doi.org/10.1016/j.abb.2005.10.018.

[245] Brodribb TJ, McAdam SAM. Passive origins of stomatal control in vascular plants. Science 2011;331:582−5. Available from: https://doi.org/10.1126/science.1197985.

[246] Batool A, Akram NA, Cheng Z-G, Lv G-C, Ashraf M, Afzal M, et al. Physiological and biochemical responses of two spring wheat genotypes to non-hydraulic root-to-shoot signalling of partial and full root-zone drought stress. Plant Physiol Biochem 2019;139:11−20. Available from: https://doi.org/10.1016/j.plaphy.2019.03.001.

[247] Castro P, Puertolas J, Dodd IC. Stem girdling uncouples soybean stomatal conductance from leaf water potential by enhancing leaf xylem ABA concentration. Environ Exp Bot 2019;159:149−56. Available from: https://doi.org/10.1016/j.envexpbot.2018.12.020.

[248] Gui YW, Sheteiwy MS, Zhu SG, Batool A, Xiong YC. Differentiate effects of non-hydraulic and hydraulic root signaling on yield and water use efficiency in diploid and tetraploid wheat under drought stress. Environ Exp Bot 2021;181:104287. Available from: https://doi.org/10.1016/j.envexpbot.2020.104287.

[249] Khaleghi A, Naderi R, Brunetti C, Maserti BE, Salami SA, Babalar M. Morphological, physiochemical and antioxidant responses of *Maclura Pomifera* to drought stress. Sci Rep 2019;9:19250. Available from: https://doi.org/10.1038/s41598-019-55889-y.

[250] Brunner I, Herzog C, Dawes MA, Arend M, Sperisen C. How tree roots respond to drought. Front Plant Sci 2015;6. Available from: https://doi.org/10.3389/fpls.2015.00547.

[251] Chinnusamy V. Molecular genetic perspectives on cross-talk and specificity in abiotic stress signalling in plants. J Exp Bot 2003;55:225−36. Available from: https://doi.org/10.1093/jxb/erh005.

[252] Tuteja N. Abscisic acid and abiotic stress signaling. Plant Signal Behav 2007;2:135−8. Available from: https://doi.org/10.4161/psb.2.3.4156.

[253] Zhu JK. Salt and drought stress signal transduction in plants. Annu Rev Plant Biol 2002;53:247−73. Available from: https://doi.org/10.1146/annurev.arplant.53.091401.143329.

[254] Fujita Y, Yoshida T, Yamaguchi-Shinozaki K. Pivotal role of the AREB/ABF-SnRK2 pathway in ABRE-mediated transcription in response to osmotic stress in plants. Physiol Plant 2013;147.

[255] Kanno Y, Hanada A, Chiba Y, Ichikawa T, Nakazawa M, Matsui M, et al. Identification of an abscisic acid transporter by functional screening using the receptor complex as a sensor. Proc Natl Acad Sci USA 2012;109:9653−8. Available from: https://doi.org/10.1073/pnas.1203567109.

[256] Rodrigues A, Santiago J, Rubio S, Saez A, Osmont KS, Gadea J, et al. The short-rooted phenotype of the brevis radix mutant partly reflects root abscisic acid hypersensitivity. Plant Physiol 2020;149. Available from: https://doi.org/10.1104/pp.108.133819.

[257] Harris JM. Abscisic acid: hidden architect of root system structure. Plants 2015;4:548−72. Available from: https://doi.org/10.3390/plants4030548.

[258] Huber AE, Melcher PJ, Piñeros MA, Setter TL, Bauerle TL. Signal coordination before, during and after stomatal closure in response to drought stress. New Phytol 2019;224:675−88. Available from: https://doi.org/10.1111/nph.16082.

[259] Hu B, Cao J, Ge K, Li L. The site of water stress governs the pattern of ABA synthesis and transport in peanut. Sci Rep 2016;6:1−11. Available from: https://doi.org/10.1038/srep32143.

[260] Davies WJ, Wilkinson S, Loveys B. Stomatal control by chemical signalling and the exploitation of this mechanism to increase water use efficiency in agriculture. New Phytol 2002;153:449−60. Available from: https://doi.org/10.1046/j.0028-646X.2001.00345.x.

[261] Wilkinson S, Davies WJ. Xylem sap pH increase: a drought signal received at the apoplastic face of the guard cell that involves the suppression of saturable abscisic acid uptake by the epidermal symplast. Plant Physiol 1997;113:559−73. Available from: https://doi.org/10.1104/pp.113.2.559.

[262] Felle HH. pH: signal and messenger in plant cells. Plant Biol 2001;3.

[263] Chaffey N. Handbook of plant science. Ann Bot (Lond.) 2008;101. Available from: https://doi.org/10.1093/aob/mcm330.

[264] Wilkinson S. pH as a stress signal. Plant Growth Regul 1999;29:87−99. Available from: https://doi.org/10.1023/a:1006203715640.

[265] Daszkowska-Golec A. The role of abscisic acid in drought stress: how ABA helps plants to cope with drought stress. Drought stress tolerance in plants, vol. 2: molecular and genetic perspectives. Springer; 2016.

[266] Schurr U, Gollan T, Schulze ED. Stomatal response to drying soil in relation to changes in the xylem sap composition of *Helianthus annuus*. II. Stomatal sensitivity to abscisic acid imported from the xylem sap. Plant Cell Environ 1992;15. Available from: https://doi.org/10.1111/j.1365-3040.1992.tb01489.x.

[267] Yao L, Cheng X, Gu Z, Huang W, Li S, Wang L, et al. The AWPM-19 family protein OsPM1 mediates abscisic acid influx and drought response in rice. Plant Cell 2018;30. Available from: https://doi.org/10.1105/tpc.17.00770.

[268] Joshi-Saha A, Valon C, Leung J. Abscisic acid signal off the STARTing block. Mol Plant 2011;4:562−80. Available from: https://doi.org/10.1093/mp/ssr055.

[269] Merilo E, Jalakas P, Laanemets K, Mohammadi O, Hõrak H, Kollist H, et al. Abscisic acid transport and homeostasis in the context of stomatal regulation. Mol Plant 2015;8:1321−33. Available from: https://doi.org/10.1016/j.molp.2015.06.006.

[270] Mongrand S, Hare PD, Chua N-H. Abscisic acid. In: Henry HL, Norman AW, editors. Encyclopedia of hormones. New York: Academic Press; 2003. p. 1−10. ISBN 978-0-12-341103-7.

[271] Hamilton DWA, Hills A, Köhler B, Blatt MR. Ca^{2+} channels at the plasma membrane of stomatal guard cells are activated by hyperpolarization and abscisic acid. Proc Natl Acad Sci USA 2000;97:4967−72. Available from: https://doi.org/10.1073/pnas.080068897.

[272] Pel ZM, Murata Y, Benning G, Thomine S, Klüsener B, Allen GJ, et al. Calcium channels activated by hydrogen peroxide mediate abscisic acid signalling in guard cells. Nature 2000;406. Available from: https://doi.org/10.1038/35021067.

[273] Kwak JM, Mori IC, Pei ZM, Leonhard N, Angel Torres M, Dangl JL, et al. NADPH Oxidase AtrbohD and AtrbohF genes function in ROS-dependent ABA signaling in Arabidopsis. EMBO J 2003;22:2623−33. Available from: https://doi.org/10.1093/emboj/cdg277.

[274] MacRobbie EAC. Control of volume and turgor in stomatal guard cells. J Memb Biol 2006;210. Available from: https://doi.org/10.1007/s00232-005-0851-7.

[275] Finkelstein RR, Rock CD. Abscisic acid biosynthesis and response. Arabidopsis Book 2002;1:e0058. Available from: https://doi.org/10.1199/tab.0058.

[276] Daszkowska-Golec A. The role of abscisic acid in drought stress: how ABA helps plants to cope with drought stress. In: Hossain MA, Wani SH, Bhattacharjee S, Burritt DJ, Tran L-SP, editors. Drought stress tolerance in plants, Vol. 2. Cham: Springer International Publishing; 2016, p. 123−51. ISBN 978-3-319-32421-0.

[277] Gamper N, Shapiro MS. Target-specific PIP_2 signalling: how might it work? J Physiol 2007;582:967−75. Available from: https://doi.org/10.1113/jphysiol.2007.132787.

[278] Nataraja KN, Parvathi MS. Tolerance to drought stress in plants: unravelling the signaling networks. In: Hossain MA, Wani SH, Bhattacharjee S, Burritt DJ, Tran L-SP, editors. Drought stress tolerance in plants, vol. 2. Cham: Springer International Publishing; 2016, p. 71−90. ISBN 978-3-319-32421-0.

[279] Daloso DM, Medeiros DB, Anjos L, Yoshida T, Araújo WL, Fernie AR. Metabolism within the specialized guard cells of plants. New Phytol 2017;216:1018−33. Available from: https://doi.org/10.1111/nph.14823.

[280] Dreyer I, Uozumi N. Potassium channels in plant cells: potassium channels in plants. FEBS J 2011;278:4293−303. Available from: https://doi.org/10.1111/j.1742-4658.2011.08371.x.

[281] Rhodes D, Nadolska-Orczyk A. Plant stress physiology. in encyclopedia of life sciences. Chichester: John Wiley & Sons, Ltd; 2001, p. a0001297. ISBN 978-0-470-01617-6.

[282] Roberts SK, Snowman BN. The effects of ABA on channel-mediated K + transport across higher plant roots. J Exp Bot 2000;51:1585−94. Available from: https://doi.org/10.1093/jexbot/51.350.1585.

[283] Rodríguez-Gamir J, Ancillo G, González-Mas MC, Primo-Millo E, Iglesias DJ, Forner-Giner MA. Root signalling and modulation of stomatal closure in flooded citrus seedlings. Plant Physiol Biochem 2011;49. Available from: https://doi.org/10.1016/j.plaphy.2011.03.003.

[284] Bradford KJ, Hsiao TC. Stomatal behavior and water relations of waterlogged tomato plants. Plant Physiol 1982;70:1508−13. Available from: https://doi.org/10.1104/pp.70.5.1508.

[285] Else MA, Tiekstra AE, Croker SJ, Davies WJ, Jackson MB. Stomatal closure in flooded tomato plants involves abscisic acid and a chemically unidentified anti-transpirant in xylem sap. Plant Physiol 1996;112:239−47. Available from: https://doi.org/10.1104/pp.112.1.239.

[286] Sojka RE. Stomatal clousre in oxygen-stressed plants. Soil Sci 1992;154:269−80.

[287] Castonuay Y, Nadeau P, Simard RR. Effects of flooding on carbohydrate and ABA levels in roots and shoots of Alfalfa. Plant Cell Environ 1993;16. Available from: https://doi.org/10.1111/j.1365-3040.1993.tb00488.x.

[288] Herzog M. Mechanisms of flood tolerance in wheat and rice—the role of leaf gas films during plant submergence. University of Copenhagen; 2017.

[289] Schmidt RR, Fulda M, Paul MV, Anders M, Plum F, Weits DA, et al. Low-oxygen response is triggered by an ATP-dependent shift in Oleoyl-CoA in Arabidopsis. Proc Natl Acad Sci USA 2018;115:E312101−10. Available from: https://doi.org/10.1073/pnas.1809429115.

[290] Richmond KE, Sussman M. Got silicon? The non-essential beneficial plant nutrient. Curr Opin Plant Biol 2003;6:268−72. Available from: https://doi.org/10.1016/S1369-5266(03)00041-4.

[291] Kosmacz M, Parlanti S, Schwarzländer M, Kragler F, Licausi F, Van Dongen JT. The stability and nuclear localization of the transcription factor RAP2.12 are dynamically regulated by oxygen concentration. Plant Cell Environ 2015;38:1094−103. Available from: https://doi.org/10.1111/pce.12493.

[292] Tan X, Xu H, Khan S, Equiza MA, Lee SH, Vaziriyeganeh M, et al. Plant water transport and aquaporins in oxygen-deprived environments. J Plant Physiol 2018;227:20−30. Available from: https://doi.org/10.1016/j.jplph.2018.05.003.

[293] Tournaire-Roux C, Sutka M, Javot H, Gout E, Gerbeau P, Luu D-T, et al. Cytosolic pH regulates root water transport during anoxic stress through gating of aquaporins. Nature 2003;425. Available from: https://doi.org/10.1038/nature01853.

[294] Zhang WH, Tyerman SD. Effect of low O_2 concentration and azide on hydraulic conductivity and osmotic volume of the cortical cells of wheat roots. Aust J Plant Physiol 1991;18:603−13.

[295] Else MA, Janowiak F, Atkinson CJ, Jackson MB. Root signals and stomatal closure in relation to photosynthesis, chlorophyll a fluorescence and adventitious rooting of flooded tomato plants. Ann Bot (Lond.) 2009;103:313−23. Available from: https://doi.org/10.1093/aob/mcn208.

[296] Liu J, Equiza MA, Navarro-Rodenas A, Lee SH, Zwiazek JJ. Hydraulic adjustments in Aspen (*Populus tremuloides*) seedlings following defoliation involve root and leaf aquaporins. Planta 2014;240. Available from: https://doi.org/10.1007/s00425-014-2106-2.

[297] Yordanova RY, Popova LP. Flooding-induced changes in photosynthesis and oxidative status in maize plants. Acta Physiol Plant 2007;29. Available from: https://doi.org/10.1007/s11738-007-0064-z.

[298] Buckley TN. The control of stomata by water balance. New Phytol 2005;168:275−92. Available from: https://doi.org/10.1111/j.1469-8137.2005.01543.x.

[299] Silveira NM, Frungillo L, Marcos FCC, Pelegrino MT, Miranda MT, Seabra AB, et al. Exogenous nitric oxide improves sugarcane growth and photosynthesis under water deficit. Planta 2016;244:181−90. Available from: https://doi.org/10.1007/s00425-016-2501-y.

[300] Apel K, Hirt H. Reactive oxygen species: metabolism, oxidative stress, and signal transduction. Annu Rev Plant Biol 2004;55:373−99. Available from: https://doi.org/10.1146/annurev.arplant.55.031903.141701.

[301] Beligni MV, Lamattina L. Nitric oxide stimulates seed germination and de-etiolation, and inhibits hypocotyl elongation, three light-inducible responses in plants. Planta 2000;210:215−21. Available from: https://doi.org/10.1007/PL00008128.

[302] O'Neill KJ, Horwitz P, Lund MA. The spatial and temporal nature of changing acidity in a wetland: the case of lake Jandabup on the Swan Coastal Plain, Western Australia. SIL Proc 1922-2010 2002;28:1284−8. Available from: https://doi.org/10.1080/03680770.2001.11902663.

[303] Misra AN, Misra M, Singh R. Nitric oxide ameliorates stress responses in plants. Plant Soil Environ 2011;57:95.

[304] Einspahr KJ, Thompson GA. Transmembrane signaling via phosphatidylinositol 4,5-bisphosphate hydrolysis in plants. Plant Physiol 1990;93:361−6. Available from: https://doi.org/10.1104/pp.93.2.361.

[305] Mehta RL, McDonald BR, Aguilar MM, Ward DM. Regional citrate anticoagulation for continuous arteriovenous hemodialysis in critically ill patients. Kidney Int 1990;38:976−81. Available from: https://doi.org/10.1038/ki.1990.300.

[306] Allan AC, Fluhr R. Two distinct sources of elicited reactive oxygen species in tobacco epidermal cells. Plant Cell 1997;9:1559−72. Available from: https://doi.org/10.2307/3870443.

[307] Gao X, Du Z, Patel TB. Copper and zinc inhibit Gαs function: a nucleotide-free state of Gαs induced by Cu2$^+$ and Zn2$^+$*. J Biol Chem 2005;280:2579–86. Available from: https://doi.org/10.1074/jbc.M409791200.

[308] Alshaal T, Alsaeedi A, El-Ramady H, Almohsen M. Enhancing seed germination and seedlings development of common bean (*Phaseolus vulgaris*) by SiO$_2$ nanoparticles. Egypt J Soil Sci 2017;. Available from: https://doi.org/10.21608/ejss.2017.891.1098.

[309] Alsaeedi AH, El-Ramady H, Alshaal T, El-Garawani M, Elhawat N, Almohsen M. Engineered silica nanoparticles alleviate the detrimental effects of Na + stress on germination and growth of common bean (*Phaseolus vulgaris*). Environ Sci Pollut Res 2017;24:21917–28. Available from: https://doi.org/10.1007/s11356-017-9847-y.

[310] Kiprutto N, Rotich LK, Riungu GK. Agriculture, climate change and food security. OALib 2015;02:1–7. Available from: https://doi.org/10.4236/oalib.1101472.

[311] Yoshida T, Fujita Y, Sayama H, Kidokoro S, Maruyama K, Mizoi J, et al. AREB1, AREB2, and ABF3 are master transcription factors that cooperatively regulate ABRE-dependent ABA signaling involved in drought stress tolerance and require ABA for full activation. Plant J 2010;61:672–85. Available from: https://doi.org/10.1111/j.1365-313X.2009.04092.x.

[312] Ranganathan S, Suvarchala V, Rajesh YBRD, Srinivasa Prasad M, Padmakumari AP, Voleti SR. Effects of silicon sources on its deposition, chlorophyll content, and disease and pest resistance in rice. Biol Plant 2006;50:713–16. Available from: https://doi.org/10.1007/s10535-006-0113-2.

[313] Matichenkov VV, Calvert DV, Snyder GH. Prospective of silicon fertilization for citrus in Florida. Proc Soil Crop Sci Soc Florida 2000;59:137–41.

[314] Rains DW, Epstein E, Zasoski RJ, Aslam M. Active silicon uptake by wheat. Plant Soil 2006;280:223–8. Available from: https://doi.org/10.1007/s11104-005-3082-x.

[315] Ma JF, Yamaji N, Mitani N, Tamai K, Konishi S, Fujiwara T, et al. An efflux transporter of silicon in rice. Nature 2007;448:209–12. Available from: https://doi.org/10.1038/nature05964.

[316] Barber DA, Shone MGT. The absorption of silica from aqueous solutions by plants. J Exp Bot 1966;17:569–78.

[317] Jarvis SC. The uptake and transport of silicon by perennial ryegrass and wheat. Plant Soil 1987;97:429–37. Available from: https://doi.org/10.1007/BF02383233.

[318] Jones LHP, Handreck KA. Silica in soils, plants, and animals. In: Norman, AG, editor. Advances in agronomy. Academic Press; 1967, p. 107–49.

[319] Liang Y, Si J, Römheld V. Silicon uptake and transport is an active process in *Cucumis sativus*. New Phytol 2005;167:797–804. Available from: https://doi.org/10.1111/j.1469-8137.2005.01463.x.

[320] Marxen A, Klotzbücher T, Jahn R, Kaiser K, Nguyen VS, Schmidt A, et al. Interaction between silicon cycling and straw decomposition in a silicon deficient rice production system. Plant Soil 2016;398:153–63. Available from: https://doi.org/10.1007/s11104-015-2645-8.

[321] Rodrigues FA, McNally DJ, Datnoff LE, Jones JB, Labbé C, Benhamou N, et al. Silicon enhances the accumulation of diterpenoid phytoalexins in rice: a potential mechanism for blast resistance. Phytopathology 2004;94:177–83. Available from: https://doi.org/10.1094/PHYTO.2004.94.2.177.

[322] Datnoff LE, Deren CW, Snyder GH. Silicon fertilization for disease management of rice in Florida. Crop Prot 1997;16:525–31. Available from: https://doi.org/10.1016/S0261-2194(97)00033-1.

[323] Hattori T, Sonobe K, Inanaga S, An P, Tsuji W, Araki H, et al. Short term stomatal responses to light intensity changes and osmotic stress in sorghum seedlings raised with and without silicon. Environ Exp Bot 2007;60:177–82. Available from: https://doi.org/10.1016/j.envexpbot.2006.10.004.

[324] Ma JF, Tamai K, Yamaji N, Mitani N, Konishi S, Katsuhara M, et al. A silicon transporter in rice. Nature 2006;440:688–91. Available from: https://doi.org/10.1038/nature04590.

[325] Emamverdian A, Ding Y, Xie Y, Sangari S. Silicon mechanisms to ameliorate heavy metal stress in plants. BioMed Res Int 2018;2018:1–10. Available from: https://doi.org/10.1155/2018/8492898.

[326] Gong HJ, Chen KM, Zhao ZG, Chen GC, Zhou WJ. Effects of silicon on defense of wheat against oxidative stress under drought at different developmental stages. Biol Plant 2008;52:592–6. Available from: https://doi.org/10.1007/s10535-008-0118-0.

[327] Lux A, Luxová M, Abe J, Tanimoto E, Hattori T, Inanaga S. The dynamics of silicon deposition in the sorghum root endodermis. New Phytol 2003;158:437–41. Available from: https://doi.org/10.1046/j.1469-8137.2003.00764.x.

[328] Hossain MT, Soga K, Wakabayashi K, Kamisaka S, Fujii S, Yamamoto R, et al. Modification of chemical properties of cell walls by silicon and its role in regulation of the cell wall extensibility in oat leaves. J Plant Physiol 2007;164:385–93. Available from: https://doi.org/10.1016/j.jplph.2006.02.003.

[329] Cornelis J-T, Titeux H, Ranger J, Delvaux B. Identification and distribution of the readily soluble silicon pool in a temperate forest soil below three distinct tree species. Plant Soil 2011;342:369–78. Available from: https://doi.org/10.1007/s11104-010-0702-x.

[330] Mitani N, Ma JF. Uptake system of silicon in different plant species. J Exp Bot 2005;56:1255–61. Available from: https://doi.org/10.1093/jxb/eri121.

[331] Ma, Miyake Y, Takahashi E. Chapter 2: Silicon as a beneficial element for crop plants Silicon in agriculture; In: Datnoff, LE, Snyder, GH, Korndörfer, GH, editors. Studies in plant science. Elsevier; 2001, p. 17–39.

[332] Liang Y, Sun W, Zhu Y-G, Christie P. Mechanisms of silicon-mediated alleviation of abiotic stresses in higher plants: a review. Environ Pollut 2007;147:422–8. Available from: https://doi.org/10.1016/j.envpol.2006.06.008.

[333] Raven JA. The transport and function of silicon in plants. Biol Rev 1983;58:179–207. Available from: https://doi.org/10.1111/j.1469-185X.1983.tb00385.x.

[334] Mouahid M, Bouzoubaa K, Zouagui Z. Preparation and use of an autogenous bacterin against infectious coryza in chickens. Vet Res Commun 1991;15:413–19. Available from: https://doi.org/10.1007/BF00346536.

[335] Yamaji N, Mitatni N, Ma JF. A transporter regulating silicon distribution in rice shoots. Plant Cell 2008;20:1381–9. Available from: https://doi.org/10.1105/tpc.108.059311.

[336] Lanning FC. Nature and distribution of silica in strawberry plants. Proc Am Soc Hortic Sci 1960;76:349–58.

[337] Osuna-Canizalez FJ, De Datta SK, Bonman JM. Nitrogen form and silicon nutrition effects on resistance to blast disease of rice. Plant Soil 1991;135:223–31.

[338] Epstein E. Silicon. Annu Rev Plant Physiol Plant Mol Biol 1999;50:641—64. Available from: https://doi.org/10.1146/annurev.arplant.50.1.641.

[339] Henriet C, Draye X, Oppitz I, Swennen R, Delvaux B. Effects, distribution and uptake of silicon in banana (Musa Spp.) under controlled conditions. Plant Soil 2006;16.

[340] Casey WH, Kinrade SD, Knight CTG, Rains DW, Epstein E. Aqueous silicate complexes in wheat, Triticum aestivum L. Plant Cell Environ 2004;27:51—4. Available from: https://doi.org/10.1046/j.0016-8025.2003.01124.x.

[341] Allahmoradi P, Ghobadi M, Taherabadi S. Physiological aspects of mungbean (Vigna radiata L.) in response to drought stress; 2011.

[342] Saud S, Li X, Chen Y, Zhang L, Fahad S, Hussain S, et al. Silicon application increases drought tolerance of Kentucky bluegrass by improving plant water relations and morphophysiological functions. Sci World J 2014;2014:368694. Available from: https://doi.org/10.1155/2014/368694.

[343] Hayat R, Ali S, Amara U, Khalid R, Ahmed I. Soil beneficial bacteria and their role in plant growth promotion: a review. Ann Microbiol 2010;60:579—98. Available from: https://doi.org/10.1007/s13213-010-0117-1.

[344] Singh S, Parihar P, Singh R, Singh VP, Prasad SM. Heavy metal tolerance in plants: role of transcriptomics, proteomics, metabolomics, and ionomics. Front Plant Sci 2015;6:1143. Available from: https://doi.org/10.3389/fpls.2015.01143.

[345] Luyckx M, Hausman J-F, Lutts S, Guerriero G. Silicon and plants: current knowledge and technological perspectives. Front Plant Sci 2017;8:411. Available from: https://doi.org/10.3389/fpls.2017.00411.

[346] Alsaeedi A, El-Ramady H, Alshaal TA, Almohsen M. Enhancing seed germination and seedlings development of common bean (Phaseolus vulgaris) by SiO_2 nanoparticles. Egypt J Soil Sci 2017;57:407—15. Available from: https://doi.org/10.21608/ejss.2017.891.1098.

[347] Alsaeedi A, El-Ramady H, Alshaal T, El-Garawani M, Elhawat N, Al-Otaibi A. Exogenous nanosilica improves germination and growth of cucumber by maintaining K + /Na + ratio under elevated Na + stress. Plant Physiol Biochem 2018;125:164—71. Available from: https://doi.org/10.1016/j.plaphy.2018.02.006.

[348] Gong H, Chen K. The regulatory role of silicon on water relations, photosynthetic gas exchange, and carboxylation activities of wheat leaves in field drought conditions. Acta Physiol Plant 2012;34:1589—94. Available from: https://doi.org/10.1007/s11738-012-0954-6.

[349] Maghsoudi K, Emam Y, Ashraf M. Foliar application of silicon at different growth stages alters growth and yield of selected wheat cultivars. J Plant Nutr 2016;39:1194—203. Available from: https://doi.org/10.1080/01904167.2015.1115876.

[350] Amin M, Ahmad R, Basra SMA, Murtaza G. Silicon induced improvement in morpho-physiological traits of maize (Zea mays L.) under water deficit. Pak J Agric Res 2014;51.

[351] Shi Y, Zhang Y, Yao H, Wu J, Sun H, Gong H. Silicon improves seed germination and alleviates oxidative stress of bud seedlings in tomato under water deficit stress. Plant Physiol Biochem 2014;78:27—36. Available from: https://doi.org/10.1016/j.plaphy.2014.02.009.

[352] Ming DF, Pei ZF, Naeem MS, Gong HJ, Zhou WJ. Silicon alleviates PEG-induced water-deficit stress in upland rice seedlings by enhancing osmotic adjustment: silicon alleviates PEG-induced water-deficit stress. J Agron Crop Sci 2012;198:14—26. Available from: https://doi.org/10.1111/j.1439-037X.2011.00486.x.

[353] Asgharipour MR, Mosapour H. A foliar application silicon enchances drought tolerance in fennel. J Anim Plant Sci 2016;26.

[354] Putra ETS, Purwanto BH. Physiological responses of oil palm seedlings to the drought stress using boron and silicon applications. J Agron 2015;13.

[355] Abdalla N, Ragab MI, Fári M, El-Ramady H, Alshaal T, Elhawat N, et al. Nanobiotechnology for plants. EBSS 2019;2. Available from: https://doi.org/10.21608/jenvbs.2018.6711.1041.

[356] Emam MM, Khattab HE, Helal NM, Deraz AE. Effect of selenium and silicon on yield quality of rice plant grown under drought stress. Aust J Crop Sci 2014;8.

[357] Gunes DA, Pilbeam DJ, Inal A, Bagci EG. Influence of silicon on antioxidant mechanisms and lipid peroxidation in chickpea (Cicer arietinum L) cultivars under drought stress. J Plant Interact 2007;2.

[358] Ahmad ST, Haddad R. Study of silicon effects on antioxidant enzyme activities and osmotic adjustment of wheat under drought stress. Czech J Genet and Plant Breed 2011;47(2011):17—27. Available from: https://doi.org/10.17221/92/2010-CJGPB.

[359] Khattab HI, Emam MA, Emam MM, Helal NM, Mohamed MR. Effect of selenium and silicon on transcription factors NAC5 and DREB2A involved in drought-responsive gene expression in rice. Biol Plant 2014;58:265—73. Available from: https://doi.org/10.1007/s10535-014-0391-z.

[360] Kaya C, Tuna L, Higgs D. Effect of silicon on plant growth and mineral nutrition of maize grown under water-stress conditions. J Plant Nutr 2006;29:1469—80. Available from: https://doi.org/10.1080/01904160600837238.

[361] Tuna AL, Kaya C, Higgs D, Murillo-Amador B, Aydemir S, Girgin AR. Silicon improves salinity tolerance in wheat plants. Environ Exp Bot 2008;62:10—16. Available from: https://doi.org/10.1016/j.envexpbot.2007.06.006.

[362] Sardans J, Peñuelas J. Potassium control of plant functions: ecological and agricultural implications. Plants (Basel), 10. 2021. p. 419. Available from: http://doi.org/10.3390/plants10020419.

[363] Liang Y. Effects of silicon on enzyme activity and sodium, potassium and calcium concentration in barley under salt stress. Plant Soil 1999;209:217. Available from: https://doi.org/10.1023/A:1004526604913.

[364] Hattori T, Inanaga S, Araki H, An P, Morita S, Luxova M, et al. Application of silicon enhanced drought tolerance in Sorghum bicolor. Physiol Plant 2005;123:459—66. Available from: https://doi.org/10.1111/j.1399-3054.2005.00481.x.

[365] Nolla A, Faria RD, Korndörfer G, Roque T, da Silva B. Effect of silicon on drought tolerance of upland rice. J Food Agric Environ 2012;10.

[366] Mauad M, Crusciol CAC, Nascente AS, Grassi Filho H, Lima GPP. Effects of silicon and drought stress on biochemical characteristics of leaves of upland rice cultivars. Rev Ciênc Agron 2016;47:532—9. Available from: https://doi.org/10.5935/1806-6690.20160064.

[367] Chen W, Yao X, Cai K, Chen J. Silicon alleviates drought stress of rice plants by improving plant water status, photosynthesis and mineral nutrient absorption. Biol Trace Elem Res 2011;142:67—76. Available from: https://doi.org/10.1007/s12011-010-8742-x.

[368] Ahmad Z, Ahmad Waraich E, Barutçular C, Hossain A, Erman M, Çiğ F, et al. Enhancing drought tolerance in wheat through improving morpho-physiological and antioxidants activities of plants by the supplementation of foliar silicon. Phyton 2020;89:529—39. Available from: https://doi.org/10.32604/phyton.2020.09143.

[369] Gong H, Zhu X, Chen K, Wang S, Zhang C. Silicon alleviates oxidative damage of wheat plants in pots under drought. Plant Sci 2005;169:313—21. Available from: https://doi.org/10.1016/j.plantsci.2005.02.023.

[370] Ahmad ST, Haddad R. Effect of silicon on antioxidant enzymes activities and osmotic adjustment contents in two bread wheat genotypes under drought stress conditions. Seed Plant Prod J 2010.

[371] Ahmad M, El-Saeid MH, Akram MA, Ahmad HR, Haroon H, Hussain A. Silicon fertilization—a tool to boost up drought tolerance in wheat (*Triticum aestivum* L.) crop for better yield. J Plant Nutr 2016;39:1283—91. Available from: https://doi.org/10.1080/01904167.2015.1105262.

[372] Thorne SJ, Hartley SE, Maathuis FJM. The effect of silicon on osmotic and drought stress tolerance in wheat landraces. Plants 2021;10:814. Available from: https://doi.org/10.3390/plants10040814.

[373] Bianchini HC, Marques DJ. Gas exchange and putrescine content as drought stress indicators in corn cultivars fertilized with silicon. Aust J Crop Sci 2020;1252—8. Available from: https://doi.org/10.21475/ajcs.20.14.08.p2339.

[374] Gao X, Zou C, Wang L, Zhang F. Silicon decreases transpiration rate and conductance from stomata of maize plants. J Plant Nutr 2006;29:1637—47. Available from: https://doi.org/10.1080/01904160600851494.

[375] Gao X, Zou C, Wang L, Zhang F. Silicon improves water use efficiency in maize plants. J Plant Nutr 2005;27:1457—70. Available from: https://doi.org/10.1081/PLN-200025865.

[376] Sattar A, Sher A, Ijaz M, Ul-Allah S, Butt M, Irfan M, et al. Interactive effect of biochar and silicon on improving morpho-physiological and biochemical attributes of maize by reducing drought hazards. J Soil Sci Plant Nutr 2020;20:1819—26. Available from: https://doi.org/10.1007/s42729-020-00253-7.

[377] Abd El-Mageed TA, Shaaban A, Abd El-Mageed SA, Semida WM, Rady MOA. Silicon defensive role in maize (*Zea mays* L.) against drought stress and metals-contaminated irrigation water. Silicon 2021;13:2165—76. Available from: https://doi.org/10.1007/s12633-020-00690-0.

[378] Ning D, Qin A, Liu Z, Duan A, Xiao J, Zhang J, et al. Silicon-mediated physiological and agronomic responses of maize to drought stress imposed at the vegetative and reproductive stages. Agronomy 2020;10:1136. Available from: https://doi.org/10.3390/agronomy10081136.

[379] Gunes A, Kadioglu YK, Pilbeam DJ, Inal A, Coban S, Aksu A. Influence of silicon on sunflower cultivars under drought stress, II: Essential and nonessential element uptake determined by polarized energy dispersive X-ray fluorescence. Commun Soil Sci Plant Anal 2008;39:1904—27. Available from: https://doi.org/10.1080/00103620802134719.

[380] Gunes A, Pilbeam DJ, Inal A, Coban S. Influence of silicon on sunflower cultivars under drought stress. I. Growth, antioxidant mechanisms, and lipid peroxidation. Commun Soil Sci Plant Anal 2008;39.

[381] Habibi G, Hajiboland R. Alleviation of drought stress by silicon supplementation in pistachio (*Pistacia vera* L.) plants. Folia Hortic 2013;25:21—9. Available from: https://doi.org/10.2478/fhort-2013-0003.

[382] Helaly MN, El-Hoseiny H, El-Sheery NI, Rastogi A, Kalaji HM. Regulation and physiological role of silicon in alleviating drought stress of mango. Plant Physiol Biochem 2017;118:31—44. Available from: https://doi.org/10.1016/j.plaphy.2017.05.021.

[383] Kang J, Zhao W, Zhu X. Silicon improves photosynthesis and strengthens enzyme activities in the C3 succulent xerophyte *Zygophyllum xanthoxylum* under drought stress. J Plant Physiol 2016;199:76—86. Available from: https://doi.org/10.1016/j.jplph.2016.05.009.

[384] Biju S, Fuentes S, Gupta D. Silicon improves seed germination and alleviates drought stress in lentil crops by regulating osmolytes, hydrolytic enzymes and antioxidant defense system. Plant Physiol Biochem 2017;119:250—64. Available from: https://doi.org/10.1016/j.plaphy.2017.09.001.

[385] Patel M, Fatnani D, Parida AK. Silicon-induced mitigation of drought stress in peanut genotypes (*Arachis hypogaea* L.) through ion homeostasis, modulations of antioxidative defense system, and metabolic regulations. Plant Physiol Biochem 2021;166:290—313. Available from: https://doi.org/10.1016/j.plaphy.2021.06.003.

[386] Hasanuzzaman M, Nahar K, Anee TI, Khan MIR, Fujita M. Silicon-mediated regulation of antioxidant defense and glyoxalase systems confers drought stress tolerance in *Brassica napus* L. South Afr J Bot 2018;115:50—7. Available from: https://doi.org/10.1016/j.sajb.2017.12.006.

[387] Saja-Garbarz D, Ostrowska A, Kaczanowska K, Janowiak F. Accumulation of silicon and changes in water balance under drought stress in *Brassica napus* Var. Napus L. Plants 2021;10:280. Available from: https://doi.org/10.3390/plants10020280.

[388] Rehman A, Hamid N, Naz B. Effect of exogenous application of silicon with drought stress on protein and carbohydrate contents of edible beans (*Vigna radiate* & *Vigna unguiculata*). Pak J Chem 2012;2:99—105. Available from: https://doi.org/10.15228/2012.v02.i02.p08.

[389] Shen X, Zhou Y, Duan L, Li Z, Eneji AE, Li J. Silicon effects on photosynthesis and antioxidant parameters of soybean seedlings under drought and ultraviolet-B radiation. J Plant Physiol 2010;167:1248—52. Available from: https://doi.org/10.1016/j.jplph.2010.04.011.

[390] Cao B, Wang L, Gao S, Xia J, Xu K. Silicon-mediated changes in radial hydraulic conductivity and cell wall stability are involved in silicon-induced drought resistance in tomato. Protoplasma 2017;254:2295—304. Available from: https://doi.org/10.1007/s00709-017-1115-y.

[391] Saleh J, Najafi N, Oustan S, Aliasgharzad N, Ghassemi-Golezani K. Changes in extractable Si, Fe, and Mn as affected by silicon, salinity, and waterlogging in a sandy loam soil. Commun Soil Sci Plant Anal 2013;44:1588—98. Available from: https://doi.org/10.1080/00103624.2013.768261.

[392] Chu M, Liu M, Ding Y, Wang S, Liu Z, Tang S, et al. Effect of nitrogen and silicon on rice submerged at tillering stage. Agron J 2018;110:183—92. Available from: https://doi.org/10.2134/agronj2017.03.0156.

[393] Pan T, Zhang J, He L, Hafeez A, Ning C, Cai K. Silicon enhances plant resistance of rice against submergence stress. Plants 2021;10:767. Available from: https://doi.org/10.3390/plants10040767.

[394] Balakhnina TI, Matichenkov VV, Wlodarczyk T, Borkowska A, Nosalewicz M, Fomina IR. Effects of silicon on growth processes and adaptive potential of barley plants under optimal soil watering and flooding. Plant Growth Regul 2012;67:35—43. Available from: https://doi.org/10.1007/s10725-012-9658-6.

[395] Sayed SA, Gadallah MAA. Effects of silicon on *Zea mays* plants exposed to water and oxygen deficiency. Russ J Plant Physiol 2014;61:460—6. Available from: https://doi.org/10.1134/S1021443714040165.

[396] Włodarczyk T, Balakhnina T, Matichenkov V, Brzezińska M, Nosalewicz M, Szarlip P, et al. Effect of silicon on barley growth and N$_2$O emission under flooding. Sci Total Environ 2019;685:1—9. Available from: https://doi.org/10.1016/j.scitotenv.2019.05.410.

11

Silicon and nanosilicon mediated heat stress tolerance in plants

Abida Parveen[1], Sahar Mumtaz[2], Muhammad Hamzah Saleem[3], Iqbal Hussain[1], Shagufta Perveen[1] and Sumaira Thind[1]

[1]Department of Botany, Government College University, Faisalabad, Pakistan [2]Department of Botany, Division of Science and Technology, University of Education, Lahore, Pakistan [3]College of Plant Science and Technology, Huazhong Agricultural University, Wuhan, P.R. China

11.1 Silicon and plants

Silicon (Si) is believed to be the most plentiful component in earth's crust, but it has yet to be linked to the number of essential elements. It has been revealed by recent research that silicon is a crucial element involved in plants growth and development [1]. It plays an essential role in controlling the ultimate biochemical and physiological features of plants. Although it is a nonessential element, it is found about 30% in rocks with the majority of its existence in the form of salts [2]. Plants absorb Si very efficiently, which is then transported with water to stem and leaves through the transpiration pull [3]. Due to its eco-friendly role, it contributed significantly toward the mitigation of various biotic and abiotic stresses in plants. However, scavenging of ROS, stimulation of antioxidative defense responses, and phytohormonal signaling are some of the vital roles of Si [4,5].

11.2 Silicon dynamics and distribution in plants

Understanding the molecular motion is critical for explaining how plants absorb nutrients and water from land and distribute to various parts [6−8]. Silicon is mainly deposited in inflorescence, bract, epidermal cells of leaves, and endodermal cells of roots [9]. Silicon deposition has an effect on passive processes such as transpiration. Deposition of Si is not an arbitrary process [10]. Kumar et al. [11] identify Si deposition trends in grasses and propose three main types of Si deposition: guided para mural silicification in silica cells, uncontrolled silicon deposition in cell walls, and controlled silicon deposition in cell walls. The mechanism of Si deposition was primarily examined in aerial parts and roots [12].

11.3 Nanosilicon and plants

Nanomaterials are substances with a molecular weight of less than 100 nm. Nanoparticles' (NPs') physicochemical characteristics can differ significantly from larger particles of the same nature [13]. Nanotechnology's usage in agriculture is growing at a quick rate [14]. Due to this reason, understanding and elucidation of the functionality of nano-Si in plant growth and production are critical. For plants, nano-Si is treated as a "quasiessential" factor that controls a variety of physiological mechanisms such as plant growth, seed germination, photosynthesis, and stress resistance [15,16]. Silicon NPs are more effective than bulk particles due to their limited

Silicon and Nano-silicon in Environmental Stress Management and Crop Quality Improvement.
DOI: https://doi.org/10.1016/B978-0-323-91225-9.00001-7

surface area and reactivity. It is critical to determine the impact of nanosilicon biochemical systems [17]. The functionality of nano-Si in plants growth, photosynthesis, and stress resistance has been discussed in detail by Siddiqui et al. [18].

11.4 Use of nanosilicon to promote plant growth and heat stress tolerance

Nanobiotechnology is a powerful method for achieving long-term agricultural sustainability [19]. Some NPs with special physical and chemical features rather than functioning as nanocarriers naturally improve plant growth and heat stress resistance [20]. Nano-Si's function is determined by their physical and chemical characteristics implementation system (foliar distribution, hydroponics, and soil) and dosage applied. It is clear that nano-Si is advantageous for the growth and development of many plants when they are subjected to heat stress [21]. Under heat-stressed conditions, the use of nano-Si can enhance all facets of plant development. The usage of nano-Si assists to counteract the effects of heat stress on plant functions such as stomatal conductance [22]. Under extreme heat, the leaf dry and fresh weight and the potassium value have all decreased. If nano-Si is not applied, then it results in uncontrolled plants growth under stressful conditions. One of the potential mechanisms of nano-Si adaptation to stress conditions is that it retains plant resilience in heat stress leading to increased cell size and leaf thickness [23].

11.5 Role of silicon and nanosilicon particles in improving heat stress endurance

Plants produce organic mineralized NPs for optimal production when they are exposed to high temperatures [24]. The activation of seed with NPs of Si has revolutionized the farming sector, which credit goes to advanced research. Nanoparticles have special physical and chemical characteristics, which attract scientists from all over the world [25]. Nano-Si-coated seeds have the capability to cope with a variety of stresses [26]. Accumulation of nano-SiO_2 in maize seed results in a substantial rise of organic substances, including proteins, chlorophyll, phenols, and dry weight of plant [27]. Tomato seed germination, average germination time, germinating seeds percentage, level of seed germination, and fresh and dry weight of seedling are significantly influenced by the application of nano-SiO_2 [18]. Under heat stress, the addition of nano-SiO_2 to the nutrient media will increase the germination and growth of lentil plants [28]. The influence of heat stress is lessened by the application of nano-SiO_2 in basil (*Ocimum basilicum*), which ultimately enhanced the fresh and dry weight of leaf and proline and chlorophyll contents in plants [29]. When *Vicia faba* is subjected to heat stress, then nano-SiO_2 can improve the number of germination features (Fig. 11.1). Furthermore, nano-SiO_2 application in *V. faba* results in increased

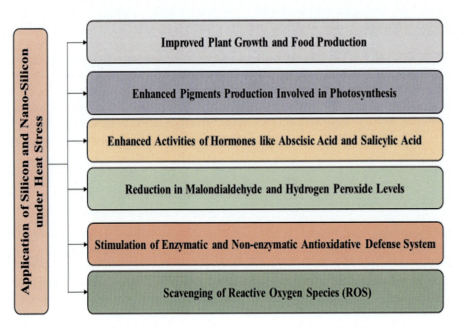

FIGURE 11.1 Effect of silicon and nanosilicon application on plants under heat stress.

fresh and dry weight and plant length under heat stress [30]. Nano-Si can be employed to increase tomato photosynthetic quantum and to reduce germination rate and plant development impairment caused by heat stress [28]. Silicon NPs are sometimes utilized as engineered oxide NPs. Plants are affected by NPs in both positive and negative ways. Nano-Si application to plants can have both positive and negative impacts on the growth of plants. Plants are able to tolerate heat stress when potassium silicate is combined with salicylic acid [31]. Nano-SiO_2 imparts a positive effect on plant growth under various stresses, which make these NPs a viable substitute for harmful fertilizers for safe agricultural activities.

11.6 Regulation of antioxidant activities by silicon in crop plants under heat stress

Silicon is the second most dominant element in soil, with its accessibility to plants ranging from 1% to 10% of the plant's overall dry weight. Many crop plants have been reported to accumulate Si in vertical pathways. From the last 20 years, research revealed that silicon application has improved plant growth and production in both natural and stressful environments [32]. Plants produce ROS like $^1O_2$, O_2^{2-}, H_2O_2, and OH $\cdot$ under stressed conditions [33–35], which trigger major harm to the structure and function of cells [36–38]. Plants have evolved a dynamic antioxidant mechanism (i.e., nonenzymatic and enzymatic) to sustain homeostasis [39–42]. Under extreme stress, exogenous implementation of Si exhibit enhanced disease resistance by controlling the production of reactive oxygen species and decreasing the absorption of harmful ions like Na^+ [1]. These findings indicate that Si plays a crucial role in antioxidant defense regulation and reducing oxidative stress in heat-sensitive plants. Many other studies get the same findings and conclude that increased the level of catalase and superoxide dismutase function as an oxidative scavenger toward reactive radicals under stress.

Nanoparticles affect positively the soybean (*Glycine max* L.) development and antioxidant activity under stress conditions [43]. Heat stress can reduce biomass of shoot and root, K^+ contents in roots and leaf, enzymatic and nonenzymatic antioxidants, phenolics, and H_2O_2, as well as lipid peroxidation and oxygen radical concentration. Treatment of plants with 0.5–1 mM nano-Si oxide can improve the growth of seedling shoot and root. A foliar usage of Si dioxide at 2 mM inhibited soybean development. Exogenously applied nano-Si reduces heat stress by increasing potassium ions levels, antioxidant activities, nonenzymatic molecules, and lipid peroxidation [44]. Elevated concentrations of lipid peroxidation caused by ROS may promote the production of malondialdehyde (MDA), which is typically seen as a key marker of abiotic stress causing harm to the lipid bilayer [45–47]. Stressed environmental conditions can substantially increase the levels of $^1O_2$ and MDA [48–50]. Furthermore, exogenously applied nano-Si hinders the development of ROS and triggers antioxidant defense mechanisms to combat heat stress [51]. Superoxide dismutase is the primary ROS scavenger, but its enzymatic behavior produces H_2O_2 by which peroxidase disintegrates through substrate oxidation into phenolic contents and antioxidants, while catalase decomposes hydrogen peroxide into simple O_2 and water molecules [52–54]. Nanosilicon was already known to stimulate disease resistance by reducing the generation of ROS, decreasing electrolytic loss and MDA materials (Table 11.1), immobilizing and minimizing uptake of harmful ions under stressed environment [66].

11.7 Mechanisms of silicon-mediated amelioration of heat stress in plants

Silicon is the second most abundant element in earth's crust, possessed several plant-friendly properties. Silicon appears to have a beneficial impact on plants under abiotic and biotic stressed conditions [32]. Adding Si to a variety of essential crops enhanced growth and production. One of the most important abiotic stresses that affects plants growth and development is heat stress. Plants are stressed by aerobic, osmotic, and ionic stresses when there is a significant concentration of heat in the growth media. Severe temperatures have a negative effect on soil, groundwater, and agricultural development [57]. Silicon helps plants to cope with heat stress in a variety of ways. Enhanced photosynthetic rate, detoxification of toxic reactive oxygen compounds utilizing antioxidants, and good nutritional regulation are involved in Si-mediated stress mitigation. Si regulates photosynthesis, root development, redox homeostasis balance, and nutritional supply in plants under extreme environmental conditions [18].

TABLE 11.1 Interactive effects of organic fertilizers on plants under abiotic stresses.

Sr. no.	Plant	Organic fertilizer applied	Growth medium	Stress	Biological effects	Reference
1	Tomato	Silicon	Soil	Heat stress	– Improvement in tomato plant biomass – Improvement in growth parameters – Enhancement in pigments production for photosynthesis – Regulation of protective mechanisms especially by antioxidants	Khan et al. [55]
2	Cucumber	Silicon	Soil	Heat	– Improvement in plant growth and food production – Improvement in production of photosynthetic pigments in leaves of cucumber – Increased capability of antioxidants action – Increased fruit yield and better quality for market	Shalaby et al. [51]
3	Rice	Soil	Silicon	Heat	– Improvement in chlorophylls of leaves – Enhancement in stomatal conductance and photosynthesis – Reduction of CO_2 conc. in intercellular spaces – Enhancement in photosynthetic rate – Decreased contents of MDA – Improvement in actions of enzymatic antioxidants	Chen-yang et al. [56]
4	Rice	Soil	Silicon	Heat, drought	– Retention of plasma membrane structure by silicon – Improvement in cellulose quantity of cell walls – Enhancement in drought stress endurance – Prevention in the leakage of electrolytes	Agarie et al. [57]
5	Date palm	Silicon	Soil	Heat	– Significantly promoted plant growth attributes – Induction of elevated levels of hormones, i.e., ABA and SA – Enhanced accumulation of antioxidants – Reduction in heat-induced oxidative stress – Decreased conc. of O_2^- and lipid peroxidation – Improvement in conc. and activity of enzymatic antioxidants	Khan et al. [58]
6	Pumpkin	Nanosilicon	Growth chamber	Salinity	– Nano-Si increased the percentage of germinated seeds – Causes reduction in leakage of electrolytes and level of H_2O_2 – Chlorophyll degradation was decreased – Enhanced photosynthetic activity – Enhanced conductance in stomata – Enhanced rate of transpiration – Enhanced WUE – Decreased injury by ROS in the presence of antioxidants	Siddiqui et al. [18]
7	Soybean	Nanosilicon	Soil	Salinity	– Nano-Si enhanced biomass of seedlings – Nanosilicon alleviated salt stress by replacing K^+ with Na^+ – Enhanced rate of action of antioxidants – Decrease in Na^+ conc. – Decrease in lipid peroxidation – Decreased generation of ROS	Farhangi-Abriz et al. [59]
8	Strawberry	Nanosilicon	Soil	Salinity	– Maintenance of epicuticle wax structure – Improvement of chlorophyll content – Improvement in carotenoid content – Less accumulation of proline – Lessen adverse structural and functional changes caused by salinity	Avestan et al. [60]
9	*Salvia splendens*	Silicon	Soil	Heat	– Increased activities of enzymatic antioxidants – Decreased rate of action of catalase	Soundararajan et al. [61]
10	Pea	Silicon	Soil	Heavy metals	– Enhanced biomass production of plants – Improvement of photosynthetic pigments – Increased activities of enzymatic antioxidants – Decreased rate of action of APX	
11	Wheat	Silicon	Soil	Drought	– Improvement in dry biomass of plants – Improvement in chlorophyll contents	Gong et al. [62]

(Continued)

TABLE 11.1   (Continued)

Sr. no.	Plant	Organic fertilizer applied	Growth medium	Stress	Biological effects	Reference
12	Sweet pepper	Nanosilicon	Sandy soil	Salinity	− Beneficial effects on N and P uptake − Improvement in nutritional status − Absorption of nutrients increased − Boosted height of plants − Root length increased − Enhanced plant weights (fresh and dry) − Enhanced yield attributes of pepper − n-Si is more valuable and efficient in comparison with regular silicon	Tantawy et al. [63]
13	Canola	Silicon	Soil	Salinity	− Reduction in the buildup of ions like Na^+ and Cl^- − Enhanced rate of action of antioxidants − Reduced lipid peroxidation and H_2O_2	Farshidi et al. [64]
14	Sorghum	Silicon	Soil	Drought	− Improvement in RWC − Enhanced rate of transpiration − Accelerated internal physiological processes of plants	Yin et al. [65]

ABA, abscisic acid, *SA,* salicylic acid, *WUE,* water use efficiency

11.8 Silicon and nanosilicon against several plant diseases

According to recent research, Si boosts plant tolerance to a variety of plant pathogens including crumbly mildew Septoria and eyespot, as well as a variety of insect pests. Plants absorb Si as silicic acid from roots. Under water-deficit conditions, Si has a beneficial impact on biomass yield. Activation of defensive enzymes and controlling the vast system of signaling channels are part of Si-induced metabolic tolerance throughout plant−pathogen encounters [67]. Silicon can function efficiently at the genetic level during stress conditions to control genes functionality involved in defense mechanisms. Knowing how Si regulates plant−microbe relationships may aid in the efficient employment of this element to enhance fruit yield and improve sensitivity caused by heat stress [21]. Silicon enhances resilience to a number of diseases serving as a physical shield centered on predefense frameworks prior to pathogen infection. The most popular faba bean illness in Egypt is chocolate spot disease, which is caused by *Botrytis fabae.* Nano-Si has been used as a resistance inducer in faba bean against *B. fabae.* It can be concluded that applying nano-Si to *V. faba* plants improved their resistance to infectious *B. fabae* [30]. Due to this reason, nano-Si is suggested as an option for important fungicides to overcome infections of *B. fabae.* When wheat leaves contaminated with *Pyricularia oryzae* are treated with Si, then it results in limited hyphae entrance to upper epidermal cells. These hyphae entered in other cells where no Si was applied [68]. When Si was applied on the tissue surface of rice contaminated with *Pyricularia grisea* and *Rhizoctonia solani,* then decline in lesions of leaf blade was observed [51].

Reference

[1] Javed MT, Saleem MH, Aslam S, et al. Elucidating silicon-mediated distinct morpho-physio-biochemical attributes and organic acid exudation patterns of cadmium stressed Ajwain *Trachyspermum ammi* L. Plant Physiol Biochem 2020;157.

[2] Kamran M, Parveen A, Ahmar S, et al. An overview of hazardous impacts of soil salinity in crops, tolerance mechanisms, and amelioration through selenium supplementation. Int J Mol Sci 2019;21(1):148.

[3] Ahmad P, Ahanger MA, Alam P, et al. Silicon (Si) supplementation alleviates NaCl toxicity in mung bean [*Vigna radiata* (L.) Wilczek] through the modifications of physio-biochemical attributes and key antioxidant enzymes. J Plant Growth Regul 2019;38(1):70−82.

[4] Kamran M, Danish M, Saleem MH, et al. Application of abscisic acid and 6-benzylaminopurine modulated morpho-physiological and antioxidative defense responses of tomato (*Solanum lycopersicum* L.) by minimizing cobalt uptake. Chemosphere 2020;263:128169.

[5] Wu J, Geilfus C-M, Pitann B, et al. Silicon-enhanced oxalate exudation contributes to alleviation of cadmium toxicity in wheat. Environ Exp Bot 2016;131:10−18.

[6] Zaheer IE, Ali S, Saleem MH, et al. Zinc-lysine supplementation mitigates oxidative stress in rapeseed (*Brassica napus* L.) by preventing phytotoxicity of chromium, when irrigated with tannery wastewater. Plants 2020;9(9):1145.

[7] Afzal J, Saleem MH, Batool F, et al. Role of ferrous sulfate ($FeSO_4$) in resistance to cadmium stress in two rice (*Oryza sativa* L.) genotypes. Biomolecules 2020;10(12):1693.

[8] Saleem MH, Ali S, Kamran M, et al. Ethylenediaminetetraacetic acid (EDTA) mitigates the toxic effect of excessive copper concentrations on growth, gaseous exchange and chloroplast ultrastructure of *Corchorus capsularis* L. and improves copper accumulation capabilities. Plants 2020;9(6):756.

[9] Zeng F-R, Zhao F-S, Qiu B-Y, et al. Alleviation of chromium toxicity by silicon addition in rice plants. Agric Sci China 2011;10(8):1188−96.

[10] Fan X, Wen X, Huang F, et al. Effects of silicon on morphology, ultrastructure and exudates of rice root under heavy metal stress. Acta Physiol Plant 2016;38(8):197.

[11] Kumar A, Bieri M, Reindl T, et al. Economic viability analysis of silicon solar cell manufacturing: Al-BSF vs PERC. Energy Procedia 2017;130:43−9.

[12] Tang H, Liu Y, Gong X, et al. Effects of selenium and silicon on enhancing antioxidative capacity in ramie (*Boehmeria nivea* (L.) Gaud.) under cadmium stress. Environ Sci Pollut Res 2015;22(13):9999−10008.

[13] Rizwan M, Ali S, Ali B, et al. Zinc and iron oxide nanoparticles improved the plant growth and reduced the oxidative stress and cadmium concentration in wheat. Chemosphere 2019;214:269−77.

[14] Ali Q, Ahmar S, Sohail MA, et al. Research advances and applications of biosensing technology for the diagnosis of pathogens in sustainable agriculture. Environ Sci Pollut Res 2021;28:1−18.

[15] Saleem MH, Ali S, Seleiman MF, et al. Assessing the correlations between different traits in copper-sensitive and copper-resistant varieties of jute (*Corchorus capsularis* L.). Plants 2019;8(12):545.

[16] Rehman M, Liu L, Wang Q, et al. Copper environmental toxicology, recent advances, and future outlook: a review. Environ Sci Pollut Res 2019;26:1−14.

[17] Jia-Wen W, Yu S, Yong-Xing Z, et al. Mechanisms of enhanced heavy metal tolerance in plants by silicon: a review. Pedosphere 2013;23(6):815−25.

[18] Siddiqui H, Ahmed KBM, Sami F, et al. Silicon nanoparticles and plants: current knowledge and future perspectives. Sustain Agric Rev 2020;41:129−42.

[19] Faizan M, Faraz A, Mir AR, et al. Role of zinc oxide nanoparticles in countering negative effects generated by cadmium in *Lycopersicon esculentum*. J Plant Growth Regul 2021;40(1):101−15.

[20] Nair PMG, Chung IM. Study on the correlation between copper oxide nanoparticles induced growth suppression and enhanced lignification in Indian mustard (*Brassica juncea* L.). Ecotoxicol Environ Saf 2015;113:302−13.

[21] Sako K, Nguyen HM, Seki M. Advances in chemical priming to enhance abiotic stress tolerance in plants. Plant Cell Physiol 2020;61(12):1995−2003.

[22] Chen B, Chu S, Cai R, et al. The effect of diffusion induced fatigue stress on capacity loss in nano silicon particle electrodes during cycling. J Electrochem Soc 2016;163(13):A2592.

[23] Cheng B, Chen F, Wang C, et al. The molecular mechanisms of silica nanomaterials enhancing the rice (*Oryza sativa* L.) resistance to planthoppers (*Nilaparvata lugens* Stal). Sci Total Environ 2021;767:144967.

[24] Yildiztugay A, Ozfidan-Konakci C, Yildiztugay E, et al. Biochar triggers systemic tolerance against cobalt stress in wheat leaves through regulation of water status and antioxidant metabolism. J Soil Sci Plant Nutr 2019;19(4):935−47.

[25] Hussain A, Ali S, Rizwan M, et al. Responses of wheat (*Triticum aestivum*) plants grown in a Cd contaminated soil to the application of iron oxide nanoparticles. Ecotoxicol Environ Saf 2019;173:156−64.

[26] Hameed A, Sheikh MA, Jamil A, et al. Seed priming with sodium silicate enhances seed germination and seedling growth in wheat (*Triticum aestivum* L.) under water deficit stress induced by polyethylene glycol. Pak J Life Soc Sci 2013;11(1):19−24.

[27] Suriyaprabha R, Karunakaran G, Yuvakkumar R, et al. Growth and physiological responses of maize (*Zea mays* L.) to porous silica nanoparticles in soil. J Nanopart Res 2012;14(12):1−14.

[28] Sabaghnia N, Janmohammadi M. Graphic analysis of nano-silicon by salinity stress interaction on germination properties of lentil using the biplot method. Agric For 2014;60(3).

[29] Kalteh M, Alipour ZT, Ashraf S, et al. Effect of silica nanoparticles on basil (*Ocimum basilicum*) under salinity stress. J Chem Health Risks 2018;4(3).

[30] Galal OA, Thabet AF, Tuda M, et al. RAPD Analysis of genotoxic effects of nano-scale SiO$_2$ and TiO$_2$ on broad bean (*Vicia faba* L.). J Fac Agric Kyushu Univ 2020;65:57−63.

[31] Hussein M, Abou-Baker NH. Growth and mineral status of moringa plants as affected by silicate and salicylic acid under salt stress. Int J Plant Soil. Sci 2013;3:163−77.

[32] Hasanuzzaman M, Alam MM, Nahar K, et al. Silicon-induced antioxidant defense and methylglyoxal detoxification works coordinately in alleviating nickel toxicity in *Oryza sativa* L. Ecotoxicology 2019;28(3):261−76.

[33] Alam H, Khattak JZK, Ksiksi TS, et al. Negative impact of long-term exposure of salinity and drought stress on native *Tetraena mandavillei* L. Physiol Plant 2020;172.

[34] Saleem M, Ali S, Rehman M, et al. Influence of phosphorus on copper phytoextraction via modulating cellular organelles in two jute (*Corchorus capsularis* L.) varieties grown in a copper mining soil of Hubei Province, China. Chemosphere 2020;248.

[35] Ali M, Kamran M, Abbasi GH, et al. Melatonin-induced salinity tolerance by ameliorating osmotic and oxidative stress in the seedlings of two tomato (*Solanum lycopersicum* L.) cultivars. J Plant Growth Regul 2020;40.

[36] Saleem MH, Kamran M, Zhou Y, et al. Appraising growth, oxidative stress and copper phytoextraction potential of flax (*Linum usitatissimum* L.) grown in soil differentially spiked with copper. J Environ Manag 2020;257:109994.

[37] Saleem MH, Fahad S, Khan SU, et al. Morpho-physiological traits, gaseous exchange attributes, and phytoremediation potential of jute (*Corchorus capsularis* L.) grown in different concentrations of copper-contaminated soil. Ecotoxicol Environ Saf 2020;189:109915.

[38] Saleem MH, Fahad S, Rehman M, et al. Morpho-physiological traits, biochemical response and phytoextraction potential of short-term copper stress on kenaf (*Hibiscus cannabinus* L.) seedlings. Peer J 2020;8:e8321.

[39] Yasmin H, Bano A, Wilson NL, et al. Drought tolerant *Pseudomonas* sp. showed differential expression of stress-responsive genes and induced drought tolerance in *Arabidopsis thaliana*. Physiol Plant 2021. Available from: https://doi.org/10.1111/ppl.13497.

[40] Mumtaz S, Saleem MH, Hameed M, et al. Anatomical adaptations and ionic homeostasis in aquatic halophyte *Cyperus laevigatus* L. under high salinities. Saudi J Biol Sci 2021;28.

[41] Nawaz M, Wang X, Saleem MH, et al. Deciphering Plantago ovata forsk leaf extract mediated distinct germination, growth and physio-biochemical improvements under water stress in maize (*Zea mays* L.) at early growth stage. Agronomy 2021;11(7):1404.

[42] Hameed A, Akram NA, Saleem MH, et al. Seed treatment with α-tocopherol regulates growth and key physio-biochemical attributes in carrot (*Daucus carota* L.) plants under water limited regimes. Agronomy 2021;11(3):469.

[43] Cao Z, Rossi L, Stowers C, et al. The impact of cerium oxide nanoparticles on the physiology of soybean (*Glycine max* (L.) Merr.) under different soil moisture conditions. Environ Sci Pollut Res 2018;25(1):930−9.

[44] Badr NB, Al-Qahtani KM, Mahmoud AED. Factorial experimental design for optimizing selenium sorption on *Cyperus laevigatus* biomass and green-synthesized nano-silver. Alex Eng J 2020;.

[45] Javed MT, Tanwir K, Abbas S, et al. Chromium retention potential of two contrasting *Solanum lycopersicum* Mill. cultivars as deciphered by altered pH dynamics, growth, and organic acid exudation under Cr stress. Environ Sci Pollut Res 2021;28:1−13.

[46] Deng G, Yang M, Saleem MH, et al. Nitrogen fertilizer ameliorate the remedial capacity of industrial hemp (*Cannabis sativa* L.) grown in lead contaminated soil. J Plant Nutr 2021;44:1−9.

[47] Ghafar MA, Akram NA, Saleem MH, et al. Ecotypic morphological and physio-biochemical responses of two differentially adapted forage grasses, *Cenchrus ciliaris* L. and *Cyperus arenarius* Retz. to drought stress. Sustainability 2021;13(14):8069.

[48] Saleem MH, Fahad S, Khan SU, et al. Copper-induced oxidative stress, initiation of antioxidants and phytoremediation potential of flax (*Linum usitatissimum* L.) seedlings grown under the mixing of two different soils of China. Environ Sci Pollut Res 2020;27(5):5211−21.

[49] Saleem MH, Fahad S, Adnan M, et al. Foliar application of gibberellic acid endorsed phytoextraction of copper and alleviates oxidative stress in jute (*Corchorus capsularis* L.) plant grown in highly copper-contaminated soil of China. Environ Sci Pollut Res 2020;27.

[50] Zaheer IE, Ali S, Saleem MH, et al. Interactive role of zinc and iron lysine on *Spinacia oleracea* L. growth, photosynthesis and antioxidant capacity irrigated with tannery wastewater. Physiol Mol Biol Plants 2020;26.

[51] Shalaby TA, Abd-Alkarim E, El-Aidy F, et al. Nano-selenium, silicon and H_2O_2 boost growth and productivity of cucumber under combined salinity and heat stress. Ecotoxicol Environ Saf 2021;212:111962.

[52] Zaheer IE, Ali S, Saleem MH, et al. Role of iron−lysine on morpho-physiological traits and combating chromium toxicity in rapeseed (*Brassica napus* L.) plants irrigated with different levels of tannery wastewater. Plant Physiol Biochem 2020;155:70−84.

[53] Saleem MH, Ali S, Irshad S, et al. Copper uptake and accumulation, ultra-structural alteration, and bast fibre yield and quality of fibrous jute (*Corchorus capsularis* L.) plants grown under two different soils of China. Plants 2020;9(3):404.

[54] Parveen A, Saleem MH, Kamran M, et al. Effect of citric acid on growth, ecophysiology, chloroplast ultrastructure, and phytoremediation potential of jute (*Corchorus capsularis* L.) seedlings exposed to copper stress. Biomolecules 2020;10(4):592.

[55] Khan A, Khan AL, Imran M, et al. Silicon-induced thermotolerance in *Solanum lycopersicum* L. via activation of antioxidant system, heat shock proteins, and endogenous phytohormones. BMC Plant Biol 2020;20:1−18.

[56] Chen-yang W, Dan C, Hai-wei L, et al. Effects of exogenous silicon on the pollination and fertility characteristics of hybrid rice under heat stress during anthesis. Yingyong Shengtai Xuebao 2013;24(11).

[57] Agarie S, Hanaoka N, Ueno O, et al. Effects of silicon on tolerance to water deficit and heat stress in rice plants (*Oryza sativa* L.), monitored by electrolyte leakage. Plant Prod Sci 1998;1(2):96−103.

[58] Khan A, Bilal S, Khan AL, et al. Silicon and gibberellins: synergistic function in harnessing ABA signaling and heat stress tolerance in date palm (*Phoenix dactylifera* L.). Plants 2020;9(5):620.

[59] Farhangi-Abriz S, Torabian S. Nano-silicon alters antioxidant activities of soybean seedlings under salt toxicity. Protoplasma 2018;255 (3):953−62.

[60] Avestan S, Ghasemnezhad M, Esfahani M, et al. Application of nano-silicon dioxide improves salt stress tolerance in strawberry plants. Agronomy 2019;9(5):246.

[61] Soundararajan P, Sivanesan I, Jana S, et al. Influence of silicon supplementation on the growth and tolerance to high temperature in *Salvia splendens*. Hortic Environ Biotechnol 2014;55(4):271−9.

[62] Gong H, Zhu X, Chen K, et al. Silicon alleviates oxidative damage of wheat plants in pots under drought. Plant Sci 2005;169(2):313−21.

[63] Tantawy A, Salama Y, El-Nemr M, et al. Nano silicon application improves salinity tolerance of sweet pepper plants. Int J ChemTech Res 2015;8(10):11−17.

[64] Farshidi M, Abdolzadeh A, Sadeghipour HR. Silicon nutrition alleviates physiological disorders imposed by salinity in hydroponically grown canola (*Brassica napus* L.) plants. Acta Physiol Plant 2012;34(5):1779−88.

[65] Yin L, Wang S, Liu P, et al. Silicon-mediated changes in polyamine and 1-aminocyclopropane-1-carboxylic acid are involved in silicon-induced drought resistance in *Sorghum bicolor* L. Plant Physiol Biochem 2014;80:268−77.

[66] Fauteux F, Chain F, Belzile F, et al. The protective role of silicon in the Arabidopsis-powdery mildew pathosystem. Proc Natl Acad Sci 2006;103(46):17554−9.

[67] Pereira de Souza Junior J, de Mello Prado R, Machado dos Santos Sarah M, et al. Silicon mitigates boron deficiency and toxicity in cotton cultivated in nutrient solution. J Plant Nutr Soil Sci 2019;182(5):805−14.

[68] Fidalgo F, Azenha M, Silva AF, et al. Copper-induced stress in *Solanum nigrum* L. and antioxidant defense system responses. Food Energy Sec 2013;2(1):70−80.

CHAPTER

12

Silicon-mediated cold stress tolerance in plants

Roghieh Hajiboland

Department of Plant Sciences, University of Tabriz, Tabriz, Iran

12.1 Introduction

Low temperature is an environmental constraint limiting the geographical locations and the extension of the cultivating areas of crop species. Low temperature includes chilling ($0°C-15°C$) and freezing ($<0°C$) temperatures and is known to be one of the major factors limiting the growth and productivity of plants at high latitudes or altitudes. Crop species such as rice, maize, bean, tomato, and potato are known to be cold sensitive with a low ability to cope with freezing temperatures [1]. In contrast, winter cereals, spinach, and cabbage grow well at chilling temperatures and show a high capacity to develop frost resistance [1]. Although plants are encountering cold stress mainly during the vegetative growth stage, chilling also severely hampers the reproductive growth of plants, inhibits flowering or postpones it, and thus reduces seed, grain, and fruit production. In rice and maize as warm-climate crops, low temperature at the time of anthesis results in abnormalities in the reproductive organs and prevents fertilization or causes premature abortion of grains [2].

12.1.1 Chilling injury in plants

It is hard to determine the primary function and the most severely affected processes under low-temperature conditions. This is due to complex indirect effects generated by chilling temperatures on various aspects of plant metabolism. From the thermodynamic point of view, reduction of temperatures reduces the kinetics of metabolic reactions, shifts the thermodynamic equilibrium [3], and directly affects the stability and the solubility of proteins leading to changes in the metabolic regulations [4].

Low temperatures induce rigidification of the membranes. The temperature at which a membrane changes from semifluid to semicrystalline state, for example, transition temperature, is mainly related to the relative proportion of unsaturated fatty acids [5]. Chilling sensitive plants usually have a higher proportion of saturated fatty acids and, therefore, a higher transition temperature while chilling resistant species are marked by a lower proportion of the saturated fatty acids and correspondingly a lower transition temperature [6]. Membrane rigidification leads to the disturbance of all membrane processes such as gating of the ion channels and electron transport processes [7].

Low temperature is also accompanied by the accumulation of reactive oxygen species (ROS). Due to the overreduction of the chloroplast electron transfer chain, ROS generation is elevated while the activities of the scavenging enzymes decreased under low temperatures [8]. The accumulation of ROS exerts damaging effects, especially on membranes, and causes leakage of inorganic and organic solutes from the cells [9]. In addition, the low temperatures favor the formation and stabilize the secondary structures in RNA, thus affecting gene and protein expression [10]. Ultimately, chilling-stressed plants show reduced leaf expansion and exhibit injury symptoms such as wilting, chlorosis, and necrosis [5].

12.1.2 Freezing injury in plants

Freezing temperatures have even more damaging effects [11]. Freezing injury in plants is not restricted to special geographic regions and many crop and horticultural plants are being encountered with frost damage in late spring or early autumn [12]. Injury to the tissue may result from ice formation, either in the extracellular space (apoplast, see later) or within the symplast [13]. Except for the tropical and subtropical species, ice formation in plants begins in the apoplastic space because of its relatively lower solute concentration than the cytosol [14]. Due to the difference in chemical potential created by a growing ice crystal, cellular water migrates to the extracellular space, resulting in cell dehydration and shrinkage [15]. The extent of dehydration increases in parallel with a reduction in temperature, and thus freezing tolerance is dependent on the difference in tolerance to dehydration [16]. The process of water moving out of the cells leads to the enlargement of existing ice crystals and causes a mechanical constraint on the cell wall (CW) and the plasma membrane, leading to cell rupture [16]. Ultimately, ice can penetrate the symplast and causes deterioration of the intracellular structures and cell death [13]. Under freezing temperatures, the membranes are the most damaged part of the cell mainly due to dehydration that results in expansion-induced cell lyses, fracture lesions, and phase transition [12].

12.1.3 Cold acclimation

Plants native to temperate climates are often chilling-tolerant and have also developed a mechanism for increasing the ability to withstand freezing temperatures. Preexposure to low but nonfreezing temperatures improves resistance to below-zero temperatures that called cold acclimation or cold-hardiness [17,18]. In temperate latitudes, cold acclimation is established in the autumn, when the temperatures are low but still positive [15].

Cold acclimation is associated with several cellular and biochemical changes that result in the acquisition of tolerance to low negative temperatures. An increase in the proportion of unsaturated fatty acids and, thereby, a drop in transition temperature occurs during cold acclimation that prevents the expansion-induced cell lyses and formation of hexagonal II phase lipids [19,20]. In addition, enhancement of the antioxidant mechanisms increased intercellular sugar levels and other cryoprotectants; induction of the genes encoding molecular chaperones, dehydrins, lipid transfer proteins, and the late-embryogenesis-abundant (LEA) proteins [15] and the posttranslational activation of cold-signaling molecules [21] occur upon preexposure to positive near zero temperatures. Although plants acquire frost tolerance through the acclimation process, continued chilling stress hampers growth and reduces dry matter production in crop species [19].

12.1.4 Cold sensing and signaling

The process of perception of low temperature as a physical parameter and its conversion into biochemical parameters followed by subsequent propagation of the signal through various intermediate molecules, that is, signal transduction pathway, can ultimately cause the physiological responses and modify the expression of genes involved in the cold response or tolerance.

12.1.4.1 Cold sensing

Perception of cold temperature does not rely on a single molecule or cell structure but is mediated by several cellular or subcellular components. Cold stress is sensed through reduction of membrane fluidity (rigidification), depolymerization of microtubules and microfilaments, changes in the conformation of DNA-binding domain of transcription factors, modifications in the activity of enzymes, and finally, through changes in the degree of DNA supercoiling and RNA secondary structures [22,23]. Based on the criteria for a temperature sensor in animal cells, COLD1/RGA1 complex has been described as a plant cold sensor in rice [24]. Regardless of the mechanism involved, cold sensing processes lead to the generation of signaling molecules (ROS, Ca^{2+}) or change the activity of enzymes of signaling cascades (phospholipase, kinases), thus modifying the gene expression and protein synthesis involved in response to cold stress [25]. Cold stress can also be sensed in all organelles through their membranes and the cellular response to a low temperature is in fact an integration of different pathways initiated in each organelle [26]. Since the time of response is different among these components; for example., membrane rigidification is instantaneous while cytoskeleton destabilization takes a few minutes, or hours, a succession of switches are turned on sequentially [27].

12.1.4.2 Cold signaling

Numerous genes involved in cold response, cold acclimation, and freezing tolerance have been identified in plants (Table 12.1). These genes belong to the membrane-associated receptors, several kinases, transcription factors, and posttranslational modifiers including proteins responsible for ubiquitination and sumoylation. There are some overlaps between cold signaling pathways and the response mechanisms to drought and salt stress. For example, one of the largest classes of the cold-responsive genes (*COR*, Cold-Regulated) encodes LEA-like proteins that are induced by dehydration stress with a well-known membrane and protein-stabilizing function [15].

TABLE 12.1 Molecular components of the cold signaling pathway in plants.

Gene	Description	Function	Ref
COR	**C**old **R**esponsive **G**enes	COR genes have a central role in the cold acclimation-induced freezing tolerance; they contain DRE (dehydration responsive elements) or CRT (C-repeats) or ABRE (ABA-responsive element) in the promoter regions	Liu et al. [28]
CBF	**C**R**T** **b**inding **f**actor	*CBF3* encodes transcription factor; is induced in response to low-temperature stress (4°C), and increases freezing tolerance; it interacts with BTF3 (Basic transcription factor 3) leading to its stability through preventing ubiquitin-mediated degradation under cold stress	Liu et al. [28] Guo et al. [29]
BTF3	**B**asic **T**ranscription **F**actor3	A positive regulator of CBF3 is phosphorylated by OST1 that enhances its interaction with CBF proteins and prevents CBF proteins from ubiquitin-mediated degradation under cold stress	Ding et al. [30]
ICE	**I**nducer of **C**BF **E**xpression 1	*ICE1* encodes a transcription factor that specifically recognized MYC sequence on the *CBF3* promoter; it is positively regulated by SIZ1 but negatively by HOS1	Miura Furumoto. [31]
HOS1	**H**igh Expression of **O**smotically Responsive Genes 1	*HOS1* encodes a ubiquitin E3 ligase with a RING finger domain; it directly interacts with ICE1 and causes its ubiquitination and degradation; is constitutively expressed but is immediately downregulated under cold stress; is a negative regulator of *COR* genes through reduction of the expression level of CBFs	Lee et al. [32]
SIZ1	**S**AP and **MIZ**-finger domain	SIZ1 encodes a SUMO E3 ligase and is responsible for the sumoylation of ICE1, which increases its stability under cold stress, resulting in increased *CBF* expression and enhanced freezing tolerance	Miura & Hasegawa 2008 [33]
COLD1	A cold sensor	*COLD1* is a G-protein regulator, localized on the plasma membrane and ER and interacts with the RGA1 (a GTPase) to enhance its enzymatic activity; COLD1-RGA1 complex facilitates the cold-induced Ca^{2+} spiking	Ma et al. [24]
CRPK1	**C**old-**R**esponsive **P**rotein **K**inase1	CRPK1 is a receptor-like cytoplasmic kinase lacking a transmembrane domain and forms complex with a cold-activated receptor-like kinase (RLK); cold-activated CRPK1 phosphorylates 14-3-3 proteins, promotes their translocation to the nucleus where they interact with CBF1/CBF3, and facilitates their degradation after ubiquitination; mutations in *CRPK1* confer an enhanced freezing tolerance	Liu et al. [34]
OST1	**O**pen **S**tomata 1	OST1, a member of the SNF1-related protein kinase family (SnRK), is a positive regulator of freezing tolerance; cold-activated OST1 phosphorylates ICE1, enhances its stability by preventing its interaction with HOS1; OST1 also phosphorylates BTF3 and facilitates their interaction with CBF proteins, and thus stabilizes CBF proteins under cold stress	Ding et al. [35]
LOS2	**L**ow Expression of **O**smotically Responsive 2	*LOS2* encodes bi-functional enolase; involved in cold-responsive gene expression; mutation in *LOS2* (*los2*) blocks cold induction of *COR* but not *CBF* genes and repress *STZ*	Lee et al. [36]
STZ/ZAT10	**S**alt **T**olerance **Z**inc Finger	*STZ* encodes a zinc-finger transcription factor, is upregulated during chilling and freezing treatments, overexpressing lines have higher cold tolerance; *ZAT10* expression is repressed in the*cbf1/2/3* triple mutant, its overexpression induces the expression of *COR* genes without need to cold treatment	Lee et al. [36]; Chinnusamy et al. [37]
ESK1	A negative regulator	*ESK1* encodes a negative regulator; in the absence of cold acclimation, the mutant (*esk1*) is more freeze-tolerant than the wild type; *ESK1* acts likely through a COR-independent pathway	Xin [38]
SFR	**S**ensitive to **F**reezing	*sfr* mutants (*sfr 2*, *sfr 3*, *sfr 4*, *sfr 5*, *sfr 7*) were identified in the genetic screening for high frost susceptibility in *Arabidopsis*, only the function of some has been known	Warren et al. [39]

Some of the *COR* genes contain DRE (dehydration responsive) or CRT (C-repeats), and some contain ABRE (ABA-responsive element) in the promoter regions [28,29,40–42] (Fig. 12.1).

12.2 Mitigation of low-temperature stress by Si

The beneficial effect of added Si has been reported for Si-accumulator species such as rice, wheat, barley, and strawberry under unstressed conditions [43–45]. In nonaccumulator species such as tomato and tobacco, cultivation of plants in low-Si substrates could reveal the effect of added Si on the improvement of biomass [46,47]. The most prominent effect of Si supplementation could be observed under various biotic and abiotic stresses such as drought, salinity, and heavy metal toxicity as amelioration of stress effects and improvement of biomass

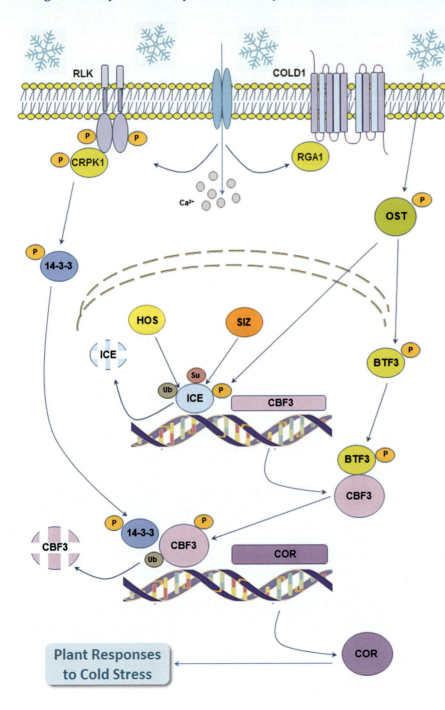

FIGURE 12.1 An illustration showing the most important components of the cold signaling pathway in plants. Cold is sensed by the membrane, Ca²⁺ channels or specific receptors, COLD1/RGA1 and CRPK1-RLK. In addition, activated OST contributes to the BTF3 phosphorylation, which, in turn, phosphorylates ICE and CBF3 leading to enhanced transcription of *CBF3* and *COR* genes, respectively. Although CRPK1-RLK is activated by the low-temperature stress, it influences negatively CBF3 through phosphorylated 14-3-3 proteins that interact with CBFs and promote their degradation via the ubiquitination pathway. HOS as a negative regulator of ICE increases ubiquitination and degradation of this transcription factor while SIZ activates ICE through sumoylation.

production [48]. Relatively little attention, however, has been paid to the effect of Si under cold stress conditions and the involving mechanisms have less been explored compared with other stresses either in accumulator or in nonaccumulator species.

In the following sections, we will summarize the reports on the ameliorative effects of Si in cold-stressed plants with an emphasis on the involving physiological and biochemical mechanisms. In the absence of data in cold-stressed plants, evidence will be provided from other stresses or unstressed plants for inferring its effect under cold stress conditions and highlighting the definite contribution of Si to the basic processes of plant growth and stress tolerance.

12.2.1 Water relations and photosynthesis under cold stress affected by Si

Water supply from the roots is restricted under low root-zone temperature [49], that is, at least partly, because of reduced root hydraulic conductivity in cold-stressed plants [50]. Thus a rapid stomatal closure under cold stress that prevents transpirational water loss is an important criterion for cold tolerance [51]. Cold-sensitive species exhibits low leaf water potentials under cold conditions while tolerant species maintain water potentials due to a rapid stomatal closure and preventing transpiration [5].

Low temperatures affect different aspects of photosynthesis [52]. Low stomatal and mesophyll conductance, impairment of chloroplasts development, degradation of chlorophyll, and restricted metabolite transport all contribute to the low photosynthetic rates under cold stress [53]. Chilling reduces thermodynamically the activities of Calvin cycle enzymes; thus a poor CO_2 fixation rate restricts the $NADP^+$ availability for accepting electrons from the electron transport chain [54]. This leads, in turn, to the generation of excess excitation energy and ROS accumulation and degradation of stromal enzymes such as Rubisco [55]. The changes in the redox status of chloroplasts result in the lower activity of thioredoxin-related Calvin cycle enzymes and additional inhibition of photosynthesis [54]. Furthermore, because an optimum rate of photosynthesis requires an appropriate balance between the rates of CO_2 fixation and sucrose synthesis, inhibition of sucrose synthesis under cold stress leads to the accumulation of phosphorylated intermediates and limitation of photosynthesis [56]. During cold acclimation, sucrose synthesis is stimulated through both posttranslational activation of sucrose phosphate synthase (SPS) and an increase in the expression of cytosolic fructose-1,6-bisphosphatase (cFBPase) and SPS. In leaves that develop at low temperature, cold acclimation causes constitutively higher expression of Calvin cycle and sucrose synthesis enzymes [57,58].

The effects of Si on the photosynthetic parameters have been reported in unstressed plants. Si supplementation of nonaccumulators tobacco [46,59] and pistachio [60] and Si accumulators strawberry [43,61], wheat [62], and barley [63] resulted in higher leaf chlorophyll and elevated rates of net CO_2 assimilation but without the effect on the leaf photochemical reactions. In rice, the study of photosynthetic parameters in the low Si mutant lacking a functional Lsi (lsi1) showed that Si improved the source capacity and sink demand associated with increased mesophyll conductance and higher nitrogen use efficiency [45]. In Phyllostachys praecox (bamboo) grown at 25°C, increasing Si amendment levels to the soil results in significant improvement of the leaf photosynthesis rate without any effect on the stomatal conductance but with lower transpiration rate leading to higher water use efficiency [64]. Higher leaf Si was associated with higher chlorophyll a and b content in rice lines overexpressing Si transporter, Lsi [65].

Si-mediated improvement of the leaf photosynthesis rate was also observed in cold-stressed barley [63] and rice [66]. The expression levels of genes associated with chlorophyll metabolism, light-harvesting complex, photosystem II reaction center proteins, cytochrome b$_6$f complex, iron-sulfur subunit, ATP synthase subunits, and ferredoxin all were higher in the rice line overexpressing Lsi in comparison to the wild-type plants [66].

12.2.2 Cold stress and ROS metabolism affected by Si

ROS are generated in biological pathways as byproducts or signaling agents; however, an excess of ROS causes oxidative stress and damage membranes and other cell structures [67]. Under cold conditions, an excessive ROS production as a result of impaired photosynthetic electron transport and CO_2 fixation and/or inefficient scavenging by antioxidant systems ultimately results in oxidative stress and severe injury [8]. Oxidative damage by ROS has been considered the main reason for freezing injury [12]. Peroxidation of unsaturated fatty acids in membranes, denaturation of proteins, breaking up of polysaccharides, and scission of DNA strands are the common deleterious consequences of ROS accumulation [6,68].

To protect plants from injury, there is an efficient antioxidant defense system comprising enzymes, for example, catalase (CAT), superoxide dismutase (SOD), peroxidase (POD), ascorbate peroxidase (APX), monodehydroascorbate reductase (MDHAR), and glutathione reductase (GR), and antioxidant metabolites, for example, glutathione (GSH) and ascorbate (AsA). An upregulation of genes encoding antioxidant enzymes and those responsible for the synthesis of metabolites is necessary for counteracting oxidative stress induced by ROS accumulation under stress conditions [8].

One of the functions of the cold acclimation process is its priming effect on the ROS defense system. During cold acclimation, relatively weak oxidative stress as an elevated H_2O_2 production induces more active antioxidant systems and alleviates freezing injury by effective scavenging of the ROS generated during the freeze-thaw stress [15,69,70]. The capacity for ROS scavenging in plants is a determining factor for resistance to cold stress as was confirmed by elevated stress tolerance in the plants overexpressing various antioxidant enzymes [71,72].

Exogenous Si has been frequently reported to be an effective treatment leading to activation of antioxidant defense in plants exposed to various environmental stresses [73]. Even under unstressed conditions, enhancement in the ratios of fatty acid unsaturation in the glycolipid and phospholipid fractions [43] and reduction of ROS and MDA, the indicator of lipid peroxidation, and activation of POD, APX, SOD, and GR have been observed in rice, and strawberry as Si-accumulator species and in tomato and tobacco as nonaccumulators [44,46,47]. Higher amounts of unsaturated membrane lipids confer higher stability for cell membranes under cold stress [5]; thus Si may play an important role in maintaining the integrity, stability, and function of the membrane [74,75]. Nevertheless, the effect of Si on the plasma membrane fatty acid composition has not yet been analyzed in cold-stressed plants.

Under cold stress, similarly, this process is one of the main reasons for mitigation of stress by Si supplementation. In wheat, activation of antioxidant defense (SOD and CAT) and concentration of nonenzymatic antioxidants (GSH and AsA) were enhanced by Si in freezing-stressed plants associated with reduction of MDA [76]. In cucumber, Si supplementation of chilling-stressed plants reduced the percentage of withered second leaves and enhanced the activities of SOD, APX, MDHAR, GR, and the concentration of AsA and GSH associated with lower ROS accumulation and MDA [77]. In seashore paspalum (*Paspalum vaginatum* Swartz) turf., the activities of SOD and CAT decreased under cold stress, resulting in higher MDA concentration while Si addition maintained significantly higher activities of SOD, POD, CAT, and reduced MDA under cold conditions [78]. *Phyllostachys praecox* a clump-forming evergreen bamboo with high susceptibility to chilling and freezing stresses showed increased activity of SOD, POD, and CAT with increasing Si amendment levels to the soil associated with lower MDA, higher membrane stability, and growth of plants under both chilling and freezing stresses [64]. A similar response to Si supplementation was observed in soil-grown pistachio (*Pistacia vera*), a tree species adapted to drought and salt but is susceptible to cold. Leaf Si spray activated ROS scavenging system in this species and improved leaf water status and protects membranes from freezing as indicated by significantly smaller leaf necrotic lesions [79]. The rice *Lsi* overexpressing line growing under chilling conditions showed higher SOD, POD, and CAT activities compared with wild-type plants [65] and are better protected against cold-induced chloroplast damages [66].

12.2.3 Accumulation of the low-molecular weight compounds under cold stress affected by Si

Synthesis and accumulation of low molecular weight protecting molecules (LMWs) such as soluble sugars, proline, and glycine betaine is another mechanism that helps plants to retain their growth or survival under cold stress. These compatible solutes are produced in response to many stresses and stabilize proteins and membranes from desiccation [80–82]. Under cold stress, these compounds increase the osmotic potential and prevent excessive cell dehydration, counterbalancing the osmotic effect of ice formation in apoplastic space (see later). In addition, the LMWs minimize the risk of cavitation during freezing by increasing the tensile strength of cellular water [83,84].

12.2.3.1 Soluble sugars

There is a clear correlation between sugar concentrations and frost resistance [85,86]. In transformed *Arabidopsis* plants expressing the antisense gene of cytosolic fructose-1,6-bisphosphatase (cFBPase) and SPS, or in the lines overexpressing SPS, the sugar concentrations are correlated with the acquired freezing tolerance upon cold acclimation [87]. Similarly, the constitutive increases in the frost tolerance of the *eskimo1* mutant [87] and in transformants overexpressing the transcriptional activator CBF3 are associated with higher levels of sucrose, raffinose, glucose, and fructose [88]. The *sensitive-to-freezing4* (*sfr4*) mutant that is impaired in cold acclimation has lower sugar levels than the wild-type plants [89]. The cold-mediated sugar accumulation depends, at least partly, on the CBF transcriptional [86]. In addition to the osmotic effect, soluble sugars are important for the protection of membranes under cold stress. Sucrose interacts with the phosphate groups in the membrane lipids and decreases membrane permeability [86]. Fructans are fructose-based oligo- and polysaccharides. Fructans and raffinose family oligosaccharides insert directly between polar head groups and increase the stability of mono- and bilayers [90]. An efficient scavenging activity against the highly deleterious hydroxyl radical (•OH) that could not be removed by enzymatic antioxidants has also been observed for soluble sugars [91].

Higher soluble carbohydrates level has been reported in the roots (but not in the leaves) of Si-treated strawberry under unstressed conditions [92]. In maize exposed to cold stress, Si treatment as seed coating results in higher total soluble carbohydrates level in the seedlings [93].

Trehalose, as a disaccharide together with its precursor trehalose-6-phosphate, plays an important role in plant development, regulation of metabolism, and abiotic stress signaling [94,95]. Trehalose serves as an energy source, osmolyte, or protein/membrane protectant under various environmental stress conditions [96]. After exposure to cold stress, coordinated changes have been observed in transcript levels of the enzymes involved in the trehalose metabolism [97]. In *Arabidopsis*, overexpression of trehalose 6-phosphate synthase gene enhances plants cold tolerance [98]. In the only work undertaken on the effect of Si on the biosynthesis and cellular level of trehalose, it has been observed that Si treatment caused an enhancement of trehalose concentration associated with higher activities of synthesizing enzymes, trehalose-6-P synthase and trehalose-6-P phosphatase and reduction of the activity of the catabolizing enzyme, trehalase in pigeon pea in both unstressed and Zn- and Cd-stressed plants [99]. Further studies are necessary for exploring the effect of Si on biosynthesis and accumulation of trehalose under other stresses including cold stress.

The metabolism of nitrogenous compounds is also responsive to low-temperature stress [101]. Gamma-aminobutyric acid (GABA) is an important amine-containing metabolite associated with cryoprotection in barley and wheat [100]. The expression of glutamate decarboxylase responsible for GABA synthesis is rapidly upregulated by the imposition of cold stress [101]. The effect of Si on the cellular levels or synthesis of GABA has not yet been studied in cold-stressed plants. However, in unstressed maize plants [102], or in tomatoes under osmotic stress [103], Si-supplemented plants showed marginally or significantly higher GABA levels compared with plants without Si treatment.

12.2.3.2 Proline

A high level of proline accumulation has been frequently reported in cold-stressed plants. In *Arabidopsis*, proline is accumulated up to 130-fold after 4 days of cold treatment [101]. A high accumulation of proline is a characteristic feature of *Arabidopsis*, *eskimo1* mutant that shows a high freezing tolerance in the absence of cold acclimation [38]. Finally, the *Arabidopsis* plants with reduced expression of proline-degrading enzyme (*AtProDH*) show higher tolerance to freezing than wild-type plants [104]. Proline protects enzymes from denaturation, stabilizes the machinery of protein synthesis, regulates the cytosolic acidity, increases water-binding capacity, and acts as a reservoir of carbon and nitrogen sources [80,105]. Proline is also able to induce the expression of specific genes containing the relevant responsive element [106].

In plants exposed to salt or drought stress, the increase in the levels of proline mediated by Si has been frequently reported and has been considered as one of the main mechanisms for mitigation of these stresses by Si [107]. Even in unstressed plants, Si-treated plants showed higher proline in the leaves and roots [59,108]. In cold-stressed plants, similarly, the ameliorative effect of Si is associated with elevated levels of proline. In seashore paspalum (*Paspalumvaginatum* Swartz), enhanced growth status and lower LT_{50} (the low temperature required to cause 50% electrolyte leakage) in Si-treated plants were associated with a significant increase in proline concentrations [78]. Similar results have been observed in wheat [76] and maize [93] plants.

12.2.3.3 Glycine betaine

Glycine betaine is synthesized and accumulated in some plant species from Poaceae and Chenopodiaceae under various environmental stresses [109]. In glycine betaine accumulator species, its amount increased during cold acclimation and the extent of its accumulation is correlated with the freezing resistance (estimated via the LT_{50}) of the genotypes [110]. In plants that do not normally accumulate glycine betaine, its role in response to cold has been studied by facilitating glycine betaine accumulation through transgenic approaches. These studies showed alleviation of the effect of chilling stress on reduction of growth and PSII activity and improvement of freezing tolerance compared with wild-type plants [111]. Glycine betaine stabilizes the transcriptional and translational machinery under low temperatures [112,113]. In addition, glycine betaine is involved in the production of H_2O_2 as a signaling molecule and thus mediates activation of cold signaling pathways leading to chilling tolerance [114].

There is no report on the effect of Si on the synthesis or accumulation of glycine betaine in cold-stressed plants. However, in the young seedlings of drought-stressed lentils (*Lens culinaris* Medik), Si treatment rather decreased the glycine betaine concentration in parallel with that of proline that was attributed to the alleviation of stress effect in the Si-treated plants [115]. It is noteworthy that proline and glycine betaine have occasionally been considered as a biochemical marker of stress rather than a factor for mitigation of stress, and thus reduction of their cellular levels likely implied the occurrence of less stress in plants. In common bean cultivars, the highest salt- and drought-induced accumulation of proline was observed in the more-susceptible genotypes to stress [116].

12.2.3.4 Polyamines

Polyamines including spermine, putrescine, and spermidine with two or more primary amino groups contribute to plant stress response [117,118]. In rice, a spermidine synthase gene, *OsSPDS2*, is upregulated in response to low temperature [119]. SAM (S-adenosyl-l-methionine; SAM) synthase is induced under cold conditions and an increase in putrescine, ornithine, and citrulline has been reported in *Arabidopsis* exposed to cold stress [120]. Polyamines protect the photosynthetic function of cold-stressed plants through association with the thylakoid membranes and the light-harvesting complex II [121]. Inhibition of polyamine synthesis led to increased oxidative damage and higher electrolyte leakage and lower photochemical efficiency of PSII in cold-stressed plants [121]. Direct inhibition of the cold-induced activation of NADPH oxidases by spermidine was observed in cucumber [122].

The endogenous level of polyamines is modified by Si in plants exposed to salt or drought stress [102,103,123,124]. In cold-stressed plants, however, there is no report on the Si-mediated increase in the polyamines levels.

12.2.4 Hormone signaling under cold stress affected by Si

CBF signaling pathway (as discussed earlier) is not the only responsible pathway for cold acclimation, chilling, and freezing tolerance. Phytohormones also play important roles in cold tolerance and acclimation [31,41,125,126]. The cold acclimation process is started as an independent pathway; however, abscisic acid (ABA) contributes at later stages and is required for the maximum tolerance to freezing temperatures [127]. This is confirmed by the reduction of the capacity for cold acclimation and freezing tolerance in the ABA-deficient or -insensitive *Arabidopsis* mutants [128] and overlap between cold-responsive and ABA-responsive genes [129]. A link has been observed between ABA and the induction of antioxidant defense enzymes [130,131] and proline biosynthesis [80]. Exposure of maize, a cold-sensitive species, to low-temperature stress significantly decreased the levels of ABA [93]. Cold stress-induced elevation of ABA level was higher in the cold-tolerant compared with cold-sensitive *Digitaria eriantha* genotypes [132]. These results suggest likely a link between the cold-induced accumulation of ABA and the extent to which plants tolerate cold stress.

Salicylic acid and jasmonic acids that are involved in plants' response to various abiotic stresses are also contributed to the cold signaling pathway [133]. Improved cold acclimation and increased cold tolerance by exogenous applications of SA and JA and altered responsiveness to cold stress in the mutants affected in SA and JA metabolism confirmed this role [133–135]. Salicylic acid and jasmonic acids were decreased under cold conditions in maize [93] and *Digitaria eriantha* [132].

The CBF pathway caused also a reduction in the biologically active gibberellic acid (GA) through increased expression of GA2 oxidases, enzymes that convert biologically active GAs to inactive GAs which, in turn, results in an increase in DELLA proteins under cold stress [136]. The level of DELLA proteins, as the negative regulator of GA with growth-restraining effect, is also increased through the ABA pathway [136,137]. It has been demonstrated that DELLA proteins are required for the full activation of freezing tolerance in *A. thaliana* likely through reducing the levels of ROS [138]. Cold stress resulted in the reduction of GA levels in the leaves and roots of maize [93] and in the cold-sensitive but not tolerant genotype of tomato [139]. Temperature lowering caused a decrease in the free auxin levels [140].

Similar to GA and auxin, the levels of zeatin declined in cold-stressed rice, maize, and *Arabidopsis* plants [93,133]. In rice, this was associated with a significant downregulation of the genes involved in the cytokinin biosynthesis [141,142]. Accordingly, exogenous cytokinins improved cold tolerance in *Arabidopsis* [143].

There are reports on the Si effect on the levels of phytohormones in unstressed plants [47,102]. There are also pieces of evidence for the interactions between Si and hormonal cold stress signaling [144]. Si supplementation resulted in an increase in the levels of ABA, SA, and JA in cold-stressed maize plants [93]. An increase in the endogenous levels of ABA, SA, and JA that was reduced under cold stress likely boost the cold-response pathways and was at least partly the mechanism of cold stress mitigation by Si in this species. In contrast, Si restored the levels of GA, auxin, and zeatin to the levels of nonstressed plants associated with the resumption of growth under cold stress conditions [93]. More studies are needed for exploring the precise mechanisms involved in hormone signaling and expression under Si treatment.

12.2.5 Mineral nutrition of plants under cold stress affected by Si

Mineral nutrients play important roles in plants' response to stress conditions [145–147]. Cold stress as low air temperatures and particularly as low root-zone temperature hampers plants' nutrients uptake through different mechanisms. Under chilling stress, root growth is impaired and reduction of root length, low hydraulic conductance, and poor branching, all lead to reduced mineral nutrients uptake in plants [148,149]. Cold stress-mediated membrane damage causes leaching losses of nutrients in the young seedlings [93] that could impair plants' establishment under field conditions. In addition, in soil-grown plants, low temperature affects the physico-chemical and microbial properties in soils, reduces the volume of the root system for exploring the nutrients, and ultimately modifies the ability of plants to acquisition the nutrients from the soil [150]. Poor root system under chilling stress is responsible for the reduction in uptake of several nutrients, including N, P, and K [148,150,151]. There are also pieces of evidence showing that deficiency of nutrients increases the susceptibility of plants to cold stress and exacerbates the damaging effects of frost events under natural conditions [152]. The critical role of nutrients in plants' cold stress tolerance is well proved by the results of priming experiments [153]. Seed priming or supplementation of soil-grown plants with macronutrients (K) [154] or micronutrients (Zn and Mn) [93] significantly improves plants chilling tolerance. The most important component of cold stress response that is improved by the supply of nutrients is the capacity for ROS scavenging. Of particular importance is the role of Zn and K supply that contribute significantly to the reduction of NADPH-mediated ROS generation [155] and enhanced SOD-mediated ROS scavenging [93]. Ca has been considered an important nutrient for chilling tolerance and for the protection of leaves from dehydration. It has been observed that Ca is required for chilling-induced stomatal closure in chilling-tolerant plants via an ABA-independent pathway [51].

The effect of Si application on the uptake of nutrients in the soil and hydroponic medium has been observed under unstressed conditions. In general, Si affects the uptake and distribution of both macronutrients and micronutrients but in different ways depending on the nutrient and growing substrate [156–158]. Results showed that Si increased the availability of P, S, Ca, Zn, and Mn for soil-grown plants and increased uptake and root-to-shoot translocation of S, Ca, Mg, Mn, and Fe in nutrient solution-grown plants but without the effect on K in any of culture conditions [158]. In cold-stressed plants, Si improved the nutritional status of Zn and Mn in maize plants and ameliorates the chilling-induced leaching losses of these nutrients through the improvement of both membrane integrity and uptake, internal mobility, and transport of these nutrients [93].

12.2.6 Phenolics metabolism under cold stress affected by Si

Activation of plant secondary metabolism is a well-known response of plants under various environmental stresses. Phenolics as the largest group of secondary metabolites have been known to be an important component of plants' stress responses. Under drought, nutritional imbalances, heavy metal stress, and particularly under high light stress, phenylpropanoid metabolism is induced in plant tissues through the increase in phenylalanine ammonia-lyase (PAL) and chalcone synthase activities, as well as the activation of a number of genes involved in phenolic metabolism [159–161].

Under low-temperature stress, changes in the plant phenolic metabolism and in the levels of soluble phenols particularly anthocyanins are frequently reported in several plant species. Free phenolics likely act as an endogenous antioxidant and their accumulation under cold stress could be considered a cellular adaptation to stress [162]. In *Arabidopsis* and cabbage, anthocyanins accumulate in leaves in response to low temperatures [75,163]. An increase in anthocyanin and mRNA abundance in the sheaths of maize seedlings is positively related to the severity and duration of the cold [164]. In wheat (*Triticum aestivum*) grown under cold conditions, leaves showed an increase in soluble phenols while the opposite was observed in roots [165]. Oilseed rape (*Brassica napus*) plants exposed to cold showed an increase in the levels of p-coumaric acid, ferulic acid, and synaptic acid, and in the esterified soluble forms of these acids that was associated with an increase in the activity of PAL [166]. In this species, PAL activity and the rate of accumulation of phenolic compounds in leaves depend on the range of low temperature applied to the plant [167]. A higher increase in the PAL activity under cold stress was also observed in the frost-tolerant compared with sensitive genotypes of *Miscanthus sinensis*, a clump-forming warm-season grass [168]. It has been stated that the increase in the levels of esterification of the free phenolics that allow their transport to the vacuoles might be a mechanism to protect the cells from toxic concentrations of free soluble phenolics in the cytosol [162]. In *Glycine max* roots also increased PAL activity during acclimation to cold (10°C) was accompanied by a significant increase in the levels of esterified forms of ferulic, syringic, and p-hydroxybenzoic acids [169].

There are pieces of evidence on the effect of Si on the phenylpropanoid pathway, accumulation of free phenolics, and their profile in unstressed plants. In nonaccumulator tobacco, Si treatment decreased the free phenolics level [108] while in accumulator strawberry, Si-treated plants showed significantly higher phenolics associated with elevated activity of PAL [92]. Si also influenced the phenolic profile of the leaves, roots, and fruits in strawberry, gallic and caffeic acids, and quercetin increased by Si treatments, while ellagic and p-coumaric acids decreased [92]. These data suggested an interaction between Si and the enzymatic components of the phenylpropanoid pathway with an unknown significance.

The effect of Si on the synthesis and accumulation of secondary metabolites has been studied mainly under biotic stresses [170]. Unfortunately, the effect of Si on the phenolics metabolism and accumulation in plants grown under stress conditions including cold stress has not been investigated so far and remained to be unraveled in the future.

12.2.7 Modifications in cell wall properties under cold stress affected by Si

The chemical and physical properties of CW are modified by environmental factors such as light, level of nutrients supply, osmotic stress, and temperature [171]. Resistance of plants to cold stress depends not only on the response of cell membranes, but also on the mechanical properties of CW. During extracellular ice formation, which leads to cell dehydration and may result in cell collapse, increased CW rigidity protects cells against intracellular freezing [83]. The mechanical characteristics of CW, and pectins, in particular, regulate water flow, the range of dehydration, and ice propagation [172]. In trees, pectins prevent water loss and ice nucleation within the cells [173].

During cold acclimation, significant modifications occur in the CW structure and function, as increase in its strength, reduction of pore size [83] and changes in the polysaccharide composition and activities of the CW modifying enzymes [172]. Since the gelation and rigidity of CW relies on the calcium cross-linkages [174], the activity of pectin methylesterase (PME) that catalyzes the de-esterification of pectin methyl groups is important for Ca^{2+} binding to the carboxylate ions and regulating the extent of calcium cross-linkages [174]. CW-bound PMEs are temperature sensitive [175]. Cold acclimation induces a significant increase in pectin content in winter oilseed rape associated with high activity of PME and stiffening leaf lamina [172]. In the tolerant pea genotype, acclimation is accompanied by increases in the pectin content, suggesting a protective role for pectin as part of the frost-tolerance mechanism [176]. In maize, as a cold-sensitive species, in contrast, low temperature caused a reduction of pectin content and PME activity in leaves [177].

There are increasing reports showing that Si interacts with CW components, such as cellulose, hemicelluloses, callose, ferulic acid, and lignin by forming a silicon-organic complex [178–181]. In Si accumulators that preferentially incorporate Si as part of CW, Si impacts considerably CW chemistry. In the low-silicon 1 (*Bdlsi1*−1) mutant of *Brachypodium distachyon*, several alterations in noncellulosic polysaccharides and lignin were recognized, indicating the effect of Si in the types of linkages of noncellulosic polymers and lignin and in the three-dimensional organization of the CW network [182]. A low concentration of Si in the straw of *B. distachyon* triggers compositional changes of CW network and structural rearrangements of polymers, rather than affecting CW thickness [182]. In *Sorghum bicolor*, substantial differences were observed in the autofluorescence of CW between −Si and + Si plants that confirmed the impact of Si on CW chemistry and suggests that in + Si roots, silica polymerized on the modified lignin and altered its structure [183].

It has been demonstrated that CW components, for example., cellulose and other polysaccharides, lignin, and proteins, are involved in the SiO_2 deposition (biosilicification) and its templating likely through a sequestration mechanism through which it hinders auto condensation of $Si(OH)_4$ and formation of cytotoxic silica [184]. In rice cell suspension cultures, Si is firmly associated with CW components (mainly hemicellulose) via Si-O-C bonds

[179] and improves the mechanical properties of CW by increasing the length of cellulose filaments and the density of cellulose microfibrils [179]. SiO_2 interacts also with the aromatic ring or phenolic acid of the lignocarbohydrate complex in rice and other species [185]. Although the Si-mediated changes in the physical properties of CW have not been directly analyzed in cold-stressed plants, these pieces of evidence show that Si may be effective in the improvement of the cold tolerance through modifications in the mechanical properties of CW due to the improvement of stiffness of CW. More studies are also required for the possible effect of Si on the modification of CW in nonaccumulator species exposed to cold stress [186].

Cold treatment also affects pectin-associated receptors or WAKs (wall-associated kinases) that are located at the interface between CW and plasma membrane and are required for cell elongation and development [14]. An upregulation of WAK genes was observed in rice seedling and in a cold-tolerant maize genotype under low-temperature stress [187]. WAKs could bind either oligogalacturonides or pectin in CW in response to cold stress, indicating a potential role of CW as a cold stress sensor [187]. Since the action of Si is mainly restricted in the space between CW and plasma membrane, it has been speculated that the priming effect of Si is, at least partly, due to the activation of CW-localized signaling events [188]. It is likely that interaction of Si with CW components is mediated by WAKs that are colocalized with the deposition site of amorphous Si [186], which leads, in turn, to triggering of the response to CW-localized stresses.

12.2.8 Lignification under cold stress affected by Si

Lignin, a complex phenolic polymer, plays an important role in the growth and development of plants. Lignin enhances plant CW rigidity and is an important barrier that protects cells against the deleterious effects of various biotic and abiotic factors [189]. Lignin metabolism response to various environmental factors and the activity of involving mechanisms are altered by various stress factors [190].

A positive correlation between lignification and the annual average temperature was observed in the latewood of *Picea abies* [191]. Similarly, an increase in the amounts of lignin, cellulose, and hemicellulose was observed with increasing temperature in monocots under temperate climate conditions [192]. These observations could be explained by reduction of photosynthesis and providing carbon skeleton for lignin synthesis pathway. In contrast, reports on the cold-mediated increase in the lignification are relatively common in various plant species. Ex vitro poplar (*Populus tremula* × *P. tremuloides* L.) seedlings grown at 10°C showed an increase in lignin content [193]. During acclimation to cold, an increase in the expression of the gene coding for p-coumarate 3-hydroxylase (C3H) involved in the biosynthesis of lignin has been reported in *Rhododendron* [194]. A higher increase in hydroxycinnamyl alcohol dehydrogenase (CAD) activity under cold stress was observed in the frost-tolerant compared with frost-sensitive genotypes of *Miscanthus sinensis* [168]. In tobacco, the expression of *PAL*, hydroxycinnamoyltransferase (*HCT*), and *CAD* is upregulated under chilling stress leading to an increase in lignin synthesis [195]. Several other genes involved in the lignin biosynthesis such as cinnamoyl coenzyme A reductase (*CCR*), caffeate O-methyltransferase (*COMT*), and caffeoylcoenzyme A, O-methyltransferase (*CCoAOMT*) are induced in response to cold [162]. Changes in the composition of lignin or enhancement of its content lead to CW strengthening, thus preventing freezing damage and cell collapse [14,161].

Reports on the effect of Si on the level of lignification of CW in the Si-accumulator rice, are controversial. Si treatment caused a significant increase in lignin and suberin in the roots of rice [196] associated with significantly higher transcription of PAL and 4CL, leading to elevated levels of monolignols. Higher amounts of ABC transporter proteins and POD activity that may facilitate the transport of the monolignols to the apoplast and their subsequent polymerization are also coincide with higher lignification [196]. The same was observed in the rice straw using wild-type and mutants that are affected in CW composition [180]. In contrast, distinctly higher lignin in the leaves of low-Si mutants (*lsi*) grown under paddy field conditions [197] suggests the impairment of lignin synthesis by Si. Similarly, wild-type rice grown under −Si condition showed higher lignin content in the leaf CW associated with upregulation of the involving enzymes including *OsPAL*, *OsCCR1*, and *OsCAD6* [198]. Higher lignin was also observed in the low-Si mutant of another grass species, *B. distachyon* grown in soil [182]. These contrastive results could be explained likely through the effect of different "Si status" of leaves on lignin synthesis. Extremely low leaf Si that could be found in the *lsi* mutant cultivated in soil may lead to higher lignin synthesis for compensating the reduced structural strange in the absence of Si. In the presence of marginal amounts of tissue Si that could be found in the −Si plants grown in hydroponic medium, lignin synthesis may respond positively to further supplementation with Si. This suggestion is confirmed by the positive effect of Si at low but negative effect at higher Si supply observed in rice [197] and soybean [199]. This may emphasize the importance of reporting the leaf Si concentration and its impurity in the medium of rice cultivation systems. In barley, Si influenced the lignin concentration depending on genotype and plant organ and affects also the lignin composition [200].

In nongrass species (nonaccumulators), studies on the effect of Si on lignin synthesis and deposition have mainly focused on the stressed plants. In tobacco leaves exposed to mechanical pressure, the activities of PAL, polyphenol oxidase (PPO), and cytosolic and covalently bound PODs increased in parallel with elevated lignin deposition, and Si treatment augmented this effect [108]. In soybean grown under shading conditions, lignin concentration increased by Si addition accompanied by increased activity of POD, 4-coumarate:CoA ligase (4CL), CAD, and PAL [199]. These results suggest that Si in stressed plants triggers signaling pathways that target, in turn, the structural strength of CW and thus lead to the intensification of the effect of stress treatments on the lignin content.

Under optimum growth conditions, in contrast, Si decreased lignin concentration in tobacco associated with reduction of activities of PAL, PPO, and soluble and covalently bound PODs, as well as the content of CW-bound phenolics and lignin [108]. Lower lignin in the Si-treated plants in the absence of any stress shows likely the Si role in enhancing the mechanical strength of CW and may reflect a trade-off between lignin and Si accumulation. Si resembles lignin in terms of structural strength and is a cheaper mechanism for providing plant mechanical support than lignin [201,202]. In accordance with this hypothetical role for Si, there is a strong negative correlation between Si and carbon in wheat straw [203]. Recently, this effect of Si has been highlighted particularly in the legumes [204].

In sum, despite numerous reports on the effect of low temperatures on CW lignification and the effect of Si on this process summarized earlier, the effect of Si in cold-stressed plants on the lignin concentration has not been studied so far and there is no information directly showing the contribution of Si-mediated lignin modification to plants chilling and freezing tolerance.

12.2.9 Contribution of apoplast to the cold tolerance affected by Si

The apoplast is an intercellular space in plant tissue that is outside the plasma membrane and connects cells to cells and the external environment [171]. The apoplast is the first compartment encountering environmental signals and contributes to the perception and transduction of these signals in parallel with the function of the plasma membrane [205–207]. The apoplast contributes also to the cold acclimation process and the biochemical activities in this compartment are markedly modified after exposure to above-freezing temperatures [206]. One of the most critical factors for frost survival is avoidance of ice formation in the intracellular space. The apoplast acts as a barrier and prevents the penetration of ice into the intracellular space. In addition, extracellular freezing that causes dehydration helps cells to avoid intracellular ice formation and increases survival under freezing temperatures [208]. Some specific proteins purified from the apoplastic region of winter cereals after cold acclimation exhibit antifreeze activity [209,210].

Under cold stress-induced dehydration, membranes are protected by soluble sugars in both cytosol and apoplast. Soluble sugars in the apoplast prevent adhesion of ice to the membrane [13], reduce ice nucleation, and block ice propagation from the apoplast to the symplast spaces and thus prevent the lethal frost damage [206]. Cold acclimation results in elevated levels of soluble sugars in the leaf apoplast in winter and spring barley cultivars [63] and in winter oat [211]. The significance of apoplastic sugars in plants' cold tolerance has been confirmed by elevated levels of constitutive and acclimation-induced cold tolerance in the transformant line of *Solanum tuberosum* L. expressing the yeast apoplastic invertase [212].

The antioxidant enzymes such as APX, POD, SOD, and CAT are also present in the leaf apoplast [63,213]. Stress factors are also able to induce ROS generation in apoplastic space that results in the activation of the enzymatic scavenging system in this compartment [214]. Differential activity of antioxidant enzymes in the apoplast and symplast implied that the apoplastic ROS scavenging system provides the first line of defense against the environmental stresses [215].

Under cold stress, the apoplastic antioxidant enzymes contribute significantly to the freezing tolerance and increased activities of antioxidant enzymes is an indication of cold hardiness [216]. In barley, cold-acclimated plants showed higher SOD and CAT activities in the leaf apoplast compared with nonacclimated plants [63]. In strawberries, the activities of apoplastic CAT, POD, and APX varied significantly depending on the cold-acclimation stage and the cold-hardiness level of cultivars [216]. Low temperature significantly increased the activities of apoplastic CAT, POD, and SOD in winter but not in spring wheat cultivars [213]. Finally, the ameliorative effect of exogenous salicylic acid on cold stress in wheat was mediated by the regulation of the activities of the apoplastic antioxidant enzymes [217]. It has been stated that the existence of considerable activities of CAT, POD, and SOD in the leaf apoplast play important roles in the tolerance to various stresses during winter [214].

There are pieces of evidence suggesting an apoplast-linked mode of action for Si [186]. It is highly probable that, in addition to the physical properties of CW (as mentioned previously), Si would influence the biochemical properties of the apoplast [218]. Studies on the effect of Si on the biochemical properties of apoplast in cold-stressed plants are rare. In the leaves of barley, soluble carbohydrate and protein concentrations were increased by Si under unstressed conditions. Under cold stress, Si exacerbated the effect of cold acclimation on the accumulation of soluble carbohydrates and on the activities of ROS scavenging enzymes [63]. The Si-mediated modifications in the biochemical properties of the leaf apoplast were associated with increased plants' survival and decreased LT_{50} so that the effect of cold acclimation was completely substituted by Si [63]. Substitution of cold acclimation treatment by Si for the amelioration of frost damage suggests that Si and acclimation treatments trigger likely similar pathways leading to freezing resistance.

The priming role of Si and its effect on the activation of plants' defense against biotic and abiotic stresses could not be explained through a solely mechanical and passive role, that is, a physical barrier. There are pieces of evidence suggesting that Si likely binds to the hydroxyl groups of proteins, acts as a second messenger, and/or exerts a signaling role in plant cells and is involved in the modification of gene expression in plants [186,219]. In

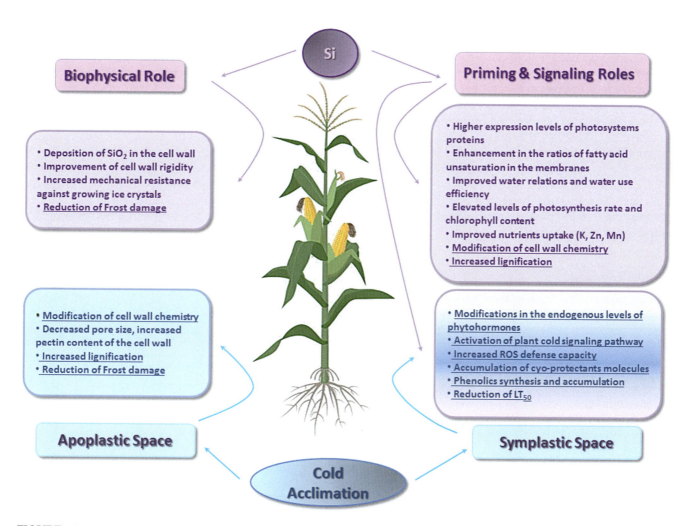

FIGURE 12.2 A schematic presentation of the mechanisms for Si-mediated mitigation of cold stress based on the nature of mechanisms: biophysical versus priming & signaling roles. The effect of cold acclimation was demonstrated depending on where they occur: symplast versus apoplast. The processes that are commonly activated under both Si and cold acclimation processes are underlined.

the "apoplasmic obstruction hypothesis" suggested by Coskun et al. [35], it has been proposed that the extracellular Si first interacts with plant cells in the apoplast and influences its function through a cascading effect.

12.3 Concluding remarks

Supplementation of plants with Si alleviates the adverse effects of both chilling and freezing temperatures and completely substitutes for the cold acclimation process in barley and likely in the grass family. Several mechanisms are involved in the mitigation of cold stress by Si ranging from structural modification of CW to the up- or down-regulation of gene expression. Depending on the cellular space where the Si action is confined to, the apoplastic and symplastic modes of action could be recognized. The leaf apoplast is likely the initial compartment in that Si exerts its effect through modification of CW structure and the biochemical properties of this extracellular space leading to the alterations in the downstream biochemical and molecular events in the symplast.

Based on the nature of these effects, however, the mechanisms of Si action could be classified into two distinct modes. The first action mode is the biophysical mechanisms through which deposition of Si in CW improves its rigidity and thus protects CW from the mechanical injury created by ice formation in the extracellular space and cell collapse under frost events. It seems that Si-accumulator species are particularly efficient to use this mode of action for alleviation of freezing stress because of the higher ability for Si uptake and deposition that is most likely active and is patterned by CW structure. The second mode of Si action is priming and signaling role via the mechanisms that are completely unknown except some pieces of evidence existing on direct Si—protein interactions. This mode is characterized by the activation of a wide spectrum of biochemical and molecular events ranging from ROS defense system, synthesis of cryo-protectants to the changes in the lignification of CW. Although we do not know about the effect of Si on the components of CBF signaling pathway so far, data from the literature suggest that Si starts a cross-talk with this pathway through alterations in the endogenous levels of phytohormones. It does not seem that Si accumulator and nonaccumulator species are different in this mode. An overview of the interactive effects of cold signaling and Si is presented in Fig. 12.2.

Although there is no clear evidence on the difference between Si accumulator and nonaccumulator species in the extent of cold stress alleviation, pieces of evidence from the overexpressing lines or disruptive mutants of *Lsi* and contrastive results on the effect of exogenous Si on lignin synthesis (mentioned in the text) suggest that Si effect is partly concentration-dependent. It is likely that the first action mode could be fulfilled only under adequately higher "Si status" that is achieved in the accumulators while the second one could also be triggered under low "Si status" in the tissues and thus in both accumulator and nonaccumulators. Thus providing data on the "Si status" of the plant tissues and information on the cultivation medium and Si impurities is important for comparison across species.

Acknowledgment

This work was supported by University of Tabriz.

References

[1] Snyder RL, Melo-Abreu JD. Frost protection: fundamentals, practice and economic, vol 1. Rome: Food and Agriculture Organization of the United Nations; 2005.

[2] Thakur P, Kumar S, Malik JA, et al. Cold stress effects on reproductive development in grain crops: an overview. Environ Exp Bot 2010;67:429—43.

[3] Siddiqui KS, Cavicchioli R. Cold-adapted enzymes. Annu Rev Biochem 2006;75:403—33.

[4] Yadav SK. Cold stress tolerance mechanisms in plants. A review. Agron Sustain Dev 2010;30:515—27.

[5] Lukatkin AS, Brazaityte A, Bobinas C, et al. Chilling injury in chilling-sensitive plants: a review. Agriculture 2012;99:111—24.

[6] Zhang H, Dong J, Zhao X, et al. Research progress in membrane lipid metabolism and molecular mechanism in peanut cold tolerance. Front Plant Sci 2019;10:838.

[7] Takahashi D, Li B, Nakayama T, et al. Plant plasma membrane proteomics for improving cold tolerance. Front Plant Sci 2013;4:90.

[8] Suzuki N, Mittler R. Reactive oxygen species and temperature stresses: a delicate balance between signaling and destruction. Physiol Plant 2006;126:45—51.

[9] Demidchik V, Straltsova D, Medvedev SS, et al. Stress-induced electrolyte leakage: the role of K^+-permeable channels and involvement in programmed cell death and metabolic adjustment. J Exp Bot 2014;65:1259—70.

[10] Melencion SM, Chi YH, Pham TT, et al. RNA chaperone function of a universal stress protein in *Arabidopsis* confers enhanced cold stress tolerance in plants. Int J Mol Sci 2017;18:2546.

[11] Beck EH, Heim R, Hansen J. Plant resistance to cold stress: mechanisms and environmental signals triggering frost hardening and dehardening. J Biosci 2004;29:449—59.

[12] Pearce RS. Plant freezing and damage. Ann Bot 2001;87:417—24.

[13] Gusta LV, Wisniewski M, Nesbitt NT, et al. The effect of water, sugars, and proteins on the pattern of ice nucleation and propagation in acclimated and nonacclimated canola leaves. Plant Physiol 2004;135:1642—53.

[14] Le Gall H, Philippe F, Domon JM, et al. Cell wall metabolism in response to abiotic stress. Plants 2015;4:112—66.

[15] Ruelland E, Vaultier MN, Zachowski A, et al. Cold signalling and cold acclimation in plants. Adv Bot Res 2009;49:35—150.

[16] Steponkus PL. Role of the plasma membrane in freezing injury and cold acclimation. Annu Rev Plant Physiol 1984;35:543—84.

[17] Puhakainen T. Physiological and molecular analyses of cold acclimation of plants [dissertation]. Finland: University of Helsinki; 2004.

[18] Janská A, Maršík P, Zelenková S, et al. Cold stress and acclimation—what is important for metabolic adjustment? Plant Biol 2010;12:395—405.

[19] Xin Z, Browse J. Cold comfort farm: the acclimation of plants to freezing temperatures. Plant Cell Environ 2000;23:893—902.

[20] Gusta LV, Wisniewski M. Understanding plant cold hardiness: an opinion. Physiol Plant 2013;147:4—14.

[21] Barrero-Gil J, Salinas J. Post-translational regulation of cold acclimation response. Plant Sci 2013;205:48—54.

[22] Minorsky PV. Temperature sensing by plants: a review and hypothesis. Plant Cell Environ 1989;12:119—35.

[23] Ruelland E, Zachowski A. How plants sense temperature. Environ Exp Bot 2010;69:225—32.

[24] Ma Y, Dai X, Xu Y, et al. COLD1 confers chilling tolerance in rice. Cell 2015;160:1209—21.

[25] Plieth C. Temperature sensing by plants: calcium-permeable channels as primary sensors—a model. J Membr Biol 1999;1999(172):121—7.

[26] Orvar BL, Sangwan V, Omann F, et al. Early steps in cold sensing by plant cells: the role of actin cytoskeleton and membrane fluidity. Plant J 2000;23:785—94.

[27] Samaj J, Baluska F, Hirt H. From signal to cell polarity: mitogen-activated protein kinases as sensors and effectors of cytoskeleton dynamicity. J Exp Bot 2004;55:189—98.

[28] Liu Y, Dang P, Liu L, et al. Cold acclimation by the CBF–COR pathway in a changing climate: Lessons from *Arabidopsis thaliana*. Plant Cell Rep 2019;38:511—19.

[29] Guo X, Liu D, Chong K. Cold signaling in plants: Insights into mechanisms and regulation. J Integr Plant Biol 2018;60:745—56.

[30] Ding Y, Jia Y, Shi Y, et al. OST1-mediated BTF3L phosphorylation positively regulates CBFs during plant cold responses. EMBO J 2018;37:e98228.

[31] Miura K, Furumoto T. Cold signaling and cold response in plants. Int J Mol Sci 2013;14:5312—37.

[32] Lee H, Xiong L, Gong Z, et al. The Arabidopsis *HOS1* gene negatively regulates cold signal transduction and encodes a RING finger protein that displays cold-regulated nucleo—cytoplasmic partitioning. Genes Dev 2001;15:912—24.

[33] Miura K, Hasegawa PM. Regulation of cold signaling by sumoylation of ICE1. Plant Signal Behav 2008;3:52—3.

[34] Liu Z, Jia Y, Ding Y, et al. Plasma membrane CRPK1-mediated phosphorylation of 14-3-3 proteins induces their nuclear import to fine-tune CBF signaling during cold response. Mo. Cell 2017;66:117—28.

[35] Coskun D, Deshmukh R, Sonah H, et al. The controversies of silicon's role in plant biology. New Phytol 2019;221:67—85.

[36] Lee H, Guo Y, Ohta M, et al. *LOS2*, a genetic locus required for cold-responsive gene transcription encodes a bi-functional enolase. EMBO J 2002;21:2692—702.

[37] Chinnusamy V, Zhu J, Zhu JK. Gene regulation during cold acclimation in plants. Physiol Plant 2006;126:52—61.

[38] Xin Z. *eskimo1* mutants of *Arabidopsis* are constitutively freezing-tolerant. Proc Natl Acad Sci USA 1998;95:7799—804.

[39] Warren G, McKown R, Marin A, et al. Isolation of mutations affecting the development of freezing tolerance in *Arabidopsis thaliana* (L.) Heynh. Plant Physiol 1996;111:1011—19.

[40] Thomashow MF. Plant cold acclimation: freezing tolerance genes and regulatory mechanisms. Annu Rev Plant Biol 1999;50:571—99.

[41] Shi Y, Ding Y, Yang S. Cold signal transduction and its interplay with phytohormones during cold acclimation. Plant Cell Physiol 2015;56:7—15.

[42] Ding Y, Shi Y, Yang S. Advances and challenges in uncovering cold tolerance regulatory mechanisms in plants. New Phytol 2019;222:1690—16704.

[43] Wang SY, Galletta GJ. Foliar application of potassium silicate induces metabolic changes in strawberry plants. J Plant Nutr 1998;21:157—67.

[44] Kim YH, Khan AL, Kim DH, et al. Silicon mitigates heavy metal stress by regulating P-type heavy metal ATPases, *Oryza sativa* low silicon genes, and endogenous phytohormones. BMC Plant Biol 2014;14:1—3.

[45] Detmann KC, Araújo WL, Martins SC, et al. Silicon nutrition increases grain yield, which, in turn, exerts a feed-forward stimulation of photosynthetic rates via enhanced mesophyll conductance and alters primary metabolism in rice. New Phytol 2012;196:752—62.

[46] Hajiboland R, Cheraghvareh L, Poschenrieder C. Improvement of drought tolerance in tobacco (*Nicotiana rustica* L.) plants by silicon. J Plant Nutr 2017;40:1661—76.

[47] Khan A, Khan AL, Imran M, et al. Silicon-induced thermotolerance in *Solanum lycopersicum* L. via activation of antioxidant system, heat shock proteins, and endogenous phytohormones. BMC Plant Biol 2020;20:1—8.

[48] Coskun D, Britto DT, Huynh WQ, et al. The role of silicon in higher plants under salinity and drought stress. Front Plant Sci 2016;7:1072.

[49] Wolfe DW. Low temperature effects on early vegetative growth, leaf gas exchange and water potential of chilling-sensitive and chilling-tolerant crop species. Ann Bot 1991;67:205—12.

[50] Aroca R, Vernieri P, Irigoyen JJ, et al. Involvement of abscisic acid in leaf and root of maize (*Zea mays* L.) in avoiding chilling-induced water stress. Plant Sci 2003;165:671—9.

[51] Wilkinson S, Clephan AL, Davies WJ. Rapid low temperature-induced stomatal closure occurs in cold-tolerant *Commelina communis* leaves but not in cold-sensitive tobacco leaves, via a mechanism that involves apoplastic calcium but not abscisic acid. Plant Physiol 2001;126:1566—78.

[52] Adam S, Murthy SD. Effect of cold stress on photosynthesis of plants and possible protection mechanisms. In: Gaur RK, Sharma P, editors. Approaches to plant stress and their management. New Delhi: Springer; 2014. p. 219—26.

[53] Mahajan S, Tuteja N. Cold, salinity and drought stresses: an overview. Arch Biochem Biophys 2005;444:139–58.

[54] Wise RR. Chilling-enhanced photooxidation: the production, action and study of reactive oxygen species produced during chilling in the light. Photosynth Res 1995;45:79–97.

[55] Sharma A, Kumar V, Shahzad B, et al. Photosynthetic response of plants under different abiotic stresses: a review. J Plant Growth Regul 2019;19:1–23.

[56] Paul MJ, Pellny TK. Carbon metabolite feedback regulation of leaf photosynthesis and development. J Exp Bot 2003;54:539–47.

[57] Huber SC, Huber JL. Role and regulation of sucrose-phosphate synthase in higher plants. Annu Rev Plant Biol 1996;47:431–44.

[58] Paul MJ, Foyer CH. Sink regulation of photosynthesis. J Exp Bot 2001;52:1383–400.

[59] Hajiboland R, Cheraghvareh L. Influence of Si supplementation on growth and some physiological and biochemical parameters in salt-stressed tobacco (Nicotiana rustica L.) plants. J Sci IR Iran 2014;25:205–17.

[60] Habibi G, Norouzi F, Hajiboland R. Silicon alleviates salt stress in pistachio plants. Prog Biol Sci 2014;4:189–202.

[61] Hajiboland R, Moradtalab N, Aliasgharzad N, et al. Silicon influences growth and mycorrhizal responsiveness in strawberry plants. Physiol Mol Biol Plants 2018;24:1103–15.

[62] Hajiboland R, Cherghvareh L, Dashtebani F. Effect of silicon supplementation on wheat plants under salt stress. J Plant Process Func 2016;5:1–2.

[63] Joudmand A, Hajiboland R. Silicon mitigates cold stress in barley plants via modifying the activity of apoplasmic enzymes and concentration of metabolites. Acta Physiol Plant. 2019;41:29.

[64] Qian ZZ, Zhuang SY, Li Q, et al. Soil silicon amendment increases Phyllostachys praecox cold tolerance in a pot experiment. Forests 2019;10:405.

[65] Azeem S, Li Z, Zheng H, et al. Quantitative proteomics study on Lsi1 in regulation of rice (Oryza sativa L.) cold resistance. Plant Growth Regul 2016;78:307–23.

[66] Fang C, Zhang P, Jian X, et al. Overexpression of Lsi1 in cold-sensitive rice mediates transcriptional regulatory networks and enhances resistance to chilling stress. Plant Sci 2017;262:115–26.

[67] Choudhury S, Panda P, Sahoo L, et al. Reactive oxygen species signaling in plants under abiotic stress. Plant Signal Behav 2013;8:e23681.

[68] Liang SM, Kuang JF, Ji SJ, et al. The membrane lipid metabolism in horticultural products suffering chilling injury. Food Qual Saf 2020;4:9–14.

[69] Shao HB, Chu LY, Shao MA, et al. Higher plant antioxidants and redox signaling under environmental stresses. Compt Rend Biol 2008;331:433–41.

[70] Dai F, Huang Y, Zhou M, Zhang G. The influence of cold acclimation on antioxidative enzymes and antioxidants in sensitive and tolerant barley cultivars. Biol Plant 2009;53:257–62.

[71] Lin KH, Sei SC, Su YH, et al. Overexpression of the Arabidopsis and winter squash superoxide dismutase genes enhances chilling tolerance via ABA-sensitive transcriptional regulation in transgenic Arabidopsis. Plant Signal Behav 2019;14:1685728.

[72] Che Y, Zhang N, Zhu X, et al. Enhanced tolerance of the transgenic potato plants over-expressing Cu/Zn superoxide dismutase to low temperature. Sci Hort 2020;261:108949.

[73] Kim YH, Khan AL, Waqas M, et al. Silicon regulates antioxidant activities of crop plants under abiotic-induced oxidative stress: a review. Front Plant Sci 2017;8:510.

[74] Agarie S, Hanaoka N, Ueno O, et al. Effects of silicon on tolerance to water deficit and heat stress in rice plants (Oryza sativa L.), monitored by electrolyte leakage. Plant Prod Sci 1998;1:96–103.

[75] He Q, Ren Y, Zhao W, et al. Low temperature promotes anthocyanin biosynthesis and related gene expression in the seedlings of purple head Chinese cabbage (Brassica rapa L.). Genes 2020;11:81.

[76] Liang Y, Zhu J, Li Z, et al. Role of silicon in enhancing resistance to freezing stress in two contrasting winter wheat cultivars. Environ Exp Bot 2008;64:286–94.

[77] Liu JJ, Lin SH, Xu PL, et al. Effects of exogenous silicon on the activities of antioxidant enzymes and lipid peroxidation in chilling-stressed cucumber leaves. Agric Sci China 2009;8:1075–86.

[78] He Y, Xiao H, Wang H, et al. Effect of silicon on chilling-induced changes of solutes, antioxidants, and membrane stability in seashore paspalum turf grass. Acta Physiol Plant 2010;32:487–94.

[79] Habibi G. Exogenous silicon leads to increased antioxidant capacity in freezing-stressed pistachio leaves. Acta Agric Slov 2015;105:43–52.

[80] Szabados L, Savouré A. Proline: a multifunctional amino acid. Trends Plant Sci 2010;15:89–97.

[81] Krasensky J, Jonak C. Drought, salt, and temperature stress-induced metabolic rearrangements and regulatory networks. J Exp Bot 2012;63:1593–608.

[82] Bhandari K, Nayyar H. Low temperature stress in plants: an overview of roles of cryoprotectants in defense. In: Ahmad P, Wani M, editors. Physiological mechanisms and adaptation strategies in plants under changing environment. New York: Springer; 2014. p. 193–265.

[83] Rajashekar CB, Lafta A. Cell-wall changes and cell tension in response to cold acclimation and exogenous abscisic acid in leaves and cell cultures. Plant Physiol 1996;111:605–12.

[84] Rajashekar CB, Burke MJ. Freezing characteristics of rigid plant tissues (development of cell tension during extracellular freezing). Plant Physiol 1996;111:597–603.

[85] Yuanyuan M, Yali Z, Jiang L, et al. Roles of plant soluble sugars and their responses to plant cold stress. Afr J Biotechnol 2009;8:2004–10.

[86] Tarkowski ŁP, Van den Ende W. Cold tolerance triggered by soluble sugars: a multifaceted countermeasure. Front Plant Sci 2015;6:203.

[87] Stitt M, Hurry V. A plant for all seasons: alterations in photosynthetic carbon metabolism during cold acclimation in Arabidopsis. Curr Opin Plant Biol 2002;5:199–206.

[88] Gilmour SJ, Sebolt AM, Salazar MP, et al. Overexpression of the Arabidopsis CBF3 transcriptional activator mimics multiple biochemical changes associated with cold acclimation. Plant Physiol 2000;124:1854–65.

[89] Uemura M, Warren G, Steponkus PL. Freezing sensitivity in the sfr4 mutant of Arabidopsis is due to low sugar content and is manifested by loss of osmotic responsiveness. Plant Physiol 2003;131:1800–7.

[90] Valluru R, Van den Ende W. Plant fructans in stress environments: emerging concepts and future prospects. J Exp Bot 2008;59:2905—16.

[91] Matros A, Peshev D, Peukert M, et al. Sugars as hydroxyl radical scavengers: proof-of-concept by studying the fate of sucralose in Arabidopsis. Plant J 2015;82:822—39.

[92] Hajiboland R, Moradtalab N, Eshaghi Z, et al. Effect of silicon supplementation on growth and metabolism of strawberry plants at three developmental stages. New Zealand J Crop Hortic Sci 2018;46:144—1461.

[93] Moradtalab N, Weinmann M, Walker F, et al. Silicon improves chilling tolerance during early growth of maize by effects on micronutrient homeostasis and hormonal balances. Front Plant Sci 2018;9:420.

[94] Fernandez O, Béthencourt L, Quero A, et al. Trehalose and plant stress responses: friend or foe? Trends Plant Sci 2010;15:409—17.

[95] Ponnu J, Wahl V, Schmid M. Trehalose-6-phosphate: connecting plant metabolism and development. Front Plant Sci 2011;2:70.

[96] Almeida AM, Cardoso LA, Santos DM, et al. Trehalose and its applications in plant biotechnology. Vitro Cell Dev Biol Plant 2007;43:167—77.

[97] Iordachescu M, Imai R. Trehalose biosynthesis in response to abiotic stresses. J Integr Plant Biol 2008;50:1223—9.

[98] Liu X, Fu L, Qin P, et al. Overexpression of the wheat trehalose 6-phosphate synthase 11 gene enhances cold tolerance in Arabidopsis thaliana. Gene 2019;710:210—17.

[99] Garg N, Singh S. Mycorrhizal inoculations and silicon fortifications improve rhizobial symbiosis, antioxidant defense, trehalose turnover in pigeon pea genotypes under cadmium and zinc stress. Plant Growth Regul 2018;86:105—19.

[100] Mazzucotelli E, Tartari A, Cattivelli L, et al. Metabolism of γ-aminobutyric acid during cold acclimation and freezing and its relationship to frost tolerance in barley and wheat. J Exp Bot 2006;57:3755—66.

[101] Kaplan F, Kopka J, Haskell DW, et al. Exploring the temperature-stress metabolome of Arabidopsis. Plant Physiol 2004;136:4159—68.

[102] Hosseini SA, Naseri Rad S, Ali N, et al. The ameliorative effect of silicon on maize plants grown in Mg-deficient conditions. Int J Mol Sci 2019;20:969.

[103] Ali N, Schwarzenberg A, Yvin JC, et al. Regulatory role of silicon in mediating differential stress tolerance responses in two contrasting tomato genotypes under osmotic stress. Front Plant Sci 2018;9:1475.

[104] Nanjo T, Kobayashi M, Yoshiba Y, et al. Antisense suppression of proline degradation improves tolerance to freezing and salinity in Arabidopsis thaliana. FEBS Lett 1999;461:205—10.

[105] Hayat S, Hayat Q, Alyemeni MN, et al. Role of proline under changing environments: a review. Plant Signal Behav 2012;7:1456—66.

[106] Satoh R, Nakashima K, Seki M, et al. ACTCAT, a novel cis-acting element for proline-and hypoosmolarity-responsive expression of the ProDH gene encoding proline dehydrogenase in Arabidopsis. Plant Physiol 2002;130:709—19.

[107] Zhu Y, Gong H. Beneficial effects of silicon on salt and drought tolerance in plants. Agron Sustain Dev 2014;34:455—72.

[108] Hajiboland R, Bahrami-Rad S, Poschenrieder C. Silicon modifies both a local response and a systemic response to mechanical stress in tobacco leaves. Biol Plant 2017;61:187—91.

[109] Annunziata MG, Ciarmiello LF, Woodrow P, et al. Spatial and temporal profile of glycine betaine accumulation in plants under abiotic stresses. Front Plant Sci 2019;10:230.

[110] Kamata T, Uemura M. Solute accumulation in wheat seedlings during cold acclimation: contribution to increased freezing tolerance. Cryo Lett 2004;25:311—22.

[111] Khan MS, Yu X, Kikuchi A, et al. Genetic engineering of glycine betaine biosynthesis to enhance abiotic stress tolerance in plants. Plant Biotechnol 2009;26:125—34.

[112] Chen TH, Murata N. Glycinebetaine protects plants against abiotic stress: mechanisms and biotechnological applications. Plant Cell Environ 2011;34:1—20.

[113] Giri J. Glycinebetaine and abiotic stress tolerance in plants. Plant Signal Behav 2011;6:1746—51.

[114] Park EJ, Jeknic Z, Chen TH. Exogenous application of glycine betaine increases chilling tolerance in tomato plants. Plant Cell Physiol 2006;47:706—14.

[115] Biju S, Fuentes S, Gupta D. Silicon improves seed germination and alleviates drought stress in lentil crops by regulating osmolytes, hydrolytic enzymes and antioxidant defense system. Plant Physiol Biochem 2017;119:250—64.

[116] Arteaga S, Yabor L, Díez MJ, et al. The use of proline in screening for tolerance to drought and salinity in common bean (Phaseolus vulgaris L.) genotypes. Agronomy 2020;10:817.

[117] Alcázar R, Marco F, Cuevas JC, et al. Involvement of polyamines in plant response to abiotic stress. Biotechnol Lett 2006;28:1867—76.

[118] Gupta K, Dey A, Gupta B. Plant polyamines in abiotic stress responses. Acta Physiol Plant 2013;35:2015—36.

[119] Imai R, Ali A, Pramanik MH, et al. A distinctive class of spermidine synthase is involved in chilling response in rice. J Plant Physiol 2004;161:883—6.

[120] Cook D, Fowler S, Fiehn O, et al. A prominent rolefor the CBF cold response pathway in configuring the low-temperature metabolome of Arabidopsis. Proc. Natl Acad. Sci. 2004;101:15243—8.

[121] Kotzabasis K, Fotinou C, Roubelakis-Angelakis KA, et al. Polyamines in the photosynthetic apparatus. Photosynth Res 1993;38:83—8.

[122] Shen W, Nada K, Tachibana S. Involvement of polyamines in the chilling tolerance of cucumber cultivars. Plant Physiol 2000;124:431—40.

[123] Yin L, Wang S, Liu P, et al. Silicon-mediated changes in polyamine and 1-aminocyclopropane-1-carboxylic acid are involved in silicon-induced drought resistance in Sorghum bicolor L. Plant Physiol Biochem 2014;80:268—77.

[124] Yin J, Jia J, Lian Z, et al. Silicon enhances the salt tolerance of cucumber through increasing polyamine accumulation and decreasing oxidative damage. Ecotoxicol Environ Saf 2019;169:8—17.

[125] Penfield S. Temperature perception and signal transduction in plants. New Phytol 2008;179:615—28.

[126] Thomashow MF. Molecular basis of plant cold acclimation: insights gained from studying the CBF cold response pathway. Plant Physiol 2010;154:571—7.

[127] Gusta LV, Trischuk R, Weiser CJ. Plant cold acclimation: the role of abscisic acid. J Plant Growth Regul 2005;24:308—18.

[128] Tamminen I, Makela P, Heino P, et al. Ectopic expression of ABI3 gene enhances freezing tolerance in response to abscisic acid and low temperature in Arabidopsis thaliana. Plant J 2001;25:1—8.

[129] Rabbani MA, Maruyama K, Abe H, et al. Monitoring expression profiles of rice genes under cold, drought, and high-salinity stresses and abscisic acid application using cDNA microarray and RNA gel-blot analyses. Plant Physiol 2003;133:1755–67.

[130] Kumar S, Kaur G, Nayyar H. Exogenous application of abscisic acid improves cold tolerance in chickpea (Cicer arietinum L.). J Agron Crop Sci 2008;194:449–56.

[131] Li HY, Zhang WZ. Abscisic acid-induced chilling tolerance in maize seedlings is mediated by nitric oxide and associated with antioxidant system. In: Gan B, Gan Y, Yu Y, editors. Advanced materials research, vol. 378. Trans Tech Publications Ltd; 2012. p. 423–7.

[132] Garbero M, Pedranzani H, Zirulnik F, et al. Short-term cold stress in two cultivars of Digitaria eriantha: effects on stress-related hormones and antioxidant defense system. Acta Physiol Plant 2011;33:497–507.

[133] Eremina M, Rozhon W, Poppenberger B. Hormonal control of cold stress responses in plants. Cell Mol Life Sci 2016;73:797–810.

[134] Horváth E, Szalai G, Janda T. Induction of abiotic stress tolerance by salicylic acid signaling. J Plant Growth Regul 2007;26:290–300.

[135] Hu Y, Jiang Y, Han X, et al. Jasmonate regulates leaf senescence and tolerance to cold stress: crosstalk with other phytohormones. J Exp Bot 2017;68:1361–9.

[136] Hu Y, Han X, Yang M, et al. The transcription factor inducer of CBF expression1 interacts with abscisic acid insensitive5 and DELLA proteins to fine-tune abscisic acid signaling during seed germination in Arabidopsis. Plant Cell 2019;31:1520–38.

[137] Harberd NP, Belfield E, Yasumura Y. The angiosperm gibberellin-GID1-DELLA growth regulatory mechanism: how an "inhibitor of an inhibitor" enables flexible response to fluctuating environments. Plant Cell 2009;21:1328–39.

[138] Achard P, Renou JP, Berthome R, et al. Plant DELLAs restrain growth and promote survival of adversity by reducing the levels of reactive oxygen species. Curr Biol 2008;18:656–60.

[139] Heidari P, Entazari M, Ebrahimi A, et al. Exogenous EBR ameliorates endogenous hormone contents in tomato species under low-temperature stress. Horticulturae 2021;7:84.

[140] Lee BH, Henderson DA, Zhu JK. The Arabidopsis cold-responsive transcriptome and its regulation by ICE1. Plant Cell 2005;17:3155–75.

[141] Xia J, Zhao H, Liu W, et al. Role of cytokinin and salicylic acid in plant growth at low temperatures. Plant Growth Regul 2009;57:211–21.

[142] Maruyama K, Urano K, Yoshiwara K, et al. Integrated analysis of the effects of cold and dehydration on rice metabolites, phytohormones, and gene transcripts. Plant Physiol 2014;164:1759–71.

[143] Jeon J, Kim NY, Kim S, et al. A subset of cytokinin two-component signaling system plays a role in cold temperature stress response in Arabidopsis. J Biol Chem 2010;285:23371–86.

[144] Tripathi DK, Vishwakarma K, Singh VP, et al. Silicon crosstalk with reactive oxygen species, phytohormones and other signaling molecules. J Hazard Mater 2020;13:124820.

[145] Cakmak I. Role of mineral nutrients in tolerance of crop plants to environmental stress factors. Proceedings of international symposium on fertigation—optimizing the utilization of water nutrients. International Potash Institute; 2005. p. 35–48.

[146] Waraich EA, Ahmad R, Halim A, et al. Alleviation of temperature stress by nutrient management in crop plants: a review. J Soil Sci Plant Nutr 2012;12:221–44.

[147] Hajiboland R. Effect of micronutrient deficiencies on plants' stress responses. In: Ahmad P, Prasad M, editors. Abiotic stress responses in plants. New York: Springer; 2012. p. 283–329.

[148] Hund A, Richner W, Soldati A, et al. Root morphology and photosynthetic performance of maize inbred lines at low temperature. Eur J Agron 2007;27:52–61.

[149] Yan QY, Duan ZQ, Mao JD. Low root zone temperature limits nutrient effects on cucumber seedling growth and induces adversity physiological response. J. Integr Agric 2013;12:1450–60.

[150] Domisch T, Finér L, Lehto T, et al. Effect of soil temperature on nutrient allocation and mycorrhizas in Scots pine seedlings. Plant Soil 2002;239:173–85.

[151] Yan Q, Duan Z, Mao J, et al. Effects of root-zone temperature and N, P, and K supplies on nutrient uptake of cucumber (Cucumis sativus L.) seedlings in hydroponics. Soil Sci Plant Nutr 2012;58:707–17.

[152] Grewal JS, Singh SN. Effect of potassium nutrition on frost damage and yield of potato plants on alluvial soils of Punjab (India). Plant Soil 1980;57:105–10.

[153] Farooq M, Wahid A, Siddique KHM. Micronutrient application through seed treatments—a review. J Soil Sci Plant Nutr 2012;12:125–42.

[154] Hakerlerler H, Oktay M, Eryüce N, et al. Effect of potassium sources on the chilling tolerance of some vegetable seedlings grown in hotbeds. In: Johnston AE, editor. Food security in the WANA region, the essential need for balanced fertilization. Basel: International Potash Institute; 1997. p. 317–27.

[155] Cakmak I. The role of potassium in alleviating detrimental effects of abiotic stresses in plants. J Plant Nutr Soil Sci 2005;168:521–30.

[156] Hernandez-Apaolaza L. Can silicon partially alleviate micronutrient deficiency in plants? A review. Planta 2014;240:447–58.

[157] Pontigo S, Ribera A, Gianfreda L, et al. Silicon in vascular plants: uptake, transport and its influence on mineral stress under acidic conditions. Planta 2015;242:23–37.

[158] Greger M, Landberg T, Vaculík M. Silicon influences soil availability and accumulation of mineral nutrients in various plant species. Plants 2018;7:41.

[159] Chalker-Scott L, Fuchigami LH. The role of phenolic compounds in plant stress responses. In: Li PH, editor. Low temperature stress physiology in crops. CRC Press; 2018. p. 67–80.

[160] Sharma A, Shahzad B, Rehman A, et al. Response of phenylpropanoid pathway and the role of polyphenols in plants under abiotic stress. Molecules 2019;24:2452.

[161] Naikoo MI, Dar MI, Raghib F, et al. Role and regulation of plants phenolics in abiotic stress tolerance: an overview. In: Khan MI, Reddy PS, Ferrante A, Khan N, editors. Plant signaling molecules. Amsterdam: Elsevier; 2019. p. 157–68.

[162] Moura JC, Bonine CA, de Oliveira FernandesViana J, et al. Abiotic and biotic stresses and changes in the lignin content and composition in plants. J Integr Plant Biol 2010;52:360–76.

[163] Leyva A, Jarillo JA, Salinas J, et al. Low temperature induces the accumulation of phenylalanine ammonia-lyase and chalcone synthase mRNAs of Arabidopsis thaliana in a light-dependent manner. Plant Physiol 1995;108:39–46.

[164] Christie PJ, Alfenito MR, Walbot V. Impact of low-temperature stress on general phenylpropanoid and anthocyanin pathways: enhancement of transcript abundance and anthocyanin pigmentation in maize seedlings. Planta 1994;194:541–9.

[165] Olenichenko N, Zagoskina N. Response of winter wheat to cold: production of phenolic compounds and l-phenylalanine ammonialyase activity. Appl Biochem Microbiol 2005;41:600–3.

[166] Solecka D, Boudet AM, Kacperska A. Phenylpropanoid and anthocyanin changes in low-temperature treated winter oilseed rape leaves. Plant Physiol Biochem 1999;37:491–6.

[167] Solecka D, Kacperska A. Phenylalanine ammonia-lyase activity in leaves of winter oilseed rape plants as affected by acclimation of plants to low-temperature. Plant Physiol Biochem 1995;33:585–91.

[168] Domon JM, Baldwin L, Acket S, et al. Cell wall compositional modifications of *Miscanthus* ecotypes in response to cold acclimation. Phytochemistry 2013;85:51–61.

[169] Janas KM, Cvikrová M, Pałągiewicz A, et al. Alterations in phenylpropanoid content in soybean roots during low temperature acclimation. Plant Physiol Biochem 2000;38:587–93.

[170] Ahanger MA, Bhat JA, Siddiqui MH, et al. Integration of silicon and secondary metabolites in plants: a significant association in stress tolerance. J Exp Bot 2020;71:6758–74.

[171] Sattelmacher B. The apoplast and its significance for plant mineral nutrition. New Phytol 2001;149:167–92.

[172] Solecka D, Żebrowski J, Kacperska A. Are pectins involved in cold acclimation and de-acclimation of winter oil-seed rape plants? Ann Bot 2008;101:521–30.

[173] Wisniewski M, Davis G. Immunogold localization of pectins and glycoproteins in tissues of peach with reference to deep supercooling. Trees Struct Funct 1995;9:253–60.

[174] Pelloux J, Rusterucci C, Mellerowicz EJ. New insights into pectin methyl-esterase structure and function. Trends Plant Sci 2007;12:267–77.

[175] Anthon GE, Barrett DM. Characterization of the temperature activation of pectin methylesterase in green beans and tomatoes. J Agric Food Chem 2006;54:204–11.

[176] Baldwin L, Domon JM, Klimek JF, et al. Structural alteration of cell wall pectins accompanies pea development in response to cold. Phytochemistry 2014;104:37–47.

[177] Bilska-Kos A, Solecka D, Dziewulska A, et al. Low temperature caused modifications in the arrangement of cell wall pectins due to changes of osmotic potential of cells of maize leaves (*Zea mays* L.). Protoplasma 2017;254:713–24.

[178] He C, Wang L, Liu J, et al. Evidence for 'silicon' within the cell walls of suspension-cultured rice cells. New Phytol 2013;200:700–9.

[179] He C, Ma J, Wang L. A hemicellulose-bound form of silicon with potential to improve the mechanical properties and regeneration of the cell wall of rice. New Phytol 2015;206:1051–62.

[180] Zhang J, Zou W, Li Y, et al. Silica distinctively affects cell wall features and lignocellulosicsaccharification with large enhancement on biomass production in rice. Plant Sci 2015;239:84–91.

[181] Soukup M, Rodriguez Zancajo VM, Kneipp J, et al. Formation of root silica aggregates in sorghum is an active process of the endodermis. J Exp Bot 2020;71:6807–17.

[182] Głazowska S, Baldwin L, Mravec J, et al. The impact of silicon on cell wall composition and enzymatic saccharification of *Brachypodium distachyon*. Biotechnol Biofuels 2018;11:1–8.

[183] Zexer N, Elbaum R. Unique lignin modifications pattern the nucleation of silica in sorghum endodermis. J Exp Bot 2020;71:6818–29.

[184] Exley C. A possible mechanism of biological silicification in plants. Front Plant Sci 2015;6:853.

[185] Watteau F, Villemin G. Ultrastructural study of the biogeochemical cycle of silicon in the soil and litter of a temperate forest. Eur J Soil Sci 2001;52:385–96.

[186] Guerriero G, Hausman JF, Legay S. Silicon and the plant extracellular matrix. Front Plant Sci 2016;7:463.

[187] Zhang F, Huang L, Wang W, et al. Genome-wide gene expression profiling of introgressedindica rice alleles associated with seedling cold tolerance improvement in a japonica rice background. BMC Genomics 2012;13:461.

[188] Hamann T. The plant cell wall integrity maintenance mechanism-concepts for organization and mode of action. Plant Cell Physiol 2015;56:215–23.

[189] Liu Q, Luo L, Zheng L. Lignins: biosynthesis and biological functions in plants. Int J Mol Sci 2018;19:335.

[190] Baxter HL, Stewart Jr CN. Effects of altered lignin biosynthesis on phenylpropanoid metabolism and plant stress. Biofuels 2013;4:635–50.

[191] Gindl W, Grabner M, Wimmer R. The influence of temperature on latewood lignin content in treeline Norway spruce compared with maximum density and ring width. Trees Struct Funct. 2000;14:409–14.

[192] Ford CW, Morrison IM, Wilson JR. Temperature effects on lignin, hemicellulose and cellulose in tropical and temperate grasses. Austr J Agric Res. 1979;30:621–33.

[193] Hausman JF, Evers D, Thiellement H, et al. Compared responses of poplar cuttings and *in vitro* raised shoots to short-term chilling treatments. Plant Cell Rep 2000;19:954–60.

[194] Wei HU, Dhanaraj AL, Arora R, et al. Identification of cold acclimation-responsive Rhododendron genes for lipid metabolism, membrane transport and lignin biosynthesis: importance of moderately abundant ESTs in genomic studies. Plant Cell Environ 2006;29:558–70.

[195] Zhou P, Li Q, Liu G, et al. Integrated analysis of transcriptomic and metabolomic data reveals critical metabolic pathways involved in polyphenol biosynthesis in *Nicotianatabacum* under chilling stress. Funct Plant Biol 2018;46:30–43.

[196] Fleck AT, Nye T, Repenning C, et al. Silicon enhances suberization and lignification in roots of rice (*Oryza sativa*). J Exp Bot 2011;62:2001–11.

[197] Suzuki S, Ma JF, Yamamoto N, et al. Silicon deficiency promotes lignin accumulation in rice. Plant Biotechnol 2012;29:391–4.

[198] Yamamoto T, Nakamura A, Iwai H, et al. Effect of silicon deficiency on secondary cell wall synthesis in rice leaf. J Plant Res 2012;125:771–9.

[199] Hussain S, Shuxian L, Mumtaz M, et al. Foliar application of silicon improves stem strength under low light stress by regulating lignin biosynthesis genes in soybean (*Glycine max* (L.) Merr.). J Hazard Mater 2021;401:123256.

[200] Vega I, Rumpel C, Ruíz A, et al. Silicon modulates the production and composition of phenols in barley under aluminum stress. Agronomy 2020;10:1138.

[201] Raven JA. The transport and function of silicon in plants. Biol Rev 1983;58:179–207.

[202] Klotzbücher T, Klotzbücher A, Kaiser K, et al. Variable silicon accumulation in plants affects terrestrial carbon cycling by controlling lignin synthesis. Glob Change Biol 2018;24:e183–9.

[203] Neu S, Schaller J, Dudel EG. Silicon availability modifies nutrient use efficiency and content, C: N: P stoichiometry, and productivity of winter wheat (*Triticum aestivum* L.). Sci Rep 2017;7:1–8.

[204] Putra R, Powell JR, Hartley SE, Johnson SN. Is it time to include legumes in plant silicon research? Funct Ecol 2020;34:1142–57.

[205] Hoson T. Apoplast as the site of response to environmental signals. J Plant Res 1998;111:167–77.

[206] Liu J. Temperature-mediated alterations of the plant apoplast as a mechanism of intracellular freezing stress avoidance [dissertation]. Canada: University of Saskatchewan; 2015.

[207] Farvardin A, González-Hernández AI, Llorens E, et al. The apoplast: a key player in plant survival. Antioxidants 2020;9:604.

[208] Sakai A, Larcher W. Frost survival of plants. Responses and adaptation to freezing stress. New York: Springer-Verlag; 1987.

[209] Atıcı Ö, Nalbantoğlu B. Effect of apoplastic proteins on freezing tolerance in leaves. Phytochemistry 1999;50:755–61.

[210] Griffith M, Yaish MW. Antifreeze proteins in overwintering plants: a tale of two activities. Trends Plant Sci 2004;2004(9):399–405.

[211] Livingston DP, Henson CA. Apoplastic sugars, fructans, fructanexohydrolase and invertase in winter oat: responses to second-phase cold-acclimation. Plant Physiol 1998;116:403–8.

[212] Deryabin AN, Burakhanova EA, Trunova TI. The involvement of apoplasticinvertase in the formation of resistance of cold-tolerant plants to hypothermia. Biol Bull 2016;43:26–33.

[213] Cakmak TU, Atici O. Effects of putrescine and low temperature on the apoplastic antioxidant enzymes in the leaves of two wheat cultivars. Plant Soil Environ 2009;55:320–6.

[214] Baek SH, Kwon IS, Park TI, et al. Activities and isozyme profiles of antioxidant enzymes in intercellular compartment of overwintering barley leaves. J Biochem Mol Biol 2000;33:385–90.

[215] Atıcı Ö, Nalbantolu B. Antifreeze proteins in higher plants. Phytochemistry 2003;64:1187–96.

[216] Turhan E, Aydogan C, Baykul A, et al. Apoplastic antioxidant enzymes in the leaves of two strawberry cultivars and their relationship to cold-hardiness. Not. Bot. Hortic Agrobot Cluj-Napoca 2012;40:114–22.

[217] Taşgın E, Atıcı Ö, Nalbantoğlu B, et al. Effects of salicylic acid and cold treatments on protein levels and on the activities of antioxidant enzymes in the apoplast of winter wheat leaves. Phytochemistry 2006;67:710–15.

[218] Wiese H, Nikolic M, Römheld V. Silicon in plant nutrition. In: Sattelmacher B, Horst WJ, editors. The apoplast of higher plants: compartment of storage, transport and reactions. Dordrecht: Springer; 2007. p. 33–47.

[219] Luyckx M, Hausman JF, Lutts S, et al. Silicon and plants: current knowledge and technological perspectives. Front Plant Sci 2017;8:411.

CHAPTER

13

Silicon and nano-silicon mediated heavy metal stress tolerance in plants

Seyed Majid Mousavi

Agricultural Research, Education and Extension Organization (AREEO), Soil and Water Research Institute (SWRI), Department of Soil Fertility and Plant Nutrition, Karaj, Iran

13.1 Introduction

Numerous biotic and abiotic stresses have continuously affected the growth and development of plants. As one of the abiotic stresses, heavy metal stresses have received more attention over the last century. The metals and metalloids with an atomic density greater than $4 \, g \, cm^{-3}$ are defined as heavy metals [1]. Furthermore, based on this definition, there are 53 heavy metals [2]. The majority of heavy metals such as lead, cadmium, arsenic, platinum, and silver do not have any essential roles in plants. Although heavy metals are naturally present in the soil, but their concentrations in the soil and environment will be increased due to geologic and anthropogenic activities [3–7]. These potentially toxic metals, through different mechanisms and reactions, disturb plant growth [8]. Also, these elements can be moved by the food chain into living creatures and consequently threaten human/public health [9]. Different strategies have been studied and recommended for mitigating the stress of toxic metal in plants in the last few decades. Nutritional management of the plants growing in heavy metal-contaminated soils has received considerable attention over the last century, and silicon (Si), because of its ability to ameliorate toxic metal stress [10,11], has been studied from different aspects.

Based on the report by Wang et al. [12], the potential role of Si in toxic metals detoxification is related to its effects on change of plant cellular mechanisms and biochemical interactions with the external growth medium. Depending on the crop species, type of the heavy metal, and its concentration level, the useful functions of Si are changed and are generally more obvious in the plants that uptake high concentrations of Si [1,13]. Even though Si has high availability in soils, but its bioavailable form is often inadequate in most types of soils. Plants absorb Si as silicic acid (H_4SiO_4) and its concentration in plants ranges from 1 to $100 \, g \, kg^{-1}$; based on the data [14], it is the largest range of mineral elements that refer to the unique abilities of plant species to absorb and translocate of Si, through certain carriers of Si [15]. This chapter will try to evaluate and discuss the Si-induced mechanism for ameliorating toxic metals stress in plants.

13.2 Heavy metals: Functions, effects, and classification based on necessity

Soils and environmental contaminations with potentially toxic metals have been known as a serious concern because of their undesirable ecological effects. Agricultural activities, such as fertilizers, pesticides, and herbicides applications, have been shown to cause soil pollution with the excess concentrations of potentially toxic metals, such as Cd, Pb, and Zn [3–7]. The agricultural soils that have been exposed to the excess concentrations of these metals for long times were reported to have undesirable effects on plant growth and metabolism, soil quality, and public health (plant, animal, and human health) [16,17].

Silicon and Nano-silicon in Environmental Stress Management and Crop Quality Improvement.
DOI: https://doi.org/10.1016/B978-0-323-91225-9.00012-1

181

Heavy metals are categorized as nonessential (Pd, Cd, Cr, As, Hg, and Ag) and essential elements (Fe, Zn, Cu, Mn, Mo, Ni, and Co). Nonessential heavy metals are potentially toxic to plants/animals/human, while essential heavy metals are vital for plant growth and development, and they are known as micronutrients [18]. Generally, a plant usually grows if the supply of a particular nutrient covers the plant's requirements. The existence of both essential and nonessential heavy metals in excess can result in the mediation, reduction, or inhibition of plant growth through structural, anatomical, and biochemical alterations [19]. Heavy metals also have interactions with essential nutrients, which are known as antagonism and synergism reactions, and especially in higher concentrations, heavy metals significantly affect the phyto-availability, uptake, and generally, the status of essential nutrients in plants [17]. Some toxicity symptoms of these metals on different plants are presented in Table 13.1.

Heavy metals toxicities on plants are raised from different pathways due to direct and indirect effects. Physiological pathways are considered as direct effects of heavy metals, while their indirect effects on plant growth and productivity are caused by biochemical and molecular pathways [2] (Table 13.2).

TABLE 13.1 Some adverse effects of heavy metals toxicities on different plant species.

Crop	Heavy metal	Result	Reference
Pigeon pea	Ni	• Reduction of chlorophyll content • Reduction of stomatal conductance • Reduction of enzyme activity, affecting Calvin cycle and CO_2 fixation	Sheoran et al. [20]
Wheat	Cr	• Reduction of the shoot and root growth	Panda and Patra [21]
Mung beans	Co	• Reduction of Mn and Fe concentrations in root and leaves • Inhibition of seedling growth • Inhibition of plant growth	Liu et al. [22]
Maize	Hg	• Reduction of elongation of primary roots as well as an inhibition of the gravimetric response of the seedlings	Patra et al. [23]
Rice	Ni	• Inhibition of root growth	Lin and Kao [24]
Raddish	Co	• Decrease in shoot length, root length, and total leaf area • Reduction of chlorophyll content • Decrease in plant nutrient contents and antioxidant enzyme activities; Reduction of plant sugar, amino acid, and protein contents	Jayakumar et al. [25]
Barley, oilseed rape, and tomato	Co	• Adverse effect on shoot growth and biomass	Li et al. [26]
Portia tree	Pb	• Decrease in number of leaves and leaf area • Reduction in plant height	Kabir et al. [27]
Maize	Ni	• Reduction of chlorophyll content • Increase in K^+ efflux and carbohydrate leakage from roots • Increase in the level of ROS generation	Ghasemi et al. [28]
Onion	Cr	• Inhibition of germination process • Reduction in plant biomass	Nematshahi et al. [29]
Maize	Pb	• Reduction in germination percentage • Reduction in growth • Reduction in plant biomass and protein content	Hussain et al. [30]
Maize genotypes	Co and Ni	• Reduction of maize seed germination	Ebru [31]
Phaseolus vulgaris	Cu	• Reduction of length and fresh and dry weights • Reduction of growth • Increase in albumin and globulin content	Karmous et al. [32]
Cotton	Cd	• Reduction in antioxidant enzyme activities in both leaves and roots	Farooq et al. [33]
Tomato	Cd, Co, Ni, Pb	• Increase in phenols, flavonoids contents, negative effects on yield and marketable yield	Hashem et al. [34]
Sunflower	Zn, Pb	• Reduction in growth and dry biomass yield	Mousavi et al. [10,11]
Maize	Cd	• Reduction in plant height, fresh weight, transpiration rate, and photosynthetic activity	Liu et al. [35]
Tomato	Cd	• Reduction in seed germination, root elongation, and plant growth	Baruah et al. [36]
Flax	Cu	• Inhibition of plant growth • Reduction in yield attributes and total chlorophyll content	El-Beltagi et al. [37]

TABLE 13.2   Direct and indirect effects of heavy metals on plants from different pathways.

Direct effects	Indirect effects	
Physiological pathways	Biochemical pathways	Molecular pathways
Decrease in plant biomass	Increase in the production of ROS	Chromosomal aberrations
Decrease in protein biosynthesis	Increase in peroxidation of lipid	Damage to DNA
Decrease in respiration	Increase in inactivation of the enzyme	Abnormal mitosis and meiosis
Decrease in water status	Increase in injury/death of the cell	
Decrease in pigment content		
Decrease in photosynthesis		

To mitigate the heavy metal toxicities in plants, different strategies have been recommended [7,38] one of which is plant nutrition management by using some beneficial elements like silicon (Si), [7,10,11] and it will be discussed from different aspects in the following sections.

13.3 Silicon/nano-silicon plays a vital role in the alleviation of heavy metals toxicity in plants

The influence of silicon (Si) on the promotion of plant tolerance to the toxicity of heavy metals is well documented. Silicon can play this role by both external and internal plant mechanisms [39]. The external mechanism (growth media) of mitigating heavy metal toxicity is mostly due to reduction in soil acidity (increase in soil pH) affected by application of silicate and there with precipitation of the heavy metal as silicate compounds and consequently reduction of its bioavailability [40]. Also, Si plays an important role in the translocation and distribution of elements in different plant tissues and therefore increases their tolerance to potentially toxic metals [41]. Plant species have different abilities to uptake Si, and it has shown that plants with more abilities to uptake Si will usually have more ability to tolerate heavy metals stress [42,43]. Based on the report by Gu et al. [44], the application of Si compounds in a multimetal (Cd, Zn, Cu, and Pb) contaminated acidic soil reduced heavy metals accumulation in plant tissues and increased plant growth. Shi et al. [21] observed that Si significantly reduces Cd toxicity in rice seedlings by lowering the uptake and also reducing the Cd translocation from root to shoot. Also, reduction of lipid peroxidation and fatty acid desaturation in plant tissues under toxic metals contamination were reported [9]. The positive roles of Si in alleviating Al toxicity in peanut and rice seedlings [45,46] sorghum, tomato, soybean, and maize [39], Zn and Pb in sunflower [10,11] were reported.

Nanoparticles (NPs) are usually defined as materials with at least one dimension smaller than 100 nm [10,11]. NPs have considerable positive points compared to usual materials, such as high surface activity, further surface reaction sites, considerable catalytic efficiency, and effective optical and magnetic properties [47,48]. NPs have been already effectively used in agriculture (as nano-fertilizers and nano-pesticides) and also for environmental objectives [49–52]. The advances in nanotechnology offer a new path for the development of soil remediation [53]. Due to the unique nature and characteristics, NPs not only have a better role than traditional materials but also can cause some new functions [54]. Studies have shown that nano-Si is more efficient than common Si in lowering toxic metals concentrations of plant tissues and ameliorating their toxicities to plants' growth and yield [55–57]. These materials can be applied as soil application and foliar application to ameliorate heavy metals stress in plants. The results of Hussain et al. [51] and Khan et al. [58] showed that the application of nano-Si ameliorated Pb and Cd toxicities in rice and wheat. Based on the report by Hussain et al. [51], foliar application of nano-Si successfully reduced the Cd and Pb concentrations in grains and improved the yield.

Considering the advantages of Si, in both agriculture and environmental problems, and also a significant decrease in Si pools in agricultural soils due to the continuous removal of Si during the crop harvest, taking into account of Si application management is an undeniable fact to compensate the soil depletion and efficient use of the unique functions of this element.

13.3.1 Silicon/nano-silicon mechanisms to ameliorate potentially toxic metals stress in plants

An accurate understanding of the Si/nano-Si-induced mechanisms for toxic metals detoxification in different plant species is vital for Si applications. These mechanisms have been evaluated from different aspects over the last few decades. The nature of these mechanisms at the soil level is entirely different from their nature at plant levels. The immobilization of heavy metals by Si is the main mechanism at the soil level. While increasing the activities of antioxidant enzymes, structural modifications in different plant tissues, distribution/compartmentation of toxic metals into metabolically inactive parts, coprecipitation with toxic metals, chelation, and alteration of gene expression are the vital mechanisms of Si at plant levels (Tables 13.3 and 13.4). It will be tried to discuss the nature, properties, condition, and efficiency of these mechanisms in detail in the following sections.

13.3.1.1 Silicon: Heavy metals immobilization in soil

Silicon, through affecting the soil properties, significantly affects the bioavailability of the heavy metals [75]. Soil acidity and soil organic matter are two of the most important properties of soils that significantly affect metal bioavailability. It has been reported that Si through reducing the soil acidity, reduces the metal bioavailability [44]. Results of Li et al. [76], Rizwan et al. [75], and Lu et al. [77] showed that availability of Pb, Cd, and Cr is decreased by Si application, and they stated that reduction of soil acidity in the treated soil is the main reason. However, some studies reported a decrease in bioavailability of Cd by calcium silicate without affecting soil acidity [78,79].

The role of Si in speciation of heavy metals in soil solution was reported by Putwattana et al. [80], and the formation of silicate complexes plays the main role. Silicon application significantly reduced the proportion of exchangeable Cr in the soil by increasing the proportion of precipitation-bound and organic matter-bound Cr fraction. Also, it was reported that there is more concentration of Cd as the form of adsorbed Fe-Mn oxides bound and organic matter fractions in the soil treated with Si [81,82]. The influence of common Si and nano-Si on change of Al, Pb, and Zn fractions was also reported by Hara et al. [83] and Mousavi et al. [10,11]. Based on these research studies, more available fractions of Al, Pb, and Zn shifted to the more stable fractions with low mobility affected by Si and nano-Si applications.

Patrícia Vieira da Cunha et al. [79] reported that Cd and Zn distribution in the soil is affected by Si application, and they were observed in the more stable form such as complexed with organic matter and crystalline iron oxides. Silicon compounds immobilize heavy metals (Cu, Cd, and Zn) in multimetal contaminated soil by decreasing soil acidity. Based on the report by Gu et al. [44], these metals are mainly precipitated as their silicates, phosphates, and hydroxides in the Si treatments. Silicon accumulation in the shoots and roots protects plants against heavy metal toxicity. Emamverdian et al. [74] stated that the formation of Si-metal ion complexes may be responsible for lowering the Cd concentration in roots and leaves of bamboo seedling. Though, the adsorption of Cd ions onto SiO_2 NPs could reduce heavy metal accumulations in roots and leaves.

Based on the above studies, it is concluded that Si application through an external interaction with heavy metals, such as alteration of soil pH and heavy metals fractionation/speciation, can immobilize potentially heavy metals. However, there is still a necessity to evaluate the effect of Si nutrition on the immobilization of heavy metals from different aspects in agricultural soils.

13.3.1.2 Silicon: Improvement and promotion of the antioxidant defense system

Plants have an advanced antioxidant defense, including enzymatic and nonenzymatic antioxidant systems [84]. Antioxidants will be stimulated under heavy metals stress which is a strategy for increasing toxic metals tolerance in plants. It has been reported that antioxidant enzyme activities decrease under higher toxic metal concentrations [85]. However, some studies showed that antioxidant enzyme activities are reduced by Si application under heavy metals stress. The increasing and reducing effects of Si on antioxidant activities in some plants have been shown in Table 13.5.

TABLE 13.3 Categorization of silicon/nano-silicon-induced mechanisms for mitigating heavy metal toxicities.

Internal mechanisms	External mechanisms
• Influence on antioxidants (both enzymatic and nonenzymatic)	• Decrease in availability by lowering the growth media acidity
• Influence on transportation and accumulation of toxic metals into vacuoles	• Coprecipitation with toxic metals in the growth media
• Chelation of heavy metals in plants	• Influence on phenolics production, this compound chelates heavy metals ions in vitro
• Influence on gene expression in plants	• Influence on activities of metal carriers
• Coprecipitation of toxic metals in plants	
• Decrease or inhibition in translocation of toxic metals from root to shoot	
• Structural changes in plants	

TABLE 13.4 Common silicon and nano-silicon induced mechanisms for improving heavy metals tolerance in plants.

Heavy metal	Plant	Mechanism	Reference
Common silicon			
Pb	Rice, cotton	• Inhibiting Pb transport from roots to shoots • Increasing antioxidant enzyme activities and preventing membrane oxidative damage of plant tissue • Elevation of photosynthesis and antioxidant enzymes in cotton affected by Si suppresses Pb uptake and oxidative stress	Yamaji et al. [59], Shi et al. [21], Bharwana et al. [60]
Cd	Wheat, rice	• Formation of apoplasmic barriers in endodermis closer to the root apex • Reduction of Cd concentration in shoots by the distribution of Cd in the root cell walls • Coprecipitation with silicates in cell walls • Reduction of metal absorption and translocation	Barcelo et al. [61], Rea [62], Dragišic Maksimovic et al. [63], Huang et al. [64]
Cr	Barley	• Increase in plant height, number of tillers, root length, and leaf size	Khandekar and Leisner [65]
Al	Maize, Barley	• Formation of hydroxyaluminosilicates in the apoplast of the root apex • Deposition of Si at the epidermis	Rogalla and Romheld [66], Li et al. [67]
Zn	Maize, *Minuartiaverna*, rice	• Formation of the zinc silicates compound in the cytoplasm, which it has low solubility • Coprecipitation as Zn silicates in the cell walls • Limitation of metal uptake and transport	Collin et al. [68], Bharwana et al. [69], Huang et al. [64]
Mn	Cucumber	• Decrease in lipid peroxidation and increase in antioxidants levels • Increase in Mn in the cell wall of shoots and therefore reduce in the symplast	Ma et al. [70], Sonah et al. [71]
Cu	Flax	• Upregulation of antioxidant defense and secondary metabolites	El-Beltagi et al. [37]
Nano-silicon			
Pb	Rice	• Inhibition of Pb transfer from roots to shoots	Liu et al. [56]
Cd	Rice	• Reduction of Cd uptake and translocation from root to shoot • Increase in antioxidant capacity	Wang et al. [72]
As	Maize	• Reduction in oxidative stress and increase in components of the ascorbate-glutathione cycle (AsA-GSH cycle)	Tripathi et al. [57]
Cd	Rice	• Reduction of Cd concentrations in rice roots, shoots, and grains	Chu et al. [55]
Al	Maize	• Activation of the antioxidant defiance systems	de Sousa et al. [73]
Cd	Moso bamboo (*Phyllostachys edulis*)	• Increase in growth and germination parameters • Stimulation of the antioxidant activities • Formation of Si-metal ion complexations	Emamverdian et al. [74]

13.3.1.3 Silicon: Distribution/compartmentation of potentially toxic metals in different plant tissues

One of the vital detoxification mechanisms of toxic metals by Si is distribution/compartmentation in different plant tissues (roots, shoots, grain). Williams and Vlamis were the first scientists who reported this mechanism [96]. They observed that Mn is uniformly distributed in barley leaves instead of discrete necrotic spots, affected by Si application, and consequently, Mn toxicity was decreased. In recent years, Rizwan et al. [75] and Naeem et al. [78] observed that Si nutrition increases toxic metals concentration in plants roots and decreases their translocation to shoots and grain. A similar result was observed by Emamverdian et al. [74]. They showed that nano-Si application significantly decreased Cd concentration in the leaves of bamboo, resulting from the formation of Si-Cd complexes in the root and soil. Inhibition of Cd translocation from roots to shoot in wheat affected by nano-Si application is another case that was reported by Ali et al. [69]. Silicon deposition in endodermis may play a vital role in reducing heavy metal translocation [91]. Based on different studies [41,97–99], it is concluded that heavy metals sequestration in metabolically less active cell compartments, such as the cell walls, might be a vital mechanism to ameliorate the stress of the heavy metal in plants by Si. Some studies showed that Si-mediated alterations in heavy metals compartmentation in the root tips may also be a vital strategy for Si-induced toxic metals tolerance [100]. The Cu localization in roots of wheat under Si application was studied by Keller et al. [101]. They reported that in the control treatment (without Si), Cu was mainly concentrated in the central cylinder of the root while in the Si treatment,

TABLE 13.5 Variable responses of antioxidant defense system (stimulation or reduction), by common silicon and nano-silicon under heavy metals stress in different plants.

Plant	Heavy metal	Result	Reference
Cucumber (Negin cultivar)	Cd	Increase in CAT activity	Khodarahmi et al. [86]
Pakchoi and Rice	Cr	Increase in activities of POD, SOD, and CAT	Zeng et al. [87], Zhang et al. [88]
Cotton and Banana	Pb	Increase in activities of POD, SOD, and CAT	Bharwana [69]
Rice and Cotton	Zn and Fe	Increase in antioxidant enzyme activities	Chalmardi et al. [89], Anwaar et al. [90]
Rice, Cucumber, and Pakchoi	Mn and Cd	Increase in the activities of nonenzymatic antioxidants [glutathione (GSH), nonprotein thiols (NPT), and ascorbic acid (AsA)]	Shi et al. [91], Song et al. [92], Li et al. [76], Wang et al. [47]
Solanum nigrum L.	Cd	Decrease in SOD, CAT, POD, and APX activities	Liu et al. [93]
Rice	Cd	Decrease in CAT and POD activities	Wang et al. [47]
Peanut and Maize	Al and Zn	Decrease in CAT, SOD, and POD activities in roots	Bokor et al. [94], Shen et al. [45]
Sorghum bicolor and Rice	Zn, As	Decrease in SOD and POD activities	Hu et al. [95]
Flax	Cu	Increase in SOD, POD, APX, and CAT activities	El-Beltagi et al. [37]
Moso bamboo	Cd	Stimulation of the antioxidant activities	Emamverdian et al. [74]

Cu was mainly concentrated in the vicinity of root epidermis while Si was concentrated mainly in the endodermis irrespective Cu treatment. Other researchers reported that Zn and Si are colocalized in the cell wall of metabolically less active tissues, mostly in sclerenchyma of roots. Therefore Si localization in the root endodermis might act as a barrier to block toxic metals' entrance into cells.

13.3.1.4 Silicon: Coprecipitation with potentially toxic metals

Many researches confirm that coprecipitation of toxic metals with Si induces mitigation of toxic metals stress in plants. Cocker et al. [40] and Barcelo et al. [61] reported that Si application under Al stress resulted in the formation of aluminosilicates and hydroxyaluminosilicates in the apoplast of the plant root apex and consequently, detoxification of Al occurred. The coprecipitation of Si with Cd in the stem of rice and therefore reduction of Cd concentration in leaves was reported by Gu et al. [44]. Shi et al. [91] found that in rice roots Si and Cd were mainly precipitated in the endodermis and decreased the cell wall porosity of inner root tissues. They hypothesized the coprecipitation of Si with Cd. However, there are controversial reports in the literature related to Si and heavy metals coprecipitation in plants. The microscopic findings of Collin et al. [68] did not approve coprecipitation of Si and Cu at the root level in bamboo plants. Also, based on the report by Dresler et al. [102], there were no Si−Cd complexes in maize plants grown in hydroponics under Cd stress with increasing Si levels. Considering the above literature reviews, it is concluded that coprecipitation of Si with potentially toxic metals may not be a vital strategy for heavy metals detoxification. However, more studies may be required to clarify this strategy in plant species.

13.3.1.5 Silicon: Chelation of potentially toxic metals

The strategies by which Si alleviates toxic metals stress in plants might be due to the chelation of flavonoid phenolics or organic acids with these metals. An increase in production of these compounds affected by Si nutrition in plants under toxic metals stress was reported by Barcelo et al. [61], Keller et al. [101], and Collin et al. [68]. The findings of Keller et al. [101] showed that under high concentrations of Cu, Si application enhanced citrate, malate, and aconitate concentrations in roots of wheat seedlings compared to the control treatment (without any Si applications). They concluded that Cu could form a complex with organic acids and decrease translocation of Cu from roots to shoots affected by Si. Based on the above information, it is concluded that the production of chelates, which can be complex heavy metals, is increased by Si. It might be an important strategy that could decrease the harmful effects of potentially toxic metals in plants up to certain levels along with other strategies. Nevertheless, further researches are required to clarify the role of Si-mediated chelation of toxic metals in soils and plants.

13.3.1.6 Silicon: Correlation with gene expression

The Si-mediated alleviation of toxic metals stress is also correlated to its role in modifying gene expression. In a study, Li et al. [66] reported that Si nutrition under Cu stress activated the genes responsible for the production of metallothioneins in *Arabidopsis thaliana* that can chelate heavy metals. In maize plants under higher concentrations of Zn was also reported that Si application caused downregulation of *ZmLsi1* and *ZmLsi2* genes in roots and upregulated the expression level of *ZmLsi6* gene in the first leaf and reduced expression of *ZmLsi6* gene in the second leaf [103]. Similarly, in rice grown in contaminated soil with Cd, upregulation of the gene expression of *OsLsi1* [encoding for Si transport NIP-III (Nodulin 26-like intrinsic proteins-III) Aquaporin] and downregulation of the gene expression of *Nramp5*, a gene involved in the Cd transport, affected by Si application were reported by Ma et al. [70]. Si application was shown to significantly upregulate the expression of genes responsible for Si transport (*OsLSi1* and *OsLSi2*) and downregulate the expression of genes encoding heavy metal transporters (*OsHMA2* and *OsHMA3*) in rice plants (Kim et al., 2014a, b). Until now, the Si-mediated mechanisms for the mitigation of potentially toxic metals are less known at the genetic and molecular levels and more studies are needed to clarify linkage relationships between Si and toxic metals stress to study the expression level of genes related to transport, deposition, and translocation of heavy metals and Si in the plant.

13.3.1.7 Silicon: Structural alterations, a beneficial mechanism under toxic metals stress

Anatomical and morphological features of plants are altered by Si application that has positive effects on mitigating toxic metals stress on plants. Increase in plant height, number and size of leaves, and root length are some typical responses of plants to Si application under Pb, Cd, and Zn stress [33,60,69]. This status was studied by some researchers in different plant species (Table 13.6). Considering these literature reviews, it is concluded that structural modifications induced by Si under toxic metals stress may clarify the mitigation of the stress of the heavy metal. Though, the relevant anatomical/morphological mechanisms still remained to be studied.

TABLE 13.6 Morphological and anatomical responses of plants to common silicon/nano-silicon application under heavy metals stress.

Plant	Heavy metal	Morphological and anatomical response	Reference
Maize	Cd, Zn	Increase in xylem diameter, epidermis, mesophyll, and the transverse area of collenchymas and midvein	Patrícia Vieira da Cunha and Williams Araújo do Nascimento [104], Patrícia Vieira da Cunha et al. [79]
Maize	Mn	Increase in leaf-epidermal-layer thickness	Doncheva et al. [105]
Wheat	Cd	Formation of apoplastic barriers in the endodermis closer to the root apex	Greger et al. [106]
Rapeseed and Indian mustard	Cd	Increase in growth of suberin lamellae in the endodermis particularly near the root tips	Vatehova et al. [107]
Maize	Cd	Increase in development of Casparian bands, suberin lamellae, and root vascular tissues	Lukacova et al. [108], Vaculík et al. [99]
Maize	Cd	Increase in Casparian bands, suberin lamellae, and root vascular tissues	Vaculik et al. [108], Lukacova et al. [99]
A. *marina*	Cd	Increase in the development of the apoplasmic barrier in the roots	Zhang et al. [41]
Barley	Cr	Increase in plant height, the number of tillers, root length, and leaf size	Ali et al. [69]
Peanut	Al	Decrease in root diameter and increase in root surface area	Shen et al. [45]
Wheat	Cu	Increase in root length	Keller et al. [101]
Wheat	Cr	Increase in development of suberization and lignification and also xylem and phloem were well arranged	Tripathi et al. [109]
Moso bamboo (*Phyllostachys edulis*)	Cd	Increase in growth and germination parameters	Emamverdian et al. [74]

13.4 Conclusion

The adverse effects of heavy metals on public health have been well studied and understood by different scientists worldwide. Also, as far as was reported, different natural and anthropogenic activities have resulted in a higher concentration of these potentially toxic metals in the environment. Scientists have been studied and evaluated different strategies for ameliorating the toxicity effects of these metals on plant growth and yield, human and generally, public health, seriously since the last century. In this regard, one of the environmentally friendly methods to alleviate the adverse effects of toxic metals on plants is plants nutrition management by applying beneficial and essential nutrients in the growth media. Silicon nutrition/application has been introduced as a practical option for lowering phytotoxicity and the concentration of potentially toxic metals in plants. Numerous studies have been evaluated the useful effects of Si application, as ordinary and nano form, under heavy metals stress and have been explained its possible mechanisms for mitigating toxic metals stress from different aspects, although further experiments and evaluations are needed to validate them. Most of these studies reported that Si application as NPs is superior to common silicon to ameliorate heavy metals stress on plants. Silicon compounds increase plant tolerance to heavy metals through different mechanisms, which are varied depending on the type of the heavy metal, plant cultivar, and concentration of the metals exposed to the plant. Silicon is commonly used in the plant defense system against potentially toxic metals in two main mechanisms: (1) avoidance and (2) tolerance.

In the avoidance strategy, Si reduces heavy metal bioavailability and uptake by decreasing the soil acidity, stimulating the chelation of toxic metals with root exudates, or lowering the heavy metals translocation from root to the shoots. While, in the tolerance strategy, Si mitigates the heavy metals stress by some vital mechanisms, such as distribution/compartmentation of toxic metals into cell walls and vacuoles, an increase in the antioxidant activity (both enzyme and nonenzyme antioxidants), reduction/prevention of the heavy metals' transportation in plants, coprecipitation with heavy metals, chelation of the heavy metals in plants and consequently reduction of their translocation in plants, regulation of the gene expression, and structural alterations. The efficiency of each of these mechanisms varies depending on the type of heavy metal and plant variety as well as the level of contamination. Numerous studies claim that the application of Si-based fertilizers constitutes a promising approach to mitigate toxic metals stress, and further field studies should be performed to determine the precision of the positive Si-derived functions.

References

[1] Hosseini SA, Naseri Rad S, Ali N, Yvin JC. The ameliorative effect of silicon on maize plants grown in Mg-deficient conditions. Int J Mol Sci 2019;20:969.

[2] Akhter Bhat J, Shivaraj SM, Singh P, Navadagi DB, Tripathi DK, Dash PK, et al. Role of silicon in mitigation of heavy metal stresses in crop plants. Plants 2019;8(3):71. Available from: https://doi.org/10.3390/plants8030071.

[3] Mousavi SM, Bahmanyar MA, Pirdashti H. Lead and cadmium availability and uptake by rice plant in response to different biosolids and inorganic fertilizers. Am J Agric Biol Sci 2010;5(1):25–31.

[4] Mousavi SM, Bahmanyar MA, Pirdashti H, Gilani SS. Trace metals distribution and uptake in soil and rice grown on a 3-year vermicompost amended soil. Afr J Biotechnol 2010;9(25):3780–5.

[5] Mousavi SM, Bahmanyar MA, Pirdashti H. Phytoavailability of some micronutrients (Zn and Cu), heavy metals (Pb, Cd), and yield of rice affected by sewage sludge perennial application. Commun soil Sci plant Anal 2013;44(22):3246–58.

[6] Mousavi SM, Bahmanyar MA, Pirdashti H, Moradi S. Nutritional (Fe, Mn, Ni, and Cr) and growth responses of rice plant affected by perennial application of two bio-solids. Environ Monit Assess 2017;189(7):1–10.

[7] Mousavi SM, Moshiri F, Moradi S. Mobility of heavy metals in sandy soil after application of composts produced from maize straw, sewage sludge and biochar: discussion of Gondeketal (2018). J Environ Manag 2018;222:132–4.

[8] Kim YH, Khan AL, Waqas M, Shim JK, Kim DH, Lee KY, et al. Silicon application to rice root zone influenced the phytohormonal and antioxidant responses under salinity stress. J Plant Growth Regul 2014;33:137–49.

[9] Nagajyoti PC, Lee KD, Sreekanth T. Heavy metals, occurrence and toxicity for plants: a review. Environ Chem Lett 2010;8:199–216.

[10] Mousavi SM, Motesharezadeh B, Hosseini HM, Alikhani H, Zolfaghari AA. Root-induced changes of Zn and Pb dynamics in the rhizosphere of sunflower with different plant growth promoting treatments in a heavily contaminated soil. Ecotoxicol Environ Saf 2018;147:206–16.

[11] Mousavi SM, Motesharezadeh B, Hosseini HM, Alikhani H, Zolfaghari AA. Geochemical fractions and phytoavailability of zinc in a contaminated calcareous soil affected by biotic and abiotic amendments. Environ Geochem Health 2018;40(4):1221–35.

[12] Wang Y, Stass A, Horst WJ. Apoplastic binding of aluminum is involved in silicon-induced amelioration of aluminum toxicity in maize. Plant Physiol 2004;136:3762–70.

[13] Ma JF, Yamaji N. Silicon uptake and accumulation in higher plants. Trends Plant Sci 2006;11:392–7.

[14] Deshmukh RK, Ma JF, Bélanger RR. Role of silicon in plants. Front Plant Sci 2017. Available from: https://doi.org/10.3389/fpls.2017.01858.

[15] Deshmukh R, Bélanger RR. Molecular evolution of aquaporins and silicon influx in plants. Funct Ecol 2016;30:1277–85.

[16] Foucault Y, Lévèque T, Xiong T, Schreck E, Austruy A, Shahid M, et al. Green manure plants for remediation of soils polluted by metals and metalloids: ecotoxicity and human bioavailability assessment. Chemosphere 2013;93:1430—5.

[17] Zia-ur-Rehman M, Sabir M, Nadeem M. Remediating cadmium-contaminated soils by growing grain crops using inorganic amendments. Soil remediation and plants: prospects and challenges. Amsterdam: Elsevier Inc., Academic Press; 2015. p. 367—96.

[18] Kalaivanan D, Ganeshamurthy AN. Mechanisms of heavy metal toxicity in plants. Abiotic stress physiology of horticultural crops. Berlin: Springer. 2016. p. 85—102.

[19] Ali S, Bai P, Zeng F, Cai S, Shamsi IH, Qiu B, et al. The ecotoxicological and interactive effects of chromium and aluminum on growth, oxidative damage and antioxidant enzymes on two barley genotypes differing in Al tolerance. Environ Exp Bot 2011;70:185—91.

[20] Sheoran IS, Singal HR, Singh R. Effect of cadmium and nickel on photosynthesis and the enzymes of the photosynthetic carbon reduction cycle in pigeon pea (Cajanus cajan L.). Photosynth Res 1990;23(3):345—51.

[21] Shi Q, Bao Z, Zhu Z, He Y, Qian Q, Yu J. Silicon-mediated alleviation of Mn toxicity in Cucumis sativus in relation to activities of superoxide dismutase and ascorbate peroxidase. Phytochemistry 2005;66:1551—9.

[22] Liu J, Reid RJ, Smith FA. The mechanism of cobalt toxicity in mung beans. Physiol Plant 2000;110:104—10. Available from: https://doi.org/10.1034/j.1399-3054.2000.110114.x.

[23] Patra M, Bhowmik N, Bandopadhyay B, Sharma A. Comparison of mercury, lead and arsenic with respect togenotoxic effects on plant systems and the development of genetic tolerance. Environ Exp Bot 2004;52(3):199—223.

[24] Lin YC, Kao CH. Nickel toxicity of rice seedlings: cell wall peroxidase, lignin, and NiSO$_4$-inhibited root growth. Crop Environ Bioinf 2005;2:131—6.

[25] Jayakumar K, Jaleel CA, Vijayarengan P. Changes in growth, biochemical constituents, and antioxidant potentials in radish (Raphanus sativus L.) under cobalt stress. Turkish J Biol 2007;31(3):127—36.

[26] Li HF, Gray C, Mico C, Zhao FJ, McGrath SP. Phytotoxicity and bioavailability of cobalt to plants in a range of soils. Chemosphere 2009;75:979—86.

[27] Kabir M, Iqbal MZ, Shafiq M. Effects of lead on seedling growth of Thespesia populnea L. Adv Environ Biol 2009;3(2):184—90.

[28] Ghasemi F, Heidari R, Jameii R, Purakbar L. Effects of Ni^{2+}toxicity on Hill reaction and membrane functionality in maize. J Stress Physiol Biochem 2012;8:55—61.

[29] Nematshahi N, Lahouti M, Ganjeali A. Accumulation of chromium and its effect on growth of (Allium cepacv. Hybrid). Eur J Exp Biol 2012;2(4):969—74.

[30] Hussain A, Abbas N, Arshad F. Effects of diverse doses of lead (Pb) on different growth attributes of Zea mays L. Agric Sci 2013;4(5):262—5.

[31] Ebru OG. Nickel and cobalt effects on maize germination. In: 2nd International conference on agriculture and biotechnology, vol. 79. IACSIT Press; 2014. Available from: https://doi.org/10.7763/IPCBEE.2014.v79.12.

[32] Karmous I, Bellani LM, Chaoui A, El Ferjani E, Muccifora S. Effects of copper on reserve mobilization in embryo of Phaseolus vulgaris L. Environ. Sci Pollut Res 2015;22:10159—65. Available from: https://doi.org/10.1007/s11356-015-4208-1.

[33] Farooq MA, Ali S, Hameed A, Ishaque W, Mahmood K, Iqbal Z. Alleviation of cadmium toxicity by silicon is related to elevated photosynthesis, antioxidant enzymes; suppressed cadmium uptake and oxidative stress in cotton. Ecotoxicol Environ Saf 2013;96:242—9.

[34] Hashem HA, Shouman AI, Hassanein RA. Physico-biochemical properties of tomato (Solanum lycopersicum) grown in heavy-metal contaminated soil. Acta Agric Scand Sect B Soil Plant Sci 2017;68(4).

[35] Liu L, Li JW, Yue FX, Yan XW, Wang FY, Bloszies S, et al. Effects of arbuscular mycorrhizal inoculation and biochar amendment on maize growth, cadmium uptake and soil cadmium speciation in Cd-contaminated soil. Chemosphere 2018;194:495—503.

[36] Baruah N, Subham CM, Farooq M, Gogoi N. Influence of heavy metals on seed germination and seedling growth of wheat, pea, and tomato. Water Air Soil Pollut 2019;230:273—88.

[37] El-Beltagi HS, Sofy MR, Aldaej MI, Mohamed HI. Silicon alleviates copper toxicity in flax plants by up-regulating antioxidant defense and secondary metabolites and decreasing oxidative damage. Sustainability 2020;12:4732. Available from: https://doi.org/10.3390/su12114732.

[38] Moshiri F, Ebrahimi H, Ardakani MR, Rejali F, Mousavi SM. Biogeochemical distribution of Pb and Zn forms in two calcareous soils affected by mycorrhizal symbiosis and alfalfa rhizosphere. Ecotoxicol Environ Saf 2019;179:241—8.

[39] Sahebi M, Hanafi MM, Siti Nor Akmar A, Rafii MY, Azizi P, Tengoua F, et al. Importance of silicon and mechanisms of biosilica formation in plants. BioMed Res Int 2015. Available from: https://doi.org/10.1155/2015/396010.

[40] Cocker KM, Evans DE, Hodson MJ. The amelioration of aluminium toxicity by silicon in higher plants: solution chemistry or an in planta mechanism? Physiol Plant 1998;104:608—14.

[41] Zhang C, Wang L, Nie Q, Zhang W, Zhang F. Long-term effects of exogenous silicon on cadmium translocation and toxicity in rice (Oryza sativa L.). Environ Exp Bot 2008;62:300—7.

[42] Greger M, Landberg T, Vaculík M. Silicon influences soil availability and accumulation of mineral nutrients in various plant species. Plants 2018;7(2):41.

[43] Shi Z, Yang S, Han D, Zhou Z, Li X, Liu Y, et al. Silicon alleviates cadmium toxicity in wheat seedlings (Triticum aestivum L.) by reducing cadmium ion uptake and enhancing antioxidative capacity. Environ Sci Pollut Res 2018;25:7638—46.

[44] Gu HH, Qiu H, Tian T, Zhan SS, Deng THB, Chaney RL, et al. Mitigation effects of silicon rich amendments on heavy metal accumulation in rice (Oryza sativa L.) planted on multi-metal contaminated acidic soil. Chemosphere 2011;83:1234—40.

[45] Shen X, Xiao X, Dong Z, Chen Y. Silicon effects on antioxidative enzymes and lipid peroxidation in leaves and roots of peanut under aluminum stress. Acta Physiol Plant 2014;36:3063—9.

[46] Singh VP, Tripathi DK, Kumar D, Chauhan DK. Influence of exogenous silicon addition on aluminium tolerance in rice seedlings. Biol Trace Elem Res 2011;44:1260—74.

[47] Wang S, Wang F, Gao S. Foliar application with nano-silicon alleviates Cd toxicity in rice seedlings. Environ Sci Pollut Res 2014;1—9.

[48] Yang J, Jiang F, Ma C, Rui Y, Rui M, Adeel M, et al. Alteration of crop yield and quality of wheat upon exposure to silver nanoparticles in a life cycle study. J Agric Food Chem 2018;66:2589—97.

[49] Adeel M, Farooq T, White J, Hao Y, He Z, Ru Y. Carbon-based nanomaterials suppress Tobacco Mosaic Virus (TMV) infection and induce resistance in Nicotiana benthamian. J Hazard Mater 2020;404.

[50] Hao Y, Fang P, Ma C, White JC, Xiang Z, Wang H, et al. Engineered nanomaterials inhibit *Podosphaerapannosa* infection on rose leaves by regulating phytohormones. Environ Res 2019;170:1−6.

[51] Hussain B, Lin Q, Hamid Y, Sanaullah M, Di L, Hashmi MLUR, et al. Foliage application of selenium and silicon nanoparticles alleviates Cd and Pb toxicity in rice (*Oryza sativa* L.). Sci Total Environ 2020;712:136497.

[52] Li M, Adeel M, Peng Z, Yukui R. Physiological impacts of zero valent iron, Fe_3O_4 and Fe_2O_3 nanoparticles in rice plants and their potential as Fe fertilizers. Environ Pollut 2020;269.

[53] Liu Y, Wu T, White JC, Lin D. A new strategy using nanoscale zero-valent iron to simultaneously promote remediation and safe crop production in contaminated soil. Nat Nanotechnol 2021;16(2):1−9.

[54] Moharem M, Elkhatib E, Mesalem M. Remediation of chromium and mercury polluted calcareous soils using nanoparticles: sorption-desorption kinetics, speciation and fractionation. Environ Res 2019;170:366−73.

[55] Chu K, Liu Y, Shen R, Liu J. Effects of nano-silicon and common silicon on the growth and cadmium concentrations in different rice cultivars. J Nanosci Nanoeng 2019;5(1):7−11.

[56] Liu J, Cai H, Mei C, Wang M. Effects of nano-silicon and common silicon on lead uptake and translocation in two rice cultivars. Front Environ Sci Eng 2015;9:905−11. Available from: https://doi.org/10.1007/s11783-015-0786-x.

[57] Tripathi DK, Singh S, Singh VP, Prasad SM, Chauhan DK, Dubey NK. Silicon nanoparticles more efficiently alleviate arsenate toxicity than silicon in maize cultiver and hybrid differing in arsenate tolerance. Front Environ Sci 2016;4. Available from: https://doi.org/10.3389/fenvs.2016.00046.

[58] Khan ZS, Rizwan M, Hafeez M, Ali S, Adrees M, Qayyum MF, et al. Effects of silicon nanoparticles on growth and physiology of wheat in cadmium contaminated soil under different soil moisture levels. Environ Sci Pollut Res 2020;27:4958−68.

[59] Yamaji N, Mitatni N, Ma JF. A transporter regulating silicon distribution in rice shoots. Plant Cell 2008;20:1381−9.

[60] Bharwana S, Ali S, Farooq M, Iqbal N, Abbas F, Ahmad M. Alleviation of lead toxicity by silicon is related to elevated photosynthesis, antioxidant enzymes suppressed lead uptake and oxidative stress in cotton. J Bioremed Biodeg 2013;4(4). Available from: https://doi.org/10.4172/2155-6199.1000187.

[61] Barcelo J, Guevara P, Poschenrieder C. Silicon amelioration of aluminium toxicity in teosinte (*Zea mays* L. ssp. mexicana). Plant Soil 1993;154:249−55.

[62] Rea PA. Phytochelatin synthase: of a protease a peptide polymerase made. Physiol Plant 2012;145:154−64.

[63] Dragišic Maksimovic J, Mojovic M, Maksimovic V, Römheld V, Nikolic M. Silicon ameliorates manganese toxicity in cucumber by decreasing hydroxyl radical accumulation in the leaf apoplast. J Exp Bot 2012;63:2411−20.

[64] Huang F, Wen XH, Cai YX, Cai KZ. Silicon-mediated enhancement of heavy metal tolerance in rice at different growth stages. Int J Environ Res Public Health 2018;15:2193. Available from: https://doi.org/10.3390/ijerph15102193.

[65] Khandekar S, Leisner S. Soluble silicon modulates expression of *Arabidopsis thaliana* genes involved in copper stress. J Plant Physiol 2011;168:699−705.

[66] Li J, Leisner M, Frantz J. Alleviation of copper toxicity in *Arabidopsis thaliana* by silicon addition to hydroponic solutions. J Am Soc Hortic Sci 2008;133:670−7.

[67] Rogalla H, Römheld V. Role of leaf apoplast in silicon-mediated manganese tolerance of *Cucumis sativus* L. Plant Cell Environ 2002;25:549−55.

[68] Collin B, Doelsch E, Keller C, Cazevieille P, Tella M, Chaurand P, et al. Evidence of sulfur-bound reduced copper in bamboo exposed to high silicon and copper concentrations. Environ Pollut 2014;187:22−30.

[69] Ali S, Farooq MA, Yasmeen T, Hussain S, Arif MS, Abbas F, et al. The influence of silicon on barley growth, photosynthesis and ultrastructure under chromium stress. Ecotoxicol Environ Saf 2013;89:66−72.

[70] Ma J, Cai H, He C, Zhang W, Wang L. A hemicellulose-bound form of silicon inhibits cadmium ion uptake in rice (Oryza sativa) cells. New Phytol 2015;206:1063−74.

[71] Sonah H, Deshmukh RK, Labbé C, Bélanger RR. Analysis of aquaporins in *Brassicaceae* species reveals high-level of conservation and dynamic role against biotic and abiotic stress in canola. Sci Rep 2017;7(1).

[72] Wang S, Wang F, Gao S. Foliar application with nano-silicon alleviates Cd toxicity in rice seedlings. Environ Sci Pollut Res 2015;22(4):2837−45.

[73] de Sousa A, Saleh AM, Habeeb TH, Hassan YM, Zrieq R, Wadaan MAM, et al. Silicon dioxide nanoparticles ameliorate the phytotoxic hazards of aluminum in maize grown on acidic soil. Sci Total Environ 2019;693.

[74] Emamverdian A, Ding Y, Mokhberdoran F, Ahmad Z, Xie Y. The Effect of silicon nanoparticles on the seed germination and seedling growth of moso bamboo (*Phyllostachys edulis*) under cadmium stress. Pol J Environ Stud 2021;30(4):1−10.

[75] Rizwan M, Meunier JD, Miche H, Keller C. Effect of silicon on reducing cadmium toxicity in durum wheat (*Triticum turgidum* L. cv. Claudio W.) grown in a soil with aged contamination. J Hazard Mater 2012;209−210:326−34.

[76] Li L, Zheng C, Fu Y, Wu D, Yang X, Shen H. Silicate-mediated alleviation of Pb toxicity in banana grown in Pb-contaminated soil. Biol Trace Elem Res 2012;145:101−8.

[77] Lu HP, Zhuang P, Li ZA, Tai YP, Zou B, Li YW, et al. Contrasting effects of silicates on cadmium uptake by three dicotyledonous crops grown in contaminated soil. Environ Sci Pollut Res 2014;21:9921−30.

[78] Naeem A, Ghafoor A, Farooq M. Suppression of cadmium concentration in wheat grains by silicon is related to its application rate and cadmium accumulating abilities of cultivars. J Sci Food Agric 2014;95(12):2467−72. Available from: https://doi.org/10.1002/jsfa.6976.

[79] Patrícia Vieira da Cunha K, Williams Araújo do Nascimento C. Silicon effects on metal tolerance and structural changes in maize (*Zea mays* L.) grown on a cadmium and zinc enriched soil. Water Air Soil Pollut 2009;197(323).

[80] Putwattana N, Kruatrachue M, Pokethitiyook P, Chaiyarat R. Immobilization of cadmium in soil by cow manure and silicate fertilizer, and reduced accumulation of cadmium in sweet basil (*Ocimum basilicum*). Sci Asia 2010;36:349−54.

[81] Chen HM, Zheng CR, Tu C, She ZG. Chemical methods and phytoremediation of soil contaminated with heavy metals. Chemosphere 2000;41:229−34.

[82] Liang YC, Wong JWC, Long W. Silicon-mediated enhancement of cadmium tolerance in maize (*Zea mays* L.) grown in cadmium contaminated soil. Chemosphere 2005;58:475−83.

[83] Hara T, Gu MH, Koyama H. Ameliorative effect of silicon on aluminum injury in the rice plant. Soil Sci Plant Nutr 1999;45:929−36.

[84] Gill RA, Zang L, Ali B, Farooq MA, Cui P, Yang S, et al. Chromium-induced physiochemical and ultrastructural changes in four cultivars of Brassica napus L. Chemosphere 2015;120:154−64.

[85] Adrees M, Ali S, Rizwan M, Ibrahim M, Abbas F, Farid M, et al. The effect of excess copper on growth and physiology of important food crops: a review. Environ Sci Pollut Res 2015;22:8148−62.

[86] Khodarahmi S, Khoshgoftarmanesh AH, Mobli M. Effect of silicon nutrition on alleviating cadmium toxicity-induced damage on cucumber (Cucumis sativus L.) at vegetative stage. J Sci Technol 2012;3:103−10.

[87] Zeng FR, Zhao FS, Qiu BY, Ouyang YN, Wu FB, Zhang GP. Alleviation of chromium toxicity by silicon addition in rice plants. Agric Sci China 2011;10:1188−96.

[88] Zhang S, Li S, Ding X, Li F, Liu C, Liao X, et al. Silicon mediated the detoxification of Cr on pakchoi (Brassica chinensis L.) in Cr-contaminated soil. J Food Agric Environ 2013;11:814−19.

[89] Chalmardi ZK, Abdolzadeh A, Sadeghipour HR. Silicon nutrition potentiates the antioxidant metabolism of rice plants under iron toxicity. Acta Physiol Plant 2014;36:493−502.

[90] Anwaar SA, Ali S, Ali S, Ishaque W, Farid M, Farooq MA, et al. Silicon (Si) alleviates cotton (Gossypium hirsutum L.) from zinc (Zn) toxicity stress by limiting Zn uptake and oxidative damage. Environ Sci Pollut Res 2014;22:3441−50.

[91] Shi X, Zhang C, Wang H, Zhang F. Effect of Si on the distribution of Cd in rice seedlings. Plant Soil 2005;272:53−60.

[92] Song AL, Li ZJ, Zhang J, Xue GF, Fan FL, Liang YC. Silicon-enhanced resistance to cadmium toxicity in Brassica chinensis L. is attributed to Si suppressed cadmium uptake and transport and Si-enhanced antioxidant defense capacity. J Hazard Mater 2009;172:74−83.

[93] Liu J, Zhang H, Zhang Y, Chai T. Silicon attenuates cadmium toxicity in Solanum nigrum L. by reducing cadmium uptake and oxidative stress. Plant Physiol Biochem 2013;68:1−7.

[94] Bokor B, Bokorová S, Ondoš S, Švubová R, Lukačová Z, Hýblová M, et al. Ionome and expression level of Si transporter genes (Lsi1, Lsi2, and Lsi6) affected by Zn and Si interaction in maize. Environ Sci Pollut Res 2014;. Available from: https://doi.org/10.1007/s11356-014-3876-6.

[95] Hu H, Zhang J, Wang H, Li R, Pan F, Wu J, et al. Effect of silicate supplementation on the alleviation of arsenite toxicity in 93-11 (Oryza sativa L. indica). Environ Sci Pollut Res 2013;20:8579−89.

[96] Williams ED, Vlamis J. The effect of silicon on yield and manganese uptake and distribution in the leaves of barley plants grown in culture solutions. Plant Physiol 1957;32:404−9.

[97] Liu C, Li F, Luo C, Liu X, Wang S, Liu T, et al. Foliar application of two silica sols reduced cadmium accumulation in rice grains. J Hazard Mater 2009;161:1466−72.

[98] Maksimović JD, Mojović M, Maksimović V, Römheld V, Nikolic M. Silicon ameliorates manganese toxicity in cucumber by decreasing hydroxyl radical accumulation in the leaf apoplast. J Exp Bot 2012;63:2411−20.

[99] Vaculik M, Landberg T, Greger M, Luxova M, Stolarikova M, Lux A. Silicon modifies root anatomy, and uptake and subcellular distribution of cadmium in young maize. Ann Bot 2012;110:433−43.

[100] Zhang Q, Yan C, Liu J, Lu H, Duan H, Du J, et al. Silicon alleviation of cadmium toxicity in mangrove (Avicennia marina) in relation to cadmium compartmentation. J Plant Growth Regul 2014;33:233−42.

[101] Keller C, Rizwan M, Davidian JC, Pokrovsky OS, Bovet N, Chaurand P, et al. Effect of silicon on wheat seedlings (Triticum turgidum L.) grown in hydroponics and exposed to 0 to 30mM Cu. Planta 2015;241:847−60.

[102] Dresler S, Wójcik M, Bednarek W, Hanaka A, Tukiendorf A. The effect of silicon on maize growth under cadmium stress. Russ J Plant Physiol 2015;62:86−92.

[103] Bokor B, Vaculik M, Slováková L, Masarovič D, Lux A. Silicon does not Always mitigate zinc toxicity in maize. Acta Physiol Plant 2014;36:733−43.

[104] Patrícia Vieira da Cunha K, Williams Araújo do Nascimento C, José da Silva A. Silicon alleviates the toxicity of cadmium and zinc for maize (Zea mays L.) grown on a contaminated soil. J Plant Nutr Soil Sci 2008;171:849−53.

[105] Doncheva S, Poschenrieder C, Stoyanova Z, Georgieva K, Velichkova M, Barceló J. Silicon amelioration of manganese toxicity in Mn-sensitive and Mn-tolerant maize varieties. Environ Exp Bot 2009;65:189−97.

[106] Greger M, Landberg T, Vaculik M, Lux A. Silicon influences nutrient status in plants. In Proceedings of the 5th international conference on silicon in agriculture, Beijing, China, 13−18 September 2011. Beijing: The Organizing Committee of the 5th Silicon in Agriculture Conference:, 2011.

[107] Vatehová Z, Kollárová K, Zelko I, Richterová-Kučerová D, Bujdoš M, Lišková D. Interaction of silicon and cadmium in Brassica juncea and Brassica napus. Biologia 2012;67:498−504.

[108] Lukacova Z, Svubova R, Kohanova J, Lux A. Silicon mitigates the Cd toxicity in maize in relation to cadmium translocation, cell distribution, antioxidant enzymes stimulation and enhanced endodermal apoplasmic barrier development. Plant Growth Regul 2013;70:89−103.

[109] Tripathi DK, Singh VP, Prasad SM, Chauhan DK, Dubey NK, Rai AK. Silicon-mediated alleviation of Cr (VI) toxicity in wheat seedlings as evidenced by chlorophyll florescence, laser induced breakdown spectroscopy and anatomical changes. Ecotoxicol Environ Saf 2015;113:133−44.

14

Silicon- and nanosilicon-mediated disease resistance in crop plants

Kaisar Ahmad Bhat[1,2], *Aneesa Batool*[2,3,4], *Madeeha Mansoor*[2],
Madhiya Manzoor[2], *Zaffar Bashir*[5], *Momina Nazir*[3] and *Sajad Majeed Zargar*[2]

[1]School of Biosciences and Biotechnology, Baba Ghulam Shah Badshah University, Rajouri, India [2]Proteomics
Laboratory, Division of Plant Biotechnology, Sher-e-Kashmir University of Agricultural Sciences and Technology
Kashmir, Srinagar, India [3]Department of Chemistry, Govt. College for Women, Cluster University Srinagar, Srinagar,
India [4]Department of Chemistry, Baghwant University of Ajmer, Ajmer, India [5]Centre of Research for Development
and PG Microbiology, University of Kashmir, Srinagar, India

14.1 Introduction

Silicon (Si) is the second most prevalent element found in earth's crust measuring up to 70% of mass of soil [1]. While Si was initially anticipated as beneficial to plant development and production, it was not earlier known as an essential component for higher plants. Due to the fact that Si absorption capacity varies among plant species, it accumulates variably in different plant species. The most frequent type of Si in soil is silica (SiO_2), which is not available to plants. Si is also present in the plant accessible form as silicic acid [$Si(OH)_4$] or monosilic acid [H_4SiO_4], which is absorbed by root cells via Si-specific transporters viz. low silicon 1 (*Lsi1*), an influx transporter and low silicon 2 (*Lsi2*), a diffusion and efflux transporter [2,3]. Si in silicic acid form is absorbed by roots and is transported to shoots where it agglomerates into hydrated and amorphous silica and accumulates on the cell walls. It has been observed that various transporters help in the absorption and translocation of Si in plants (Fig. 14.1). Si is not considered as a necessary element; however, its addition in fertilizer formulations increases stress tolerance [4–6]. The quantity of Si added to soil is related to the amount of Si accessible for human consumption. Si is abundant in soil and is found in all life forms, including plants and humans. Si compounds can be found in a variety of soil fractions, including solid, liquid, and adsorbed phases (Fig. 14.1). Various research findings have established that Si accumulates in plants [7] and has a range of beneficial effects on various families like that of Gramineous plants, which include sugarcane and rice, as well as other species from Cyperaceous family. Apart from being present in cell wall, absorbed Si also has a role in stress signaling pathways [8]. As a result of Si's mechanical and physiological improvements, plants are better equipped to deal with environmental stresses [9]. Many plant species benefit from the presence of Si, which helps them resist disease and pests caused by bacteria, fungi, and pests [10]. It can also help them cope with different abiotic stresses, which include drought, salinity, and prevention of water logging and metal toxicity [11,12]. Research findings have examined how Si influences the interaction of plants with microbes [13] and how it works to increase resistance to a range of diseases caused by microbes, thus triggering defensive responses in several studies [14]. Nevertheless, the molecular foundation and mechanisms controlling Si-mediated disease resistance remain a mystery. More research is needed to understand how Si regulates plant–microbe interactions in plants. Nanosilicon, also known as SiO_2-nanoparticles (Si-NPs), is a valuable supply of Si that may be used to help plants withstand a variety of stresses [15]. These Si-NPs have been used in a variety of applications, including phytoremediation [16];

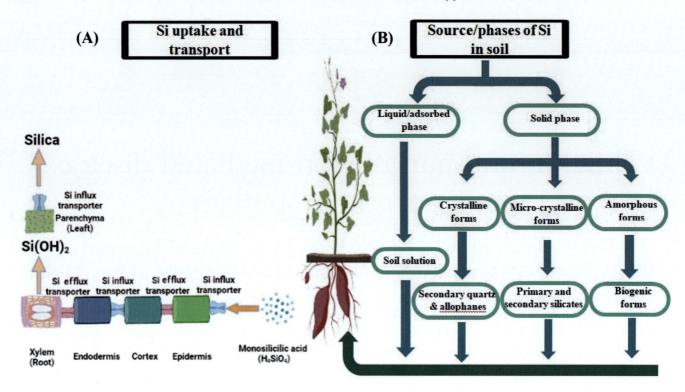

FIGURE 14.1   (A) Silicon uptake and transport in plants and (B) sources of silicon in soil.

nanofertilizer, nanopesticide, and nanoherbicide in agricultural practices [17,18]; food processing [19]; industrial applications [20]; biomedical issues [21]; biosensors [15]; and improving crop yield during various stress conditions [22]. Many studies have confirmed the role of Si nano-fertilizers in increasing biomass and productivity of various plant species like maize under salinity stress [23], rice under salinity and fluoride stress [24,25], and common bean under metal toxicity [22]. Plant diseases endanger agricultural productivity by compromising crop yield and quality. Si has been demonstrated in numerous studies to be useful in decreasing diseases caused by fungal and bacterial infections in a range of plants [26]. Si benefits plant—pathogen interactions by increasing resistance in plants to disease caused by viruses, nematodes, bacteria, and fungi. Plant defense response, changes in phytohormone homeostasis, and defense signaling networking components are all possible mechanisms triggered by Si priming. This book chapter attempts to discuss the role of Si and nano-Si to plant development and growth with a special focus on mitigating biotic stresses and to understand the mechanism for mediating disease resistance in crop plants.

14.2 Role of Si and nano-Si in mitigating plant stresses

The effect of environmental stresses both biotic and abiotic on various plant families is a major concern across the globe as it affects the plants causing loss to crops, decreasing crop yield, and overall development of plants [27]. The function of many micronutrients in mitigating various stresses in different species of plants has been recorded [28,29]. Si and applicability of different Si-NPS have shown remarkable results for mitigating various biotic as well as abiotic stress in plants and thus protecting plants from different pathogens and other adversities [30]. Several putative processes linked to Si and nano-Si have been observed under different biotic stress tolerance mechanisms in varied plant families (Table 14.1).

14.2.1 Role of Si in alleviating biotic stress

Plants show tolerance to other organisms ranging from microorganisms to mammalia in the presence of Si [26,39—42]. Deposition of Si as phytoliths increases the immunity of plants and acts as an obstacle for fungal

TABLE 14.1 Role of Si and nanosilica (Si-NPs) in biotic stress mitigation in various plant families.

Silicon				
Disease	Plant species	Family	Possible mechanism	Reference(s)
Brown spot fungus	*Oryiza sativa* (Rice)	Poaceae	Regulation of signaling pathways (ethylene signaling)	Han et al. [31]
Leaf folder	*Oryiza sativa* (Rice)	Poaceae	Increased activity of enzymes related to plant defense	Han et al. [31]
Powdery mildew	*Cucumis melo* (Melon)	Cucurbitaceae	Reduction in oxidative stress damage	Sakr [32]
Bacterial Blight	*Oryiza sativa* (Rice)	Poaceae	Enables the deposition of phenolics as well as lignin in various tissues, as well as elevated enzyme activity (PPO and PAL)	Song et al. [33], Webb et al. [34]
Fruit blotch	*Cucumis melo* (Melon)	Cucurbitaceae	Combating different disease manifestations	Ferreira et al. [35]
Rice blast	*Oryiza sativa* (Rice)	Poaceae	Regulation of phenylalanine ammonia lyase and lipoxygenase activities	Rahman et al. [36]

Nanosilica				
Disease	Size of nanosilicon (nm)	Plant species	Mode of application	Reference(s)
Papaya ring spot virus	Not defined	*Cucumber*	Soil	Elsharkawy and Mousa, [37]
Pyricularia oryzae (Blast disease)	32.5 ± 13.8	*Oryza sativa* L. (Rice)	Foliar	Tuan et al. [38]
Botrytis fabae (Chocolate spot Disease)	60	*Vicia faba* (Broad bean)	Foliar	Hasan et al. [39]
Orobanchae (Broomrape infection)	5–15	*Lycopersicon esculentum* L. (Tomato)	Seed priming	Madany et al. [40]
Colletotrichum sp. (anthracnose disease)	30–50	*Capsicum frutescens* (Chilli)	Foliar	Nguyen et al. [41]
Cucumber mosaic virus (CMV)	Not defined	*Nicotiana tabacum cv. Xanthi-nc* (Tobacco)	Soil	Elsharkawy et al. [42]
Neoscytalidium dimidiatum (Brown Spot Disease)	45	*Hylocereus undatus* (Dragon fruit)	Foliar	Tuan et al. [26]

invasion [43,44] and also shuts either outwears feeding parts of insects [45] or decreases their appetite [40,44]. The application of Si is proved to be inhibiting various diseases like that of rotting of stem leaf and neck blast, sheath blight, leaf scald and bacterial leaf blight [45], fungal disease of tomato [46], and sweet pepper [47]. In mango plants, Si depositions in epidermal tissues have been found to hinder the access of *Pseudomonas syringae* [48]. Moreover, it is ascertained that Si supplementation helps in increasing the hardness of cuticles of fruits [27] and also increased hardness in the canes protects them from the borer attack [49] and suppresses the infection by nematodes in cucumber plants [50] and also inhibits powdery mildew root rot in cucumber and wheat, leaf spotting in Bermuda grass (*Cynodon dactylon*), rust in cowpea, ring spot in sugarcane [51,52], rusting (*Hemileia vastatrix*), leaf spot (*Cercospora coffeicola*), and *Phoma ascochyta* leaf spot (*Phomatarda*) [53–57]. The studies have shown that Si supplementation also inhibits bacterial growth, for example, bacterial wilt of tomato [13]. In tobacco plants, Si has counteracted viral infection [58]. Rice blast disease that emerged in 1917 is the most worst fungal disease caused by *Magnaporthe grisea* affecting the shoots of rice plants [59]. Application of Si to soils of Florida deficient in Si content is considered as a fungicide supplement combating the rice blast [60]. Si also resists *Sphaerotheca juliginea*, which causes powdery mildew disease. In wheat and barley, the deficiency of Si makes them more prone to powdery mildew disease [61]. In cucumber, muskmelon, and grape leaves, the foliar application of Si has proved to be an efficient way of combating powdery mildew disease [62].

14.2.2 Role of nano-Si in alleviating biotic stress

Nano-Si has got antifungal and antibacterial properties, hence can combat various plant diseases [63] like resistance shown to *Aspergillus* and *Fusarium* in maize plants [64]; *Botrytis fabae* in *Vicia faba* [65], bacterial (*Pectobacterium betavasculorum*), parasitic (*Meloidogyne incognita*), and fungal (*Rhizoctonia solani*) diseases in beetroot [66]. The foliar spray of mesoporous nanosilica on tomato plants decreased their early blight [67]. Nanosilica induced the activation of salicylic acid (SA)- and jasmonic acid (JA)-sensitive genes in cucumber plants, which aided in the defense against the papaya ring spot virus (PRSV) [37] and also decrease in the tobacco mosaic virus has been observed in tobacco plants [68]. The reason behind the defensive role of nanosilica is the inducement of certain structural and some biochemical changes, which block the entry of the pathogens and hence their subsequent growth. The nano-Si supplementation brings various changes like thickening of cuticle, modifications of the primary and secondary cell wall by organic acids, and papilla formation [1]. Furthermore the bioavailability and uptake of silica are enhanced by nanosilica [69]. The hybrid forms of nanosilica or their composites have also shown alleviation of biotic stresses, for example, in watermelon against *Fusarium oxysporum* f. sp. *Niveum* [70] and decline in the extent of leaf blast of rice due to *Pyricularia oryzae* and bacterial blight in rice due to *Xanthomonas oryzae* [38].

14.3 Disease resistance modulation by Si

Different diseases attacking crop plants cause biotic stresses, which are mainly caused by bacteria, fungi, yeast, nematodes, insects, or arachnids. Plants have a defense system that helps them to deal with biotic stress. Plants enhance cell lignification in response to pathogen attack. This defense mechanism prevents the invasion of parasites, thus lowering the vulnerability of the host. The different examples of biotic stress defense include morphological barriers, chemical substances, proteins, and enzymes particularly pathogen-related (PR) proteins [71]. By preserving products and giving them strength and stiffness, these defense strategies confer tolerance to biotic stressors. Many of these studies have been compiled to validate the mechanism of action of Si in alleviating biotic stress (Fig. 14.2). In biotic stress signaling, a number of signal transduction pathways, plant hormones, and different transcription factors play an important role against stress mitigation. Si plays an important role in mediating disease resistance and mitigating stress in a number of ways and mechanisms.

14.3.1 Physical mechanisms

The positive role of Si on the growth of plants is ascribed to increased mechanical strength and formation of an exterior protective layer [72]. Many studies have been published on the defense mechanism of action of Si in plants against a variety of biotic stress conditions. For efficiently infecting the host plants, pathogens must overcome physical barriers like cell wall wax and cuticle [73,74]. Si resistance is allied with the formation of the papilla, double cuticular layer, silica thick layer just beneath the cuticle, the thickened membrane of Si and cellulose, and complexes in epidermal cells formed with organic compounds in cell walls that make plants stronger [75]. Physical obstacles keep pathogens out and protect plant cells from enzymatic damage caused by fungal pathogens [76]. In order to limit pathogen infiltration and disease occurrence, Si can create a cuticle-Si double layer beneath the cuticle [9,77]. The mechanical strength and enhancement of regeneration in cell wall is promoted by cross-linking of Si with hemi-cellulose [78,79]. The constituents of cell wall like pectins and polyphenols interact with Si to increase the cell wall flexibility during initial cell wall extension development. The severity of blast disease is lessened in rice after the epidermal cell wall is strengthened with silica [43]. The use of Si therapy on wheat leaves infested with *P. oryzae* reduced hyphael invasion of the first invading epidermal cell, whereas no Si treatment resulted in hyphae invading several adjoining leaf cells [80]. In the wheat (*Bipolaris sorokiniana*) pathosystem, the pathogen invasion into cells of epidermis and fungal colonization of foliar tissue decreased with increased Si supply [81]. Si deposition on tissue surfaces of rice diseased with *R. solani* and *Pyricularia grisea* was associated with a decrease in leaf blade at longer incubation time [82]. Furthermore, rice treated with Si showed a lower frequency of efficient penetrative appressorial sites for *P. oryzae*, suggesting that

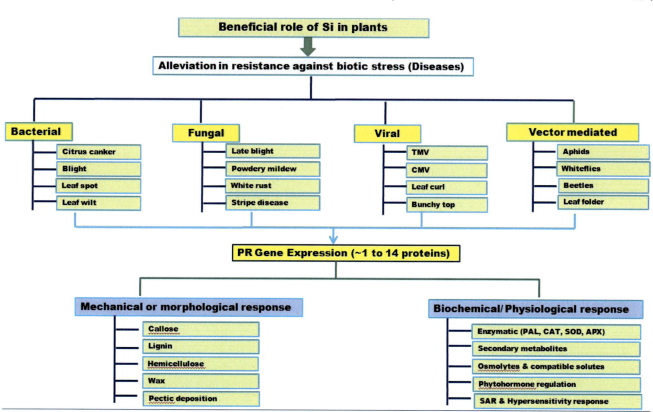

FIGURE 14.2 Schematic representation of the positive impact of silicon on plants in alleviating biotic stress.

the denser Si layer aided in pathogen delaying or inhibition [83]. In addition to strengthening cell wall, Si has been found to promote the formation of papillae upon pathogen infestation. The pathogen resistance in haustorial neck, collar regions, and papillae of fungi was aided by the deposition of Si [84]. Zeyen and coworkers (1993) [85] confirmed that under Si treatment, the epidermal cells of barley in response to *Blumeria graminis* infection can produce papillae. In reaction to *Podosphaera pannosa* infection, augmenting Si supply improved the papillae number in leaf cells, culminating in a similar consequence [86]. Si absorption into root symplasts is responsible for Si's capacity to control fungal infection in root apices. Furthermore, *Pythium aphanidermatum* growth in tomato or bitter gourd roots is not hampered by the buildup of Si on root cell walls [87]. Treatment of roots with Si confers systemic resistance on cucumber plants, and Si foliar spray confers physical barrier and osmotic resistance on cucumber plants [88]. The presence of Si in the cell wall, cuticle, wax, and papillae contributes to a partial increase in physical resistance to pathogen entry. Nevertheless, biochemical pathogen resistance, as regulated by Si, has been suggested to be more complex than physical resistance solely.

14.3.2 Si-mediated biochemical resistance mechanism

Si-mediated biochemical resistance is linked to a variety of factors, including (1) a spike in the functioning of various enzymes like polyphenol oxidase (PPO), phenylalanine ammonia lyase (PAL) gene, and peroxidase (POD), which form an important part of defense strategy in plants; (2) recruitment of antimicrobial compounds like flavonoids, phenolics, and PR proteins in proteins; and (3) regulation of various signaling pathways, which include JA pathway, SA pathway, and ET pathway [76,89].

14.3.2.1 Pathogen-related enzymes and antimicrobial defense response

Disease resistance has been demonstrated to be strongly associated with defense-related enzymes, and during plant—pathogen interactions, Si is believed to increase the action of these enzymes [76]. Si plays a pivotal role in disease resistance by triggering these enzymes such as superoxide dismutase, PAL, PODs, 1,3-glucanase, glutathione reductase, glucanase, catalase, and lipoxygenase [90]. The net content of soluble phenolics and derivatives of lignin thioglycol acid surged in the leaves of coffee and banana plants after Si incorporation [89]. Another oxidative enzyme of phenolic nature viz, PPO, is involved in the synthesis of lignin, which enhances antibacterial activity [90] and thus increases disease resistance of plants [33]. The administration of Si is thought to upregulate the activity of POD and chitinase (CHT). The former is engaged in the strengthening and cross-linking of the cell wall, whereas CHT has a role in cell wall hydrolysis of several plant pathogenic fungi, thereby participating in host—pathogen interactions [91]. Si-induced defense associated enzyme activities may also regulate gene expression encoding enzymes as is the case with the expression pattern of genes encoding for lipoxygenase (LOXa), PALa, and PALb, which substantially increased in Si-treated perennial ryegrass plants, affiliated with gray leaf spot suppression [36]. Application of Si may knock up functioning of polyphenol and oxidase POD (kind of defense-related enzymes) in rice by increasing JA-inducible herbivory responses [14]. The studies on the pathosystems of pea, cucumber, melon, wheat, rice, and soya bean revealed the positive effects of Si on reducing pathogen infections through overexpression of defense-related enzymes.

Antimicrobial compound production and accrual in plants following pathogen infiltration are an important response to enzymes for defense. Following the application of Si, there was a decrease in disease prevalence in plants due to an increase in activity of enzymes for defense, which elicit different antimicrobial compounds in plants like that of flavonoids, phytoalexins, phenols, and PR proteins [92]. As a result, it leads to an increase in defense-related enzyme activity. PAL and PPO produce lignin-associated compounds of polyphenolic nature or antimicrobial phenols during invasion of pathogens [36]. Si induction increases PAL activity, which is considered to be responsible for enhanced lignin and flavonoid production [93]. Secondary phenolic and lignin metabolism contribute to plant-mediated disease resistance because Si is engaged in the metabolism of phenolic compounds and the synthesis of lignin in plant cell walls [10]. Furthermore it also boosts plant resistance to blast disease by increasing the amount of lignin—carbohydrate complexes and lignin in rice epidermal cell walls. Plant disease resistance is induced by an overall increase in the amount of soluble phenolics in host plants [89,92]. Flavonoids are another phenolic molecule generated by Si that aid rose plants in their battle against *P. pannosa* and wheat in its fight against *P. oryzae* [86,94]. Phytoalexins are widely known to have a critical role in defense of plants against pathogen incursion by boosting phytoalexin production in powdery mildew occurred due to *Podosphaera xanthii* in cucumber plants, and rice blast manifested by *M. grisea* can be decreased. Infestation with *P. xanthii* has been linked to an increase in flavonoid phytoalexin buildup in cucumber plant species. The synthesis of phytoalexins in rice after boosting Si exhibited similar results [95,96]. Increased expression of PAL and LOX genes as well as chlorogenic acid and flavonoids were associated with increased resistance to gray leaf spot disease in perennial ryegrass (*Magnaporthe oryzae*) pathosystems [36].

14.3.2.2 *Plant systemic signal transduction*

Plants have systemic signaling networks that permit the translation of perceptions of local stressors into plant-wide responses. Even though the information may transmit by various substances including hormones and RNAs that move through the bulk flow or transpiration stream in phloem. These signals prompt distant reactions that lead the plant's unchallenged tissues for a more effective defense or stress response. Acquired acclimatization and stress-induced systemic signaling are required for plant viability during environmental stress events. To adapt to adverse environmental circumstances, plants developed sophisticated sensing, greater signaling, and enhanced acclimation systems that enable them to thrive under various stress situations [97]. The success of the process of acclimatization of plants to stress conditions, on the other hand, needs an efficient, rapid, and coordinated response involving the majority of the plant's components and tissues [98]. Plants have evolved various systemic signaling channels that allow them to transmit different stress signals from the plant part (local tissue) that senses the stress to the entire plant (systemic tissue) within few minutes [98—101]. Whenever plant's systemic tissues receive these signals, they initiate a process called systemic acquired acclimation, which permits these tissues to resist stress even though they have not experienced it earlier [99,101,102]. To defend themselves against plant invasion, plants have evolved a sophisticated immune system with numerous layers of constitutive and induced defensive mechanisms controlled by a complex network of signal transduction channels [103]. JA, SA, and ET all play critical roles in immunity boosting networks and are implicated in regulating plant defensive responses [104]. SA is mostly involved in the fight against infections of biotrophic and hemibiotrophic nature. JA and ET, on the other hand, are largely used to combat necrotrophic diseases [105]. Various studies have shown that Si regulates hormonal balance and signaling molecules, among other processes, to impact stress response in plants [13,42,106,107]. Phytohormones rack up in plants treated with Si in reaction to incursion of pathogen, cutting, or herbivore attack [14,108,109]. For example, Si-mediated defense against insect in rice plants via accumulation of JA and regulated wound-induced biosynthesis of JA [14,108]. The synthesis of SA, JA, and ET was increased in leaves of *Arabidopsis* treated with Si and infested with powdery mildew pathogens, resulting in better resistance than control plants [109]. Additionally, infection of *Ralstonia solanacearum* in *Lysopercicum* plants demonstrated that Si activates the ET and JA signaling pathways [13,110]. In rice infected with *M. oryzae*, the stimulatory effects of Si on the JA and ET signaling pathways demonstrate that the Si-mediated signaling system is crucial for rice blast resistance [106,111,112]. According to Van and coworkers, resistance to *Cochliobolus miyabeanus* in rice after induction of Si is regulated by fungal ET activation rather than SA or JA [113]. Despite the fact that Si increases the gene expression during SA pathway enzymes, resistant phenotypes generate less SA and express fewer defense genes than vulnerable controls, indicating that Si-induced resistance is mediated by mechanisms which are different from SA-dependent defense responses [114]. Nonexpressor of pathogenesis-related genes 1 (NPR1) is required to stimulate expression of PR gene in response to SA, and SA-inducible WRKY transcription factors positively control NPR1. In response to Si, transcription factor WRKY1 was enhanced during *R. solanacearum* infection in tomato [13]. SA-dependent pathway members are characterized by a rise in endogenous SA levels and the consequent synthesis of PRs is activated in response to silica exposure [115]. Si promotes silicification of leaves and phytolith-bearing silica cells, as well as the maturation of defense-related enzymes and proteins, along with transcriptional encoding proteins involved in JA signaling [14,109]. In transgenic events stifling the expression of either coronatine insensitive 1 (OsCOI1) or allene oxide synthase, significant decreases in Si deposition and an apparent loss of Si-induced leaf folder (LF) resistance were detected, both of which are implicated in JA biosynthesis, during LF caterpillar attack on rice plants [14]. The protein ligase enzyme (ubiquitin) is considered to have a role in upgrading the overall response of JA. Silencing ubiquitin-protein ligase might help plants to establish their defensive response signals after pathogen invasion [116]. Ethylene marker genes include *TSRF1*, *JERF3*, and *ACCO*, in which *TSRF1* acts as an ethylene-responsive transcription factor and *JERF3* is a transcription factor that responds to ET and JA signaling [117]. The Si-mediated resistance through ET and JA signaling pathways offered increased expression of *TSRF1*, *JERF3*, and *ACCO* genes when tomato plants were challenged with *R. solanacearum* [13]. When a pathogen is detected, ET and JA interact with other proteins during pathogen manifestation in order to monitor the expression of genes involved in defense, for example, PDF1.2. After infection with *Botrytis cinerea*, Si increased PDF1.2 expression in *Arabidopsis*, thus acting to regulate signaling pathways in response to fungal attack [105,118]. Silver thiosulfate that usually works as Si and ET signaling blocker had very little cumulative effect on pathosystems of rice—*C. miyabeanus*, showing that Si particularly targets the ET signaling pathway to defend resistance [113]. Three different types of active defensive mechanisms have been discovered during application of Si regulation of plant—pathogen interactions. The first response takes place in pathogen-infected cells; the secondary response is regulated by inducers and is restricted to cells at the infection site, but the systemic acquired response is hormonally translocated to all infected plant tissues nearby [69].

14.3.3 Gene alteration (molecular mechanisms)

Cooke and Leishman performed a meta-analysis on Si's ability to alleviate stress and observed that the bulk of research has relied on one species and one stress condition [119]. Numerous mechanisms have been reported as contributing to the mitigation of stress in higher plants, including better structural texture and photosynthetic rate [36,120—122], improved stomatal conductance shifts, and increased water uptake efficiency [40,123]. Among the several methods through which Si alleviates stress, one of the most important stress-alleviating techniques is to enhance their photosynthetic rate by plants. The main enzymes in electron transport chain (ETC) and Calvin Cycle are distressed by oxidative stress, which has a negative effect on photosynthesis [120]. Si functions as a major stimulus, regulating the activity of intracellular signaling networks that control the expression of cell wall structural modification—related defense genes, hormone synthesis, hypersensitivity responses, PR proteins, and antimicrobial compounds synthesis [124]. The defensive responses in various pathosystems by Si have been characterized by different transcriptomic and proteomic analysis [125—127]. Si increases gene expression of defense and stress-related responses, including ferritin, disease resistance response protein, WRKY1 transcription factor, trehalose phosphatase, and late embryogenesis abundant protein, which may offer tomato resistance to *R. solanacearum* [13]. Si supplementation during normal conditions lowered the expression of key housekeeping genes in rice but augmented the expression of housekeeping genes to sustain cellular functioning under pathogen attack [106]. Likewise, Ghareeb et al. reported an increase in the expression of actin, TBA, and phosphoglycerate kinase (housekeeping genes) in tomatoes infested with *R solanacearum* [13]. Jarosch and coworkers (2005) asserted that the actin cytoskeleton conferred basal resistance to *R. solanacearum*. As a result, the upregulation of actin in plants of tomato resulted in the development of host resistance after Si treatment [13]. Aquaporins are central to the transport of proteins that regulate the absorption and transport of water molecules through cell membranes, especially in the presence of abiotic stress. Nevertheless, numerous variables like abscisic acid, calcium ion concentration, free radicals, and ethylene have been involved in the activity of aquaporin [128]. Even in water stress conditions, the number of aquaporin genes in the roots regulates the plants' water absorption. After inoculation with *R. solanacearum*, tomato stems treated with rhizobacteria or Si showed strong effects, with the majority of upregulated genes associated with transduction, protein synthesis, metabolism, and defense. A large number of downregulated genes, on the other hand, were linked to photosynthesis, lipid metabolism, and transcription [50,115]. Si can reverse several transcriptional alterations exacerbated by pathogen infection. For example, when *Arabidopsis* is afflicted with the *Erysiphe cichoracearum* (fungi), the expression of nearly 4000 genes is altered. However, the number of upregulated genes implicated in defense remains unchanged over Si treatment, but there was a decrease in the number of downregulated genes involved in primary metabolism compared with control [109]. Around 900 genes responsive to pathogen infection were damaged in control plants of wheat plants attacked by *B. graminis*, but the pathogen affected just a few genes in Si-treated plants, implying that Si nearly or completely reduced the stress produced by invasion of pathogens [127]. Brunings et al. (2009) found that Si inoculation had a similar effect on the transcriptome of rice after *M. oryzae* inoculation. Si seems to reduce the impact of pathogen invasion on host plant transcriptomes, most likely by preventing pathogen virulence factors from being activated, as well as increasing resistance through reprogramming of transcriptional genes involved in defense [113].

14.3.4 Nanosilicon mediated mechanisms for disease resistance

The application of nanosilica has also shown resistance against various biotic stresses, which is due to the alleviated concentration of phenols and some protective enzymes [64,129]. Nanosilica leads to various physical changes by elevating the bioavailability of silica like the thickening of cuticle and modification of primary and secondary cell wall, which ultimately exist as barriers in the path of pathogen entry [1]. In beetroot, nanosilica has prevented the fungal, bacterial, and parasitic growth by restricting the entry of pathogens [66]. The increment in the activity of PODs and PPOs after applying nanosilica has declined the growth of *Cephalosporium maydis* in maize plants. The nanosilica supplementation has decreased the serious effects of chocolate spot disease in *V. faba* owing to the enhanced activity of antioxidant enzymes, phenols and proline [65]. The studies have shown that nanosilicon supplementation has improved the action of protective enzymes like PAL and polyphenol which have increased protection against the phythogens [130,131]. The release of enzymatic and nonenzymatic oxidants as well as an increase in photosynthesis due to Si-NPs in tomato plants protected them against the stress caused by *Orobanche* [132]. The disintegration of cell wall and disruption of ETC of fungus *A. solani* in tomato plants have been studied following foliar application of mesoporous nanosilica [67]. The disruption of DNA copying in pathogens can also be impacted by nanosilica [67]. Cucumber plants could show resistance against the PRSV due to nanosilica bringing SA and JA reactive genes into play [37]. In *Arabidopsis*, the proteins (PR1, PR2, and PR5) linked with invoking resistance against pathogens have shown excellent expression incited by nanosized silver silica [128]. In the root system of maize, nanosilica has increased the concentration of phenols during biotic stress in comparison to silica in bulk [133].

14.4 Conclusion and future perspective

Since the designation of Si as quasielement, great strides in the research of plant Si have taken place. As this element is excluded from most of the formulation of plant growth media, but its beneficial effects on prominent agricultural crops are proved day by day. The urgency of crop improvement by Si application is due to the modern-day threats faced by the plants viz, population growth rate, climate change, and increasing biotic and abiotic stresses. With the advent of nano-biotechnology, Si-NPs have proved to be another cap to stress mitigation in both biotic and abiotic stresses. Based on the current research a vast array of functions attributed to Si and nano-Si has been reported in combating different levels of biotic stress. Si and nanosilica regulate genes involved in the pathogen interaction, stress bursting pathways, photosynthesis, polyamine biosynthesis, and secondary metabolism.

Still there is a need to explore the biochemical mechanisms and signaling pathways involved in the alleviation of biotic stresses by silica nanoparticles and how these pathways or signals differ from the bulk form of silica. While these benefits are strictly counted in terms of enhancement during stress but the growing research indicates the role of Si in improving growth under relatively benign conditions. Under stressful environment circumstances, the physical, biochemical, and molecular changes induced by Si supplementation correspond to an overall increase in plant growth and stress recovery. The cellular, metabolic, and physiological activities of plants are influenced by Si application. The antioxidant machinery, ion homeostasis, and heavy metal chelation are among the primary changes induced by Si. It also affects the plant cell wall and controls the gene expression in a variety of pathways, enhancing the mechanism for tolerating the stress. Si has been found to stimulate ET, JA, and SA signaling in response to biotic signals. All of these processes, however, are influenced by plant species, genotypes, type of stress, and growth circumstances. With the emergence of omics technology, the study on the molecular pathways of biotic and abiotic stress tolerance mediated by Si has been expanded to systemic levels, such as the transcriptomic and proteomic levels. Understanding these mechanisms should serve as the foundation for our future insights, which will expand our understanding of plant adaptability to adverse environmental circumstances and may be exploited to boost agricultural output. To assess the function of Si in plant biology and get in-depth information to know and understand the mechanisms of Si-mediated stress alleviation, a systematic study is required that could help to design agricultural systems that are both cost-effective and ecologically friendly.

References

[1] Wang M, Gao L, Dong S, Sun Y, Shen Q, Guo S. Role of Si on plant—pathogen interactions. Front Plant Sci 2017;8. Available from: https://doi.org/10.3389/fpls.2017.00701.

[2] Katz O, Puppe D, Kaczorek D, Prakash N, Schaller J. Silicon in the soil—plant continuum: intricate feedback mechanisms within ecosystems. Plants 2021;10(4):652 10.3390/plants10040652.

[3] Ma JF, Tamai K, Yamaji N, Mitani N, Konishi S, Katsuhara M, et al. A silicon transporter in rice. Nature. 2006;440:688—91.

[4] Luyckx M, Hausman JF, Lutts S, Guerriero G. Silicon and plants: current knowledge and technological perspectives. Front Plant Sci 2017;8:411.

[5] Rizwan M, Ali S, Ibrahim M, Farid M, Adrees M, Bharwana SA. Mechanisms of silicon-mediated alleviation of drought and salt stress in plants: a review. Environ Sci Pollut 2015;22:15416—31. Available from: https://doi.org/10.1007/s11356-015-5305-x 2015 Res.

[6] Adrees M, Ali S, Rizwan M, Zia-ur-Rehman M, Ibrahim M, Abbas F, et al. Mechanisms of silicon-mediated alleviation of heavy metal toxicity in plants: a review. Ecotoxicol Environ Saf 2015;119:186—97.

[7] Liang YC, Wong JW, Wei L. Silicon-mediated enhancement of cadmium tolerance in maize (Zea mays L.) grown in cadmium contaminated soil. Chemosphere. 2005;5:475—83. Available from: https://doi.org/10.1016/j.chemosphere.2004.09.034.

[8] Rajput VD, Minkina T, Feizi M, Kumari A, Khan M, Mandzhieva S, et al. Effects of silicon and silicon-based nanoparticles on rhizosphere microbiome, plant stress and growth. Biology. 2021;10(8):791.

[9] Ma JF, Yamaji N. Silicon uptake and accumulation in higher plants. Trends Plant Sci 2006;11(8):392—7.

[10] Marschner P. Marschner's mineral nutrition of higher plants. London: Academic Press; 2012.

[11] Coskun D, Britto DT, Huynh WQ, Kronzucker HJ. The role of silicon in higher plants under salinity and drought stress. Front Plant Sci 2016;7:1072.

[12] Liu P, Yin L, Deng X, Wang S, Tanaka K, Zhang S. Aquaporin-mediated increase in root hydraulic conductance is involved in silicon-induced improved root water uptake under osmotic stress in Sorghum bicolor L. J Exp Bot 2014;65(17):4747—56.

[13] Ghareeb H, Bozso Z, Ott PG, Repenning C, Stahl F, Wydra K. Transcriptome of silicon-induced resistance against Ralstonia solanacearumin the silicon non-accumulator tomato implicates priming effect. Physiol Mol Plant Pathol 2011;75(3):83—9.

[14] Ye M, Song YY, Long J, Wang RL, Baerson SR, Pan ZQ, et al. Priming of jasmonate-mediated antiherbivore defense responses in rice by silicon. Proc Natl Acad Sci USA 2013;110(3631—3639) 10.1073/pnas.1305848110.

[15] Cai K, Gao D, Luo S, Zeng R, Yang J, Zhu X. Physiological and cytological mechanisms of silicon-induced resistance in rice against blast disease. Physiol Plant 2008;134:324—33. Available from: https://doi.org/10.1111/j.1399-3054.2008.01140.x.

[16] Zuo R, Liu H, Xi Y, Gu Y, Ren D, Yaun X, et al. Nano-SiO$_2$ combined with a surfactant enhanced phenanthrene phytoremediation by Erigeron annuus (L.) Pers. Environ Sci Pollut Res 2020;27(16):20538—2054.

[17] El-Nagga ME, Abdelsalam NR, Fouda MM, et al. Soil application of nano silica on maize yield and its insecticidal activity against some stored insects after the post-harvest. Nanomaterials 2020;10:739. Available from: https://doi.org/10.3390/nano10040739.

[18] Peerzada JG, Chidambaram R. A statistical approach for biogenic synthesis of nano-silica from different agro-wastes. Silicon 2021;13. Available from: https://doi.org/10.1007/s12633-020-00629-5.

[19] Mittal D, Kaur G, Singh P, Yadav K, et al. Nanoparticle-based sustainable agriculture and food science: recent advances and future outlook. Front Nanotechnol 2020;2:579954. Available from: https://doi.org/10.3389/fnano.2020.579954.

[20] AlKhatib A, Maslehuddin M, Al-Dulaijan SU. Development of high-performance concrete using industrial waste materials and nano-silica. J Mater Res Technol 2020;9(3):6696—711. Available from: https://doi.org/10.1016/j.jmrt.2020.04.067.

[21] Selvarajan V, Obuobi SE. Silica nanoparticles: a versatile tool for the treatment of bacterial infections. Front Chem 2020;8:602. Available from: https://doi.org/10.3389/fchem.2020.00602.

[22] El-Saadony MT, Desoky EM, Saad AM, et al. Biological silicon nanoparticles improve Phaseolus vulgaris L. yield and minimize its contaminant contents on a heavy metals-contaminated saline soil. J Environ Sci (China) 2021;106:1—14. Available from: https://doi.org/10.1016/j.jes.2021.01.012.

[23] Naguib DM, Abdalla H. Metabolic status during germination of nano silica primed *Zea mays* seeds under salinity stress. J Crop Sci Biotechnol 2019;22(5):415–23. Available from: https://doi.org/10.1007/s12892-019-0168-0.

[24] Banerjee A, Singh A, Sudarshan M, et al. Silicon nanoparticle-pulsing mitigates fluoride stress in rice by finetuning the ionomic and metabolomic balance and refining agronomic traits. Phytology 2021;221:67–85.

[25] Abdul-Haliem MEF, Hegazy HS, Hassan NS, et al. Effect of silica ions and nano silica on rice plants under salinity stress. Ecol Eng 2020;99:282–9.

[26] Rodrigues FA, Dallagnol LJ, Duarte HSS, Datnoff LE. Silicon control of foliar diseases in monocots and dicots. In: Rodrigues FA, DatnoffL E, editors. Silicon and plant diseases. Cham: Springer; 2015. p. 67–108.

[27] Zargar SM, Mahajan R, Bhat JA, Nazir M, Deshmukh R. Role of silicon in plant stress tolerance: opportunities to achieve a sustainable cropping system. 3 Biotech 2019;9:73. Available from: https://doi.org/10.1007/s13205-019-1613-z.

[28] Bradacova K, Weber NF, Morad-Talab N, Asim M, Imran M, Weinmann M, et al. Micronutrients (Zn/Mn), seaweed extracts, and plant growth-promoting bacteria as cold-stress protectants in maize. Chem Biol Technol Agric 2016;3:19.

[29] Vanderschuren H, Boycheva S, Li KT, Szydlowski N, GruissemW, Fitzpatrick TB. Strategies for vitamin B6 biofortification of plants: a dual role as a micronutrient and a stress protectant. Front Plant Sci 2013;4:143.

[30] Ma JF. Role of silicon in enhancing the resistance of plants to biotic and abiotic stresses. Soil Sci Plant Nutr 2004;5:11–18.

[31] Han Y, Li P, Gong S, Yang L, Wen L, Hou M. Defense responses in rice induced by silicon amendment against infestation by the leaf folder *Cnaphalocrocis medinalis*. PLoS One 2016;11(4):e0153918. Available from: https://doi.org/10.1371/journal.pone.0153918.

[32] Sakr N. The role of silicon (Si) in increasing plant resistance against fungal diseases. Hell Plant Prot J 2016;9(1):1–15. Available from: https://doi.org/10.1515/hppj-2016-0001.

[33] Song A, Xue G, Cui P, Fan F, Liu H, Chang Y, et al. The role of silicon in enhancing resistance to bacterial blight of hydroponic- and soil-cultured rice. Sci Rep 2016;62:4640. Available from: https://doi.org/10.1038/srep24640.

[34] Webb KM, Garcia E, Cruz CMV, Leach JE. Influence of rice development on the function of bacterial blight resistance genes. Eur J Plant Pathol 2010;128:399–407 https://.org/10.1007/s10658-010-9668-z.

[35] Ferreira HA, Nascimento CWA, Datnoff LE, Nunes GHS, Preston W, Souza EB, et al. Effects of silicon on resistance to bacterial fruit blotch and growth of melon. Crop Prot 2015;1807–8622. Available from: https://doi.org/10.1016/j.cropro.2015.09.025.

[36] Rahman A, Wallis CM, Uddin W. Si-induced systemic defense responses in perennial ryegrass against infection by *Magnaporthe oryzae*. Phytopathology 2015;105:748–57.

[37] Elsharkawy MM, Mousa KM. Induction of systemic resistance against 816 Papaya ring spot virus (PRSV) and its vector *Myzus persicae* by *Penicillium simplicissimum* GP17-2 and silica (SiO$_2$) nanopowder. Int J Pest Manag 2015;61:353–8. Available from: https://doi.org/10.1080/09670874.2015.1070930.

[38] Tuan LNA, Du BD, Ha LD, Dzung LTK, Van Phu D, Hien NQ. Induction of chitinase and brown spot disease resistance by oligochitosan-nand nanosilica–oligochitosan in dragon fruit plants. Agric Res 2019;8:184–90. Available from: https://doi.org/10.1007/s40003-018-0384-9.

[39] Farooq MA, Dietz KJ. Silicon as versatile player in plant and human biology: overlooked and poorly understood. Front Plant Sci 2015;6:994. Available from: https://doi.org/10.3389/fpls.2015.00994.

[40] Frew A, Weston LA, Reynolds OL, Gurr GM. The role of Si in plant biology: a paradigm shift in research approach. Ann Bot 2018;121 (7):1265–73. Available from: https://doi.org/10.1093/aob/mcy009.

[41] Nguyen NT, Nguyen DH, Pham DD, Dang VP, Nguyen QH, Hoang DQ. New oligochitosan-nanosilica hybrid materials: preparation and application on chili plants for resistance to anthracnose disease and growth enhancement. Polym J 2017;49:861–9. Available from: https://doi.org/10.1038/pj.2017.58.

[42] Reynolds OL, Padula MP, Zen RS, Gurr GM. Silicon: potential to promote direct and indirect effects on plant defense against arthropod pests in agriculture. Front Plant Sci 2016;7:744. Available from: https://doi.org/10.3389/fpls.2016.00744.

[43] Kim SG, Kim KW, Park EW, Choi D. Silicon-induced cell wall fortification of rice leaves: a possible cellular mechanism of enhanced host resistance to blast. Phytopathology 2002;92:1095–103. Available from: https://doi.org/10.1094/PHYTO.2002.92.10.1095.

[44] Massey FP, Ennos AR, Hartley SE. Grasses and the resource availabilityhypothesis: the importance of silica-based defences. J Ecol 2007;95:414–24.

[45] Jeer M, Telugu UM, Voleti SR, Padmakumari A. Soil application of siliconreduces yellow stem borer, *Scirpophaga incertulas* (Walker) damage in rice. J Appl Entomol 2017;141:189–201.

[46] Somapala K, Weerahewa D, Thrikawala S. Silicon rich rice hull amended soil enhances anthracnose resistance in tomato. Proc Food Sci 2016;6:190–3.

[47] Jayawardana HK, Weerahewa HLD, Saparamadu MD. Effect of root or foliar application of soluble silicon on plant growth, fruit quality and anthracnose development of capsicum. Trop Agric Res 2014;26(1):74–81.

[48] Frew A, Powell JR, Sallam N, Allsopp PG, Johnson SN. Trade-offs between silicon and phenolic defenses may explain enhanced performance of root herbivores on phenolic-rich plants. J Chem Ecol 2016;42:768–71.

[49] Rao SDV. Hardness of sugarcane varieties in relation to shoot borer infestation. Andhra Agric J 1967;14:99–105.

[50] Silva RV, Oliveria RDL, Nascimento KJT, Rodrigues FA. Biochemical responses of coffee resistance against *Meloidogyne exigua* mediated by silicon. Plant Pathol 2010;59:586–93.

[51] Fawe A, Menzies JG, Chérif M, Bélanger RR. Silicon and disease resistance in dicotyledons. In: Datnoff LE, Snyder GH, Korndörfer GH, editors. Silicon in agriculture. Studies in plant science. Florida: Elsevier; 2001. p. 159–69. 2001.

[52] Belanger RR, Benhamou N, Menzies JG. Cytological evidence of an active role of silicon in wheat resistance to powdery mildew (*Blumeria graminis* f. sptritici). Phytopathology 2003;93:402–12.

[53] Pozza AA, Alves E, Pozza EA, Carvalho JG, Montanari M, Guimaraes PTG, et al. Effect of silicon on the control of brown eye spot in three coffee cultivars. Fitopatol Bras 2004;29:185–8.

[54] Botelho DMS, Pozza EA, Pozza A, Carvalho JG. Effect of silicon doses and sources on the intensity of the brown eye spot of coffee seedlings. Fitopatol Bras 2005;30:582–8.

[55] Reis THP, Figueiredo FC, Guimaraes PTG, Botrel PP, Rodrigues CR. Efeito da associaçaosilíciolíquidosoluvel com fungicida no controlefitossanitário do cafeeiro. Coffee Sci 2008;3:76—80.

[56] Carre-Missio V, Rodrigues FA, Schurt DA, Rezende DC, Moreira WR, Korndorfer GH, et al. Componentesepidemiologicos da ferrugem do cafeeiroafetadospelaaplicaçao foliar de silicato de potássio. Trop Plant Pathol 2012;37:50—6.

[57] Carre-Missio V, Rodrigues FA, Schur DA, Resende RS, Souza NFA, Rezende DC, et al. Effect of foliar-applied potassium silicate on coffee leaf infection by *Hemileia vastatrix*. Ann Appl Biol 2014;164:396—403.

[58] Thakral V, Bhat FA, Kumar N, Myaka B, et al. Role of silicon under contrasting biotic and abiotic stress conditions provides benefits for climate smart cropping. Environ Exp Bot 2021;189:104545.

[59] Onodera I. Chemical studies on rice blast (I). J Sci Agric Soc 1917;180:606—17.

[60] Datnoff LE, Deren CW, Snyder GH. Silicon fertilization for disease management of rice in Florida. Crop Prot 1997;16:525—31.

[61] Sathe AP, Amit K, Rushi K, et al. Role of silicon in elevating resistance against sheath blight and blast diseases in rice (*Oryza sativa* L.). Plant Physiol Biochem 2021;166:128—39.

[62] Zeyen RJ. Silicon in plant cell defenses against cereal powdery mildew disease. Abstract of Second Silicon in Agriculture Conference. 2002. p. 15—21.

[63] Rajwade JM, Chikte RG, Paknikar KM. Nanomaterials: new weapons in a crusade against phytopathogens. Appl Microbiol Biotechnol 2020;104:1437—61. Available from: https://doi.org/10.1007/s00253-019-10334-y.

[64] Hoffmann J, Berni R, Hausman JF, Guerriero G. A review on the beneficial role of silicon against salinity in non-accumulator crops: tomato as a model. Biomolecules. 2020;. Available from: https://doi.org/10.3390/biom10091284.

[65] Hasan KA, Soliman H, Baka Z, Shabana YM. Efficacy of nano-silicon in the control of chocolate spot disease of *Vicia faba* L. caused by *Botrytis fabae*. Egypt J Basic Appl Sci 2020;7:53—66. Available from: https://doi.org/10.1080/2314808x.2020.1727627.

[66] Khan MR, Siddiqui ZA. Use of silicon dioxide nanoparticles for the management of *Meloidogyne incognita, Pectobacterium betavasculorum* and *Rhizoctonia solani* disease complex of beetroot (*Beta vulgaris* L.). Sci Hortic 2020;265:109211.

[67] Derbalah A, Shenashen M, Hamza A, Mohamed A, El Safty S. Antifungal activity of fabricated mesoporous silica nanoparticles against early blight of tomato. Egypt J Basic Appl Sci 2018;5:145—50. Available from: https://doi.org/10.1016/j.ejbas.2018.05.002.

[68] Elsharkawy MM, Suga H, Shimizu M. Systemic resistance induced by Phomasp. GS8-3 and nanosilica against cucumber mosaic virus. Environ Sci Pollut Res 2018;27:16. Available from: https://doi.org/10.1007/s11356-018-3321-3.

[69] Mathur P, Roy S. Nanosilica facilitates silica uptake, growth and stress tolerance in plants. Plant Physiol Biochem 2020;157:114—27. Available from: https://doi.org/10.1016/j.plaphy.2020.10.011.

[70] Buchman JT, Elmer WH, Ma C, Landy KM, White JC, Haynes CL. Chitosan-coated mesoporous silica nanoparticle treatment of *Citrullus lanatus* (watermelon): enhanced fungal disease suppression and modulated expression of stress related genes. ACS Sustain Chem Eng 2019;7:19649—59.

[71] Madani B, Mirshekari A, Imahori Y. Physiological responses to stress. In: Elhadi MYahia, editor. Postharvest physiology and biochemistry of fruits and vegetables. Woodhead Publishing; 2019. p. 405—23.

[72] Sun W, Zhang J, Fan Q, Xue G, Li Z, Liang Y. Silicon-enhanced resistance to rice blast is attributed to silicon-mediated defence resistance and its role as physical barrier. Eur J plant Pathol 2010;128(1):39—49.

[73] Łazniewska J, Macioszek VK, Kononowicz AK. Plant-fungus´interface: the role of surface structures in plant resistance and susceptibility to pathogenic fungi. Physiol Mol Plant Pathol 2012;78:24—30. Available from: https://doi.org/10.1016/j.pmpp.2012.01.004.

[74] Nawrath C. Unraveling the complex network of cuticular structure and function. Curr Opin Plant Biol 2006;9:281—7. Available from: https://doi.org/10.1016/j.pbi.2006.03.001.

[75] Datnoff LE, Elmer WH, Huber DM. Mineral nutrition and plant disease. St. Paul, MN: The American Phytopathological Society; 2007.

[76] Van, BJ, De Vleesschauwer D, Höfte M. Towards establishing broad-spectrum disease resistance in plants: silicon leads the way. J Exp Bot 2013;64:1281—93. Available from: https://doi.org/10.1093/jxb/ers329.

[77] Ma JF, Yamaji N. Functions and transport of silicon in plants. Cell Mol Life Sci 2008;65:3049—57. Available from: https://doi.org/10.1007/s00018-008-7580-x.

[78] Guerriero G, Hausman JF, Legay S. Silicon and the plant extracellular matrix. Front Plant Sci 2016;7:463. Available from: https://doi.org/10.3389/Fpls.2016.00463.

[79] He CW, Ma J, Wang LJ. A hemicellulose-bound form of silicon with potential to improve the mechanical properties and regeneration of the cell wall of rice. New Phytol 2015;206:1051—62. Available from: https://doi.org/10.1111/nph.13282.

[80] Liu X, Yin L, Deng X, Gong D, Du S, Wang S, et al. Combined application of silicon and nitric oxide jointly alleviated cadmium accumulation and toxicity in maize. J Hazard Mater 2020;395:122679. Available from: https://doi.org/10.1016/j.jhazmat.2020.122679.

[81] Domiciano GP, Rodrigues FA, Guerra AMN, Vale FXR. Infection process of *Bipolaris sorokiniana* on wheat leaves is affected by silicon. Trop Plant Pathol 2013;38:258—63. Available from: https://doi.org/10.1590/S1982-56762013005000006.

[82] Ma JF, Yamaji N, Mitani N, et al. An efflux transporter of silicon in rice. Nature. 2007;448(7150):209—12.

[83] Hayasaka T, Fujii H, Ishiguro K. The role of silicon in preventing appressorial penetration by the rice blast fungus. Phytopathology. 2008;98:1038—44. Available from: https://doi.org/10.1094/PHYTO-98-9-1038.

[84] Ahanger MA, Bhat JA, Siddiqui MH, Rinklebe J, Ahmad P. Integration of Si and secondary methylene abolites in plants: a significant association in stress tolerance. J Exp Bot 2020;71:6758—74.

[85] Ma D, Sun D, Wang C, Qin H, Ding H, Li Y, et al. Silicon application alleviates drought stress in wheat through transcriptional regulation of multiple antioxidant defense pathways. J Plant Growth Regul 2006;. Available from: https://doi.org/10.1007/s00344-015-9500-2.

[86] Shetty R, Jensen B, Shetty NP, Hanse M, Hansen CW, Starkey KR, et al. Silicon induced resistance against powdery mildew of roses caused by *Podosphaera pannosa*. Plant Pathol 2012;6:120—31. Available from: https://doi.org/10.1111/j.1365-3059.2011.02493.x.

[87] Heine G, Tikum G, Horst WJ. The effect of silicon on the infection by and spread of *Pythium aphanidermatum* in single roots of tomato and bitter gourd. J Exp Bot 2007;58:569—77. Available from: https://doi.org/10.1093/jxb/erl232.

[88] Liang YC, Sun W, Si J, Römheld V. Effects of foliar-and rootapplied silicon on the enhancement of induced resistance to powdery mildew in *Cucumis sativus*. Plant Pathol 2005;54:678—85. Available from: https://doi.org/10.1111/j.1365-3059.2005.01246.x.

[89] Fortunato AA, Rodrigues F, Do-Nascimento KJ. Physiological and biochemical aspects of the resistance of banana plants to Fusarium wilt potentiated by silicon. Phytopathology. 2012;102:957−66. Available from: https://doi.org/10.1094/PHYTO-02-12-0037.

[90] Quarta A, Mita G, Durante M, Arlorio M, De PA. Isolation of a polyphenol oxidase (PPO) cDNA from artichoke and expression analysis in wounded artichoke heads. Plant Physiol Biochem 2013;68:52−60. Available from: https://doi.org/10.1016/j.plaphy.2013.03.020.

[91] Weerahewa D, Somapala K. Role of Si on enhancing disease resistance in tropical fruits and vegetables: a review. OUSL J 2016;7:43−59.

[92] Mamaeva V, Sahlgren C, Lindén M, MesoNguyen NT, Nguyen DH, Pham DD, et al. New oligochitosan-nanosilica hybrid materials: preparation and application on chili plants for resistance to anthracnose disease and growth enhancement. Polym J 2017;49:861−9. Available from: https://doi.org/10.1038/pj.2017.58.

[93] Hao Z, Wang L, He Y, Liang J, Tao R. Expression of defense genes and activities of antioxidant enzymes in rice resistance to rice stripe virus and small brown planthopper. Plant Physiol Biochem 2011;49:744−51. Available from: https://doi.org/10.1016/j.plaphy.2011.01.014.

[94] Silva WLD, Cruz MFA, Fortunato AA, Rodrigues F. Histochemical aspects of wheat resistance to leaf blast mediated by silicon. Sci Agric 2015;72:322−7. Available from: https://doi.org/10.1590/0103-9016-2014-0221.

[95] Rodrigues FA, Mcnally DJ, Datnoff LE, Jones JB, Labbé C, Benhamou N, et al. Silicon enhances the accumulation of diterpenoid phytoalexins in rice: a potential mechanism for blast resistance. Phytopathology 2004;94:177−83. Available from: https://doi.org/10.1094/PHYTO.2004.94.2.177.

[96] Rodrigues FA, Jurick WM, Datnoff LE, Jones JB, Rollins JA. Silicon influences cytological and molecular events in compatible and incompatible rice—Magnaporthe grisea interactions. Physiol Mol Plant Pathol 2005;66:144−59. Available from: https://doi.org/10.1016/j.pmpp.2005.06.002.

[97] Bailey-Serres JE, Parker EA, Ainsworth GED, Oldroyd JI, Schroeder. Genetic strategies for improving crop yields. Nature 2019;575:109−18.

[98] Kollist H, et al. Rapid responses to abiotic stress: priming the landscape for the signal transduction network. Trends Plant Sci 2019;24:25−37.

[99] Zandalinas SIS, Sengupta D, Burks RK, Azad R. Mittler. Identification and characterization of a core set of ROS wave-associated transcripts involved in the systemic acquired acclimation response of Arabidopsis to excess light. Plant J 2019;98:126−41.

[100] Toyota M, et al. Glutamate triggers long-distance, calcium-based plant defense signaling. Science 2018;361:1112−15.

[101] Suzuki N, et al. Temporal-spatial interaction between reactive oxygen species and abscisic acid regulates rapid systemic acclimation in plants. Plant Cell 2013;25:3553−69.

[102] Devireddy AR, Zandalinas SI, Gómez C, Blumwald E, Mittler R. Co-ordinating the overall stomatal response of plants: rapid leaf-to-leaf communication during light stress. Sci Signal 2018;11 eaam9514.

[103] Grant MR, Kazan K, Manners JM. Exploiting pathogens tricks of the trade for engineering of plant disease resistance: challenges and opportunities. Microb Biotechnol 2013;6:212−22. Available from: https://doi.org/10.1111/1751-7915.12017.

[104] Devadas SK, Enyedi A, Raina R. The Arabidopsis hrl1 mutation reveals novel overlapping roles for salicylic acid, jasmonic acid and ethylene signalling in cell death and defence against pathogens. Plant J 2002;30:467−80. Available from: https://doi.org/10.1046/j.1365-313X.2002.01300.

[105] Pei ZF, Ming DF, Liu D, Wan GL, Geng XX, Gong HJ, et al. Silicon improves the tolerance to water-deficit stress induced by polyethylene glycol in wheat (Triticum aestivum L.) seedlings. J Plant Growth Regul 2010;29:106−15.

[106] Brunings AM, Datnoff LE, Ma JF, Mitani N, Nagamura Y, Rathinasabapathi B, et al. Differential gene expression of rice in response to silicon and rice blast fungus Magnaporthe oryzae. Ann Appl Biol 2009;155:161−70. Available from: https://doi.org/10.1111/j.1744-7348.2009.00347.x.

[107] De Vleesschauwer D, Djavaheri M, Bakker P, Hofte M. Pseudomonas fluorescens WCS374r-induced systemic resistance in rice against Magnaporthe oryzae is based on pseudobactin-mediated priming for a salicylic acid-repressible multifaceted defense response. Plant Physiol 2008;148:1996−2012. Available from: https://doi.org/10.1104/pp.108.127878.

[108] Kim YH, Khan AL, Kim DH, Lee SY, Kim KM, Waqas M, et al. Silicon mitigates heavy metal stress by regulating P-type heavy metal ATPases, Oryza sativa low silicon genes, and endogenous phytohormones. BMC Plant Biol 2014;14:13. Available from: https://doi.org/10.1186/1471-2229-14-13.

[109] Fauteux F, Chain F, Belzile F, Menzies JG, Belanger RR. The protective role of silicon in the Arabidopsis-powdery mildew pathosystem. Proc Natl Acad Sci USA 2006;103:17554−9. Available from: https://doi.org/10.1073/pnas.0606330103.

[110] Chen YY, Lin YM, Chao TC, Wang JF, Liu AC, Ho FI, et al. Virus-induced gene silencing reveals the involvement of ethylene, salicylic acid and mitogen-activated protein kinase-related defense pathways in the resistance of tomato to bacterial wilt. Physiol Plant 2009;136:324−35. Available from: https://doi.org/10.1111/j.1399-3054.2009.01226.x.

[111] Vleesschauwer D. Primary metabolism plays a central role in moulding silicon- inducible brown spot resistance in rice. Mol Plant Pathol 2015;16:811−24.

[112] Iwai T, Miyasaka A, Seo S, Ohashi Y. Contribution of ethylene biosynthesis for resistance to blast fungus infection in young rice plants. Plant Physiol 2006;142:1202. Available from: https://doi.org/10.1104/pp.106.085258.

[113] Van BJ, Spíchal L, Novák O, Strnad M, Asano T, Kikuchi S, et al. Silicon induces resistance to the brown spot fungus Cochliobolus miyabeanus by preventing the pathogen from hijacking the rice ethylene pathway. New Phytol 2015;206:761−73. Available from: https://doi.org/10.1111/nph.13270.

[114] Vivancos J, Labbe C, Menzies JG, Belanger RR. Silicon-mediated resistance of Arabidopsis against powdery mildew involves mechanisms other than the salicylic acid (SA)-dependent defence pathway. Mol Plant Pathol 2015;16:572−82. Available from: https://doi.org/10.1111/mpp.12213.

[115] Kurabachew H, Stahl F, Wydra K. Global gene expression of rhizobacteria-silicon mediated induced systemic resistance in tomato (Solanum lycopersicum) against Ralstonia solanacearum. Physiol Mol Plant Pathol 2013;84:44−52. Available from: https://doi.org/10.1016/j.pmpp.2013.06.004.

[116] Dreher K, Callis J. Ubiquitin, hormones and biotic stress in plants. Ann Bot 2007;99:787−822. Available from: https://doi.org/10.1093/aob/mcl255.

[117] Pirrello J, Prasad BN, Zhang W, Chen K, Mila I, Zouine M, et al. Functional analysis and binding affinity of tomato ethylene response factors provide insight on the molecular bases of plant differential responses to ethylene. BMC Plant Biol 2012;12:190. Available from: https://doi.org/10.1186/1471-2229-12-190.

[118] Cabot C, Gallego B, Martos S, Barceló J, Poschenrieder C. Signal cross talk in Arabidopsis exposed to cadmium, silicon, and *Botrytis cinerea*. Planta 2013;237:337−49. Available from: https://doi.org/10.1007/s00425-012-1779-7.

[119] Cooke J, Leishman MR. Consistent alleviation of abiotic stress with silicon addition: a meta-analysis. Funct Ecol 2016;30(8):1340−57.

[120] Muneer S, Park YG, Kim S, Jeong BR. Foliar or subirrigation silicon supply mitigates high temperature stress in strawberry by maintaining photosynthetic and stress-responsive proteins. J Plant Growth Regul 2017;36:836−45.

[121] Rodrigues FÁ, Vale FXR, Korndörfer GH, Prabhu AS, Datnoff LE, Oliveira AMA, et al. Influence of silicon on sheath blight of rice in Brazil. Crop Prot 2003;22:23−9.

[122] Sanglard LMVP, Martins SCV, Detmann KC, et al. Si nutrition alleviates the negative impacts of arsenic on the photosynthetic apparatus of rice leaves: an analysis of the key limitations of photosynthesis. Physiol Plant 2014;152:355−66.

[123] Kurdali F, Al-Chammaa M. Growth and nitrogen fixation in Si and/or potassium fed chickpeas grown under drought and well watered conditions. J Stress Physiol Biochem 2013;9:385−406.

[124] Fauteux F, Remus-Borel W, Menzies JG, Belanger RR. Silicon and plant disease resistance against pathogenic fungi. FEMS Microbiol Lett 2005;249:1−6. Available from: https://doi.org/10.1016/j.femsle.2005.06.034.

[125] Nwugo CC, Huerta AJ. The effect of silicon on the leaf proteome of rice (*Oryza sativa* L.) plants under cadmium-stress. J Proteome Res 2011;10:518−28. Available from: https://doi.org/10.1021/pr100716h.

[126] Zargar SM, Nazir M, Kumar A, Kim DW, Rakwal R. Silicon in plant tolerance against environmental stressors: towards crop improvement using omics approaches. Curr Proteom 2010;7:135−43. Available from: https://doi.org/10.2174/157016410791330507.

[127] Chain F, Côté-Beaulieu C, Belzile F, Menzies J, Bélanger R. A comprehensive transcriptomic analysis of the effect of silicon on wheat plants under control and pathogen stress conditions. Mol Plant Microbe Interact 2009;22:1323−30. Available from: https://doi.org/10.1094/MPMI-22-11-1323.

[128] Boursiac Y, Boudet J, Postaire O, Luu DT, Tournaire-Roux C, Maurel C. Stimulus-induced downregulation of root water transport involves reactive oxygen species-activated cell signalling and plasma membrane intrinsic protein internalization. Plant J 2008;56 (2):207−18.

[129] Hamza AM, El-Kot GA, El-Moghazy S. Non-traditional methods for controlling maize late wilt disease caused by *Cephalosporium maydis*. Egypt J Biol Pest Control 2013;23:87−93.

[130] Mathur P, Sharma E, Singh SD, Bhatnagar AK, Singh VP, Kapoor R. Effect of elevated CO_2 on infection of three foliar diseases in oilseed (*Brassica juncea*). J Plant Pathol 2013;95:135−44. Available from: https://doi.org/10.4454/JPP.V95I1.013.

[131] Dickinson M. Molecular plant pathology. London: BIOS Scientific Publishers; 2003.

[132] Madany M, Saleh A, Habeeb T, Hozien W, Elgawad H. Silicon dioxide nanoparticles alleviate the threats of broomrape infection in tomato by inducing cell wall fortification and modulating ROS homeostasis. Environ Sci Nano 2020;7:5.

[133] Rangaraj S, Gopalu K, Muthusamy P, Rathinam Y, Venkatachalam R, Narayanasamy K. Augmented biocontrol action of silica nanoparticles and *Pseudomonas fluorescens* bioformulant in maize (*Zea mays* L.). RSC Adv 2014;4:8461.

15

Silicon and nanosilicon mitigate nutrient deficiency under stress for sustainable crop improvement

Krishan K. Verma[1], Xiu-Peng Song[1], Zhong-Liang Chen[1,2], Dan-Dan Tian[3], Vishnu D. Rajput[4], Munna Singh[5], Tatiana Minkina[4] and Yang-Rui Li[1]

[1]Key Laboratory of Sugarcane Biotechnology and Genetic Improvement (Guangxi), Ministry of Agriculture and Rural Affairs/Guangxi Key Laboratory of Sugarcane Genetic Improvement/Sugarcane Research Institute, Guangxi Academy of Agricultural Sciences/Sugarcane Research Center, Chinese Academy of Agricultural Sciences, Nanning, P.R. China [2]College of Agriculture, Guangxi University, Nanning, P.R. China [3]Institute of Biotechnology, Guangxi Academy of Agricultural Sciences, Nanning, P.R. China [4]Academy of Biology and Biotechnology, Southern Federal University, Rostov-on-Don, Russia [5]Department of Botany, University of Lucknow, Lucknow, India

15.1 Introduction

Silicon (Si) is the second common element after oxygen, found in the earth's crust as a component of clay minerals [1,2]. The examples of silica minerals are quartzite, tridymite, metamorphic rock, cristobalite, and their polymorphs. The combination of Si and O_2 is known as silicate, which makes about 90% of the earth's crust [3], may be determined quantitatively by biogeochemical cycling. It is commonly found in soil as silica (SiO_2), may be up to 50−70% [4,5], also in soil solution as monosilicic acid [$Si(OH)_4$], taken up by plant roots (pH 9.0). It has been discovered to improve plant development and yield, even though it is not currently considered an essential element for physiological fitness in plants [6−13]. Silicon depletion in the soil has occurred in the agriculture system extensively and indiscriminate use of commercial fertilizers [5], and promoted its regular use as a fertilizer. It protects plants from biotic and abiotic stresses [7,8,14−16]. Plants can be classified as active, passive, or rejective Si accumulators based on the capacity to absorb Si from rhizosphere and transport to plant organs [1,17]. Plants can absorb Si in an average mass of 0.1−10% of their dry mass, depending on the plant species [2,4,9,14,15,18,19]. Silicon has been placed between essential and nonessential elements for plants as it is not required to complete plants' life cycle. However, under normal and stressful conditions, it provides some benefits to plants [20,21] due to its involvement in various intrinsic cellular activities against environmental stressors. Consequently, it has been shown to improve plant resistance to heavy metal toxicity, elevated temperature, drought, salinity, soil flooding, UV radiation, cold, and other environmental factors [1,9,21−23] with profound importance in crop improvement.

The combined effect of SiO_2 and organic fertilizers has boosted the overall plant performance [8,24−26]. Moreover, Si NPs' mesoporous nature makes them appropriate nanocarriers for various compounds used in agricultural systems. Eventually, nanosensors and nanozeolites, which are made up of Si NPs, have been successfully employed in agriculture to enhance soil moisture-water retention [20]. Silicon fertilizer is a high-quality fertilizer that promotes environmental friendly agriculture practices. Nanosilicon has tunable physical, chemical, optical, and mechanical properties and found quick usage in agroindustries, medical biology, supercapacitors, batteries, optical fibers, and concrete materials [3]. The beneficial effects of Si and nSi are found most obvious in treated plants as compared to control [1,2,27,28]. An overview presented in this chapter for Si and nSi have considerable and noticeable positive effects to mitigate plants' nutritional deficiency to achieve the optimum potential of crop productivity to ensure food security in future (Table 15.1).

15.2 Silicon and nanosilicon application in soil and plants

Silicon may be found in various forms in the soil, and most of them are detected slightly soluble. Monosilicic acid [Si $(OH)_4$] is more water-dissolve form of Si, even though it can be converted to $H_3SiO_4^-$ (pH 9) and $H_2SiO_4^{2-}$ (pH 11) [1,19,29]. Accessible Si in soil is made up of monosilicic acid in soil solution (0.1−0.6 mM), and portions of silicate may also easily be transferred into monosilicic acids, that is, polymerized silicic acid, exchangeable silicates, and colloidal silicates. Silicon with limited availability may be found in the soil solution (pH 8−9) due to significant adsorption of $H_3SiO_4^-$ to soil colloids via interactions with iron and aluminum oxides [1,29]. The impact of pH on Si availability is most complicated and depends on a variety of other parameters such as soil properties, temperature, organic matter content, and other ions [30]. The ability of Si accumulation varies depending upon the plant species or cultivars [9,28,31−33]. Equisetales, Cyperales, and Poales families have the best Si-accumulator plants, may acquire up to 10% Si on dry biomass basis [5] while 2−4% Si may be found in Cucurbitales and Urticales [34], whereas others almost exclude [35]. The effective uptake of Si in plants from the soil solution is found to be linked with the expression of Si transporters in roots [31].

The effective transport of Si in rice plants allows its accumulation around 90% in stem [31]. The aquaporins are also associated with facilitating the accumulation buildup of Si in certain crop plants. The uptake of Si by roots occurs as monosilicic acid and gets precipitated as amorphous silica within the epidermal cell wall in plants [36]. This process protects plants from mineral shortage or toxicity, tolerance to temperature (low and high), water deprivation, salinity, heavy metals, cold, light, UV radiation, and nutritional inadequacy during abiotic stressors [1,2,11,37,38]. The addition of Si to growth media boosts Zn content of plants, which is crucial for Indole acetic acid (IAA) synthesis and functioning of terminal oxidase enzyme in mitochondria to sustain respiration. During metal poisoning, Si raises the IAA content via increasing the Zn content in plants. Si acts as a physical barrier in plants, but also plays an important function in their morphological and physiological functions during stress [1,2,15,28,39]. Si has an indirect effect on biological processes by allowing other elements to reach plants and promoting physiological fitness of plants to ensure crop production [3,19,28]. The repeated cultivation of crops and the use of synthetic fertilizers like nitrogen, phosphorous, and potassium have reduced Si availability in soil [40], recognized as a limiting factor in crop productivity. Hence, application of Si in irrigation water to soil may enhance soil fertility, nutrient availability to plants, soil cation exchange capacity, and phosphorus uptake in plants with reducing nutritional deficiency [28,41−43]. Recently, several studies have investigated the action of Si in nutritional insufficiency mitigation to acquire physiological fitness of crop plants [1,44−47].

15.3 Silicon/nano-Si and micronutrients

15.3.1 Iron (Fe)

15.3.1.1 Role of Fe and Fe deficiency in plants

Iron (Fe) is a crucial element for plants throughout their life cycle. A variety of cellular processes, viz., respiration, biosynthesis of green pigments, operation of photosynthetic electron transport (PET) chain, and hexose production [48,49]. The morphological and physiological responses of plants may get altered during iron deficiency [50]. The chlorosis of young leaves, stunting of leaves and roots, and lower yields are common indications of Fe deficiency in plants [51−53]. The chlorosis of the leaves, characterized by a dramatic drop in photosynthetic pigments, is the first indication of Fe deficiency [54]. Leaf chlorosis usually starts with pale leaf color, followed by a yellowing of the interveinal portions with the main veins remaining green. The entire leaf may get chlorotic and acquire necrotic areas if the problem is severe. In severe cases, iron chlorosis causes plants to grow slowly and eventually forced them to die (Table 15.1). Fe is the more abundant element in the earth's crust, even though oxides and hydroxide compounds make it less soluble in calcareous soils. The bicarbonate (HCO_3^-) is abundant in calcareous soils, causing an increase in soil pH which decreases Fe availability [55]. The solubility of Fe is pH dependent in soil, decreases 1000-fold for each unit increase in pH [56], prevalent in calcareous dry soils. Inadequate aeration causes CO_2 accumulation with an increase in HCO_3^- found to be conducive for Fe shortage in plants [57].

Fe efficient plants use two different strategies to overcome the iron deficit. Fe (III) reduction capability mediated by an enzyme Fe reductase through acidification of the rhizosphere mediated by proton excretion via plasmalemma H^+-ATPase. Dicots and monocots enhance accessible Fe in the rhizosphere by increasing the production of Fe (II) transporter and Fe(III) chelate reductase (step I). Acidification of the rhizosphere and the production of phenolic compounds were also mentioned to increase the amount of Fe available in the rhizosphere (step II) [58]. Fe is absorbed in *A. thaliana* in three ways—rhizosphere acidification via H^+-ATPases, ferric reductase oxidase 2 mediated reduction of Fe (III) to Fe (II), and iron-regulated transporter-mediated intake of Fe (II). Step II plants, which include graminaceous plants, get Fe from the soil via secretion of high-affinity Fe (III) chelator phytosiderophores [58]. Monocot plants solubilize Fe from soils to release rhizosphere-specific Fe chelating molecules called Phyto siderophores [58]. Insoluble Fe pools have been found in the apoplast of plants [59]. The cultivars with bigger iron apoplastic pools are less likely to acquire iron chlorosis symptoms [60]. The remobilization of stored iron pools from the concerned cells in A. *thaliana* balances its level in stem [61−63]. The effect of phenolics on the remobilization of the Fe pool in *Trifolium pretense* roots was assessed [61], and it was discovered that it delayed Fe (III) chelate reductase activation process which reduces rhizospheric acidification [61] during iron remobilization (Table 15.1).

15.3.1.2 Si/nSi-mediated alleviation of Fe deficiency in plants

Si and nSi addition appears to enhance Fe storage in the root apoplast as pools [64]. Fe remobilization appears to be induced by Si during Fe deficiency and also prevents chlorophyll loss in plants during stress [65−67], shown in Table 15.1. There are two possible strategies to improve chlorophyll by Si. Primarily, the action of Si and nSi on stronger cell walls may lead to green leaves in a better position to intercept light for photosynthetic response [11,47,65,66]. Second, Si and nSi also contribute to prevent the damage of photosynthetic pigments, enzymatically [46,67−69].

The effects of Si and nSi on the restoration of Fe deficiency in soybean (*Glycine max* L.)—an iron-inefficient and chlorosis-prone plant, and cucumber (*Cucumis sativus* L.)—an Fe-deficient stressed plant were grown in a water-culture system using sodium silicate hydrate as a source of Si. Silicon prevented chlorophyll loss and preserved leaf Fe content in the iron-inefficient soybean. Cucumber deficiency symptoms were improved when silicon was added to the nutritional solution. Zn and Mn deficiency symptoms were also partially reduced [44,46]. The effects of Si were found linked with increased Fe distribution and accumulation of Fe mobilizing chemicals in the apical meristem of shoot. The mechanism that regulates the accumulation of Si was found to be strongly integrated with transpiration rates of the plants. Iron may follow Si movement intrinsically at the cellular level in plants and be efficiently remobilized from the source to the sink. Silicon increased rice root oxidation efficiency and increased Fe oxidation to insoluble ferric compounds, forming Fe plaques [70,71]. The transfer of Si from external to the cortical cells was revealed to be mediated by a comparable energy-dependent transporter in different cultivars, implying that differences in Si accumulation between plant cultivars are attributable to variable densities of the transporter [35].

15.3.2 Zinc (Zn)

15.3.2.1 Role of Zn and Zn deficiency in plants

Zinc is required for physiological functions in plants. It is a component of several necessary enzymes involved in energy transfer, protein synthesis, nitrogen metabolism, and catabolism of proteins, carbohydrates, lipids, and nucleic acids [57,72]. Zn is a plasma membrane stabilizer and is required for the stability of biological membranes. Consequently, membrane lipids and proteins are protected from oxidative degradation [73,74]. Zn is necessary for the healthy establishment of plants. In most crops, the amount of Zn in leaf is nearly 15−20 mg kg^{-1} dry weight [75]. Zn is essential not only for the plant development and metabolism but also for animal and human health. As a result, Zn deficiency has a significant impact on both plant yield and productivity, as well as human health. Nearly, three billion people worldwide are badly affected by Fe and Zn insufficiency, particularly those who rely on cereal-based meals, as Zn deficiency is the most frequent mineral disease in cereals [76−78]. More than half of the world's cultivated land is Zn deficient [76]. Zinc is found across the globe. However, its availability to plants is restricted. Majorly, weathering of rocks, soil-forming elements such as climate variables, and weathering intensity influence the occurrence of Zn in soil [79]. Despite this, high pH values, high $CaCO_3$ levels, low organic matter, decreased clay contents, and increased phosphorous treatments contribute to this element's limited availability in plants [80]. Thus Zn shortage is expected in calcareous, sandy, and peat soils and the soils with high-phosphorus concentrations (Table 15.1).

Zinc is essential for plant fitness and is also involved in different enzymatic functions. So, its scarcity has a negative impact on crop productivity. The lower growth, leaf chlorosis, decreased chlorophyll index—pigment contents of the leaf, extended maturity period, fewer tillers, reduced leaf-area expansion, and decreased crop quality symptoms are found to be linked with Zn deficiency [46,81]. In Zn-deficient plants, oxidative stress is caused by low superoxide dismutase (SOD) activity to enhance the production of reactive oxygen species (ROS). These ROS severely harm constituents of the cellular membranes, viz., lipids proteins, enzymes, and green pigments that impair photosynthetic activity [77]. Therefore enough Zn must be supplied to ensure optimal plant growth [82,83]. Zn shortage also have an impact on vegetative and reproductive growth in rice plants along with reduction of dry mass, grain-seed weight, and plant productivity [84]. The stunting of plants, reduced petioles/internodes, necrotic patches, leaves cupping upward and developing interveinal chlorosis, bronzing, and rosetting of leaves are all common deficiency symptoms caused by Zn [85] with dark brown necrotic lesions in interveinal areas of mature plants [46,86].

15.3.2.2 Si/nSi-mediated alleviation of Zn deficiency in plants

Cucumber plants cultivated in a Zn-free solution showed no significant changes in chlorophyll content, but their leaves developed necrotic patches. Si-mediated antioxidative defense helps to avoid the development of necrotic patches induced by hydroxyl radicals and superoxide anions, as seen in cucumber plants during Zn stress [46]. The presence of Si reduces the number of necrotic patches on the leaves. The application of Si increased fumarate levels in Zn-stressed plants [46]. Zn deficiency causes overproduction of ROS while Si mediates plants' antioxidant defense mechanism during soil salinity [87−89], cold [90], and water deficiency [91]. The availability of Si in soil supports Zn distribution at higher concentrations [92−94]. Si strengthens the root cell wall that prevents Zn from moving roots to shoots [95−98]. Zn may get precipitated around the root epidermis as zinc silicate, which obstructs its translocation in the xylem [92,99]. The binding sites for Zn^{2+} increase as silicate precipitates on the cell wall [96,100] and prevent its movement across the plant [99,101,102]. Upon fracture of zinc silicate, Zn gets retained in cell vacuoles in an unknown form [103]. The application of Si increases the binding of Zn in the rice plant mainly in stem, root, leaves, and sheath [93]. The use of Si increases citrate content that promotes Zn redistribution [104]. Si application also boosts vegetative and reproductive growth in rice plants in case co-cultivated by using Zn ($1-100\,g\,L^{-1}$) while $10\,g\,L^{-1}$ concentration of treatment resulted in the most significant growth of the rice plant. Si had a favorable effect by reducing the toxicity higher dose of Zn, that is, $100\,g\,Zn\,L^{-1}$. Interestingly, Si generated an increase in K^+ and Fe levels, which could have helped to alleviate Zn toxicity (Table 15.1) [84].

15.3.3 Manganese (Mn)

15.3.3.1 Role of Mn and Mn deficiency in plants

Mn is the 12th more significant nutrient in the entire world which is found about 0.098% in nature [105], viz., soil, water, and sediments, necessary for humans and plants. It can be found in various forms such as oxides, silicates, phosphates, and borates [106,107]. Manganite, pyrolusite, and rhodonite are the most frequent manganese-bearing minerals. Mn is released in the soil as a result of mineral weathering and atmospheric deposition, and it comes from natural and anthropogenic sources. Mn in the soil can be found in three different oxidation states, viz., Mn (II), Mn (III), and Mn (IV). In the soil solution, only the divalent ions are stable, whereas Mn (III) and Mn (IV) are only stable in the solid state [108]. The mobility of Mn in the soil is mainly based on the soil structure. Mn solubility is governed by redox potential and pH of the soil. Mn binds with organic materials, oxides, and silicates at pH over 6.0. Thus Mn availability and solubility are low at higher pH and organic matter level. However, Mn availability and solubility are maximum in acidic soils with the lower level of organic matter. Apart from this, Mn has a high solubility under anaerobic and aerobic circumstances at pH over 6 and below 5.5, respectively [108,109].

Mn is a key element in plants that helps them to grow and develop by activating certain enzyme activities and acting as a cofactor for other enzymes [110]. It is important for functionality of photosynthetic apparatus to onset the photolysis of water oxidizing complex [111] to liberate electrons, protons, and molecular oxygen essential for the operation of PET chain in all green plants. Mn is also associated with SOD enzyme, that is, Mn-SOD that protects plants from ROS, such as singlet oxygen and hydrogen peroxide by converting them into H_2O_2 and then water, eventually. The vulnerability of plant species and cultivars within species to Mn deficiency varies greatly. Mn shortage causes severe losses in cereals such as barley and wheat that may result in the complete loss of their harvests. It also impairs dry matter production, leaf gas exchange, and photosynthetic pigments quickly. Under Mn deficit conditions, cell organelles and chlorophyll get degraded, as reflected by the formation of interveinal leaf chlorosis, considered the most distinctive indication of Mn deficiency (Table 15.1) [112]. The main visual signs in cereals are greenish-gray patches on the basal leaves. Sandy, calcareous soils with a maximum amount of soil organic matter may enhance soluble Mn^{2+} to plant, while MnO_2 may not be available [113].

15.3.3.2 Si/nSi-mediated alleviation of Mn deficiency in plants

The interplay of Mn, Si, and nSi in a range of plants cultivated in the water-culture system for a variety of plants [70,114−117] has been well documented. $MnSO_4$ is commonly applied as a source of Mn fertilizer in solutions or in irrigation water [56]. In normal soil above pH 4, oxidation of Mn^{2+} to Mn^{4+} is preferred thermodynamically at the expenditure of large activation energy [118]. Si promotes distribution Mn uniformly in leaves [114−116,119], although the mechanism that governs the process is yet unknown. In contrast to this, Si application did not affect Mn absorption in pumpkin, but stimulated Mn buildup at the base of trichomes [117]. Accordingly, Si-treated cucumber plants could acquire c. 90% of the Mn bonded to the cell wall [120,121]. Mn is found as a precipitate in cell walls of cucumber plants that may use it to compensate related deficiencies. The positive role of Si in cucumber can be explored by its role in reducing lipid peroxidation by enhancing antioxidative activities [119] and by lowering OH^- in the apoplast of the leaves [122].

15.3.4 Copper (Cu)

15.3.4.1 Role of Cu and Cu deficiency in plants

Copper (Cu) is a necessary component for plant fitness. Cu level in the rhizospheric soil is detected in the range from 3 to 100 mg kg^{-1}. Nearly, 1−20% found free and easily accessible and the remaining gets linked with organic matter. Cu is relatively immobile between plant cell organs. Therefore signs of shortage emerge first in freshly produced younger cells and reproductive organs. This trace element is involved in many proteins, including those associated with the photosynthetic machinery and respiratory electron transport chains (etc). It is an integral element for plastocyanin protein, which acts as an immediate electron donor to PSI during the operation of PET chain, using PSII and PSI. It plays a key role in photosynthetic CO_2 assimilation and ATP generation [123]. Cu uptake and accumulation from the soil to the root system are influenced by the total amount of Cu in the rhizospheric soil and the ability of plants to bridge the soil−root interface [124]. Cu is listed as micronutrient, necessary for plants, but it may also be harmful. The folding up and wilting of leaves, as well as impaired PET, slower respiration rate, and diminished plant growth due to anomalies in apical meristems, are all signs of Mn deficiency [57].

15.3.4.2 Si/nSi-mediated alleviation of Cu deficiency in plants

Fewer research has looked at the Si−Cu relationship in plants [125,126]. Cu toxicity signs, that is, leaf chlorosis and loss in plant morphological and yield capacity were observed in *Arabidopsis thaliana* when Si was added to the nutrient solution [126]. Similar results were shown in *Triticum aestivum* [127]. However, there was no significant variation in the Cu level in plant leaves due to Si application, implying that Si altered the uptake and accumulation of Cu within plant leaves during Cu toxicity [128]. The creation of Si deposits on the cell wall improved Cu-binding sites. It prevented the effect of excess Cu toxicity on plant cells [15,40,120,125] while excess Cu may cause plant mortality. The application of Si lowers the level of Cu accumulation up to 32% in the leaves while its level in roots may enhance (Table 15.1). Furthermore, Si deposits found in phytolith leaves connected with Cu, Ca, K, and P. Cu transport from roots to shoots was found to be impaired in case coapplied with Si, which induced Cu resistance. The studies favored the impact of Si on Cu toxicity in *T. turgidum*, in case exposed to 0−30 M in the water-culture system, evidently proved that Si may reduce Cu toxicity as shown in Table 15.1 [129].

15.4 Si/nSi-mediated alleviation of heavy metal stress in plants

The several studies conducted so far elucidated the relationship between Si, nSi, and micronutrients. Pavlovic et al. [45,130] presented the first evidence that Si mitigates Fe shortage in cucumber by mobilizing Fe toward

TABLE 15.1 Impact of silicon and nanosilicon on crop plants subjected to nutritional deficiency.

Plant	Nutrient	Source	Application method	Impact	Source
Sorghum (*Sorghum bicolor* L.)	Zn	Si	Foliar application	Increased photosynthetic leaf gas exchange, chlorophylls, biomass, protein, Zn accumulation, and reduced cell membrane damage	[150]
Soybean (*Glycine max* L.)	Zn	Si	Hydroponic culture	Enhanced plant growth, physiological, biochemical parameters, Zn and Si content in leaves	[151]
Soybean (*Glycine max* L.)	Hg	nSi	Hydroponic culture	Increased plant biomass, reduced Hg accumulation in plant organs, and maintained cellular ultrastructure	[47]
Soybean (*Glycine max* L.) and cucumber (*Cucumis sativus* L.)	Fe	Si	Hydroponic culture	Increased photosynthetic pigments, soil plant analysis development (SPAD) content, growth traits, photosynthetic activities, improved/maintained accumulation of Fe. The observed response to Si application in Fe deficiency was plant-specific, probably correlated with Fe efficiency strategies	[44]
Rice (*Oryza sativa* L.)	Fe	Si	Hydroponic culture	Silicon reduced the Fe content in rice shoots and increased Casparian bands and upregulated Fe homeostasis-related genes. Improved plant growth, development, and yield capacity during Fe toxicity	[152]
Barley (*Hordeum vulgare* L.)	Fe	Si	Soil irrigation	Silicon mitigated Fe toxicity in barley plants, diminishing photosynthetic pigments and reduction in biomass, and upgrading the antioxidant enzyme activities, resulting in lowered reactive oxidative species (ROS) uptake in the photosynthetically matured leaves. The gene express upregulated in barley leaves and roots during Fe stress with Si application	[131]
Cowpea (*Vigna unguiculata* L), soybean (*Glycine max* L.), sunflower (*Helianthus annuus* L.), and white lupin (*Lupinus albus* L.)	Mn	Si	Water-culture system	Si enhances/balances apoplastic sorption of Mn-stressed plant leaves, thereby reducing free Mn^{2+} uptake in the apoplast/cytoplasm.	[153]
Maize (*Zea mays* L.)	Mn	Si	Soil application	The application of Si decreased the density of brown spots/leaf area as well as MDA content and enhanced plant performance as well as biomass subjected to Mn-stressed plants	[154]
Cucumber (*Cucumis sativus* L.)	Mn	Si	Water culture system	Morphological changes are absent in Si-applied plants with excess Mn level. POD activity enhanced and the accumulation of ·OH level in leaf apoplast was reduced in Mn-stressed plants with Si application. Overall maintain plant performance during stress with Si	[155]
Rice (*Oryza sativa* L.)	Pb	nSi and Si	Soil irrigation	nSi and Si improved plant performance and productivity, reduced Pb uptake and accumulation in plant parts and grains during stress condition. The translocation factor of Pb reduced from roots-shoots-grains	[156]
Soybean (*Glycine max* L.) and sorghum (*Sorghum bicolor* L.)	Al	Si	Water-culture system	Silicon enhanced root elongation rate during Al-toxic element. Si decreases the Al toxicity in planta through the formation of Al–Si complexes	
Tobacco (*Nicotiana tabacum* L.)	Cu	Si	Soil culture	Silicon reduced the Cu accumulation/uptake in roots subjected to Cu-stressed plants. Si-mediated mitigation of Cu stress correlated with enhanced ethylene biosynthetic expression of genes	[158]
Cucumber (*Cucumis sativus* L.)	Cu	Si	Soil irrigation	Si-mitigated Cu toxicity in cucumber plants with enhanced enzymatic activities and biomass	[159]

younger leaves and induces synthesis of citrate and malate in roots. Si also increased the iron-related expression of genes, which enhanced nicotianamine contents and made iron loading and unloading in the phloem. Si is also found to reduce Fe deficiency in cucumber and barley by raising iron levels in mature leaves and redistribution of some metals, viz., copper and zinc [131,132]. Plant-specific Si-mitigated iron stress was discovered to be pH-dependent and plant-specific [44,132]. All studies, as mentioned earlier, conducted in water culture while the scenario in soil may differ based on a variety of circumstances.

Fe and Al hydrous oxides may adsorb silicate on soil colloid surfaces. Aluminum oxides were found most suitable than iron due to a large surface area (Table 15.1) [133]. The effects of pH on Si located in the rhizospheric soil are still unresolved [134]. The impact of Si during high boron (B) in soil-cultivated crops has also been investigated in the past few years [135–138]. Si mitigated the adverse impact of B stress in cotton [139]. The authors demonstrated foliar Si administration that improved photosynthetic responses during B deficiency, whereas root Si application inhibited B transfer into shoots during B toxicity [135]. Silicon also reduces B stress by lowering oxidative stress and modifying antioxidant enzymes with loss in leaf B status [135,140]. The application of Si raises soil pH, which inhibited Zn accumulation in root by transferring to shoot [141].

The cadmium (Cd)-induced stress in *O. sativa* was reported downregulated to extend the benefit to affected crops from Si application. Rice plants exposed to Cd stress acquired lower Cd deposition in roots and shoots with increased leaf gas exchange capacity after using Si [142,143]. Rice proteome changed after Si treatment during Cd stress, regulated by the photosynthetic system, redox homeostasis, synthesis of protein, chaperon responses, and pathogen activity [144]. In *Z. mays*, Si deposition in the endodermis and pericycle of the root and subsequent variations in cell extensibility helped to reduce Cd stress. The positive effects of Si on antioxidant enzyme activities have also been correlated to reducing oxidative stress during Cd poisoning [142,145].

O. sativa has been the most common cultivar used to investigate the connection between Si and arsenic (As) in Zn- and Cd-stressed plants. Si has also been shown to inhibit As absorption and translocation in the shoot. Guo et al. [146] demonstrated that Si and As employ the same transport system (Lsi1) in rice and Si inhibited As accumulation. Notably, Si did not influence As levels in tomato plant organs, implying that the accumulation of heavy metal seems to be cultivar/species dependent [147]. A connection between As uptake and radial oxygen loss (ROL) was discovered in rice plants with high ROL accumulation and minimum arsenic content [148], as shown in Table 15.1.

15.5 Conclusion and future prospective

The studies have been conducted to reveal the influence of Si, nSi, and micronutrients to mitigate nutritional deficiency to extend physiological fitness for crop plants. It has enabled plants to uptake of nutrients and water from the rhizosphere followed by its translocation to the needed plant parts to safeguard the expression of their optimal genetic potential under abiotic adversities. Primarily, it may be sustained through ecofriendly behavior of soil and metal toxicity, critical for future agriculture under climate change. Undoubtedly, Si biochemistry seems to be capable of acquiring positive interventions as it affects pH, texture, and organic matter of the rhizosphere [1,149], home to the hidden half of the plants that significantly influence the upper canopy of plants located in the biosphere linked with plant productivity, biomass, and economical yield. Hence, Si application, uptake, translocation, and its utilization by the cellular system must be visualized through precision studies based on proteomics and genomics to enhance nutritional qualities during stress. Serious research efforts must also be made to explore the role of Si and nSi in symbiotic nitrogen-fixing bacteria in times to come to design them suitably as plant growth-promoting rhizobacteria to establish its favorable interaction on plant growth characteristics to mitigate abiotic stressors for crop improvement [10].

Acknowledgments

We are thankful to the Guangxi Academy of Agricultural Sciences, Nanning, Guangxi, China for providing the necessary facilities for this study. This study was financially supported by the National Natural Science Foundation of China (31760415), Fund for Guangxi Innovation Teams of Modern Agriculture Technology (nycytxgxcxtd-2021-03), Fund of Guangxi Academy of Agricultural Sciences (2021YT011), and Youth Program of National Natural Science Foundation of China (31901594).

Conflict of Interest

The authors declare there are no competing financial interests.

References

[1] Coskun D, Deshmukh R, Sonah H, Menzies JG, Reynolds O, Ma JF, et al. The controversies of silicon's role in plant biology. New Phytol 2019;221:67–85. Available from: https://doi.org/10.1111/nph.15343.

[2] Pavlovic J, Kostic L, Bosnic P, Kirkby EA, Nikolic M. Interactions of silicon with essential and beneficial elements in plants. Front Plant Sci 2021;12:697592. Available from: https://doi.org/10.3389/fpls.2021.697592.

[3] Prabha S, Durgalakshmi D, Rajendran S, Lichtfouse E. Plant-derived silica nanoparticles and composites for biosensors, bioimaging, drug delivery and supercapacitors: a review. Environ Chem Lett 2020;. Available from: https://doi.org/10.1007/s10311-020-01123-5.

[4] Epstein E. The anomaly of silicon in plant biology. Proc Natl Acad Sci 1994;91:11–17.

[5] Ma JF, Tamai K, Yamaji N, Mitani N, Konishi S, Katsuhara M, et al. A silicon transporter in rice. Nature 2006;440:688–91.

[6] Adrees M, Ali S, Rizwan M, Zia-ur-Rehman M, Ibrahim M, Abbas F, et al. Mechanisms of silicon-mediated alleviation of heavy metal toxicity in plants: a review. Ecotoxicol Environ Saf 2015;119:186–97.

[7] Rizwan M, Ali S, Ibrahim M, Farid M, Adrees M, Bharwana SA, et al. Mechanisms of silicon-mediated alleviation of drought and salt stress in plants: a review. Environ Sci Pollut Res 2015;22:15416–31.

[8] Wang M, Gao L, Dong S, Sun Y, Shen Q, Guo S. Role of silicon on plant pathogen interactions. Front Plant Sci 2017;8.

[9] Chen D, Wang S, Yin L, Deng X. How does silicon mediate plant water uptake and loss under water deficiency? Front Plant Sci 2018;9:281. Available from: https://doi.org/10.3389/fpls.2018.00281.

[10] Verma KK, Song X-P, Li D-M, Singh M, Rajput VD, Malviya MK, et al. Interactive role of silicon and plant–rhizobacteria mitigating abiotic stresses: a new approach for sustainable agriculture and climate change. Plants 2020;9:1055.

[11] Verma KK, Song X-P, Lin B, Guo D-J, Singh M, Rajput VD, et al. Silicon induced drought tolerance in crop plants: physiological adaptation strategies. Silicon 2021;. Available from: https://doi.org/10.1007/s12633-021-01071-x.

[12] Verma KK, Song X-P, Verma CL, Chen Z-L, Rajput VD, Wu K-C, et al. Functional relationship between photosynthetic leaf gas exchange in response to silicon application and water stress mitigation in sugarcane. Biol Res 2021;54:15.

[13] Verma KK, Song X-P, Verma CL, Malviya MK, Guo D-J, Rajput VD, et al. Predication of photosynthetic leaf gas exchange of sugarcane (*Saccharum* spp.) leaves in response to leaf positions to foliar spray of potassium salt of active phosphorus under limited water irrigation. ACS Omega 2021;6:2396–409.

[14] Epstein E. Silicon. Ann Rev Plant Physiol Plant Mol Biol 1999;50:641–64.

[15] Liang Y, Sun W, Zhu Y-G, Christie P. Mechanisms of silicon-mediated alleviation of abiotic stresses in higher plants: a review. Environ Pollut 2007;147:422–8.

[16] Guntzer F, Keller C, Meunier J-D. Benefits of plant silicon for crops: a review. Agron Sustain Dev 2012;32:201–13.

[17] Stoyanova Z, Zozikova E, Poschenrieder C, Barcelo J, Doncheva S. The effect of silicon on the symptoms of manganese toxicity in maize plants. Acta Biol Hung 2008;59:479–87.

[18] Ma JF, Yamaji N. Functions and transport of silicon in plants. Cell Mol Life Sci 2008;65:3049–57.

[19] Etesami H, Jeong BR. Silicon (Si): review and future prospects on the action mechanisms in alleviating biotic and abiotic stresses in plants. Ecotoxicol Environ Saf 2018;147:881–96.

[20] Rastogi A, Tripathi DK, Yadav S, Chauhan DK, Zivcak M, Ghorbanpour M, et al. Application of silicon nanoparticles in agriculture. 3 Biotech 2019;9:1–11.

[21] Seleiman MF, Refay Y, Al-Suhaibani N, Al-Ashkar I, El-Hendawy S, Hafez E, et al. Integrative effects of rice-straw biochar and silicon on oil and seed quality, yield and physiological traits of *Helianthus annuus* L. grown under water deficit stress. Agronomy 2019;9:637.

[22] Verma KK, Song X-P, Zeng Y, Guo D-J, Singh M, Rajput VD, et al. Foliar application of silicon boosts growth, photosynthetic leaf gas exchange, antioxidative response and resistance to limited water irrigation in sugarcane (*Saccharum officinarum* L.). Plant Physiol Biochem 2021;166:582.

[23] Verma KK, Song XP, Tian DD, Singh M, Verma CL, Rajput VD, et al. Investigation of defensive role of silicon during drought stress induced by irrigation capacity in sugarcane: physiological and biochemical characteristics. ACS Omega 2021;6:19811–21.

[24] Cooke J, Leishman MR. Is plant ecology more siliceous than we realize? Trends Plant Sci 2011;16:61–8.

[25] Janmohammadi M, Amanzadeh T, Sabaghnia N, Ion V. Effect of nano-silicon foliar application on safflower growth under organic and inorganic fertilizer regimes. Bot Lith 2016;22:53–64.

[26] Seleiman MF, Ali S, Refay Y, Rizwan M, Alhammad BA, El-Hendawy SE. Chromium resistant microbes and melatonin reduced Cr uptake and toxicity, improved physio-biochemical traits and yield of wheat in contaminated soil. Chemosphere 2020;250:126239.

[27] Cooke J, Leishman MR. Consistent alleviation of abiotic stress with silicon addition: a *meta*-analysis. Funct Ecol 2016;30:1340–57.

[28] Frew A, Weston LA, Reynolds OL, Gurr GM. The role of silicon in plant biology: a paradigm shift in research approach. Ann Bot 2018;121:1265–73.

[29] Liang Y, Nikolic M, Bélanger R, Gong H, Song A. Silicon in agriculture. Dordrecht: Springer; 2015. ISBN 9789401799775.

[30] Imtiaz M, Rizwan MS, Mushtaq MA, Ashraf M, Shahzad SM, Yousaf B, et al. Silicon occurrence, uptake, transport and mechanisms of heavy metals, minerals and salinity enhanced tolerance in plants with future prospects: a review. J Environ Manag 2016;183:521–9.

[31] Ma JF, Yamaji N, Mitani N, Tamai K, Konishi S, Fujiwara T, et al. An efflux transporter of silicon in rice. Nature 2007;448:209–12.

[32] Nable RO, Lance RCM, Cartwright B. Uptake of boron and silicon by barley genotypes with differing susceptibilities to boron toxicity. Ann Bot 1990;66:83–90.

[33] Greger M, Landberg T, Vaculík M. Silicon influences soil availability and accumulation of mineral nutrients in various plant species. Plants 2018;7:41.

[34] Hodson MJ, White PJ, Mead A, Broadley MR. Phylogenetic variation in the silicon composition of plants. Ann Bot 2005;96:1027—46.

[35] Mitani N, Ma JF. Uptake system of silicon in different plant species. J Exp Bot 2005;56:1255—61.

[36] Prychid CJ, Rudall PJ, Gregory M. Systematics and biology of silica bodies in monocotyledons. Bot Rev 2003;69:377—440.

[37] Belanger RR, Benhamou N, Menzies JD. Cytological evidence of an active role of silicon in wheat resistance to powdery mildew (*Blumeria graminis* f. sp. tritici). Phytopathology 2003;93:402—12.

[38] Cote-Beaulieu CF, Chain JG, Menzie SD, Kinrade, Belanger RR. Absorption of aqueous inorganic and organic silicon compounds by wheat and their effect on growth and powdery mildew control. Environ Exp Bot 2009;65:155—61.

[39] Datnoff LE, Rodrigues FA, Seebold KW. Silicon and plant disease. In: Datnoff LE, Elmer WH, Huber DM, editors. Mineral nutrition and plant disease. Paul, MN: APS Press, St.; 2007. p. 233—46.

[40] Ma JF, Yamaji N. Silicon uptake and accumulation in higher plants. Trends Plant Sci 2006;11:392—7.

[41] Cocker KM, Evans DE, Hodson MJ. The amelioration of aluminium toxicity by silicon in higher plants: solution chemistry or an in planta mechanism? Physiol Plant 1998;104:608—14.

[42] Matichenkov VV, Bocharnikova EA. The relationship between silicon and soil physical and chemical properties. In: Datnoff LE, Snyder GH, Korndorfer GH, editors. Silicon in agriculture: studies in plant science. Amsterdam: Elsevier Science; 2001. p. 209—19.

[43] Meena VD, Dotaniya ML, Coumar V, Rajendiran S, Kundu S, Rao AS. A case for silicon fertilization to improve crop yields in tropical soils. Proc Indian Natl Sci Acad B Biol Sci 2014;84:505—18.

[44] Gonzalo MJ, Lucena JJ, Hernández-Apaolaza L. Effect of silicon addition on soybean (*Glycine max*) and cucumber (*Cucumis sativus*) plants grown under iron deficiency. Plant Physiol Biochem 2013;70:455—61.

[45] Pavlovic J, Samardzic J, Maksimovic V, Timotijevic G, Stevic N, Laursen KH, et al. Silicon alleviates iron deficiency in cucumber by promoting mobilization of iron in the root apoplast. New Phytol 2013;198:1096—107.

[46] Bityutskii N, Pavlovic J, Yakkonen K, Maksimović V, Nikolic M. Contrasting effect of silicon on iron, zinc and manganese status and accumulation of metal-mobilizing compounds in micronutrient-deficient cucumber. Plant Physiol Biochem 2014;74:205—11.

[47] Li Y, Zhu N, Liang X, Bai X, Zheng L, Zhao J, et al. Silica nanoparticles alleviate mercury toxicity via immobilization and inactivation of Hg(II) in soybean (*Glycine max*). Environ Sci Nano 2020;7:1807—17.

[48] Guerinot ML. Improving rice yields—ironing out the details. Nat Biotechnol 2000;19:417—18.

[49] Briat JF, Vert G. Acquisition et gestion du fer par les plantes. Cah d'e/tudes et de recherché francophones agriculture, vol. 13. 2004. p. 183—201.

[50] Briat JF. Iron dynamics in plants. In: Kader JC, Delseny M, editors. Advances in botanical research:incorporating advances in plant pathology. London: Academic Press; 2007. p. 138—69.

[51] Welch RM, Graham RD. Breeding for micronutrients in staple food crops from a human nutrition perspective. J Exp Bot 2004;55:353—64.

[52] Christin H, Petty P, Ouertani K, Burgado S, Lawrence C, Kassem MA. Influence of iron, potassium, magnesium, and nitrogen deficiencies on the growth and development of sorghum (*Sorghum bicolor* L.) and sunflower (*Helianthus annuus* L.) seedlings. J Biotechnol Res 2009;1:64—71.

[53] Kabir AH, Paltridge NG, Able AJ, Paull JG, Stangoulis JCR. Natural variation for Fe-efficiency is associated with up regulation of strategy I mechanisms and enhanced citrate and ethylene synthesis in *Pisum sativum* L. Planta 2012;235:1409—19.

[54] Pestana M, de Varennes A, Abadia J, Faria EA. Differential tolerance to iron deficiency of citrus rootstocks grown in nutrient solution. Sci Hortic 2005;104:25—36.

[55] Romheld V, Marschner H. Evidence of a specific uptake system for iron phyto-siderophores in roots of grasses. Plant Physiol 1986;78:175—80.

[56] Lindsay WL. Chemical equilibria in soils. New York: Wiley; 1979.

[57] Marschner H. Mineral nutrition of higher plants. 2nd ed. London: Academic Press; 1995.

[58] Hindt MN, Guerinot ML. Getting sense for signals. Regulation of the plant iron deficiency response. Biochim Biophys Acta 2012;1823:1521—30.

[59] Briat JF, Fobis-Loisy, Grignon N, Lobreaux S, Pascal N, Savino G, et al. Cellular and molecular aspects of iron metabolism in plants. Biol Cell 1995;84:69—81.

[60] Longnecker N, Welch RM. Accumulation of apoplastic iron in plant roots. Plant Physiol 1990;92:17—22.

[61] Jin CW, You GY, He YF, Tang C, Wu P, Zheng SJ. Iron deficiency-induced secretion of phenolics facilitates the reutilization of root apoplastic iron in red clover. Plant Physiol 2007;144:278—85.

[62] Baxter IR, Vitek O, Lahner B, Muthukumar B, Borghi M, Morrissey J, et al. The leaf ionome as a multivariable system to detect plant's physiological status. Proc Natl Acad Sci USA 2008;105:12081—6.

[63] Garcia-Mina JM, Bacaicoa E, Fuentes M, Casanova E. Fine regulation of leaf iron use efficiency and iron root uptake under limited iron bioavailability. Plant Sci 2013;198:39—45.

[64] Nazaralian S, Majid A, Irian S, Najafi F, Ghahremaninejad F, Landberg T, et al. Comparison of silicon nanoparticles and silicate treatments in fenugreek. Plant Physiol Biochem 2017;115:25—33.

[65] Al-aghabary K, Zhu Z, Shi Q. Influence of silicon supply on chlorophyll content, chlorophyll fluorescence, and antioxidative enzyme activities in tomato plants under salt stress. J Plant Nutr 2004;27:2101—15.

[66] Feng J, Shi Q, Wang X, Wei M, Yang F, Xu H. Silicon supplementation ameliorated the inhibition of photosynthesis and nitrate metabolism by cadmium (Cd) toxicity in *Cucumis sativus* L. Sci Hortic 2010;123:521—30.

[67] Gottardi S, Iacuzzo F, Tomasi N, Cortella G, Manzocco L, Pinton R, et al. Beneficial effects of silicon on hydroponically grown corn salad (*Valerianella locusta* (L.) Laterr) plants. Plant Physiol Biochem 2012;56:14—23.

[68] Bybordi A. Effect of ascorbic acid and silicium on photosynthesis, antioxidant enzyme activity, and fatty acid contents in canola exposure to salt stress. J Integr Agric 2012;11:1610—20.

[69] Emamverdian A, Ding Y, Xiu Y, Sangari S. Silicon mechanisms to ameliorate heavy metal stress in plants. BioMed Res Int 2018;8492898. Available from: https://doi.org/10.1155/2018/8492898.

[70] Okuda A, Takahashi E. Effect of silicon supply on the injuries due to excessive amounts of Fe, Mn, Cu, As, AI, Co of barley and rice plant. Jpn J Soil Sci Plant Nutr 1962;33:1—8.

[71] Fu YQ, Shen H, Wu DM, Cai KZ. Silicon-mediated amelioration of Fe^{2+} toxicity in rice (*Oryza sativa* L.) roots. Pedosphere 2012;22:795−802.

[72] Pedas P, Schjoerring JK, Husted S. Identification and characterization of zinc-starvation-induced ZIP transporters from barley roots. Plant Physiol Biochem 2009;47:377−83.

[73] Aravind P, Prassad MNV. Zinc protects chloroplasts and associated photochemical functions in cadmium exposed *Ceratophyllum demersum* L., a fresh water macrophyte. Plant Sci 2004;166:1321−7.

[74] Zhao ZQ, Zhu YG, Kneer R, Smith SE. Effect of zinc on cadmium toxicity induced oxidative stressing winter wheat seedlings. J Plant Nutr 2005;28:1947−59.

[75] Broadley MR, White PJ, Hammond JP, Zelko I, Lux A. Zinc in plants. New Phytol 2007;173:677−702.

[76] Cakmak I, Kalayci M, Ekiz H, Braun HJ, Yilmaz A. Zinc deficiency as an actual problem in plant and human nutrition in Turkey: a NATO-science for stability project. Field Crop Res 1999;60:175−88.

[77] Cakmak I. Plant nutrition research: priorities to meet human needs for food in sustainable ways. Plant Soil 2002;247:3−24.

[78] Graham RD, Welch RM, Bouis HE. Addressing micronutrients malnutrition through enhancing the nutritional quality of staple foods principles, perspectives and knowledge gaps. Adv Agron 2001;70:77−142.

[79] Saeed M, Fox RL. Relation between suspension pH and Zn solubility in acid and calcareous soils. Soil Sci 1977;124:199−204.

[80] Imtiaz M. Zn deficiency in cereals (Ph.D. thesis), Reading University, Reading, UK, 1999.

[81] Hafeez FY, Abaid-Ullah M, Hassan MN. Plant growth promoting rhizobacteria as zinc mobilizers: a promising approach for cereals biofortification. Bacteria in agrobiology: crop productivity. New York: Springer; 2013. p. 217−35.

[82] Marschner H, Kirkby EA, Cakmak I. Effect of nutrition mineral status on shoot-root partitioning of photo assimilates and cycling of mineral nutrient. J Exp Bot 1996;47:1255−63.

[83] White PJ, Broadley MR. Biofortification of crops with seven mineral elements often lacking in human diets − iron, zinc, copper, calcium, magnesium, selenium and iodine. New Phytol 2009;182:49−84.

[84] Mehrabanjoubani P, Abdolzadeh A, Sadeghipour HR, Aghdasi M. Impacts of silicon nutrition on growth and nutrient status of rice plants grown under varying zinc regimes. Theor. Exp Plant Physiol 2014;27:19−29.

[85] Snowball K, Robson AD. Symptoms of nutrient deficiencies: lupins. Nedlands: University of Western Australia Press; 1986.

[86] Benton JJ. Agronomic handbook: management of crops, soils and their fertility. Boca Raton, FL: CRC Press; 2003.

[87] Liang YC. Effects of silicon on enzyme activity, and sodium, potassium and calcium concentration in barley under salt stress. Plant Soil 1999;209:217−24.

[88] Liang YC, Chen Q, Liu Q, Zhang WH, Ding RX. Exogenous silicon (Si) increases antioxidant enzyme activity and reduces lipid peroxidation in roots of salt-stressed barley (*Hordeum vulgare* L.). J Plant Physiol 2003;160:1157−64.

[89] Zhu ZJ, Wei GQ, Li J, Qian QQ, Yu JQ. Silicon alleviates salt stress and increases antioxidant enzymes activity in leaves of salt-stressed cucumber (*Cucumis sativus* L.). Plant Sci 2004;167:527−33.

[90] Liang YC, Zhu J, Li ZJ, Chu GX, Ding YF, Zhang J, et al. Role of silicon in enhancing resistance to freezing stress in two contrasting winter wheat cultivars. Environ Exp Bot 2008;64:286−94.

[91] Gong HJ, Zhu XY, Chen KM, Wang SM, Zhang CL. Silicon alleviates oxidative damage of wheat plants in pots under drought. Plant Sci 2005;169:313−21.

[92] Gu HH, Qiu H, Tian T, Zhan SS, Deng THB, Chaney RL, et al. Mitigation effects of silicon rich amendments on heavy metal accumulation in rice (*Oryza sativa* L.) planted on multimetal contaminated acidic soil. Chemosphere 2011;83:1234−40.

[93] Gu HH, Zhan SS, Wang SZ, Tang YT, Chaney RL, Fang XH, et al. Silicon mediated amelioration of zinc toxicity in rice (*Oryza sativa* L.) seedlings. Plant Soil 2012;350:193−204.

[94] Song A, Li P, Fan F, Li Z, Liang Y. The effect of silicon on photosynthesis and expression of its relevant genes in rice (*Oryza sativa* L.) under high-zinc stress. PLoS One 2014;9:e113782.

[95] Gong HJ, Randall DP, Flowers TJ. Silicon deposition in the root reduces sodium uptake in rice (*Oryza sativa* L.) seedlings by reducing bypass flow. Plant Cell Environ 2006;29:1970−9.

[96] Currie HA, Perry CC. Silica in plant biological, biochemical and chemical studies. Ann Bot 2007;100:1383−9.

[97] Huang CF, Yamaji N, Nishimura M, Tajima S, Ma JF. A rice mutant sensitive to Al toxicity is defective in the specification of root outer cell layers. Plant Cell Physiol 2009;50:976−85.

[98] Peleg ZY, Saranga T, Fahima A, Aharoni, Elbaum R. Genetic control over silica deposition in wheat awns. Physiol Plant 2010;140:10−20.

[99] da Cunha KPV, do Nascimento CWA. Silicon effects on metal tolerance and structural changes in maize (*Zea mays* L.) grown on a cadmium and zinc enriched soil. Water Air Soil Pollut 2009;197:323−30.

[100] Wang LJ, Wang YH, Chen Q, Cao WD, Li M, Zhan FS. Silicon induced cadmium tolerance of rice seedlings. J Plant Nutr 2000;23:1397−406.

[101] Hodson MJ, Sangster AG. Aluminium/silicon interactions in conifers. J Inorg Biochem 1999;76:89−98.

[102] Shi X, Zhang C, Wang H, Zhang F. Effect of Si on the distribution of Cd in rice seedlings. Plant Soil 2005;272:53−60.

[103] Neumann D, Zur Nieden U. Silicon and heavy metal tolerance of higher plants. Phytochemistry 2001;56:685−92.

[104] Hernandez-Apaolaza L. Can silicon partially alleviate micronutrient deficiency in plants? A review. Planta 2014;240:447−58.

[105] Siegel A, Siegel H. Metal ions in biological systems: manganese and its role in biological processes. Boca Raton, FL: CRC Press; 2000. p. 37.

[106] Gerber GB, Leonard A, Hantson P. Carcinogenicity, mutagenicity and teratogenicity of manganese compounds. Crit Rev Oncol Hematol 2002;42:25−34.

[107] Howe P, Malcolm H, Dobson S. Manganese and its compounds: environmental aspects. Geneva: World Health Organization; 2004.

[108] Mcbride MB. Environmental chemistry of soils. 1st ed. Oxford: Oxford University Press; 1994.

[109] Kabata-Pendias A, Pendias H. Trace elements in soils and plants. 3rd ed. Boca Raton, FL: CRC Press; 2001.

[110] Mukhopadhyay MJ, Sharma A. Manganese in cell metabolism of higher plants. Bot Rev 1991;57:117−49.

[111] Li P, Song A, Li Z, Fan F, Liang Y. Silicon ameliorates manganese toxicity by regulating both physiological processes and expression of genes associated with photosynthesis in rice (*Oryza sativa* L.). Plant Soil 2015;397:289−301.

[112] Papadakis IE, Giannakoula A, Therios IN, Bosabalidis AM, Moustakas M, Nastou A. Mn-induced changes in leaf structure and chloroplast ultrastructure of *Citrus volkameriana* L. plants. J Plant Physiol 2007;164:100−3.

[113] Husted S, Thomsen MU, Mattsson M, Schjoerring J. Influence of nitrogen and sulphur form on manganese acquisition by barley (*Hordeum vulgare*). Plant Soil 2005;268:309−17.

[114] Williams DE, Vlamis J. The effect of silicon on yield and manganese-54 uptake and distribution in the leaves of barley grown in culture solutions. Plant Physiol 1957;32:404−9.

[115] Horiguchi T, Morita S. Mechanism of manganese toxicity and tolerance of plants. IV. Effect of silicon on alleviation of manganese toxicity of barley. J Plant Nutr 1987;10:2299−310.

[116] Horst WJ, Marschner H. Effect of silicon on manganese tolerance of bean plants (*Phaseolus vulgaris* L.). Plant Soil 1978;50:287−303.

[117] Iwasaki K, Matsumura A. Effect of silicon on alleviation of manganese toxicity in pumpkin (*Cucurbita moschata* Duch cv. Shintosa). Soil Sci Plant Nutr 1999;45:909−20.

[118] Gilkes RJ, McKenzie RM. Geochemistry and mineralogy of manganese in soils. In: Graham R, Hannan RJ, Uren NC, editors. Manganese in soils and plants. Dordrecht: Kluwer Academic Publishers; 1988. p. 23−35.

[119] Shi Q, Bao Z, Zhu Z, He Y, Qian Q, Yu J. Silicon mediated alleviation of Mn toxicity in *Cucumis sativus* in relation to activities of superoxide dismutase and ascorbate peroxidase. Phytochemistry 2005;66:1551−9.

[120] Rogalla H, Romheld V. Role of leaf apoplast in silicon mediated manganese tolerance of *Cucumis sativus* L. Plant Cell Environ 2002;25:549−55.

[121] Wiese H, Nikolic M, Romheld V. Silicon in plant nutrition. Effect of zinc, manganese and boron leaf concentrations and compartmentation. In: Sattelmacher B, Horst WJ, editors. The apoplast of higher plants: compartment of storage, transport and reactions. Dordrecht: Springer; 2007. p. 33−47.

[122] Dragisic J, Bogdanovic J, Maksimovic V, Nikolic M. Silicon modulates the metabolism and utilization of phenolic compounds in cucumber (*Cucumis sativus* L.) grown at excess manganese. J Plant Nutr Soil Sci 2007;170:739−44.

[123] Demirevska-kepova K, Simova-Stoilova L, Stoyanova Z, Holzer R, Feller U. Biochemical changes in barley plants after excessive supply of copper and manganese. Environ Exp Bot 2004;52:253−66.

[124] Agata F, Ernest B. *Meta*-metal interactions in accumulation of V5+, Ni2+, Mo6+, Mn2+ and Cu2+ in under and above ground parts of *Sinapis alba*. Chemosphere 1998;36:1305−17.

[125] Frantz JM, Khandekar S, Leisner S. Silicon differentially influences copper toxicity response in silicon-accumulator and non-accumulator species. J Am Soc Hortic Sci 2011;136:329−38.

[126] Khandekar S, Leisner S. Soluble silicon modulates expression of *Arabidopsis thaliana* genes involved in copper stress. J Plant Physiol 2011;168:699−705.

[127] Nowakowski W, Nowakowska J. Silicon and copper interaction in growth of spring wheat seedlings. Biol Plant 1997;39(3):463−6.

[128] Li J, Leisner SM. Alleviation of copper toxicity in *Arabidopsis thaliana* by silicon addition to hydroponic solutions. J Am Soc Hortic Sci 2008;133(5):670−7.

[129] Keller C, Rizwan M, Davidian JC, Pokrovsky OS, Bovet N, Chaurand P, et al. Effect of silicon on wheat seedlings (*Triticum turgidum* L.) grown in hydroponics and exposed to 0 to 30 μM Cu. Planta 2015;241:847−60.

[130] Pavlovic J, Samardzic J, Kostic L, Laursen KH, Natic M, Timotijevic G, et al. Silicon enhances leaf remobilization of iron in cucumber under limited iron conditions. Ann Bot 2016;118:271−80.

[131] Nikolic DB, Nesic S, Bosnic D, Kostic L, Nikolic M, Samardzic JT. Silicon alleviates iron deficiency in barley by enhancing expression of strategy II genes and metal redistribution. Front Plant Sci 2019;10:416. Available from: https://doi.org/10.3389/fpls.2019.00416.

[132] Bityutskii NP, Yakkonen KL, Petrova AI, Lukina KA, Shavarda AL. Silicon ameliorates iron deficiency of cucumber in a pH-dependent manner. J Plant Physiol 2018;231:364−73.

[133] Jones LHP, Handreck KA. Silica in soils, plants, and animals. Adv Agron 1967;19:107−49.

[134] Haynes RJ. What effect does liming have on silicon availability in agricultural soils? Geoderma 2019;337:375−83.

[135] Farooq MA, Saqib ZA, Akhtar J, Bakhat HF, Pasala RK, Dietz KJ. Protective role of silicon (Si) against combined stress of salinity and boron (B) toxicity by improving antioxidant enzymes activity in rice. Silicon 2019;11:2193−7.

[136] Gunes A, Inal A, Bagci EG, Coban S, Sahin O. Silicon increases boron tolerance and reduces oxidative damage of wheat grown in soil with excess boron. Biol Plant 2007;51:571−4.

[137] Inal A, Pilbeam DJ, Gunes A. Silicon increases tolerance to boron toxicity and reduces oxidative damage in barley. J Plant Nutr 2009;32:112−28.

[138] Liu C, Lu W, Ma Q, Ma C. Effect of silicon on the alleviation of boron toxicity in wheat growth, boron accumulation, photosynthesis activities, and oxidative responses. J Plant Nutr 2017;40:2458−67.

[139] De Souza Junior JP, de Mello Prado R, dos Santos Sarah MM, Felisberto G. Silicon mitigates boron deficiency and toxicity in cotton cultivated in nutrient solution. J Plant Nutr Soil Sci 2019;182:805−14.

[140] Akcay UC, Erkan IE. Silicon induced antioxidative responses and expression of BOR2 and two PIP family aquaporin genes in barley grown under boron toxicity. Plant Mol Biol Rep 2016;34:318−26.

[141] Zajaczkowska A, Korzeniowska J, Sienkiewicz-Cholewa U. Effect of soil and foliar silicon application on the reduction of zinc toxicity in wheat. Agriculture 2020;10:522.

[142] Liu J, Ma J, He C, Li X, Zhang W, Xu F, et al. Inhibition of cadmium ion uptake in rice (*Oryza sativa*) cells by a wall-bound form of silicon. New Phytol 2013;200:691−9.

[143] Li L, Ai S, Li Y, Wang Y, Tang M. exogenous silicon mediates alleviation of cadmium stress by promoting photosynthetic activity and activities of antioxidative enzymes in rice. J Plant Growth Regul 2017;37:602−11.

[144] Nwugo CC, Huerta AJ. The effect of silicon on the leaf proteome of rice (*Oryza sativa* L.) plants under cadmium-stress. J Proteome Res 2011;10:518−28.

[145] Wu J, Geilfus CM, Pitann B, Mühling KH. Silicon-enhanced oxalate exudation contributes to alleviation of cadmium toxicity in wheat. Environ Exp Bot 2016;131:10−18.

[146] Guo W, Zhang J, Teng M, Wang LH. Arsenic uptake is suppressed in a rice mutant defective in silicon uptake. J Plant Nutr Soil Sci 2009;172:867—74.

[147] Marmiroli M, Pigoni V, Savo-Sardaro ML, Marmiroli N. The e_ect of silicon on the uptake and translocation of arsenic in tomato (*Solanum lycopersicum* L.). Environ Exp Bot 2014;99:9—17.

[148] Wu C, Zou Q, Xue S, Mo J, Pan W, Lou L, et al. Effects of silicon (Si) on arsenic (As) accumulation and speciation in rice (*Oryza sativa* L.) genotypes with different radial oxygen loss (ROL). Chemosphere 2015;138:447—53.

[149] Exley C. A possible mechanism of biological silicification in plants. Front Plant Sci 2015;6:853. Available from: https://doi.org/10.3389/fpls.2015.00853.

[150] de Farias Guedes VH, de Mello Prado R, Frazão JJ, et al. Foliar-applied silicon in sorghum (*Sorghum bicolor* L.) alleviate zinc deficiency. Silicon 2020;. Available from: https://doi.org/10.1007/s12633-020-00825-3.

[151] Pascual MB, Echevarria V, Gonzalo MJ, Hernández-Apaolaza L. Silicon addition to soybean (*Glycine max* L.) plants alleviate zinc deficiency. Plant Physiol Biochem 2016;108:132—8.

[152] Becker M, Ngo NS, Schenk MKA. Silicon reduces the iron uptake in rice and induces iron homeostasis related genes. Sci Rep 2020;10:5079. Available from: https://doi.org/10.1038/s41598-020-61718-4.

[153] Blamey FPC, McKenna BA, Li C, Cheng M, Tang C, Jiang H, et al. Manganese distribution and speciation help to explain the effects of silicate and phosphate on manganese toxicity in four crop species. New Phytol 2018;217:1146—60.

[154] Song AL, Li P, Li ZJ, Fan FL, Nikolic M, Liang YC. The alleviation of zinc toxicity by silicon is related to zinc transport and antioxidative reactions in rice. Plant Soil 2011;344:319—33.

[155] Maksimovic JD, Mojovic M, Maksimovic V, Romheld V, Nikolic M. Silicon ameliorates manganese toxicity in cucumber by decreasing hydroxyl radical accumulation in the leaf apoplast. J Exp Bot 2012;63:2411—20.

[156] Liu J, Cai H, Mei C, Wang M. Effects of nano-silicon and common silicon on lead uptake and translocation in two rice cultivars. Front Environ Sci Eng 2015;9:905—11.

[157] Kopittke PM, Gianoncelli A, Kourousias G, Green K, McKenna BA. Alleviation of Al toxicity by Si is associated with the formation of Al—Si complexes in root tissues of sorghum. Front Plant Sci 2017;8:2189. Available from: https://doi.org/10.3389/fpls.2017.02189.

[158] Flora C, Khandekar S, Boldt J, Leisner S. Silicon modulates expression of pathogen defense-related genes during alleviation of copper toxicity in *Nicotiana tabacum*. J Plant Nutr 2021;44:723—33.

[159] Bosnic D, Bosnic P, Nikolic D, Nikolic M, Samardzic J. Silicon and iron differently alleviate copper toxicity in cucumber leaves. Plants 2019;8:554. Available from: https://doi.org/10.3390/plants8120554.

CHAPTER

16

Silicon as a natural plant guard against insect pests

C.M. Kalleshwaraswamy[1], M. Kannan[2] and N.B. Prakash[3]

[1]Department of Agricultural Entomology, College of Agriculture, University of Agricultural and Horticultural Sciences, Shivamogga, India [2]Department of Nano Science and Technology, Tamil Nadu Agricultural University, Coimbatore, India [3]Department of Soil Science and Agricultural Chemistry, College of Agriculture, University of Agricultural Sciences, GKVK, Bangalore, India

16.1 Introduction

Like any other living things, the plants should take nutrients from soil, and invest energy to produce off-springs. As herbivores consume plants, they remove the nutrients from plants that affect the plant performances including offspring production. Hence plants have developed an array of mechanisms to prevent feeding or resist feeding [1]. However, insects have evolved to feed on plants and thus there is a constant warfare between plants defenses and insect herbivory. Plant nutrients have one or more roles to play in plant biology. Few essential and nonessential nutrients have an effect on plant traits that induce resistance or tolerance to herbivory. One such amendment that has been widely studied in the recent past around the world is the use of silicon against biotic and abiotic stress [2]. Silicon (Si) is the second most abundant element in the earth's crust, including plant tissues, accounting for around 28% (wt/wt) of the total [3]. Despite the fact that silica is a nonessential element for plant growth and development, evidence revealed that under unfavorable circumstances it facilitates induced resistance in plants against biotic and abiotic stress [4].

Because of the significant reliance on insecticides and their detrimental consequences around the world, interest in host plant resistance has increased [5]. The search for chemical elicitors to activate host plant defense against insect pests was never-ending [6]. Signaling pathways control the defensive systems, with external supplementation of plant growth regulators and other modifications playing a key role [7]. Exogenous application of Si reduces insect growth and development by interfering with their feeding and biology [8]. Silicon promotes plant resistance in two ways: first, through physical resistance, and second, through chemical defense [9]. In the recent past, a lot of research is being aimed at the utilization of Si as a soil amendment or foliar application, and its mechanism of uptake, expression, effects on herbivory and also on tritropic interactions [10]. McColloch and Salmon (1923) [11] were the first to show that silica plays a role in hessian fly, *Mayetiola destructor* (Say) (Cecidomyiidae: Diptera) resistance in maize. Further, involvement of Si resistance in the sorghum shoot fly, *Atherigona indica infuscata* Emden (Diptera: Muscidae) [12] and the rice stem borer, Chilo simplex Butler (Lepidoptera: Pyralidae) [13] was reported. Recent research findings revealed increased plant resistance to insect herbivores in a variety of Si-fertilized crops, with a rising focus on discovering the underlying mechanisms [14]. The range of insects affected due to increased Si in plants includes folivores, phloem-feeding, and xylem-feeding [15]. This chapter summarizes the impact of Si as a plant guard against insect pests and the mechanisms by which herbivory is affected. The detailed summary of methods of Si application to plants and its effects is represented in Fig. 16.1 and are discussed under the following headings.

Silicon and Nano-silicon in Environmental Stress Management and Crop Quality Improvement.
DOI: https://doi.org/10.1016/B978-0-323-91225-9.00004-2

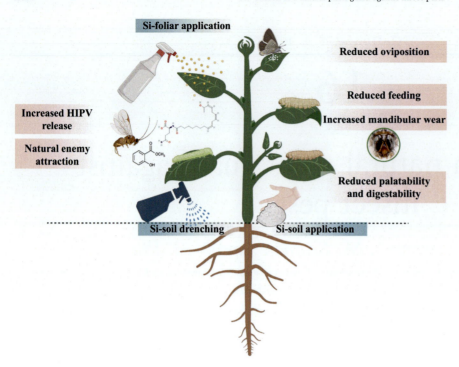

FIGURE 16.1 Methods of Si application to plants and its effect of induced resistance against herbivory. For details see the text.

16.2 Effect of Si on host plant selection for oviposition and feeding

Silicon application is known to reduce the preference of host plants by herbivory [10]. *Schizaphis graminum* (Rondani) (Hemiptera: Aphididae) grown on sodium silicate–treated plants, for example, demonstrated a decrease in feeding preference [16]. Similarly, lowered host preference and reduced aphid colonization were observed due to sodium silicate application. In another report, the number of corn leaf aphids, *Rhopalosiphum maidis* (Aphididae) was significantly reduced on corn leaf treated with either foliar Si sprays or soil application of Si [17].

In the case of *Bemisia tabaci* (Aleyrodidae), oviposition preference was significantly reduced due to Si application [18]. Foliar silicon application was less preferred for oviposition (75.05 eggs) compared to soil drenching (87.08 eggs). Even the source of Si also had varied effects on oviposition preference. Among Si sources, plants treated with SiO_2 had a significantly lower number of oviposited eggs (77.03) compared to K_2SiO_3-treated cotton plants (85.10). Similarly, significantly less oviposition of *B. tabaci* was recorded on plants treated with 200 (84.16 eggs) and 400 (51.83 eggs) ppm Si concentrations as compared to untreated control (107.22 eggs) [18]. Further studies reported that Si applications increase the developmental period and reduction of oviposition preference of *B. tabaci*, resulting in a significant decrease in the pest population, number of generations, and ultimately yield losses in cotton and cucumber crops, respectively [18–20]. A similar type of Si-influenced resistance has been reported in *Zinnia elegans* to *Myzus persicae* Sulzer [21]. High Si addition extended the planthopper, *Nilaparvata lugens* (Stal) stylet pathway and the time it took to reach the first phloem puncture and duration of phloem ingestion thus reducing the quantity of phloem ingestion [22]. In general, at this juncture, it is not clear whether insects can detect the presence of volatiles produced by plants rich in Si. The information generated in this line may be helpful to understand whether herbivores help their offsprings indirectly by avoiding oviposition so that offsprings get benefitted.

16.3 Si physical defense against herbivores

In the contexts of plant defense against herbivory, the chemical defense mechanism has received much more attention than the physical defense [23]. However, Si physical defense mechanism impact is little known [24]. Leaf toughness, hairs, spines, and other structures can act as a physical defense against herbivore feeding [25]. However, the mechanism involved may vary from crop to crop. In grasses, physical defenses induced by silica are considered to be the primary mode of defense than other defenses in herbivory deterrence [26].

The effects of silica on herbivores are thought to be caused by three pathways, both of which are currently lacking in experimental support. The first is the increased abrasiveness of leaves containing silica, which increases mouthpart wear. Second, silica has been shown to reduce herbivores' ability to digest grass leaves. Finally, it acts as a physical barrier within the leaves, restricting nitrogen access [27]. The plants with less nitrogen could be preferred less by insects. Silica may also physically damage the digestive system of insects thus reducing food conversion efficiency [27].

Silica can cause the wearing of mandibles in insects [28]. It is very commonly reported in rice with differing silica contents against stem borers and defoliators [29]. Silica influenced resistance to sugarcane borer, *Eldana saccharina* (Walker) or *Chilo suppressalis* Walker (both belong to Crambidae: Lepidoptera), on rice leaf and stalk silicification which affected delayed stalk penetration by larvae [30,31]. In *Spodoptera exempta*, exposure to silica-rich diets resulted in increased mandible wear [32]. A rapid and instar-wise effect was reported by reducing feeding and growth rates of the larva. These effects on larval growth and feeding efficiency are nonreversible but continued to act even after the larvae moved to different diets The increased exposure of *S. exempta* to silica impacted more adverse effects on the growth and development [33]. A similar type of effect has been reported on one of the serious pests of rice, yellow stem borer (YSB), *Scirpophaga incertulas* (Walker) in India [34]. Thus herbivores cannot fast evolve Si-induced physical defenses of plants, suggesting that this form of defense will have a major effect on herbivore fitness. This was evident in grasshopper, *Oxya grandis* (Acrididae) wherein Mir et al. [35] reported deformation of the incisor teeth through scanning electron microscopy (SEM) micrographs. However, Santos et al. [36] reported no changes in the mandible morphology tomato pinworm larva, *Tuta absoluta* (Meyrick) (Gelechiidae) fed on silicon compared to the control group. This indicates that the effect of silicon may vary across crops and herbivores involved.

Grass family is known to accumulate more Si than other crops. On a dry weight basis, rice shoots have been shown to accumulate Si up to 10%. Silica content of many species of grasses was increased up to seven times due to high silica treatment [37]. This higher content of Si in the shoot or other parts of the plant is known to help in physical defense against herbivory. The uptake of Si is due to the presence and involvement of Si transporter genes [38]. Different Si transporter genes are expressed differently in different parts of the plant. Silicon transporter genes such as *EcLsi1*, *EcLsi2*, and *EcLsi6* are involved in the Si uptake process in Ragi plants [39] inducing resistance to pink stem borer, *S. inferens*. Amendment with Si to Ragi plant significantly increased accumulation in leaf, stem, and root. This increased accumulation led to a significant negative effect on the length of tunnel bored and the boring success of larvae. Same authors identified Si transporter genes in root and stem tissues of finger millet, which include *EcLsi1*, *EcLsi2*, and *EcLsi6* responsible for increased Si uptake followed by its exogenous application. However, in rice, other transporter genes such as *Lsi1* and *Lsi2* were identified as Si transporter genes [40,41]. The mouthparts of many groups of herbivores might be an evolution for grass feeding. There is a need to understand whether larvae of insects can evolve their mandible in response to the Si abrasive effect.

16.4 Effect of Si on palatability and digestibility

Plants are believed to have increased Si accumulation to prevent herbivory in the form of phytoliths through evolutionary adaptation [42,43]. Phytoliths are microscopic bodies of amorphous silica found in cells or as silicified tissue sections toughening the plant tissues leading to averting food intake and digestion [44]. Silicon deposition causes plant tissues to become more abrasive, lowering herbivores' palatability and digestibility [33]. Using scanning electron microscopy with energy-dispersive X-ray spectroscopy (SEM-EDX), Hartley et al. [45] demonstrated that the morphology of the phytolith within the tissues has more influence on the abrasivity than the actual concentration of Si. Keeping et al. [46] used the same technique to show the pattern of Si deposition in sugarcane responsible for increased resistance to *E. saccharina*. A similar type of report suggests that the rescue grass *B. catharticus* seemed to discourage herbivore feeding in silicon-rich diets by making leaves less palatable [35]. Marwat and Baloch [47] observed a negative correlation between Si content in rice plants and YSB damage. Silicon possibly damaged the basement membrane of the alimentary canal and impaired its normal digestive activity.

Si reduces the effect on palatability and digestibility of leaves in herbivores, many additional factors known to increase the effectiveness of Si [33]. One such factor is plant nutrient status and herbivore compensatory ability [48]. In the case of true armyworm, *Pseudeletia unipuncta* Haworth (Lepidoptera: Noctuidae), Si addition resulted in poorly developed mandibles with poor ingestion of food leading to high larval mortality compared to control [49]. However, larvae fed on plants treated with both Si and N lived longer than larvae fed on Si-only plants but the pupal weight did not differ across treatments. Increased postingestion feeding physiology, rather than compensatory food intake, could have accounted for the lack of Si effects on pupal weight [48]. This study enlightens the complex interaction of nutrient dynamics and their effect on herbivore feeding. Some Si-accumulating plants, for example, have lower phenolic compound contents [50]. A similar study of the relationship between root phenolics content and Si concentration is demonstrated in sugarcane in Australia. Frew et al. [51] observed a negative correlation between sugarcane root feeding grub fitness and high Si content in the root. Taking into consideration of many previous studies [50–52], the effect of phenolics is pronounced under low Si content in the host plant parts. Hence, there is a need to evaluate Si-induced defense against insects under combined nutrient experiments.

16.5 Effect of Si on biology, feeding behavior, and performance of insects

The defense mechanisms induced by Si are due to changes in the trichome morphology and increased production and accumulation of secondary metabolites [52,53]. Similarly, Si-treated plants can significantly increase antioxidant enzyme activities and the production of antifungal compounds such as phenolic metabolism product, phytoalexins, and pathogenesis-related proteins. Molecular and biochemical detections show that Si can activate the expression of defense-related genes and may play an important role in the transduction of plant stress signals such as salicylic acid, jasmonic acid, and ethylene [54]. In addition, Si improves plant vigor that can resist extraneous stress [55]. The protective action of Si was initially attributed to physical barriers that strengthen the cell wall. However, recent research has revealed that Si's function on plants is significantly more complex, involving interactions with cell components and metabolism [54]. It is also now known that phytoliths deposited in leaf epidermal cells at definite intervals would be often encountered by chewing larvae of Orthopterans and Lepidoptera, rather than phloem-feeding Hemipterans, which probe from veins and hence phytoliths may not be fed [44]. There are numerous literature reports indicating the Si-induced changes in plant morphology and physiology that have a negative impact on biology, feeding behavior, and fitness of herbivorous insects [56]. But the effect of Si supplementation is higher in grasses than in the plants of other families.

Among defoliating insects, the Si-induced effect has been reported in ortopteran and lepidopteran insects. Massey and Hartley [33] reported the effect of *S. exempta* growth performance on different grasses. *S. exempta* larvae had 40%–66% reduction in the relative growth rate fed on plants with high silica compared to plants of different species with low silica. This growth rate reduction resulted in lower pupal weight. Similarly, the relative growth rate of grasshopper, *Schistocerca gregaria* nymphs was reduced to an extent of 17%–33% on high silica grasses compared with low silica grasses. Variation in growth performance was observed across different species of grasses tested indicating silica uptake varies between plant species and herbivores [33]. Silica addition led to feeding deterrence and brought down the performance of two leaf-feeding chewing insects investigated, but had no negative impacts on the feeding preference or population growth of *S. avenae*, a phloem-feeding aphid. In another study, increasing Si supply reduced ultimate larval survival of *Sesamia calamistis* Hampson in maize from 26% (control) to 4.0% at 0.56 g Si/plant [57].

Similarly, silicon inhibited the *E. saccharina* Walker, both directly and indirectly [58]. The direct effect was by delaying stalk penetration and feeding damage and thus it indirectly led to a prolonged exposure time of young larvae to natural enemies [58]. Sugarcane borer larvae, *Diatraea saccharalis*, had significantly lower relative growth rates on Si-treated plants than on nontreated plants [59]. However, no effects of Si treatment on the lengths of larval and pupal phases, larval and pupal survival, or sex ratio were found in the fall armyworm (FAW), *Spodoptera frugiperda* (J.E. Smith), feeding on maize [60]. FAW larvae-fed leaves from Si-treated plants had lower leaf consumption, larval and pupal weights, male and female longevity, egg quantity, and egg viability. Higher rice Si content was found to be associated with unfavorable consequences. This demonstrated that the foliar application of Si is efficient in negatively effecting *S. frugiperda* development [60,61].

The effect of Si has also been reported against hemipterans [62]. The use of sodium silicate resulted in wheat plant resistance against green aphid *S. graminum* by reducing preference, lifespan, and the number of generations [56,62]. Another aphid pest *M. persicae* was not affected much with Si application. The silicon fertilization did not affect the preference but it reduced fecundity and the rate of population growth of this species of aphid [63].

16.6 Effect of Si on natural enemies and tritrophic interaction

It is a fairly well-established fact that Si directly induces defense against herbivory, very little is targeted on tritrophic level interactions involving natural enemies of insect pests [8]. No clear evidence of Si application on host-herbivore-parasitoids interactions [64]. A few investigations on Si and parasitoids suggest that supplementing with Si increased parasite attraction by altering volatile emissions [54,65] and increased parasitism efficiency [66,67]. Using a Y-tube olfactometer, Kvedaras et al. [68] studied the response of the generalist predatory red and blue beetle, *Dicranolaius bellulus* (Coleoptera: Melyridae) to *Helicoverpa armigera*. A significantly high number of *D. bellulus* chose the olfactometer arm connected to Si$^+$ plants but only fewer predators attracted to the Si$^-$ treatment plants on which *H. armigera* larvae feeding. This indicates the selective preference of predatory beetle to *H. armigera* that is feeding and producing herbivore-induced plant volatiles due to high Si concentration in the plant parts [69]. In the field experiment, a significantly higher level of removal of *H. armigera* eggs where larvae were feeding Si-treated cucumbers than from Si nontreated *H. armigera* feeding plants, indicating a differential attraction of the predators toward cucumber from the lucerne crop. Silicon promotes biological control by increasing natural enemy attraction to pest-infested plants. Postinsect infestation, the activity of natural enemies increases in Si$^+$ plants, which results in efficient biocontrol in the field [68]. It is also demonstrated in one of the serious pests of maize, FAW, *Sp frugiperda* that SI supplementation increases the activity of predatory anthocorid bug, *Orius insidiosus* [70] Contradictorily, because of the significantly larger detrimental impacts on top—down biocontrol, Hall et al. [71] strongly warned that using Si to defend against aphid pests is potentially unwise. External supply of Si affected parasitoid fitness by altering leaf alkaloid concentrations and reducing aphid mummy size [71]. Hence, there is a need of a strong research base, to generalize any such reports and use Si in pest control looking into the tropic levels operating in any cropping ecosystem.

Silicon is also known to increase the activity of entomopathogenic fungi [72]. In a study, the fungus *Beauveria bassiana* and the potassium silicate were added to nutrient solutions and applied to plant roots 7 days after inoculation with spider mite, *Tetranychus urticae* Koch. The pest mites were not killed by potassium silicate alone, but when it was applied at greater rates, corresponding to 80 and 160 mg of pure silicon per liter, pest mortality was induced by *B. bassiana*, which was up to 92%. The authors of the study emphasized biochemical defenses post-Si application which interfered with the mites feeding leading to susceptibility to the entomopathogen [72]. However, there is no clear evidence of Si defense on other groups of biocontrol agents such as entomopathogenic nematodes, viruses, and bacteria widely used in pest control.

16.7 Commercial sources of Si and their induced resistance against herbivory

Exogeneous application of Si is commonly done with soil or as a foliar spray to the crops. Sources of agricultural Si range from chemicals to natural minerals, as well as byproducts from the steel and steel industries [73]. Water-soluble potassium or sodium or calcium silicates' foliar fertilizers are widely used but are expensive [74]. Slow-release Si-containing potassium or rich in potassium that are made with feldspar as raw materials are highly profitable, cheap, and ecofriendly [75]. Another source of Si, the Diatomaceous earth (DE) was also effectively used in field crops. However, there is a need of a strong research base to clarify the effect of DE against biotic stress [76]. Fly ash, another source of silicon, induces resistance against stem borer larvae, grasshopper, leaf folder and blue beetle on rice due to high uptake of Si that caused the wearing of mandibles of chewing insects [77]. Silicon amendment to plants can induce enhanced resistance against insect pests [22]. High silica content of 18 rice cultivars had a negative role with respect to food consumption by surface grasshopper, *Oxya nitidula* (Walker) [78]. Silicon amendment suppressed the *C. suppressalis* through reduced larval performance and feeding damage and indirectly through increased exposure time of larvae to natural enemies and other control tactics [79]. Jeer et al. [34] investigated the effect of rice husk ash, a byproduct of rice milling as a cheap renewable source of Si and commercially available imidazole (a Si solubilizer and carrier) through soil application against YSB, *S. incertulas*, and its damage in five rice varieties. In comparison to the untreated control, all of the soil treatments reduced YSB damage at the vegetative and reproductive phases in all five cultivars tested.

With the application of calcium silicate slag, damage intensity of sugarcane borer, *D. saccharalis* was reduced in five cultivars [80]. Similarly, Keeping and Meyer [81] demonstrated calcium silicate treatment considerably increased resistance at a faster rate than the control against sugarcane borer. The larval weight of *E. saccharina* (Walker) was reduced by 19.8% and stalk length bored by 24.4%. Vilela et al. [82] studied Si-induced resistance in two sugarcane cultivars RB 72454 (moderately resistant) and SP 801842 (susceptible) to *D. saccharalis* (Fabricius). The susceptible cultivar produced similar numbers of holes in Si-treated soil as the moderately resistant variety. Si treatment led to cuticular thickening and crystal formation on stomata of leaves. The addition of Si to the maize plants increased their resistance to the corn leaf aphid *R. maidis* [17]. Si dose in rice at 150 and 100 mg kg^{-1} recorded 59.3% and 49.6% damage, respectively, when compared to the control treatment, which recorded 84.8% damage. This was due to the lower preference of leaves and straw by larvae because of the presence of higher silicon content [83].

According to Keeping and Kvedaras [15], Si is likely to be just as significant in phloem and xylem feeding insect resistance as in folivore or borer resistance. The mechanisms of Si-mediated resistance appear to include one or a mix of constitutive and induced mechanical and chemical defenses for all feeding guilds. Chandramani et al. [84] investigated the effect of induced resistance on rice planthopper in pot culture and in the field. Higher levels of phenol (3.5 and 2.85 mg g^{-1} in stem and leaf), tannin (5.65 and 4.50 mg g^{-1} in stem and leaf), and silica (6.20 and 6.46 mg g^{-1} in stem and leaf) in the treatments conferred induced resistance to rice hoppers via an antibiosis mechanism.

Silicon also regulates the production and accumulation of secondary metabolic compounds, which influence the protective and detoxification enzyme activity of leaf folder larvae, ultimately affected larval survival on rice [85]. Through stimulating the feeding stress defense mechanism, Si increased rice resistance in the leaf folder. Silicon treatment boosts the expression of defense-related genes, which activates the plant defensive enzymes, resulting in higher amounts of defensive chemicals including phenolics, phytoalexins, and momilactones [64].

16.8 Combined effect of Si with other amendments and plant growth regulators

Plant growth regulators (PGRs) and Si both have separately demonstrated capacity to induce resistance against herbivory, whereas a few studies have recorded their combined effect [61]. Foliar silicic acid (FSA) application to maize plants adversely affects the FAW, *Sp. frugiperda* larval growth. Gibberellic acid (GA3) had no effect on mean larval weight when used alone, but when combined with FSA, mean larval weight decreased [61]. Following the treatment of FSA, GA3, and jasmonic acid (JA), the biological parameters of *Sp. frugiperda* were adversely linked with increases in Si, phenols and tannins, and potassium concentration in leaves JA [61]. Gibberellic acid can modify the vegetative properties and Si intake of maize plants, resulting in a decrease in *Sp. frugiperda* larval consumption and female moth oviposition [86]. In wheat, the combination of Si and potassium reduced pink stem borer damage and increased photosynthesis, yield, and other metrics [87]. Si's beneficial impacts on crop growth, yield, and quality have been thoroughly proven [50]. Silicon has boosted plant growth and yield by enhancing biotic and abiotic stress resistance, regulating pH, and obtaining macronutrients and micronutrients with silicate fertilizers [54]. It has also been discovered that when plants are exposed to abiotic and biotic stresses, the beneficial benefits of Si become more apparent [76]. Combining Si and PGRs can be a powerful and long-lasting method for improving plant growth, evaluating the combined use of Si and PGRs on plants suffering from abiotic and biotic stresses could be potential areas of research.

16.9 Conclusions and future prospects

It is now understood beyond doubt that Si application induces resistance against insect pests because of various combined effects. There is a wealth of information and literature accumulated over the years with respect to Si-induced resistance in insect pests. The combined effects on insects such as reduced oviposition, feeding, digestibility, the fitness of insects, and increased activity of biological control agents are well apparent. Further, Si is compatible with other amendments. Although few physical and chemical mechanisms of Si-induced resistance against herbivory are understood, there is a lack of understanding with respect to molecular mechanisms. There is a further need for standardizing dosages, effect of soil type on resistance, method of applications for maximum output, and understanding mode of actions on different groups of feeding guilds.

References

[1] War AR, Paulraj MG, Ahmad T, et al. Mechanisms of plant defense against insect herbivores. Plant Signal Behav 2012;7(10):1306—20. Available from: https://doi.org/10.4161/psb.21663.

[2] Dordas C. Role of nutrients in controlling plant diseases in sustainable agriculture. A review. Agron Sustain Dev 2008;28:33—46. Available from: https://doi.org/10.1051/agro:2007051.

[3] Epstein E. Silicon. Annu Rev Plant Biol 1999;50:641—64.

[4] Luyckx M, Hausman JF, Lutts S, et al. Silicon and plants: current knowledge and technological perspectives. Front. Plant. Sci. 2017;8:411. Available from: https://doi.org/10.3389/fpls.2017.00411.

[5] Dent D. Insect pest management. Wallingford: CAB International; 1991. p. 583.

[6] Hamm JC, Stout MJ, Riggio RM. Herbivore-and elicitor-induced resistance in rice to the rice water weevil (*Lissorhoptrus oryzophilus* Kuschel) in the laboratory and field. J Chem Ecol 2010;36:192—9.

[7] Ruan J, Zhou Y, Zhou M, et al. Jasmonic acid signaling pathway in plants. Int J Mol Sci 2019;20(10):2479.

[8] Leroy N, de Tombeur F, Walgraffe Y, et al. Silicon and plant natural defenses against insect pests: impact on plant volatile organic compounds and cascade effects on multitrophic interactions. Plants 2019;8:444. Available from: https://doi.org/10.3390/plants8110444.

[9] Alhousari F, Greger M. Silicon and mechanisms of plant resistance to insect pests. Plants 2018;7(2):33. Available from: https://doi.org/10.3390/plants7020033.

[10] Reynolds OL, Keeping MG, Meyer JH. Silicon-augmented resistance of plants to herbivorous insects: a review. Ann. Appl. Biol. 2009;155:171−86. Available from: https://doi.org/10.1111/j.1744-7348.2009.00348.x.

[11] McColloch JW, Salmon S. The resistance of wheat to the hessian fly − a progress report. J Econ Entomol 1923;16:293−8.

[12] Ponnaiya BWX. Studies on the genus Sorghum. II. The cause of resistance in sorghum to the insect pest *Atherigona indica* M. Madras Univ J 1953;21:203−17.

[13] Sasamoto K. Studies on the relation between insect pests and silica content in rice plant (II). On the injury of the second generation larvae of rice stem borer. Oyo Kontyu 1953;9:108−10.

[14] Fernández V, Eichert T. Uptake of hydrophilic solutes through plant leaves: current state of knowledge and perspectives of foliar fertilization. Crit Rev Plant Sci 2009;28:36−68. Available from: https://doi.org/10.1080/07352680902743069.

[15] Keeping MG, Kvedaras OL. Silicon as a plant defence against insect herbivory: response to Massey, Ennos and Hartley. J Anim Ecol 2008;77(3):631−3. Available from: https://doi.org/10.1111/j.1365-2656.2008.01380.x.

[16] Carvalho SP, Moraes JC, Carvalho JG. Silica effect on the resistance of *Sorghum bicolor* (L.) Moench to the greenbug *Schizaphis graminum* (Rond.) (Homoptera: Aphididae). An Soc Entomol Bras 1999;28:505−10.

[17] Moraes JC, Goussain MM, Carvalho GA, et al. Feeding non-preference of the corn leaf aphid *Rhopalosiphum maidis* (Fitch, 1856) (Hemiptera: Aphididae) to corn plants (*Zea mays* L.) treated with silicon. Cienc Agrotec 2005;29:761−6.

[18] Abbasi A, Sufyan M, Arif MJ, et al. Effect of silicon on oviposition preference and biology of *Bemisia tabaci* (Gennadius) (Homoptera: Aleyrodidae) feeding on *Gossypium hirsutum* (Linnaeus). Inter J Pest Manage 2020;. Available from: https://doi.org/10.1080/09670874.2020.1802084.

[19] Peixoto ML, Moraes JC, Silva AA, et al. Effect of silicon on the oviposition preference of *Bemisia tabaci* biotype B (Genn.)(Hemiptera: Aleyrodidae) on bean (*Phaseolus vulgaris* L.) plants. Cienc Agrotec 2011;35(3):478−81.

[20] Correa RS, Moraes JC, Auad AM, et al. Silicon and acibenzolar-S-methyl as resistance inducers in cucumber, against the whitefly *Bemisia tabaci* (Gennadius) (Hemiptera: Aleyrodidae) biotype B. Neotrop Entomol 2005;34:429−33.

[21] Ranger CM, Singh AP, Frantz JM, et al. Influence of silicon on resistance of *Zinnia elegans* to *Myzus persicae* (Hemiptera: Aphididae). Environ Entomol 2009;38:129−36.

[22] Yang L, Han Y, Li P, et al. Silicon amendment to rice plants impairs sucking behaviors and population growth in the phloem feeder *Nilaparvata lugens* (Hemiptera: Delphacidae). Sci Rep 2017;7(1):1−7. Available from: https://doi.org/10.1038/s41598-017-01060-4.

[23] Hochuli DF. Does silica defend grasses against invertebrate herbivory? Trends Ecol Evol 1993;8:418−19.

[24] Hochuli DF. The ecology of plant/insect interactions: implications of digestive strategy for feeding by phytophagous insects. Oikos 1996;75:133−41.

[25] Hanley ME, Lamont BB, Fairbanks MM, et al. Plant structural traits and their role in anti-herbivore defence. Perspect Plant Ecol Evol Syst 2007;8:157−78.

[26] Massey FP, Ennos AR, Hartley SE. Silica in grasses as defence against insect herbivores: contrasting effects on folivores and phloem feeder. J Anim Ecol 2007;75:595−603.

[27] Vicari M, Bazely DR. Do grasses fight back − the case for antiherbivore defenses. Trends Ecol Evol 1993;8:137−41.

[28] Drave EH, Lauge G. Study of the action of silica on the wearing of mandibles of the pyralid of rice: *Chilo suppressalis* (F. Walker). Bull Entomol Soc Fr 1978;83:159−62.

[29] Ramachandran R, Khan ZR. Mechanisms of resistance in wild-rice *Oryza brachyantha* to rice leaf folder *Cnaphalocrocis medinalis* (Guenee) (Lepidoptera, Pyralidae). J Chem Ecol 1991;17:41−65.

[30] Kvedaras OL, Byrne MJ, Coombes NE, et al. Influence of plant silicon and sugarcane cultivar on mandibular wear in the stalk borer *Eldana saccharina*. Agric For Entomol 2009;11(3):301−6. Available from: https://doi.org/10.1111/j.1461-9563.2009.00430.x.

[31] Djamin A, Pathak MD. Role of silica in resistance to Asiatic rice borer, *Chilo suppressalis* Walker, in rice varieties. J Econ Entomol 1967;60:347−51.

[32] Massey FP, Hartley SE. Physical defences wear you down: progressive and irreversible impacts of silica on insect herbivores. J Anim Ecol 2009;78(1):281−91.

[33] Massey FP, Hartley SE. Experimental demonstration of the antiherbivore effects of silica in grasses: impacts on foliage digestibility and vole growth rates. Proc R Soc B Biol Sci 2006;273:2299−304.

[34] Jeer M, Telugu UM, Voleti SR, et al. Soil application of silicon reduces yellow stem borer, *Scirpophaga incertulas* (Walker) damage in rice. J Appl Entomol 2017;141(3):189−201.

[35] Mir SH, Rashid I, Hussain B, et al. Silicon supplementation of rescue grass reduces herbivory by a grasshopper. Front Plant Sci 2019;10:671. Available from: https://doi.org/10.3389/fpls.2019.00671.

[36] Santos MC, Junqueira AMR, Mendes VG, et al. Effect of silicon on the morphology of the midgut and mandible of tomato leafminer *Tuta absoluta* (Lepidoptera: Gelechiidae) larvae. Invertebr Surviv J 2015;12:158−65.

[37] Massey FP, Ennos AR, Hartley SE. Silica in grasses as defence against insect herbivores: contrasting effects on folivores and phloem feeder. J Anim Ecol 2006;75:595−603.

[38] Kaur H, Greger M. A review on Si uptake and transport system. Plants 2019;8(4):81. Available from: https://doi.org/10.3390/plants8040081.

[39] Jadhao KR, Bansal A, Rout GR. Silicon amendment induces synergistic plant defense mechanism against pink stem borer (*Sesamia inferens* Walker.) in finger millet (*Eleusine coracana* Gaertn.). Sci Rep 2020;10:4229. Available from: https://doi.org/10.1038/s41598-020-61182-0.

[40] Ma JF, Yamaji NA. Cooperative system of silicon transport in plants. Trends Plant Sci 2015;20:435−42.

[41] Wangkaew B, Prom-u-thai CT, Jamjod S, et al. Silicon concentration and expression of silicon transport genes in two Thai rice varieties. CMU J Nat Sci 2019;18:358−72.

[42] McNaughton SJ, Tarrants JL, McNaughton MM, et al. Silica as a defense against herbivory and a growth promotor in African grasses. Ecology 1985;66:528–35. Available from: https://doi.org/10.2307/1940401.

[43] Endara MJ, Coley PD, Ghabash G, et al. Herbivores "chase" hosts based on plant defenses. Proc Natl Acad Sci USA 2017;114:7499–505. Available from: https://doi.org/10.1073/pnas.1707727114.

[44] Strömberg CAE, Di Stilio VS, Song Z. Functions of phytoliths in vascular plants: an evolutionary perspective. Funct Ecol 2016;30:1286–97. Available from: https://doi.org/10.1111/1365-2435.12692.

[45] Hartley SE, Fitt RN, McLamon EL, et al. Defending the leaf surface: intra- and inter-specifipeci-ing the leaf surface: intra- and nts: an evolutionse to damage and silicon supply. Front Plant Sci 2015;6:35.

[46] Keeping MG, Meyer JH, Sewpersad C. Soil silicon amendments increase resistance of sugarcane to stalk borer Eldana saccharina Walker (Lepidoptera: Pyralidae) under field conditions. Plants Soil 2013;363(1/2):297–318.

[47] Marwat NK, Baloch UK. Varietal resistance in rice to Tryporyza species of stem borers and its association with plant moisture, total ashes and silica contents. Pak J Agric Res 1985;6(4):278–81.

[48] Moise ERD, McNeil JN, Hartley SE, et al. Plant silicon effects on insect feeding dynamics are influenced by plant nitrogen availability. Entomol Exp Appl 2019;167:91–7. Available from: https://doi.org/10.1111/eea.12750.

[49] Frew A, Weston LA, Reynolds OL, et al. The role of silicon in plant biology: a paradigm shift in research approach. Ann Bot 2018;121:1265–73.

[50] Cooke J, Leishman MR. Trade-offs between foliar silicon and carbon-based defences: evidence from vegetation communities of contrasting soil types. Oikos 2012;121:2052–60.

[51] Frew A, Allsopp PG, Gherlenda AN, et al. Increased root herbivory under elevated atmospheric carbon dioxide concentrations is reversed by silicon-based plant defences. J Appl Ecol 2017;54:1310–19.

[52] Schaller J, Brackhage C, Dudel EG. Silicon availability changes structural carbon ratio and phenol content of grasses. Environ Exp Bot 2012;77:283–7.

[53] Samuels AL, Glass ADM, Ehret DL, et al. The effects of silicon supplementation on cucumber fruit: changes in surface characteristics. Ann Bot 1993;72:433–40.

[54] Liu J, Zhu J, Zhang P, et al. Silicon supplementation alters the composition of herbivore induced plant volatiles and enhances attraction of parasitoids to infested rice plants. Front Plant Sci 2017;8:1265. Available from: https://doi.org/10.3389/fpls.2017.01265.

[55] Fauteux F, Rémus-Borel W, Menzies JG, et al. Silicon and plant disease resistance against pathogenic fungi. FEMS Microbiol Lett 2005;249:1–6.

[56] Goussain MM, Prado E, Moraes JC. Effect of silicon applied to wheat plants on the biology and probing behaviour of the greenbug Schizaphis graminum (Rond.) (Hemiptera: Aphididae). Neotrop Entomol 2005;34(5):807–13. Available from: https://doi.org/10.1590/S1519-566X2005000500013.

[57] Sétamou MF, Schulthess F, Bosque-Perez NA, et al. Effect of plant N and Si on the bionomics of Sesamia calamistis Hampson (Lepidoptera: Noctuidae). Bull Entomol Res 1993;83:405–11.

[58] Kvedaras OL, Keeping MG. Silicon impedes stalk penetration by the borer Eldana saccharina in sugarcane. Entomol Exp Appl 2007;125:103–10.

[59] Sidhu JK, Stout MJ, Blouin DC, et al. Effect of silicon soil amendment on performance of sugarcane borer, Diatraea saccharalis (Lepidoptera: Crambidae) on rice. Bull Entomol Res 2013;103(6):656–64.

[60] Nascimento AM, Assis FA, Moraes JC, et al. Silicon application promotes rice growth and negatively affects development of Spodoptera frugiperda (J E Smith). J Appl Entomol 2017;142(1-2):241–9.

[61] Nagaratna W, Kalleshwaraswamy CM, Dhananjaya BC, et al. Effect of silicon and plant growth regulators on the biology and fitness of fall armyworm, Spodoptera frugiperda, a recently invaded pest of Maize in India. Silicon 2021;. Available from: https://doi.org/10.1007/s12633-020-00901-8.

[62] Basagli MA, Moraes JC, Carvalho GA, et al. Effect of sodium silicate application on the resistance of wheat plants to the green-aphids, Schizaphis graminum (Rond.) (Hemiptera: Aphididae). Neotrop Entomol 2003;32:659–63.

[63] Gomes FB, Moraes JCD, Santos CDD, et al. Resistance induction in wheat plants by silicon and aphids. Sci Agric 2005;62(6):547–51.

[64] Reynolds OL, Padula MP, Rensen Z, Gurr GM, et al. Silicon: potential to promote direct and indirect effects on plant defense against arthropod pests in agriculture. Front Plant Sci 2016;7:744. Available from: https://doi.org/10.3389/fpls.2016.00744.

[65] Moraes JC, Goussain MM, Basagli MAB, et al. Silicon influence on the tritrophic interaction: wheat plants, the greenbug Schizaphis graminum (Rondani) (Hemiptera: Aphididae), and its natural enemies, Chrysoperla externa (Hagen) (Neuroptera: Chrysopidae) and Aphidius colemani Viereck (Hymenoptera: Aphidiidae). Neotrop Entomol 2004;33:619–24.

[66] de Oliveira RS, Peñaflor MFGV, Gonçalves FG, et al. Silicon-induced changes in plant volatiles reduce attractiveness of wheat to the bird cherry-oat aphid Rhopalosiphum padi and attract the parasitoid Lysiphlebus testaceipes. PLoS One 2020;15:e0231005. Available from: https://doi.org/10.1371/journal.pone.0231005.

[67] Nikpay A. Improving biological control of stalk borers in sugarcane by applying silicon as a soil amendment. J Plant Prot Res 2016;56:394–401. Available from: https://doi.org/10.1515/jppr-2016-0058.

[68] Kvedaras OL, An M, Choi YS, et al. Silicon enhances natural enemy attraction and biological control through induced plant defences. Bull Entomol Res 2010;100:367–71. Available from: https://doi.org/10.1017/S0007485309990265.

[69] Haftay GG, Nakamuta K. Responses of a predatory bug to a mixture of herbivore-induced plant volatiles from multiple plant species. Arthropod Plant Interact 2016;10:429–44.

[70] Pereira P, Nascimento AM, de Souza BH, et al. Silicon supplementation of maize implacts fall armyworm colonization and increases predator attraction. Neotrop Entomol 2021;50:654–61. Available from: https://doi.org/10.1007/s13744-021-00891-1.

[71] Hall CR, Rowe RC, Mikhael M, et al. Plant silicon application alters leaf alkaloid concentrations and impacts parasitoids more adversely than their aphid hosts. Oecologia 2021;196:145–54. Available from: https://doi.org/10.1007/s00442-021-04902-1.

[72] Gatarayiha MC, Laing MD, Miller RM. Combining applications of potassium silicate and Beauveria bassiana to four crops to control two spotted spider mite, Tetranychus urticae Koch. Int J Pest Manag 2010;56:291–7. Available from: https://doi.org/10.1080/09670874.2010.495794.

[73] Branca TA, Colla V, Algermissen D, et al. Reuse and recycling of by-products in the steel sector: recent achievements paving the way to circular economy and industrial symbiosis in Europe. Metals 2020;10:345. Available from: https://doi.org/10.3390/met10030345.

[74] Meena VD, Dotaniya ML, Coumar V, et al. A case for silicon fertilization to improve crop yields in tropical soils. Proc Natl Acad Sci India Sect B Biol Sci 2014;84:505−18.

[75] Wu L, Liu M. Slow-release potassium silicate fertilizer with the function of superabsorbent and water retention. Ind Eng Chem Res 2007;46(20):6494−500.

[76] Liang Y, Nikolic M, Bélanger R, et al. Silicon in agriculture from theory to practice. Dordrecht: Springer; 2015.

[77] Vijayakumar N, Narayanasamy P. Use of fly ash as a carrier in insecticide formulation In: Fly ash in agriculture. Proceedings og the national seminar on use of lignite fly ash in agriculture, Annamalainagar, India, 28-30; 1995.

[78] Chand DS, Muralirangan MC. Silica of rice cultivars, *Oryza sativa* (Linn.) vs feeding by *Oxya nitidula* (Wlk.). Uttar Pradesh J Zool 2000;20:29−35.

[79] Hou M, Han Y. Silicon-mediated rice plant resistance to the Asiatic rice borer (Lepidoptera: Crambidae): effects of silicon amendment and rice varietal resistance. J Econ Entomol 2010;103(4):1412−19.

[80] Anderson DL, Sosa OJ. Effect of silicon on expression of resistance to sugarcane borer (*Diatraea saccharalis*). J Am Soc Sugar Cane Technol 2001;21:43−50.

[81] Keeping MG, Meyer JH. Calcium silicate enhances resistance of sugarcane to the African stalk borer *Eldana saccharina* Walker (Lepidoptera: Pyralidae). Agric For Entomol 2002;4:265−74.

[82] Vilela M, Moraes JC, Alves E, et al. Induced resistance to *Diatraea saccharalis* (Lepidoptera: Crambidae) *via* silicon application in sugarcane. Rev Colomb Entomol 2014;40(1):44−8.

[83] Ranganathan S, Suvarchala V, et al. Effects of silicon sources on its deposition, chlorophyll content, and disease and pest resistance in rice. Biol Plant 2006;50(4):713−16.

[84] Chandramani P, Rajendra NR, Sivasubramania NP, et al. Management of hoppers in rice through host nutrition. J Biopestic 2009;2(1):99−106.

[85] Han Y, Lei W, Wen L, et al. Silicon-mediated resistance in a susceptible rice variety to the rice leaf folder, Cnaphalocrocis medinalis Guenée (Lepidoptera: Pyralidae). PLoS One 2015;10(4):e0120557. Available from: https://doi.org/10.1371/journal.pone.0120557.

[86] Alvarenga R, Moraes JC, Auad AM, et al. Induction of resistance of corn plants to *Spodoptera frugiperda* (J. E. Smith, 1797) (Lepidoptera: Noctuidae) by application of silicon and gibberellic acid. Bull Entomol Res 2017;107:527−33.

[87] Jeer M, Yele Y, Sharma KC, et al. Exogenous application of different silicon sources and potassium reduces pink stem borer damage and improves photosynthesis, yield and related parameters in wheat. Silicon 2020;13:901−10. Available from: https://doi.org/10.1007/s12633-020-00481-7.

17

Recent developments in silica-nanoparticles mediated insect pest management in agricultural crops

Mallikarjuna Jeer

Entomology, ICAR-National Institute of Biotic Stress Management, Raipur, India

17.1 Introduction

Worldwide, insect pests are the major constraints in agricultural production and productivity. Insect pests cause direct damage to crop plants by feeding on them and indirectly by transmitting plant viruses leading to major yield losses. The global economic losses due to insect pests in different agricultural crops are well documented [1]. The estimated crop losses due to insect pests and other biotic stresses are stood around 32.1% globally [1]. Successful management of insect pests requires continuous application of harmful synthetic pesticides, which creates environmental pollution, severe health hazards, pest resurgence, resistance development, and also increases the cost of cultivation. There is a need to develop alternative, sustainable management practices to combat insect pests and their losses [2].

Silicon (Si) nutrition is one such eco-friendly management practice for the sustainable management of insect pests [3]. Si is the second most abundant element in the earth's crust, with soils containing approximately 32% Si by weight [4]. Si is considered to be a quasi-essential nutrient element for plants as it is not required for plant growth and development; however, it has some beneficial effects on plants [5]. Many researchers across the disciplines proved significant beneficial effects of Si nutrition to plants [6–9]. Si's role in alleviating both biotic and abiotic stresses is well documented and appreciated [9–11]. Si can be supplemented in various forms as the various Si sources are listed in Table 17.1.

Nanotechnology is one such emerging field of science where we can explore better application and utilization of Si as a nutrient source to plants. Nanoparticles exhibit unique physical and chemical properties compared to their bulk material and are more advantageous because of those properties [24]. Similarly, silica nanoparticles (SiNPs) have a diverse role in various fields of agriculture (Fig. 17.1) and exhibit unique physical and chemical properties than other Si sources. Because of their unique properties, SiNPs have a great potential to be used in agriculture for alleviating biotic and abiotic stresses [25,26]. Apart from being a potential Si source, SiNPs can also be used as insecticides for controlling many insect pests without any pesticide hazards like the residual problem, environmental pollution, and resistance development [27]. SiNPs can also act as carriers for pesticides and fertilizers for precise delivery into the plant system [28]. In this chapter, recent developments in nano-silica mediated tolerance against insect pests were thoroughly discussed.

17.2 Synthesis of SiNPs

SiNPs are synthesized mainly by two methods; top-down approach and bottom-up approach [29]. SiNPs can be synthesized from biological resources (biological synthesis) and/or from chemical materials (chemical synthesis). The size of the SiNPs is reduced from starting size through some modern techniques in the top-down

Silicon and Nano-silicon in Environmental Stress Management and Crop Quality Improvement.
DOI: **https://doi.org/10.1016/B978-0-323-91225-9.00016-9**

TABLE 17.1   Different sources of silicon used in various studies and their application method.

Sl. No	Source	Plant available silicon and application type	References
1	Calcium silicate (Ca_2SiO_4)	10%−39% soil application	[12−18]
2	Calcium *meta*-silicate ($CaSiO_3$) (wollastonite)	12% soil drenching	
3	Bagasse furnace ash or fly ash	10% soil application	[19]
4	Blast-furnace slag (Slagment)	18% soil application	
5	Sodium silicate (Na_2SiO_3)	Soil drench and foliar sprays (0.5% or 1% SiO_2) in combination with the soil drench	[20,21]
6	Potassium silicate (K_2SiO_3)	Soil and foliar application	[22,23]
7	Imidazole	A Si solubilizer	[8]
8	Rice husk ash	Soil application	[7,8]
9	Diatomaceous earth	10%−12%	[6]
10	Orthosilicic acid	—	

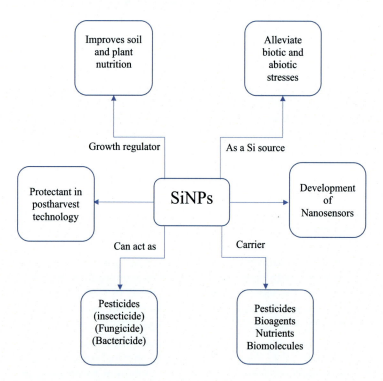

FIGURE 17.1   The different roles of SiNPs in agriculture.

approach; whereas in the bottom-up approach, the synthesis of SiNPs is carried out from atomic or molecular scale [29]. In this regard, the SiNPs can be synthesized through the widely accepted sol−gel process (Fig. 17.2) [31] or microemulsion [32], or flame synthesis [33].

17.2.1 Chemical synthesis

SiNPs can be synthesized from nonbiological materials as reported by many researchers [34−39]. Sodium silicate can be a very good precursor for SiNPs synthesis [34]. Initially, sodium silicate is dissolved in aquadest and stirred at room temperature by adding hydrochloric acid. Later, the SiNPs are recovered by filtration followed by cleaning with distilled water [34]. SiNPs are produced from silicon alkoxide using the modified sol−gel

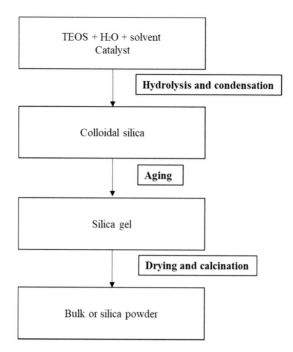

FIGURE 17.2   Flowchart of the typical sol—gel process of SiNPs synthesis [30].

method/Stober method [35]. In this method, the aqueous solution of alcoholic (ethanolic or methanolic) silicon alkoxide is subjected to ultrasonication for 10 min. Further, tetraethyl orthosilicate or tetramethyl orthosilicate is added to an aqueous solution depending on the type of alcohol used and then sonicated again for 10 min. Sodium hydroxide is used as a catalyst to promote condensation [35]. Many workers used nonbiological materials such as bentonite clay, river sand, coal fly ash, and silica fumes for the chemical synthesis of SiNPs [36—39].

17.2.2 Biological synthesis

As the chemical synthesis of SiNPs involves high toxic chemicals, heat energy, and high temperature, the resultant SiNPs have a greater negative impact on soil-plant-human health and environment [40]. Hence, the green synthesis of SiNPs, that is, synthesis from biological raw materials such as agricultural wastes and microbes, is required, which has great potential applications in different fields of agriculture [41].

Due to the wider application of silica nanoparticles in agriculture and allied sectors, it is important to produce SiNPs in large quantities from different sources. In agriculture, waste management is one of the complex and cost-intensive practices, which need to be optimized. Nanotechnology as a science gives us a very good opportunity for utilizing these biological wastes into useful materials such as synthesis of SiNPs from biological/agricultural wastes. SiNPs can be synthesized from various biological materials as reported by many researchers [42,43]. Rice husk can be a good source of biological material for SiNPs synthesis [42,43]. SiNPs can be synthesized from rice husk through an alternative microwave heating process [42]. During the process, acid (hydrochloric acid) leached rice husk is washed and calcinated at 700°C for obtaining the SiNPs (Fig. 17.3). In an alternate method, rice husk is pretreated initially with acid to remove inorganic impurities and hydrolysis of organic substances under pressurized conditions, and later the residues from acid pretreatment are calcinated at different temperatures for obtaining SiNPs [43]. SiNPs can also be prepared from incinerated paddy straw through hydrolysis by acid and alkali treatments [45]. Sugarcane bagasse and sugar beet bagasse can also be a very good source of SiNPs and can be synthesized through the sol—gel method [46,47]. SiNPs are synthesized using cassava periderm and maize stalk as precursors through the modified sol—gel method [48,49]. Some of the agricultural wastes like corn cob straw, ground nutshell, bamboo leaves, and sedge weed can also be used as precursors for obtaining SiNPs [50—53].

Bamboo is one of the good, cheapest biological raw materials from which SiNPs can be synthesized [44]. Initially, after harvesting bamboo, bamboo culms were separated and chop into small pieces of 5 mm size, and

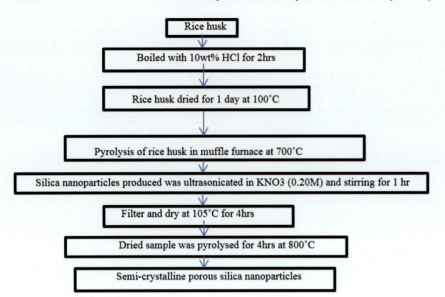

then the small pieces were subjected to pyrolysis at 1250°C temperature in argon atmosphere. The resultant charcoal-bio templates will be cooled at 200°C and inert material created by argon may be removed and SiNPs are synthesized [52].

Microbial agents can be a very good source for the biological synthesis of SiNPs. Many uni- and multicellular microbes produce either organic or inorganic materials intracellularly or extracellularly, which of nanoscale dimensions [54]. Many researchers reported the possible synthesis of SiNPs from microbial agents such as bacteria, fungi, actinomycetes, yeast, and algae [41,55,56].

17.3 Uptake and deposition of SiNPs

The uptake and deposition of Si in crop plants and its efficacy against various biotic stresses is well documented [3]. The plants absorb Si from the solution in the form of mono-silicic acid (H_4SiO_4), which has no electrical charge and is not mobile in the plant [57,58]. The uptake of H_4SiO_4 is nonselective and energetically passive and its transport from root to shoot is in the transpiration stream in the xylem [59]. The uptake of Si takes place via influx and efflux transporter $Ls1$ and $Ls2$, respectively, present in the roots [60]. The uptake mechanisms of SiNPs are not clearly understood but they can be absorbed in relatively more quantity than other silicates [61]. It has been proposed that movement and uptake of SINPs will follow the apoplastic route, as these are less affected by Si transporters [62]. However, very few researchers studied the uptake mechanisms of SiNPs in crop plants [63].

The FITC-labeled SiNPs uptake by the rice roots was well demonstrated with microscopic studies and the uptake was correlated to enhanced germination [64,65]. It was demonstrated that FITC-labeled SiNPs can able to infiltrate even the chloroplast and thereby enhancing the photosynthetic rate of wheat and lupin plants. It was also noticed that the application of mesoporous SiNPs has a tremendous effect in enhancing the growth, silicon, and chlorophyll content in maize plants than the application of bulk Si [66]. The uptake and deposition of SiNPs in the leaves of rice plants was confirmed with scanning electron microscopic studies [67].

17.4 SiNPs versus conventional insecticides in insect pest management

The farming community heavily rely on the use of insecticides for the control of insect pests of various crops due to their quick action and easy application methods [68]. However, there are numerous disadvantages in using insecticides such as environmental pollution, human and animal health hazards, residual toxicity in food chain and soil ecosystem, loss of insecticides as drift, resistance development in insect against insecticides [69], and ban on usage of insecticides in organic farming and so on.

Most of the newer generation pesticides are not water soluble or poorly water soluble in nature and most of the formulations are either emulsifiable concentrate (EC) or oil-in water emulsions or similar kinds [70]. EC formulations consist of organic solvents, which are toxic, flammable, and expensive in nature and oil-in water formulations overcome these drawbacks but they require high energy [71]. To overcome the disadvantages of insecticides, encapsulation of active ingredients of insecticides using nanoparticles and/or developments of nanoemulsions is one of the best solutions [72]. As Si is known for mitigating various biotic and abiotic stresses [11], SiNPs become the first choice for encapsulation of insecticides or carriers [73]. Some of the examples of utilizing SiNPs as insecticide carriers are summarized in Table 17.2.

The role of silica in managing a wide variety of insect pests is well documented [9]. Silicon is known to affect insect pests physically, mechanically, and physiologically [3]. SiNPs have a great advantage over other nanoparticles due to their unique physical and chemical properties. They can also act as a pesticide and can be a good source of Si. SiNPs also act as carrier material for various pesticides for precise delivery [74,75]. However, the role of SiNPs in managing insect pests is limited to storage pests and a few field pests that affect various crops [29]. The major mechanisms of SiNPs are cuticular disruption and dehydration when applied to insects [2]. The other possible mechanism might be blockage of spiracles and trachea and kill insects by adsorption and abrasion [74]. Two different SiNPs were tested against pulse beetle, *Callosobruchus maculatus* at different concentrations and found reduction in the oviposition, larval weight, poor growth, and development [76]. Increased mortality and reduced fecundity were observed in saw-toothed grain beetle, *Oryzaephilus surinamensis* and groundnut bruchid, *Caryedon serratus* with increased dosage and exposure time of SiNPs [77,78]. Fumigation with SiNPs at different concentrations was tested against stored pests like *Sitophilus oryzae*, *Rhizopertha dominica*, *Tribolium castaneum*, and *O. surinamenisis* and noticed increased mortality of adult insects [79]. Further, it was found that adults were desiccated to death due to the adsorption of SiNPs into cuticular lipids.

Some of the insect pests (foliage feeders) of field crops that were recently managed by SiNPs successfully are *Spodoptera littoralis* [80,81], *S. litura*, *Achaea janata* [82], *Mythimna separate* [83], and *Plutella xylostella* [2,84]. The sucking insect pests were also managed by SiNPs, namely, the aphids *Lipaphis pseudobrassicae* [85], *Aphis gossypii*, and the mealybug *Phenacoccus solenopsis* [86]. SiNPs were also tested against a field pest tomato leaf miner, *Liriomyza trifolii* at different concentrations and noticed a significantly lesser number of mines, low survival rate, and puparial weight [27]. Further, it was mentioned that antioxidant enzymes like catalase and superoxide dismutase were upregulated at low concentrations of SiNPs. Increased mortality was observed in diamond back moth, *P. xylostella*, a pest of cruciferous crops due to cuticular abrasion, spiracular obstruction, and desiccation [2]. However, in cotton leaf worm *S. littoralis* increased mortality rate was noticed due to SiNPs but had no significant changes concerning reproduction and development [87] (Table 17.3).

Some of the workers reported physiological alterations and disruption of midgut epithelial cells due to the application of SiNPs without any external symptoms of damage [87]. The cellular proliferation was induced in *S. frugiperda* Sf9 cells when applied with positively charged SiNPs and it was varying with concentrations [88].

TABLE 17.2 Examples for utilization of SiNPs as an insecticide carrier for insect pest control [24].

Insecticide	Crop	Target insect pest	References
Chlorfenapyr	*Brassica* chinese	Cotton bollworm (*H. armigera*) and *P. xylostella*	[89]
a-pinene and Linalool	Castor leaf disks	Tobacco cutworm (*S. litura*) and Castor semi-looper (*A. janata*)	[82]
Abamectin	—	—	[90]
Fipronil	—	Termites	[33]
Avermectin	—	—	[91−93]
	Brassica oleraceae	*P. xylostella* larva	[94]
Chlorpyriphos	Storage insect pests	*Rhyzopertha dominica* F. and *Tribolium confusum*	[95]
Deltamethrin, Pyriproxyfen, and Chlorpyrifos	Storage insect pests	*Trogoderma granarium*	[75]
Buprofezin	Rice	*Nilaprvata lugens*	[96]

TABLE 17.3 Mortality percentage among neonate larvae of *Spodoptera littoralis* tested with nanosilica after 15 days post-emergence [87].

Nanosilica (ppm)	Mortality (%) among neonate larvae of *Spodoptera littoralis* tested with nanosilica after 15 days post-emergence					Mean (%)
	R1	R2	R3	R4	R5	
Control	0.00 (0/148)	0.00 (0/148)	0.00 (0/150)	0.00 (0/146)	0.00 (0/148)	0.00
100 ppm	15.75 (23/146)	12.16 (18/148)	20.40 (30.147)	19.46 (29/149)	12.16 (18/148)	15.98
150 ppm	16.89 (25/148)	21.23 (31/146)	19.33 (29/150)	14.76 (22/149)	25.00 (37/148)	19.44
200 ppm	20.54 (30/146)	29.33 (44/150)	36.66 (55/150)	25.00 (37/148)	30.13 (44/146)	28.33
250 ppm	66.43 (97/146)	54.73 (81/148)	64.86 (96/148)	67.78 (101/149)	67.11 (100/149)	64.18
300 ppm	60.00 (90/150)	50.00 (74/148)	80.00 (120/150)	87.33 (131/150)	67.33 (101/150)	68.93
350 ppm	97.88 (139/142)	98.00 (147/150)	100.00 (150/150)	97.33 (146/150)	97.98 (146/149)	98.24

SiNPs are known to increase the production of glandular trichomes, which are known for defensive responses against insect pests. In tomatoes, the rapid increase in glandular trichome density was observed at an increasing rate of foliar application of SiNPs [97]. However, the types of glandular trichomes that were increased were not studied. This is one area where we need to explore the effect of SiNPs on the production of different types of glandular trichomes and their defensive secretions against insect pests.

Some of the thermosensitive pesticide-controlled release formulations were developed with the help of SiNPs [96]. Buprofezin was loaded into the thermosensitive mesoporous SiNPs via absorption and tested against brown planthopper (BPH) in rice and observed that long-duration effect on BPH and better growth of rice plants. Further, it was noticed that no toxic effects of these thermosensitive SiNPs on zebrafish and human pneumonocyte cells [96].

17.4.1 SiNPs and biocontrol agents

Entomopathogens are widely used for the biocontrol of various insect pests of crops and are regular components in integrated pest management strategies [98]. Entomopathogens include bacteria, fungi, viruses, actinomycetes, and nematodes. The effectiveness of entomopathogens increases under congenial environmental conditions and is mostly species specific [98]. However, there are some bottlenecks in utilizing these entomopathogens for successful management of insect pests such as low shelf life, host specific, environmental sensitive, incompatibility with other management practices like insecticides application, and slow in action [99].

SiNPs by virtue of their stability, sustainability, chemical flexibility, compatibility, and low cost can become a good carrier for microbial control agents. SiNPs can be explored for delivering these biological control agents for insect pest management practices and to overcome disadvantages [100]. Different concentrations of *Penicillium* sp. and SiNPs were evaluated for controlling major potato pest, *Myzus persicae*, and observed increased adult mortality of insects in comparison to control (no SiNPs and *Penicillium* sp.) [34]. It was proved that SiNPs can act synergistically with microbial biocontrol agents and enhance their efficacy [34]. Similarly, the synergetic activity of SiNPs and *Bacillus thuringiensis* was reported in controlling *Caenorhabditis elegans* [101].

There were attempts to develop nanoparticles of *B. thuringiensis* (spores and crystal proteins) and to use as a nanopesticide by reducing its size and some properties [102].

17.5 SiNPs in tri-trophic interactions

As such no harmful effects have been reported by the direct application of bulk Si and very few studies were carried out in this aspect [29]. Soil application of Si enhanced the attraction of predator beetle, *Dicranolaius bellulus* to *Helicoverpa armigera* in cucumber plants. Further, it was elaborated that soil applied Si enhanced herbivore-induced plant volatiles that attracted the predator beetle [103]. Moderately harmful effects of SiNPs were observed on the emergence potential of eggs of green lacewings and egg parasitoid, *Trichogramma chilonis* [104]. It was observed that decreasing rate of emergence of eggs with increasing dosage of SiNPs (Table 17.4). Severe toxicity of SiNPs was

TABLE 17.4 Effect of silica nanoparticles on egg parasitoids, *Trichogramma chilonis* Ishii, and green lacewing, *Chrysoperla zastrowi* Sillemi [104].

Dose (ppm)	Egg parasitoids, *Trichogramma chilonis*			Green lacewing, *Chrysoperla zastrowi sillemi*		Category[a]
	Emergence of eggs (%)	Reduction over control (%)	Emergence of eggs (%)	Reduction over control (%)	Emergence of eggs (%)	
20,000	16.25	19.78	83.20	29.41	69.53	Moderately harmful
15,000	30.41	43.72	68.55	49.07	49.15	Slightly harmful
10,000	52.56	53.92	45.64	52.54	45.56	Harmless
5000	63.92	76.38	33.89	58.75	39.13	Harmless
4500	67.97	79.44	29.70	71.95	25.45	Harmless
1000	76.43	82.65	20.96	76.32	20.93	Harmless
500	78.86	87.53	18.44	76.81	20.41	Harmless
100	84.95	91.34	12.14	77.36	19.85	Harmless
50	93.10	90.38	3.72	93.15	3.48	Harmless
Control	96.69	93.48	—	96.51	—	Harmless

[a] < 50 mortality: harmless; 50–79: slightly harmful; 80–90: moderately harmful; >90: harmful.

reported against bumble bee, *Bombus terrestris* [105]. In a series of toxicological studies, significantly lower mortality or foraging activity of bees was observed at lower concentrations of baited SiNPs. However, at higher concentrations, the damage of midgut epithelial cells of worker bees was observed. By virtue of their smaller size, the foliar application of SiNPs might pose a threat of spiracular blockage and physio-sorption of beneficial arthropods [29]. Hence, the acceptable concentrations of SiNPs for field application must be thoroughly assessed to avoid any harmful effects on beneficial insects. The studies on the effect of long-term exposure to SiNPs are very scarce and need to be assessed for their bioaccumulation process. The effect of either foliar or soil application of SiNPs on beneficial insects like parasitoids, predators, and pollinators both above and below ground needs to be assessed.

17.6 SiNPs and genetic engineering

With the advent of nanotechnology, the use of nanoparticles for transfer/delivery of biomolecules like DNA, RNA, proteins, and/or genetic engineering of crop plants against insect pests is also increasing. The physical and chemical properties of nanoparticles are advantageous in the accurate delivery of biomolecules into the crop systems [106]. In comparison to traditional genetic transformation techniques like *Agrobacterium*-mediated transformation, electroporation, and biolistics, nanoparticle-mediated transformations are more superior due to their more adaptability in a wide range of crops and because of their stable and transient transformation [107]. Among the nanoparticles, SiNPs look promising in targeted biomolecules delivery into the plant cellular systems because of their unique properties like stability, biocompatibility, biodegradability, and less toxic to plants [108]. SiNP-mediated gene transformation method is most successful in comparison to traditional methods due to less cellular/biomolecular damage and faster gene function assessment as in the case of pea (*Pisum sativum* L.) and Arabidopsis (*Arabidopsis thaliana*) [109,110].

17.7 Toxicity of SiNPs to crop plants

The quantity of SiNPs supplementation decides its fate in crop plants. The toxicity of SiNPs mainly depends on the particle size, concentration, method of synthesis, and raw materials used for synthesis [111,112]. It was observed that the supply of SiNPs in higher concentrations affected the growth and development of wheat plants and noticeable alterations were found in biochemical parameters such as chlorophyll content, lipid peroxidation, and higher activity of antioxidant enzymes [113]. Similarly, the decreased root and shoot growth was observed in *Bt*-cotton when supplemented with high dosage of SiNPs [114]. Genomic level toxicity was observed in some of the crops due to higher dosage of SiNPs. In *Allium cepa*, the aberrations in chromosomes along with reduced meiotic index with higher dosage of SiNPs were observed [115]. Similar things were observed in broad bean (*Vicia faba*) due to the application of SiNPs at the rate of 50 and 75 mg L^{-1} [27].

On contrary, some of the workers reported the nontoxic effects on crop plants upon application of SiNPs at higher doses. In rice, the nonsignificant effect of SiNPs dosage on growth and development was observed [116]. Similarly, in the case of potatoes, the five different concentrations of SiNPs had no significant effect [117].

There are some reports of toxicity of SiNPs on soil microflora apart from phytotoxicity. The higher concentrations of SiNPs were found toxic to some of the bacteria like *Bacillus subtilis*, *Escherichia coli*, and *Pseudomonas fluorescens* [118]. However, the toxic effect of SiNPs on soil microbial community is very limited and needs to be explored further.

17.8 SiNPs: Advantages and disadvantages

As the SiNPs have a tremendous potential role in insect pest management owing to their unique characteristics, there are some disadvantages in utilizing these particles. The list of advantages and disadvantages is summarized below.

Advantages	Disadvantages
They are unique in their physical and chemical properties like size, shape, surface tension, porosity, hydrophilic and hydrophobic, solubility	SiNPs synthesized by chemical methods have toxic nature and require high cost
They are cheap, can be synthesized from agricultural wastes, and are easy to use	Nanoformulations have low active ingredient content, nonavailability of surfactants
They can be used as nanovehicles for safe of carrying biomolecules into plant cellular systems for genetic transformation studies due to bioavailability	Lack of supply and production units
They can be used as pesticide carriers for targeted and controlled delivery due to high surface area:volume ratio	They are not insect specific and they can harm beneficial insects also by blocking the tracheal system
They can be used in developing nanoformulations due to tank mix compatibility, low flammability, low drift, and high efficiency	
They are high hydrophobicity, resistance to pH and environmental changes, and have multifunction	
They are also used in DNA detection, purification, and separation	

17.9 Conclusions and future line of work

SiNPs were proved to be an excellent material with diverse applications in the field of agriculture, in general, and in the field of plant protection, in particular. In plant protection, the SiNPs play a potential role as these are environmental-friendly in nature as compared to harmful pesticides. With growing awareness about organic food without any pesticide load, environmental hazards, and climate change, SiNPs may be one of the alternate strategies for insect pest management in emerging production systems like organic farming. The applied aspects of SiNPs are carried out by many researchers like carriers for pesticides, microbial biocontrol agents, plant nutrients, and fertilizers and in genomic engineering. Very limited studies are available on the application of SiNPs as Si sources. The following points need to be taken on priority during any studies on SiNPs:

- The effect of SiNPs on both insect pests and beneficial insects needs to be assessed under field conditions and dosage of application should be critically established.
- The ecological fate of SiNPs needs to be studied after field application with respect to biodegradation, bioaccumulation, and biomagnification.

- The application of SiNPs for insect pest management in emerging production systems like organic farming, protected cultivation, and conservation agriculture needs immediate attention.
- Encapsulation of microbial biocontrol agents using SiNPs and their efficacy under field conditions need to be assessed as they pave the way for efficient biocontrol of insect pests.
- The tri-trophic interactions of SiNPs need to be addressed in insect pest management strategies.
- The application of SiNPs in the field of pesticides should be thoroughly studied in all aspects before commercialization.
- Marketing and easy availability of SiNPs to farmers should be strengthened as these have a potential role in the field of agriculture.

References

[1] Oerke EC. Crop losses to pests. J Agric Sci 2006;144:31–43.

[2] Shoaib A, Elabasy A, Waqas M, et al. Entomotoxic effect of silicon dioxide nanoparticles on *Plutella xylostella* (L.) (Lepidoptera: Plutellidae) under laboratory conditions. Toxicol Environ Chem 2018;100:80–91.

[3] Reynolds OL, Padula MP, Zeng R, Gurr GM. Silicon: potential to promote direct and indirect effects on plant defense against arthropod pests in agriculture. Front Plant Sci 2016;7:744.

[4] Lindsay WL. Chemical equilibria in soil. New York: John Wiley & Sons; 1979.

[5] Epstein E. The anomaly of silicon in plant biology. Proc Natl Acad Sci USA 1994;91(I):1–17.

[6] Jeer M, Yele Y, Sharma KC, Prakash NB. Exogenous application of different silicon sources and potassium reduces pink stem borer damage and improves photosynthesis, yield and related parameters in wheat. Silicon 2021;13:901–10.

[7] Jeer M, Suman K, Telugu UM, Voleti SR, Padmakumari AP. Rice husk ash and imidazole application enhances silicon availability to rice plants and reduces yellow stem borer damage. Field Crop Res 2018;224:60–6.

[8] Jeer M, Telugu UM, Voleti SR, Padmakumari AP. Soil application of silicon reduces yellow stem borer, *Scirpophaga incertulas* (Walker) damage in rice. J Appl Entomol 2017;141:189–201.

[9] Islam W, Tayyab M, Khalil F, Hua Z, Huang Z, Chena HUH. Silicon-mediated plant defense against pathogens and insect pests. Pestic Biochem Physiol 2020;168:104641.

[10] Bakhat FZ, Bibi N, Zia Z, Abbas S, Hammad HM, Fahad S, et al. Silicon mitigates biotic stresses in crop plants: a review. Crop Prot 2008;104:21–34.

[11] Ma JF. Role of silicon in enhancing the resistance of plants to biotic and abiotic stresses. Soil Sci Plant Nutr 2004;50:11–18.

[12] Anderson DL, Sosa OJ. Effect of silicon on expression of resistance to sugarcane borer (*Diatraea saccharalis*). J Am Soc Sugar Cane Technol 2001;21:43–50.

[13] Keeping MG, Meyer JH. Calcium silicate enhances resistance of sugarcane to the African stalk borer, *Eldana saccharina* Walker (Lepidoptera: Pyralidae). Agric For Entomol 2002;4:265–74.

[14] Kvedaras OL, Keeping MG, Goebel FR, Byrne MJ. Larval performance of the pyralid borer, *Eldana saccharina* Walker and stalk damage in sugarcane: influence of plant silicon, cultivar and feeding site. Int J Pest Manag 2007;53:183–94.

[15] Goussain MM, Prado E, Moraes JC. Effect of silicon applied to wheat plants on the biology and probing behaviour of the greenbug *Schizaphis graminum* (Rond.) (Hemiptera: Aphididae). Neotrop Entomol 2005;34:807–13.

[16] Korndorfer AP, Cherry R, Nagata R. Effect of calcium silicate on feeding and development of tropicalsod webworms (Lepidoptera: Pyralidae). Flo Entomol 2004;87:393–5.

[17] Correa RS, Moraes JC, Auad AM, Carvalho GA. Silicon and acibenzolar-S-methyl as resistance inducersin cucumber, against the whitefly *Bemisia tabaci* (Gennadius) (Hemiptera: Aleyrodidae) biotype B. Neotrop Entomol 2005;34:429–33.

[18] Redmond CT, Potter DA. Silicon fertilization does not enhance creeping bentgrass resistance to cutworms and white grubs. USGA Turfgrass Environ Res 2007;6:1–7.

[19] Keeping MG, Meyer JH. Silicon-mediated resistance of sugarcane to *Eldana saccharina* Walker (Lepidoptera: Pyralidae): effects of silicon source and cultivar. J Appl Entomol 2006;130:410–20.

[20] Basagli MA, Moraes JC, Carvalh GA, Ecole CC, Goncalves-Gervasio R, de CR. Effects of sodium silicate application on the resistance of wheat plants tothe green-aphid *Schizaphis graminum* (Rond.) (Hemiptera: Aphididae). Neotrop Entomol 2003;32:659–63.

[21] Moraes JC, Goussain MM, Basagli MAB, Carvalho GA, Ecole CC, Sampaio MV. Silicon influence on the tritrophic interaction: wheat plants, the greenbug *Schizaphis graminum* (Rondani) (Hemiptera: Aphididae), and its natural enemies, *Chrysoperla externa* (Hagen) (Neuroptera: Chrysopidae) and *Aphidius colemani* Viereck (Hymenoptera: Aphidiidae). Neotrop Entomol 2004;33:619–24.

[22] Parrella MP, Costamagna TP, Kaspi R. The addition of potassium silicate to the fertilizer mix to suppress *Liriomyza* leafminers attacking chrysanthemums. Acta Hortic 2007;747:365–9.

[23] Subbarao DV, Perraju A. Resistance in some rice strains to first-instar larvae of *Tryporyza incertulas* (Walker) in relation to plant nutrients and anatomical structure of the plants. Int Rice Res Newsl 1976;1:14–15.

[24] Worrall EA, Hamid A, Mody KT, Mitter N, Pappu HR. Nanotechnology for plant disease management. Agronomy 2018;8:285.

[25] Tripathi DK, Singh S, Singh VP, Prasad SM, Dubey NK, Chauhan DK. Silicon nanoparticles more effectively alleviated UV-B stress than silicon in wheat (*Triticum aestivum*) seedlings. Plant Physiol Biochem 2017;110:70–81.

[26] Cui J, Liu T, Li F, Yi J, Liu C, Yu H. Silica nanoparticles alleviate cadmium toxicity in rice cells: mechanisms and size effects. Environ Pollut 2017;228:363–9.

[27] Thabet AF, Galal OA, Tuda M, El–Samahy MF, Fujita R, Hino M. Silica nanoparticle effect on population parameters and gene expression of an internal feeder, American serpentine leafminer. Preprints 2020;100369.

[28] Panpatte DG, Jhala YK. Nanotechnology for agriculture: crop production & protection. Singapore: Springer Nature; 2019.

[29] Cáceres M, Vassena CV, Dolores BM, Garcerá Pablo L, Santo-Orihuela D. Silica nanoparticles for insect pest control. Curr Pharm Des 2019;25:1–9.

[30] Rahman IA, Padavettan V. Synthesis of silica nanoparticles by sol-gel: size-dependent properties, surface modification, and applications in silica-polymer nanocomposites—a review. J Nanomater 2012;132424.

[31] Klabunde KJ. Nanoscale materials in chemistry. New York: Wiley-Interscience; 2001.

[32] Heinemann S, Coradin T, Desimone MF. Bio-inspired silica collagen materials: applications and perspectives in the medical field. Biomater Sci 2013;1:688–702.

[33] Wibowo D, Zhao CX, Peters BC, Middelberg AP. Sustained release of fipronil insecticide in vitro and in vivo from biocompatible silica nanocapsules. J Agric Food Chem 2014;62:12504–11.

[34] Hersanti H, Hidayat S, Susanto A, Virgiawan R, Joni IM. The effectiveness of *Penicillium* sp. mixed with silica nanoparticles in controlling *Myzus persicae*. AIP Conf Proc 2018;1927.

[35] Debnath N, Das S, Seth D, Chandra R, Bhattacharya SC, Goswami A. Entomotoxic effect of silica nanoparticles against *Sitophilus oryzae* (L.). J Pest Sci 2011;84:99–105.

[36] Zulfiqar U, Subhani T, Husain SW. Synthesis and characterization of silica nanoparticles from clay. J Asian Ceram Soc 2016;4:91–6.

[37] Yan F, Jiang J, Tian S, Liu Z, Shi J, Li K, et al. A green and facile synthesis of ordered mesoporous nanosilica using coal fly ash. ACS Sustain Chem Eng 2016;4:4654–61.

[38] Jafari V, Allahverdi A, Vafaei M. Ultrasound-assisted synthesis of colloidal nanosilica from silica fume: effect of sonication time on the properties of product. Adv Powder Technol 2014;25:1571–7.

[39] Kuddus A, Islam R, Tabassum S, Abu AB. Synthesis of Si NPs from river sand using the mechanochemical process and its applications in metal oxide heterojunction solar cells. Silicon 2019;12:1723–33.

[40] Naik BS. Nanoparticles from endophytic fungi and their efficacy in biological control. In: Thangadurai D, et al., editors. Nanotechnology for food, agriculture, and environment, nanotechnology in the life sciences. Switzerland: Springer Nature; 2020.

[41] Margarita S. Pesticides formulations, effects, fate. In: Tech Janeza Trdine 9, 51000. Rijeka, Croatia; 2013.

[42] Peres EC, Slaviero JC, Cunha AM, Hosseini-Bandegharaei A, Dotto GL. Microwave synthesis of silica nanoparticles and its application for methylene blue adsorption. J Environ Chem Eng 2018;6:649–59.

[43] Vaiyapuri JA, Periasamy S, Khalid MA, AlatiahAli A, Alshatwi A. Synthesis of biogenic silica nanoparticles from rice husks for biomedical applications. Ceram Int 2015;41:275–81.

[44] Snehal S, Lohani P. Silica nanoparticles: its green synthesis and importance in agriculture. J Pharmacogn Phytochem 2018;7:3383–93.

[45] Uda MNA, Gopinath SCB, Uda H, Halim NH, Parmin NA, Uda MNA, et al. Production and characterization of silica nanoparticles from fly ash: conversion of agro-waste into resource. Prep Biochem Biotechnol 2021;51:86–95.

[46] San NO, Kurşungoz C, Tümtas Y, Yas O, Ortaç B, Tekinay T. Novel onestep synthesis of silica nanoparticles from sugarbeet bagasse by laser ablation and their effects on the growth of freshwater algae culture. Particuology 2014;17:29–35.

[47] Falk G, Shinhe GP, Teixeira LB, Moraes EG, Novaesde Oliveira AP. Synthesis of silica nanoparticles from sugarcane bagasse ash and nano-silicon via magnesiothermic reactions. Ceram Int 2019;45:21618–24.

[48] Adebisi JA, Agunsoye JO, Bello SA, Haris M, Ramakokovhu MM, Daramola MO, et al. Green production of silica nanoparticles from maize stalk. Part Sci Technol 2020;38:667–75.

[49] Adebisi JA, Agunsoye JO, Bello SA, Haris M, Ramakokovhu MM, Daramola MO, et al. Extraction of silica from cassava periderm using modified sol-gel method. Niger J Tech Dev 2018;15:57–65.

[50] Rovani S, Santos JJ, Corio P, Fungaro DA. Highly pure silica nanoparticles with high adsorption capacity obtained from sugarcane waste ash. ACS Omega 2018;3:2618–27.

[51] Ghorbani F, Younesi H, Mehraban Z, Çelik MS, Ghoreyshi AA, Anbia M. Preparation and characterization of highly pure silica from sedge as agricultural waste and its utilization in the synthesis of mesoporous silica MCM-41. J Taiwan Inst Chem Eng 2013;44:821–8.

[52] Kumar V, Tiwari P, Krishnia L, Kumari R, Singh A, Ghosh A, et al. Green route synthesis of silicon/silicon oxide from bamboo. Adv Mater 2016;7:271–6.

[53] Vaibhav V, Vijayalakshmi U, Roopan M. Agricultural waste as a source for the production of silica nanoparticles. Spectrochim Acta Part A Mol Biomol Spectrosc 2015;139:515–20.

[54] Adetunji CO, Ugbenyen MA. Mechanism of action of nanopesticide derived from microorganism for the alleviation of abiotic and biotic stress affecting crop productivity. In: Panpatte DG, Jhala YK, editors. Nanotechnology for agriculture: crop production & protection. Singapore: Springer Nature; 2019.

[55] Rai M, Yadav A, Gade A. Current trends in photosynthesis of metal nanoparticles. Crit Rev Biotechnol 2008;28:277–84.

[56] Thakkar KN, Mhatre SS, Parikh RY. Biological synthesis of metallic nanoparticles. Nanomedicine 2010;6:257–62.

[57] Savant NK, Datnoff LE, Snyder GH. Depletion of plant available silicon in soils: a possible cause of declining rice yields. Commun Soil Sci Plant Anal 1997;28:1245–52.

[58] Yoshida S. The physiology of silicon in rice. Technical bulletin No. 25. Taipei, Taiwan: Food Fertilization Technology Centre; 1975.

[59] Jones LHP, Handreck KA. Silica in soils, plants and animals. Adv Agron 1967;19:107–47.

[60] Ma JF, Tamai K, Yamaji N, Mitani N, Konishi S, Katsuhara M, et al. A silicon transporter in rice. Nature 2006;440:688–91.

[61] Asgari F, Majd A, Jonoubi P, Najafi F. Effects of silicon nanoparticles on molecular, chemical, structural and ultrastructural characteristics of oat (*Avena sativa* L.). Plant Physiol Biochem 2018;127:152–60.

[62] Nazaralian S, Majd A, Irian S, Najafi F, Ghahremaninejad F, Landberg T, et al. Comparison of silicon nanoparticles and silicate treatments in fenugreek. Plant physiology and biochemistry. Elsevier Masson SAS; 2017.

[63] Mathur P, Roy S. Nanosilica facilitates silica uptake, growth and stress tolerance in plants. Plant Physiol Biochem 2020;157:114–27.

[64] Nair R, Poulose AC, Nagaoka Y, Yoshida Y, Maekawa T, Kumar DS. Uptake of FITC labeled silica nanoparticles and quantum dots by rice seedlings: effects on seed germination and their potential as biolabels for plants. J Fluoresc 2011;21:2057–68.

[65] Sun D, Hussain HI, Yi Z, Siegele R, Cresswell T, Kong L, et al. Uptake and cellular distribution, in four plant species, of fluorescently labeled mesoporous silica nanoparticles. Plant Cell Rep 2014;33:1389–402.

[66] Suriyaprabha R, Karunakaran G, Yuvakkumar R, Prabu P, Rajendran V, Kannan N. Application of silica nanoparticles for increased silica availability in maize. AIP Conf Proc 2013;1512:424–5.

[67] Félix Alvarez R, Prado R, Felisberto G, Fernandes Deus AC, Lima De Oliveira RL. Effects of soluble silicate and nanosilica application on rice nutrition in an oxisol. Pedosphere 2018;28:597–606.

[68] Lamichhane JR, Dachbrodt-Saaydeh S, Kudsk P, Messéan A. Toward a reduced reliance on conventional pesticides in European agriculture. Plant Dis 2016;100:10–24.

[69] Dey D. Impact of indiscriminate use of insecticide on environmental pollution. Int J Plant Prot 2016;9:264–7.

[70] Knowles A. Global trends in pesticide formulation technology: the development of safer formulations in China. Outlooks Pest Manag 2009;20:165–70.

[71] Kah M, Beulke S, Tiede K, Hofmann T. Nanopesticides: state of knowledge, environmental fate, and exposure modeling. Crit Rev Environ Sci Technol 2013;43:1823–67.

[72] Song SL, Liu XH, Jiang JH, Qian YH, Zhang N, Wu QH. Stability of triazophos in self-nanoemulsifying pesticide delivery system. Colloid Surf A 2009;350:57–62.

[73] Barik TK, Sahu B, Swain V. Nanosilica—from medicine to pest control. Parasitol Res 2008;103:253–8.

[74] Rastogi A, Tripathi DK, Yadav S, Chauhan DK, Zivcak M, Ghorbanpour M, et al. Application of silicon nanoparticles in agriculture. 3 Biotech 2019;9:1–11.

[75] Ziaee M, Babamir-Satehi A. Characterization of nanostructured silica as carrier for insecticides deltamethrin, pyriproxyfen, and chlorpyrifos and testing the insecticidal efficacy against *Trogoderma granarium* (Coleoptera: Dermestidae) larvae. J Econ Entomol 2019;20:1–7.

[76] Irani RY, Karimpour Y, Ziaee M. Oviposition deterrence, progeny reduction and weight loss by *Callosobruchus maculatus* (F.) in pulses treated with two nanosilica formulations. J Plant Prot 2019;41:1–15.

[77] Diagne A, Diop BN, Ndiaye PM, Andreazza C, Sembene M. Efficacy of silica nanoparticles on groundnut bruchid, *Caryedon serratus* (Olivier) (Coleoptera, Bruchidae). Afr Crop Sci J 2019;27:229–35.

[78] Zahran NF, Sayed RM. Protective effect of nanosilica on irradiated dates against saw toothed grain beetle, *Oryzaephilus surinamensis* (Coleoptera: Silvanidae) adults. J Stored Prod Res 2021;92:101799.

[79] El-Naggar ME, Abdelsalam NR, Fouda MMG, Mackled MI, Al-Jaddadi MAM, Ali HM, et al. Soil application of nano silica on maize yield and its insecticidal activity against some stored insects after the post-harvest. Nanomaterials 2020;10:739.

[80] El-Samahy MFM, Khafagy IF, El-Ghobary AMA. Efficiency of silica nanoparticles, two bioinsecticides, peppermint extract and insecticide in controlling cotton leafworm, *Spodoptera littoralis* Boisd and their effects on some associated natural enemies in sugar beet fields. Mansoura J Plant Prot Pathol 2015;6:1221–30.

[81] Ayoub HA, Khairy M, Rashwan FA, Abdel-Hafez HF. Synthesis and characterization of silica nanostructures for cotton leaf worm control. J Nanostruct Chem 2017;7:91–100.

[82] Rani PU, Madhusudhanamurthy J, Sreedhar B. Dynamic adsorption of a-pinene and linalool on silica nanoparticles for enhanced antifeedant activity against agricultural pests. J Pest Sci 2014;87:191–200.

[83] Mousa KM, Elsharkawy MM, Khodeir IA, El-Dakhakhni TN, Youssef AE. Growth perturbation, abnormalities and mortality of oriental armyworm *Mythimna separata* (Walker) (Lepidoptera: Noctuidae) caused by silica nanoparticles and *Bacillus thuringiensis* toxin. Egypt J Biol Pest Control 2014;24:283–7.

[84] Bilal M, Xu C, Cao L, Zhao P, Cao C, Li F, et al. Indoxacarb-loaded fluorescent mesoporous silica nanoparticles for effective control of *Plutella xylostella* L. with decreased detoxification enzymes activities. Pest Manag Sci 2020;76:3749–58.

[85] Debnath N, Das S, Brahmachary RL, Chandra R, Sudan S, Goswami A. Entomotoxicity assay of silica, zinc oxide, titanium dioxide, aluminium oxide nanoparticles on *Lipaphis pseudobrassicae*. AIP Conf Proc 2010;1276:307–10.

[86] Pavitra G, Sushila N, Sreenivas AG, Ashok J, Sharanagouda H. Biosynthesis of green silica nanoparticles and its effect on cotton aphid, *Aphis gossypii* Glover and mealybug, *Phenacoccus solenopsis* Tinsley. Int J Curr Microbiol Appl Sci 2018;7:1450–60.

[87] El-bendary HM, El-Helaly AA. First record nanotechnology in agricultural: Silica nanoparticles a potential new insecticide for pest control. Appl Sci Rep 2013;4:241–6.

[88] Santo-Orihuela PL, Foglia ML, Targovnik AM, Miranda MV, Desimone MF. Nanotoxicological effects of SiO$_2$ nanoparticles on *Spodoptera frugiperda* Sf9 cells. Curr Pharm Biotechnol 2016;17:465–70.

[89] Song MR, Cui SM, Gao F, Liu YR, Fan CL, Lei TQ, et al. Dispersible silica nanoparticles as carrier for enhanced bioactivity of chlorfenapyr. J Pestic Sci 2012;37:258–60.

[90] Wang Y, Cui H, Sun C, Zhao X, Cui B. Construction and evaluation of controlled-release delivery system of Abamectin using porous silica nanoparticles as carriers. Nanoscale Res Lett 2014;9:2490.

[91] Wen LX, Li ZZ, Zou HK, Liu AQ, Chen JF. Controlled release of avermectin from porous hollow silica nanoparticles. Pest Manag Sci 2005;61:583–90.

[92] Li ZZ, Xu SA, Wen LX, Liu F, Liu AQ, Wang Q, et al. Controlled release of avermectin from porous hollow silica nanoparticles: influence of shell thickness on loading efficiency, UV-shielding property and release. J Control Rel 2006;111:81–8.

[93] Li ZZ, Chen JF, Liu F, Liu AQ, Wang Q, Sun HY, et al. Study of UV-shielding properties of novel porous hollow silica nanoparticle carriers for avermectin. Pest Manag Sci 2007;63:241–6.

[94] Kaziem AE, Gao Y, Zhang Y, Qin X, Xiao Y, Zhang Y, et al. Amylase triggered carriers based on cyclodextrin anchored hollow mesoporous silica for enhancing insecticidal activity of avermectin against *Plutella xylostella*. J Hazard Mater 2018;359:213–21.

[95] Satehi AB, Ziaee M, Ashrafi A. Silica nanoparticles: a potential carrier of chlorpyrifos in slurries to control two insect pests of stored products. Entomol Gen 2018;37:77–91.

[96] Yang J, Feng J, He K, Chen Z, Chen W, Cao H, et al. Preparation of thermosensitive buprofezin-loaded mesoporous silica nanoparticles by the sol–gel method and their application in pest control. Pest Manag Sci 2021;77:4627–37.

[97] Ahmad B, Masroor M, Khan A, Jaleel H, Shabbir A, Sadiq Y, et al. Silicon nanoparticles mediated increase in glandular trichomes and regulation of photosynthetic and quality attributes in *Mentha piperita* L. J Plant Growth Regul 2020;39:346–57.

[98] Lacey LA, Grzywacz D, Shapiro-Ilan DI, Frutos R, Brownbridge M, Goettel MS. Insect pathogens as biological control agents: back to the future. J Invertebr Pathol 2015;132:1—41.

[99] Lacey LA. Microbial control of insects. In: Capinera JL, editor. Encyclopedia of entomology. Dordrecht: Springer; 2017.

[100] Vimala Devi PS, Duraimurugan P, Chandrika KSVP. *Bacillus thuringiensis*-based nanopesticides for crop protection. In: Opender K, editor. Nano-biopesticides today and future perspectives. Academic Press; 2019.

[101] Qin X, Xiang X, Sun X, Ni H, Li L. Preparation of nanoscale *Bacillus thuringiensis* chitinases using silica nanoparticles for nematicide delivery. Int J Biol Macromol 2016;82:13—21.

[102] Mahadeva Swamy HM, Asokan R. *Bacillus thuringiensis* as 'Nanoparticles'—a perspective for crop protection. Nanosci Nanotechnol Asia 2013;3:102—5.

[103] Kvedaras OL, An M, Choi YS, Gurr GM. Silicon enhances natural enemy attraction and biological control through induced plant defences. Bull Entomol Res 2010;100:367—71.

[104] Kannan M, Elango K. Effect of silica nano particles to egg parasitoids, *Trichogramma chilonis* Ishii and green lacewing, *Chrysoperla zastrowi sillemi* (Esben-Peterson) (unpublished data); 2019.

[105] Mommaerts V, Jodko K, Thomassen LCJ, Martens JA, Kirsch- Volders M, Smagghe G. Assessment of side-effects by Ludox TMA silica nanoparticles following a dietary exposure on the bumblebee *Bombus terrestris*. Nanotoxicology 2012;6:554—61.

[106] Li Y, et al. Combinatorial library of light-cleavable lipidoid nanoparticles for intracellular drug delivery. ACS Biomater Sci Eng 2019;5:2391—8.

[107] Serag MF, et al. Nanobiotechnology meets plant cell biology: carbon nanotubes as organelle targeting nanocarriers. RSC Adv 2013;3:4856—62.

[108] Hussain HI, et al. Mesoporous silica nanoparticles as a biomolecule delivery vehicle in plants. J Nanopart Res 1676;2013:15.

[109] Yi Z, et al. Functionalized mesoporous silica nanoparticles with redox-responsive short-chain gatekeepers for agrochemical delivery. ACS Appl Mater Interfaces 2015;7:9937—46.

[110] Tripathi DK, et al. Silicon nanoparticles (SiNP) alleviate chromium (VI) phytotoxicity in *Pisum sativum* (L.) seedlings. Plant Physiol Biochem 2015;96:189—98.

[111] Rizwan M, Ali S, Qayyum MF, Ok YS, Adrees M, Ibrahim M, et al. Effect of metal and metal oxide nanoparticles on growth and physiology of globally important food crops: a critical review. J Hazard Mater 2017;322:2—16.

[112] Yanga J, Cao W, Rui Y. Interactions between nanoparticles and plants: phytotoxicity and defense mechanisms. J Plant Interact 2017;12:158—69.

[113] Karimi J, Mohsenzadeh S. Effects of silicon oxide nanoparticles on growth and physiology of wheat seedlings. Russ J Plant Physiol 2016;63:119—23.

[114] Le VN, Rui Y, Gui X, Li X, Liu S, Han Y. Uptake, transport, distribution and bio-effects of SiO$_2$ nanoparticles in *Bt*-transgenic cotton. J Nanobiotechnol 2014;12:1—15.

[115] Silva GH, Monteiro RTR. Toxicity assessment of silica nanoparticles on *Allium cepa*. Ecotoxicol Environ Contam 2017;12:25—31.

[116] Adhikari T, Kundu S, Rao AS. Impact of SiO$_2$ and Mo nano particles on seed germination of rice (*Oryza sativa* L.). Int J Agric Food Sci Technol 2013;4:2249—3050.

[117] Mushinskiy AA, Aminova EV, Korotkova AM. Evaluation of tolerance of tubers *Solanum tuberosum* to silica nanoparticles. Environ Sci Pollut Res 2018;25:34559—69.

[118] Jiang W, Mashayekhi H, Xing B. Bacterial toxicity comparison between nanoand micro-scaled oxide particles. Environ Pollut 2009;157:1619—25.

CHAPTER

18

The combined use of silicon/nanosilicon and arbuscular mycorrhiza for effective management of stressed agriculture: Action mechanisms and future prospects

Hassan Etesami[1], Ehsan Shokri[2] and Byoung Ryong Jeong[3]

[1]Soil Science Department, College of Agriculture and Natural Resources, University of Tehran, Karaj, Iran
[2]Department of Nanotechnology, Agricultural Biotechnology Research Institute of Iran (ABRII), Karaj, Iran
[3]Department of Horticulture, Division of Applied Life Science (BK21 Four), Graduate School, Gyeongsang National University, Jinju, Republic of Korea

18.1 Introduction

The global population grows by 80 million people annually, and food shortage in this century is a growing concern. An appropriate solution to meet the world's ever-growing needs is greatly needed, as food security is a hallmark of sustainable development [1]. The per capita arable land is on the decline, and thus maximizing production per unit area is the only strategy to meet the increasing food needs. To do so, new tools should be utilized to increase plant fertility and agricultural productivity. In this regard, microorganisms associated with promoting plant growth play a beneficial role in improving plant yield via various mechanisms in the rhizosphere [2–4]. In fact, such microorganisms are new biological sources and tools with a high potential for root colonization and growth stimulation, to ultimately lead to an increased yield for various crops. Arbuscular mycorrhizal fungi (AMF) are the most common among such microorganisms [5]. These fungi stimulate plant growth by improving water relations, increasing tolerance against abiotic stresses, providing protection against soil-borne pathogens, and enhancing mineral uptake, particularly of phosphorous [5–14].

Silicon is now generally regarded as a beneficial element, helping a variety of plants and crops mitigate stresses, which include drought, herbivory, extreme temperatures, salinity, imbalanced nutrition, metal toxicity, and UV radiation [15–20]. It has been observed that silicon promotes antipathogenic [19,21] and antioxidative activities [22,23], root growth in drought [24], regulation of certain stress-tolerance-related gene expressions [25,26], and stress signaling [19]. After oxygen, silicon is the second most common constituent of soils, at 28%. It has a strong affinity with O_2, and thus it always exists as silica (SiO_2) or silicate (SiO_4^{4-}). Silicon is found as a compound with many minerals in rocks rather than in its elemental form in nature [27]. Despite its abundance, most silicon in soils is not readily available for plants to absorb [27]. Plants uptake silicon in the form of monosilicic acid or orthosilicic acid H_4SiO_4, whose concentration in soils is between 0.1 and 0.6 mM [28]. Plants take up silicon actively, passively, or rejectively, along with water [29]. At solution pH levels lower than 9, roots take up silicic acid $[Si(OH)_4]$, an uncharged monomeric molecule [30], which is then polymerized into silica gel (SiO_2. nH_2O) in the shoots. Different plants have differing silicon accumulating capacities, which can vary between

0.1% and 10.0% [30,31]. This difference in the absorption capacity is attributed to silicon uptake capacity of the roots [32].

In the last few years, nanobiotechnology has received considerable momentum for plant sciences. Various metal oxides as nanoparticles have been applied to improve growth and productivity and protect crops. Nanosilica has especially been prominent in promoting plant growth and enhancing plant resistance to different biotic and abiotic stresses [33].

Because plants have a limited capacity to uptake silicon and nanosilica, their benefits are significant but limited [34]. This is because most plant species cannot benefit from silicon due to the lack of an efficient silicon transport system [35]. To benefit from silicon/nanosilica, plants must transport it from the soil solution to the different tissues. In addition, it is known that silicon must be accumulated in large amounts to lend plants maximum benefits for tolerance against environmental stresses. However, some plants (e.g., legumes that are low to moderate Si accumulators) [36] require some external agent to optimize the silicon uptake. Ma and Yamaji [37] suggested that dicots, which cannot accumulate sufficient silicon for the element to be beneficial. Genetic manipulation of the roots might help plants increase their capacity to take up and accumulate silicon and thus improve their aptitude to overcome abiotic and biotic stresses. However, genetic manipulation of roots is usually time- and labor-consuming.

In agricultural fields, silicon contents are low due to the frequent postharvest silicon-rich litter elimination [38]. Biological and chemical pathways are used for silicon release into the soil [39]. Minerals containing silicon are resistant to decomposition and weathering processes, leading to a low-silicon level in the soil solution. The soil silicon content can considerably vary, from 0.1% to 45% of the dry weight [40], and plants can only absorb silicon present as silicic acid. This results in a low concentration of absorbable silicon in the soil and a low-silicon absorption by plants. According to previous studies, it has been shown that AMF present an alternative option to absorb, translocate, and transfer silicon to the root cells from the environment to the root cells, and as such, may represent another potential mechanism of mycorrhizal enhancement of plant resistance to stresses [34,41,42]. It has been proven that AMF enable plants to exploit the rhizosphere more efficiently and increase nutrient absorption [6]. The different parameters that affect mycorrhizal colonization effectiveness should be strategically investigated, along with implementing systematic approaches to improve the quality and productivity of crops [43]. Silicon has been shown to increase the effectiveness of mycorrhizal colonization in plants [5]. Therefore the main objective of this chapter is to highlight the importance of AMF, silicon, nanosilica, and their combined use for plant stress alleviation. According to the studies reviewed in this chapter, the combined use of silicon and AMF seems to improve plant growth more effectively than their respective isolated applications do.

18.2 Silicon-mediated plant stress alleviation

Silicon fertilization was proposed to stimulate plant growth and help mitigate environmental stresses in plants in an environmentally friendly, ecologically compatible manner [15,16]. Silicon's significant involvement in plant processes was first noted several years ago, as it was observed to be involved with multiple critical plant processes [44]. Silicon is classified as a quasiessential, but not an essential, element for plant development and growth at present. However, an increasing number of evidence is found in the literature that supports that silicon is beneficial to plants, especially under different biotic and abiotic stresses [16]. Silicon's benefits are seen more clearly when plants are stressed, as silicon provides protection against a range of biotic and abiotic stresses [15]. Silicon was observed to increase the photosynthetic pigment contents, photosystem II activities, and the gas exchange in unstressed plants [45–47].

Silicon application has also been reported to improve the yield and quality of agricultural crops [48,49] and enhance the pathogenic resistance in plants [49–51]. Most benefits of silicon application are attributed to the element's accumulation in the cell walls of leaves, hulls, roots, and stems. Silicon accumulation in the roots, for example, provides binding sites for metals and reduces the apoplastic bypass flow, which results in a decreased uptake and translocation of toxic salts and roots to the shoots from the roots. Silicon deposition strengthens the cell walls and reduces transpiration from the cuticles, which increases plant resistance to stresses such as extreme temperatures, radiation, drought, UV irradiation, and lodging. Silicon application to crops is also known to modulate the antioxidant defense system in plants and upregulate their components [22,52,53].

Many researchers have validated that silicon is a vital agent to mitigate stresses in plants [15,16,54–59]. There exists a comprehensive review in the literature of how silicon helps plants tolerate biotic and abiotic stresses [15,16,54,59,60]. A few recent studies are presented further.

Stressed plants produce a higher level of reactive oxygen species (ROS), which leads to ion leakage and protein/lipid peroxidation. It has been observed that silicon helps plants maintain the normal form of membranes, decrease their permeability, and reduce the malondialdehyde (MDA) levels, the end product of lipid peroxidation in environmentally stressed plants [52,61]. Enzymes such as superoxide dismutase (SOD, EC 1.15.1.1), monodehydroascorbate reductase, peroxidase (POD, EC 1.11.1.7), dehydroascorbate reductase, ascorbate peroxidase (APX, EC 1.11.1.11), ascorbate (AsA), glutathione reductase (GR, EC 1.8.1.7), glutathione S-transferase (glyoxalase I and II and redox pools of reduced glutathione and ascorbate), catalase (CAT, EC 1.11.1.6), and total glutathione (GSH) [62–65] are involved in the antioxidant defense systems of plants under various stresses.

Plant growth regulators (PGRs) such as abscisic acid (ABA), methyl jasmonate (MeJ), salicylic acid (SA), indole-3-acetic acid (IAA), and gibberellins (GAs) improve plant tolerance of abiotic stresses in stressful environments [66,67]. For example, in response to salinity, soybean triggered GA production [68]. The regulatory interactions of PGRs and silicon are well-coordinated to mitigate environmental stresses in plants effectively. Silicon influences endogenous phytohormones, which are commonly monitored in observing plant responses to different stresses. SA and MeJ biosynthesis were reduced with silicon application in *Oryza sativa* plants exposed to heavy metal stress, whereas the opposite trend was observed for ABA, confirming the inverse relationship between jasmonic acid (JA)/SA and ABA biosynthesis [69,70]. It was also observed that silicon negatively affects the JA levels in response to wounding [71]. Silicon was observed to mediate phytohormone homeostasis in barley plants and potentially help plant stress tolerance in potassium deficiency when additional stresses like osmotic stress are also involved, as silicon regulates hormone/metabolite homeostasis.

Many studies in the literature report that silicon modifies the concentration of different solutes such as glycine betaine [72], proline [68,73], carbohydrates [74], polyols, antioxidant compounds like total phenolics [75], total free amino acids [76,77], and total soluble sugars, to minimize the osmotic shock created by salinity and the corresponding Na^+ and Cl^- ion toxicity and leads to increased tolerance of drought and salinity stresses. Compatible solutes may also scavenge oxygen radicals [62]. Increased osmolytes and the corresponding improved osmotic adjustment potential following silicon application [78] could explain the increased growth and photosynthetic ability of silicon-treated plants exposed to salinity.

Silicon's help in alleviating drought and salinity stresses have been associated with an increased antioxidant defense capacity [22,52,53]. Silicon benefits activities of enzymes involved in oxidative stress signaling and those that control vital plant activities like starch and sucrose metabolism. The activities of enzymes such as sucrose synthase, sucrose phosphate synthase, sucrose invertase, and sucrose synthase are regulated differently by silicon both the leaves and roots of *Cucumis sativus* L. seedlings in salinity because silicon regulated the carbohydrate metabolism enzyme activities to decrease the soluble sugar levels in the leaves [79].

Maillard, Ali [80] found that silicon altered genes related to ABA metabolisms, such as ADP-ribosylation factor (ADP-RF1), cyclophilin, glyceraldehydes-3-phosphate dehydrogenase, and the sulfur transporter to mitigate sulfur deficiency and osmotic stresses efficiently. Similarly, silicon affects the starch-metabolic, glycolytic, and tricarboxylic acid pathways to help plants deal with osmotic stresses [81].

Carbon assimilation and photosynthesis are responsible for a large part of the growth and biomass accumulation of plants. Silicon application has been widely observed to improve the water use efficiency and net photosynthetic rate [45,82,83], and exogenous silicon application increases the expression of vital photosynthesis-related genes, regardless of the plant stress levels [84,85]. The feedforward stimulation of silicon on the net assimilation rate of CO_2 is basically associated with the enhanced mesophyll conductance, which leads to maximum carboxylation rate and increased chloroplastic CO_2 levels [45].

Silicon application can also improve shoot and root remobilization of amino acids and carbohydrates [45,79]. The soluble carbohydrate content in cucumber leaves decreased in drought and salinity after silicon application [79,86]. Detmann et al. [45] found that the flag leaf glucose, fructose, and sucrose contents in low-silicon mutant rice decreased in response to silicon application, and the flag leaf content of several amino acids in flag leaves of wild type rice decreased following silicon application as well. It has been widely reported that increased silicon accumulation in crops generally results in increased biomass accumulation and yield [87–89]. Silicon can thus accumulate in crops and stimulate higher translocation rates of photoassimilates, resulting in a strengthened carbon sink. Hence increased silicon storage in crops exerts a feedforward effect on immobilizing carbon in the plant biomass. To summarize, silicon accumulation can stimulate photosynthetic pigments and enhance the net photosynthetic rate in crops. From a bigger perspective, applying silicon in agriculture helps immobilize the atmospheric CO_2 as photoassimilates.

It is well established that silicon helps environmentally stressed plants take up nutrients [16]. Previous studies have explored how silicon influences micronutrient (e.g., Zn) uptake of plants in a drought. Stimulated root growth [90] that increases space from which plants may absorb Zn [91] or enhanced concentrations of organic compounds with low molecular weights (e.g., citrate) that contributes to nutrient uptake and root-to-shoot transport to alleviate deficiency symptoms [92] are the likely pathways for how silicon affects micronutrient uptake in plants. Silicon's influence on Zn transporters is also expected to be a factor in the increased Zn uptake following silicon application. Silicon was observed to increase the Fe transporter (IRT1 and IRT2) expression levels [93], which are members of the ZIP (Zrt/IRT-like protein) family that also include Zn transporters. Applying silicon to the soil medium also decreased soil fixation of phosphorous and thus enhanced the soil phosphorous solubility, which resulted in a higher phosphorous uptake by plants [94]. Silicon has also been reported to upregulate the transcription of certain aquaporin genes to increase the root hydraulic conductance [95].

Silicon can also inhibit toxic metal uptake by the roots [96] and improve the acquisition of beneficial and essential minerals such as phosphorous [97–99]. Silicon can be bound to toxic metals in the shoots to neutralize them and transport them to specialized compartments [100,101], maintain the cell wall and cell membrane stability under stresses [23,102], and mediate the apoplastic and symplastic osmotic gradients [17]. Hasanuzzaman et al. [65] discovered that exogenous silicon application attenuated cadmium-induced oxidative stress in *Brassica napus* L. plants by modulating the Asada–Halliwell pathway (AsA-GSH) and enzymes of the glyoxalase system. It was concluded that the activities of the enzymes in the glyoxalase systems and the AsA-GSH pathway were modulated by silicon to maintain the redox state of GSH and AsA, which ameliorated the damages of cadmium-induced oxidative stress.

In addition to helping plants deal with abiotic stresses, silicon helps form strong barriers in the aboveground and belowground parts of plants to protect them against different biotic and abiotic stresses [103]. The cell wall is considered the first target barrier site to reduce the cascade of stresses on plants [79]. It is thought that the inflow into the cortex of excess freely available ions in the soil solution is slowed by components of the cell wall [104]. Silicon has been observed to facilitate the development of cell wall components in many crops, such as *Allium cepa, O. sativa, Guizotia abyssinica, Tradescantia virginiana,* and *Zea mays* [105,106]. Silicon enhances lignification, silicification, and suberization to support certain plant divisions and plays a prominent role in strengthening cell walls [17]. Silicon bonded to hemicellulose has been observed to enhance the structural stability and rigidity [102] to benefit and protect plants from environmental stresses, including osmotic stresses. Through a process known as biosilification, silicon forms an amorphous silica barrier as the silicic acid polymerizes with the cell wall [107] and helps plants resist stresses like penetration of pathogens, metals, and metalloids [25,106,108]. It was observed that silicon increased the number of sodium ions to bind to the cell walls in the roots of wheat plants [82,109,110]. The study by Hinrichs et al. [105] with rice plants elaborates on silicon's role in cell wall synthesis. Silicon-induced Casparian strip formations in rice plants, which help reduce radial oxygen losses. The findings were further confirmed with transcriptomic techniques and identified genes that are potentially involved in silicon's contribution to forming the cell wall components, such as ATP-binding cassette (ABC) transporter, class III POXs, transferases, and ligases. The part that the ABC transporter (*OsABCG25*) plays in the cell wall formation of the exodermis was further investigated by using overexpression and knockout mutants of the aforementioned genes. This was also clear in the expression of other *OsABCG25* transporters in the silicon-induced formation of cell wall genes, which corresponds to 4-coumarate-CoA ligase, phenyl alanine-ammonia-lyase, and diacylglycerol O-acyltransferase of the phenylpropanoid pathway, while the different developments of various cell wall components in mutants and the silicon supply also regulated the cell wall exodermis functions.

18.3 Nanosilica-mediated plant stress alleviation

Nanosilica or Si nanoparticles (Si-NPs) are particles generally smaller than 100 nm synthesized from bulk silica [111,112]. Bulk silica and Si-NPs possess different physiochemical properties [113]. Plant responses to Si-NPs are a function of the application method, size, and shape of the particles, and physiochemical properties [114]. Mesoporous Si-NPs are gaining increasing attention for use in agricultural sciences [115,116]. The higher surface area to volume ratio, better permeability in plant cells, unique thermal and electrical properties put Si-NPs at an advantage compared to their bulk counterpart [111]. Si-NP entry into the plant cells is speculated to be relatively simple silicon transporters (*lsi1*) are less sensitive to Si-NPs, and hence it is proposed that Si-NPs are taken up by the roots mainly through the apoplastic route [117]. The uptake mechanism has yet to be clearly understood, but plants uptake Si-NPs at a relatively higher rate than they do other silicates [118,119].

There are fewer studies investigating how Si-NPs improve plant growth compared to such studies regarding bulk silicon [33]. It is known that Si-NPs can (1) facilitate silica uptake, which increases the expression of the stress-regulating OsNAC proteins, and proline and sugar accumulation; (2) improve stress tolerance and growth of plants [33,120]; and (3) increase seed germination by enhancing nutrient availability. Under stressful conditions (e.g., salinity and drought), Si-NPs improved photosynthesis, stomatal conductance, and leaf pigment (chlorophyll content and carotenoid) contents, blocked the Na^+ and Cl^- uptake to protect plants from toxic salinity, enhanced various growth parameters (plant height, fresh and dry root and shoot weights) and physiological parameters (K, P, and N contents), stimulated enzymatic antioxidant activities (e.g., that of catalase, peroxidase, polyphenol oxidase, and superoxide dismutase, as well as that of other enzymes like lipases and amylases), and enhanced the activities of gibberellins and respiratory enzymes that contribute to a better cell metabolic status [112,121−134]. The increased growth and yield, as well as pigment contents, following Si-NP application in plants under drought and salinity, may be attributed to increased levels of nitrogen and potassium levels in plant tissues [128]. For example, for wheat plants in drought, the application of Si-NPs at concentrations up to 200 mg kg^{-1} increased the uptake of nitrogen, potassium, and a majority of the other minerals [135]. Foliar Si-NP sprays (at 0, 15, 30, 60, and 120 mg L^{-1}) to cucumber (*Cucumis sativus* L.) enhanced the chemical compositions, growth, and yield [121]. Foliar Si-NP applications to rice grown in saline-sodic soils of the Nile delta also improved the yield parameters, such as the chlorophyll content, number of grains per panicle, panicle length, grain yield, and plant height [136]. In another study, Si-NPs reduced Na^+ accumulation in plant tissues, which was crucial to inducing salinity tolerance to bean plants (*Phaseolus vulgaris* L.) [137]. Si-NP applications at concentrations up to 200 mg kg^{-1} to cucumber seedlings in drought and salinity increased the K^+/Na^+ ratio. Si-NPs can also increase the turgor pressure by improving the relative water content and water use efficiency [127,138]. For example, it was demonstrated that Si-NP application relieved drought stress in experiments with *Prunus* sp. [125]. One gram of 7-nm Si-NPs has an approximate absorption surface area of 400 m^2; this large surface area to weight ratio renders Si-NP application effective for improved water translocation and ultimately water use efficiency [139]. Farhangi-Abriz and Torabian [140] reported that a 1 mM solution of Si-NPs applied under salinity stress increased the seedling root and shoot lengths. A controlled release of Si-NP-containing fertilizers may help plants alleviate drought stresses by helping the soil retain water and gradually releasing nutrients [141].

Applying Si-NPs has been found to enhance growth and biomass accumulation in plants exposed to heavy metals [142]. This may be because Si-NPs help alleviates heavy metal stresses in plants [143]. Other mechanisms with which Si-NPs may also help plants grow are by protecting the photosynthetic apparatus from heavy metal toxicity; increasing antioxidant enzyme activities to decrease ROS content and protect cell membrane integrity; upregulating gene expression for heavy metal (e.g., Cd) transport to vacuoles (*OsHMA3*) and silicon uptake (*OsLsi1*); downregulating the genes associated with heavy metal (e.g., Cd) uptake (*OsLCT1* and *OsNRAMP5*); upregulating the expression of osNAC proteins associated with upregulating genes for redox homeostasis, proline synthesis, stress tolerance, and soluble sugar biosynthesis; and promoting nutrient availability to plants [142,144−148]. For example, Si-NPs were observed to simultaneously decrease Cd ion accumulation in rice grains by 30%−60% and increased K, Mg, and Fe translocations [149]. Liu et al. [150] also conducted a trial to investigate how Si-NPs and common silicon affected the uptake and translocation of lead (Pb) and reported a marked reduction of both in rice cultivars. Application of Si-NPs and common silicon increased the biomass yields of rice plants from 3.3% to 11.8% and 1.8% to 5.2%, respectively, in soils treated with 500 and 1000 mg kg^{-1}, compared to the control (without silicon). Moreover, grains accumulated the lowest amount of Pb when treated with silicon salts. The application of Si-NPs and common Si significantly reduced the root-to-shoot and shoot-to-grain Pb translocation. Generally, Si-NPs efficiently remediated the toxic effects of Pb on rice growth. Bharwana et al. [151] also carried out an experiment to explore how Si-NPs alleviated Pb toxicity in cotton and indicated that Si application significantly reduced the Pb uptake, electrolyte leakage, as well as the MDA and hydrogen peroxide (H_2O_2) contents in cotton plants.

Si-NPs give rise to antibacterial and antifungal activities and thus play a significant role in plant disease management [152−155]. For example, applying Si-NPs led to increased resistance of maize plants against *Fusarium* and *Aspergillus*. Increased phenol and defense enzyme concentrations were responsible for this increased resistance [156]. Si-NP fertilization of *Vicia faba* infected with *Botrytis fabae* similarly increased the antioxidant enzyme activities, total phenol, and proline concentrations and led to lower severity of chocolate spot disease [157]. Tuan et al. [158] found that Si-NPs/chitosan hybrid application led to decreased *Pyricularia oryzae*-induced leaf blast index and *Xanthomonas oryzae*-driven bacterial blight disease index in rice. Chitosan-silica composites have been applied as a fungicide to preserve grapes postharvest [159]. Leaf tissue deposition of Si-NPs leads to enhanced plant defense against pathogens [33]. Si-NPs form a physical barrier that limits pathogen entry and protects hydrolyzing enzymes secreted by pathogens from enzymatically degrading cell walls [160]. In addition, Si-NPs affect DNA replication of pathogens, express JA and SA genes to induce systematic defense, increase the growth, lignification of the roots, and the photosynthetic rate, increase the various defense enzyme activities, induce expression of pathogenesis-related proteins (PR1, PR2, and PR5) involved in the systematically acquired resistance pathways in plants, and enhance the production nonenzymatic and enzymatic oxidants [33,153,154,159,161,162]. A dissolving fungal cell membrane, and a disrupted fungal cell wall that affects the electron transport chain that increases silica accumulation and causes cell lysis were also responsible for the decreased pathogen growth [153].

Recently, Si-NP application was also shown to effectively reduce the lately rising fluoride stress in rice plants caused by uncontrolled groundwater extraction [163]. Si-NP application was observed to enhance nonenzymatic and enzymatic antioxidant activities, help plants maintain their ionic balance, and reduce fluoride bioaccumulation and tissue injury.

It is noteworthy that both soil drench and foliar spray of Si-NPs effectively increased the yield and growth of plants and improved the physiological parameters of stressed plants [142,164]. Future studies should explore the stress-mitigating and growth-promoting attributes of Si-NPs for different horticultural crops [120].

18.4 Arbuscular mycorrhizal fungi-mediated plant stress alleviation

The associations between roots and fungi are called mycorrhizae, which are classified as arbuscular, arbutoid, ectoericoid, monoptropoid, and orchid, depending on their morphological characteristics [165]. AMF are the most common and also regarded as the oldest obligate symbionts, which colonizes nearly 80% of the root cortex of the terrestrial plant order biographically and whose identity is 400 million years old [166]. These fungi are from the phylum Glomeromycota, order Glomales [167], describing around 340 species [168]. Parniske [169] defined arbuscular mycorrhizal (AM) symbiosis in natural ecosystems as "the mother of plant root endosymbiosis." AM symbiosis helps the fungi get as much as 20% of the fixed carbon from plants, while the plants receive an improved supply of nutrients like nitrogen and phosphorous [43,170]. AMF have been recognized as biocontrol, bioenhancers, and biostimulant agents [171]. AM symbiosis is one of the most common ways with which plants mitigate biotic and abiotic stresses. The symbiosis has traditionally been associated with an improved acquisition of nutrients and water acquisition, but recent research shed light on a more complex picture of this symbiosis. For example, it has been reported that AMF are involved in the absorption, translocation, use efficiency, and cycling of various nutrients (e.g., Cu, Fe, K, N, P, Zn, etc.), enzymatic activities, photosynthesis, respiration, and plant metabolism, and help the host plant to enhance the resistance against diseases and pathogenic organisms [14,166,172−177].

AMF help plants maintain normal growth in stressful environments by mitigating the negative impacts of different stresses. Some mechanisms with which AMF help plants deal with stresses are accumulating compatible osmolytes and secondary metabolites, modifying the mineral uptake, mineral assimilation, and the phytohormone profiles, producing phytochelatins, expressing proteins, compartmentalizing and sequestrating toxic ions, and upregulating the antioxidant system [6,166]. AMF facilitate carbon sequestration and increases the carbon content of soils via aggregation, and prevents decomposition of organic carbon [14]. AM inoculation improves the net assimilation rates by increasing stomatal conductance and improving photosystem II [178].

AMF can also improve crop quality, in addition to improving the nutritional status of crops. For instance, AMF-colonized strawberries exhibited improved antioxidant properties due to enhanced levels of secondary metabolites [179]. AMF modified the production of carotenoids and some volatile compounds to improve the dietary quality of crops [180]. It was observed that AMF benefited the quality of tomatoes [181]. Mycorrhizal symbiosis is known to improve the accumulation of chlorophylls, anthocyanins, carotenoids, tocopherols, total soluble phenolics, and numerous mineral nutrients [182]. AMF can also increase the biosynthesis of valuable phytochemicals in edible plants and help construct a healthy food production chain [183,184]. AMF inoculation can also play an essential role in improving the concentration of essential oils (terpenoids) in medicinal plants [43]. AMF also interact with different soil microorganisms, plant growth-promoting rhizobacteria, and mycorrhiza helper bacteria, which significantly affect agriculture [14].

AMF are widely accepted to help plants tolerate various stresses that include heavy metal toxicity, nutritional imbalance, extreme temperatures, salinity, and drought. Simultaneous exposure of plants to drought and salinity, for example, increases ROS production, which can be highly injurious to plants [185,186]. ROS is commonly detoxified by enzymes such as CAT, SOD, GR, and POD [187]. AM colonization in plants showed increased antioxidant enzyme (e.g. POD and SOD) activities, contributing to plant productivity [188]. Improved vegetative growth, total dry weight, and fresh weight along with increased photosynthesis, stomatal conductance, relative leaf water content, nutrient, and water uptake along with reduced electrolyte leakage have been found in AM-inoculated plants [188–190]. The enhanced nutrient absorption and water status of mycorrhizal plants may be due to the ability of the fungi to uptake nutrients and water more efficiently and the increased absorption surfaces of the soil-growing hyphae [191–193]. The plant–water relations are affected in several different ways by AMF, which include effective absorption of soil water, hormonal changes, enhanced soil-root contact, regulation of osmotic adjustment, improvement of the gas exchange, and direct water uptake from the soil through the mycelium and transport to the plant [191]. The AM mycelium can penetrate soil pores that plant root hairs cannot and might be able to transport water that is not available directly to roots both internally and externally through the hyphal surfaces [194–196]. AM *Lactuca sativa* plants inoculated with *Glomus claroideum*, *Glomus coronatum*, *Glomus intraradices*, and *Glomus mosseae* have received a 3–4.75 mL plant^{-1} greater daily water flow than noncolonized plants have, with the increase influenced by the root colonization frequency and extraradical mycelium production, although the different AMF exhibited differing abilities to enhance the water uptake [196].

Modifying root characteristics such as hydraulic conductivity can considerably improve the osmotic stress tolerance levels [197]. Aquaporins (AQPs) are essential for regulating the plant water flow under osmotic stresses. AQPs come from a multifunctional family that branches from a major intrinsic protein superfamily that is important in osmoregulation and acts as water channels [198]. The passive water transport is facilitated by AQPs, following a potential gradient [199]. Furthermore, AQPs can transport low-molecular-weight molecules like ammonium, CO_2, and glycerol [199,200]. The first aquaporin from an AM fungus (*GintAQP1*) was cloned by Aroca, Bago [201]. The authors found evidence that supports that fungal AQPs can compensate for the drought-induced downregulation of AQPs in the host plant, although the functionality of *GintAQP1* could not be demonstrated. The authors also observed that osmotically unstressed parts of the mycelium saw upregulation of *GintAQP1* expression, while other parts of the mycelium were stressed. This suggests that unstressed and stressed mycelium may communicate with each other. *GintAQPF1* and *GintAQPF2* are two other characterized functional genes that encode for AQPs present in the AMF *Rhizophagus intraradices* [202]. These two genes are overexpressed under osmotic stresses to help alleviate the stress for the fungus and potentially increase the water supply for the host plant [203].

AMF may also improve the soil characteristics to indirectly reduce osmotic stresses for plants and biochemically and physically increase the water retention capacity of soils [204]. The physical improvements are mainly through the interaction between the soil particles and the extraradical mycelium to form stable aggregates that contribute to water retention and organic matter protection in the soil [205]. Biochemically, AMF are released to the soil organic products such as hydrophobins, glomalin (operationally measured as "glomalin-related soil protein" GRSP), mucilage, and polysaccharides [206]. Such compounds participate in carbon sequestration and the formulation of stable aggregates [207]. These stable aggregates improve infiltration and water retention, retarding the rhizospheric drought effects [208,209]. Zhang et al. [210] observed in a recent study that in the rhizosphere of *Poncirus trifoliate* plants growing in saline conditions (irrigated with 100 mM NaCl), inoculation with *Diversispora versiform* results in a higher GRSP amount than that in noncolonized plants, which was also associated with a greater amount of water-stable aggregates. Zou et al. [211] additionally found a high correlation between the amounts of GSRP and of water-stable aggregates and between the amounts of GRSP and available water, which highlights the important role of GSRP for plants to cope with osmotic stresses in the soil.

Drought generally causes reduced uptake in plants of major minerals such as calcium, iron, zinc, copper, manganese, silicon, etc. In AM-colonized plants under osmotic stresses, especially in soils where drought affects nutrient diffusion, plants uptake an increased amount of mineral nutrients [212]. Safir et al. [213,214] first demonstrated that AM improve plants' nutritional status and improves tolerance to osmotic stresses. Phosphorous is studied heavily in AM because it is rapidly absorbed by plants but is slow to be diffused in the soil, which leads to depletion zones to be formed around the roots [166,215]. AM formation leads to improved phosphorous absorption in plants growing under limiting conditions [166,215,216]. Several studies have observed that direct phosphorous uptake pathway of roots in some plants may be reduced in some plants, while it is completely suppressed in other species by AM colonization [217,218]. AM may be responsible for 100% of the phosphorous uptake in tomato and flax plants. The extraradical mycelium absorbs phosphates, which are then polymerized into polyphosphates, which accumulate in the vacuoles of the extraradical mycelium [219].

The high mobilities of NO_3^- and NH_4^+ are reduced by osmotic stresses. AMF increase the absorption of NO_3^-, NH_4^+, and some organic nitrogen sources in such conditions [220–222]. Bago et al. [223] proposed that a process associated with the polyphosphate transport and the urea cycle occur for the transfer and the subsequent absorption of nitrogen by plants, which was later confirmed by Govindarajulu et al. [222]. According to Bago, Vierheilig [224], proton symport is responsible for the NO_3^- absorption by the extraradical mycelium.

Subsequently, NO_3^- assimilation in the AM mycelium involves nitrate reductase converting it to NH_4^+ and inclusion in the glutamine synthetase cycle, followed by transformation of glutamate synthase into arginine and transported to the intrarradical mycelium associated with the polyphosphate transport [222,223]. Extraradical hyphal NH_4^+ absorption involves the same cycle.

Similar to drought stress, osmotic stresses (mainly in saline soils) also lead to mineral nutrient deficiency and decreased Ca^{2+}, K^+, and Mg^{2+} levels in plant tissues (ionic imbalances) because Na^+ hinders absorption of the aforementioned ions [225,226]. AMF act as the primary barrier that selects ions, therefore improving the absorption transport of certain mineral nutrients [227]. AM reduces the Na^+ translocation to plant tissues and thus prevents toxic Na^+ levels. This is because AM can store such ions in their vacuoles and retain them in structures like vesicles and intraradical mycelium [191,228]. Studies with *R. intraradices* further demonstrated that AMF selectively absorbed mineral nutrients while preventing Na^+ entry into the mycorrhizal structures and maintaining high K^+, Ca^{2+}, and Mg^{2+} levels with respect to the Na^+ level [229]. Lower leaf Na^+ concentrations were observed in AM-colonized plants compared to noncolonized plants of *Acacia nilotica* [230], *Olea europea* [231], and *Triticum aestivum* [232] grown in saline conditions. Four gene sequences have been identified in *Rhizophagus irregularis* that encode K^+ transporter systems; three encode SKC (small conductance calcium-activated potassium channels) ion channels, while one encodes transporter family type HAK/KT/KUP [233] that selectively can modify the K^+:Na^+ ratio.

Proline is a widely researched free amino acid that acts as an osmoprotectant and is highly accumulated in plants placed under abiotic stresses, such as heavy metal accumulation [234], water constraints [235,236], low temperature [237], salinity [238], and especially osmotic stresses. Proline is manufactured in the chloroplast and the cytoplasm and is accumulated in the vacuoles. It is synthesized from ornithine under normal growing conditions and from glutamate in stressed growing conditions [239–241]. Proline is also a variable amino acid necessary for scavenging ROS under stresses and in determining the membrane and protein structures [242]. Proline is also a nutritional source, in addition to being an osmotolerant. Proline is utilized as a source for both carbon and nitrogen by fungi. AMF are reported to be able to stimulate mycorrhizal plants to produce proline when under stress [215,243–245]. This proline production depends on the plant species and the specific plant-fungus combination's symbiotic efficiency [246,247]. For example, maize colonized by *Funneliformis mosseae* exposed to 0.5 and 1 g NaCl kg^{-1} soil produced higher proline levels than noncolonized plants did [248]. Mycorrhizal lettuce accumulated higher proline levels than noncolonized lettuce did when in drought [249]. These examples suggest that mycorrhizal plants had the highest capacity for osmotic adjustments, as plants with AM had the highest root tissue proline content, which allowed them to tolerate the low soil water potential and maintain a favorable water potential gradient for root water supply. Mycorrhizal plants have relatively greater proline contents when in low water potentials and hence experience improved root water absorption, which leads to enhanced root hydraulic conductivity and a better osmotic balance [197].

Sugars are essential for plants in dealing with stresses because they act as osmoprotectants, like proline, which strongly helps with the osmotic adjustment of plants, contributing up to 50% of a plant's osmotic potential, and also serves as a source of carbon [248,250]. Osmotically stressed mycorrhizal plants increase the accumulation of total soluble sugars as a defense mechanism [251,252]. Mycorrhizal maize had significantly higher levels of soluble and reduced sugars in comparison to noncolonized plants [248]. The increased sugar accumulation in mycorrhizal plants under osmotic stress is due to the increased photosynthetic capacity [253,254]. Organic acids also play an essential role in unfavorable conditions as active metabolites are involved in plant vacuoles' osmotic adjustment, countering Cl^- in cells, and regulating the cytosol's pH [255,256]. Under saline conditions of 0, 0.5, and 1 g NaCl kg^{-1} soil, the total organic acid concentration was, respectively, 31%, 24%, and 8% higher in mycorrhizal maize than in their noncolonized plants [248]. Mycorrhizal plants had the highest concentrations of acetic, citric, fumaric malic, oxalic, propionic, and valeric acids [248,257].

AM mainly increase the tolerance to heavy metals by reducing their availability to plants, by forming toxic ion complexes with the hyphal cell components (chitin, cellulose, etc.) and their exudates, compartmentalizing metals in vesicles, sequestering metals in the extraradicular hyphal structures, and influencing metal(loid) speciation through the alteration of the microenvironment in the mycorrhizosphere. Furthermore, the glomalin secreted by the extracellular hyphae, a glycoprotein, binds to metalloids to immobilize them in the soil and reduce their availability and uptake by plants [258–264]. Metal dilution in plant tissues could result from the increased growth and uptake exclusion by precipitation or chelation in the rhizosphere [259,265]. For example, in a field trial study, it was found that AM fungi, together with Fe-bearing phyllosilicate amendments, could help plants survive in metal-contaminated soils [266].

It is also known that mycorrhizae ameliorate harmfully impacts of biotic stresses. AMF are known to compete with soil-borne pathogens more efficiently than fungicide treatments do, and therefore they can substitute chemical fungicides [267]. They enhance plants' growth attributes and reduce their susceptibility to diseases to protect plants from pathogens [268–270]. For example, an AM fungus (*G. mosseae*) effectively reduced root disease symptoms in tomato plants [271]. In another study, tomato plants colonized with *F. mosseae* and *R. irregularis* were observed to increase the accumulation of folic acid, riboflavin, etc. to modulate the oxylipin pathway, change the activities of chitinase, lipoxygenase, β-1,3-glucanase, and phenylalanine ammonia-lyase, and lead to an enhanced resistance of against *Botrytis cinerea* and early blight disease [272]. Moreover, the authors have also demonstrated that jasmonate signaling also mediates the tolerance to pathogen attacks [273]. Complex crosstalk of hormones like SA, JA, increased expressions of genes involved in the biosynthesis of JA (e.g., OPR), genes responsive to SA (e.g., PR1), wound-inducible polypeptide prosystemin, and the enzyme levels involved in the biosynthesis of JA and SA are mechanisms with which mycorrhiza enhance stress resistance in plants [274]. Inoculation of potato plants with *R. irregularis* MUCL 41833 has recently been demonstrated to enhance the plants' defense response against *Rhizoctonia solani* mediated by ERF3, which suggests that the ethylene signaling pathway is involved [275]. Accumulation of such genes as *OsAP2*, *OsEREBP*, *OsJAmyb*, and *OsNPR1*, as well as genes involved in the calcium-mediated signaling process, such as *OsCaM*, *OsCBP*, and *OsCML4*, and genes from the signaling pathway like *OsMPK6* and *OsDUF26* induced by mycorrhizal inoculation has been proven to be effective in protecting rice plants from pathogen attacks [43]. Much of the studies in the literature emphasize the role AM fungi play in plant nutrition, as the plant-AMF symbiosis has its basis on nutrient transfer. Indeed, the influence of AMF on plant nutrition can further affect the plant's chemical defenses [276].

18.5 Plant stress alleviation mediated by the combined use of silicon and arbuscular mycorrhizal fungi

It has been found that the combined use of AMF and silicon better improves plant growth in comparison with the use of either of them by itself [5]. In previous studies, the potential application of mycorrhizal inoculations and silicon fortifications in improving plant growth under unstressed and stressed conditions has been studied [34,276–280]. For example, in a study with well-watered (unstressed) strawberry plants, a significant increase in the shoot growth was observed with silicon treatment. The highest shoot biomass production was when AMF and silicon were applied together. The application of AMF and silicon also enhanced the net photosynthetic rate by inducing a higher stomatal conductance [281]. In a study [277], rapid production of phytochelatins under cadmium and zinc stresses progressively depleted glutathione, resulting in diminished phytochelatins synthesis. Inoculation with *R. irregularis* and/or silicon amendment boosted glutathione production, which further increased the synthesis of phytochelatins. AM fungi have been reported to enhance phytochelatin biosynthesis in *Solanum lycopersicum* through overexpressing genes responsible for phytochelatin synthesis [282]. The higher tolerance to heavy metals displayed by pigeon pea Pusa 2002 (a metal-tolerant genotype) could be correlated with its higher production of phytochelatins induced by AMF and silicon compared to that of pigeon pea Pusa 991 (a metal-sensitive genotype). Heavy metals induce ROS formation through the consumption of glutathione and its derivatives, which are necessary to scavenge these reactive species and synthesize phytochelatins responsible for the vacuolar compartmentalization of heavy metals [283]. In the study of Garg and Singh [277], AM inoculation and silicon supplementation modulated the proline biosynthesis under cadmium and zinc stresses to affect the physiological processes and nutrient uptake in plants. Moreover, the increase in the total proline content may be attributed to enhanced activities of glutamate dehydrogenase and pyrroline-5-carboxylate synthetase, as well as the reduced proline dehydrogenase activity induced by silicon supplementation and AM colonization. This demonstrates the importance of the balance between the catabolic and anabolic biosynthesis-related enzymes. Under cadmium and zinc stresses, applying silicon to AM-inoculated plants resulted in enhanced nutrient status, growth, water status, and ultimately a better harvest index as the proline biosynthesis was promoted and metal uptake was limited. AM played an important role in enhancing the uptake of nutrients (iron, magnesium, nitrogen, and phosphorous) and the root biomass, and Si significantly contributed to the enhanced calcium and potassium contents and the improved shoot biomass. The benefits of AM colonization and silicon supplementation are important on their own and become cumulative as silicon uptake is enhanced with mycorrhization. The genotype PUSA 2002 is able to more easily form effective mycorrhizal symbioses, and observed a better silicon uptake compared to the genotype PUSA 991. Negative effects of zinc were completely canceled and the cadmium levels in PUSA 2002 were much lower due to the independent and cumulative benefits of silicon supplementation and mycorrhization. In addition, Garg and Kashyap [284] observed that silicon fortification and AM inoculations restored plant growth and productivity in two pigeon pea genotypes subjected to As (V) and As (III) stresses by reducing the As uptake and increasing the silicon contents in the plant tissues. The detoxification mechanisms employed by AMF and silicon include metalloid immobilization in the soil, metal sequestration, or complexation by exuding organic compounds from the roots [285]. In another study, chickpea plants had a significantly increased ROS generation in both the roots and leaves in response to salinity, where the salt-sensitive genotype exhibited higher levels of metabolites associated with stresses than did the salt-tolerant genotype. AM inoculation and silicon application both effectively mitigated the adverse effects of salinity in chickpea plants and helped them adapt to stresses. AM was more efficient than silicon was in enhancing the enzymatic and nonenzymatic activities, and shifted the redox state from the oxidized (CSSG, DHA) to the reduced forms (GSH, AsA) to ensure redox homeostasis, while silicon was more effective than AM in reducing the accumulation of the oxidative metabolites. Furthermore, a combined application of silicon and AM exhibited cumulative effects that provided a better platform for chickpea to adapt to salinity stress. The vitality of the AsA-GSH cycle was evaluated to be better in the salinity-tolerant genotype and played an essential role in lowering the oxidative damages by scavenging the excess ROS compared to the salinity-sensitive genotype. This is congruous with the relatively better recovery of the growth reductions observed in the salinity-tolerant genotype. Therefore the study can summarize AM inoculation and silicon fertilization as important tools in effectively mitigating the harms of salinity in chickpea plants [278]. Combined applications of AMF and silicon reduced the concentrations of toxic ions and metabolites related to stresses—H_2O_2 and MDA—by enhancing APX, CAT, and SOD activities and reversing the salt-induced growth retardations [278,279]. Low silicon accumulating chickpea genotypes colonized with *F. mosseae* under salinity stress mitigated the stress symptoms through a significantly enhanced silicon uptake [278,286].

Garg and Bhandari [278] reported increased salinity stress tolerance in chickpea with silicon fertilization and/or AMF colonization. An essential mechanism with which mycorrhizal colonization enhances salinity tolerance is maintaining a high cytosolic K^+/Na^+ ratio in AM plants [287]. AM fungi enhance the K^+ absorption in saline conditions by preventing Na^+ from translocating to the shoot tissues [252,288,289] thereby maintaining the ionic balance. AM plants have been observed to maintain relatively lower Na^+ levels [231,290]. Higher Ca^{2+}, Mg^{2+}, and K^+ concentrations in contrast to lower Na^+ levels are observed in the *R. irregularis* hyphae and spores. This indicates that AMF act as the barrier that selects the ions for entry and preselects nutrients to prevent toxic salt ions from entering the plant and result in mitigating salinity stress in plants [229,289]. The authors attributed this selective absorption to the Na^+/H^+ antiporter SOS1 activities located in the plasma membrane [291]. Estrada et al. [292] also demonstrated that the Na^+ and K^+ transporters are differentially expressed in plants putatively involved in Na^+/K^+ homeostasis in roots during AM colonization.

Reducing the uptake of toxic ions is one of the foremost mechanisms by which silicon alleviates stresses. Adding silicon to the rooting medium has been validated to reduce the Na^+ and Cl^- translocation to the shoots and roots for plants grown in saline soils [293]. This is attributed to H^+-ATPase stimulation by silicon in the root plasma membrane [294]. Increased H^+-ATPase activities in the plasma membrane may accelerate the K^+ import into the cell and Na^+ export from the cell, helping maintain a constant ionic ratio. Silicon enhanced the H^+-activity to improve the K^+ uptake in both soil-based and hydroponic experiments, according to Mali et al. [295]. It has recently been suggested that the voltage-dependent K^+ inward rectifying channels are activated by the H^+ gradient induced by silicon application, which increases the K^+ uptake, increasing the turgor pressure and enabling the cell walls to expand, to ultimately lead to increased plant growth [296]. The findings indicate that the combined Si and AM application to plants growing in saline conditions can improve the endogenous nutrition, growth attributes, and yield parameters. Garg and Bhandari [278] reported that silicon is better than AMF at improving the K^+/Na^+ ratio, while AMF are more efficient than silicon at enhancing growth and productivity. Mycorrhization substantially improves the silicon uptake, and therefore silicon supplementation with mycorrhization significantly decreases the Na^+ levels, enhances the nutrient uptake, improves growth and yield, prevents chlorophyll damage, and increases the rubisco activity. The study highlights the AM inoculation and silicon fertilization as important, sustainable methodologies for effectively mitigating salinity's harms on plants [278]. *R. irregularis* and silicon application to *Cajanus cajan* L. Millsp. (pigeon pea) genotypes place under cadmium and zinc stresses led to modulated proline biosynthesis and yield [277]. Under greenhouse and field conductions, AM fungi and potassium silicate at low concentrations effectively reduced the occurrence and severity of white rot and increased the growth of onion plants [297]. In another study, a combined application of AMF, *Enterobacter* sp., and silicon had the potential to suppress *Rigidoporus microporus* and improve the growth of rubber seedlings in glasshouse conditions [298].

As previously mentioned, the antioxidant defense enzymes are activated by stresses in the roots and leaves of plants. But the antioxidant defense enzymes are not sufficient on their own, as reflected by the increasing MDA concentration in stressed plants for protection against ROS. Silicon and AMF have been reported to similarly increase antioxidant defense enzyme activities, particularly that of SOD. It has been observed that silicon and AMF also lead to a decline in the stress metabolite (MDA and H_2O_2) levels. This may indicate that AMF and silicon help plants deal with oxidative damage by reducing the production of stress metabolites, in addition to elevating the capacity of the antioxidant defense system. Higher enzymatic and nonenzymatic antioxidant defense system activities have frequently been observed in plants with a higher root AMF colonization [299,300] than in non-AMF-inoculated plants. It has been argued that silicon deposition in the cell membrane is how the antioxidant defense mechanisms are biochemically enhanced [301]. Several researchers argue that silicon is involved in plant metabolism because silicon gives rise to increased antioxidant enzyme activities and the levels of nonenzymatic antioxidant substances in abiotically stressed plants [56,63,302]. As a common response in plants under drought, organic osmolyte accumulation (e.g., of proline, soluble sugars, etc.) is known to lead to an osmotic gradient with the environment [303], that is naturally associated with a reduced osmotic potential. In a study, silicon and AMF application increased the root concentration of organic osmolytes, which indicated that the water economy of the roots might be controlled through a different strategy involving AMF and silicon [281]. According to the results of various studies, the combined use of silicon and AMF are more beneficial to plant growth than their standalone applications are.

As mentioned before, silicon nutrition can impart significant benefits to plants, but the extent to which it can is limited because its uptake by plants is limited [34]. In addition to increasing P, Zn, Cu, Ca, and N uptake by the plant, mycorrhizae have been consistently reported to increase the silicon uptake [34,42,279,284,304,305], and this enhanced silicon uptake induced by AM may be utilized to improve growth and productivity [278], improving nitrogen fixation efficiency in legumes [277], and imparting stress tolerance in plants such as Z. mays [41,305], Glycine max [304,306], soybean [306], pigeon pea and chickpea [278,279,284,307], and banana [34]. Silicon is reported to accumulate into the spores and hyphae of mycorrhizal fungi such as Glomus etunicatum, G. coronatum, Glomus versiform, R. irregularis (= G. intraradices), F. mosseae (= G. mosseae), and Rhizophagus clarus (= Glomus clarum) and can be transferred to the host roots [34,229,278]. Hammer et al. [229] found that hyphae and spores of AMF in saline soils exhibited a higher silicon buildup compared to hyphae and spores of AMF in regular growth environments by quantifying the elemental composition in spores and hyphae of AMF collected from two saline sites using PIXE at high salinity. AM is known to enhance lateral root formation and therefore modify the root system architecture [308], and Ma et al. [309] reported that silicon was mainly absorbed by the secondary roots. Clark and Zeto [305] observed an elevated Si concentration in AM-colonized plants growing in acidic soils, in saline soils [278], in low-silicon soils [276], and in metal-contaminated soils [277,306,310].

It remains unknown how AMF increase the silicon uptake of plants. Transporters at the extraradical hyphae at the soil-fungus and plant-fungus interface for the silicon uptake and transfer across the periarbuscular interface in the plant cells through active transport, respectively, are likely involved in the increased silicon uptake induced by AMF [34,229,278,304]. The studies mentioned above highlight that AMF inoculation is a sustainable method to effectively enhance plants' silicon uptake. Therefore it is of great future research interest to investigate how AM symbiosis enhances the silicon uptake in the host plant and how AM and silicon work together to enhance plant resistance to biotic and abiotic stresses.

It is known that silicon increases mycorrhizal effectiveness, which is defined as the growth difference between plants with and without mycorrhizae [5,311]. Different factors influence mycorrhizal effectiveness, such as the fungal species, soil conditions, plant species, and genotype [312]. In contrast to the wide range of studies investigating phosphorous availability as a soil chemical factor on the mycorrhizal effectiveness, silicon's effects on the mycorrhizal effectiveness have yet to be widely explored. Silicon applications to strawberry inoculated with the AMF G. versiform, R. clarus, and R. intraradices were observed to increase the mycorrhizal effectiveness compared to the mycorrhizal effectiveness of AM plants not treated with silicon in recent studies [280,281]. Some known synergistic mechanisms between silicon and AMF effectiveness include: (1) stimulated root growth of AM plants and enhanced uptake and transfer of nutrients, which may promote AMF colonization [280]. In a study, silicon enhanced the nutrient uptake by restricting the absorption of both Zn and Cd, which might be due to either immobilization or sequestration of heavy metals in the soil or root vacuoles [313]. It has also been well known that salinity stress and drought diminish the nutritional status of plants, resulting in Mn, Cu, N, P, K, Zn, and Fe deficiencies. AMF and Si applications improve the nutritional status, equaling or even exceeding the critical deficiency thresholds of the different macronutrients and micronutrients. Pavlovic et al. [93] and Dragišić et al. [314] observed that silicon application increased the iron and zinc uptake at low concentrations on the rhizoplane. Plants with a higher root AMF colonization more efficient uptake and translocate macronutrients and micronutrients to the shoot than noninoculated plants do [184,315]; (2) increased photosynthetic rate that provides an increased carbon source for the fungi, for example, by increasing the leaf chlorophyll contents, stomatal conductance, and photosynthetic enzymes activities [316,317] and improving the physical stability of the leaves, leading to a more horizontal orientation of the leaves [318]. It is known that mycorrhizal association is completely dependent on the organic carbon supply from their plant partner, as 4%−20% of the carbon fixed through photosynthesis is transferred to the AMF [6]. The hyphal absorption capacity and the photosynthetic rate (or organic carbon supply) are positively correlated with the formation of arbuscules [6,34,281]. Silicon was observed to increase the formation of arbuscules in a study that may have resulted from enhanced uptake and transfer of nutrients within the plant, improved photosynthetic rate, and enhanced root growth [5,34]. The hyphal absorption capacity is positively correlated with the photosynthetic rate [166]; (3) reduced lignin synthesis and polymerization and/or modified metabolic pathways of phenolics in the AM host plant [280,319,320], which can influence how the host plant interacts with the AMF. Silicon's influence on the metabolizing phenolic compounds has been reported [321,322]. Plant-produced phenolic compounds are identified to be important in initiating AM symbiosis. Still, root penetration and AM symbiosis establishment depend on the host plant and its interaction with fungi. Experiments with the AM fungus Glomus on sorghum and clover, using phenolic compounds such as p-hydroxybenzoic acid, p-coumaric acid, or quercetin as the growth simulator, have validated this observation. Strigolactones are found in the root exudates of various plants and have been observed to influence the AMF establishment, direct AMF growth, and affect the hyphae branching to positively influence AMF symbiosis [323,324]. In another study, silicon was reported to help enhance the phenolic compound (flavonoid-type phenolics) metabolism [319]. Phenolic compounds like flavonoids may also help facilitate the AMF-host plant interactions [320] and positively influence the fungal growth parameters, such as spore germination, hyphal growth and branching, root colonization [325,326], and secondary spore formations. Moreover, they play a part in the fungal invasion and arbuscule formation inside the roots [327]. Some flavonoids have been observed to stimulate the AMF-plant interaction, which becomes more apparent in the presence of rhizospheric CO_2. In addition, it has been shown that the alteration in the profile of flavonoids in the root extracts is obtained through changes in the expression of genes involved in flavonoid, phenylpropanoid, and isoflavoid metabolic pathways [320]. The expression of genes involved in the metabolic pathways of flavonoids, isoflavoids, and phenylpropanoids have been observed to alter the profile of flavonoids in the root extracts [43]. Strigolactones have recently been identified as the host-recognition signals for AMF, but this raises questions on how flavonoids generally act as signaling molecules in AMF-plant interactions [328]; and (4) increased root pool of soluble sugars, which is important for supporting AMF entry, and further establishment in the roots are other probable mechanisms [281]. For example, it has been observed that drought triggers sugar accumulation and leads to an adjustment in the photosynthetic rate [329]. This drought-triggered accumulation of soluble sugars in turn causes an impaired plant metabolism by changing either the translocation or composition of sugars in the leaves [86]. It has been observed that the soluble sugar concentrations in the leaves of stressed plants, decrease with AMF and silicon application because of the resumed growth and carbohydrate consumption necessary for biomass production. This indicates that Si and AMF may mediate the accumulation of soluble sugars in drought-stressed leaves through a negative feedback mechanism. In contrast to the leaves of stressed plants, it has been found that the soluble sugars concentration in the roots of stressed plants increases with AMF and Si application. This increase may be due to an improved photosynthate allocations to the roots and/or net CO_2 assimilation, and may in turn help stimulate root growth in stressed plants. Considering the osmotic effects of soluble carbohydrates, an elevated pool of soluble sugars may also improve the root water uptake capacity from a dry substrate [281].

18.6 Conclusions and future perspectives

Environmental stresses, especially drought and salinity, are one of the major agricultural problems reducing crop yield worldwide. The use of silicon and AMF is known as one of the most effective and economical ways to enhance plant tolerance to various environmental stresses. Based on what is known about them, the strategy of combining silicon and AMF application may be highly useful in improving plant tolerance to various environmental stresses, compared to their isolated applications. It is known that silicon increases mycorrhizal effectiveness, and AMF have also been reported to consistently enhance the silicon uptake in plants (Fig. 18.1). The mechanisms with which AMF increase silicon uptake in plants remain unknown. An understanding of how plant silicon/nanosilicon concentrations and the AM symbiosis can impact the plant tolerance of biotic and abiotic stresses could better inform stress management for increased efficacy of biocontrol/biofertilizer strategies in the field. Therefore, for future research, detailed investigations are needed on the role of AMF in silicon/nanosilicon

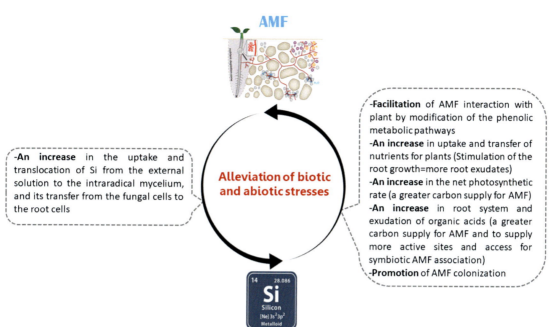

FIGURE 18.1 Some known interactions between arbuscular mycorrhizal fungi (AMF) and silicon (Si) in alleviating biotic and abiotic stresses in plant.

uptake and the subsequent impacts in alleviating environmental stresses, especially at the intraspecific level. Interactions at the metabolic levels and molecular mechanisms for the synergistic effects of silicon/nanosilicon on the AMF effectiveness also need further research. Compared to the studies done on AMF-mediated silicon uptake, studies on the impact of AMF-mediated nanosilicon uptake and their role in improving plant growth under environmental stresses are scarce. Further studies are also required to validate the AMF performance in combination with suitable insoluble silicate sources under various field conditions and different ecosystems.

Acknowledgments

We express our gratitude to the University of Tehran for providing the facilities necessary for this research.

References

[1] FAO. Agriculture: key to achieving the 2030 agenda for sustainable development. Rome: Food and Agriculture Organization of the United Nations. 2016.

[2] Etesami H, Maheshwari DK. Use of plant growth promoting rhizobacteria (PGPRs) with multiple plant growth promoting traits in stress agriculture: action mechanisms and future prospects. Ecotoxicol Environ Saf 2018;156:225–46. Available from: https://doi.org/10.1016/j.ecoenv.2018.03.013.

[3] Etesami H, Adl SM. Plant growth-promoting rhizobacteria (PGPR) and their action mechanisms in availability of nutrients to plants. Phyto-microbiome in stress regulation. Singapore: Springer; 2020. p. 147–203.

[4] Etesami H, Beattie GA. Plant-microbe interactions in adaptation of agricultural crops to abiotic stress conditions. Probiotics and plant health. Singapore: Springer; 2017. p. 163–200.

[5] Etesami H, Jeong BR, Glick BR. Contribution of arbuscular mycorrhizal fungi, phosphate-solubilizing bacteria, and silicon to P uptake by plant. Front Plant Sci 2021;12(1355). Available from: https://doi.org/10.3389/fpls.2021.699618.

[6] Smith SE, Read DJ. Mycorrhizal symbiosis. New York: Academic Press; 2010.

[7] Ismail Y, McCormick S, Hijri M. The arbuscular mycorrhizal fungus, *Glomus irregulare*, controls the mycotoxin production of *Fusarium sambucinum* in the pathogenesis of potato. FEMS Microbiol Lett 2013;348(1):46–51.

[8] Nadeem SM, Ahmad M, Zahir ZA, Javaid A, Ashraf M. The role of mycorrhizae and plant growth promoting rhizobacteria (PGPR) in improving crop productivity under stressful environments. Biotechnol Adv 2014;32(2):429–48.

[9] Miransari M. Contribution of arbuscular mycorrhizal symbiosis to plant growth under different types of soil stress. Plant Biol 2010;12(4):563–9.

[10] Porcel R, Aroca R, Ruiz-Lozano JM. Salinity stress alleviation using arbuscular mycorrhizal fungi. A review. Agron Sustain Dev 2012;32(1):181–200.

[11] Novair SB, Hosseini HMS, Etesami H, Razavipour T, Pirmoradian N. The role of arbuscular mycorrhizal fungal community in paddy soil. Agriculturally important fungi for sustainable agriculture. Cham: Springer; 2020. p. 61–88.

[12] Bahraminia M, Zarei M, Ronaghi A, Sepehri M, Etesami H. Ionomic and biochemical responses of maize plant (*Zea mays* L.) inoculated with *Funneliformis mosseae* to water-deficit stress. Rhizosphere 2020;16:100269.

[13] Chen M, Arato M, Borghi L, Nouri E, Reinhardt D. Beneficial services of arbuscular mycorrhizal fungi—from ecology to application. Front Plant Sci 2018;9:1270. Available from: https://doi.org/10.3389/fpls.2018.01270.

[14] Malhi GS, Kaur M, Kaushik P, Alyemeni MN, Alsahli AA, Ahmad P. Arbuscular mycorrhiza in combating abiotic stresses in vegetables: an eco-friendly approach Saudi J Biol Sci 2021;28(2):1465–76. Available from: https://doi.org/10.1016/j.sjbs.2020.12.001 Epub 2021/02/23. Available from: 33613074.

[15] Etesami H, Jeong BR. Silicon (Si): review and future prospects on the action mechanisms in alleviating biotic and abiotic stresses in plants. Ecotoxicol Environ Saf 2018;147:881–96.

[16] Etesami H, Jeong BR, Rizwan M. The use of silicon in stressed agriculture management: action mechanisms and future prospects. Metalloids in plants: advances and future prospects. Hoboken, NJ: Wiley; 2020. p. 381–431.

[17] Coskun D, Britto DT, Huynh WQ, Kronzucker HJ. The role of silicon in higher plants under salinity and drought stress. Front Plant Sci 2016;7:1072.

[18] Liang Y, Sun W, Zhu Y-G, Christie P. Mechanisms of silicon-mediated alleviation of abiotic stresses in higher plants: a review. Environ Pollut 2007;147(2):422–8.

[19] Van Bockhaven J, De Vleesschauwer D, Höfte M. Towards establishing broad-spectrum disease resistance in plants: silicon leads the way. J Exp Bot 2013;64(5):1281–93.

[20] Bityutskii NP, Yakkonen KL, Petrova AI, Shavarda AL. Interactions between aluminium, iron and silicon in *Cucumber sativus* L. grown under acidic conditions. J Plant Physiol 2017;218:100–8.

[21] Cai K, Gao D, Luo S, Zeng R, Yang J, Zhu X. Physiological and cytological mechanisms of silicon-induced resistance in rice against blast disease. Physiol Plant 2008;134(2):324–33.

[22] Gong H, Zhu X, Chen K, Wang S, Zhang C. Silicon alleviates oxidative damage of wheat plants in pots under drought. Plant Sci 2005;169(2):313–21.

[23] He Y, Xiao H, Wang H, Chen Y, Yu M. Effect of silicon on chilling-induced changes of solutes, antioxidants, and membrane stability in seashore paspalum turfgrass. Acta Physiol Plant 2010;32(3):487–94.

[24] Hattori T, Inanaga S, Araki H, An P, Morita S, Luxová M, et al. Application of silicon enhanced drought tolerance in *Sorghum bicolor*. Physiol Plant. 2005;123(4):459–66. Available from: https://doi.org/10.1111/j.1399-3054.2005.00481.x.

[25] Fauteux F, Chain F, Belzile F, Menzies JG, Bélanger RR. The protective role of silicon in the Arabidopsis—powdery mildew pathosystem. Proc Natl Acad Sci U S A 2006;103(46):17554–9.

[26] Rios JJ, Martínez-Ballesta MC, Ruiz JM, Blasco B, Carvajal M. Silicon-mediated improvement in plant salinity tolerance: the role of aquaporins. Front Plant Sci 2017;8:948.

[27] Epstein E. Silicon. Annu Rev Plant Biol 1999;50(1):641–64.

[28] Epstein E. The anomaly of silicon in plant biology. Proc Natl Acad Sci U S A 1994;91(1):11–17.

[29] Mitani N, Ma JF. Uptake system of silicon in different plant species. J Exp Bot 2005;56(414):1255–61.

[30] Ma JF, Takahashi E. Soil, fertilizer, and plant silicon research in Japan. Amsterdam; Boston: Elsevier; 2002.

[31] Epstein E. Silicon: its manifold roles in plants. Ann Appl Biol 2009;155(2):155–60.

[32] Takahashi E, Ma JF, Miyake Y. The possibility of silicon as an essential element for higher plants. Comments Agric Food Chem 1990;2(2):99–102.

[33] Mathur P, Roy S. Nanosilica facilitates silica uptake, growth and stress tolerance in plants. Plant Physiol Biochem 2020;157:114–27.

[34] Anda CCO, Opfergelt S, Declerck S. Silicon acquisition by bananas (cV Grande Naine) is increased in presence of the arbuscular mycorrhizal fungus *Rhizophagus irregularis* MUCL 41833. Plant Soil 2016;409(1-2):77–85.

[35] Montpetit J, Vivancos J, Mitani-Ueno N, Yamaji N, Rémus-Borel W, Belzile F, et al. Cloning, functional characterization and heterologous expression of TaLsi1, a wheat silicon transporter gene Plant Mol Biol 2012;79(1-2):35–46. Available from: https://doi.org/10.1007/s11103-012-9892-3 Epub 2012/02/22. Available from: 22351076.

[36] Guerriero G, Hausman J-F, Legay S. Silicon and the plant extracellular matrix. Front Plant Sci 2016;7:463.

[37] Ma JF, Yamaji N. Silicon uptake and accumulation in higher plants. Trends Plant Sci 2006;11(8):392–7.

[38] Vandevenne FI, Barão L, Ronchi B, Govers G, Meire P, Kelly EF, et al. Silicon pools in human impacted soils of temperate zones. Glob Biogeochem Cy 2015;29(9):1439–50. Available from: https://doi.org/10.1002/2014GB005049.

[39] Bennett PC, Roger JR, Choi WJ, Hiebert FK. Silicates, silicate weathering, and microbial ecology. Geomicrobiol J 2001;18(1):3–19. Available from: https://doi.org/10.1080/01490450151079734.

[40] Sommer M, Kaczorek D, Kuzyakov Y, Breuer J. Silicon pools and fluxes in soils and landscapes—a review. J Plant Nutr Soil Sci 2006;169(3):310–29.

[41] Kothari SK, Marschner H, Römheld V. Direct and indirect effects of VA mycorrhizal fungi and rhizosphere microorganisms on acquisition of mineral nutrients by maize (*Zea mays* L.) in a calcareous soil. New Phytol 1990;116(4):637–45. Available from: https://doi.org/10.1111/j.1469-8137.1990.tb00549.x.

[42] Clark RB, Zeto SK. Mineral acquisition by mycorrhizal maize grown on acid and alkaline soil. Soil Biol Biochem 1996;28(10):1495–503. Available from: https://doi.org/10.1016/S0038-0717(96)00163-0.

[43] Basu S, Rabara RC, Negi S. AMF: The future prospect for sustainable agriculture. Physiol Mol Plant Pathol 2018;102:36–45.

[44] Cooke J, Leishman MR. Silicon concentration and leaf longevity: is silicon a player in the leaf dry mass spectrum? Funct Ecol 2011;25(6):1181–8.

[45] Detmann KC, Araújo WL, Martins SCV, Sanglard LMVP, Reis JV, Detmann E, et al. Silicon nutrition increases grain yield, which, in turn, exerts a feed-forward stimulation of photosynthetic rates via enhanced mesophyll conductance and alters primary metabolism in rice. New Phytol 2012;196(3):752–62.

[46] Pilon C, Soratto RP, Moreno LA. Effects of soil and foliar application of soluble silicon on mineral nutrition, gas exchange, and growth of potato plants. Crop Sci 2013;53(4):1605–14.

[47] Harizanova A, Zlatev Z, UniversityPlovdiv LKA. Effect of silicon on activity of antioxidant enzymes and photosynthesis in leaves of cucumber plants (*Cucumis sativus* L.). Türk Tarım ve Doğa Bilimleri Derg 2014;1(Özel Sayı-2):1812–17.

[48] Korndörfer GH, Snyder GH, Ulloa M, Powell G, Datnoff LE. Calibration of soil and plant silicon analysis for rice production. J Plant Nutr 2001;24(7):1071–84.

[49] Richmond KE, Sussman M. Got silicon? The non-essential beneficial plant nutrient. Curr Opin Plant Biol 2003;6(3):268–72.

[50] Reynolds OL, Keeping MG, Meyer JH. Silicon-augmented resistance of plants to herbivorous insects: a review. Ann Appl Biol 2009;155(2):171–86.

[51] Meyer JH, Keeping MG. Impact of silicon in alleviating biotic stress in sugarcane in South Africa. Sugar Cane Int 2005;23:14–18.

[52] Liang Y, Chen Q, Liu Q, Zhang W, Ding R. Exogenous silicon (Si) increases antioxidant enzyme activity and reduces lipid peroxidation in roots of salt-stressed barley (*Hordeum vulgare* L.). J Plant Physiol 2003;160(10):1157–64. Available from: https://doi.org/10.1078/0176-1617-01065.

[53] Zhu Z, Wei G, Li J, Qian Q, Yu J. Silicon alleviates salt stress and increases antioxidant enzymes activity in leaves of salt-stressed cucumber (*Cucumis sativus* L.). Plant Sci 2004;167(3):527–33.

[54] Etesami H, Jeong BR. Importance of silicon in fruit nutrition: agronomic and physiological implications. Fruit crops. Amsterdam: Elsevier; 2020. p. 255–77.

[55] Rizwan M, Ali S, Ibrahim M, Farid M, Adrees M, Bharwana SA, et al. Mechanisms of silicon-mediated alleviation of drought and salt stress in plants: a review. Environ Sci Pollut Res 2015;22(20):15416–31.

[56] Zhu Y, Gong H. Beneficial effects of silicon on salt and drought tolerance in plants. Agron Sustain Dev 2014;34(2):455–72.

[57] Kim Y-H, Khan AL, Waqas M, Lee I-J. Silicon regulates antioxidant activities of crop plants under abiotic-induced oxidative stress: a review. Front Plant Sci 2017;8:510.

[58] Tripathi DK, Singh S, Singh VP, Prasad SM, Dubey NK, Chauhan DK. Silicon nanoparticles more effectively alleviated UV-B stress than silicon in wheat (*Triticum aestivum*) seedlings. Plant Physiol Biochem 2017;110:70–81. Available from: https://doi.org/10.1016/j.plaphy.2016.06.026.

[59] Ranjan A, Sinha R, Bala M, Pareek A, Singla-Pareek SL, Singh AK. Silicon-mediated abiotic and biotic stress mitigation in plants: underlying mechanisms and potential for stress resilient agriculture. Plant Physiol Biochem 2021;163:15–25. Available from: https://doi.org/10.1016/j.plaphy.2021.03.044.

[60] Bakhat HF, Bibi N, Zia Z, Abbas S, Hammad HM, Fahad S, et al. Silicon mitigates biotic stresses in crop plants: a review. Crop Prot 2018;104:21–34.

[61] Xu L, Islam F, Ali B, Pei Z, Li J, Ghani MA, et al. Silicon and water-deficit stress differentially modulate physiology and ultrastructure in wheat (*Triticum aestivum* L.). 3 Biotech. 2017;7(4):1–13.

[62] Abbas T, Balal RM, Shahid MA, Pervez MA, Ayyub CM, Aqueel MA, et al. Silicon-induced alleviation of NaCl toxicity in okra (*Abelmoschus esculentus*) is associated with enhanced photosynthesis, osmoprotectants and antioxidant metabolism. Acta Physiol Plant 2015;37(2):6.

[63] Shi Y, Zhang Y, Han W, Feng R, Hu Y, Guo J, et al. Silicon enhances water stress tolerance by improving root hydraulic conductance in *Solanum lycopersicum* L. Front Plant Sci 2016;7:196.

[64] Li Y-T, Zhang W-J, Cui J-J, Lang D-Y, Li M, Zhao Q-P, et al. Silicon nutrition alleviates the lipid peroxidation and ion imbalance of *Glycyrrhiza uralensis* seedlings under salt stress. Acta Physiol Plant 2016;38(4):96.

[65] Hasanuzzaman M, Nahar K, Anee TI, Fujita M. Exogenous silicon attenuates cadmium-induced oxidative stress in *Brassica napus* L. by modulating AsA-GSH pathway and glyoxalase system. Front Plant Sci 2017;8:1061.

[66] Fahad S, Hussain S, Matloob A, Khan FA, Khaliq A, Saud S, et al. Phytohormones and plant responses to salinity stress: a review. Plant Growth Regul 2015;75(2):391–404.

[67] Wani SH, Kumar V, Shriram V, Sah SK. Phytohormones and their metabolic engineering for abiotic stress tolerance in crop plants. Crop J 2016;4(3):162–76.

[68] Lee SK, Sohn EY, Hamayun M, Yoon JY, Lee IJ. Effect of silicon on growth and salinity stress of soybean plant grown under hydroponic system. Agrofor Syst 2010;80(3):333–40.

[69] Kim Y-H, Khan AL, Kim D-H, Lee S-Y, Kim K-M, Waqas M, et al. Silicon mitigates heavy metal stress by regulating P-type heavy metal ATPases, *Oryza sativa* low silicon genes, and endogenous phytohormones. BMC Plant Biol 2014;14(1):1–13.

[70] Kim YH, Khan AL, Waqas M, Shim JK, Kim DH, Lee KY, et al. Silicon application to rice root zone influenced the phytohormonal and antioxidant responses under salinity stress. J Plant Growth Regul 2014;33(2):137–49.

[71] Kim Y-H, Khan AL, Hamayun M, Kang SM, Beom YJ, Lee I-J. Influence of short-term silicon application on endogenous physiohormonal levels of *Oryza sativa* L. under wounding stress. Biol Trace Elem Res 2011;144(1):1175–85.

[72] Torabi F, Majd A, Enteshari S. The effect of silicon on alleviation of salt stress in borage (*Borago officinalis* L.). Soil Sci Plant Nutr 2015;61(5):788–98.

[73] Yin L, Wang S, Li J, Tanaka K, Oka M. Application of silicon improves salt tolerance through ameliorating osmotic and ionic stresses in the seedling of Sorghum bicolor. Acta Physiol Plant 2013;35(11):3099–107.

[74] Ming DF, Pei ZF, Naeem MS, Gong HJ, Zhou WJ. Silicon alleviates PEG-induced water-deficit stress in upland rice seedlings by enhancing osmotic adjustment. J Agron Crop Sci 2012;198(1):14–26.

[75] Hashemi A, Abdolzadeh A, Sadeghipour HR. Beneficial effects of silicon nutrition in alleviating salinity stress in hydroponically grown canola, *Brassica napus* L., plants. Soil Sci Plant Nutr 2010;56(2):244–53. Available from: https://doi.org/10.1111/j.1747-0765.2009.00443.x.

[76] Hajiboland R, Cherghvareh L, Dashtebani F. Effect of silicon supplementation on wheat plants under salt stress. J Plant Process Funct 2016;5(18):1–12.

[77] Sonobe K, Hattori T, An P, Tsuji W, Eneji AE, Kobayashi S, et al. Effect of silicon application on sorghum root responses to water stress. J Plant Nutr 2010;34(1):71–82.

[78] Pereira TS, da Silva Lobato AK, Tan DKY, da Costa DV, Uchoa EB, do Nascimento Ferreira R, et al. Positive interference of silicon on water relations, nitrogen metabolism, and osmotic adjustment in two pepper ('*Capsicum annuum*') cultivars under water deficit. Aust J Crop Sci 2013;7(8):1064.

[79] Zhu Y, Guo J, Feng R, Jia J, Han W, Gong H. The regulatory role of silicon on carbohydrate metabolism in *Cucumis sativus* L. under salt stress. Plant Soil 2016;406(1):231–49.

[80] Maillard A, Ali N, Schwarzenberg A, Jamois F, Yvin J-C, Hosseini SA. Silicon transcriptionally regulates sulfur and ABA metabolism and delays leaf senescence in barley under combined sulfur deficiency and osmotic stress. Environ Exp Bot 2018;155:394–410.

[81] Hosseini SA, Maillard A, Hajirezaei MR, Ali N, Schwarzenberg A, Jamois F, et al. Induction of barley silicon transporter HvLsi1 and HvLsi2, increased silicon concentration in the shoot and regulated starch and ABA homeostasis under osmotic stress and concomitant potassium deficiency. Front Plant Sci 2017;8:1359.

[82] Gong HJ, Randall DP, Flowers TJ. Silicon deposition in the root reduces sodium uptake in rice (*Oryza sativa* L.) seedlings by reducing bypass flow Plant Cell Env 2006;29(10):1970–9. Available from: https://doi.org/10.1111/j.1365-3040.2006.01572.x Epub 2006/08/26. Available from: 16930322.

[83] Ouzounidou G, Giannakoula A, Ilias I, Zamanidis P. Alleviation of drought and salinity stresses on growth, physiology, biochemistry and quality of two *Cucumis sativus* L. cultivars by Si application. Braz J Bot 2016;39(2):531–9.

[84] Ashfaque F, Inam A, Iqbal S, Sahay S. Response of silicon on metal accumulation, photosynthetic inhibition and oxidative stress in chromium-induced mustard (*Brassica juncea* L.). South Afr J Bot 2017;111:153–60.

[85] Song A, Li P, Fan F, Li Z, Liang Y. The effect of silicon on photosynthesis and expression of its relevant genes in rice (*Oryza sativa* L.) under high-zinc stress. PLoS One 2014;9(11):e113782.

[86] Silva ON, Lobato AKS, Ávila FW, Costa RCL, Neto CFO, Santos Filho BG, et al. Silicon-induced increase in chlorophyll is modulated by the leaf water potential in two water-deficient tomato cultivars. Plant Soil Environ 2012;58(11):481–6.

[87] Tuna AL, Kaya C, Higgs D, Murillo-Amador B, Aydemir S, Girgin AR. Silicon improves salinity tolerance in wheat plants. Environ Exp Bot 2008;62(1):10–16.

[88] Kaya C, Tuna L, Higgs D. Effect of silicon on plant growth and mineral nutrition of maize grown under water-stress conditions. J Plant Nutr 2006;29(8):1469–80. Available from: https://doi.org/10.1080/01904160600837238.

[89] Anderson DL. Soil and leaf nutrient interactions following application of calcium silicate slag to sugarcane. Fertil Res 1991;30(1):9–18. Available from: https://doi.org/10.1007/BF01048822.

[90] Hattori T, Inanaga S, Tanimoto E, Lux A, Luxová M, Sugimoto Y. Silicon-induced changes in viscoelastic properties of sorghum root cell walls Plant Cell Physiol 2003;44(7):743–9. Available from: https://doi.org/10.1093/pcp/pcg090 Epub 2003/07/26. Available from: 12881502.

[91] Rengel Z. Availability of Mn, Zn and Fe in the rhizosphere. J Soil Sci Plant Nutr 2015;15(2):397–409.

[92] Hernandez-Apaolaza L. Can silicon partially alleviate micronutrient deficiency in plants? A review. Planta 2014;240(3):447–58.

[93] Pavlovic J, Samardzic J, Maksimović V, Timotijevic G, Stevic N, Laursen KH, et al. Silicon alleviates iron deficiency in cucumber by promoting mobilization of iron in the root apoplast. New Phytol 2013;198(4):1096–107.

[94] Owino-Gerroh C, Gascho GJ. Effect of silicon on low pH soil phosphorus sorption and on uptake and growth of maize. Commun Soil Sci Plant Anal 2005;35(15-16):2369–78. Available from: https://doi.org/10.1081/LCSS-200030686.

[95] Liu P, Yin L, Deng X, Wang S, Tanaka K, Zhang S. Aquaporin-mediated increase in root hydraulic conductance is involved in silicon-induced improved root water uptake under osmotic stress in *Sorghum bicolor* L. J Exp Bot 2014;65(17):4747–56.

[96] Ma J, Cai H, He C, Zhang W, Wang L. A hemicellulose-bound form of silicon inhibits cadmium ion uptake in rice (*Oryza sativa*) cells New Phytol 2015;206(3):1063–74. Available from: https://doi.org/10.1111/nph.13276 Epub 2015/02/04. Available from: 25645894.

[97] Ma J, Takahashi E. Effect of silicon on the growth and phosphorus uptake of rice. Plant Soil 1990;126(1):115–19.

[98] Gao D, Cai K, Chen J, Luo S, Zeng R, Yang J, et al. Silicon enhances photochemical efficiency and adjusts mineral nutrient absorption in *Magnaporthe oryzae* infected rice plants. Acta Physiol Plant 2011;33(3):675–82.

[99] Neu S, Schaller J, Dudel EG. Silicon availability modifies nutrient use efficiency and content, C:N:P stoichiometry, and productivity of winter wheat (*Triticum aestivum* L.). Sci Rep 2017;7(1):1–8.

[100] Cocker KM, Evans DE, Hodson MJ. The amelioration of aluminium toxicity by silicon in wheat (*Triticum aestivum* L.): malate exudation as evidence for an in planta mechanism. Planta 1998;204(3):318–23.

[101] Wang Y, Stass A, Horst WJ. Apoplastic binding of aluminum is involved in silicon-induced amelioration of aluminum toxicity in maize Plant Physiol 2004;136(3):3762–70. Available from: https://doi.org/10.1104/pp.104.045005 Epub 10/22. Available from: 15502015.

[102] He C, Ma J, Wang L. A hemicellulose-bound form of silicon with potential to improve the mechanical properties and regeneration of the cell wall of rice. New Phytol 2015;206(3):1051–62.

[103] Cai K, Gao D, Chen J, Luo S. Probing the mechanisms of silicon-mediated pathogen resistance. Plant Signal Behav 2009;4(1):1–3.

[104] Faiyue B, Al-Azzawi MJ, Flowers TJ. The role of lateral roots in bypass flow in rice (*Oryza sativa* L.). Plant Cell Environ 2010;33 (5):702–16.

[105] Hinrichs M, Fleck AT, Biedermann E, Ngo NS, Schreiber L, Schenk MK, et al. An ABC transporter is involved in the silicon-induced formation of Casparian bands in the exodermis of rice. Front Plant Sci 2017;8:671.

[106] Fleck AT, Schulze S, Hinrichs M, Specht A, Waßmann F, Schreiber L, et al. Silicon promotes exodermal casparian band formation in Si-accumulating and Si-excluding species by forming phenol complexes. PLoS One 2015;10(9):e0138555.

[107] Exley C. A possible mechanism of biological silicification in plants. Front Plant Sci 2015;6:853.

[108] Fauteux F, Rémus-Borel W, Menzies JG, Bélanger RR. Silicon and plant disease resistance against pathogenic fungi. FEMS Microbiol Lett 2005;249(1):1–6.

[109] Saqib M, Zörb C, Schubert S. Silicon-mediated improvement in the salt resistance of wheat (*Triticum aestivum*) results from increased sodium exclusion and resistance to oxidative stress. Funct Plant Biol 2008;35(7):633–9.

[110] Krishnamurthy P, Ranathunge K, Nayak S, Schreiber L, Mathew MK. Root apoplastic barriers block Na^+ transport to shoots in rice (*Oryza sativa* L.). J Exp Bot 2011;62(12):4215–28.

[111] Jeelani PG, Mulay P, Venkat R, Ramalingam C. Multifaceted application of silica nanoparticles. A review. Silicon 2020;12(6):1337–54.

[112] Yuvakkumar R, Elango V, Rajendran V, Kannan NS, Prabu P. Influence of nanosilica powder on the growth of maize crop (*Zea mays* L.). Int J Green Nanotechnol 2011;3(3):180−90.

[113] O'Farrell N, Houlton A, Horrocks BR. Silicon nanoparticles: applications in cell biology and medicine. Int J Nanomed 2006;1(4):451.

[114] Rastogi A, Zivcak M, Sytar O, Kalaji HM, He X, Mbarki S, et al. Impact of metal and metal oxide nanoparticles on plant: a critical review. Front Chem 2017;5(78). Available from: https://doi.org/10.3389/fchem.2017.00078.

[115] Mamaeva V, Sahlgren C, Lindén M. Mesoporous silica nanoparticles in medicine—recent advances. Adv Drug Deliv Rev 2013;65 (5):689−702.

[116] Moradipour M, Saberi-Riseh R, Mohammadinejad R, Hosseini A. Nano-encapsulation of plant growth-promoting Rhizobacteria and their metabolites using alginate-silica nanoparticles and carbon nanotube improves UCB1 Pistachio micropropagation. J Microbiol Biotechnol 2019;29(7):1096−103.

[117] Nazaralian S, Majd A, Irian S, Najafi F, Ghahremaninejad F, Landberg T, et al. Comparison of silicon nanoparticles and silicate treatments in fenugreek. Plant Physiol Biochem 2017;115:25−33.

[118] Schaller J, Brackhage C, Bäucker E, Dudel EG. UV-screening of grasses by plant silica layer? J Biosci 2013;38(2):413−16.

[119] Asgari F, Majd A, Jonoubi P, Najafi F. Effects of silicon nanoparticles on molecular, chemical, structural and ultrastructural characteristics of oat (*Avena sativa* L.). Plant Physiol Biochem 2018;127:152−60.

[120] Siddiqui H, Ahmed KBM, Sami F, Hayat S. Silicon nanoparticles and plants: current knowledge and future perspectives. Sustainable agriculture reviews, 41. New York: Springer; 2020. p. 129−42.

[121] Yassen A, Abdallah E, Gaballah M, Zaghloul S. Role of silicon dioxide nano fertilizer in mitigating salt stress on growth, yield and chemical composition of cucumber (*Cucumis sativus* L.). Int J Agric Res 2017;22:130−5.

[122] Tantawy AS, Salama YAM, El-Nemr MA, Abdel-Mawgoud AMR. Nano silicon application improves salinity tolerance of sweet pepper plants. Int J ChemTech Res 2015;8(10):11−17.

[123] Kalteh M, Alipour ZT, Ashraf S, Marashi Aliabadi M, Falah, Nosratabadi A. Effect of silica nanoparticles on basil (*Ocimum basilicum*) under salinity stress. J Chem Health Risks 2018;4(3):49−55.

[124] Sabaghnia N, Janmohammadi M. Graphic analysis of nano-silicon by salinity stress interaction on germination properties of lentil using the biplot method. Agric For 2014;60(3):29−40.

[125] Ashkavand P, Tabari M, Zarafshar M, Tomásková I, Struve D. Effect of SiO_2 nanoparticles on drought resistance in hawthorn seedlings. Leśne Prace Badawcze 2015;76(4):350−9.

[126] Pei ZF, Ming DF, Liu D, Wan GL, Geng XX, Gong HJ, et al. Silicon improves the tolerance to water-deficit stress induced by polyethylene glycol in wheat (*Triticum aestivum* L.) seedlings. J Plant Growth Regul 2010;29(1):106−15.

[127] Haghighi M, Pessarakli M. Influence of silicon and nano-silicon on salinity tolerance of cherry tomatoes (*Solanum lycopersicum* L.) at early growth stage. Sci Hortic 2013;161:111−17.

[128] Alsaeedi A, El-Ramady H, Alshaal T, El-Garawany M, Elhawat N, Al-Otaibi A. Silica nanoparticles boost growth and productivity of cucumber under water deficit and salinity stresses by balancing nutrients uptake. Plant Physiol Biochem 2019;139:1−10.

[129] Avestan S, Ghasemnezhad M, Esfahani M, Byrt CS. Application of nano-silicon dioxide improves salt stress tolerance in strawberry plants. Agronomy. 2019;9(5):246.

[130] Mushtaq A, Jamil N, Riaz M, Hornyak GL, Ahmed N, Ahmed SS, et al. Synthesis of silica nanoparticles and their effect on priming of wheat (*Triticum aestivum* L.) under salinity stress. Biol Forum Inter J 2017;9:150−7.

[131] Naguib DM, Abdalla H. Metabolic status during germination of nano silica primed *Zea mays* seeds under salinity stress. J Crop Sci Biotechnol 2019;22(5):415−23.

[132] Abbasi Khalaki M, Ghorbani A, Moameri M. Effects of silica and silver nanoparticles on seed germination traits of *Thymus kotschyanus* in laboratory conditions. J Rangel Sci 2016;6(3):221−31.

[133] Janmohammadi M, Sabaghnia N. Effect of pre-sowing seed treatments with silicon nanoparticles on germinability of sunflower ( ). Botanica 2015;21(1):13−21. Available from: https://doi.org/10.1515/botlit-2015-0002.

[134] Suciaty T, Purnomo D, Sakya AT. Supriyadi. The effect of nano-silica fertilizer concentration and rice hull ash doses on soybean (*Glycine max* (L.) Merrill) growth and yield. IOP Conf Ser Earth Environ Sci 2018;129:012009. Available from: https://doi.org/10.1088/1755-1315/129/1/012009.

[135] Aqaei P, Weisany W, Diyanat M, Razmi J, Struik PC. Response of maize (*Zea mays* L.) to potassium nano-silica application under drought stress. J Plant Nutr 2020;43(9):1205−16. Available from: https://doi.org/10.1080/01904167.2020.1727508.

[136] Kheir AMS, Abouelsoud HM, Hafez EM, Ali OAM. Integrated effect of nano-Zn, nano-Si, and drainage using crop straw-filled ditches on saline sodic soil properties and rice productivity. Arab J Geosci 2019;12(15):1−8.

[137] Alsaeedi AH, El-Ramady H, Alshaal T, El-Garawani M, Elhawat N, Almohsen M. Engineered silica nanoparticles alleviate the detrimental effects of Na + stress on germination and growth of common bean (*Phaseolus vulgaris*). Environ Sci Pollut Res 2017;24(27):21917−28.

[138] Rawson HM, Long MJ, Munns R. Growth and development in NaCl-treated plants. I. Leaf Na^+ and Cl^- concentrations do not determine gas exchange of leaf blades in barley. Funct Plant Biol 1988;15(4):519−27.

[139] Wang J, Naser N. Improved performance of carbon paste amperometric biosensors through the incorporation of fumed silica. Electroanalysis 1994;6(7):571−5.

[140] Farhangi-Abriz S, Torabian S. Nano-silicon alters antioxidant activities of soybean seedlings under salt toxicity. Protoplasma 2018;255 (3):953−62.

[141] Mushtaq A, Jamil N, Rizwan S, Mandokhel F, Riaz M, Hornyak GL, et al. Engineered silica nanoparticles and silica nanoparticles containing controlled release fertilizer for drought and saline areas. IOP Conf Ser Mater Sci Eng 2018;414(1):012029.

[142] Hussain A, Rizwan M, Ali Q, Ali S. Seed priming with silicon nanoparticles improved the biomass and yield while reduced the oxidative stress and cadmium concentration in wheat grains. Environ Sci Pollut Res 2019;26(8):7579−88.

[143] Tripathi DK, Singh VP, Prasad SM, Chauhan DK, Kishore Dubey N, Rai AK. Silicon-mediated alleviation of Cr(VI) toxicity in wheat seedlings as evidenced by chlorophyll florescence, laser induced breakdown spectroscopy and anatomical changes. Ecotoxicol Environ Saf 2015;113:133−44. Available from: https://doi.org/10.1016/j.ecoenv.2014.09.029.

[144] Gao M, Zhou J, Liu H, Zhang W, Hu Y, Liang J, et al. Foliar spraying with silicon and selenium reduces cadmium uptake and mitigates cadmium toxicity in rice. Sci Total Environ 2018;631:1100—8.

[145] Cui J, Liu T, Li F, Yi J, Liu C, Yu H. Silica nanoparticles alleviate cadmium toxicity in rice cells: mechanisms and size effects Environ Pollut 2017;228:363—9. Available from: https://doi.org/10.1016/j.envpol.2017.05.014 Epub 2017/05/30. Available from: 28551566.

[146] Tripathi DK, Singh S, Singh VP, Prasad SM, Chauhan DK, Dubey NK. Silicon nanoparticles more efficiently alleviate arsenate toxicity than silicon in maize cultiver and hybrid differing in arsenate tolerance. Front Environ Sci 2016;4:46.

[147] Tripathi DK, Singh VP, Prasad SM, Chauhan DK, Dubey NK. Silicon nanoparticles (SiNp) alleviate chromium (VI) phytotoxicity in Pisum sativum (L.) seedlings. Plant Physiol Biochem 2015;96:189—98. Available from: https://doi.org/10.1016/j.plaphy.2015.07.026.

[148] Manivannan A, Ahn Y-K. Silicon regulates potential genes involved in major physiological processes in plants to combat stress. Front Plant Sci 2017;8:1346.

[149] Chen R, Zhang C, Zhao Y, Huang Y, Liu Z. Foliar application with nano-silicon reduced cadmium accumulation in grains by inhibiting cadmium translocation in rice plants Environ Sci Pollut Res Int 2018;25(3):2361—8. Available from: https://doi.org/10.1007/s11356-017-0681-z Epub 2017/11/11. Available from: 29124638.

[150] Liu J, Cai H, Mei C, Wang M. Effects of nano-silicon and common silicon on lead uptake and translocation in two rice cultivars. Front Environ Sci Eng 2015;9(5):905—11. Available from: https://doi.org/10.1007/s11783-015-0786-x.

[151] Bharwana SA, Ali S, Farooq MA, Iqbal N, Abbas F, Ahmad MSA. Alleviation of lead toxicity by silicon is related to elevated photosynthesis, antioxidant enzymes suppressed lead uptake and oxidative stress in cotton. J Bioremed Biodeg 2013;4(4):187.

[152] Rajwade JM, Chikte RG, Paknikar KM. Nanomaterials: new weapons in a crusade against phytopathogens Appl Microbiol Biotechnol 2020;104(4):1437—61. Available from: https://doi.org/10.1007/s00253-019-10334-y Epub 2020/01/05. Available from: 31900560.

[153] Derbalah A, Shenashen M, Hamza A, Mohamed A, El Safty S. Antifungal activity of fabricated mesoporous silica nanoparticles against early blight of tomato. Egypt J Basic Appl Sci 2018;5(2):145—50. Available from: https://doi.org/10.1016/j.ejbas.2018.05.002.

[154] Madany MMY, Saleh AM, Habeeb TH, Hozzein WN, AbdElgawad H. Silicon dioxide nanoparticles alleviate the threats of broomrape infection in tomato by inducing cell wall fortification and modulating ROS homeostasis. Environ Sci Nano 2020;7(5):1415—30. Available from: https://doi.org/10.1039/C9EN01255A.

[155] Buchman JT, Elmer WH, Ma C, Landy KM, White JC, Haynes CL. Chitosan-coated mesoporous silica nanoparticle treatment of Citrullus lanatus (Watermelon): Enhanced fungal disease suppression and modulated expression of stress-related genes. ACS Sustain Chem Eng 2019;7(24):19649—59. Available from: https://doi.org/10.1021/acssuschemeng.9b04800.

[156] Suriyaprabha R, Karunakaran G, Yuvakkumar R, Rajendran V, Kannan N. Foliar application of silica nanoparticles on the phytochemical responses of maize (Zea mays L.) and its toxicological behavior. Synth React Inorg Met Org Nano-Metal Chem 2014;44(8):1128—31. Available from: https://doi.org/10.1080/15533174.2013.799197.

[157] Hasan KA, Soliman H, Baka Z, Shabana YM. Efficacy of nano-silicon in the control of chocolate spot disease of Vicia faba L. caused by Botrytis fabae. Egypt J Basic Appl Sci 2020;7(1):53—66. Available from: https://doi.org/10.1080/2314808X.2020.1727627.

[158] Tuan LNA, Du BD, Ha LDT, Dzung LTK, Van Phu D, Hien NQ. Induction of chitinase and brown spot disease resistance by oligochitosan and nanosilica—oligochitosan in Dragon fruit plants. Agric Res 2019;8(2):184—90. Available from: https://doi.org/10.1007/s40003-018-0384-9.

[159] Youssef K, de Oliveira AG, Tischer CA, Hussain I, Roberto SR. Synergistic effect of a novel chitosan/silica nanocomposites-based formulation against gray mold of table grapes and its possible mode of action. Int J Biol Macromol 2019;141:247—58. Available from: https://doi.org/10.1016/j.ijbiomac.2019.08.249.

[160] Khan MR, Siddiqui ZA. Use of silicon dioxide nanoparticles for the management of Meloidogyne incognita, Pectobacterium betavasculorum and Rhizoctonia solani disease complex of beetroot (Beta vulgaris L.). Sci Horticult 2020;265:109211. Available from: https://doi.org/10.1016/j.scienta.2020.109211.

[161] Elsharkawy MM, Suga H, Shimizu M. Systemic resistance induced by Phoma sp. GS8-3 and nanosilica against Cucumber mosaic virus. Environ Sci Pollut Res 2020;27(16):19029—37. Available from: https://doi.org/10.1007/s11356-018-3321-3.

[162] Chu H, Kim H-J, Su Kim J, Kim M-S, Yoon B-D, Park H-J, et al. A nanosized Ag—silica hybrid complex prepared by γ-irradiation activates the defense response in Arabidopsis. Radiat Phys Chem 2012;81(2):180—4. Available from: https://doi.org/10.1016/j.radphyschem.2011.10.004.

[163] Banerjee A, Singh A, Sudarshan M, Roychoudhury A. Silicon nanoparticle-pulsing mitigates fluoride stress in rice by fine-tuning the ionomic and metabolomic balance and refining agronomic traits Chemosphere 2021;262:127826. Available from: https://doi.org/10.1016/j.chemosphere.2020.127826 Epub 2020/11/14. Available from: 33182120.

[164] Behboudi F, Tahmasebi Sarvestani Z, Kassaee MZ, Modares Sanavi SAM, Sorooshzadeh A. Improving growth and yield of wheat under drought stress via application of SiO_2 nanoparticles. J Agric Sci Technol 2018;20(7):1479—92.

[165] Wang B, Qiu YL. Phylogenetic distribution and evolution of mycorrhizas in land plants. Mycorrhiza 2006;16(5):299—363. Available from: https://doi.org/10.1007/s00572-005-0033-6.

[166] Smith SE, Read DJ. Mycorrhizal symbiosis. London: Academic Press; 2008.

[167] Schüßler A, Schwarzott D, Walker C. A new fungal phylum, the Glomeromycota: phylogeny and evolution* *Dedicated to Manfred Kluge (Technische Universität Darmstadt) on the occasion of his retirement. Mycol Res 2001;105(12):1413—21. Available from: https://doi.org/10.1017/S0953756201005196.

[168] Bonfante P, Desirò A. Arbuscular mycorrhizas: the lives of beneficial fungi and their plant hosts. Principles of plant-microbe interactions. Leiden: Springer; 2015. p. 235—45.

[169] Parniske M. Arbuscular mycorrhiza: the mother of plant root endosymbioses. Nat Rev Microbiol 2008;6(10):763—75. Available from: https://doi.org/10.1038/nrmicro1987.

[170] Keymer A, Pimprikar P, Wewer V, Huber C, Brands M, Bucerius SL, et al. Lipid transfer from plants to arbuscular mycorrhiza fungi. Elife. 2017;6:e29107.

[171] Pereira Junior P, Rezende PM, Malfitano SC, Lima RK, Corrêa LVT, Carvalho ER. Effects of doses of silicon in the yield and agronomic characteristics of soybean [Glycine max (L.) Merrill]. Ciência e Agrotecnologia 2010;34(4):908—13.

[172] Fay P, Mitchell DT, Osborne BA. Photosynthesis and nutrient-use efficiency of barley in response to low arbuscular mycorrhizal colonization and addition of phosphorus New Phytol 1996;132(3):425–33. Available from: https://doi.org/10.1111/j.1469-8137.1996.tb01862.x Epub 1996/03/01. Available from: 26763638.

[173] Romero-Munar A, Del-Saz NF, Ribas-Carbó M, Flexas J, Baraza E, Florez-Sarasa I, et al. Arbuscular mycorrhizal symbiosis with *Arundo donax* decreases root respiration and increases both photosynthesis and plant biomass accumulation Plant Cell Env 2017;40(7):1115–26. Available from: https://doi.org/10.1111/pce.12902 Epub 2017/01/07. Available from: 28060998.

[174] Kumar A, Verma JP. Does plant-microbe interaction confer stress tolerance in plants: a review? Microbiol Res 2018;207:41–52. Available from: https://doi.org/10.1016/j.micres.2017.11.004.

[175] Ortas I, Sari N, Akpinar Ç, Yetisir H. Screening mycorrhiza species for plant growth, P and Zn uptake in pepper seedling grown under greenhouse conditions. Sci Horticult 2011;128(2):92–8. Available from: https://doi.org/10.1016/j.scienta.2010.12.014.

[176] Latef AAHA, Hashem A, Rasool S, Abd_Allah EF, Alqarawi AA, Egamberdieva D, et al. Arbuscular mycorrhizal symbiosis and abiotic stress in plants: a review. J Plant Biol 2016;59(5):407–26. Available from: https://doi.org/10.1007/s12374-016-0237-7.

[177] Hashem A, Alqarawi AA, Radhakrishnan R, Al-Arjani A-BF, Aldehaish HA, Egamberdieva D, et al. Arbuscular mycorrhizal fungi regulate the oxidative system, hormones and ionic equilibrium to trigger salt stress tolerance in *Cucumis sativus* L. Saudi J Biol Sci 2018;25 (6):1102–14.

[178] Vicente-Sánchez J, Nicolás E, Pedrero F, Alarcón JJ, Maestre-Valero JF, Fernández F. Arbuscular mycorrhizal symbiosis alleviates detrimental effects of saline reclaimed water in lettuce plants Mycorrhiza 2014;24(5):339–48. Available from: https://doi.org/10.1007/s00572-013-0542-7 Epub 2013/11/30. Available from: 24287607.

[179] Castellanos-Morales V, Villegas J, Wendelin S, Vierheilig H, Eder R, Cárdenas-Navarro R. Root colonisation by the arbuscular mycorrhizal fungus *Glomus intraradices* alters the quality of strawberry fruits (Fragaria x ananassa Duch.) at different nitrogen levels J Sci Food Agric 2010;90(11):1774–82. Available from: https://doi.org/10.1002/jsfa.3998 Epub 2010/06/24. Available from: 20572056.

[180] Hart M, Ehret DL, Krumbein A, Leung C, Murch S, Turi C, et al. Inoculation with arbuscular mycorrhizal fungi improves the nutritional value of tomatoes. Mycorrhiza. 2015;25(5):359–76. Available from: https://doi.org/10.1007/s00572-014-0617-0.

[181] Bona E, Cantamessa S, Massa N, Manassero P, Marsano F, Copetta A, et al. Arbuscular mycorrhizal fungi and plant growth-promoting pseudomonads improve yield, quality and nutritional value of tomato: a field study Mycorrhiza. 2017;27(1):1–11. Available from: https://doi.org/10.1007/s00572-016-0727-y Epub 2016/08/20. Available from: 27539491.

[182] Baslam M, Garmendia I, Goicoechea N. Arbuscular mycorrhizal fungi (AMF) improved growth and nutritional quality of greenhouse-grown lettuce. J Agric Food Chem 2011;59(10):5504–15. Available from: https://doi.org/10.1021/jf200501c.

[183] Sbrana C, Avio L, Giovannetti M. Beneficial mycorrhizal symbionts affecting the production of health-promoting phytochemicals Electrophoresis 2014;35(11):1535–46. Available from: https://doi.org/10.1002/elps.201300568 Epub 2014/07/16. Available from: 25025092.

[184] Rouphael Y, Franken P, Schneider C, Schwarz D, Giovannetti M, Agnolucci M, et al. Arbuscular mycorrhizal fungi act as biostimulants in horticultural crops. Sci Horticult 2015;196:91–108. Available from: https://doi.org/10.1016/j.scienta.2015.09.002.

[185] Bauddh K, Singh RP. Growth, tolerance efficiency and phytoremediation potential of *Ricinus communis* (L.) and *Brassica juncea* (L.) in salinity and drought affected cadmium contaminated soil Ecotoxicol Env Saf 2012;85:13–22. Available from: https://doi.org/10.1016/j.ecoenv.2012.08.019 Epub 2012/09/11. Available from: 22959315.

[186] Oztekin GB, Tuzel Y, Tuzel IH. Does mycorrhiza improve salinity tolerance in grafted plants? Sci Horticult 2013;149:55–60. Available from: https://doi.org/10.1016/j.scienta.2012.02.033.

[187] Ahanger MA, Agarwal RM. Potassium up-regulates antioxidant metabolism and alleviates growth inhibition under water and osmotic stress in wheat (*Triticum aestivum* L). Protoplasma. 2017;254(4):1471–86. Available from: https://doi.org/10.1007/s00709-016-1037-0.

[188] Hegazi AM, El-Shraiy AM, Ghoname AA. Mitigation of salt stress negative effects on sweet pepper using arbuscular mycorrhizal fungi (AMF), *Bacillus megaterium* and Brassinosteroids (BRs). Gesunde Pflanz 2017;69(2):91–102. Available from: https://doi.org/10.1007/s10343-017-0393-9.

[189] Augé RM, Toler HD, Saxton AM. Mycorrhizal stimulation of leaf gas exchange in relation to root colonization, shoot size, leaf phosphorus and nitrogen: a quantitative analysis of the literature using meta-regression. Front Plant Sci 2016;7:1084. Available from: https://doi.org/10.3389/fpls.2016.01084.

[190] Hashem A, Kumar A, Al-Dbass AM, Alqarawi AA, Al-Arjani A-BF, Singh G, et al. Arbuscular mycorrhizal fungi and biochar improves drought tolerance in chickpea. Saudi J Biol Sci 2019;26(3):614–24. Available from: https://doi.org/10.1016/j.sjbs.2018.11.005.

[191] Augé RM. Water relations, drought and vesicular-arbuscular mycorrhizal symbiosis. Mycorrhiza 2001;11(1):3–42. Available from: https://doi.org/10.1007/s005720100097.

[192] Ruiz-Lozano JM. Arbuscular mycorrhizal symbiosis and alleviation of osmotic stress. New perspectives for molecular studies. Mycorrhiza 2003;13(6):309–17. Available from: https://doi.org/10.1007/s00572-003-0237-6.

[193] Bárzana G, Aroca R, Paz JA, Chaumont F, Martinez-Ballesta MC, Carvajal M, et al. Arbuscular mycorrhizal symbiosis increases relative apoplastic water flow in roots of the host plant under both well-watered and drought stress conditions Ann Bot 2012;109(5):1009–17. Available from: https://doi.org/10.1093/aob/mcs007 Epub 01/31. Available from: 22294476.

[194] Allen MF. Influence of vesicular-arbuscular mycorrhizae on water movement through bouteloua gracilis (h.b.k.) lag ex steud*. New Phytol 1982;91(2):191–6. Available from: https://doi.org/10.1111/j.1469-8137.1982.tb03305.x.

[195] Hardie KAY. The effect of removal of extraradical hyphae on water uptake by vesicular-arbuscular mycorrhizal plants. New Phytol 1985;101(4):677–84. Available from: https://doi.org/10.1111/j.1469-8137.1985.tb02873.x.

[196] Marulanda A, Azcón R, Ruiz-Lozano JM. Contribution of six arbuscular mycorrhizal fungal isolates to water uptake by *Lactuca sativa* plants under drought stress. Physiol Plant 2003;119(4):526–33. Available from: https://doi.org/10.1046/j.1399-3054.2003.00196.x.

[197] Evelin H, Kapoor R, Giri B. Arbuscular mycorrhizal fungi in alleviation of salt stress: a review. Ann Bot 2009;104(7):1263–80.

[198] Maurel C, Verdoucq L, Luu DT, Santoni V. Plant aquaporins: membrane channels with multiple integrated functions Annu Rev Plant Biol 2008;59:595–624. Available from: https://doi.org/10.1146/annurev.arplant.59.032607.092734 Epub 2008/05/01. Available from: 18444909.

[199] Kruse E, Uehlein N, Kaldenhoff R. The aquaporins. Genome Biol 2006;7(2):206. Available from: https://doi.org/10.1186/gb-2006-7-2-206.

[200] Maurel C, Santoni V, Luu D-T, Wudick MM, Verdoucq L. The cellular dynamics of plant aquaporin expression and functions. Curr Opin Plant Biol 2009;12(6):690–8. Available from: https://doi.org/10.1016/j.pbi.2009.09.002.

[201] Aroca R, Bago A, Sutka M, Paz JA, Cano C, Amodeo G, et al. Expression analysis of the first arbuscular mycorrhizal fungi aquaporin described reveals concerted gene expression between salt-stressed and nonstressed mycelium Mol Plant Microbe Interact 2009;22 (9):1169–78. Available from: https://doi.org/10.1094/mpmi-22-9-1169 Epub 2009/08/07. Available from: 19656051.

[202] Li T, Hu YJ, Hao ZP, Li H, Wang YS, Chen BD. First cloning and characterization of two functional aquaporin genes from an arbuscular mycorrhizal fungus Glomus intraradices New Phytol 2013;197(2):617–30. Available from: https://doi.org/10.1111/nph.12011 Epub 2012/11/20. Available from: 23157494.

[203] Li T, Hu YJ, Hao ZP, Li H, Chen BD. Aquaporin genes GintAQPF1 and GintAQPF2 from Glomus intraradices contribute to plant drought tolerance Plant Signal Behav 2013;8(5):e24030. Available from: https://doi.org/10.4161/psb.24030 Epub 2013/02/26. Available from: 23435173.

[204] Martin SL, Mooney SJ, Dickinson MJ, West HM. The effects of simultaneous root colonisation by three Glomus species on soil pore characteristics. Soil Biol Biochem 2012;49:167–73. Available from: https://doi.org/10.1016/j.soilbio.2012.02.036.

[205] Mardhiah U, Caruso T, Gurnell A, Rillig MC. Arbuscular mycorrhizal fungal hyphae reduce soil erosion by surface water flow in a greenhouse experiment. Appl Soil Ecol 2016;99:137–40. Available from: https://doi.org/10.1016/j.apsoil.2015.11.027.

[206] Singh PK, Singh M, Tripathi BN. Glomalin: an arbuscular mycorrhizal fungal soil protein. Protoplasma 2013;250(3):663–9. Available from: https://doi.org/10.1007/s00709-012-0453-z.

[207] Tisdall JM, Oades JM. Organic matter and water-stable aggregates in soils. J Soil Sci 1982;33(2):141–63. Available from: https://doi.org/10.1111/j.1365-2389.1982.tb01755.x.

[208] Rillig MC, Mummey DL. Mycorrhizas and soil structure. New Phytol 2006;171(1):41–53. Available from: https://doi.org/10.1111/j.1469-8137.2006.01750.x.

[209] Manoharan PT, Shanmugaiah V, Balasubramanian N, Gomathinayagam S, Sharma MP, Muthuchelian K. Influence of AM fungi on the growth and physiological status of Erythrina variegata Linn. grown under different water stress conditions. Eur J Soil Biol 2010;46 (2):151–6. Available from: https://doi.org/10.1016/j.ejsobi.2010.01.001.

[210] Zhang Y-C, Wang P, Wu Q-H, Zou Y-N, Bao Q, Wu Q-S. Arbuscular mycorrhizas improve plant growth and soil structure in trifoliate orange under salt stress. Arch Agron Soil Sci 2017;63(4):491–500. Available from: https://doi.org/10.1080/03650340.2016.1222609.

[211] Zou Y-N, Srivastava AK, Wu Q-S, Huang Y-M. Glomalin-related soil protein and water relations in mycorrhizal citrus (Citrus tangerina) during soil water deficit. Arch Agron Soil Sci 2014;60(8):1103–14. Available from: https://doi.org/10.1080/03650340.2013.867950.

[212] Marschner P, Crowley DE, Higashi RM. Root exudation and physiological status of a root-colonizing fluorescent pseudomonad in mycorrhizal and non-mycorrhizal pepper (Capsicum annuum L.). Plant Soil 1997;189(1):11–20. Available from: https://doi.org/10.1023/A:1004266907442.

[213] Safir GR, Boyer JS, Gerdemann JW. Mycorrhizal enhancement of water transport in soybean Science 1971;172(3983):581–3. Available from: https://doi.org/10.1126/science.172.3983.581 Epub 1971/05/07. Available from: 17802222.

[214] Safir GR, Boyer JS, Gerdemann JW. Nutrient status and mycorrhizal enhancement of water transport in soybean Plant Physiol 1972;49 (5):700–3. Available from: https://doi.org/10.1104/pp.49.5.700 Epub 1972/05/01. Available from: 16658032.

[215] Garg N, Manchanda G. Role of arbuscular mycorrhizae in the alleviation of ionic, osmotic and oxidative stresses induced by salinity in Cajanus cajan (L.) Millsp. (pigeonpea). J Agron Crop Sci 2009;195(2):110–23. Available from: https://doi.org/10.1111/j.1439-037X.2008.00349.x.

[216] Bowles TM, Barrios-Masias FH, Carlisle EA, Cavagnaro TR, Jackson LE. Effects of arbuscular mycorrhizae on tomato yield, nutrient uptake, water relations, and soil carbon dynamics under deficit irrigation in field conditions. Sci Total Environ 2016;566-567:1223–34. Available from: https://doi.org/10.1016/j.scitotenv.2016.05.178.

[217] Liu C, Muchhal US, Uthappa M, Kononowicz AK, Raghothama KG. Tomato phosphate transporter genes are differentially regulated in plant tissues by phosphorus Plant Physiol 1998;116(1):91–9. Available from: https://doi.org/10.1104/pp.116.1.91 Epub 1998/02/05. Available from: 9449838.

[218] Smith SE, Smith FA, Jakobsen I. Functional diversity in arbuscular mycorrhizal (AM) symbioses: the contribution of the mycorrhizal P uptake pathway is not correlated with mycorrhizal responses in growth or total P uptake. New Phytol 2004;162(2):511–24. Available from: https://doi.org/10.1111/j.1469-8137.2004.01039.x.

[219] Viereck N, Hansen PE, Jakobsen I. Phosphate pool dynamics in the arbuscular mycorrhizal fungus Glomus intraradices studied by in vivo31P NMR spectroscopy. New Phytol 2004;162(3):783–94. Available from: https://doi.org/10.1111/j.1469-8137.2004.01048.x.

[220] Hodge A, Campbell CD, Fitter AH. An arbuscular mycorrhizal fungus accelerates decomposition and acquires nitrogen directly from organic material. Nature. 2001;413(6853):297–9. Available from: https://doi.org/10.1038/35095041.

[221] Hodge A, Helgason T, Fitter AH. Nutritional ecology of arbuscular mycorrhizal fungi. Fungal Ecol 2010;3(4):267–73. Available from: https://doi.org/10.1016/j.funeco.2010.02.002.

[222] Govindarajulu M, Pfeffer PE, Jin H, Abubaker J, Douds DD, Allen JW, et al. Nitrogen transfer in the arbuscular mycorrhizal symbiosis. Nature 2005;435(7043):819–23. Available from: https://doi.org/10.1038/nature03610.

[223] Bago B, Pfeffer P, Shachar-Hill Y. Could the urea cycle be translocating nitrogen in the arbuscular mycorrhizal symbiosis? New Phytol 2001;149(1):4–8. Available from: https://doi.org/10.1046/j.1469-8137.2001.00016.x.

[224] Bago B, Vierheilig H, Piché Y, AzcÓN-Aguilar C. Nitrate depletion and pH changes induced by the extraradical mycelium of the arbuscular mycorrhizal fungus Glomus intraradices grown in monoxenic culture. New Phytol 1996;133(2):273–80. Available from: https://doi.org/10.1111/j.1469-8137.1996.tb01894.x.

[225] Adiku SGK, Renger M, Wessolek G, Facklam M, Hecht-Bucholtz C. Simulation of the dry matter production and seed yield of common beans under varying soil water and salinity conditions. Agric Water Manag 2001;47(1):55–68. Available from: https://doi.org/10.1016/S0378-3774(00)00094-9.

[226] Hu Y, Schmidhalter U. Drought and salinity: a comparison of their effects on mineral nutrition of plants. J Plant Nutr Soil Sci 2005;168(4):541–9.

[227] Daei G, Ardekani MR, Rejali F, Teimuri S, Miransari M. Alleviation of salinity stress on wheat yield, yield components, and nutrient uptake using arbuscular mycorrhizal fungi under field conditions. J Plant Physiol 2009;166(6):617—25. Available from: https://doi.org/10.1016/j.jplph.2008.09.013.

[228] Mardukhi B, Rejali F, Daei G, Ardakani MR, Malakouti MJ, Miransari M. Arbuscular mycorrhizas enhance nutrient uptake in different wheat genotypes at high salinity levels under field and greenhouse conditions. C R Biol 2011;334(7):564—71.

[229] Hammer EC, Nasr H, Pallon J, Olsson PA, Wallander H. Elemental composition of arbuscular mycorrhizal fungi at high salinity Mycorrhiza 2011;21(2):117—29. Available from: https://doi.org/10.1007/s00572-010-0316-4 Epub 2010/05/26. Available from: 20499112.

[230] Giri B, Kapoor R, Mukerji KG. Improved tolerance of *Acacia nilotica* to salt stress by *Arbuscular mycorrhiza, Glomus fasciculatum* may be partly related to elevated K/Na ratios in root and shoot tissues Microb Ecol 2007;54(4):753—60. Available from: https://doi.org/10.1007/s00248-007-9239-9 Epub 2007/03/21. Available from: 17372663.

[231] Porras-Soriano A, Soriano-Martín ML, Porras-Piedra A, Azcón R. Arbuscular mycorrhizal fungi increased growth, nutrient uptake and tolerance to salinity in olive trees under nursery conditions. J Plant Physiol 2009;166(13):1350—9. Available from: https://doi.org/10.1016/j.jplph.2009.02.010.

[232] Talaat NB, Shawky BT. Modulation of nutrient acquisition and polyamine pool in salt-stressed wheat (*Triticum aestivum* L.) plants inoculated with arbuscular mycorrhizal fungi. Acta Physiol Plant 2013;35(8):2601—10. Available from: https://doi.org/10.1007/s11738-013-1295-9.

[233] Casieri L, Ait Lahmidi N, Doidy J, Veneault-Fourrey C, Migeon A, Bonneau L, et al. Biotrophic transportome in mutualistic plant—fungal interactions. Mycorrhiza. 2013;23(8):597—625. Available from: https://doi.org/10.1007/s00572-013-0496-9.

[234] Sharma SS, Dietz KJ. The significance of amino acids and amino acid-derived molecules in plant responses and adaptation to heavy metal stress J Exp Bot 2006;57(4):711—26. Available from: https://doi.org/10.1093/jxb/erj073 Epub 2006/02/14. Available from: 16473893.

[235] Hare PD, Cress WA, Van Staden J. Dissecting the roles of osmolyte accumulation during stress. Plant Cell Environ 1998;21(6):535—53. Available from: https://doi.org/10.1046/j.1365-3040.1998.00309.x.

[236] Shirmohammadi E, Alikhani HA, Pourbabaei AA, Etesami H. Improved phosphorus (P) uptake and yield of rainfed wheat fed with P fertilizer by drought-tolerant phosphate-solubilizing fluorescent pseudomonads strains: a field study in drylands. J Soil Sci Plant Nutr 2020;20(4):2195—211.

[237] Naidu BP, Paleg LG, Aspinall D, Jennings AC, Jones GP. Amino acid and glycine betaine accumulation in cold-stressed wheat seedlings. Phytochemistry 1991;30(2):407—9. Available from: https://doi.org/10.1016/0031-9422(91)83693-F.

[238] Munns R. Genes and salt tolerance: bringing them together. New Phytol 2005;167(3):645—63. Available from: https://doi.org/10.1111/j.1469-8137.2005.01487.x.

[239] Pérez-Pérez JG, Robles JM, Tovar JC, Botía P. Response to drought and salt stress of lemon 'Fino 49' under field conditions: water relations, osmotic adjustment and gas exchange. Sci Horticult 2009;122(1):83—90. Available from: https://doi.org/10.1016/j.scienta.2009.04.009.

[240] Lehmann S, Funck D, Szabados L, Rentsch D. Proline metabolism and transport in plant development Amino Acids 2010;39(4):949—62. Available from: https://doi.org/10.1007/s00726-010-0525-3 Epub 2010/03/06. Available from: 20204435.

[241] Kavi Kishor PB, Sreenivasulu N. Is proline accumulation per se correlated with stress tolerance or is proline homeostasis a more critical issue? Plant Cell Environ 2014;37(2):300—11. Available from: https://doi.org/10.1111/pce.12157 Epub 2013/06/25. Available from: 23790054.

[242] Ashraf M, Foolad MR. Roles of glycine betaine and proline in improving plant abiotic stress resistance. Environ Exp Bot 2007;59 (2):206—16. Available from: https://doi.org/10.1016/j.envexpbot.2005.12.006.

[243] Yooyongwech S, Phaukinsang N, Cha-um S, Supaibulwatana K. Arbuscular mycorrhiza improved growth performance in *Macadamia tetraphylla* L. grown under water deficit stress involves soluble sugar and proline accumulation. Plant Growth Regul 2013;69(3):285—93. Available from: https://doi.org/10.1007/s10725-012-9771-6.

[244] Evelin H, Giri B, Kapoor R. Ultrastructural evidence for AMF mediated salt stress mitigation in *Trigonella foenum-graecum*. Mycorrhiza. 2013;23(1):71—86. Available from: https://doi.org/10.1007/s00572-012-0449-8.

[245] Mo Y, Wang Y, Yang R, Zheng J, Liu C, Li H, et al. Regulation of plant growth, photosynthesis, antioxidation and osmosis by an arbuscular mycorrhizal fungus in watermelon seedlings under well-watered and drought conditions Front Plant Sci 2016;7:644. Available from: https://doi.org/10.3389/fpls.2016.00644 Epub 2016/06/01. Available from: 27242845.

[246] Ruiz-Lozano JM, Porcel R, Azcón C, Aroca R. Regulation by arbuscular mycorrhizae of the integrated physiological response to salinity in plants: new challenges in physiological and molecular studies J Exp Bot 2012;63(11):4033—44. Available from: https://doi.org/10.1093/jxb/ers126 Epub 2012/05/04. Available from: 22553287.

[247] Ruiz-Lozano JM, Azcón R. Hyphal contribution to water uptake in mycorrhizal plants as affected by the fungal species and water status. Physiol Plant 1995;95(3):472—8. Available from: https://doi.org/10.1111/j.1399-3054.1995.tb00865.x.

[248] Sheng M, Tang M, Zhang F, Huang Y. Influence of arbuscular mycorrhiza on organic solutes in maize leaves under salt stress Mycorrhiza 2011;21(5):423—30. Available from: https://doi.org/10.1007/s00572-010-0353-z Epub 2010/12/31. Available from: 21191619.

[249] Ruíz-Lozano JM, del Carmen Perálvarez M, Aroca R, Azcón R. The application of a treated sugar beet waste residue to soil modifies the responses of mycorrhizal and non mycorrhizal lettuce plants to drought stress. Plant Soil 2011;346(1):153. Available from: https://doi.org/10.1007/s11104-011-0805-z.

[250] Abdel Latef AAH, Chaoxing H. Does inoculation with *Glomus mosseae* improve salt tolerance in pepper plants? J Plant Growth Regul 2014;33(3):644—53. Available from: https://doi.org/10.1007/s00344-014-9414-4.

[251] Porcel R, Ruiz-Lozano JM. Arbuscular mycorrhizal influence on leaf water potential, solute accumulation, and oxidative stress in soybean plants subjected to drought stress J Exp Bot 2004;55(403):1743—50. Available from: https://doi.org/10.1093/jxb/erh188 Epub 2004/06/23. Available from: 15208335.

[252] Talaat NB, Shawky BT. Influence of arbuscular mycorrhizae on yield, nutrients, organic solutes, and antioxidant enzymes of two wheat cultivars under salt stress. J Plant Nutr Soil Sci 2011;174(2):283—91. Available from: https://doi.org/10.1002/jpln.201000051.

[253] Sheng M, Tang M, Chen H, Yang B, Zhang F, Huang Y. Influence of arbuscular mycorrhizae on photosynthesis and water status of maize plants under salt stress Mycorrhiza 2008;18(6-7):287—96. Available from: https://doi.org/10.1007/s00572-008-0180-7 Epub 2008/06/28. Available from: 18584217.

[254] Wu Q-S, Zou Y-N, He X-H. Contributions of arbuscular mycorrhizal fungi to growth, photosynthesis, root morphology and ionic balance of citrus seedlings under salt stress. Acta Physiol Plant 2010;32(2):297−304. Available from: https://doi.org/10.1007/s11738-009-0407-z.

[255] Hasegawa PM, Bressan RA, Zhu J-K, Bohnert HJ. Plant cellular and molecular responses to high salinity. Annu Rev Plant Physiol Plant Mol Biol 2000;51(1):463−99. Available from: https://doi.org/10.1146/annurev.arplant.51.1.463.

[256] Yang C, Chong J, Li C, Kim C, Shi D, Wang D. Osmotic adjustment and ion balance traits of an alkali resistant halophyte *Kochia sieversiana* during adaptation to salt and alkali conditions. Plant Soil 2007;294(1):263−76. Available from: https://doi.org/10.1007/s11104-007-9251-3.

[257] Rozpądek P, Rąpała-Kozik M, Wężowicz K, Grandin A, Karlsson S, Ważny R, et al. Arbuscular mycorrhiza improves yield and nutritional properties of onion (*Allium cepa*). Plant Physiol Biochem 2016;107:264−72. Available from: https://doi.org/10.1016/j.plaphy.2016.06.006.

[258] Bano SA, Ashfaq D. Role of mycorrhiza to reduce heavy metal stress. Nat Sci 2013;2013.

[259] Kapoor R, Evelin H, Mathur P, Giri B. Arbuscular mycorrhiza: approaches for abiotic stress tolerance in crop plants for sustainable agriculture. Plant acclimation to environmental stress. New York: Springer; 2013. p. 359−401.

[260] Vodnik D, Grčman H, Maček I, Van Elteren JT, Kovačevič M. The contribution of glomalin-related soil protein to Pb and Zn sequestration in polluted soil. Sci Total Environ 2008;392(1):130−6.

[261] Chen BD, Zhu YG, Duan J, Xiao XY, Smith SE. Effects of the arbuscular mycorrhizal fungus *Glomus mosseae* on growth and metal uptake by four plant species in copper mine tailings. Environ Pollut 2007;147(2):374−80.

[262] Hutchinson JJ, Young SD, Black CR, West HM. Determining uptake of radio-labile soil cadmium by arbuscular mycorrhizal hyphae using isotopic dilution in a compartmented-pot system. New Phytol 2004;477−84.

[263] Subramanian KS, Tenshia V, Jayalakshmi K, Ramachandran V. Biochemical changes and zinc fractions in arbuscular mycorrhizal fungus (*Glomus intraradices*) inoculated and uninoculated soils under differential zinc fertilization. Appl Soil Ecol 2009;43(1):32−9.

[264] Wu S, Zhang X, Sun Y, Wu Z, Li T, Hu Y, et al. Transformation and immobilization of chromium by arbuscular mycorrhizal fungi as revealed by SEM−EDS, TEM−EDS, and XAFS. Environ Sci Technol 2015;49(24):14036−47.

[265] Audet P. Arbuscular mycorrhizal fungi and metal phytoremediation: ecophysiological complementarity in relation to environmental stress. Emerging technologies and management of crop stress tolerance. San Diego, CA: Academic Press; 2014. p. 133−60.

[266] Sprocati AR, Alisi C, Pinto V, Montereali MR, Marconi P, Tasso F, et al. Assessment of the applicability of a "toolbox" designed for microbially assisted phytoremediation: the case study at Ingurtosu mining site (Italy). Environ Sci Pollut Res 2014;21(11):6939−51.

[267] Eid KE, Abbas MHH, Mekawi EM, ElNagar MM, Abdelhafez AA, Amin BH, et al. Arbuscular mycorrhiza and environmentally biochemicals enhance the nutritional status of *Helianthus tuberosus* and induce its resistance against *Sclerotium rolfsii* Ecotoxicol Environ Saf 2019;186:109783. Available from: https://doi.org/10.1016/j.ecoenv.2019.109783 Epub 2019/10/20. Available from: 31629192.

[268] Caradonia F, Francia E, Morcia C, Ghizzoni R, Moulin L, Terzi V, et al. Arbuscular mycorrhizal fungi and plant growth promoting Rhizobacteria avoid processing tomato leaf damage during chilling stress. Agronomy. 2019;9(6):299. Available from: https://doi.org/10.3390/agronomy9060299.

[269] Song YY, Ye M, Li CY, Wang RL, Wei XC, Luo SM, et al. Priming of anti-herbivore defense in tomato by arbuscular mycorrhizal fungus and involvement of the jasmonate pathway J Chem Ecol 2013;39(7):1036−44. Available from: https://doi.org/10.1007/s10886-013-0312-1 Epub 2013/06/26. Available from: 23797931.

[270] Song YY, Cao M, Xie LJ, Liang XT, Zeng RS, Su YJ, et al. Induction of DIMBOA accumulation and systemic defense responses as a mechanism of enhanced resistance of mycorrhizal corn (*Zea mays* L.) to sheath blight Mycorrhiza. 2011;21(8):721−31. Available from: https://doi.org/10.1007/s00572-011-0380-4 Epub 2011/04/13. Available from: 21484338.

[271] Pozo MJ, Cordier C, Dumas-Gaudot E, Gianinazzi S, Barea JM, Azcón-Aguilar C. Localized versus systemic effect of arbuscular mycorrhizal fungi on defence responses to Phytophthora infection in tomato plants. J Exp Bot 2002;53(368):525−34. Available from: https://doi.org/10.1093/jexbot/53.368.525.

[272] Sanchez-Bel P, Troncho P, Gamir J, Pozo MJ, Camañes G, Cerezo M, et al. The nitrogen availability interferes with mycorrhiza-induced resistance against *Botrytis cinerea* in tomato. Front Microbiol 2016;7(1598). Available from: https://doi.org/10.3389/fmicb.2016.01598.

[273] Song Y, Chen D, Lu K, Sun Z, Zeng R. Enhanced tomato disease resistance primed by arbuscular mycorrhizal fungus. Front Plant Sci 2015;6(786). Available from: https://doi.org/10.3389/fpls.2015.00786.

[274] Nair A, Kolet SP, Thulasiram HV, Bhargava S. Systemic jasmonic acid modulation in mycorrhizal tomato plants and its role in induced resistance against *Alternaria alternata* Plant Biol (Stuttgart, Ger.) 2015;17(3):625−31. Available from: https://doi.org/10.1111/plb.12277 Epub 2014/10/21. Available from: 25327848.

[275] Velivelli SL, Lojan P, Cranenbrouck S, de Boulois HD, Suarez JP, Declerck S, et al. The induction of ethylene response factor 3 (ERF3) in potato as a result of co-inoculation with Pseudomonas sp. R41805 and *Rhizophagus irregularis* MUCL 41833—a possible role in plant defense Plant Signal Behav 2015;10(2):e988076. Available from: https://doi.org/10.4161/15592324.2014.988076 Epub 2015/02/28. Available from: 25723847.

[276] Frew A, Powell JR, Allsopp PG, Sallam N, Johnson SN. Arbuscular mycorrhizal fungi promote silicon accumulation in plant roots, reducing the impacts of root herbivory. Plant Soil 2017;419(1):423−33. Available from: https://doi.org/10.1007/s11104-017-3357-z.

[277] Garg N, Singh S. Arbuscular mycorrhiza *Rhizophagus irregularis* and silicon modulate growth, proline biosynthesis and yield in *Cajanus cajan* L. Millsp. (pigeonpea) genotypes under cadmium and zinc stress. J Plant Growth Regul 2018;37(1):46−63.

[278] Garg N, Bhandari P. Silicon nutrition and mycorrhizal inoculations improve growth, nutrient status, K + /Na + ratio and yield of *Cicer arietinum* L. genotypes under salinity stress. Plant Growth Regul 2016;78(3):371−87.

[279] Garg N, Bhandari P. Interactive effects of silicon and arbuscular mycorrhiza in modulating ascorbate-glutathione cycle and antioxidant scavenging capacity in differentially salt-tolerant *Cicer arietinum* L. genotypes subjected to long-term salinity. Protoplasma. 2016;253(5):1325−45.

[280] Hajiboland R, Moradtalab N, Aliasgharzad N, Eshaghi Z, Feizy J. Silicon influences growth and mycorrhizal responsiveness in strawberry plants. Physiol Mol Biol Plants 2018;24(6):1103−15.

[281] Moradtalab N, Hajiboland R, Aliasgharzad N, Hartmann TE, Neumann G. Silicon and the Association with an arbuscular-mycorrhizal fungus (*Rhizophagus clarus*) mitigate the adverse effects of drought stress on strawberry. Agronomy. 2019;9(1):41. Available from: https://doi.org/10.3390/agronomy9010041 PubMed PMID.

[282] Fuentes A, Almonacid L, Ocampo JA, Arriagada C. Synergistic interactions between a saprophytic fungal consortium and *Rhizophagus irregularis* alleviate oxidative stress in plants grown in heavy metal contaminated soil. Plant Soil 2016;407(1):355–66.

[283] Sharma P, Jha AB, Dubey RS, Pessarakli M. Reactive oxygen species, oxidative damage, and antioxidative defense mechanism in plants under stressful conditions. J Bot 2012;2012:217037. Available from: https://doi.org/10.1155/2012/217037.

[284] Garg N, Kashyap L. Silicon and *Rhizophagus irregularis*: potential candidates for ameliorating negative impacts of arsenate and arsenite stress on growth, nutrient acquisition and productivity in *Cajanus cajan* (L.) Millsp. genotypes. Environ Sci Pollut Res 2017;24(22):18520–35.

[285] Emamverdian A, Ding Y, Mokhberdoran F, Xie Y. Heavy metal stress and some mechanisms of plant defense response. Sci World J 2015;2015.

[286] Garg N, Pandey R. High effectiveness of exotic arbuscular mycorrhizal fungi is reflected in improved rhizobial symbiosis and trehalose turnover in *Cajanus cajan* genotypes grown under salinity stress. Fungal Ecol 2016;21:57–67.

[287] Chinnusamy V, Jagendorf A, Zhu JK. Understanding and improving salt tolerance in plants. Crop Sci 2005;45(2):437–48.

[288] Evelin H, Giri B, Kapoor R. Contribution of *Glomus intraradices* inoculation to nutrient acquisition and mitigation of ionic imbalance in NaCl-stressed *Trigonella foenum-graecum*. Mycorrhiza 2012;22(3):203–17.

[289] Hajiboland R. Role of arbuscular mycorrhiza in amelioration of salinity. Salt stress in plants. New York: Springer; 2013. p. 301–54.

[290] Garg N, Pandey R. Effectiveness of native and exotic arbuscular mycorrhizal fungi on nutrient uptake and ion homeostasis in salt-stressed *Cajanus cajan* L.(Millsp.) genotypes. Mycorrhiza. 2015;25(3):165–80.

[291] Miransari M. Mycorrhizal fungi to alleviate salinity stress on plant growth. Use of microbes for the alleviation of soil stresses. New York: Springer; 2014. p. 77–86.

[292] Estrada B, Aroca R, Barea JM, Ruiz-Lozano JM. Native arbuscular mycorrhizal fungi isolated from a saline habitat improved maize anti-oxidant systems and plant tolerance to salinity. Plant Sci 2013;201:42–51.

[293] Gunes A, Inal A, Bagci EG, Coban S, Pilbeam DJ. Silicon mediates changes to some physiological and enzymatic parameters symptomatic for oxidative stress in spinach (*Spinacia oleracea* L.) grown under B toxicity. Sci Horticult 2007;113(2):113–19.

[294] Liang Y, Zhang W, Chen Q, Ding R. Effects of silicon on H + -ATPase and H + -PPase activity, fatty acid composition and fluidity of tonoplast vesicles from roots of salt-stressed barley (*Hordeum vulgare* L.). Environ Exp Bot 2005;53(1):29–37.

[295] Mali M, Aery, Naresh C. Silicon effects on nodule growth, dry-matter production, and mineral nutrition of cowpea (*Vigna unguiculata*). J Plant Nutr Soil Sci 2008;171(6):835–40.

[296] Azeem M, Iqbal N, Kausar S, Javed MT, Akram MS, Sajid MA. Efficacy of silicon priming and fertigation to modulate seedling's vigor and ion homeostasis of wheat (*Triticum aestivum* L.) under saline environment. Environ Sci Pollut Res 2015;22(18):14367–71.

[297] Abd-Elbaky A, El-Abeid S, Osman N. Effect of integration between vascular arbuscular mycorrhizal fungi and potassium silicate supplementation on controlling onion white rot. Egypt J Phytopathol 2018;46(1):125–42.

[298] Shabbir I, Abd Samad MY, Othman R, Wong M-Y, Sulaiman Z, Jaafar NM, et al. Silicate solubilizing bacteria UPMSSB7, a potential biocontrol agent against white root rot disease pathogen of rubber tree. J Rubber Res 2020;23(3):227–35.

[299] Hajiboland R, Aliasgharzadeh N, Laiegh SF, Poschenrieder C. Colonization with arbuscular mycorrhizal fungi improves salinity tolerance of tomato (*Solanum lycopersicum* L.) plants. Plant Soil 2010;331(1):313–27.

[300] Wu Q-S, Srivastava AK, Zou Y-N. AMF-induced tolerance to drought stress in citrus: a review. Sci Horticult 2013;164:77–87.

[301] Savvas D, Ntatsi G. Biostimulant activity of silicon in horticulture. Sci Horticult 2015;196:66–81. Available from: https://doi.org/10.1016/j.scienta.2015.09.010.

[302] Zhu Y-X, Xu X-B, Hu Y-H, Han W-H, Yin J-L, Li H-L, et al. Silicon improves salt tolerance by increasing root water uptake in *Cucumis sativus* L. Plant Cell Rep 2015;34(9):1629–46.

[303] Ashraf M, Akram NA, Al-Qurainy F, Foolad MR. Drought tolerance: roles of organic osmolytes, growth regulators, and mineral nutrients. Advances in agronomy, vol. 111. Amsterdam: Elsevier; 2011. p. 249–96.

[304] Yost RS, Fox RL. Influence of mycorrhizae on the mineral contents of cowpea and soybean grown in an oxisol 1. Agron J 1982;74(3): 475–81.

[305] Clark RB, Zeto SK. Mineral acquisition by arbuscular mycorrhizal plants. J Plant Nutr 2000;23(7):867–902.

[306] Nogueira MA, Cardoso E, Hampp R. Manganese toxicity and callose deposition in leaves are attenuated in mycorrhizal soybean. Plant Soil 2002;246(1):1–10.

[307] Bhandari P, Garg N. Arbuscular mycorrhizal symbiosis: a promising approach for imparting abiotic stress tolerance in crop plants. Plant-microbe interactions in agro-ecological perspectives. New York: Springer; 2017. p. 377–402.

[308] Paszkowski U, Gutjahr C. Multiple control levels of root system remodeling in arbuscular mycorrhizal symbiosis. Front Plant Sci 2013;4:204.

[309] Ma JF, Goto S, Tamai K, Ichii M. Role of root hairs and lateral roots in silicon uptake by rice. Plant Physiol 2001;127(4):1773–80.

[310] Turnau K, Henriques FS, Anielska T, Renker C, Buscot F. Metal uptake and detoxification mechanisms in *Erica andevalensis* growing in a pyrite mine tailing. Environ Exp Bot 2007;61(2):117–23.

[311] Janos DP. Plant responsiveness to mycorrhizas differs from dependence upon mycorrhizas. Mycorrhiza. 2007;17(2):75–91.

[312] Tawaraya K. Arbuscular mycorrhizal dependency of different plant species and cultivars. Soil Sci Plant Nutr 2003;49(5):655–68.

[313] Miransari M. Arbuscular mycorrhizal fungi and heavy metal tolerance in plants. Arbuscular mycorrhizas and stress tolerance of plants. Singapore: Springer; 2017. p. 147–61.

[314] Dragišić Maksimović J, Mojović M, Maksimović V, Römheld V, Nikolic M. Silicon ameliorates manganese toxicity in cucumber by decreasing hydroxyl radical accumulation in the leaf apoplast. J Exp Bot 2012;63(7):2411–20.

[315] Cakmak I, Marschner H, Bangerth F. Effect of zinc nutritional status on growth, protein metabolism and levels of indole-3-acetic acid and other phytohormones in bean (*Phaseolus vulgaris* L.). J Exp Bot 1989;40(3):405–12.

[316] Guntzer F, Keller C, Meunier J-D. Benefits of plant silicon for crops: a review. Agron Sustain Dev 2012;32(1):201–13. Available from: https://doi.org/10.1007/s13593-011-0039-8.

[317] Hajiboland R. Effect of micronutrient deficiencies on plants stress responses. Abiotic stress responses in plants. New York: Springer; 2012. p. 283–329.

[318] Botta A, Rodrigues FA, Sierras N, Marin C, Cerda JM, Brossa R, editors. Evaluation of Armurox®(complex of peptides with soluble silicon) on mechanical and biotic stresses in gramineae. In: 6th International conference on silicon in agriculture, Sweden, 2014.

[319] Rodrigues FÁ, McNally DJ, Datnoff LE, Jones JB, Labbé C, Benhamou N, et al. Silicon enhances the accumulation of diterpenoid phytoalexins in rice: a potential mechanism for blast resistance. Phytopathology. 2004;94(2):177–83.

[320] Mandal SM, Chakraborty D, Dey S. Phenolic acids act as signaling molecules in plant-microbe symbioses Plant Signal Behav 2010;5 (4):359–68. Available from: https://doi.org/10.4161/psb.5.4.10871 Epub 2010/04/20. Available from: 20400851.

[321] Dragišić Maksimović J, Bogdanović J, Maksimović V, Nikolic M. Silicon modulates the metabolism and utilization of phenolic compounds in cucumber (*Cucumis sativus* L.) grown at excess manganese. J Plant Nutr Soil Sci 2007;170(6):739–44.

[322] Hajiboland R, Bahrami-Rad S, Poschenrieder C. Silicon modifies both a local response and a systemic response to mechanical stress in tobacco leaves. Biol Plant 2017;61(1):187–91.

[323] Akiyama K, Hayashi H. Strigolactones: chemical signals for fungal symbionts and parasitic weeds in plant roots Ann Bot 2006;97 (6):925–31. Available from: https://doi.org/10.1093/aob/mcl063 Epub 03/30. Available from: 16574693.

[324] Bouwmeester HJ, Matusova R, Zhongkui S, Beale MH. Secondary metabolite signalling in host–parasitic plant interactions. Curr Opin Plant Biol 2003;6(4):358–64. Available from: https://doi.org/10.1016/S1369-5266(03)00065-7.

[325] Tsai SM, Phillips DA. Flavonoids released naturally from alfalfa promote development of symbiotic glomus spores in vitro Appl Environ Microbiol 1991;57(5):1485–8. Available from: https://doi.org/10.1128/aem.57.5.1485-1488.1991 Epub 1991/05/01. Available from: 16348488.

[326] Steinkellner S, Lendzemo V, Langer I, Schweiger P, Khaosaad T, Toussaint JP, et al. Flavonoids and strigolactones in root exudates as signals in symbiotic and pathogenic plant-fungus interactions Molecules. 2007;12(7):1290–306. Available from: https://doi.org/10.3390/12071290 Epub 2007/10/03. Available from: 17909485.

[327] Hassan S, Mathesius U. The role of flavonoids in root-rhizosphere signalling: opportunities and challenges for improving plant-microbe interactions J Exp Bot 2012;63(9):3429–44. Available from: https://doi.org/10.1093/jxb/err430 Epub 2012/01/04. Available from: 22213816.

[328] Abdel-Lateif K, Bogusz D, Hocher V. The role of flavonoids in the establishment of plant roots endosymbioses with arbuscular mycorrhiza fungi, rhizobia and Frankia bacteria Plant Signal Behav 2012;7(6):636–41. Available from: https://doi.org/10.4161/psb.20039 Epub 05/14. Available from: 22580697.

[329] McCormick AJ, Cramer MD, Watt DA. Regulation of photosynthesis by sugars in sugarcane leaves. J Plant Physiol 2008;165(17): 1817–29. Available from: https://doi.org/10.1016/j.jplph.2008.01.008.

19

Biodissolution of silica by rhizospheric silicate-solubilizing bacteria

Hassan Etesami[1] and Byoung Ryong Jeong[2]

[1]Soil Science Department, College of Agriculture and Natural Resources, University of Tehran, Karaj, Iran
[2]Department of Horticulture, Division of Applied Life Science (BK21 Four), Graduate School, Gyeongsang National University, Jinju, Republic of Korea

19.1 Introduction

One way to improve the food security to support the growing world population is to increase the production quantity per unit area. The most important factor related to crop production is proper nutrition, which is crucial to increasing the growth and yield of crops. On the other hand, various abiotic and biotic stresses affect crops and reduce their productivity, act as a main obstacle against achieving the maximum yield potential of different crops. Results of several studies indicate that mineral nutrition helps reduce damages from stresses [1–3]. Certain elements are also known to benefit plants and help some plant species grow and develop under various stresses [4]. Among the different elements, silicon has been observed in many studies to benefit the growth, yield and stress tolerances of plants [5–8]. Various mechanisms are responsible for how silicon improves the growth and yield of biotically and abiotically stressed plants [5,6] (Fig. 19.1). Numerous laboratory, greenhouse and field studies carried out since 1840 demonstrate the favorable effects of silicon-containing fertilizers on various plants, including barley, rice, maize, sugarcane, and wheat [6,9]. Silicon is also known to benefit (as a functional element), the development and growth of some plant species, especially Gramineae (Poaceae) and cereals [9–13].

Although feeding plants with silicon-containing fertilizers indicates increased yield and improved crop quality, the beneficial effects of silicon under favorable conditions do not appear to be significant. According to most published studies, benefits of silicon become more pronounced when plants stressed [6,7]. Many plants are able to absorb silicon and the amount of the absorption varies between 0.1%–10% of the plant biomass, depending on the species [12].

Despite comprising 27% of the earth's crust with the status of the second most copious element [10], their poor solubility renders most forms of silicon not absorbable by plants [10,14]. Silicon has a strong affinity to oxygen; therefore it usually exists as silicates (SiO_3) in nature, which is not absorbable by plants. Silicates include aluminosilicates (clay, feldspar, mica, etc.), ferromagnesian silicates (amphiboles, olivine, pyroxenes, etc.), silicon dioxides (quartz, amorphous silica, etc.), and silicates of calcium, sodium, potassium, or iron [15]. More than 90% of the earth's crust comprises of silicates, and most metamorphic, igneous, and sedimentary also contain silicates [16]. Silicon also exists as silicic acid [$Si(OH)_4$], depending on the soil pH [10].

Plants are unable to absorb silicon before weathering processes or dissolution into the soil release silicon [17]. Plants absorb the solubilized form of silicon, orthosilicic acid (H_4SiO_4), along with water. Irrigation, silicon fertilizers, weathering of silicon-containing minerals, and desorption from the soil fertilizers give rise to monosilicic acid [18].

Repeated cultivation and limited dissolution of silicon resources decrease the available silicon levels for the plants. For instance, wheat harvest was observed to take 50–150 kg of silicon from the soil [19]. Fertilizers that can provide calcium/potassium/sodium silicates, or fine silica, are needed to maximize crop productivity [12,20]. Contrary to conventional fertilizers, only limited amounts of silicon fertilizers are available, which are

Silicon and Nano-silicon in Environmental Stress Management and Crop Quality Improvement.
DOI: **https://doi.org/10.1016/B978-0-323-91225-9.00020-0**

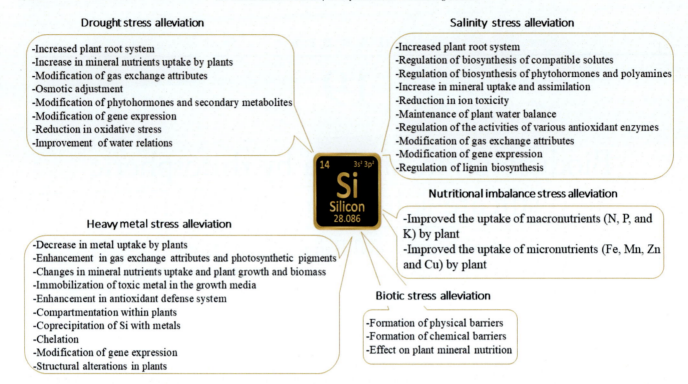

FIGURE 19.1   The multiple action mechanisms of silicon (Si) in alleviating biotic and abiotic stresses in plants. For details, see Etesami and Jeong [6].

often prohibitively expensive for most agricultural practitioners [21]. Therefore silicon fertilizers are rarely adopted, especially in developing countries. Generally, silicate fertilizers are contain (1) silicon-rich slags or industrial byproducts, whose use may contaminate soils; (2) bentonite, feldspars, micas, and diatomaceous earth, which are biological/mineral silicon fertilizers that re poorly soluble but have high application rates; and (3) potassium silicates which are highly soluble but very costly [22]. Economically friendly silicate fertilizers can be produced by recycling silicon-rich crop residues, mining, containing minerals and metals, construction and demolition wastes containing silicates of aluminum, potassium and calcium.

The primary and secondary mineral solubilities are a main factor that affect the silicon concentration in the soil solution [23]. These materials can be solubilized at an accelerated rate via biochemical and physicochemical interventions for their soil applications [24], but microbial activities for biochemical action is deemed more important than others [25]. It is well known that microorganisms dissolve and mobilize minerals in the soil [26–28]. It has also been documented in numerous studies that microbes isolated from silicate mineral surfaces weather the different silicates [29–31]. This indicates the important role silicate-solubilizing microorganisms (SSM) play as a biofertilizer to solubilize silicates and phosphates [32,33]. Microorganisms are abundant in soils, but only a small minority can solubilize the insoluble silicates [34]. However, it is known the plants and the microflora produce chelating ligands, alter the soil physical properties, and produce chelating ligands to affect the dissolution and mobilization of the soil silicate minerals [35]. Among microorganisms, plant-associated bacteria have been reported to help dissolve silicates and release silicon more quickly in to the plant-soil system [27,28,36–38] via bioweathering processes [18].

The interactions in the soil, plant, microbe system have received increasing interest in recent decades. Many microorganisms inhabit the soil, especially the rhizosphere. Numerous bacterial and fungal species are well-known to constitute a holistic system with plants. They can easily multiply in the rhizosphere to enhance plant growth and yield [39]. The different types of microorganisms in the rhizosphere (e.g., plant growth promoting rhizobacteria, PGPR) are important for improving the nutrient availability to plants [40]. According to previous studies, several PGPR have been reported to help plants under different stresses better uptake nutrients [17,33,41]. One such PGPR is the silica-solubilizing bacteria (SSB), which solubilize silica, which solubilize insoluble silicate forms to increase the supply of available Si for plants [17]. SSB have recently garnered great interest, as solubilizing silicates in rhizospheric soils (silicate-rich zones) leads to improved silicon and potassium uptake, which lessens the need for potash fertilizers [42]. Silicon is released into the rhizosphere when SSB degrade silicates [17,43]. Furthermore, it has been observed that SSB improves the soil structure and nutrition, and suppresses pathogens to benefit the growth of plants [17,44−46]. However, research on SSB are still scarce compared to those regarding other PGPR [28,29,47]. Accordingly the current chapter discusses the various soil SSB that can solubilize silicates and be potentially used as biofertilizers.

19.2 Plant growth-promoting rhizosphere bacteria

The rhizosphere is a microbial reservoir located in the soil around the plant roots, where the roots influence the biological and chemical properties of the soil [48]. Soil microbes and plants interact in many different ways in the rhizosphere, which occur as signals released in as simple chemical molecules are perceived [49]. Rhizodeposits are released/secreted by plants to influence the surrounding soil, which are mainly composed of amino acids, carbohydrates, and organic acids of secondary metabolites [50,51]. Mesotrophic soils are rhizospheric soils that favor the growth of microbial populations. Three zones are distinguished for the rhizosphere; endorhizosphere, which describes the space around the apoplastic space between cells, endodermis, and the root cortex; the rhizoplane, which is around the root surface; and the ectorhizosphere, which describes the zone between the bulk soil and the rhizoplane [52]. Various microbial groups inhabiting the rhizosphere perform many functions and affect the plant growth in many ways. These microbial groups are involved in protecting plants from phytopathogens, biotic stresses, and abiotic stresses, nutrient cycling, and plant pathogenesis. These rhizospheric microorganisms that directly and indirectly benefit plants are known as plant growth-promoting rhizosphere microorganisms. *Bacillus, Pseudomonas, Streptomyces, Burkholderia, Klebsiella, Azospirillium, Rhizobium, Trichoderma, Penicillium*, and *Aspergillus* are some of the most common microorganisms promoting the growth of plants. These microorganisms have been found to be associated with various crops and have been isolated from various sources, including plant surfaces and soils [53,54]. Among these microorganisms, plant growth-promoting rhizosphere bacteria (PGPR) are the most widely studied, most effective soil microorganisms that enhance plant performance [55]. About 2%−5% of rhizobacteria are classified as PGPR, which extend benefits to the plant growth when reintroduced to soils containing competitive microflora [56]. The use of PGPR has increased over the past few decades due to the sustainability and security of agriculture around the world. Various mechanisms are involved in how PGPR directly and indirectly affect the growth of plants. PGPR improve plant growth by improving the soil availability of nutrients (e.g., by phosphate solubilization, potassium solubilization, production of siderophores, nitrogen fixation, nitrification, etc.), producing various plant growth regulators like hormones (e.g., auxin), protecting plants against pathogens (e.g., by antifungal metabolites like phenazines, 2,4-diacetylphloroglucinol, pyrrolnitrin, viscosinamide, pyoluteorin, tensin, and HCN), helping exudation of soluble compounds, nutrient mineralization, mobilization, storage and release and soil organic matter decomposition, improving soil structures, bioremediating soils contaminated with toxic heavy metals, and decomposing xenobiotic compounds [53,55,57−59]. These rhizospheric microbial activities can also change the quantity, quality, and composition of the root exudates and thus affect the microbial components [60]. The role of PGPR in improving the availability of nutrients for plants is important and necessary for crop production. Plant-associated bacteria are also known to be important in dissolving minerals [61−63]. Bacterial strains from the diverse PGPR genera were observed to be able to weather minerals [28]. Silicate solubilizing bacteria (SSB) refer to those that can solubilize silicates, which can convert mineral structural or insoluble silicon compounds into forms in the soil that are available for plants to uptake.

19.3 Silicate-solubilizing bacteria

Soil microbes contribute to releasing crucial nutrients for their own and plants' nutrition, from primary minerals, and therefore play an essential role in the environment [28]. SSB as a biofertilizer on its own or in conjunction with silicates has proven effective for dissolving silicates [29,45]. These bacteria have the potential to release soluble silica from insoluble inorganic silicates of aluminum, calcium, magnesium, and potassium, from silicate minerals like biotite and feldspar [31], and biogenic materials like siliceous earth, diatomaceous earth, rice straw and rice husk [38]. The soil silicon depends on the element's sorption and desorption [64]. The silicate solubilization rates of microbial consortia vary according to the soil properties and the type of parent material [41,65]. For example, the silicate solubilization potential of the six isolates of eight minerals differed depending on the mineral and the isolate in a study [66]. As the knowledge of silicon's benefits to plants increased, the rhizospheric soils have been explored for new SSB [17,25]. Various bacteria in the soil were observed to solubilize silicate minerals [67]. The different bacteria were isolated from different habitats, including rice field soil samples [68], rhizosphere of rice plants [17,38], weathered rock surfaces [69], weathered purple siltstone surfaces [70], surroundings of the Quercus petreae oak mycorrhizal roots [71], weathered feldspar surfaces [27], potassium mine tailings [72], and river water, pond sediments, soils, talc mineral [73], and weathered rocks [31]. The capability to depolymerize crystalline silicates (silicon-solubilizing capacity) has been reported in various Gram-negative and Gram-positive bacteria (i.e., *Aeromonas* sp., *Bacillus* sp., *Burkholderia eburnea* CS4−2, *Proteus* sp., *Bacillus mucilaginosus*, *Bacillus globisporus*, *Bacillus mucilaginosus*, *Bacillus flexus*, *B. mucilaginosus*, *B. megaterium*, *B. edaphicus*, *Pseudomonas fluorescens*, *Burkholderia susongensis* sp., *Rhizobium* sp., *Rhizobium yantingense*, *Rhizobium tropici*, *Sphingobacterium* sp., *Enterobacter* sp., and *Pseudomonas stutzeri*) [17,25,29,31,33,38,66,68−70,73−81].

19.3.1 Isolating and screening of silicate-solubilizing bacteria

Previous research has demonstrated that SSB release potassium, calcium, iron, and phosphates from silicate minerals in the soil, in addition to solubilizing silica. Therefore SSB has been garnering the attention of scientists, to advocate their use as potassium-mobilizing biofertilizers [82]. In the absence of any specific recommended media, the investigators selected a medium of their choice and supplemented it with insoluble silicate minerals like magnesium trisilicate, illite, muscovite, biotite, talc, and feldspar for isolation [17,29,66,83−86]. Invariably Bunt and Rovira, Aleksandrov or mineral media were used and supplemented with the aforementioned insoluble silicates [82]. SSB were previously isolated from the soil and water in Bunt and Rovira media with 0.25% magnesium trisilicate [83]. However, there is no specifically recommended medium for either enumeration and screening or isolation. Vasanthi et al. [82] compared the Bunt and Rovira medium, nutrient agar, glucose agar, and soil extract agar, each with 0.25% magnesium trisilicate, for their suitability for isolation, enumeration and screening in a study. Based on the growth and clarity of the dissolution zone, it was observed that soil extract agar medium with 0.25% magnesium trisilicate was the most ideal for enumeration. The plain glucose medium with 0.25% magnesium trisilicate was more ideal for screening the isolates, as the solubilization is rapid and a larger zone is cleared. In another study, Bist et al. [44] also proposed an efficient screening medium for microbes that solubilize silicon (NBRISSM), which more clearly discerns potassium and silicon solubilization compared to the media mentioned previously. NBRISSM is composed of (in gL^{-1}): hydroxyapatite (2.5), glucose (2.5), $CaCl_2$ (1.25), $MgNO_3$ (1.25), $Mg_2O_8Si_3$ (0.1), and $(NH_4)_2SO_4$ (0.1). The potassium source for the medium makes it more appropriate for identifying silicon-solubilizing microbes, and eliminates phosphate interference as a result. These researchers also have demonstrated how silicon solubilization is functionally correlated with organic (maleic, fumaric, succinic, gluconic, and tartaric) acid production and acidic phosphatase activities. Silicon solubilization is a phenomenon based on production of organic acids, and leads to changed medium color (Fig. 19.2). Glucose consumption from the growth medium, that leads to organic acid production, may be responsible for the essence of this acidification [87]. Although it has been observed that all organic acids are involved in silicate dissolution, gluconic acid was identified as the most effective agent [29].

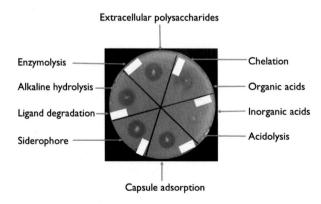

FIGURE 19.2   Silicate solubilization by various japonica rice (*Oryza sativa* L. cv. Dongjin) rhizosphere bacterial isolates after 7 days. For details, see Kang et al. [17].

Vasanthi et al. [82] have also demonstrated the difference in the abilities of SSB to produce acids. In this study, the SSB that produced relatively lower acidity exhibited a larger zone in the glucose medium. It is likely that acidolysis, which is considered to be important in silicate solubilization, may operate in one bacterium but other mechanisms may play a role in other SSB [82]. According to Vasanthi et al. [82], it is always better to test the isolates in more than one media to assess the growth and solubilization so as to avoid discarding the likely best isolates as the solubilization potential differs according to the media.

19.3.2 Silicate-solubilizing bacteria action mechanisms for the silicon availability for plants

Although research has lent much knowledge on SSB over the past decades, much has yet to be uncovered regarding biogenic weathering, and the role microbial communities in soils play [88,89]. Currently, little is known how SSB solubilize silicates. SSB weather silicate minerals in the soil to release potassium and silica, and hence are used as a biofertilizer for crops [90]. Silicate weathering provides access to minerals containing potassium, such as apatites, and therefore is also related to phosphorus solubilization [28,91]. Organic acid production, for example, is known to solubilize mineral potassium, silicon, and inorganic phosphorus (TCP) [29,92,93].

Accordingly, the same mechanisms for potassium and phosphorus solubilization are responsible for the biogenic silicate weathering [28,91,92,94]. It has been observed that alkaline hydrolysis, acidolysis, ligand degradation, capsule adsorption, enzymolysis, extracellular polysaccharides, chelation, and redox contribute to microbial silicate dissolution. Acidolysis is the most prominent and widely known mechanism of silicate mineral weathering [95]. SSB release soluble silica from insoluble silicates via various mechanisms, which are summarized below.

19.3.2.1 Organic acid production

As SSB solubilize insoluble tri-calcium silicates and other insoluble nutrient sources, they also produce various organic acids [96], including citric, malic, acetic, 2-keto-gluconic, tartaric, gluconic, oxalic, hexadecanoic, propionic, succinic, malonic, phthalic, lactic, heptadecanoic, formic, hydroxypropionic, and oleic acids [25,94,97,98]. Sheng et al. [29] observed that Bacillus globisporus Q12, a silicate-solubilizing bacterial strain isolated from the weathered feldspar surfaces, solubilized silicon and potassium from silicate minerals like biotite, muscovite, and feldspar via organic acids such as acetic and gluconic acids (which most actively solubilized the silicate minerals). Sheng and He [27] reported in another study that SSB-driven feldspar and illite are attributed to the production of organic acids such as 2-ketogluconic, tartaric, gluconic, citric, oxalic, succinic, and malic acids. Tartaric acid appears to be the most commonly found silicate or potassium solubilizing agent [99—101]. Different organisms have different types of various organic acids produced by SSB [29]. It is known that dissolution rates of silicates depend on the strength of the bond between the metal ion and the ligand (mono-, di-, and tricarboxylic acids), and the number of organic acid functional groups (-COOH) that is able to react with surfaces of different minerals [61]. In addition, the net rate of silicon solubilization is significantly related (positively) to the organic acid carboxyl groups [61,102]. The organic acids have metal-complexing properties (as chelating agents) and may chelate (bind to) the iron and aluminum in silicates, rendering the silicates soluble, and also protonating (providing protons, H^+) for silicate hydrolyzes [103—106]. The SSB-induced pH decrease and organic acid release primarily control the rhizospheric silicon dissolution [61,103,107]. However, as the ability to decrease the pH did not necessarily correlate with the ability to solubilize silicon and potassium from minerals, acidification does not appear to be the sole mechanism for solubilization [99,108]. Moreover, it has been shown that adding 0.05 M EDTA to the medium was just as effective in solubilization as inoculation with Penicilum bilaii was, indicating that the chelating capacity of organic acids is also important [27,78]. In a study, acidulous dissolution of the crystal network and aluminum chelation by SSB enabled them to weather phlogopite [109].

It is known that silicon, potassium, and phosphorus solubilization from silicate rocks is accelerated by potassium and phosphate-solubilizing bacteria (KSB & PSB, respectively) [61]. Different organic acids released by these bacteria affect silicon, potassium and phosphorus solubilization in three main ways: (1) binding (complexation) metal (Al, Ca, Fe) cations to the hydroxyl (-OH) and carboxylic (-COOH) groups; (2) exchanging absorbed silicon, potassium, and phosphorus, with organic compounds (ligand exchange); and (3) acidifying the soil solution to facilitate the pH-dependent dissolution of apatites and silicates [110—112]. High silicon concentrations in soil solution have been correlated to high phosphorus levels as well. This is likely due to the fact that phosphorus and silicon compete for binding sites on mineral surfaces, as supported by researchers who demonstrated that high silicon concentrations accompanied a higher level of phosphorus dissolution in permafrost soils and peat [102,113]. The aforementioned results suggest that organic acids either served as substituents of silicon (ligand exchange) or effectively complexed metal cations present in the crystal lattice (i.e., Al, Ca, FE, and Mg), to promote solubilization [33,78,114,115]. The amount of silicon dissolved in the soil solution is increased as a result of the formation of organic ligands and stable metal ion complexes [65,116,117]. In general, the synthesis and discharge of organic acids by the SSB, PSB, and KSB into their surroundings leads to acidification of their cells and the environment, which results in potassium, phosphorus, and silicon ions being released from the silicate minerals, and ultimately increases the bioavailability of these nutrients.

19.3.2.2 Production of inorganic acids

Certain bacterial isolates produce inorganic acids that solubilize insoluble minerals like silicates [90]. It has been reported that bacterial inorganic acids can convert insoluble forms of phosphorus and silicon (biotite feldspar, mica, and muscovite) into their corresponding soluble forms, with the net result being an increased nutrient availability to plants [61,103,107]. The bacterial production of inorganic acids in the rhizosphere can originate from (1) *Thiobacillus* oxidizing sulphur and reduced sulphide into sulphuric acid; (2) *Nitrosomonas* oxidizing ammonia into nitrates, and *Nitrobacter* converting nitrate into nitric acid; and (3) carbonic anhydrase synthesis and discharge, that catalyze the interconversion of water and soil-microbe-produced carbon dioxide into the dissociated ions of carbonic acid [118], which facilitates silicate mineral conversions by microbes, as observed in orthoclase being degraded to kaolinite [119]. Furthermore, CO_2 sequestration in basaltic aquifers and the associated mineralization of carbonate is reported to potentially provide a suitable environment for silicate minerals to be dissolved in [120–122].

19.3.2.3 Production of siderophores

Siderophores are organic chelators with low molecular weights that have a very high and specific affinity to Fe(III). Siderophore biosynthesis is controlled by the iron levels, and siderophores mediate the microbial cell iron uptake [123]. Bacteria (e.g., cyanobacteria), fungi, and plants (phytosiderophores) growing at a low-Fe^{3+} concentration produce siderophores [124]. SSB that produce siderophores are able to solubilize silicon by scavenging iron from silicate minerals (Fe(III)–siderophore complex formation), as is the case for the degradation of hornblende [125]. It is also known that phosphate-solubilizing microorganisms are capable of solubilizing silicates by secreting different siderophores that may affect silicon phosphorus solubilization from rocks [110–112].

19.3.2.4 Production of extracellular polysaccharides

Extracellular polysaccharides, also known as exopolysaccharides (EPSs), are sugar-based polymers with high molecular weights many microorganisms synthesize and secrete [126]. SSB are known to solubilize silicon by producing EPSs [17,127]. Extracellular polysaccharides are involved in silicate breakdown and rock weathering, due to their drying and wetting properties, and by acting as a sorbent of metal ions (binding silicates, and affecting the fluid-mineral phase equilibrium to render the silicates soluble). Extracellular polysaccharides could disturb the homeostasis of organic acids or H^+ involved in the silicon and phosphorus solubilization processes by holding the free silicon and phosphorus in the medium, consequently inducing higher release of silicon and phosphorus from phosphorus minerals [128] and silicate ones. Extracellular polysaccharides are a known major component of mature biofilm structures [126]. The biofilm formation also solubilizes the microenvironmental silicates [79]. However, more research is necessary to understand how EPSs synergistically solubilize silicon.

19.3.2.5 Acidolysis and alkaline hydrolysis

Acidolysis is mineral dissolution that occurs when the medium acidifies [28] and is the most commonly occurring mechanism involved in silicate mineral dissolution [17,38,95]. Rhizospheric SSB-produced organic acid acidolysis can either directly dissolve the mineral potassium by slowly releasing exchangeable potassium and readily available exchangeable potassium, or chelate potassium mineral-associated aluminum and silicon ions [129]. SSB are known to be able to solubilize silicates by shifting the environmental pH towards alkalinity through fixing nitrogen and decomposing organic matter, subsequently forming amines and ammonia [25,130].

SSB can also adsorb and bind the inorganic silicate ions to surfaces of bacteria, as they have ionizable lipopolysaccharide-carboxylates and phosphates in Gram-negative bacteria and peptidoglycan, teichoic acids, and teichuronic acids in Gram-positive bacteria, and their high ion reactivity, rendering silicon dissolution [131]. In addition, these bacteria solubilize silicates by reducing sulphates and producing H_2S, which react with cations (e.g., Ca, Fe, etc.) of silicate minerals to form sulphides [132].

19.4 Plant growth-promoting effects of silicate-solubilizing bacteria

The combined application of silicate bacteria and organo-minerals was first reported to enhance the growth and yield of maize and wheat by Aleksandrov [133]. Microbe-produced organic acids solubilizing insoluble silicon and phosphates are known to enhance the availability of the minerals to plants [134]. SSB isolated from soil minerals and plant roots could also increase the silicon uptake and subsequently the silicon concentration in plants [17,25,38,135]. Researchers suggested that the promoted plant growth was associated with organic acid release by the silicate-solubilizing bacterial strains, as well as silicate solubilization [17]. A number of bacterial strains of the genus *Pseudomonas, Bacillus, Proteus, Burkholderia, Rhizobia,* and *Enterobacter* are known to degrade silicates to release silicon and therefore promote plant growth [17,33,38,74–76]. Kang et al. [17] inoculated japonica rice plants with the silicate-solubilizing bacterial strain *Burkholderia eburnea* CS4–2 and found that the silicon level in plants grown on the plant growth substrate including insoluble silicates increased. Furthermore, attributes of plant growth (chlorophyll content, root and shoot lengths, root and shoot fresh weights, etc.) also increased when compared to the control and those treated with insoluble silica. When applied in conjunction with insoluble silica, CS4–2 significantly promoted the growth of rice plants [17]. In another study [38], *Rhizobium* sp. IIRR-1, an SSB strain isolated from rhizospheric soils of rice, was found to be able to colonize and grow on all insoluble silicates, which resulted in an increased (12.45%–60.15% over that of the control) release of silica into the culture media. *Rhizobium* sp. IIRR-1 also effectively colonized the roots of rice seedlings and increased the seedling vigor by 29.18%, compared to that of the uninoculated control. In addition to providing silicon to plants, silicate-bound potassium, calcium, iron, magnesium, phosphorus, and zinc can be solubilized and provided to plants by SSB such as *B. mucilaginosus* [25,33,38,66,67,77,78,82,136]. For example, silicon uptake enhanced the phosphorus availability, which decreased the iron and manganese availability for plants in a study [137]. In another study, mica and soil minerals were dissolved, while SiO_2 and K^+ were simultaneously released

TABLE 19.1 Effect of silicate-solubilizing bacteria with multiple plant growth promoting traits on the growth of plants.

Silicate-solubilizing bacterial isolate	Silicon source	Other PGP traits	Observed effect	Reference
Enterobacter sp.	Insoluble magnesium trisilicate and calcium silicate	Phosphate solubilization, potassium solubilization, IAA production, siderophore prodction, nitrogen fixation, and antagonistic activity against pathogens	The isolate UPMSSB7 (*Enterobacter* sp.)showed the highly significant percent inhibition of radial growth (57.24%) against the pathogen *Rigidoporus microporus*	Shabbir et al. [45]
Burkholderia eburnea CS4–2	Magnesium trisilicate	IAA production	When combined with silica fertilization, soil inoculation with CS4–2 promoted all rice growth attributes over those of the water-treated (control) and insoluble silica-fertilized plants.	Kang et al. [17]
Bacillus globisporus Q12	Feldspar, muscovite, and biotite	–	Solubilization of potassium and silicon from the silicate minerals by the action of organic acids.	Sheng et al. [29]
Bacillus mucilaginosus	Mica and feldspar	–	B. mucilaginosus was found to dissolve soil minerals and mica and simultaneously release K^+ and SiO_2 from the crystal lattices. B. mucilaginosus also produced organic acids and polysaccharides during growth.	Liu et al. [78]
Unidentified	Magnesium trisilicate	Phosphate solubilization and potassium solubilization	35 bacterial isolates were capable of solubilizing either silicate, phosphate or potassiumand could inhibit the growth of plant pathogenic fungi Magnaporthae grisae, Rhizoctonia solani, Altarnaria alternata and Macrophomina pheasolina	Naureen et al. [34]
Bacillus edaphicus	Calcium silicate	–	SSB and calcium silicate significantly produced higher cane yields in fresh plantings and also in ratoon crop	Brindavathy et al. [81]

from the crystal lattices by *B. mucilaginosus* [78]. The extent of SSB-induced K solubilization was 4.90 mg L^{-1} at pH 6.5—8.0 as observed in a study (Badr [138]. *Enterobacter* sp. (GAK2) was able produce indole-3-acetic acid, in addition to solubilizing the insoluble phosphates and silicates [33]. Thus ecofriendly crop production and reduced use of agrochemicals like potassium fertilizers can be brought about with SSB application for biofertilization. SSB is also known to help decompose crop residues, organic matter, etc. [46], and provide plants with nutrients. SSB can also provide a biological defense system against pathogenic fungi for plants [8,34,45,139]. In addition to their silicon-solubilizing capacity, SSB can also fix nitrogen, produce plant growth hormones and siderophores, and solubilize phosphorus (Table 19.1).

Results of the studies mentioned above demonstrate that SSB utilization may potentially improve silicon solubilization, to increase the silicon uptake by plants and resultantly enhance growth and health of plants. In contrast to the vast array of studies conducted on PSB and other PGPB, little research currently exists to isolate how SSB act to promote plant growth in the rhizosphere [17,25]. The proportion of SSB to the total bacteria present in silicate minerals and soils is low, which points to their uniqueness [25].

19.5 Conclusion and future perspectives

Silicon is plentiful in the earth's crust and is important for agriculture due to its benefits for plants. Biological and chemical soil reactions generally govern the accessibility of silicon for plant roots. Silicon can be converted from its insoluble forms to its soluble forms only by rock weathering or the biological activities of roots and microorganisms. The microbe-driven transformation of polymerized silica into monomeric forms is important for the biogeochemical silicate cycles in nature. SSB have recently garnered great research interest in this regard, since silicate solubilization in silicate-rich rhizospheric soils promotes the uptake of potassium and silicon, thus reducing the need for potash fertilization. Inoculation of soils with SSB has been shown effective for dissolving insoluble silicates, that lead to a higher agricultural performance. In addition to their silicon-solubilizing abilities, SSB can also produce siderophores and plant growth hormones, fix nitrogen, solubilize potassium and phosphorus, and fight plant pathogens. In general, SSB-driven silicate solubilization can profitably be utilized to enhance crop production on its own, or in conjunction with silicate materials, since silica is also agronomically beneficial, and its mobilization is always accompanied by the release of other macro-and micronutrients that are bound to silicate minerals.

Silicate solubilization is complex and is influenced by numerous factors. Environmental factors, the type of SSB used, the type, size, and quantity of minerals and the nutritional status of the soil all affect SSB-driven silicate solubilization. Moreover, SSB stability after soil inoculations is also important in solubilizing silicon to benefit the growth and development of crops. Therefore more research is necessary in order to understand how to develop an efficient, indigenous silicate-solubilizing microbial consortium to benefit the growth and yield of crops. Another significant concern is commercially propagating, preserving, and transporting to the crop fields of the silicate-solubilizing consortium for practical agricultural applications. SSB usually solubilize silicon (and potassium) by (1) lowering the pH, or (2) by enhancing chelation of the silicon-bound cations, and (3) acidolysis of the area surrounding the microorganisms. However, a molecular-level understanding of the mechanisms involved in the weathering of silicate minerals weathering mechanism with these SSB remains unclear, and future research should be pursued in this regard. Future research should also focus on determining the optimum concentration of SSB to be used for inoculating different carriers, and evaluating the benefits of SSB application in different field conditions.

Acknowledgments

We wish to thank University of Tehran for providing the necessary facilities for this study

References

[1] Cakmak I. (ed.) Role of mineral nutrients in tolerance of crop plants to environmental stress factors; 2005.

[2] Waraich EA, Ahmad R, Ashraf MY. Role of mineral nutrition in alleviation of drought stress in plants. Australian J Crop Sci 2011;5 (6):764—77.

[3] Waraich EA, Ahmad R, Halim A, Aziz T. Alleviation of temperature stress by nutrient management in crop plants: a review. J soil Sci plant Nutr 2012;12(2):221—44.

[4] Marschner H. Beneficial mineral elements. Miner Nutr High plants 1995;2:405—34.

[5] Etesami H, Jeong BR, Rizwan M. The use of silicon in stressed agriculture management: action mechanisms and future prospects. Metalloids Plants Adv Future Prospect 2020;381−431.

[6] Etesami H, Jeong BR. Silicon (Si): review and future prospects on the action mechanisms in alleviating biotic and abiotic stresses in plants. Ecotoxicol Environ Saf 2018;147:881−96.

[7] Etesami H, Jeong BR. Importance of silicon in fruit nutrition: agronomic and physiological implications. Fruit crops. Elsevier; 2020. p. 255−77.

[8] Fauteux F, Rémus-Borel W, Menzies JG, Bélanger RR. Silicon and plant disease resistance against pathogenic fungi. FEMS Microbiology Lett 2005;249(1):1−6.

[9] Hodson MJ, White PJ, Mead A, Broadley MR. Phylogenetic variation in the silicon composition of plants. Ann Botany 2005;96 (6):1027−46.

[10] Epstein E. Silicon. Annu Rev plant Biol 1999;50(1):641−64.

[11] Broadley M, Brown P, Cakmak I, Ma JF, Rengel Z, Zhao F. Chapter 8-beneficial elements M. Petra Marschner's mineral nutrition of higher plants. San Diego: Academic Press; 2012.

[12] Liang Y, Nikolic M, Bélanger R, Gong H, Song A. doi Silicon in agriculture, 10. Dordrecht: Springer; 2015. p. 978−94.

[13] Cooke J, Leishman MR. Is plant ecology more siliceous than we realise? Trends Plant Sci 2011;16(2):61−8.

[14] Vasanthi N, Saleena LM, Raj SA. Silicon in day today life. World Appl Sci J 2012;17(11):1425−40.

[15] Ma JF, Takahashi E. Soil, fertilizer, and plant silicon research in Japan. Elsevier; 2002.

[16] White AF, Brantley SL. Weathering rates of silicate minerals. Chem Weathering Rates Silicate Minerals, Rev Mineral 1995;31:1−22.

[17] Kang S-M, Waqas M, Shahzad R, You Y-H, Asaf S, Khan MA, et al. Isolation and characterization of a novel silicate-solubilizing bacterial strain Burkholderia eburnea CS4-2 that promotes growth of japonica rice (Oryza sativa L. cv. Dongjin). Soil Sci Plant Nutr 2017;63 (3):233−41.

[18] Klotzbücher T, Marxen A, Vetterlein D, Schneiker J, Türke M, Van Sinh N, et al. Plant-available silicon in paddy soils as a key factor for sustainable rice production in Southeast Asia. Basic Appl Ecol 2015;16(8):665−73.

[19] Tubana BS, Babu T, Datnoff LE. A review of silicon in soils and plants and its role in US agriculture: history and future perspectives. Soil Sci 2016;181(9/10):393−411.

[20] Datnoff LE, Rodrigues FA. The role of silicon in suppressing rice diseases:. American Phytopathological Society; 2005.

[21] Meena VD, Dotaniya ML, Coumar V, Rajendiran S, Kundu S, Rao AS. A case for silicon fertilization to improve crop yields in tropical soils. Proc Natl Acad Sci, India Sect B: Biol Sciences 2014;84(3):505−18.

[22] Datnoff LE, Snyder GH, Korndörfer GH. Silicon in agriculture:. Elsevier; 2001.

[23] Sommer M, Kaczorek D, Kuzyakov Y, Breuer J. Silicon pools and fluxes in soils and landscapes—a review. J Plant Nutr Soil Sci 2006;169 (3):310−29.

[24] Bin L, Ye C, Lijun ZHU, Ruidong Y. Effect of microbial weathering on carbonate rocks. Earth Sci Front 2008;15(6):90−9.

[25] Vasanthi N, Saleena LM, Raj SA. Silica solubilization potential of certain bacterial species in the presence of different silicate minerals. Silicon 2018;10(2):267−75.

[26] Calvaruso C, Turpault M-P, Frey-Klett P. Root-associated bacteria contribute to mineral weathering and to mineral nutrition in trees: a budgeting analysis. Appl Environ Microbiol 2006;72(2):1258−66.

[27] Sheng XF, He LY. Solubilization of potassium-bearing minerals by a wild-type strain of Bacillus edaphicus and its mutants and increased potassium uptake by wheat. Can J microbiol 2006;52(1):66−72.

[28] Uroz S, Calvaruso C, Turpault M-P, Frey-Klett P. Mineral weathering by bacteria: ecology, actors and mechanisms. Trends Microbiol 2009;17(8):378−87.

[29] Sheng XF, Zhao F, He LY, Qiu G, Chen L. Isolation and characterization of silicate mineral-solubilizing Bacillus globisporus Q12 from the surfaces of weathered feldspar. Can J Microbiol 2008;54(12):1064−8.

[30] Lapanje A, Wimmersberger C, Furrer G, Brunner I, Frey B. Pattern of elemental release during the granite dissolution can be changed by aerobic heterotrophic bacterial strains isolated from Damma Glacier (central Alps) deglaciated granite sand. Microb Ecol 2012;63 (4):865−82.

[31] Wang RR, Wang Q, He LY, Qiu G, Sheng XF. Isolation and the interaction between a mineral-weathering Rhizobium tropici Q34 and silicate minerals. World J Microbiol Biotechnol 2015;31(5):747−53.

[32] Chen W, Yang F, Zhang L, Wang J. Organic acid secretion and phosphate solubilizing efficiency of Pseudomonas sp. PSB12: effects of phosphorus forms and carbon sources. Geomicrobiol J 2016;33(10):870−7.

[33] Lee K-E, Adhikari A, Kang S-M, You Y-H, Joo G-J, Kim J-H, et al. Isolation and characterization of the high silicate and phosphate solubilizing novel strain Enterobacter ludwigii GAK2 that promotes growth in rice plants. Agronomy. 2019;9(3):144.

[34] Naureen Z, Aqeel M, Hassan MN, Gilani SA, Bouqellah N, Mabood F, et al. Isolation and screening of silicate bacteria from various habitats for biological control of phytopathogenic fungi. Am J Plant Sci 2015;6(18):2850.

[35] Cornelis JT, Delvaux B, Georg R, Lucas Y, Ranger J, Opfergelt S. Tracing the origin of dissolved silicon transferred from various soil-plant systems towards rivers: a review. 2011.

[36] Savant NK, Snyder GH, Datnoff LE. Silicon management and sustainable rice production 58 Advances in agronomy. Elsevier; 1996. p. 151−99.

[37] Hutchens E, Valsami-Jones E, McEldowney S, Gaze W, McLean J. The role of heterotrophic bacteria in feldspar dissolution−an experimental approach. Mineral Mag 2003;67(6):1157−70.

[38] Chandrakala C, Voleti SR, Bandeppa S, Kumar NS, Latha PC. Silicate solubilization and plant growth promoting potential of Rhizobium sp. isolated from rice rhizosphere. Silicon 2019;11(6):2895−906.

[39] Vessey JK. Plant growth promoting rhizobacteria as biofertilizers. Plant Soil 2003;255(2):571−86.

[40] Etesami H, Adl SM. Plant growth-promoting rhizobacteria (PGPR) and their action mechanisms in availability of nutrients to plants. Phyto-microbiome in stress regulation. Springer; 2020. p. 147−203.

[41] Rogers JR, Bennett PC. Mineral stimulation of subsurface microorganisms: release of limiting nutrients from silicates. Chem Geol 2004;203(1−2):91−108.

[42] Sheng XF. Growth promotion and increased potassium uptake of cotton and rape by a potassium releasing strain of Bacillus edaphicus. Soil Biol Biochem 2005;37(10):1918−22.

[43] Hutchens E, Valsami-Jones E, McEldowney S, Gaze W, McLean J. The role of heterotrophic bacteria in feldspar dissolution—an experimental approach. Mineral Mag 2003;67(6):1157−70.

[44] Bist V, Niranjan A, Ranjan M, Lehri A, Seem K, Srivastava S. Silicon-solubilizing media and its implication for characterization of bacteria to mitigate biotic stress. Front Plant Sci 2020;11:28.

[45] Shabbir I, Abd Samad MY, Othman R, Wong M-Y, Sulaiman Z, Jaafar NM, et al. Silicate solubilizing bacteria UPMSSB7, a potential biocontrol agent against white root rot disease pathogen of rubber tree. J Rubber Res 2020;23(3):227−35.

[46] Aleksandrov VG, Blagodyr RN, Ilev IP. Liberation of phosphoric acid from apatite by silicate bacteria. Mikrobiol Z 1967;29(11):1.

[47] Vasanthi N, Saleena LM, Raj SA. Concurrent release of secondary and micronutrient by a Bacillus sp. Am Eurasian J Agric Env Sci 2012;2:1061−4.

[48] Hiltner L. Uber nevere Erfahrungen und Probleme auf dem Gebiet der Boden Bakteriologie und unter besonderer Beurchsichtigung der Grundungung und Broche. Arb Deut Landw Ges Berl 1904;98:59−78.

[49] Dakora FD, Matiru VN, Kanu AS. Rhizosphere ecology of lumichrome and riboflavin, two bacterial signal molecules eliciting developmental changes in plants. Front plant Sci 2015;6:700. Available from: https://doi.org/10.3389/fpls.2015.00700 PubMed PMID: 26442016.

[50] Marschner H. Marschner's mineral nutrition of higher plants. Academic press; 2011.

[51] Dakora FD, Phillips DA. Root exudates as mediators of mineral acquisition in low-nutrient environments. Food security nutrient-stressed environments: exploiting plants' genet capabilities, 2002, 201-213.

[52] McNear Jr DH. The rhizosphere-roots, soil and everything in between. Nat Educ Knowl 2013;4(3):1.

[53] Singh R, Kumar A, Singh M, Pandey KD. PGPR amelioration in sustainable agriculture. Elsevier; 2019.

[54] Kumar M, Etesami H, Kumar V. Saline soil-based agriculture by halotolerant microorganisms. Springer; 2019.

[55] Etesami H, Maheshwari DK. Use of plant growth promoting rhizobacteria (PGPRs) with multiple plant growth promoting traits in stress agriculture: action mechanisms and future prospects. Ecotoxicol Environ Saf 2018;156:225−46. Available from: https://doi.org/10.1016/j.ecoenv.2018.03.013.

[56] Kloepper JW. editor. Plant growth-promoting rhizobacteria on radishes; 1978.

[57] Etesami H. Plant−microbe interactions in plants and stress tolerance. Plant life under changing environment. Elsevier; 2020. p. 355−96.

[58] Glick BR. Plant growth-promoting bacteria: mechanisms and applications. Scientifica 2012;2012.

[59] Ahemad M, Kibret M. Mechanisms and applications of plant growth promoting rhizobacteria: Current perspective. J King Saud Univ - Sci 2014;26(1):1−20. Available from: https://doi.org/10.1016/j.jksus.2013.05.001.

[60] Philippot L, Raaijmakers JM, Lemanceau P, Van Der Putten WH. Going back to the roots: the microbial ecology of the rhizosphere. Nat Rev Microbiol 2013;11(11):789−99.

[61] Pastore G, Kernchen S, Spohn M. Microbial solubilization of silicon and phosphorus from bedrock in relation to abundance of phosphorus-solubilizing bacteria in temperate forest soils. Soil Biol Biochem 2020;151:108050. Available from: https://doi.org/10.1016/j.soilbio.2020.108050.

[62] Banfield JF, Barker WW, Welch SA, Taunton A. Biological impact on mineral dissolution: application of the lichen model to understanding mineral weathering in the rhizosphere. Proc Natl Acad Sci 1999;96(7):3404−11.

[63] Vorhies JS, Gaines RR. Microbial dissolution of clay minerals as a source of iron and silica in marine sediments. Nat Geosci 2009;2(3):221−5.

[64] Haynes RJ, Zhou Y-F. Silicate sorption and desorption by a Si-deficient soil−Effects of pH and period of contact. Geoderma. 2020;365:114204.

[65] Vandevivere P, Welch SA, Ullman WJ, Kirchman DL. Enhanced dissolution of silicate minerals by bacteria at near-neutral pH. Microb Ecol 1994;27(3):241−51.

[66] Vasanthi N, Saleena L, Raj SA. Silica solubilization potential of certain bacterial species in the presence of different silicate minerals. Silicon 2018;10(2):267−75.

[67] Sheng X-F, He L-y, Huang W-y. The conditions of releasing potassium by a silicate-dissolving bacterial strain NBT. Agric Sci China 2002;1:662−6.

[68] Vasanthi N, Saleena LM, Anthoni AR. Evaluation of media for isolation and screening of silicate solubilising bacteria. Int J Curr Res 2013;5(2):406−8.

[69] Gu J-Y, Zang S-G, Sheng X-F, He L-Y, Huang Z, Wang Q. Burkholderia susongensis sp. nov., a mineral-weathering bacterium isolated from weathered rock surface. Int J Syst Evolut Microbiol 2015;65(3):1031−7.

[70] Chen W, Sheng X-F, He L-Y, Huang Z. Rhizobium yantingense sp. nov., a mineral-weathering bacterium. Int J Syst Evolut Microbiol 2015;65(2):412−17.

[71] Calvaruso C, Turpault M-P, Leclerc E, Ranger J, Garbaye J, Uroz S, et al. Influence of forest trees on the distribution of mineral weathering-associated bacterial communities of the Scleroderma citrinum mycorrhizosphere. Appl Env Microbiol 2010;76(14):4780−7.

[72] Huang Z, He L, Sheng X, He Z. Weathering of potash feldspar by Bacillus sp. L11. Wei Sheng Wu Xue Bao = Acta Microbiologica Sin 2013;53(11):1172−8.

[73] Umamaheswari T, Srimeena N, Vasanthi N, Cibichakravarthy B, Anthoniraj S, Karthikeyan S. Silica as biologically transmutated source for bacterial growth similar to carbon. Matters Archive 2016;2(3) e201511000005.

[74] Meena V, Dotaniya M, Coumar V, Rajendiran S, Kundu S, Rao AS. A case for silicon fertilization to improve crop yields in tropical soils. Proc Natl Acad Sci, India Sect B Biol Sci 2014;84(3):505−18.

[75] Wang M, Gao L, Dong S, Sun Y, Shen Q, Guo S. Role of silicon on plant−pathogen interactions. Front Plant Sci 2017;8:701.

[76] Kumawat N, Kumar R, Kumar S, Meena VS. Nutrient solubilizing microbes (NSMs): its role in sustainable crop production. Agriculturally important microbes for sustainable agriculture. Springer; 2017. p. 25−61.

[77] Lin Q-M, Rao Z-H, Sun Y-X, Yao J, Xing L-J. Identification and practical application of silicate-dissolving bacteria. Agric Sci China 2002;1(1):81−5.

[78] Liu W, Xu X, Wu X, Yang Q, Luo Y, Christie P. Decomposition of silicate minerals by Bacillus mucilaginosus in liquid culture. Environ Geochem Health 2006;28(1):133−40.

[79] Malinovskaya IM, Kosenko LV, Votselko SK, Podgorskii VS. Role of Bacillus mucilaginosus polysaccharide in degradation of silicate minerals. Microbiology 1990;59(1):49−55.

[80] Cruz JA. Silicate-solubilizing bacteria in Louisiana soils: identification, profiling, and functions in crop production. LSU (Doctoral dissertations). 5473; 2021.

[81] Brindavathy R, Dhara N, Rajasundari K. Biodissolution of silica by silicon bacteria in sugarcane rhizosphere. Res J Agr Sci 2012;3:1042−4.

[82] Vasanthi N, Saleena LM, Raj SA. Evaluation of media for isolation and screening of silicate solubilising bacteria. Int J Curr Res 2013;5 (2):406−8.

[83] Kannan NM, Raj SA. Occurrence of silicate solubilizing bacteria in rice ecosystem. Madras Aric J 1998;85:47−9.

[84] Zhou H-b, Zeng X-x, Liu F-F, Qiu G-z, Hu Y-h. Screening, identification and desilication of a silicate bacterium. J Cent South Univ Technol 2006;13(4):337−41.

[85] Archana DS. Studies on potassium solubilizing bacteria. Mestrado—Dharwad, Karnataka, Índia: University Of Agricultural Sciences. 2007.

[86] Purushothaman A. Distribution of silicate dissolving bacteria in vellar estuary. 1974.

[87] Mardad I, Serrano A, Soukri A. Solubilization of inorganic phosphate and production of organic acids by bacteria isolated from a Moroccan mineral phosphate deposit. Afr J Microbiol Res 2013;7(8):626−35.

[88] White AF, Brantley SL. The effect of time on the weathering of silicate minerals: why do weathering rates differ in the laboratory and field? Chem Geol 2003;202(3−4):479−506.

[89] Brucker E, Kernchen S, Spohn M. Release of phosphorus and silicon from minerals by soil microorganisms depends on the availability of organic carbon. Soil Biol Biochem 2020;143:107737.

[90] Meena VS, Maurya BR, Verma JP. Does a rhizospheric microorganism enhance K+ availability in agricultural soils? Microbiol Res 2014;169(5):337−47. Available from: https://doi.org/10.1016/j.micres.2013.09.003.

[91] Gorbushina AA, Broughton WJ. Microbiology of the atmosphere-rock interface: how biological interactions and physical stresses modulate a sophisticated microbial ecosystem. Annu Rev Microbiol 2009;63:431−50.

[92] Etesami H. Enhanced phosphorus fertilizer use efficiency with microorganisms. Nutrient dynamics for sustainable crop production. Springer; 2020. p. 215−45.

[93] Etesami H, Emami S, Alikhani HA. Potassium solubilizing bacteria (KSB): mechanisms, promotion of plant growth, and future prospects: a review. J Soil Sci Plant Nutr 2017;17(4):897−911.

[94] Vassilev N, Vassileva M, Nikolaeva I. Simultaneous P-solubilizing and biocontrol activity of microorganisms: potentials and future trends. Appl Microbiol Biotechnol 2006;71(2):137−44.

[95] Jongmans AG, Van Breemen N, Lundström U, Van Hees PAW, Finlay RD, Srinivasan M, et al. Rock-eating fungi. Nature 1997;389 (6652):682−3.

[96] Park KH, Lee CY, Son HJ. Mechanism of insoluble phosphate solubilization by Pseudomonas fluorescens RAF15 isolated from ginseng rhizosphere and its plant growth-promoting activities. Lett Appl Microbiol 2009;49(2):222−8.

[97] Joseph M, Dhargave T, Deshpande C, Srivastava A. Microbial solubilisation of phosphate: Pseudomonas vs Trichoderma. Ann Plant Soil Res 2015;17(3):227−32.

[98] Wu SC, Cao ZH, Li ZG, Cheung KC, Wong MH. Effects of biofertilizer containing N-fixer, P and K solubilizers and AM fungi on maize growth: a greenhouse trial. Geoderma 2005;125(1−2):155−66.

[99] Keshavarz Zarjani J, Aliasgharzad N, Oustan S, Emadi M, Ahmadi A. Isolation and characterization of potassium solubilizing bacteria in some Iranian soils. Arch Agron Soil Sci 2013;59(12):1713−23.

[100] Prajapati K, Sharma MC, Modi HA. Isolation of two potassium solubilizing fungi from ceramic industry soils. Life Sci Leafl 2012;5:71−5.

[101] Prajapati KB, Modi HA. Isolation and characterization of potassium solubilizing bacteria from ceramic industry soil. CIBTech J Microbiol 2012;1(2−3):8−14.

[102] Hömberg A, Obst M, Knorr K-H, Kalbitz K, Schaller J. Increased silicon concentration in fen peat leads to a release of iron and phosphate and changes in the composition of dissolved organic matter. Geoderma 2020;374:114422.

[103] Harley AD, Gilkes RJ. Factors influencing the release of plant nutrient elements from silicate rock powders: a geochemical overview. Nutrient Cycl Agroecosyst 2000;56(1):11−36.

[104] Avakyan ZA, Pivovarova TA, Karavaiko GI. Properties of a new species Bacillus mucilaginosus. Microbiology. 1986;55(3):369−74.

[105] Drever JI, Stillings LL. The role of organic acids in mineral weathering. Colloids Surf A Physicocheml Eng Asp 1997;120 (1−3):167−81.

[106] Duff RB, Webley DM. 2-Ketogluconic acid as a natural chelator produced by soil bacteria. Chem Ind 1959;1376−7.

[107] Cama J, Ganor J. The effects of organic acids on the dissolution of silicate minerals: a case study of oxalate catalysis of kaolinite dissolution. Geochim Cosmochim Acta 2006;70(9):2191−209.

[108] Rosa-Magri MM, Avansini SH, Lopes-Assad ML, Tauk-Tornisielo SM, Ceccato-Antonini SR. Release of potassium from rock powder by the yeast Torulaspora globosa. Braz Arch Biol Technol 2012;55(4):577−82.

[109] Abou-el-Seoud II, Abdel-Megeed A. Impact of rock materials and biofertilizations on P and K availability for maize (Zea maize) under calcareous soil conditions. Saudi J Biol Sci 2012;19(1):55−63.

[110] Wang D, Xie Y, Jaisi DP, Jin Y. Effects of low-molecular-weight organic acids on the dissolution of hydroxyapatite nanoparticles. Environ Science: Nano 2016;3(4):768−79.

[111] Oburger E, Jones DL, Wenzel WW. Phosphorus saturation and pH differentially regulate the efficiency of organic acid anion-mediated P solubilization mechanisms in soil. Plant Soil 2011;341(1):363−82.

[112] Welch SA, Taunton AE, Banfield JF. Effect of microorganisms and microbial metabolites on apatite dissolution. Geomicrobiol J 2002;19 (3):343−67.

[113] Schaller J, Faucherre S, Joss H, Obst M, Goeckede M, Planer-Friedrich B, et al. Silicon increases the phosphorus availability of Arctic soils. Sci Rep 2019;9(1):1—11.

[114] Violante A, Cozzolino V, Perelomov L, Caporale AG, Pigna M. Mobility and bioavailability of heavy metals and metalloids in soil environments. J Soil Sci Plant Nutr 2010;10(3):268—92.

[115] Smits MM, Wallander H. Role of mycorrhizal symbiosis in mineral weathering and nutrient mining from soil parent material. Mycorrhizal mediation of soil:. Elsevier; 2017. p. 35—46.

[116] Welch SA, Ullman WJ. The effect of organic acids on plagioclase dissolution rates and stoichiometry. Geochim Cosmochim Acta 1993;57 (12):2725—36.

[117] Blume H-P, Brümmer GW, Fleige H, Horn R, Kandeler E, Kögel-Knabner I, et al. Inorganic soil components—minerals and rocks. Scheffer/schachtschabelsoil science. Springer; 2016. p. 7—53.

[118] Brucker E, Kernchen S, Spohn M. Release of phosphorus and silicon from minerals by soil microorganisms depends on the availability of organic carbon. Soil Biol Biochem 2020;107737.

[119] Waksman SA, Starkey RL. Microbioligical analysis of soil as an index of soil fertility: VII. Carbon dioxide evolution1. Soil Sci 1924;17 (2):141—62.

[120] Cornelis JT, Delvaux B. Soil processes drive the biological silicon feedback loop. Funct Ecol 2016;30(8):1298—310.

[121] Kanakiya S, Adam L, Esteban L, Rowe MC, Shane P. Dissolution and secondary mineral precipitation in basalts due to reactions with carbonic acid. J Geophys Res Solid Earth 2017;122(6):4312—27.

[122] Pokrovsky O, Shirokova L, Stockman G, Zabelina S, Bénézeth P, Gerard E, et al., editors. Quantifying the role of microorganisms in silicate mineral dissolution at the conditions of CO2 storage in basalts; 2011.

[123] Raines DJ, Sanderson TJ, Wilde EJ, Duhme-Klair AK. Siderophores. Reference module in chemistry, molecular sciences and chemical engineering. Elsevier; 2015.

[124] Řezanka T., Palyzová A., Faltýsková H., Sigler K. Chapter 5 - Siderophores: amazing metabolites of microorganisms. In: Atta ur R., (ed.) Studies in natural products chemistry. 60: Elsevier; 2019. p. 157—188.

[125] Kalinowski BE, Liermann LJ, Brantley SL, Barnes A, Pantano CG. X-ray photoelectron evidence for bacteria-enhanced dissolution of hornblende. Geochim Cosmochim Acta 2000;64(8):1331—43.

[126] Ferreira A, Silva I, Oliveira V, Cunha R, Moreira L. Insights into the role of extracellular polysaccharides in Burkholderia adaptation to different environments. Front Cell Infect Microbiol 2011;1(16). Available from: https://doi.org/10.3389/fcimb.2011.00016.

[127] Xiao B, Sun Y-F, Lian B, Chen T-M. Complete genome sequence and comparative genome analysis of the Paenibacillus mucilaginosus K02. Microb Pathogenesis 2016;93:194—203.

[128] Yi Y, Huang W, Ge Y. Exopolysaccharide: a novel important factor in the microbial dissolution of tricalcium phosphate. World J Microbiol Biotechnol 2008;24(7):1059—65.

[129] Römheld V, Kirkby EA. Research on potassium in agriculture: needs and prospects. Plant Soil 2010;335(1):155—80.

[130] Kutuzova RS. Release of silica from minerals as a result of microbial activity. Mikrobiologiya. 1969;38:596—602.

[131] Urrutia MM, Beveridge TJ. Formation of fine grained silicate minerals and metal precipitates by a bacterial surface and the implications on the global cycling of silicon. Chem Geol 1994;116:261—80.

[132] Ehrlich HL, Newman DK, Kappler A. Ehrlich's geomicrobiology. CRC press;; 2015.

[133] Aleksandrov VG. editor. Organo-mineral fertilizers and silicate bacteria; 1958.

[134] Ameen F, AlYahya SA, AlNadhari S, Alasmari H, Alhoshani F, Wainwright M. Phosphate solubilizing bacteria and fungi in desert soils: species, limitations and mechanisms. Arch Agron Soil Sci 2019;65(10):1446—59.

[135] Peera SKPG, Balasubramaniam P, Mahendran PP. Effect of fly ash and silicate solubilizing bacteria on yield and silicon uptake of rice in Cauvery Delta Zone. Environ Ecol 2016;34(4A):1966—71.

[136] Kannan N, Raj SA. Occurrence of silicate solubilizing bacteria in rice ecosystem. Madras Agric J 1998;85:47—9.

[137] Sahebi M, Hanafi MM, Siti Nor Akmar A, Rafii MY, Azizi P, Tengoua F, et al. Importance of silicon and mechanisms of biosilica formation in plants. BioMed Res Int 2015;2015.

[138] Badr MA. Efficiency of K-feldspar combined with organic materials and silicate dissolving bacteria on tomato yield. J Appl Sci Res 2006;2:1191—8.

[139] Carver TLW, Zeyen RJ, Ahlstrand GG. The relationship between insoluble silicon and success or failure of attempted primary penetration by powdery mildew (*Erysiphe graminis*) germlings on barley. Physiol Mol Plant Pathol 1987;31(1):133—48.

20

Silicon and nano-silicon in plant nutrition and crop quality

Saima Riaz[1], Iqbal Hussain[1], Abida Parveen[1], Muhammad Arslan Arshraf[1], Rizwan Rasheed[1], Saman Zulfiqar[2], Sumaira Thind[1] and Samiya Rehman[3]

[1]Department of Botany, Government College University, Faisalabad, Pakistan [2]Department of Botany, The Government Sadiq College Women University, Bahawalpur, Pakistan [3]Department of Biochemistry, University of Okara, Okara, Pakistan

20.1 Introduction

Silicon (Si) is the second most copious element found on the planet [1] and is abundant in the soil as part of the clay minerals that are the main constituents of soil. It founds between 0.1 and 1.4 mM in soil solution [2]. At a pH value below 9 (PKA = 9.8), Si uptake as silicic acid [Si (OH)$_4$] [3]. As a result, all plants grown up in this kind of soil have some Si deposits in their tissues [4]. In contrast, the cultivation and use of chemical fertilizers have reduced the availability of Si to plants [4]. In soil, Si deficiency is now documented as a restraining factor in crop production, mainly for Si-accumulator crop cultivars like rice and sugarcane, and Si-enrich fertilizers are more often used for better yield [5]. Efficient Si-accumulating crops, passive Si-accumulators, and Si-repellent cultivars are classified according to their ability to absorb Si [6].

Plants also show a disparity in their Si deposition potency from 0.1% to 10.0% Si per dry weight [7,8]. Differences in capability of the roots to absorb the Si is an account for the difference in Si accumulation between species. Since the pioneering discovery of genes (*LSi1*, *LSi2*, and *LSi6*), which demonstrated the Si uptake and transport in plants a decade ago, scientists have worked out to understand the molecular Pathway behind Si uptake and transport among cultivars [9–11]. Many kinds of research have been conducted to understand how and why Si benefits the plants' growth and development [12]? Silicon is not only thought of as essential for plant growth and development, although literature pieces of evidence have shown that this metalloid is beneficial for plant growth, specifically in stressful conditions [13,14]. Silicon mitigates the negative effects caused under abiotic stress, for example, salt stress, water deficit condition, waterlogging, and metals [15,16]. Biogenic silica also acts as a protective measure to reduce the palatability and digestion of herbivores [17–19]. Silicon also improves the survival of plants with an increase in their resistance to exogenous stress (Fig. 20.1). This is due to the physical barrier that strengthens the cell wall with SiO$_2$ precipitation and incorporation [20,21]. While the transport of Si from soil solution to cortical cells is carried out by the presence of specific Si transporters depending upon the type of plants [22].

Silicon is transported into the shoot by xylem after the roots are taken up, and significant variances are found among species during this xylem-load process [23] (Fig. 20.2). In contrast to the passive diffusion process used by cucumbers and tomatoes, Si content in rice xylem sap is 20 times greater than in cucumbers due to xylem load-mediated transporters [22–24]. As a result, xylem loading was identified as a critical stage in active accumulation of Si in rice. A lower density transporter transports the Si from the ionic medium to the cortical cells, while a defective transporter transport the Si from cortical cells to the xylem can explain the much lower Si accumulation in cucumbers and tomatoes [25]. It is most complicated to develop a mechanistic model that explains the specific

Silicon and Nano-silicon in Environmental Stress Management and Crop Quality Improvement.
DOI: https://doi.org/10.1016/B978-0-323-91225-9.00021-2

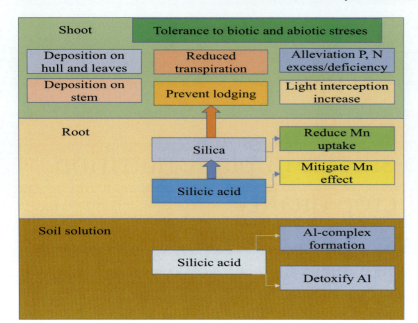

FIGURE 20.1 Schematic representation of Si effects in plant growth under biotic and abiotic stress.

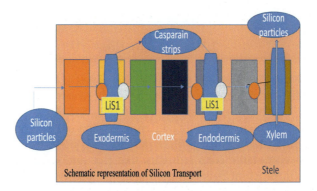

FIGURE 20.2 Schematic representation of Si transport in plants.

processes involved in Si-derived stress tolerance [9]. Many attempts were made to solve this dilemma and are still in progress, including in-depth analysis of Si application's effects on a wide range of biotic and abiotic stress variables, biophysiochemical parameters, and mineral localization with metabolomics and transcriptome responses. The current research topic brings together a variety of factors that will help researchers better understand how Si can best be used to support sustainable growth and climate-adapted cultivation?

Nanotechnology now has attained more attention due to its multilayered uses in the environment, agriculture, and related sectors [26,27]. Nanotechnology is a modern field of science that reforms technical developments in the present world. A pioneering scientific methodology has certainly more functions to monitor the environmental pollutants for the wellbeing of the ecosystem [28]. However, their extraordinary application and hyperactive discharge headed as a huge risk to the soil microbiome [28]. Nanotechnology is documented as vital for agriculture and is used to apply biofertilizers, biopesticides, and as indicators of soil quality to overcome the adverse effects of metals in agriculture [29].

However, nanoparticles (NPs) can affect different processes of photosynthesis, or even in different parts of chloroplasts. In other words, NPs may affect the morphology of photosynthetic structures, contents of photosynthetic pigments, etc. [30]. For example, NPs can increase the photosynthesis rate by enhancing the activity of rubisco's enzyme [31]. Nanoparticles can also alter the performance of photosystem II (PSII) and CO_2 harvesting [32]. Some NPs may decrease the absorption of sunlight by decreasing the Chl contents of plants [33]. As photosynthesis rate depends on the Chl contents of plants, decreased Chl contents can negatively affect the photosynthesis rate [34].

Some NPs can change the wavelength domains of photosynthesis. For example, carbon dots (CDs) nanocapsules absorb wavelengths between 200 and 700 nm, which can increase the light harvest and also electron transport chain performance in PSII, and ultimately promote photosynthesis [35–37]. Interestingly, NPs may also act as a shadow, preventing light harvesting in plants, which is referred to as the "shading effect." As a result, plants start producing more Chl to harvest more light, however, they usually cannot afford it, leading to a decrease in photosynthesis [38]. Hence plants can be affected by NPs in different ways. Due to the significance of photosynthesis for plants and human life, it is vital to investigate NPs-plant interactions and their impact on photosynthesis.

In this context, previous literature has shown an effect of various metal and metal oxide NPs on the growth, yield, and nutrient quality of essential crops [39] like silicon NPs (Si-NPs), zinc oxide (ZnO), copper oxide (CuO), silver oxide (AgO), titanium dioxide (TiO_2), and chitosan NPs as a nutrient source to maintain food [40–42].

Silicon-nanoparticles have the potential to increase plant growth. A positive impact on photosynthesis and gas exchange has been documented in Si-NPs-plant interactions [43]. For example, Si-NPs increased the total Chl contents and therefore photosynthesis in *Zea mays* L. [44]. Also, SiO_2-NPs show the same impact on plants as Si-NPs [45]. It is now acknowledged that SiO_2-NPs display a positive impact on transpiration, stomatal conductance, PSII, and electron transport chain activity as well as the photosynthesis rate [43, 45]. Elevated contents of Chl a and b were detected in sugarcane plants treated with SiO_2-NPs [46]. However, SiO_2-NPs caused no significant change in photosynthetic pigments of maize [47]. Treatment with 30 mg L^{-1} of Si quantum dots increased the Chl a and Chl b contents in *Lactuca sativa* plants [48]. SiC-NPs, at concentrations below 150 mg L^{-1}, increased the light absorption, growth, and lipid accumulation in the microalga *Scenedesmus* sp.; however, 250 mg L^{-1} of SiC-NPs decreased the growth and lipid accumulation in this algal species [49].

On a whole, Si-NPs are found as a promising tool against water stress [50], salinity stress [51], metal toxicity [52,53], and UV-B stress [16], in the agricultural sector. Silicon-nanoparticles are firstly up taken by root; transported to the stem, to get better plant growth [41]. Currently, wheat plants promote growth and development with improved biomass accumulation, seed germination, sprout elongation[42], or stimulate photosynthesis in crops [54]. Hence Si-NPs have shown the potential to increase the crop harvest for sustainable agriculture. Nanoparticles are particular in nature which make it differ from their bulk material are; their small size, larger surface-to-weight ratio, and different shapes [55,56]. Thus it is essential to understand by what means different Si-NPs interact in the environment. Their particular and unique properties make it a great potential source in agriculture and can improve the various abiotic stresses better than bulk material [53, 57] as shown in Fig. 20.3.

20.2 Silicon as micronutrient

Silicon is the next most abundantly found element on the earth's surface after carbon [1]. Still its essentiality is debatable [58] and it is not declared as vital to higher plants, although its application to various crops, including sugarcane and rice, results in better growth and yield [59,60]. Vast cultivation systems and undiscriminating uses

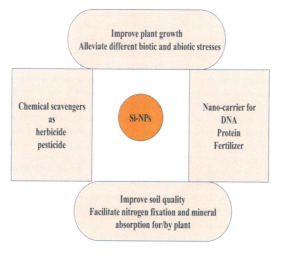

FIGURE 20.3   Schematic representation of Si-NPs benefits. *NP*, Nanoparticle.

of commercial fertilizers depleted the Si from the soil[4], urging to use it as fertilizer. Because of its role in stress tolerance, and protects the plants from growth inhibition and yield loss [7,61,62]. The mechanism of Si uptake differs from plant to plant [63]. Multiple studies have been done to examine the mechanism of Si uptake among rice, cucumber, and tomato plants. Mainly, the same transporter Lsi (channels) and Lsi2 (anion-type transporter) usually transports the Si in all of these systems; only the density of the transporters showed variation between plants [22,63,64]. Silicon is typically uptake and transport from different points present on the plasma membrane but not from the vacuoles due to lack of vacuolar membrane transporter [65].

Rice is considered among Si-accumulators and 10% of Si-contents accumulated in the shoot is much higher than the content of macronutrients such as N, P, and K, While its deficiency inhibited the rice growth increases the number of the empty grain, which leads to reduce the grain yields [66,67]. Rust brown spots were observed on the stem and shoot of rice particularly *Piricularia oryzae* under silicon deficient condition. Silicon is not considered as an essential element for plant growth since rice can ripen under Si-deficit medium with a major decrease in grain yield, though it is reported as an agronomically important for rice. Similarly, in tomato plants, the Si deficiency is visualized on the tip of meristematic tissue, however, the tissue was not found ruptured, while the young leaves near the tip were deformed, then hardened and became brittle [68,69]. The prolonged deficiency of Si causes chlorosis. A flower blossomed, but failed to pollinate, thus producing distorted fruit or no fruit [4]. Some authors [2] concluded that Si deficiency in tomatoes is happened due to Zn deficiency since in the absence of Si, more P has been taken up, and precipitated with Zn, which lead to reduce the availability of Zn internally. However, the addition of Si to growth media increases the Zn content, which is mandatory for indole acetic acid production in plants. This result improve the uptake of other nutrients in plants to counteract the different abiotic stresses like metals [70]. Silicon is also passively taken up by cucumber, and increased the actions of chitinase, polyphenol oxidase, and peroxidase enzymes [71].

Silicon indirectly influences the biological systems, provides plants with other elements, and thus improves the plant yield. Silicon was not counted in the list of essential elements for crops. Essential elements criteria given by Arnon and Stout (1939) [72] reported that a plant cannot complete its life cycle without the essential element. Though, there is still no evidence that plants cannot complete their life cycle without Si. On the other hand, it is classified as a beneficial element and has been documented for more than 50 years by those teaching and researching plant nutrition [59]. In this context, both Si-accumulating and non-accumulating plants become mature without Si addition, although their growth and grain/fruit yield were decreased exponentially. While somehow, previous literature indicated that Si could be taken as a micronutrient, and those currently available techniques are unable to exclude Si from the growth medium altogether. In addition to criteria of the essential elements, the element must be directly involved in plant metabolism. However, Si performs multiple functions, such as stimulating photosynthesis, improving tissue strength, and reducing transpiration rate in plants [73]. All these factors lead to increase the dry matter production and resistance to physical, and biochemical stress as well as optimize the fertility of soil with improved soil cation-exchange capacity [74] make feasible nutrients available to plants,[75] enhanced P uptake in plants [76], and reduced the aluminum (Al) toxicity after binding it to form hydroxyl-aluminosilicates [77,78] (Fig. 20.4). However, the vast study on silicon transport in plants and its significant results on the yield of crops due to silicon fertilization leads the International Plant Nutrition Institute (IPNI) to conclude it a "beneficial substance" in 2015 [79]. Furthermore, Si also influences the availability of different elements in soils after competing with binding on soil particles dependent on the speciation of silicic acid [80].

20.3 Direct impact of Si and Si-NPs on plants

Numerous silicon beneficial effects on the plant growth and development have been studied in both Si-accumulators and non-Si-accumulators plant cultivars. The effects are characterized by alleviating many stresses, both biotic and abiotic [79,81–83]. Likewise, plant response toward NPs is influenced by various factors, including NPs' size, shape, and application means [84]. Current literature has shown that Si-NPs influence directly plants and affect their morphological and physiological parameters to improve growth and yield [85–87]. Several studies have also reported some negative influence of Si-NPs on plants [88,89].

Silicon promotes plant growth with an increase in cell elongation and division, possibly owed to an increase in the extensibility of cell walls in rice and the apoplast barriers of endodermis [20, 90]. A recent study reported that Si is found responsible for an increase in the extension of cell wall of growing zone with decrease in the basal zone of stellar tissues lined up by endodermal lateral walls in roots of sorghum. This is concluded as Si's beneficial role in root elongation and stele protection through hardening [91].

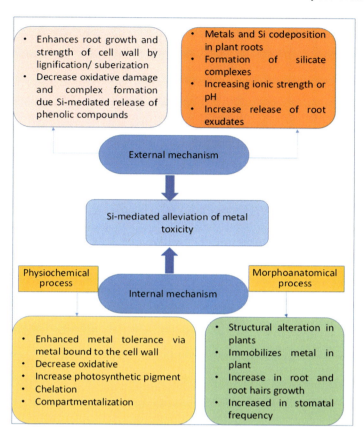

FIGURE 20.4 Schematic hypotheses for Si-mediated metal tolerance in plants [7].

Maximum positive effects of Si and nano-silicon are due to its deposition in cell walls of epidermal tissues of aerial parts. It is deposited as a polymer made of hydrated, amorphous silicon dioxide and formed silicon-cuticle or silicon cellulose double layers [92]. Silicon also gets deposited on the different cellular layers of leaves and shells. Thus the positive role of Si is quantitatively related to the concentration of Si accumulation in shoots. It is also believed that Si produces an active defense system in suppressing pathogenic infections with an increase in phytoalexin production in cucumber and rice [68,93]. Literature-based pieces of evidence proved the presence of SiO_2 precipitation as specific cell wall components [21]. Silicon bound with hemicellulose is also found in the suspension culture of rice cells [94]. Mixed-linkage glucans (MLGs) participate in SiO_2 formation reported in horsetail,[95] later confirmed in rice, where the action of hydrolase enzyme on MLGs promote silicification [96]. Recently, the function of callose in biosilicication has been studied in *Arabidopsis* under overexpression or downregulation of the callose synthase gene PMR4 [97]. Silicon showed an advantageous response in the control of biotic diseases in different plant species caused by plant pathogenic microorganisms or herbivores. Silicon improved the resistance against leaf and neck blast, sheath blight, brown spot, leaf scald, and stem rot in rice cultivars [98]. Similarly, the prevalence of powdery mildew in barley, cucumber, strawberry, and wheat, ringspot in sugarcane, and rust in cowpea [71,99]. The literature surveyed by Hartley et al., 2015 [19] demonstrated that Si deposition inside the leaf tissue increases the roughness in leaf tissues and reduced its palatability and digestibility for herbivores by X-ray spectroscopy (SEM-EDX). Keeping et al., 2009 [100] also studied this Si deposition pattern in sugarcane. Biochemical and molecular mechanisms are also strengthened by Si, which enables the plant to improve the resistance with increasing metabolites like phenols, phytoalexins, and momilactones,[101] also activating the enzymatic antioxidants metabolites such as peroxidase, polyphenol oxidase, lipoxygenase, and phenylalanine ammonia-lyase [102]. It is reported that [103] Si treatments also promote the regulation of defense-related genes.

In addition, Reynold et al., 2016 [104] reported that Si also defense the plants from predators' or parasitoids' and herbivores attack. In Si-treated plants, the phenology of the life cycle of insects slowed down making them more prone to predation [105,106]. Silicon plays a crucial function to determine the plants' reaction toward multiple environmental conditions. Two main pathways that are responsible for stress resistance are generally (1) physiomechanical protection by SiO_2 deposition and (2) a biochemical reaction that triggers the change in metabolites [107]. The distribution pattern of Si in cellular parts of the plant consents to our focus to gain additional information about its mode of action.

According to Liang et al., 2013 [108] Si deposition improves the strength and stability of stem tissues against lodging. It fetches tolerance toward ultraviolet radiation due to the shielding effect of Si accumulation on leaf epidermis [109] or by reducing UV-B-induced tissue damage [110]. Under water-stressed conditions or low humidity, Si increases the water relations [111]. It results in silica-cuticle double layer formation under the leaf epidermis with reduced water evaporation from it [112]. It also reduces stomata conductivity with improved integrity and turgidity of guard cells after Si accretion and reformed cell wall properties [61]. The improvement in drought resistance through Si application results in an increase in root elongation [113] and upregulation of the aquaporin genes [114] in drought-stressed plants. These results indicate the Si is involvement in the plant water status of rice leaves, and reduced transpiration up to 30%, which has a thin cuticle [115].

Under salt stress, Si reduces the uptake and translocation of sodium [61] and chlorine [116] from root to shoot in barley, rice, and wheat to induce salt tolerance [117]. In rice, Si amended the salt toxicity by hindering the transpiratory flow with SiO_2 deposition between exodermis and endodermis layer of cell [118]. The increased K uptake, which enables K/Na to be retained, is improved by the Si implementation, which directly stabilizes the activity of sodium-potassium pumps in salt-affected roots [119]. The Si also mitigates the damage from climatic stress like cyclones, low temperatures, and scarce sunlight in the summer time. Typically, a cyclone caused infertility in rice flowers, which results in a significant decrease in their yield. A high Si addition increases the rice shelf life [111].

Si in soil contaminated with metal influenced the bioavailability of toxic elements. Its existence in the form of sodium *meta*-silicate or alkaline Si particles results in an increase in pH of the rhizosphere, which leads to a decrease in the availability of metal concentration in soil [120] or the soluble form of Si in soil convert to produce viscous metasilicic acid (H_2SiO_3) that holds back toxic metals [121]. According to Kidd et al., 2001 [122], Si increases the contents of phenolic compounds, for example, catechin and quercetin which chelates with Al compounds to mitigate Al toxicity in several plant species like barley, maize, rice, sorghum, pea, and soybeans [123,124]. However, the presence of hydroxyl aluminum silicate in the apoplast also contributes to Al detoxification reported by Wang et al., 2004 [125]. The compartmentalization or reduced metal uptake by root with an apparent accumulation in the endodermis is significant in metal stress tolerance [23].

There are controversial data on the coprecipitation of Si with metals also available in the literature [23, 126]. The mechanism recognized for codeposition of Si with Cd in rice cell walls is through Si-Cd complexation, which results in a decrease in Cd inflow [127]. Similarly, down-regulation of the Nramp5 transporter responsible for Cd uptake with Si hemicellulose deposition is reported [128]. Down-regulation of other metal Cu/Cd linked Si transporters (*OsHMA2* and *OsHMA3*) in rice is also witnessed [129].

Further, literature surveys described that Si improved the antioxidant enzymes treated with metal [130] or an over-synthesis of endogenous antioxidants that leads to attenuation of oxidative stress,[131] maintains net photosynthesis rate after the stabilization of chloroplast structures, photosystem integrity, and an increased pigment concentration [16]. Therefore, Si may be of significant importance to trigger the plant responses, but particular molecular signals involved in this adaptation process have yet to recognize.

The Si role under phosphorus (P) deficit or excess conditions studied in various crops, containing barley and rice because excessive P levels in the soil result in the disturbance of plant development and reduced nutrition quality [132,133]. In nutrient solution having excessive P content, adding Si results in a more significant increase in the dry weight of rice [134] with down-regulation of P transporter *OsPHT1* in roots, subsequently reducing Pi uptake. While in some cases under P deficiency, it also improved accessibility of internal P by reducing the additional influx of Fe and Mn [135]. Effects of Si under Mn stress have been seen in hydroponically cultivated barley, beans, pumpkin, and rice [111,136,137] with improved antioxidant profile rather than an increase in remobilization of Mn.

It is also viewed that Si-NPs form a binary film on the epidermal cell wall, giving the plant a structural color [86]. The effect was not only limited to coloring. It is also presented that Si-NPs are used as a strengthening material that prevents mycological, microbial, and nematode-caused infections to increase the disease resistance [138]. The authors also concluded that Si-NPs formed a layer to decrease transpiration and make plants more tolerant to drought or temperature stress. Hydrophobic Si-NPs have considerable potential against plant-based ectoparasites of veterinary importance [139] It has been speculated that the mechanism by which Si-NPs fights pests after the break of protective lipid water barrier, which leads to the death of target organisms [140]. To increase the effectiveness of insecticides, it helps in release and translocation of these chemicals at the target. For the control and organized release of bio-formulations, however, mesoporous SiO_2-NPs are essential to delivering pesticides [141], and help to increase the shelf life and efficiency of commercial pesticides.

Mesoporous silica nanoparticles (MSNPs) (20 nm) were taken up by three crucial crop plants comprising of lupine, wheat, maize as well as *Arabidopsis* and, after entering the roots, relocated to the aboveground parts via the xylem bypass through symplast or apoplast routes [142]. Interestingly, this finding confirms the affinity of NPs with cell walls after the accumulation of MSNPs in the cell wall components. The versatile nature of MSPs and their particles magnitude attained after stabilizing the pH value and surfactant concentration is essential for efficient uptake by plants. They are taken up through the minute opening present in the root cells [142]. It has been shown that mesoporous Si-NPs increase the total photosynthetic rate of lupine and wheat sprouts [143]. In this study, the authors saw a prominent change in the absorbance peak of chlorophyll from 14 to 10 cm^{-1} of wheat and lupine, suggesting a change in the molecular structure of chlorophyll. It is stated that the presence of silicone in leaves is essential in plants; it is beneficial for plants because it keeps them in the upright position and keeps leaves in an expanded position so resulting in the rise of photosynthesis and total yield [144].

Silica nanoparticles (SNPs) are used to protect wheat seedlings from UV-B stress with a solid activated antioxidant defense system [145]. Remarkably, Si-NPs amended the harmful effects of UV-B stress, that is, low fresh weight, chlorophyll, and tissue impairment. Silicon NPs (50 mg L^{-1}) application to fenugreek leaves enlarged the shoots and leaves [146]. Similar to Yassen et al., 2017 [147] reported that cucumber plant yield, growth, and chemical compounds were increased with the application of silicone NPs at lower doses than the controls. At the same time, they also reported that growth parameters increased by the application of Si-NPs, the height of the plant included the weight, dry weight, and the number of leaves per plant, also including the fruit quality measures like the number of fruits and fruit bulk.

Under metal stress like Cr (VI), the enzymatic antioxidants such as superoxide dismutase (SOD) and ascorbate peroxidase (APX) rise significantly in the presence of Si-NPs, while lesser effects were observed in catalase (CAT), glutathione reductase, and dehydroascorbate reductase [16]. SNPs have also been found to improve the seedling germination in well-known Si exclusion plants such as tomatoes and improve fresh and dry weight [148]. While SiO_2-NPs promote tap root length as well as lateral root growth, and chlorophyll contents have proven valuable in the production of seedlings [85]. However, the influence of SNPs depends on the plant nature, as they significantly reduced plant growth in Bt-transgenic cotton [149]. The toxicity of SNPs is linked to pH and nutrient absorption systems [87]. The phytotoxicity of SNPs was prompted if the pH of the solution is not adjusted [88]. The alkaline pH 8 lowers the nutrient absorbance, whereas the negatively charged Si-NPs be likely to absorb nutrients. However, there is still an information gap on the properties of Si-NPs on zeta potential or pH stability [149]. These Si-NPs in agriculture is gaining more significance to make more strategic solution for agriculture sustainability. For example, nanomaterials are engineered to control the nutrient immobility in the soil [150]. It could be concluded the role of Si and nano-silicon is reported in some studies in Table 20.1.

20.4 Si-NPs as a delivering agent for fertilizers

In-plant systems, Si is a very favorable element for growth and development: its content in plants is comparable to that of macronutrients is 20%−91% of dry weight, based on the plant species [68], and it is generally used as fertilizer to increase the yield of many crops [138]. Silicon has unique physical, chemical, and structural properties, due to which its entrance into plants. After entering plants, its effects on physiological processes like photosynthesis, growth, etc. [163]. In a recent study, Fitiyani and Haryanti (2016) [164] revealed that silicon NPs were applied as fertilizers in *Solanum lycopersicum*, which cause improvement in plants height, increased the number of leaves and roots. Suciaty et al., 2018 [165] have reported in a similar study that nano-silicone particles based fertilizers positively promoted the yield of soybean plants (*Glycine max*). Silicone NPs also increase the net assimilation rate and leaf area index [165]. Silicone is essential and beneficial for plants because it keeps them in an erect position and leaves in expanded position, resulting in increasing the rate of photosynthesis and yield in plants [140]. Although, after applying silicone NPs, yield and growth increased because of increasing amino acids, protein, N, K, and P in plants [47].

Si-NPs mechanism is an efficient vehicle for target-based delivery of pesticides and fertilizers to crops [166] and carries commercial herbicides attached in a diatom fistula and quickly delivers the herbicide to the crop [41, 156].

NPs possess a certain unique set of properties like sizes in the range of 1−100 nm, lower toxicity, chelating abilities, and other higher strength thus they are strong candidates in serving as a delivery system for the various biological molecules like nucleic acids into cellular bodies. Singh (2013) [167] explained the role of nanodelivery systems in cancer therapeutics and described the role of various NPs like carbon nanotubes, fullerenes, and nanoshells in delivering siRNA. However, in plant cells the outermost boundary of the cell or cell wall hinders the

TABLE 20.1   Effect of Si and Si-NPs on growth, physiochemical, and yield attributes in different plant species.

Growth medium	Plant species (Scientific name)	Response of plant to applied silicon and nano-silicon	References
Silicon as SiO₂ Seedling, priming	*Larix olgensis*	Increase in plant height, tap root length, number of lateral roots, and chlorophyll contents	[85]
Soil, field	*Zea mays*	Root elongation, silica deposition than control plant was observed	[151]
Seeds, Petri plate	*Lens culinaris* *Lycopersicum esculentum*	Better germination index and early growth of plants under salinity stress. Improved seed germination, time, germination index, vigor index, seedling fresh, and dry weight	[148,152]
Seedling, irrigation	*Crataegus aronia*	An increase in plant growth, concentrations of photosynthetic pigments, with decrease in water potential and oxidative stress indicator, the MDA content	[153]
Si-priming Under high tunnel and field conditions	*Arabidopsis thaliana*	Reduction in powdery mildews and high yield	[17]
Nano-Si Seeds, Petri plate	*Lycopersicum esculentum*	Improved root and shoot growth	[154]
Seedlings, nutrient solution	*Solanum lycopersicum*	Rise in fresh weight, chlorophyll contents, photosynthesis rate, and leaf water potential the plant	[51]
Plant, foliar spray	*Ocimum basilicum*	NPs mitigated the effects of salinity stress	[155]
seeds, pod	*Vicia faba*	Improved flowering results high yield	[151]
SiO₂-NP fertilizer Encapsulated farmyard manure with NPK fertilizers	*Helianthus annuus*	Significantly effective at improving the reduction in the growth characteristics	[156]
Si-fertilizer: OPTYSIL Laboratory trial Small field trial	*Triticum aestivum* *Brassica napus*	Induce drought tolerance with increase in pod number, average seed yield/plant, thousand grain weight, and decrease electrolyte leakage	[157]
SAAT: a stable Silicic acid Foliar spray, Hydroponically or soil solution, vegetative stage	*Oryza sativa, Saccharum officinarum, Solanum tuberosum, Solanum lycopersicum, Triticum aestivum*, and *Eleusine coracana*	Increase in root mass, leaf area, chlorophyll content, nutrient uptake (P, Ca, K, and Si), yield with good quality	[158—160]
Elkem: as Silicon material Greenhouse, field experiment	Vegetables and *Oryza sativa*	Induce resistance against insect and its infections with increase in antioxidants, specific and nonspecific or both	[161]
Silixol: foliar fertilizer Field trial, vegetative, booting, and seed development stage	*Triticum aestivum*	Mitigate drought stress with increase relative water content, chlorophyll contents, root elongation, K and P contents of both straw and seeds	[162]

efficient delivery of NPs carrying genetic material. The removal of cell walls can lead to efficient delivery [168]. Taylor and Fauquet (2002) [169] described the NPs bombardment method to deliver the genetic material inside the cell and thus there is no need to remove cell wall.

Mesoporous Si-NPs of 2–10 nm worked as a highly efficient delivery vector for fertilizers like boron and urea [166,170]. Silicone NPs help in the highly efficient delivery of organic macromolecules, that is, proteins, nucleotides (for DNA and RNA), and chemicals. They help to improve the plant resistance and nutrient efficiency for increasing the crop yields [171]. Nanoencapsulation is a modern technique and is being an efficient way for the protected use of chemicals with no or fewer wastes released to the environment [172,173]. Mesoporous Si-NPs have stable structures and compositions to chemical and heat, while showing more significant surface areas, variable pore sizes, and several well-characterized surface properties. These adaptations made them appropriate for the delivery of macroorganic guest molecules to plants [170]. Surface-coated Si-NP is very much useful for transport of nuclear DNA and siRNA [174]. Similarly, Torney et al., 2007 [170] reported the NPs-based delivery of macromolecules, and chemical substances (with the gene and its chemical inducer) to crop or whole leaves is a highly efficient application. Another highly beneficial application of NPs is the direct delivery of a prerecombinase protein through gold-plated MSNs using a biolistic method in maize plants [175]. Martin-Ortigosa et al., 2012 [176] performed a study to comprehend the role of MSNPs in delivering plasmid DNA and protein simultaneously into plant cells. The researchers successfully interpreted the use of MSNPs as delivery agents for synergistic delivery of proteins and plasmid DNA. Therefore MSPs have established transportation materials and can be used for the development GMO (Fig. 20.5).

20.5 Effects of Si and Si-NPs on plant nutrient uptake

Plants transport nutrients through the roots into the vascular cylinder in two separate ways: apoplastic and transcellular. Meanwhile, absorption mechanism of each nutrient may differ based on plant nature and ecological factors [177]. Several barriers have been developed to promote the selective mineral absorption after controlling the apoplastic flow of solutes into the root stele. Silicon deposits in both exodermis and the endodermis, which allow the development of a more effective apoplast barrier [178]. Plant transpiration can be reduced by Si accretion in epidermal layers of cell, which reduces the apoplast nutrient uptake [179].

In current agriculture practice, Si is identified as an efficient supplement for the growth and development of various plants particularly rice and sugarcane [180,181]. According to reports, the uptake of Si in rice and sugarcane is sometimes higher than the intake of N and K [99]. The Si treatment can increase the optimal N rate, which leads to an increase the rice productivity due to the synergistic effect. In combination with the P fertilizer with Si use efficiency increased from 24% to 34% as it decreased the soil P retention capacity, resulting in a higher concentration of water-soluble P [182]. According to Haynes (2014) [4] frequent harvest and regular use of inorganic fertilizers like N, P, and K have reduced the amount of Si in soil. Now, its deficiency in the soil is counted as one of the limiting factors for

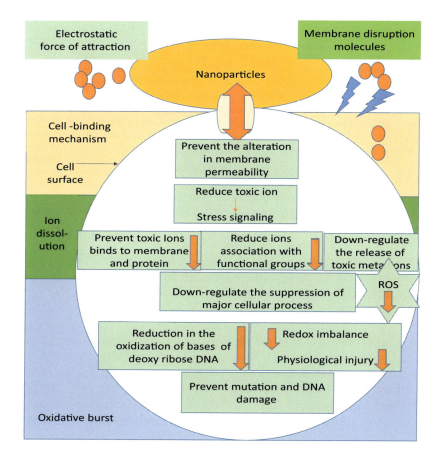

FIGURE 20.5 Schematic representation of underlying mechanisms associated with NPs. *NPs*, Nanoparticles.

better plant production. Very little information is found on Si-based nutrient deficit improvement, but research has just come to focus. Several researchers have examined the effects of Si on nutrient deficiency [136,183].

The response of Si toward nutrient uptake has widely been discussed in this literature. Iron (Fe) is a necessary micronutrient that has an important role in many cellular functions, such as respiration, chlorophyll biosynthesis, photosynthetic, etc., and hexose biosynthesis [184]. In calcareous soils, hydrogen carbonate is found in excess amounts, which leads to an increase in pH value and thus insolubility of Fe occurred [185]. The solubility of Fe in soils is strongly pH-dependent and decreases with an increase in pH value [186].

Silicon has also improved the Fe flow from root to shoot, even though Fe treatment has little effect on Si uptake and transport [187−189]. It is suggested after the addition of Si, the increase of Si transporters (FRO2, IRT1, and HA1) expression could influence the Fe uptake and translocation [190] and cucumber [183,191,192]. In addition, the redox potential of the nutrient or soil solution is essential for Fe stability in the solution. Silicon efficiency mitigates the Fe deficiency was studied by Gonzalo et al., 2013 [183] in soybean (an iron-inefficient plant) and cucumber (an iron-efficient plant) with sodium silicate used as a Si source. The addition of Si inhibited the chlorophyll degradation, slow down the growth retardation, and maintained the Fe content in Fe-inefficient soybean variety under Fe-deficient conditions. Silicon inflow delayed the reduction in shoot dry weight, shoot length, number of nodes, and Fe content of root and shoot of cucumber plant reported by Stevic et al., 2016 [193]. The beneficial aspects of Si in Fe deficiency increase the Fe content in leaf and tissue deposits with organic acids and phenolic compounds (citrate or catechin) [136]. This results in an increase in the translocation of Fe in apical apexes and buildup of Fe-mobilizing substances. Pavlovic et al., 2013 [192] concluded that the root responses to Fe deficit condition in cucumbers are indirectly influenced as a result of improved Fe contents in an entire plant through Si-mediated expression of strategy-dependent genes (FRO2, IRT1, and HA1) [194].

Zinc (Zn) is one of the essential micronutrients, a component of several essential enzymes that are being used for protein synthesis, nitrogen metabolism, and macromolecules catabolism [2,195]. It also acts as a stabilizer of the plasma membrane and protects the plants from oxidative damage [196]. The average total Zn concentration in soil is 50 mg kg^{-1} [197]. Several factors affect the Zn availability in soils as well as its toxicity in plants such as pH, physiochemical properties of soil, and the tolerance level of the crop species/variety [198]. Normally the Zn required for leaf in most cultivars is 1520 mg kg^{-1} dry weight [199]. Its deficiency is not only affecting the crop yield and production but has also shown some serious effects on humans. Around 3 billion population worldwide are under Fe and Zn deficiencies, mainly those reliant on grain foods, because Zn deficiency is the most common mineral disorder in cereals [200,201]. A deficiency in Zn leads to growth disorders, chlorosis, a longer ripening period, fewer tillers, smaller leaves, and poor quality of harvested plants [202]. In rice, where Zn deficiency leads to reducing the dry weight, grain weight per panicle, the weight of seeds, and yield [203]. Although, with Si addition, the Zn pool has become more mobile under deficient conditions, Since Si and Zn deposits in leaves with nutrient mobilization by phloem lead to a rise in Zn content of seeds and fruits, which suggests a more productive use of available Zn, especially at times of Zn scarcity [136]. Cucumber plants were sown in Zn-free solution have no significant alteration in chlorophyll, except necrotic spots on leaves. The addition of Si reduced the necrotic stains on the leaves. The citrate contents of roots decreased under Zn insufficient condition, but Si amends this response with an increase of these compounds. Likewise, the fumarate and shikimate concentration in root increases with increasing of Si in Zn-deficient plants [136]. A Si supplement increased the Zn contents in stems, roots, and shells of the rice plant [204]. The zinc bound with Si remobilized under its deficient medium since the correlation of Si supplementation with Zn remobilization has not yet been investigated [205]. Furthermore, Si decreased the expression of zinc transporter OsZIP1 in the root of rice plants to minimize the oxidative stress under Zn-induced toxicity with no effect on root to shoot Zn translocation [206].

Manganese (Mn) is the 12th most crucial element, 0.098% Mn is present in the earth biosphere [207]. It occurs in the soil and water, and sediments, which is important for living organisms. It exists in various forms containing oxides, silicates, phosphates, and borates [208,209]. Manganese is released into the soil through rock crystal weathering and atmospheric deposition from natural and human activities. Manganese movement in the soil is found sensitive to soil states like acidity, moisture, and organic matter. The solubility of soil Mn is dependent on redox potential and soil pH. The mobility of Mn is increased at low pH that favors the reduction of insoluble manganese oxides [210]. Manganese is an essential component of the SOD containing Mn (Mn-SOD), which helps the plant to scavenge ROS and convert them into hydrogen peroxide and then water. Sandy and calcareous soils usually contain little Mn, in combination with a high organic matter favors the oxidation of soluble ionic Mn to oxide state, easily taken up by the plant [211]. Foliar application of Mn^{+2} is only essential for citrus fruits and other trees or crops to remedy the yield reductions or complete loss of crops during the winter [212]. Dragišić et al. 2007 [213] report the Si is indirectly used to reduce the hydroxyl ions in leaf apoplast by reducing free apoplast Mn^{2+} to protect the plant. Williams and Vlamis (1957) [214] revealed that Si has no effect on the Mn contents of leaves, but Si caused more even distribution in leaves rather than in discrete necrotic points that were more supported by future experiments [215]. Silicon alleviates the oxidative stress caused by the deficiency of Mn in sorghum plants,[137] presenting the indirect beneficial role rather than direct response with an increase in Mn uptake and remobilization.

Copper (Cu) contents in soils are between 3 and 100 mg kg^{-1}, rarely 1%−20% are found in free, readily available state. Copper is immovable between tissues, and deficiency first appears in fresh newer cells. The tendency for Cu uptake and accumulation from soil to root system depends on the total contents of Cu available in soil and also on the ability of the plants to cross the interface between soils and roots [216]. Copper is dual in nature, it acts as a micronutrient as well as metal that can induce stress in plants. The leaves that contain Si are not only deposited with Cu, but other plant nutrients such as Ca, K, and P. The Si under Cu toxicity inhibited the Cu transport from roots to shoots with restricted mobilization [217]. Silicon positively ameliorates the Cu toxicity in wheat, *Arabidopsis*, and tobacco with a decrease in the expression of Cu transporters like *AtCOPT1*, *AtHMA5*, and *NtCOPT1* [218] or with an increase in the accumulation of Cu in root rather than root to shoot translocation in wheat [23].

20.6 Effects of Si and Si-NPs fertilizer on protein and amino acids contents

Silicon nanoparticles work for the targeted distribution of micro and macromolecules in plants [171]. Nanoplatforms and their application in different areas under *vivo* or *vitro* conditions have raised the significance of agricultural nanotechnology. This field provides controlled and regulated release of agrochemicals to enhance plant resistibility and nutrient efficiency with increased crop yields [171]. Nanoencapsulation is an effective and protected means with lower release of chemicals into environment and ensures its protection [172,173]. The uptake, capability, and impact of different NPs on plant growth, development, and biochemical processes may vary based on the plant species. Mesoporous Si-NPs are chemically and thermally stable with huge surfaces, adjustable pore sizes, and well-characterized surface properties, making them right for the uptake of molecules [166]. It has been seen that mesoporous Si-NPs increase the growth, protein, and photosynthetic rate of lupine and wheat seedlings [143] with an increase in the leucine-rich repeat receptor-like kinase (LRR-RLK) in rice,[219] a protein used in intracellular signal transduction. The recent omics study will help us to elucidate the absence of genes or proteins involved in signal transduction [220]. In fact, in rice (Si-accumulator), it has been suggested that Si acts as a second messenger by binding with OH group of proteins in signaling cascade [71]. Rice transporter (*OsLsi1*) and its homologs belong to the Noduline-26 Major Intrinsic Protein 3 (NIP3) subfamily of aquaporin have an amino acid sequence of six transmembrane domains and two Asn-Pro-Ala (NPA) motifs, found responsible for the transport of Si from the soil solution to the root [221]. Haynes (2014) [4] revealed that Si usually expresses in roots in form of *OsLsi1*, while it seems to be down-regulated. *OsLsi1* expression in root is lower in apical than basal root regions, signifying Si uptake was more obvious in mature root zone instead of root tips [9]. Lsi2 (*OsLsi2*) is first found rice Si-efflux transporter gene probably encodes for a membrane protein with 11 transmembrane protein domains and is responsible for transportation of Si from root to stele. Over 90% of the Si was uptake by roots then transferred to the stem [111] via these genes, and their expression depends on genotypic differences in Si accumulation [9]. Cucumbers are species that accumulate Si in the stem more strongly than other dicotyledons [222]. The Si contents in cucumber leaves are in a range from 1.8% to 2.9% [221] Si content especially in the xylem sap is often higher than in peripheral solutions [25]. *CmLsi1* is the first gene that transports Si in pumpkin [223]. In contrast to *OsLsi1* transporters of rice, pumpkin *CmLsi1* is localized on all root cells, like barley and maize [224]. Two Si-efflux transporters (*CmLsi2-1* and *CmLsi2-2*) were also isolated by two pumpkin varieties [223].

Silicon positively influenced the root hydraulic conductance and upregulation of the plasma membrane intrinsic proteins genes studied in sorghum seedlings [225]. This recommended that Si might be effective for maintaining the phospholipids to protein ratio and plasma membranes permeability,[226] although; the mechanism behind this is still unclear. Similar results of Si foliar application were observed in strawberries with a rise in unsaturated fatty acids and phospholipids [227] and protein contents in wheat under drought [228]. Silicon in salt-sensitive plants (barley) under salinity decreases saturated to unsaturated fatty acids. Conversely, this effect was observed intolerant plants to mitigate the salt stress [25]. However, Si stimulated the upregulation of polysaccharides contents in rice cell walls and other components like pectic acid, protein, polyphenols, and lignin [219]. Some authors [229] found no obvious effect of Si on grain quality of two wheat varieties. However, the proportion of brown rice, the proportion of milled rice, and the proportion of head rice in connection with the fatty acid content of Si-treated rice were considerably higher than in Si-untreated rice [230].

20.7 The role of Si and Si-NPs in crop quality

It is well documented that Si improves the growth and yield of economically important crops [231]. The main monocotyledons cultivars generally reported to respond positively to Si fertilization are rice, wheat, maize, barley, millet, sorghum, and sugarcane which actively absorb and accumulate a huge amount of Si in their organs, and dicotyledonous plants such as cotton and soya beans that are also able to accumulate Si through special transporters [138]. Although the increase in percentage yield after Si fertilization was seen with better food quality. Silicon fertilizers are mainly reported to improve rice grain quality, sugarcane, vegetables, and fruits. Silicon significantly improved the qualitative traits of sugarcane juice, such as commercial cane sugar (CCS), and the sugar production in either salt-sensitive or salt-tolerant sugar cultivars [232]. The Si supply to the hydroponic medium improved the fruit firmness, soluble solid contents, and ascorbic acid of tomato fruits,[233] while the fertilization of strawberries with Si enhanced tissue consistency and shelf life of the fruit after harvesting [234]. In apples, the use of Si increased the content of soluble solids, vitamin profile (ascorbic acid) with decreased tartaric acid substances in the fruit but did not affect the hardness of the fruit [235]. Previously cited literature demonstrates that Si applications increased the total content of soluble solids, sugars, and organic acids with decreased nitrate content in grapes [236]. It is also observed an increase of 5.1%−10.2% in the yield of cucumbers with Si implementation [237]. A similar rise in the weight of individual cucumbers with an improvement in the quality of sugar and vitamin C levels [225]. Thus silicon's importance for improving plant mineral nutrition is not only significant, although its role in plant growth and development with a decrease in plant-based diseases is significant [238,239]. Currently, nanotechnology is providing an auspicious space for interdisciplinary research that facilities us in several fields.

20.8 Conclusions and future perspectives

This chapter is exploring the potency of Si and Si-NPs in agriculture. It brings together the literature related to the multiple uses of NPs as insecticides, fertilizers, herbicides, gene or drug transfer agents, soil-improving substances and biosensors for soil analysis. Studies show the Si-NPs as revolutionary means used in various sectors from agricultural to medical sciences. Poor agricultural practices which include the excessive use of pesticides and fertilizers are the leading cause of deteriorating soil quality and food scarcity, is now emerging as a major challenge to mitigate global 2050 demands of food for over nine billion people [240,241]. There has been ever increase in the need for sustainable development in agriculture optimizing the yields and increment in crop biomass in an environment-friendly way. Nanotechnology holds the promise and potential in the big picture of global 2050 food demands [242]. Similarly, silica fertilizers are in practice to improve the plant-available silicon (PAS) as well the uptake of micronutrients (N, P, and K) with an increase in quality and magnitude of yield of various commercial and noncommercial crops including rice, Bt-cotton, onion, tomato, citrus, banana, pomegranate, and chili under various biotic and abiotic stress [243]. In the future, it would be beneficial to use these practices in inducing resistance to several environmental factors. These NPs thought helpful for green and eco-friendly alternatives to various chemical fertilizers devoid of harming nature and used as integrated disease management strategies to enhance host plant resistance. As researchers and growers become more aware of the potential of silicon and its NPs, it is often known as the quasiessential element that will be known as a viable means of enhancing plant health and performance.

References

[1] Singer MJ, Munns DN. Soils: an introduction. Upper Saddle River, NJ: Pearson Prentice Hall; 2006.
[2] Yan GC, Nikolic M, YE MJ, et al. Silicon acquisition and accumulation in plant and its significance for agriculture. J Integr Agric 2018;17(10):2138−50.
[3] Deshmukh RK, Vivancos J, Guérin V, et al. Identification and functional characterization of silicon transporters in soybean using comparative genomics of major intrinsic proteins in Arabidopsis and rice. Plant Mol Biol 2013;83(4−5):303−15.
[4] Haynes RJ. A contemporary overview of silicon availability in agricultural soils. J Plant Nutr Soil Sci 2014;177(6):831−44.

[5]   Ma JF, Yamaji N, Mitani N, Xu XY, et al. Transporters of arsenite in rice and their role in arsenic accumulation in rice grain. Nat Acad Sci 2006;105:9931–5.

[6]   Mitani N, Ma JF. Uptake system of silicon in different plant species. J Exp Bot 2005;56(414):1255–61.

[7]   Liang Y, Sun W, Zhu YG, Christie P. Mechanisms of silicon-mediated alleviation of abiotic stresses in higher plants: a review. Environ Pollut 2007;47:422–8.

[8]   Ma JF, Yamaji N. Functions and transport of silicon in plants. Cell Mol Life Sci 2008;65(19):3049–57.

[9]   Ma JF, Yamaji N, Mitani N, et al. An efflux transporter of silicon in rice. Nature. 2007;448:209–12.

[10] Rao GB, Susmitha P. Silicon uptake, transportation and accumulation in rice. J Pharmacogn Phytochem 2017;6:290–3.

[11] Mitani-Ueno N, Ma JF. Linking transport system of silicon with its accumulation in different plant species. J Soil Sci Plant Nutr 2021;67 (1):10–17.

[12] Luyckx M, Hausman JF, Lutts S, Guerriero G. Silicon and plants: current knowledge and technological perspectives. Front Plant Sci 2017;8:411.

[13] Yizhu L, Imtiaz M, Ditta A, Rizwan MS, et al. Response of growth, antioxidant enzymes and root exudates production towards As stress in *Pteris vittata* and in *Astragalus sinicus* colonized by arbuscular mycorrhizal fungi. Environ Sci Pollut Res 2020;27:2340–52.

[14] Rahman SU, Xuebin Q, Yatao X, et al. Silicon and its application methods improve physiological traits and antioxidants in *Triticum aestivum* (L.) under cadmium stress. J Soil Sci Plant Nutr 2020;20(3):1110–21.

[15] Singh R, Gautam N, Mishra A, Gupta R. Heavy metals and living systems: an overview. Indian J Pharmacol 2011;43:246.

[16] Tripathi DK, Singh VP, Prasad SM, et al. Silicon nanoparticles (SiNp) alleviate chromium (VI) phytotoxicity in *Pisum sativum* (L.) seedlings. Plant Physiol Biochem 2015;96:189–98.

[17] Vivancos J, Labbé C, Menzies JG, Bélanger RR. Silicon-mediated resistance of Arabidopsis against powdery mildew involves mechanisms other than the salicylic acid (SA)-dependent defence pathway. Mol Plant Pathol 2015;16(6):572–82.

[18] Massey FP, Hartley SE. Physical defences wear you down: progressive and irreversible impacts of silica on insect herbivores. J Anim Ecol 2009;78:281–91.

[19] Hartley SE, Fitt RN, McLarnon EL, Wade RN. Defending the leaf surface: intra-and inter-specific differences in silicon deposition in grasses in response to damage and silicon supply. Front Plant Sci 2015;6:35.

[20] Lukačová Z, Švubová R, Kohanová J, Lux A. Silicon mitigates the Cd toxicity in maize in relation to cadmium translocation, cell distribution, antioxidant enzymes stimulation and enhanced endodermal apoplasmic barrier development. Plant Growth Regul 2013;70 (1):89–103.

[21] Guerriero G, Hausman JF, Legay S. Silicon and the plant extracellular matrix. Front Plant Sci 2016;7:463.

[22] Mitani N, Ma JF. Uptake system of silicon in different plant species. J Exp Bot 2005;56:1255–61.

[23] Keller C, Rizwan M, Davidian JC, et al. Effect of silicon on wheat seedlings (*Triticum turgidum* L.) grown in hydroponics and exposed to 0 to 30 μM Cu. Planta. 2015;241(4):847–60.

[24] Ghareeb H, Bozsó Z, Ott PG, et al. Transcriptome of silicon-induced resistance against *Ralstonia solanacearum* in the silicon non-accumulator tomato implicates priming effect. Physiol Mol Plant Path 2011;75(3):83–9.

[25] Liang Y, Wong JW, Wei L. Silicon-mediated enhancement of cadmium tolerance in maize (*Zea mays* L.) grown in cadmium contaminated soil. Chemosphere. 2005;58:475–83.

[26] Biswas P, Wu CY. Nanoparticles and the environment. J Air Waste Manag Assoc 2005;55(6):708–46.

[27] Raffi MM, Husen A. Impact of fabricated nanoparticles on the rhizospheric microorganisms and soil environment. Nanomaterials and plant potential. Cham: Springer; 2019. p. 529–52.

[28] Khanna K, Kohli SK, Handa N, et al. Enthralling the impact of engineered nanoparticles on soil microbiome: a concentric approach towards environmental risks and cogitation. Ecotoxicol Environ Saf 2021;222:112459.

[29] Servin A, Elmer W, Mukherjee A, et al. A review of the use of engineered nanomaterials to suppress plant disease and enhance crop yield. J Nanopart Res 2015;17:1–21.

[30] Tan W, Deng C, Wang Y, et al. Interaction of nanomaterials in secondary metabolites accumulation, photosynthesis, and nitrogen fixation in plant systems. Compr Anal Chem 2019;84:55–74.

[31] Kataria S, Jain M, Rastogi A, et al. Role of nanoparticles on photosynthesis: avenues and applications. Nanomaterials in plants, algae and microorganisms. San Diego, CA: Academic Press; 2019. p. 103–27.

[32] Falco WF, Scherer MD, Oliveira SL, Wender H, Colbeck I, Lawson T, et al. Phytotoxicity of silver nanoparticles on *Vicia faba*: evaluation of particle size effects on photosynthetic performance and leaf gas exchange. Sci Total Environ 2020;701:134816.

[33] Tao X, Yu Y, Fortner JD, et al. Effects of aqueous stable fullerene nanocrystal (nC60) on *Scenedesmus obliquus*: evaluation of the sub-lethal photosynthetic responses and inhibition mechanism. Chemosphere 2015;122:162–7.

[34] Wang H, Zhang M, Song Y, et al. Carbon dots promote the growth and photosynthesis of mung bean sprouts. Carbon 2018;136:94–102.

[35] Chandra S, Pradhan S, Mitra S, et al. High throughput electron transfer from carbon dots to chloroplast: a rationale of enhanced photosynthesis. Nanoscale 2014;6(7):3647–55.

[36] Verma SK, Das AK, Gantait S, et al. Applications of carbon nanomaterials in the plant system: a perspective view on the pros and cons. Sci Total Environ 2019;667:485–99.

[37] Xu X, Mao X, Zhuang J, Lei B, et al. PVA-coated fluorescent carbon dot nanocapsules as an optical amplifier for enhanced photosynthesis of lettuce. ACS Sustain Chem Eng 2020;8(9):3938–49.

[38] Middepogu A, Hou J, Gao X, Lin D. Effect and mechanism of TiO_2 nanoparticles on the photosynthesis of *Chlorella pyrenoidosa*. Ecotoxicol Environ Saf 2018;161:497–506.

[39] Rizwan M, Ali S, ur Rehman MZ, et al. Effect of foliar applications of silicon and titanium dioxide nanoparticles on growth, oxidative stress, and cadmium accumulation by rice (*Oryza sativa*). Acta Physiol Plant 2019;41:1–2.

[40] Yang KY, Doxey S, McLean JE, et al. Remodeling of root morphology by CuO and ZnO nanoparticles: effects on drought tolerance for plants colonized by a beneficial pseudomonad. Botany 2018;96:175–86.

[41] Rastogi A, Tripathi DK, Yadav S, et al. Application of silicon nanoparticles in agriculture. 3 Biotech 2019;9:90.

[42] Ali S, Rizwan M, Noureen S, et al. Combined use of biochar and zinc oxide nanoparticle foliar spray improved the plant growth and decreased the cadmium accumulation in rice (*Oryza sativa* L.) plant. Environ Sci Pollut Res 2019;11:11288−99.

[43] Mahto R, Chatterjee N, Priya T, Singh RK. Nanotechnology and its role in agronomic crops. Agronomic Crops. Singapore: Springer; 2019. p. 605−36.

[44] Suriyaprabha R, Karunakaran G, Yuvakkumar R, et al. Growth and physiological responses of maize (*Zea mays* L.) to porous silica nanoparticles in soil. J Nanopart Res 2012;14(12):1−4.

[45] Siddiqui MH, Al-Whaibi MH, Faisal M, Al Sahli AA. Nano-silicon dioxide mitigates the adverse effects of salt stress on *Cucurbita pepo* L. Environ Toxicol Chem 2014;33(11):2429−37.

[46] Tayemeh MB, Esmailbeigi M, Shirdel I, et al. Perturbation of fatty acid composition, pigments, and growth indices of *Chlorella vulgaris* in response to silver ions and nanoparticles: a new holistic understanding of hidden ecotoxicological aspect of pollutants. Chemosphere. 2020;238:124576.

[47] Zhao L, Zhang H, White JC, et al. Metabolomics reveals that engineered nanomaterial exposure in soil alters both soil rhizosphere metabolite profiles and maize metabolic pathways. Environ Sci Nano 2019;6(6):1716−27.

[48] Li Y, Li W, Zhang H, et al. Amplified light harvesting for enhancing Italian lettuce photosynthesis using water soluble silicon quantum dots as artificial antennas. Nanoscale. 2020;12(1):155−66.

[49] Ren HY, Dai YQ, Kong F, et al. Enhanced microalgal growth and lipid accumulation by addition of different nanoparticles under xenon lamp illumination. Bioresour Technol 2020;297:122409.

[50] Jullok N, Van Hooghten R, Luis P, et al. Effect of silica nanoparticles in mixed matrix membranes for pervaporation dehydration of acetic acid aqueous solution: plant-inspired dewatering systems. J Clean Prod 2016;112:4879−89.

[51] Haghighi M, Pessarakli M. Influence of silicon and nano-silicon on salinity tolerance of cherry tomatoes (*Solanum lycopersicum* L.) at early growth stage. Sci Hortic 2013;161:111−17.

[52] Thind S, Hussain I, Rasheed R, et al. Alleviation of cadmium stress by silicon nanoparticles during different phenological stages of Ujala wheat variety. Arab J Geosci 2021;14:1−5.

[53] Cui J, Liu T, Li F, et al. Silica nanoparticles alleviate cadmium toxicity in rice cells: mechanisms and size effects. Environ Pollut 2017;228:363−9.

[54] Fatemi H, Pour BE, Rizwan M. Isolation and characterization of lead (Pb) resistant microbes and their combined use with silicon nanoparticles improved the growth, photosynthesis and antioxidant capacity of coriander (*Coriandrum sativum* L.) under Pb stress. Environ Pollut 2020;266:114982.

[55] Roduner E. Size matters: why nanomaterials are different. Chem Soc Rev 2006;35:583−92.

[56] O'Farrell N, Houlton A, Horrocks BR. Silicon nanoparticles: applications in cell biology and medicine. Int J Nanomed 2006;1:451.

[57] Abdel-Haliem ME, Hegazy HS, Hassan NS, Naguib DM. Effect of silica ions and nano silica on rice plants under salinity stress. Ecol Eng 2017;99:282−9.

[58] Hodson MJ, Evans DE. Aluminium−silicon interactions in higher plants: an update. J Exp Bot 2020;71(21):6719−29.

[59] Epstein E, Bloom AJ. Mineral nutrition of plants: principles and perspectives. Sunderland: Sinauer Associates; 2005.

[60] Debona D, Rodrigues FA, Datnoff LE. Silicon's role in abiotic and biotic plant stresses. Annu Rev Phytopathol 2017;55:85−107.

[61] Zhu Y, Gong H. Beneficial effects of silicon on salt and drought tolerance in plants. Agron Sustain Dev NZ J Crop Hortic Sci 2014;17:455−72.

[62] Van Bockhaven J, De Vleesschauwer D, Höfte M. Silicon-mediated priming results in broad spectrum resistance in rice (*Oryza sativa* L.). In: Proceedings of the abstract retrieved from abstracts of the 64th international symposium on crop protection, 2012: p. 62.

[63] Pavlovic J, Kostic L, Bosnic P, et al. Interactions of silicon with essential and beneficial elements in plants. Front Plant Sci 2021;12:1224.

[64] Noronha H, Silva A, Mitani-Ueno N, et al. The VvNIP2; 1 aquaporin is a grapevine silicon channel. J Exp Bot 2020;71(21):6789−98.

[65] Moore KL, Chen Y, van de Meene AM, et al. Combined NanoSIMS and synchrotron X-ray fluorescence reveal distinct cellular and subcellular distribution patterns of trace elements in rice tissues. New Phytol 2014;201:104−15.

[66] Réthoré E, Ali N, Yvin JC, Hosseini SA. Silicon regulates source to sink metabolic homeostasis and promotes growth of rice plants under sulfur deficiency. J Mol Sci 2020;21(10):3677.

[67] Che J, Yamaji N, Shao JF, et al. Silicon decreases both uptake and root-to-shoot translocation of manganese in rice. J Exp Bot 2016;67(5):1535−44.

[68] Ranjan A, Sinha R, Bala M, Pareek A, et al. Silicon-mediated abiotic and biotic stress mitigation in plants: underlying mechanisms and potential for stress resilient agriculture. Plant Physiol Biochem 2021;163:15−25.

[69] Ma JF, Takahashi E. Functions of silicon in plant growth. Soil, fertilizer, and plant silicon research in Japan. Amsterdam: Elsevier; 2002. p. 107−80.

[70] Tripathi DK, Singh VP, Kumar D, Chauhan DK. Impact of exogenous silicon addition on chromium uptake, growth, mineral elements, oxidative stress, antioxidant capacity, and leaf and root structures in rice seedlings exposed to hexavalent chromium. Acta Physiol Plant 2012;34:279−89.

[71] Fauteux F, Rémus-Borel W, Menzies JG, Bélanger RR. Silicon and plant disease resistance against pathogenic fungi. FEMS Microbiol Lett 2005;249:1−6.

[72] Arnon DI, Stout PR. The essentiality of certain elements in minute quantity for plants with special reference to copper. Plant Physiol 1939;14:37.

[73] Thind S, Hussain I, Ali S, Hussain S, Rasheed R, Ali B, et al. Physiological and biochemical bases of foliar silicon-induced alleviation of cadmium toxicity in wheat. J Soil Sci Plant Nutr 2020;20(4):2714−30.

[74] Smyth TJ, Sanchez PA. Effects of lime, silicate, and phosphorus applications to an Oxisol on phosphorus sorption and ion retention. Soil Sci Soc Am J 1980;44:500−5.

[75] Meena VD, Dotaniya ML, Coumar V, et al. A case for silicon fertilization to improve crop yields in tropical soils. Proc Natl Acad Sci India Sec B Biol Sci 2014;84:505−18.

[76] Matichenkov VV, Bocharnikova EA. The relationship between silicon and soil physical and chemical properties. Stud Plant Sci 2001;8:209−19.

[77] Wada SI, Wada K. Formation, composition and structure of hydroxy-aluminosilicate ions. J Soil Sci 1980;31:457−67.

[78] Cocker KM, Evans DE, Hodson MJ. The amelioration of aluminium toxicity by silicon in higher plants: solution chemistry or an in planta mechanism? Physiol Plant 1998;104:608−14.

[79] Coskun D, Deshmukh R, Sonah H, et al. The controversies of silicon's role in plant biology. New Phytol 2019;221(1):67−85.

[80] Schaller J, Puppe D, Kaczorek D, et al. Silicon cycling in soils revisited. Plants. 2021;10(2):295.

[81] Hoffmann J, Berni R, Hausman JF, Guerriero G. A review on the beneficial role of silicon against salinity in non-accumulator crops: tomato as a model. Biomolecules 2020;10(9):1284.

[82] Sun H, Duan Y, Mitani-Ueno N, et al. Tomato roots have a functional silicon influx transporter but not a functional silicon efflux transporter. Plant Cell Environ 2020;43(3):732−44.

[83] Asher CJ. Beneficial elements, functional nutrients, and possible new essential elements. Micronutr Agric 1991;4:703−23.

[84] Rastogi A, Zivcak M, Sytar O, et al. Impact of metal and metal oxide nanoparticles on plant: a critical review. Front Chem 2017;12(5):78.

[85] Bao-Shan L, Chun-hui L, Li-jun F, et al. Effect of TMS (nanostructured silicon dioxide) on growth of *Changbai larch* seedlings. J Res 2004;15:138−40.

[86] Strout G, Russell SD, Pulsifer DP, et al. Silica nanoparticles aid in structural leaf coloration in the Malaysian tropical rainforest understorey herb *Mapania caudata*. Ann Bot 2013;112:1141−8.

[87] Suriyaprabha R, Karunakaran G, Yuvakkumar R, et al. Foliar application of silica nanoparticles on the phytochemical responses of maize (*Zea mays* L.) and its toxicological behavior. Synth React Inorg Met-Org Nano-Met Chem 2014;44:1128−31.

[88] Slomberg DL, Schoenfisch MH. Silica nanoparticle phytotoxicity to *Arabidopsis thaliana*. Environ Sci Technol 2012;46:10247−54.

[89] Li P, Song A, Li Z, Fan F, Liang Y. Silicon ameliorates manganese toxicity by regulating manganese transport and antioxidant reactions in rice (*Oryza sativa* L.). Plant Soil 2012;354:407−19.

[90] Bélanger RR, Bowen PA, Ehret DL, Menzies JG. Greenhouse crops. Plant Dis 1995;79:329.

[91] Bendz G, editor. Biochemistry of silicon and related problems. Berlin: Springer Science & Business Media; 2013.

[92] Ahmad R, Zaheer SH, Ismail S. Role of silicon in salt tolerance of wheat (*Triticum aestivum* L.). Plant Sci 1992;85:43−50.

[93] Bennett WF. Nutrient deficiencies and toxicities in crop plants. St Paul, MN: The American Phytopathological Society; 1993.

[94] He C, Ma J, Wang L. A hemicellulos bound form of silicon with potential to improve the mechanical properties and regeneration of the cell wall of rice. New Phytol 2015;206:1051−62.

[95] Fry SC, Nesselrode BH, Miller JG, Mewburn BR. Mixed-linkage (1→3, 1→4)-β-d-glucan is a major hemicellulose of Equisetum (horsetail) cell walls. New Phytol 2008;179:104−15.

[96] Kido N, Yokoyama R, Yamamoto T, et al. The matrix polysaccharide (1; 3, 1; 4)-β-D-glucan is involved in silicon-dependent strengthening of rice cell wall. Plant Cell Physiol 2015;56:268−76.

[97] Brugiére T, Exley C. Callose-associated silica deposition in Arabidopsis. J Trace Elem Med Biol 2017;39:86−90.

[98] Debona D, Rodrigues FA, Datnoff LE. Silicon's role in abiotic and biotic plant stresses. Annu Rev Phytopatho 2017;55:85−107.

[99] Savant NK, Snyder GH, Datnoff LE. Silicon management and sustainable rice production. Adv Agron 1996;58:151−99.

[100] Keeping MG, Kvedaras OL, Bruton AG. Epidermal silicon in sugarcane: cultivar differences and role in resistance to sugarcane borer *Eldana saccharina*. Environ Exp Bot 2009;66:54−60.

[101] Rémus-Borel W, Menzies JG, Bélanger RR. Silicon induces antifungal compounds in powdery mildew-infected wheat. Physiol Mol Plant 2005;66:108−15.

[102] Rahman A, Wallis CM, Uddin W. Silicon-induced systemic defense responses in perennial ryegrass against infection by *Magnaporthe oryzae*. Phytopathology. 2015;105:748−57.

[103] Cai K, Gao D, Luo S, et al. Physiological and cytological mechanisms of silicon-induced resistance in rice against blast disease. Physiol Plant 2008;134(2):324−33.

[104] Reynolds OL, Padula MP, Zeng R, Gurr GM. Silicon: potential to promote direct and indirect effects on plant defense against arthropod pests in agriculture. Front Plant Sci 2016;7:744.

[105] James DG. Field evaluation of herbivore-induced plant volatiles as attractants for beneficial insects: methyl salicylate and the green lacewing, *Chrysopa nigricornis*. J Chem Ecol 2003;29:1601−9.

[106] Connick VJ. The impact of silicon fertilisation on the chemical ecology of the grapevine, *Vitis vinifera*; constitutive and induced chemical defenses against arthropod pests and their natural enemies [Doctoral dissertation]. Charles Sturt University; 2011.

[107] Tahir MA, Rahmatullah T, Aziz M, Ashraf S, Kanwal S, Maqsood MA. Beneficial effects of silicon in wheat (*Triticum aestivum* L.) under salinity stress. Pak J Bot 2006;38:1715−22.

[108] Liang SJ, Li ZQ, Li XJ. Effects of stem structural characters and silicon content on lodging resistance in rice (*Oryza sativa* L.). Res Crop 2013;14:621−36.

[109] Goto M, Ehara H, Karita S. Protective effect of silicon on phenolic biosynthesis and ultraviolet spectral stress in rice crop. Plant Sci 2003;164:349−56.

[110] Shen X, Zhou Y, Duan L, Li Z, Eneji AE, Li J. Silicon effects on photosynthesis and antioxidant parameters of soybean seedlings under drought and ultraviolet-B radiation. J Plant Physiol 2010;167:1248−52.

[111] Ma JF, Takahashi E. Soil, fertilizer, and plant silicon research in Japan. Amsterdam; Boston: Elsevier; 2002.

[112] Gong HJ, Chen KM, Chen GC, Wang SM, Zhang CL. Effects of silicon on growth of wheat under drought. J Plant Nutr 2003;26:1055−63.

[113] Hattori T, Inanaga S, Tanimoto E, et al. Silicon-induced changes in viscoelastic properties of sorghum root cell walls. Plant Cell Physiol 2003;44:743−9.

[114] Liu P, Yin L, Wang S, et al. Enhanced root hydraulic conductance by aquaporin regulation accounts for silicon alleviated salt-induced osmotic stress in *Sorghum bicolor* L. Environ Exp Bot 2015;111:42−51.

[115] Agarie S, Hanaoka N, Ueno O, et al. Effects of silicon on tolerance to water deficit and heat stress in rice plants (*Oryza sativa* L.), monitored by electrolyte leakage. Plant Prod Sci 1998;1:96−103.

[116] Shi Y, Wang Y, Flowers TJ, Gong H. Silicon decreases chloride transport in rice (*Oryza sativa* L.) in saline conditions. J Plant Physiol 2013;170:847−53.

[117] Savvas D, Ntatsi G. Biostimulant activity of silicon in horticulture. Sci Horticult 2015;196:66—81.

[118] Yeo AR, Flowers SA, Rao G, et al. Silicon reduces sodium uptake in rice (*Oryza sativa* L.) in saline conditions and this is accounted for by a reduction in the transpirational bypass flow. Plant Cell Environ 1999;22:559—65.

[119] Xu CX, Ma YP, Liu YL. Effects of silicon (Si) on growth, quality and ionic homeostasis of aloe under salt stress. S Afr J Bot 2015;98:26—36.

[120] Jia-Wen WU, Yu SH, et al. Mechanisms of enhanced heavy metal tolerance in plants by silicon: a review. Pedosphere 2013;23:815—25.

[121] Gu HH, Qiu H, Tian T, et al. Mitigation effects of silicon rich amendments on heavy metal accumulation in rice (*Oryza sativa* L.) planted on multi-metal contaminated acidic soil. Chemosphere. 2011;83:1234—40.

[122] Kidd PS, Llugany M, Poschenrieder CH, et al. The role of root exudates in aluminium resistance and silicon-induced amelioration of aluminium toxicity in three varieties of maize (*Zea mays* L.). J Exp Bot 2001;52:1339—52.

[123] Feng J, Shi Q, Wang X, et al. Silicon supplementation ameliorated the inhibition of photosynthesis and nitrate metabolism by cadmium (Cd) toxicity in *Cucumis sativus* L. Sci Hortic 2010;123(4):521—30.

[124] Kopittke PM, Gianoncelli A, Kourousias G, et al. Alleviation of Al toxicity by Si is associated with the formation of Al—Si complexes in root tissues of sorghum. Front Plant Sci 2017;8:2189.

[125] Wang Y, Stass A, Horst WJ. Apoplastic binding of aluminum is involved in silicon-induced amelioration of aluminum toxicity in maize. Plant Physiol 2004;136:3762—70.

[126] Collin B, Doelsch E, Keller CC, et al. Evidence of sulfur-bound reduced copper in bamboo exposed to high silicon and copper concentrations. Environ Pollut 2014;187:22—30.

[127] He C, Wang L, Liu J, et al. Evidence for 'silicon' within the cell walls of suspension-cultured rice cells. New Phytol 2013;200:700—9.

[128] Ma J, Cai H, He C, et al. A hemicellulose-bound form of silicon inhibits cadmium ion uptake in rice (*Oryza sativa*) cells. New Phytol 2015;206:1063—74.

[129] Kim YH, Khan AL, Kim DH, et al. Silicon mitigates heavy metal stress by regulating P-type heavy metal ATPases, *Oryza sativa* low silicon genes, and endogenous phytohormones. BMC Plant Bio 2014;14:1—3.

[130] Adrees M, Ali S, Rizwan M, Zia-ur-Rehman M, et al. Mechanisms of silicon-mediated alleviation of heavy metal toxicity in plants: a review. Ecotoxicol Environ Saf 2015;119:186—97.

[131] Imtiaz M, Rizwan MS, Mushtaq MA, et al. Silicon occurrence, uptake, transport and mechanisms of heavy metals, minerals and salinity enhanced tolerance in plants with future prospects: a review. J Environ Manage 2016;183:521—9.

[132] Zhang W, Liu D, Li C, Cui Z, Chen X, Russell Y, et al. Zinc accumulation and remobilization in winter wheat as affected by phosphorus application. Field Crop Res 2015;184:155—61.

[133] Nikolic M, Nikolic N, Kostic L, et al. The assessment of soil availability and wheat grain status of zinc and iron in Serbia: implications for human nutrition. Sci Total Environ 2016;553:141—8.

[134] Hu AY, Che J, Shao JF, et al. Silicon accumulated in the shoots results in down-regulation of phosphorus transporter gene expression and decrease of phosphorus uptake in rice. Plant Soil 2018;423(1):317—25.

[135] Soratto RP, Fernandes AM, Pilon C, Souza MR. Phosphorus and silicon effects on growth, yield, and phosphorus forms in potato plants. J Plant Nutr 2019;42(3):218—33.

[136] Bityutskii N, Pavlovic J, Yakkonen K, et al. Contrasting effect of silicon on iron, zinc and manganese status and accumulation of metal-mobilizing compounds in micronutrient-deficient cucumber. Plant Physiol Biochem 2014;74:205—11.

[137] de Oliveira RL, de Mello Prado R, Felisberto G, et al. Silicon mitigates manganese deficiency stress by regulating the physiology and activity of antioxidant enzymes in sorghum plants. J Plant Nutr Soil Sci 2019;19(3):524—34.

[138] Lesharadevi K, Parthasarathi T, Muneer S. Silicon biology in crops under abiotic stress: a paradigm shift and cross-talk between genomics and proteomics. J Biotechnol 2021;333:21—38.

[139] Ulrichs C, Mewis I, Goswami A. Crop diversification aiming nutritional security in West Bengal: biotechnology of stinging capsules in nature's water-blooms. Ann Tech Issue State Agric Technol Serv Assoc 2005;1—18.

[140] Rai M, Ingle A. Role of nanotechnology in agriculture with special reference to management of insect pests. Appl Microbiol Biotechnol 2012;94:287—93.

[141] Chen J, Wang W, Xu Y, Zhang X. Slow-release formulation of a new biological pesticide, pyoluteorin, with mesoporous silica. J Agric Food Chem 2011;59:307—11.

[142] Sun D, Hussain HI, Yi Z, et al. Uptake and cellular distribution, in four plant species, of fluorescently labeled mesoporous silica nanoparticles. Plant Cell Rep 2014;33:1389—402.

[143] Sun D, Hussain HI, Yi Z, et al. Mesoporous silica nanoparticles enhance seedling growth and photosynthesis in wheat and lupin. Chemosphere. 2016;152:81—91.

[144] Putri FM, Suedy SWA, Darmanti S. The effect of nanosilica fertilizer on number of stomata, chlorophyll content, and growth of black rice (*Oryza sativa* L. cv. Japonica). Bull Anat dan Fisiologi 2017;2(1):72—9.

[145] Tripathi DK, Singh S, Singh VP, et al. Silicon nanoparticles more effectively alleviated UV-B stress than silicon in wheat (*Triticum aestivum*) seedlings. Plant Physiol Biochem 2017;110:70—81.

[146] Ivani R, Sanaei Nejad SH, et al. Role of bulk and Nanosized SiO$_2$ to overcome salt stress during Fenugreek germination (*Trigonella foenum-graceum* L.). Plant Signal Behav 2018;13:1044190.

[147] Yassen A, Abdallah E, Gaballah M, Zaghloul S. Role of silicon dioxide nano fertilizer in mitigating salt stress on growth, yield and chemical composition of cucumber (*Cucumis sativus* L.). Int J Agric Res 2017;12:130—5.

[148] Siddiqui MH, Al-Whaibi MH. Role of nano-SiO2 in germination of tomato (*Lycopersicum esculentum* seeds Mill.). Saudi J Biol Sci 2014;21:13—17.

[149] Rui Y, Gui X, Li X, et al. Uptake, transport, distribution and bio-effects of SiO2 nanoparticles in Bt-transgenic cotton. J Nanobiotechnol 2014;12:1—5.

[150] Faceto LF, Grillo R, Gerson Ade Medeiros, et al. Nanotechnology in agriculture: Which innovation potential does it have? Front Environ Sci 2016;4:20.

[151] Roohizadeh G, Majd A, Arbabian S. The effect of sodium silicate and silica nanoparticles on seed germination and growth in the *Vicia faba* L. Trop Plant Res 2015;2:85−9.

[152] Sabaghnia N, Janmohammadi M. Effect of nano-silicon particles application on salinity tolerance in early growth of some lentil genotypes. Biologia. 2015;69:39−55.

[153] Ashkavand P, Tabari M, Zarafshar M, et al. Effect of SiO$_2$ nanoparticles on drought resistance in hawthorn seedlings. For Res Pap 2015;76:350−9.

[154] Haghighi M, Afifipour Z, Mozafarian M. The effect of N-Si on tomato seed germination under salinity levels. J Biol Environ Sci 2012;6:87−90.

[155] Kalteh M, Alipour ZT, Ashraf S, et al. Effect of silica nanoparticles on basil (*Ocimum basilicum*) under salinity stress. J Chem Health Risks 2018;4.

[156] Janmohammadi M, Amanzadeh T, Sabaghnia N, Ion V. Effect of nano-silicon foliar application on safflower growth under organic and inorganic fertilizer regimes. Botanica 2016;22:53−64.

[157] Ciecierski W, Kardasz H. Impact of silicon based fertilizer Optysil on abiotic stress reduction and yield improvement in field crops. In: Proceedings of the 6th international conference on silicon in agriculture, Stockholm, Sweden; 2014; p. 54−55.

[158] Prakash NB, Chandrashekar N, Mahendra C, et al. Effect of foliar spray of soluble silicic acid on growth and yield parameters of wetland rice in hilly and coastal zone soils of Karnataka, South India. J Plant Nutr 2011;34(12):1883−93.

[159] Laane HM. The effects of the application of foliar sprays with stabilized silicic acid: an overview of the results from 2003-2014. Silicon. 2017;9(6):803−7.

[160] du Jardin P. The science of plant bio stimulants—a bibliographic analysis, Ad hoc study report. European Commission; 2012.

[161] Matichenkov VV, Fomina IR, Biel KY. Protective role of silicon in living organisms. Complex biological systems: adaptation and tolerance to extreme environments. Beverly: Scrivener Publishing LLC; 2018. p. 175.

[162] Passala R, Jain N, Deokate PP, Rao V, Minhas VS. Assessment of silixol (OSA) efficacy on wheat physiology: growth and nutrient content under drought conditions. In: 6th International conference on silicon in agriculture. Stockholm (Sweden); 2014; p. 26−30.

[163] Wanyika H, Gatebe E, Kioni P, et al. Mesoporous silica nanoparticles carrier for urea: potential applications in agrochemical delivery systems. J Nanosci Nanotechnol 2012;12:2221−8.

[164] Fitiyani HP, dan Haryanti S. The effect of using nano-silica fertilizer on growth tomato plant (*Solanum lycopersicum*) var. Round. Bull Anat dan Fisiologi 2016;24:34−41.

[165] Suciaty T, Purnomo D, Sakya AT. The effect of nano-silica fertilizer concentration and rice hull ash doses on soybean (*Glycine max* (L.) Merrill) growth and yield. IOP Conf Ser Earth Environ Sci 2018;129:012009.

[166] Wanyika H, Gatebe E, Kioni P, et al. Mesoporous silica nanoparticles carrier for urea: potential applications in agrochemical delivery systems. J Nanosci Nanotechnol 2012;12:2221−8.

[167] Singh S. Nanomaterials as non-viral siRNA delivery agents for cancer therapy. Bioimpacts 2013;3(2):53.

[168] Ochatt S. Plant cell electrophysiology: applications in growth enhancement, somatic hybridization and gene transfer. Biotechnol Adv 2013;31(8):1237−46.

[169] Taylor NJ, Fauquet CM. Microparticle bombardment as a tool in plant science and agricultural biotechnology. DNA Cell Biol 2002;21(12):963−77.

[170] Torney F, Trewyn BG, Lin VS, Wang K. Mesoporous silica nanoparticles deliver DNA and chemicals into plants. Nat Nanotechnol 2007;2:295−300.

[171] Nair R, Varghese SH, Nair BG, et al. Nanoparticulate material delivery to plants. Plant Sci 2010;179:154−63.

[172] Tsuji K. Microencapsulation of pesticides and their improved handling safety. J Microencapsul 2001;18:137−47.

[173] Boehm AL, Martinon I, Zerrouk R, et al. Nanoprecipitation technique for the encapsulation of agrochemical active ingredients. J Microencapsul 2003;20:433−41.

[174] Xia T, Kovochich M, Liong M, et al. Polyethyleneimine coating enhances the cellular uptake of mesoporous silica nanoparticles and allows safe delivery of siRNA and DNA constructs. ACS Nano 2009;3:3273−86.

[175] Martin-Ortigosa S, Peterson DJ, Valenstein JS, et al. Mesoporous silica nanoparticle-mediated intracellular Cre protein delivery for maize genome editing via loxP site excision. Plant Physiol 2014;164:537−47.

[176] Martin-Ortigosa S, Valenstein JS, Lin VS, et al. Gold functionalized mesoporous silica nanoparticle mediated protein and DNA codelivery to plant cells via the biolistic method. Adv Funct Mater 2012;22(17):3576−82.

[177] White PJ. The pathways of calcium movement to the xylem. J Exp Bot 2001;52:891−9.

[178] Gong HJ, Randall DP, Flowers TJ. Silicon deposition in the root reduces sodium uptake in rice (*Oryza sativa* L.) seedlings by reducing bypass flow. Plant Cell Environ 2006;29:1970−9.

[179] Prychid CJ, Rudall PJ, Gregory M. Systematics and biology of silica bodies in monocotyledons. Botanical Rev 2003;69:377−440.

[180] Emanuel E. Silicon. Annu Rev Plant Physiol 1999;50:641−64.

[181] Hodson MJ, White PJ, Mead A, Broadley MR. Phylogenetic variation in the silicon composition of plants. Ann Bot 2005;96:1027−46.

[182] Singh AK, Singh R, Singh K. Growth, yield and economics of rice (*Oryza sativa*) as influenced by level and time of silicon application. Indian J Agron 2005;50:190−3.

[183] Gonzalo MJ, Lucena JJ, Hernández-Apaolaza L. Effect of silicon addition on soybean (*Glycine max*) and cucumber (*Cucumis sativus*) plants grown under iron deficiency. Plant Phsyio Biochem 2013;70:455−61.

[184] Guerinot ML. Improving rice yields—ironing out the details. Nat biotech 2001;19:417−18.

[185] Römheld V, Marschner H. Evidence for a specific uptake system for iron phytosiderophores in roots of grasses. Plant Physiol 1986;80:175−80.

[186] Lindsay WL. Chemical equilibria in soils. New York: John Wiley and Sons Ltd.; 1979.

[187] Dos Santos MS, Sanglard LM, Martins SC, et al. Silicon alleviates the impairments of iron toxicity on the rice photosynthetic performance via alterations in leaf diffusive conductance with minimal impacts on carbon metabolism. Plant Physiol Biochem 2019;143:275−85.

[188] Becker M, Ngo NS, Schenk MK. Silicon reduces the iron uptake in rice and induces iron homeostasis related genes. Sci Rep 2020;10(1):5079.

[189] Hernández-Apaolaza L, Escribano L, Zamarreño ÁM, et al. Root silicon addition induces Fe deficiency in cucumber plants, but facilitates their recovery after Fe resupply. A comparison with Si foliar sprays. Front Plant Sci 2020;11:1851.

[190] Gottardi S, Iacuzzo F, Tomasi N, et al. Beneficial effects of silicon on hydroponically grown corn salad (*Valerianella locusta* (L.) Laterr) plants. Plant Physiol Biochem 2012;56:14—23.

[191] You-Qiang FU, Hong SH, Dao-Ming WU, Kun-Zheng CA. Silicon-mediated amelioration of Fe^{2+} toxicity in rice (*Oryza sativa* L.) roots. Pedosphere 2012;22:795—802.

[192] Pavlovic J, Samardzic J, Maksimović V, et al. Silicon alleviates iron deficiency in cucumber by promoting mobilization of iron in the root apoplast. New Phytol 2013;198:1096—107.

[193] Stevic N, Korac J, Pavlovic J, Nikolic M. Binding of transition metals to monosilicic acid in aqueous and xylem (*Cucumis sativus* L.) solutions: a low-T electron paramagnetic resonance study. Biometals. 2016;29(5):945—51.

[194] Nikolic DB, Nesic S, Bosnic D, et al. Silicon alleviates iron deficiency in barley by enhancing expression of Strategy II genes and metal redistribution. Front Plant Sci 2019;10:416.

[195] Pedas P, Schjoerring JK, Husted S. Identification and characterization of zinc-starvation-induced ZIP transporters from barley roots. Plant Phsyio Biochem 2009;47:377—83.

[196] Zhao ZQ, Zhu YG, Kneer R, Smith SE. Effect of zinc on cadmium toxicity-induced oxidative stress in winter wheat seedlings. J Plant Nutr 2005;28:1947—59.

[197] Sumner ME, editor. Handbook of soil science. Boca Raton, FL: CRC Press; 1999.

[198] Garcia-Gomez C, Garcia S, Obrador AF, Gonzalez D, Babin M, Fernandez MD. Effects of aged ZnO NPs and soil type on Zn availability, accumulation and toxicity to pea and beet in a greenhouse experiment. Ecotoxicol Environ Saf 2018;160:222—30.

[199] Broadley MR, White PJ, Hammond JP, et al. Zinc in plants. New Phytol 2007;173:677—702.

[200] Cakmak I. Plant nutrition research: priorities to meet human needs for food in sustainable ways. Plant Soil 2002;247:3—24.

[201] Graham RD, Welch RM, Bouis HE. Addressing micronutrient malnutrition through enhancing the nutritional quality of staple foods: principles, perspectives and knowledge gaps. Adv Agron 2001;70:77—142.

[202] Hafeez FY, Abaid-Ullah M, Hassan MN. Plant growth-promoting rhizobacteria as zinc mobilizers: a promising approach for cereals biofortification. Bacteria agrobiology: Crop productivity. New York: Springer; 2013. p. 217—35.

[203] Mehrabanjoubani P, Abdolzadeh A, Sadeghipour HR, Aghdasi M. Impacts of silicon nutrition on growth and nutrient status of rice plants grown under varying zinc regimes. Theor Exp Plant Physiol 2015;27:19—29.

[204] Gu HH, Zhan SS, Wang SZ, et al. Silicon-mediated amelioration of zinc toxicity in rice (*Oryza sativa* L.) seedlings. Plant Soil 2012;350:193—204.

[205] Hernandez-Apaolaza L. Can silicon partially alleviate micronutrient deficiency in plants? A review. Planta. 2014;240:447—58.

[206] Huang S, Ma JF. Silicon suppresses zinc uptake through down-regulating zinc transporter gene in rice. Physiol Plant 2020;170(4):580—91.

[207] Siegel A, Siegel H. Metal ions in biological systems: manganese and its role in biological processes. New York: Marcel Dekker; 2000. p. 37.

[208] Gerber GB, Leonard A, Hantson PH. Carcinogenicity, mutagenicity and teratogenicity of manganese compounds. Crit Rev Oncol Hematol 2002;42:25—34.

[209] Howe P, Malcolm H, Dobson S. Manganese and its compounds: environmental aspects. Geneva: World Health Organization; 2004.

[210] Moore JW. Inorganic contaminants of surface water: research and monitoring priorities. New York: Springer; 2012.

[211] Husted S, Thomsen MU, Mattsson M, Schjoerring JK. Influence of nitrogen and sulphur form on manganese acquisition by barley (*Hordeum vulgare*). Plant Soil 2005;268:309—17.

[212] Papadakis IE, Giannakoula A, Therios IN, et al. Mn-induced changes in leaf structure and chloroplast ultrastructure of *Citrus volkameriana* (L.) plants. J Plant Physiol 2007;164:100—3.

[213] Dragišić Maksimović J, Bogdanović J, et al. Silicon modulates the metabolism and utilization of phenolic compounds in cucumber (*Cucumis sativus* L.) grown at excess manganese. J Plant Nutr Soil Sci 2007;170:739—44.

[214] Williams DE, Vlamis J. The effect of silicon on yield and manganese-54 uptake and distribution in the leaves of barley plants grown in culture solutions. Plant Physiol 1957;32:404.

[215] Shi X, Zhang C, Wang H, Zhang F. Effect of Si on the distribution of Cd in rice seedlings. Plant Soil 2005;272:53—60.

[216] Agata F, Ernest B. *Meta*-metal interactions in accumulation of V^{5+}, Ni^{2+}, Mo^{6+}, Mn^{2+} and Cu^{2+} in under and above ground parts of *Sinapis alba*. Chemosphere 1998;36:1305—17.

[217] Oliva SR, Mingorance MD, Leidi EO. Effects of silicon on copper toxicity in Erica and evalensis Cabezudo and Rivera: a potential species to remediate contaminated soils. J Environ Monit 2011;13:591—6.

[218] Khandekar S, Leisner S. Soluble silicon modulates expression of *Arabidopsis thaliana* genes involved in copper stress. J Plant Physiol 2011;168:699—705.

[219] Fleck AT, Nye T, Repenning C, et al. Silicon enhances suberization and lignification in roots of rice (*Oryza sativa*). J Exp Bot 2011;62:2001—11.

[220] Balmer A, Pastor V, Gamir J, et al. The 'prime-ome': towards a holistic approach to priming. Trends Plant Sci 2015;20:443—52.

[221] Ma JF, Yamaji N. Functions and transport of silicon in plants. Cell Mol life Sci 2008;65:3049—57.

[222] Wiese H, Nikolic M, Römheld V. Silicon in plant nutrition. The apoplast of higher plants: compartment of storage, transport and reactions. New York: Springer; 2007. p. 33—47.

[223] Mitani N, Yamaji N, Ago Y, et al. Isolation and functional characterization of an influx silicon transporter in two pumpkin cultivars contrasting in silicon accumulation. Plant J 2011;66:231—40.

[224] Mitani N, Yamaji N, Ma JF. Identification of maize silicon influx transporters. Plant Cell Physio 2009;50:5—12.

[225] Liu X, Zhang N, Bing T, Shangguan D. Carbon dots based dual-emission silica nanoparticles as a ratio metric nanosensor for Cu^{2+}. Anal Chem 2014;86:2289—96.

[226] Liang Y, Hua H, Zhu YG, et al. Importance of plant species and external silicon concentration to active silicon uptake and transport. New Phytol 2006;172:63–72.

[227] Wang SY, Galletta GJ. Foliar application of potassium silicate induces metabolic changes in strawberry plants. J Plant Nutr 1998;2:157–67.

[228] Gong H, Zhu X, Chen K, et al. Silicon alleviates oxidative damage of wheat plants in pots under drought. Plant Sci 2005;169:313–21.

[229] Yu L, Gao J. Effects of silicon on yield and grain quality of wheat. J Triticeae Crop 2012;32:469–73.

[230] Zhang GL, Dai QG, Wang JW, et al. Effects of silicon fertilizer rate on yield and quality of japonica rice Wuyujing 3. Chin J Rice Sci 2007;21:299–303.

[231] Guntzer F, Keller C, Meunier JD. Benefits of plant silicon for crops: a review. Agron Sustain Dev 2012;32:201–13.

[232] Ashraf M, Ahmad R, Afzal M, et al. Potassium and silicon improve yield and juice quality in sugarcane (Saccharum officinarum L.) under salt stress. J Agron Crop Sci 2009;195:284–91.

[233] Gaofeng X, Guilong Z, Yanxin S, et al. Influences of spraying two different forms of silicon on plant growth and quality of tomato in solar greenhouse. Chin Agri Sci Bull 2012;16:272–6.

[234] Babini E, Marconi S, Cozzolino S, et al. Bio-available silicon fertilization effects on strawberry shelf-life. Int Symp Postharv Technol Glob Mark 2012;934:815–18.

[235] Su XW, Wei SC, Jiang YM, Huang YY. Effects of silicon on quality of apple fruit and Mn content in plants on acid soils. Shandog Aric Sci 2011;6:018.

[236] Shi G, Cai Q, Liu C, Wu L. Silicon alleviates cadmium toxicity in peanut plants in relation to cadmium distribution and stimulation of antioxidative enzymes. Plant Growth Regul 2010;61:45–52.

[237] Wang XS, Han JG. Effects of NaCl and silicon onion distribution in the roots, shoots and leaves of two alfalfa cultivars with different salt tolerance. Soil Sci Plant Nutr 2007;53:278–85.

[238] Datnoff LE. Silicon in the life and performance of turfgrass. Appl Turfgrass Sci 2005;2:1–6.

[239] Hossain KZ, Monreal CM, Sayari A. Adsorption of urease on PE-MCM-41 and its catalytic effect on hydrolysis of urea. Colloids Surf B Biointerfaces 2008;62:42–50.

[240] Liu F, Wen LX, Li ZZ, Yu W, Sun HY, Chen JF. Porous hollow silica nanoparticles as controlled delivery system for water-soluble pesticide. Mater Res Bull 2006;41(12):2268–75.

[241] Naderi MR, Danesh-Shahraki A. Nanofertilizers and their roles in sustainable agriculture. Int J Agric Crop Sci 2013;5(19):2229–32.

[242] Vishwakarma K, Upadhyay N, Kumar N, et al. Potential applications and avenues of nanotechnology in sustainable agriculture. Nanomaterials in plants, algae, microorganisms. London: Academic Press; 2018. p. 473–500.

[243] Crooks R, Prentice P. Extensive investigation into field based responses to a silica fertiliser. Silicon. 2017;9(2):301–4.

21

Effect of silicon and nanosilicon application on rice yield and quality

Norollah Kheyri

Department of Agronomy, Gorgan Branch, Islamic Azad University, Gorgan, Iran

21.1 Introduction

Rice (*Oryza sativa* L.) is the staple food of about 3 billion people across the world, predominantly in Asia [1]. Silicon (Si) is the second most abundant element in soil (27.7% of total soil weight) after oxygen (47%) [2,3]. Plants generally take up Si in the form of monosilicic acid (H_4SiO_4) from the soil solution, which is about $150-300$ kg Si ha^{-1} in rice [4]. Rice takes up Si through an active process [5]. Si is mainly found in the aerial part and in the epidermis of leaf blade, sclerenchyma cell, vascular tissues, and vascular sheath. In addition, Si is present in the outer epidermis, along the cell wall, and root tissues of the plant, and is more abundant in older leaves than younger leaves [6]. Among the crop plants of the Poaceae family, rice has the highest capability to take up Si [7]. Rice is a typical Si-accumulator and the Si content in the shoot is more than 10% of the plant dry matter, which is higher than the content of essential macronutrients such as nitrogen (N), phosphorus (P), and potassium (K) [8]. It seems that the Si uptake by rice begins after the tillering stage or after the stem elongation [9]. Si uptake varies in different rice cultivars as well as in different plant organs [10,11]. Although Si has not been defined as an essential element in plant nutrition [12], it has positive impacts on plants physiology, especially on the rice plant [13]. Si has a dual impact on the soil-plant system, so that it improves the Si plant nutrition, plant resistance to pests, diseases, and environmental stresses, and improves soil fertility by improving water, physicochemical properties of soil and preserving nutrients in available form for the plant [14]. Si plays an important role in the metabolism, physiological and structural activities of rice plant by improving light interception, increasing photosynthesis rate, improving grain yield (GY), and alleviating damages caused by biotic (pests and diseases) and abiotic (salinity, drought, heat, chilling, ultraviolet (UV) radiation, and heavy metals [HMs] toxicity) stresses [15–21]. Application of Si increases rice GY by improving the agronomic parameters such as fertile tillers number per hill and panicle fertility [22–24]. It has been reported that the application of Si at the reproductive growth stage of rice has positive effects on growth, photosynthesis, and rice performance [25]. Rice grain fortification through the application of various sources of Si fertilizers helps to improve human health. The addition of Si fertilizers improves the uptake and transport of Si to rice grains [22]. Many studies have demonstrated that Si fertilizers significantly increase the rice GY [13,22,25,26] and improve the rice quality characters (i.e., protein content and macronutrients and micronutrients concentrations) [21,22,26–30]. Si fertilizers applied in agriculture are generally as conventional Si such as calcium silicate (Ca_2SiO_4), potassium silicate (K_2SiO_3), and sodium silicate (Na_2SiO_3). The application of Si-containing nanofertilizers for plant nutrition has attracted attention as a modern technique in agriculture in the recent years [22]. Nanoparticles (NPs) contain particles less than 100 nm in size by reducing the particle size to nanoscale [31]. This feature allows modern Si fertilizers (SiNPs) to easily penetrate into the leaves and create a thick silicate layer on the leaf surface [14]. Conventional Si fertilizers are typically of low availability, and the application of high-availability nanosilicon (nano-Si) is a more promising option for improving the accumulation of Si in rice tissue [22]. Numerous reports indicate the superiority of SiNPs over conventional Si fertilizers in terms of improving rice GY and quality [22,32,33].

Thus, in this chapter, we review the recent findings on how Si and nano-Si improve growth, yield, and quality of rice. Special emphasis is placed on the importance of applying SiNPs as a modern agronomic technique in agriculture to increase the quantitative and qualitative yield of rice. We also describe the diverse roles of Si and nano-Si in increasing yield and quality of rice under biotic and abiotic stresses.

21.2 Impacts of Si and nano-Si on rice yield and quality

The improvement in growth, GY, and quality of rice following Si application has been reported by many researchers [13,22,25,29,34]. Si deficiency reduced rice growth and yield, while Si application increased the number of filled spikelets per panicle, thereby improving GY [24]. The addition of high doses of Si to the nutrient solution improves growth and GY of rice [29,34]. Although the application of Si in both vegetative and reproductive stages has positive impacts on the growth and yield of rice, but a dramatic increase in GY and significant improvement in quality of rice occurs following Si application at the reproductive growth phase [25,35,36].

Si helps enhance the plant growth and the nutrients uptake in plants grown under nutritional imbalances [37]. Si fertilization significantly increased the nutrient uptake in rice and, as a result, led to increased growth, and higher GY and quality. Thus rice GY and quality affected by Si fertilization may be pertinent to an enhanced uptake of other nutrients such as N, P, K, Ca, Mg, Zn, Fe, and Mn [22,26,29,30].

Si also plays protective role for rice against certain stresses such as salinity [16,38], drought [17], chilling [19], heat [18], UV radiation [20,39,40], HM toxicity [21,30,41−46], pests, and diseases [15,47], thereby improving GY and quality of rice under stress.

The vital role of Si in both the conventional Si and nano-Si forms has been observed in increasing the different growth parameters, and improving qualitative and quantitative yield of rice [22,25,26,29,48]. However, some findings indicate that the Si application as NPs has better effects than conventional Si in the terms of rice yield and quality [22,32,33,49,50].

The beneficial impacts of Si on the yield and quality of rice are depicted in Table 21.1. These impacts are achieved by increasing growth parameters, improving nutrient uptake, and alleviating the damages caused by several biotic and abiotic stresses.

21.2.1 Impacts of Si and nano-Si on increasing growth, agronomic parameters, and grain yield of rice

The impacts of Si on the vegetative and reproductive growth of rice are well recognized. The addition of Si reduced the lodging index (13.7%) and increased plant height (12.2%−16.7%), pushing resistance (10.5%− 13.8%), and yield up to 15.1% [51]. Application of Si (100 mg L^{-1}) fertilizer in the forms of potassium silicate (K_2SiO_3) and sodium silicate (Na_2SiO_3) at the tillering stage increased plant height, the number of tillers, leaf area, and shoot dry weight compared with the control [14]. Also, various vegetative traits such as plant height, number of tillers, and leaf area were improved by adding Si (100 mg L^{-1} K_2SiO_3) at the tillering stage of rice [35]. Yasari et al. [52] demonstrated that the Si-treated plants showed an increase in total number of tiller plant^{-1}, number of fertile tiller plant^{-1}, and rice GY by about 11.6%, 14.2%, and 18.2%, respectively, when compared to non-Si-treated plants. Nhan et al. [53] also found positive effects of Si application in rice. Si (12 mL L^{-1} OryMax) application increased the internode diameter (17%), the internode wall thickness (13%), the chlorophyll a content (43%), the chlorophyll b content (47%), the total carotenoid content (47%), the number of panicles pot^{-1} (7%), the number of filled grains panicle^{-1} (22%), the filled grain ratio (4%), the 1000-grain weight (15%), and the yield pot^{-1} (15%). Significant increases in reproductive parameters, including panicle development, grain fertility, and diameter and number of pollen grains in stigma, were recorded in rice following Si application [18,54].

The increase in different growth traits following Si application depends on various factors, including the dose of Si [55]. Application of high dose of Si (500 kg ha^{-1} Ca_2SiO_4) to rice increased the total number of spikelets per panicle by 3.2%, filled spikelets per panicle by 5.2%, and GY by 4.9%, when compared to control plant [34]. Effects of different doses of SiO_2 (100, 200, 300, and 400 kg hm^{-2} SiO_2) were studied on yield attributes and GY of rice. The addition of Si (400 kg hm^{-2} SiO_2) to the NPK fertilizers in rice increased the tillers number hill^{-1} (28%), the number of grains panicle^{-1} (6%), the 1000-seed weight (33%), and the GY (23%), compared with nonapplication of Si [29]. In a study [56], application of 750 kg ha^{-1} Si in the form of calcium silicate (Ca_2SiO_4) increased rice GY, probably by increasing total tillers number per hill, and enhancing fertile tillers number per hill. A similar response to Si fertilization was also reported in different rice lines [13] (Table 21.2).

TABLE 21.1 Beneficial impacts of Si and nano-Si on rice yield and quality.

Si sources	Optimum dose of Si	Impact on rice	References
Na$_2$SiO$_3$	100 mg kg^{-1}	(i) Increased panicle number, filled spikelets percentage, 1000-grain weight, spikelet number plant^{-1}, grain yield plant^{-1}; and (ii) improved plant resistance to Cd toxicity	Lin et al. [13]
Na$_2$SiO$_3$	1.5 mM	(i) Increased number and diameter of pollen grains; (ii) improved anther dehiscence percentage, pollination, and fertilization; and (iii) enhanced plant resistance to heat stress	Li et al. [18]
Na$_2$SiO$_3$	2.5 mM	(i) Increased seedling height, dry biomass, and soluble protein content; (ii) reduced Cr uptake and translocation; and (iii) enhanced activities of antioxidant enzymes such as POD, CAT, SOD, and APX	Zeng et al. [45]
Na$_2$SiO$_3$	1.5 mM	Improved Si, P, K, Ca, Zn, and B concentrations	Mehrabanjoubani et al. [26]
Na$_2$SiO$_3$	3 mM	(i) Increased Si content in shoot and root; and (ii) decreased as uptake	Zia et al. [21]
Na$_2$SiO$_3$	50 mL	(i) Increased expression of Si uptake genes; (ii) enhanced activation of defense systems of enzymatic and nonenzymatic antioxidants; (iii) improved production of osmolytes; and (iv) alleviated negative effects of salinity stress	Abdel-Haliem et al. [16]
Na$_2$SiO$_3$	5 mM	(i) Improved dry weight of grains and shoots; and (ii) reduced Cd accumulation in grains and shoots	Liu et al. [41]
Na$_2$SiO$_3$	3 mM	(i) Improved growth; (ii) increased photosynthetic rate, stomatal conductance, and WUE; (iii) reduced transpiration rate; (iv) decreased transport of chloride to the shoots; and (v) reduced damages caused by salinity	Shi et al. [38]
Na$_2$SiO$_3$	100 mg kg^{-1}	(i) Increased leaf length, leaf width, stem dry mass, and SPAD values healthy; and (ii) reduced damages caused by YSB	Ranganathan et al. [47]
Na$_2$SiO$_3$	1.70 mM	(i) Increased root nutrient absorbing capability; and (ii) enhanced tolerance to chilling stress	Liang et al. [19]
Ca$_2$SiO$_4$	750 kg ha^{-1}	Increased total tillers number per hill, fertile tillers number per hill, and GY	Ghasemi Mianaei et al. [56]
Ca$_2$SiO$_4$	500 kg ha^{-1}	Increased total number of spikelets per panicle, filled spikelets per panicle, and GY	Ghanbari-Malidareh [34]
K$_2$SiO$_3$	100 mg L^{-1}	Improved plant height, number of tillers, leaf area, and content of chlorophyll a and b	Gerami et al. [35]
K$_2$SiO$_3$	1.5 mM	(i) Increased photosynthesis rate, photochemical efficiency, and WUE; (ii) improved regulation of mineral nutrient uptake such as K, Na, Ca, Mg, and Fe; and (iii) alleviated drought stress in plant	Chen et al. [17]
K$_2$SiO$_3$ and Na$_2$SiO$_3$	100 mg L^{-1}	Increased plant height, number of tillers, leaf area, and shoot dry weight	Meena et al. [14]
SiO$_2$	400 kg hm^{-2}	(i) Increased tillers number hill^{-1}, number of grains panicle^{-1}, 1000-seed weight, and GY; (ii) enhanced N, P, K, and Si concentrations in biomass; and (iii) improved protein content	Cuong et al. [29]
SiO$_2$	4 mM	(i) Increased shoot and root biomass; and (ii) decreased shoot Cd concentrations and Cd distribution ratio in shoots	Zhang et al. [46]
SiO$_2$	200 kg ha^{-1}	(i) Increased net photosynthetic rate, intercellular CO$_2$ concentration, stomatal conductivity, and WUE; (ii) decreased transpiration rate; and (iii) mitigated negative effects of elevated UV radiation on photosynthesis and transpiration	Lou et al. [20]
SiO$_2$	240 mg L^{-1}	(i) Increased plant resistance to BPH; and (ii) decreased BPH fertility, survival rate, honeydew excretion quantity, and settled insect number	He et al. [15]
SiO$_2$	20 g SiO$_2$ kg^{-1} soil	(i) Increased shoot biomass; and (ii) decreased As concentrations in shoot and root	Wu et al. [44]
H$_4$SiO$_4$	2 mM	(i) Enhanced net photosynthesis; and (ii) increased number of grains plant^{-1}, grains biomass plant^{-1}, filled grains percentage, 1000-grain weight, and harvest index	Lavinsky et al. [25]

(Continued)

TABLE 21.1 (*Continued*)

Si sources	Optimum dose of Si	Impact on rice	References
OryMax	12 ml L^{-1}	Increased internode diameter, internode wall thickness, contents of chlorophyll a, b and total carotenoid, number of panicles pot^{-1}, number of filled grains panicle^{-1}, filled grain ratio, 1000-grain weight, and yield pot^{-1}	Nhan et al. [53]
Nano-SiO$_2$, Ca$_2$SiO$_4$	50 mg L^{-1}, 392 kg Si ha^{-1}	(i) Increased number of fertile tillers hill^{-1}, number of filled grains panicle^{-1}, and GY; (ii) enhanced tissue Si and Zn concentrations; and (iii) improved protein content	Kheyri et al. [22]
Nano-Si	2.5 mM	(i) Improved rice growth; (ii) increased Zn, Fe, and Mg contents in shoot; (iii) enhanced antioxidant capacity; and (iv) reduced Cd accumulation, Cd partitioning in shoot and MDA level	Wang et al. [30]
Nano-SiO$_2$	40 mg L^{-1}	(i) Increased shoot length and shoot dry weight; and (ii) improved shoot Si uptake	Adhikari et al. [27]
Nano-Si	50 mg L^{-1}	(i) Increased tissue Si uptake; and (ii) improved protein content	Yazdpour et al. [33]
Nano-SiO$_2$	500 mg kg^{-1}	Reduced Cd and As accumulation in rice tissues	Wang et al. [65]
Nano-Si	30 mg L^{-1}	(i) Increased plant height, chlorophyll concentrations, shoot and root biomass; (ii) improved activities of CAT, POD, SOD, and APX in shoots; (iii) decreased EL, and MDA content; and (iv) reduced Cd accumulation in shoot and root	Rizwan et al. [64]
Nano-Si and Na$_2$SiO$_3$	0.0025 mol L^{-1}	(i) Increased plant biomasses; and (ii) reduced Pb content in grains	Liu et al. [43]

Si: silicon; N: nitrogen; P: phosphorus; K: potassium; Ca: calcium; Mg: magnesium; Fe: iron; Zn: zinc; B: boron; Na: sodium; Cd: cadmium; Cr: chromium; As: arsenic; Pb: lead; UV: Ultraviolet; YSB: yellow stem borer; BPH: brown planthopper; POD: peroxidase; CAT: catalase; SOD: superoxide dismutase; APX: ascorbate peroxidase; MDA: malondialdehyde; EL: electrolyte leakage; SPAD value: chlorophyll content; WUE: water use efficiency; GY: grain yield.

TABLE 21.2 Effect of Si application rate on yield components and grain yield of different rice lines (Lin et al., 2016).

Agronomic parameters	Si application rate (mg kg^{-1})	WT	Lsi1-OE line	Lsi1-RNAi line
Panicle number	0	4.41 ± 0.11	7.00 ± 0.11	6.44 ± 0.11
	10	4.43 ± 0.44	7.38 ± 0.28	6.66 ± 0.12
	100	5.02 ± 0.01	8.05 ± 0.05	6.77 ± 0.10
Filled spikelets percentage	0	66.31 ± 0.11	89.92 ± 0.12	82.91 ± 1.11
	10	66.44 ± 0.80	89.87 ± 0.43	82.57 ± 0.85
	100	70.46 ± 0.64	93.70 ± 0.72	83.62 ± 1.04
1000-grain weight	0	23.13 ± 0.23	21.23 ± 0.55	21.02 ± 0.19
	10	23.15 ± 0.08	21.52 ± 0.46	20.08 ± 0.10
	100	22.71 ± 0.53	21.40 ± 0.52	20.32 ± 0.71
Spikelet number plant^{-1}	0	234.23 ± 1.03	141.28 ± 0.90	136.94 ± 1.58
	10	241.17 ± 1.03	144.03 ± 100	139.22 ± 2.83
	100	232.68 ± 2.15	146.58 ± 3.47	132.47 ± 2.62
Grain yield plant^{-1}	0	15.32 ± 0.15	18.59 ± 0.20	15.17 ± 0.14
	10	15.59 ± 0.04	18.79 ± 0.14	15.04 ± 0.14
	100	15.37 ± 0.12	18.83 ± 0.09	15.66 ± 0.08

Si application at the reproductive growth stage has very positive impacts on rice growth and yield. Lavinsky et al. [25] investigated the effect of additions or removals of Si (0 or 2 mM, respectively, -Si e + Si) on the growth and yield of rice during three specific periods of the rice growth cycle: (1) the vegetative growth stage (V), from transplanting to panicle initiation; (2) the reproductive growth stage (R1), from panicle initiation to heading; and (3) the ripening stage (R2), from heading to maturity. Eight treatments were designed; Si was supplied (+ Si) or not (-Si) during V, R1 and R2 growth phases, as follows: -Si/-Si/-Si, -Si/-Si/+Si, -Si/+Si/-Si, -Si/+Si/+Si, +Si/-Si/-Si, +Si/-Si/+Si, +Si/+Si/-Si, and +Si/+Si/+Si. Compared with the Si supply during V or R2 growth phases, the Si supply during R1 stage was more effective in increasing yield and yield components of rice (Table 21.3). Si supply only during R1 phase produced grain biomass comparable to that during whole growth period (V + R1 + R2). The Si-treated plants during R1, irrespective of Si removal or application during the other growth phases, showed higher number of grains panicle^{-1} (49% on average), higher grains biomass plant^{-1} (77% on average), higher filled grains percentage (20% on average), higher 1000-grain weight (20% on average), higher harvest index (61% on average), and higher net photosynthesis (15% on average), when compared to Si-treated plants only in V and/or R2 phases. Ma et al. [36] reported that when Si was supplied to rice at various growth phases of vegetative, reproductive, or ripening, grain filling significantly increased by adding Si only during the reproductive phase. Plants that received Si (100 mg L^{-1} K$_2$SiO$_3$) at the flowering stage (reproductive phase), displayed higher content of chlorophyll a and b, when compared to plants that received Si at the tillering stage (vegetative phase) [35].

Application of new Si fertilizers (nano-Si) has an important role in improving rice growth and yield due to high availability and better uptake by the plant [22]. Nanofertilizers have higher use efficiency than conventional chemical fertilizers and release nutrients slowly and steadily at the critical growth stages, which facilitates nutrients uptake for plants [57,58]. In addition to providing plant nutrients and soil reviving to an organic state, nano-fertilizers are not harmful to the environment, so they have more benefits compared to conventional chemical fertilizers [50]. Although studies on the effects of SiNPs on rice are scarce, it has been proved that the application of nano-Si has more beneficial effects in improving the yield and quality of rice than conventional Si fertilizers. Amrullah et al. [32] reported that NPs foliar application of Si had more positive impacts than conventional Si fertilizer on the measured traits of rice including plant height, root length, leaf length, fresh and dry weight of root, and fresh and dry weight of rice plant canopy. By comparing the effects of various forms of Si as nano-Si and conventional Si on agronomic parameters and GY of rice, Kheyri et al. [22] concluded that the Si application by both nano-SiO$_2$ (50 mg L^{-1} or 300 g ha^{-1}) and Ca$_2$SiO$_4$ (392 kg Si ha^{-1}) resulted in a significant increase in the number of fertile tillers hill^{-1} by 11% and 8.3%, the number of filled grains panicle^{-1} by 5.2% and 4.3%, and GY by 9.6% and 6.9%, respectively, compared with control treatment (Table 21.4). However, the Si application through NPs had better effects than conventional Si in terms of yield components and GY of rice. They found that small amounts of Si applied as nanoscale fertilizers (nano-SiO$_2$) provided a benefit that was similar to or greater than large amounts of conventional fertilizer (Ca$_2$SiO$_4$), thus, the application of SiNPs is suggested to increase GY, reduce fertilizer costs and environmental pollution, and enrich rice grains [22].

TABLE 21.3 The effect of Si supply (0 or 2 mM, respectively, -Si and + Si), during the vegetative growth stage, V (from transplanting to panicle initiation), reproductive growth stage, R1 (from panicle initiation to heading) and the ripening stage, R2 (from heading to maturity), on the rice yield attributes (Lavinsky et al. [25]).

V	-Si				+ Si			
R1	-Si		+ Si		-Si		+ Si	
R2	-Si	+ Si	-Si	+ Si	-Si	+ Si	-Si	+ Si
Panicle number plant^{-1}	43.9b	42.4b	63.6a	62.8a	41.4b	43.2b	63.4a	64.1a
Grains number panicle^{-1}	304b	289b	440a	429a	287b	302b	436a	445a
Filled grain percentage (%)	63.5b	66.2b	83.3a	83.2a	75.7b	70.2b	82.9a	82.8a
1000-grain weight (g)	25.7b	26.3b	32.5a	32.5a	29.5b	26.8b	32.3a	32.9a
Grain biomass (g)	7.79b	7.69b	14.3a	14.0a	8.50b	8.08b	14.0a	14.6a
Harvest index (g g^{-1})	0.21b	0.21b	0.34a	0.35a	0.22b	0.22b	0.35a	0.35a
Net photosynthesis (μmol CO$_2$ m^{-2} s^{-1})	22.9b	23.3b	25.9a	26.8a	21.8b	22.6b	25.9a	25.8a

The means followed by a different letter differ significantly among the treatments, at the $P < .05$ level using a Scott–Knott test. $n = 6$.

TABLE 21.4 Effect of Si sources on agronomic parameters and grain yield of rice (Kheyri et al., 2019).

Si sources	Fertile tillers hill^{-1}	Number of filled grains panicle^{-1}	Grain yield (kg ha^{-1})
Control	15.4b	88.3b	3969b
Nano-SiO$_2$	17.3a	93.2a	4389a
Calcium silicate	16.8a	92.3a	4265a

Means separation in columns followed by the same letter(s) are not significantly different at $P \leq .05$.

21.2.2 Impacts of Si and nano-Si on improving nutrient uptake of rice

Application of Si improves the protein content and grain quality of rice through supplying N to the plant [29]. Rice plant displayed increased N, P, K, and Si accumulations in biomass when supplied with increasing levels of SiO$_2$ fertilizer (100–400 kg hm^{-2} SiO$_2$). The application of a high dose of Si (400 kg hm^{-2} SiO$_2$) improved the N, P, K, and Si concentrations in rice biomass by 33%, 68%, 37%, and 59%, respectively [29]. The addition of Si (1.5 mM) improved the Si, P, K, Ca, Zn, and B concentrations in rice plants supplied up to 50 μg L^{-1} Zn [26]. Zia et al. [21] found that Si (3 mM) application increased the Si concentration in the rice root and shoot by about 48% and 42%, respectively. Ma et al. [59] compared two rice genotypes, Kasalath (indica) and Nipponbare (japonica), under various doses of Si. Under the application of 0.15 mM Si(OH)$_4$, Nipponbare showed higher Si content than Kasalath, while the Si content in both genotypes was nearly the same by application of 1.5 mM Si(OH)$_4$. One reason for these changes is the different mechanisms involved in the Si accumulation in the two rice genotypes [60]. In a study, Crusciol et al. [61] investigated the effect of lime, phosphorous gypsum, and silicate on nutrient uptake of flag leaf and rice GY. The highest uptake of flag leaf Si (29 g kg^{-1}) was achieved when silicate was applied alone, while the uptake of macronutrients (N, P, K, Ca, and Mg) remained essentially unchanged. The researchers noted that the highest number of panicles per m^2 (225 panicles) and consequently the highest GY (4362 kg ha^{-1}) was obtained by the combined application of silicate and phosphorous gypsum, although there was no significant difference with the separate application of silicate.

Many reports indicated the synergistic interaction between Si and other nutrients such as Zn in rice plant. In a study, Ghasemi et al. [48] found that Zn concentration in rice increased by 21% when Si was added to the rice plant and Si content in rice improved by 24% when Zn was applied. In another study, Mehrabanjoubani et al. [26] evaluated the impacts of Si (0 and 1.5 mM as sodium silicate) nutrition on growth and nutrient status of rice plants under varying Zn supplies (1–100 μg L^{-1} Zn as zinc sulfate). The Si nutrition displayed a positive impact on Zn uptake in rice plants supplied up to 50 μg L^{-1} Zn. Thus Si could alleviate Zn deficiency of rice plants due to greater Zn uptake and facilitate the process of Zn transport from root to shoot through transpiration stream. The positive interaction between Si and Zn on the uptake of these elements in rice tissue has been confirmed by other researchers [22,30].

The beneficial role of SiNPs in improving the nutrient uptake and quality of rice has been recognized. The application of nanofertilizers can increase the germination rate of the plant, improve the plant's resistance to biotic and abiotic stresses, increase nutrient utilization efficiency, increase plant growth, and reduce environmental effects compared to traditional fertilizers [49]. Nanofertilizers release nutrients in a controlled way synchronized with plant demand, thereby improving nutrient efficiency [62]. Kheyri et al. [22] investigated the nutrient uptake and quality of rice under different sources of Si. Plants that received Si as calcium silicate (392 kg Si ha^{-1}) and nano-SiO$_2$ (300 g ha^{-1}) showed higher concentration of Si and Zn in grain as well as protein yield in grain, when compared to plants that did not receive Si. In this review, the Si application via either conventional Si or nano-Si increased the Si concentration in grain by 11.7% and 15.9%, the Zn concentration in grain by 10% and 13% and the protein yield in grain by 10% and 13%, respectively. However, the application of Si fertilizer as NPs had greater impacts than traditional Si in terms of Si, Zn, and protein content in rice tissue. The study demonstrated that the Si application in the form of NPs is more appropriate than conventional Si in increasing the rice quality (i.e., protein content, and Si and Zn concentrations) and reducing the occurrence of nutrient deficiencies in the human diet [22]. Foliar application of nano-Si significantly increased the Si and protein content in rice grain and straw, thereby enhancing the rice quality, when compared with conventional Si [33]. Wang et al. [30] observed that foliar application of 2.5 mM nano-Si improved rice grain quality, probably by increasing Zn, Fe, and Mg content in the shoot, and reducing Cd accumulation in the shoot. The effects of different doses of nano-Si (0, 10, 20, 40, 60, 80, and 100 mg L^{-1}) on the growth and seed germination of rice were studied. Si had positive effects in terms of shoot length, shoot dry weight, and increase in shoot Si uptake by application of Si at a dose of 40 mg L^{-1} [27].

21.2.3 Impacts of Si and nano-Si on ameliorating yield and quality of rice under biotic and abiotic stresses

The application of Si fertilizer to alleviate HM toxicity to rice has attracted increasing attention in recent years [30]. Si application increases rice GY and improves grain quality by lowering the HM content in grain [63], thus, it reduces the deleterious effects of HM on human health [21]. Many reports have shown the protective role of Si in rice against a range of toxic metal stresses such as cadmium (Cd), arsenic (As), lead (Pb), chromium (Cr), etc. Rice plants supplemented with Si (5 mM Na_2SiO_3) at the tillering stage and subsequently exposed to 50 μM Cd showed amelioration of the negative impacts of Cd toxicity by improvements in dry weight of grains and shoots, and alleviating toxicity and accumulation of Cd in the grains and shoots of rice. The alleviation of Cd accumulation in rice grains is probably due to the Cd sequestration in the shoot cell walls following the Si foliar application [41]. Addition of Si (2 and 4 mM) under Cd (0, 2, and 4 μM) stress decreased shoot Cd concentrations by 30%–50% and Cd distribution ratio in shoots by 25.3%–46% and increased shoot and root biomass of rice by 125%–171% and 100%–106%, respectively [46]. Foliar application of nano-Si (2.5 mM) improved rice growth parameters under Cd stress as indicated by decreased Cd accumulation, Cd partitioning in shoot and malondialdehyde (MDA) level and by increasing content of some mineral elements (Mg, Fe, and Zn) and antioxidant capacity [30]. Rice plants treated with SiNPs accumulated higher biomass against Cd stress. The significant impact of SiNPs in mitigating Cd-induced toxicity depended on the dose of NPs used. Foliar application of 5, 10, 20, and 30 mg L^{-1} SiNPs increased rice shoot dry weight by 34%, 62%, 74%, and 101% and reduced the shoot Cd concentrations by 21%, 37%, 49%, and 55% over the control plants. The reasons mentioned included the possibility that SiNPs could improve photosynthetic rate by increasing contents of chlorophyll a, chlorophyll b, and carotenoid. Application of SiNPs also decreased the electrolyte leakage and MDA content and increased the production of antioxidant enzymes including peroxidase (POD), catalase (CAT), superoxide dismutase (SOD), and ascorbate peroxidase (APX) in rice tissues under Cd stress [64]. Lin et al. [13] found that applying 100 mg kg^{-1} Si (Na_2SiO_3) to rice under Cd stress increased the plant's resistance to Cd toxicity and thus improved rice GY. Wang et al. [65] investigated the impacts of different concentrations of SiNPs (0, 150, 500, or 2000 mg kg^{-1}) on the uptake of Cd (1 mg kg^{-1}) and As (5 mg kg^{-1}) by rice seedlings under both continuous flooding (CF) and alternate wetting and drying (AWD) conditions. The effectiveness of SiNPs on rice Cd and As uptake depended on the concentration of SiNPs and the water management schemes. Simultaneous reduction of Cd and As in rice shoots was observed by application of 500 mg kg^{-1} SiNPs under AWD irrigation. The addition of SiNPs (500 mg kg^{-1}) reduced rice shoots Cd and As concentrations by 50% and 70% under AWD than CF irrigation. Supplementation of Si (20 g SiO_2 kg^{-1} soil) to the As (40 mg As kg^{-1} soil) decreased shoot and root total As concentrations in six rice genotypes and increased the shoot biomass [44]. Liu et al. [42] proposed a probable mechanism for Si-induced alleviation of As toxicity. The proposed mechanism includes a release of silicic acid from the silicate gel, which would increase the accumulation of dimethylarsinic acid (DMA) in both vegetative and reproductive tissues of rice by inhibiting DMA adsorption on the soil solid phase or by displacing adsorbed DMA. Zia et al. [21] evaluated the impact of optimum Si (3 mM) supplementation on rice seed germination, seedling growth, and P and As uptake in rice plants exposed to As (150 and 300 μM) stress under anaerobic and aerobic conditions. They observed that the aerobic conditions are more favorable to decrease As uptake in rice. As stress decreased, the seed germination up to 40%–50% as compared to control plants. As (300 μM) application also significantly decreased the seedling length and P uptake in the rice plant while Si supplementation enhanced the seedlings length and reduced the negative effect of As on P uptake on rice shoot especially in plants grown under aerobic conditions. Si (2.5 mM Na_2SiO_3) addition is documented as reducing the uptake and translocation of Cr from root to shoot of rice, and increasing the activities of antioxidant enzymes such as CAT, POD, SOD, and APX. The Si alleviated Cr (100 μM) toxicity in the roots and shoots of rice by 30% and 41%, respectively [45]. Liu et al. [43] studied the effects of different Si sources as nano-Si and common Si on Pb toxicity, uptake, translocation, and accumulation in the rice plant grown under Pb stress. When Si (0.0025 mol L^{-1}) was applied as common Si and nano-Si under Pb (500 and 1000 mg kg^{-1}) stress, it improved the plant biomasses by 1.8%–5.2% and 3.3%–11.8%, respectively, compared with the untreated control. In addition, application of common Si and nano-Si significantly reduced the Pb concentrations in rice grains by 21.3%–40.9% and 38.6%–64.8%, respectively, and the Pb translocation from shoots to grains by 8.3%–13.7% and 15.3%–21.1%, respectively, compared with control. They concluded that nano-Si is more efficient than common Si in alleviating the toxic impacts of Pb on rice growth, preventing Pb transport from roots to shoots of rice, and blocking Pb accumulation in rice grains.

The roles of Si in alleviating salt stress to plants have been observed in many plant species including rice. Application of exogenous Si (3 mM Na_2SiO_3) modulated different physiological processes (photosynthetic rate, stomatal conductance, water use efficiency, and transpiration rate) that improved the rice growth under salinity (50 mM NaCl) [38]. The alleviative effect of Si on salinity stress is related to decreased transport of chloride to the rice shoots by reducing transpirational bypass flow in the rice roots. Abdel-Haliem et al. [16] investigated the effects of Si ions (Na_2SiO_3) on the physiological and biochemical responses and the expression of the two Si uptake genes LSi1 and LSi2 in rice grown under NaCl salt concentrations (0, 100, 200, 400, and 800 mM). They revealed that Si supplementation helped rice plants mitigate the negative impact of salinity stress. Si application increased the expression of Si uptake genes (Lsi1 and Lsi2) by jasmonic acid production. This phytohormone upregulated the expression of the Si uptake genes in salt-treated plants supplied with Si ions, activated the defense systems of enzymatic (Polyphenol oxidase [PPO], SOD, POD, and CAT) and the nonenzymatic antioxidants (polyphenols and proline), and produces osmolytes such as proline and free amino acid.

Si can reduce transpiration rate by 30% by creating deposits in rice [60], thus increasing drought resistance. In rice seedlings, Si (1.5 mM K_2SiO_3) alleviates drought stress in rice plant by increasing water use efficiency (WUE) and photosynthesis rate, and by adjustment of mineral nutrient (K, Na, Ca, Mg, and Fe) uptake in rice [16].

Reports on the impacts of Si on the growth of rice under high and low temperatures (heat and chilling) stresses are scarce. However, some findings have indicated that Si mitigates the adverse effects of heat and chilling stresses. Si protects the rice plant against high temperatures by efficient maintenance of transpiration [63]. Application of Si (1.5 mM Na_2SiO_3) enhanced the number and diameter of pollen grains of rice against heat treatment [18]. Exogenous Si supplementation increased the root nutrient absorbing capability and preventing the wilting, thereby enhancing the rice resistance to chilling ($0°C−4°C$) stress [19].

Si can be used in rice production to mitigate the negative impacts of elevated UV-B radiation on the photosynthesis and transpiration parameters of rice. Si application in UV-B stress conditions, increased the net photosynthetic rate, intercellular CO_2 concentration, stomatal conductivity, and WUE by 16.9%−28.0%, 3.5%−14.3%, 16.8%−38.7%, and 29.0%−51.2%, respectively, but decreased transpiration rate by 1.9%−10.8%, compared with control rice plants [20]. Si application improves growth of rice under UV radiation, as reported in several studies. The addition of Si can increase Si deposits and decrease cinnamyl alcohol dehydrogenase activity and ferulic and p-coumaric acid contents in rice, and these responses are closely related to alterations in the UV defense system [40]. Exogenous Si application increases the uptake of Si, P, Na, and Ca in rice leaves under UV stress [39].

Si application enhances rice plants resistance and/or tolerance to pests and diseases. Addition of Si (100 mg kg^{-1} Na_2SiO_3) increased the leaf length by 25.5%, the leaf width by 2.7%, the stem dry mass by 47.6%, and the chlorophyll content (SPAD value) by 4.5%, and reduced the damage caused by the yellow stem borer in rice by about 40%, compared with control plant. The probable reason for this alleviation of pest damage was lower digestibility of the leaves and straw by the insect due to the presence of higher Si content [47]. Application of high concentrations of Si (240 mg L^{-1} SiO_2) enhanced the rice plants resistance to brown planthopper (BPH) insects compared to controls. Rice plants treated with high Si solution reduced the damages caused by BPH by significantly decreasing the BPH fertility, survival rate, honeydew excretion quantity, and settled insect number and inhibiting the weight increase of the BPH [15].

21.3 Conclusion and future perspective

The published studies indicate that the application of various sources of Si improve the growth, yield, and quality of rice, which may be related to increasing agronomic parameters, improving nutrient uptake, and mitigating stress-induced damage. Si plays a vital role in a wide spectrum of physiological processes in rice plant. For example, it helps enhance the photosynthesis rate, improve the nutrients uptake, and protect the rice plant from biotic and abiotic stresses-induced damages. However, research conducted to date on the role of SiNPs under stressful conditions is scarce. The prominent role of Si in protecting rice to alleviate damages from biotic and abiotic stresses has been published in many studies. However, the basic physiological mechanisms by which Si could protect rice plants from stressful conditions should be considered in future research. In particular, the physiological and molecular mechanisms that following the application of Si reduce the uptake of HM in the shoots and grain of rice. Despite the improvement in nutrient uptake and subsequent increase in yield and quality by the supplying Si to the rice plants, excessive uptake of nutrients can also be challenging due to the element's toxicity. Thus it is recommended to evaluate the exact amount of all nutrients including macronutrients and micronutrients in rice grain following the Si application in future studies.

Given the importance of Si nutrition in human health, new fertilizers research programs seek to improve not only yields but also grain Si uptake to address both food security and quality. The potential of application of

NPs as a modern technique in the agricultural sector that can pave the way for increasing the yield and quality of rice plants while reducing the use of chemical fertilizers and may provide an ecofriendly alternative to various chemical fertilizers without harming nature has attracted considerable attention in recent years. Findings from some field experiments confirm the superiority of SiNPs over conventional Si. The application of nano-Si has great superiority due to its advantages in nutrients uptake, high crop yield and quality, high adaptation to environment, and low cost. In addition, the involvement of nano-Si in the proper availability of key macronutrients and micronutrients to the rice plant increases the enrichment of rice grains, thereby improving human health. Many studies indicate that high yields require the high doses application of traditional Si fertilizers, which in turn increases labor and costs. On the other hand, numerous studies have shown that small amounts of Si applied as NPs provide a benefit similar to or greater than large amounts of conventional Si. Thus SiNPs application is considerably more cost-effective and more convenient compared with traditional Si. However, the exact dose of SiNPs application is still under study.

Although further investigation is needed to identify various physiological and molecular mechanisms affecting rice growth, yield, and quality, especially under stressful conditions, the findings from experimental field research studies confirm the ability of application of both Si and nano-Si to improve rice yield and quality. However, the most important issue is to apply sources of Si fertilizer that are cheaper, more effective, and more environmentally friendly. Thus the use of modern technologies and the application of SiNPs with lower cost and higher efficiency compared with conventional Si can have concrete solutions to many agricultural problems regarding yield and the quality of rice.

References

[1] Stone R. Food safety. Arsenic and paddy rice: a neglected cancer risk? Science 2008;321:184−5. Available from: https://doi.org/10.1126/science.321.5886.184.

[2] Datnoff LE, Snyder GH, Korndorfer GH. Silicon in agriculture. Amsterdam: Elsevier; 2001.

[3] Haghighi M, Pessarakli M. Influence of silicon and nano-silicon on salinity tolerance of cherry tomatoes (*Solanum lycopersicum* L.) at early growth stage. Sci Hortic 2013;161:111−17. Available from: https://doi.org/10.1016/j.scienta.2013.06.034.

[4] Bazilevich NI. The biological productivity of North Eurasian ecosystems. RAS Institute of Geography. Nauka: Moscow; 1993.

[5] Ma JF, Tamai K, Yamaji N, et al. A silicon transporter in rice. Nature 2006;440:688−91.

[6] Tanaka A, Park YD. Significance of the absorption and distribution of silica in the growth of the rice plant. Soil Sci Plant Nutr 1966;12:23−8. Available from: https://doi.org/10.1080/00380768.1966.10431957.

[7] Tamai K, Ma JF. Characterization of silicon uptake by rice roots. New Phytol 2003;158:431−6. Available from: https://doi.org/10.1046/j.1469-8137.2003.00773.x.

[8] Nakata Y, Ueno M, Kihara J, et al. Rice blast disease and susceptibility to pests in a silicon uptake-deficient mutant lsi1 of rice. Crop Protec 2008;27:865−8. Available from: https://doi.org/10.1016/j.cropro.2007.08.016.

[9] Kato N, Owa N. Dissolution mechanism of silicate slage fertilizers in paddy. Soil Sci 1990;4:609−10.

[10] Kabata-Pendias A. Trace elements in soils and plants. Boca Raton, FL: CRC Press; 1984.

[11] Windslow MD, Okada K, Correa-Victoria F. Silicon deficiency and the adaptation of tropical rice ecotypes. Plant Soil 1997;188:239−48.

[12] Epstein E, Bloom AJ. Mineral nutrition of plants: principles and perspectives. 2nd ed. Sunderland, MA: Sinauer; 2005.

[13] Lin H, Fang C, Li Y, et al. Effect of silicon on grain yield of rice under cadmium-stress. Acta Physiol Plant 2016;38:186. Available from: https://doi.org/10.1007/s11738-016-2177-8.

[14] Meena VD, Dotaniya ML, Coumar V, et al. A case for silicon fertilization to improve crop yields in tropical soils. Proc Natl Acad Sci India Sect B Biol Sci 2014;84:505−18. Available from: https://doi.org/10.1007/s40011-013-0270-y.

[15] He W, Yang M, Li Z, et al. High levels of silicon provided as a nutrient in hydroponic culture enhances rice plant resistance to brown planthopper. Crop Protec 2015;67:20−5. Available from: https://doi.org/10.1016/j.cropro.2014.09.013.

[16] Abdel-Haliem MEF, Hegazy HS, Hassan NS, et al. Effect of silica ions and nano silica on rice plants under salinity stress. Ecol Engg 2017;99:282−9. Available from: https://doi.org/10.1016/j.ecoleng.2016.11.060.

[17] Chen W, Yao X, Cai K, et al. Silicon alleviates drought stress of rice plants by improving plant water status, photosynthesis and mineral nutrient absorption. Biol Trace Elem Res 2011;142:67−76. Available from: https://doi.org/10.1007/s12011-010-8742-x.

[18] Li WB, Wang H, Zhang FS. Effects of silicon on anther dehiscence and pollen shedding in rice under high temperature stress. Acta Agron Sin 2005;31:134−6.

[19] Liang Y, Hua HX, Zhu YG, et al. Importance of plant species and external silicon concentration to active silicon uptake and transport. New Phytol 2006;172:63−72. Available from: https://doi.org/10.1111/j.1469-8137.2006.01797.x.

[20] Lou YS, Wu L, Lixuan R, et al. Effects of silicon application on diurnal variations of physiological properties of rice leaves of plants at the heading stage under elevated UV-B radiation. Int J Biometeorol 2016;60:311−18. Available from: https://doi.org/10.1007/s00484-015-1039-1.

[21] Zia Z, Bakhat HF, Saqib ZA, et al. Effect of water management and silicon on germination, growth, phosphorus and arsenic uptake in rice. Ecotoxicol Environ Safe 2017;144:11−18. Available from: https://doi.org/10.1016/j.ecoenv.2017.06.004.

[22] Kheyri N, Ajam Norouzi H, Mobasser HR, et al. Effects of silicon and zinc nanoparticles on growth, yield, and biochemical characteristics of rice. Agron J 2019;111:3084−90. Available from: https://doi.org/10.2134/agronj2019.04.0304.

[23] Tamai K, Ma JF. Reexamination of silicon effects on rice growth and production under field conditions using a low silicon mutant. Plant Soil 2008;307:21−7. Available from: https://doi.org/10.1007/s11104-008-9571-y.

[24] Mobasser H.R., Ghanbari-Malidareh A., Sedghi A.H. Effect of silicon application to nitrogen rate and splitting on agronomical characteristics of rice (*Oryza sativa* L.). In: Silicon in agriculture conference, Wild Coast Sun, South Africa; 2008.

[25] Lavinsky AO, Detmann KC, Reis JV, et al. Silicon improves rice grain yield and photosynthesis specifically when supplied during the reproductive growth stage. J Plant Physiol 2016;206:125−32. Available from: https://doi.org/10.1016/j.jplph.2016.09.010.

[26] Mehrabanjoubani P, Abdolzadeh A, Sadeghipour HR, et al. Impacts of silicon nutrition on growth and nutrient status of rice plants grown under varying zinc regimes. Theor Exp Plant Physiol 2015;27:19−29. Available from: https://doi.org/10.1007/s40626-014-0028-9.

[27] Adhikari T, Kundu S, Rao AS. Impact of SiO₂ and Mo nano particles on seed germination of rice (*Oryza sativa* L.). Int J Agric Food Sci Tech 2013;4:809−16.

[28] Ahmad A, Afzal M, Ahmad AUH, et al. Effect of foliar application of silicon on yield and quality of rice (*Oryza sativa* L.). Cercetari Agron Moldova 2013;46:21−8. Available from: https://doi.org/10.2478/v10298-012-0089-3.

[29] Cuong TX, Ullah H, Datta A, et al. Effects of silicon-based fertilizer on growth, yield and nutrient uptake of rice in tropical zone of Vietnam. Rice Sci 2017;24:283−90. Available from: https://doi.org/10.1016/j.rsci.2017.06.002.

[30] Wang S, Wang F, Gao S. Foliar application with nano-silicon alleviates Cd toxicity in rice seedlings. Environ Sci Pollut Res 2015;22:2837−45. Available from: https://doi.org/10.1007/s11356-014-3525-0.

[31] Shaw AK, Hossain Z. Impact of nano-CuO stress on rice (*Oryza sativa* L.) seedlings. Chemosphere 2013;93:906−15. Available from: https://doi.org/10.1016/j.chemosphere.2013.05.044.

[32] Amrullah, Sopandie D, Sugianta, et al. Influence of nano-silica on the growth of rice plant (*Oryza sativa* L.). Asian J Agric Res 2015;9:33−7. Available from: https://doi.org/10.3923/ajar.2015.33.37.

[33] Yazdpour H, Noormohamadi G, Madani H, et al. Role of nano-silicon and other silicon resources on straw and grain protein, phosphorus and silicon contents in Iranian rice cultivar (*Oryza sativa* cv. Tarom). Int J Biosci 2014;5:449−56. Available from: https://doi.org/10.12692/ijb/5.12.449-456.

[34] Ghanbari-Malidareh A. Silicon application and nitrogen on yield and yield components in rice (*Oryza sativa* L.) in two irrigation systems. Int J Agric Biosyst Engin 2011;5:40−7.

[35] Gerami M, Fallah A, Khatami Moghadam MR. Study of potassium and sodium silicate on the morphological and chlorophyll content on the rice plant in pot experiment. Int J Agric Crop Sci 2012;4:658−61.

[36] Ma J, Nishimura K, Takahashi E. Effect of silicon on the growth of rice plant at different growth stages. Soil Sci Plant Nutr 1989;35:347−56. Available from: https://doi.org/10.1080/00380768.1989.10434768.

[37] Etesami H, Jeong BR. Silicon (Si): review and future prospects on the action mechanisms in alleviating biotic and abiotic stresses in plants. Ecotoxicol Environ Safe 2018;147:881−96. Available from: https://doi.org/10.1016/j.ecoenv.2017.09.063.

[38] Shi Y, Wang Y, Flowers TJ, et al. Silicon decreases chloride transport in rice (*Oryza sativa* L.) in saline conditions. J Plant Physiol 2013;170:847−53. Available from: https://doi.org/10.1016/j.jplph.2013.01.018.

[39] Gao W, Zheng Y, Slusser JR, et al. Impact of enhanced ultraviolet-B irradiance on cotton growth, development, yield, and qualities under field conditions. Agr Meteorol 2003;120:241−8. Available from: https://doi.org/10.1016/j.agrformet.2003.08.019.

[40] Goto M, Ehara H, Karita S, et al. Protective effect of silicon on phenolic biosynthesis and ultraviolet spectral stress in rice crop. Plant Sci 2003;164:349−56. Available from: https://doi.org/10.1016/S0168-9452(02)00419-3.

[41] Liu C, Li F, Luo C, et al. Foliar application of two silica sols reduced cadmium accumulation in rice grains. J Hazard Mater 2009;161:1466−72. Available from: https://doi.org/10.1016/j.jhazmat.2008.04.116.

[42] Liu WJ, McGrath SP, Zhao FJ. Silicon has opposite effects on the accumulation of inorganic and methylated arsenic species in rice. Plant Soil 2014;376:423−31. Available from: https://doi.org/10.1007/s11104-013-1991-7.

[43] Liu J, Cai H, Mei C, et al. Effects of nano-silicon and common silicon on lead uptake and translocation in two rice cultivars. Front Environ Sci Engg 2015;9:905−11. Available from: https://doi.org/10.1007/s11783-015-0786-x.

[44] Wu C, Zou Q, Xue S, et al. Effects of silicon (Si) on arsenic (As) accumulation and speciation in rice (*Oryza sativa* L.) genotypes with different radial oxygen loss (Rol). Chemospher 2015;138:447−53. Available from: https://doi.org/10.1016/j.chemosphere.2015.06.081.

[45] Zeng FR, Zhao FS, Qiu BY, et al. Alleviation of chromium toxicity by silicon addition in rice plants. Agr Sci China 2011;10:1188−96. Available from: https://doi.org/10.1016/S1671-2927(11)60109-0.

[46] Zhang C, Wang L, Nie Q, et al. Long-term effects of exogenous silicon on cadmium translocation and toxicity in rice (*Oryza sativa* L.). Environ Exp Bot 2008;62:300−7. Available from: https://doi.org/10.1016/j.envexpbot.2007.10.024.

[47] Ranganathan S, Suvarchala V, Rajesh YBRD, et al. Effects of silicon sources on its deposition, chlorophyll content, and disease and pest resistance in rice. Biol Plant 2006;50:713−16. Available from: https://doi.org/10.1007/s10535-006-0113-2.

[48] Ghasemi M, Mobasser HR, Asadimanesh H, et al. Investigating the effect of potassium, zinc and silicon on grain yield, yield components and their absorption in grain rice (*Oryza sativa* L.). Elect J Soil Manage Sustain Prod 2014;4:1−24.

[49] Alharby HF, Metwali EMR, Fuller MP, et al. Impact of application of zinc oxide nanoparticles on callus induction, plant regeneration, element content and antioxidant enzyme activity in tomato (*Solanum lycopersicum* mill.) under salt stress. Arch Biol Sci 2016;68:723−35. Available from: https://doi.org/10.2298/ABS151105017A.

[50] Sabir S, Arshad M, Ghaudhari SK. Zinc oxide nanoparticles for revolutionizing agriculture: synthesis and applications. Sci World J 2014;1−8. Available from: https://doi.org/10.1155/2014/925494.

[51] Kim YH, Khan AL, Shinwari ZK, et al. Silicon treatment to rice (*Oryza sativa* L cv "Gopumbyeo") plants during different growth periods and its effects on growth and grain yield. Pak J Bot 2012;44:891−7. Available from: http://www.pakbs.org/pjbot/PDFs/44(3)/08.pdf.

[52] Yasari E, Yazdpoor H, Poor Kolhar H, et al. Effects of plant density and the application of silica on seed yield and yield components of rice (*Oryza sativa* L.). Int J Biol 2012;4:46−53. Available from: https://doi.org/10.5539/ijb.v4n4p46.

[53] Nhan PP, Dong NT, Nhan HT, et al. Effects of OryMax^SL and Siliysol^MS on growth and yield of MTL560 rice. World Appl Sci J 2012;19:704−9. Available from: https://doi.org/10.5829/idosi.wasj.2012.19.05.703.

[54] Ma JF, Yamaji N. Functions and transport of silicon in plants. Cell Mol Life Sci 2008;65:3049−57. Available from: https://doi.org/10.1007/s00018-008-7580-x.

[55] Ali A, Basra SMA, Ahmad R, et al. Optimizing silicon application to improve salinity tolerance in wheat. Soil Environ 2009;28:136−44.

[56] Ghasemi Mianaei A, Mobasser HR, Madani H, et al. Silicon and potassium application facts on lodging related characteristics and quantity yield in rice (*Oryza sativa* L.) Tarom Hashemi variety. New Find Agric 2011;5:423−35.

[57] Mazaherinia S, Astaraei A, Fotovat A, et al. Effect of nano iron oxide particles on Fe, Mn, Zn and Cu concentrations in wheat plant. World Appl Sci J 2010;7:156−62.

[58] Preetha PS, Balakrishnan N. A review of nano fertilizers and their use and functions in soil. Int J Curr Microbiol Appl Sci 2017;6:3117−33. Available from: https://doi.org/10.20546/ijcmas.2017.612.364.

[59] Ma JF, Tamai K, Ichii M, et al. A rice mutant defective in active Si uptake. Plant Physiol 2002;130:2111−17. Available from: https://doi.org/10.1104/pp.010348.

[60] Ma JF, Takahashi E. Soil, fertilizer, and plant silicon research in Japan. Amsterdam: Elsevier; 2002.

[61] Crusciol CAC, Artigiani ACCA, Arf O, et al. Soil fertility, plant nutrition, and grain yield of upland rice affected by surface application of lime, silicate, and phosphogypsum in a tropical no-till system. Catena 2016;137:87−99. Available from: https://doi.org/10.1016/j.catena.2015.09.009.

[62] DeRosa MC, Monreal C, Schnitzer M, et al. Nanotechnology in fertilizers. Nat Nanotechnol 2010;5:91.

[63] Meharg C, Meharg AA. Silicon, the silver bullet for mitigating biotic and abiotic stress, and improving grain quality, in rice? Environ Exp Bot 2015;120:8−17. Available from: https://doi.org/10.1016/j.envexpbot.2015.07.001.

[64] Rizwan M, Ali S, Rehman MZ, et al. Effect of foliar applications of silicon and titanium dioxide nanoparticles on growth, oxidative stress, and cadmium accumulation by rice (*Oryza sativa*). Acta Physiol Plant 2019;41. Available from: https://doi.org/10.1007/s11738-019-2828-7.

[65] Wang X, Jiang J, Dou F, et al. Simultaneous mitigation of arsenic and cadmium accumulation in rice (*Oryza sativa* L.) seedlings by silicon oxide nanoparticles under different water management schemes. Paddy Water Environ 2021. Available from: https://doi.org/10.1007/s10333-021-00855-6.

22

Biological impacts on silicon availability and cycling in agricultural plant-soil systems

Daniel Puppe[1], Danuta Kaczorek[1,2] and Jörg Schaller[1]

[1]Leibniz Centre for Agricultural Landscape Research (ZALF), Müncheberg, Germany [2]Department of Soil Environment Sciences, Warsaw University of Life Sciences (SGGW), Warsaw, Poland

22.1 Introduction

Silicon (Si) is considered a beneficial substance for the majority of higher plants nowadays because Si accumulation in plants increases their resistance against abiotic and biotic stress [1–3]. Thus bioavailability of Si in soils is crucial, especially for agricultural plant-soil systems. Biogenic silica (BSi) pools play an important role as a source of bioavailable Si (monomeric silicic acid, H_4SiO_4) because BSi (i.e., hydrated amorphous silica, $SiO_2 \cdot nH_2O$) is in general much more soluble compared to silicate minerals. In this context, physicochemical properties of BSi structures and residues in soils control BSi dissolution, and thus Si release rates [4,5]. Numerous prokaryotic and eukaryotic organisms have been evolutionary adapted to use bioavailable Si for the formation of siliceous structures in a process called biosilicification. Furthermore, plants and some soil-inhabiting organisms have been found to enhance the dissolution of amorphous and crystalline silica mainly by the release of acidic metabolites, a process that is called bioleaching or bioweathering [6]. In soils BSi structures and residues of bacteria, fungi, plants, animals, and protists can be found, which represent bacterial, fungal, phytogenic, zoogenic, and protistic BSi pools, respectively [1–3] (Fig. 22.1).

BSi is not only an important source for bioavailable Si but also plays a key role in the link between global Si and carbon (C) cycles. This is because BSi controls Si fluxes from terrestrial to aquatic ecosystems [7–9]. These Si fluxes control marine diatom production on a global scale, because marine diatoms need Si for frustule (cell wall) formation, and thus reproduction. Marine diatoms in turn are able to fix large quantities of carbon dioxide (CO_2) via photosynthesis—in fact, up to 54% of the biomass in the oceans is represented by these unicellular organisms [10]. Moreover, Si accumulation in plants promotes plant performance with positive effects on ecosystem functioning and C sequestration in plant biomass [11].

Humans directly influence Si cycling on a global scale by intensified land use, that is, agriculture and forestry. Si exports via crop harvesting and increased erosion rates lead to pronounced Si losses in agricultural plant-soil systems causing a depletion of bioavailable Si in agricultural soils (anthropogenic desilication) [12,13]. Aside from climate change, a growing global population, and decreasing resources, anthropogenic desilication might be one of the big challenges for agriculture in the 21st century. Actually, harvest-related Si losses of up to 100–500 kg Si ha^{-1} occur in agricultural plant-soil systems year by year depending on crops [14,15]. Recently, it was estimated that about 35% of BSi in the plant biomass worldwide is accumulated in field crops and this proportion is going to increase with increased agricultural production within the next decades [16].

This clearly emphasizes the need for a profound understanding of anthropogenic desilication of agricultural plant-soil systems and appropriate prevention strategies. In this context, biological impacts on Si availability and cycling in agricultural plant-soil systems should be the focus of research. Subsequently, we summarize recent knowledge on (1) biosilicification and bioweathering by different organisms, that is, plants (Section 22.2) and

Silicon and Nano-silicon in Environmental Stress Management and Crop Quality Improvement.
DOI: **https://doi.org/10.1016/B978-0-323-91225-9.00006-6**

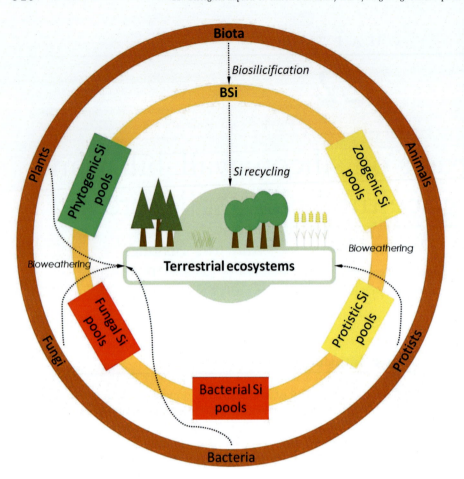

FIGURE 22.1 Schematic overview of the role of biota, biosilicification, and BSi pools in Si cycles of terrestrial ecosystems. The different colors of boxes of BSi pools indicate the corresponding level of knowledge on corresponding Si pool quantities (*green* = numerous studies available, *yellow* = only few studies available, and *red* = no studies available yet). Source: *Modified from Schaller J, Puppe D, Kaczorek D, Ellerbrock R, Sommer M. Silicon cycling in soils revisited. Plants 2021;10(2):295. https://doi.org/10.3390/plants10020295.*

further organisms like protists or fungi (Section 22.3) and (2) implications for ecosystem functioning and services of agricultural plant-soil systems (Section 22.4). The chapter ends with concluding remarks (Section 22.5) and future directions (Section 22.6).

22.2 Plants and phytogenic silica

Plants can increase mineral weathering, and thus Si bioavailability (Box 22.1) in soils mainly by (1) the production of weathering agents like CO_2, organic acids, and ligands in the rhizosphere and (2) taking up elements (e.g., Ca, K, and Si) from soil solution leading to a concentration gradient and consequential increased weathering [17,18]. In addition, the biogeochemical cycling of silica by plants has been found to play a key role in Si bioavailability in soils and Si cycling on a global scale [9,19,20]. In the following, we will focus on phytogenic silica in plants (Section 22.2.1) and soils (Section 22.2.2).

22.2.1 Phytogenic silica in plants—formation and function

Phytogenic silica or phytoliths, that is, the siliceous structures precipitated in plants, are mainly made of amorphous, hydrated silica ($SiO_2 \cdot nH_2O$), but also contain organic matter and various elements like aluminum, calcium, iron, manganese, and phosphorus [21,22]. Phytogenic silica can be found (1) in living plants within cells (i.e., in the cell wall and the cell lumen) forming relatively stable, recognizable phytoliths, that can be found in soils as plant microfossils or (2) in intercellular spaces and extracellular (cuticular) layers forming relatively fragile silica structures [23,24]. The size of Si precipitates in plants ranges from about 100 nm to 1 mm [25,26]. For plant microfossils (soil phytoliths) an international nomenclature based on phytolith morphology has been developed, which is especially used in archeological, paleoenvironmental, evolutionary, taxonomic, and climatological

BOX 22.1

Si bioavailability.

Si is very abundant in the earth's crust, which consists of more than 90 vol.% of SiO_2 and silicates. However, only dissolved Si in the form of H_4SiO_4 is available to organisms, that is, bioavailable. Mineral weathering represents the ultimate source of dissolved Si in terrestrial ecosystems on a geological time scale. Mineral weathering in turn is controlled by climate, specific physicochemical soil conditions, and vegetation [7,8]. Biosilicification denotes the incorporation of inorganic Si into living organisms in the form of biogenic silica. Biosilicification by plants and further organisms like protists has established a biological Si cycle, which controls Si bioavailability on shorter time scales. Bioavailability of Si in soils is controlled by several key factors: (1) the Si concentration in soil solution (dependent on soil pH), (2) the reserve in the solid phase as Si source (dependent on quantities of minerogenic/pedogenic, biogenic, adsorbed, and fertilizer Si pools; physicochemical properties), and (3) the Si adsorption capacity or retention capability (dependent on adsorption/desorption reactions, leaching) of the soil [19,69,85]. For the determination of bioavailable Si, several extraction methods with different extractants like calcium chloride, acetate/acetic acid, or citrate extractions have been developed [19,83].

studies for taxonomic identification of plants [27]. However, regarding Si cycling by plants and Si effects on plants' resistance total Si concentrations, that is, all forms of phytogenic silica, in plants have to be considered.

Phytogenic Si can be found in almost all plant organs, for example, in leaves, stems, and roots (Fig. 22.2).

Si contents vary considerably between plant species with values ranging from about 0.1% to 10% Si per dry mass [28]. Si absorption by plants is controlled by specific influx (called *Lsi1* and *Lsi6*) and efflux (called *Lsi2*) channels, which have been found especially in crops like rice (*Oryza sativa*), wheat (*Triticum aestivum*), or sorghum (*Sorghum bicolor*) [29]. However, it should be kept in mind that the mechanisms behind the uptake, transport, and accumulation of Si in plants (active vs. passive Si transport) are still not fully understood yet, and thus are under controversial discussion [30,31].

Based on their Si content plants have been divided into three groups, that is, (1) nonaccumulators or excluders of Si (Si content per dry mass <0.5%), (2) intermediate accumulators of Si (Si content per dry mass 0.5%–1%), and (3) accumulators of Si (Si content per dry mass >1%) [32]. Regarding field crops, especially cereal grasses of the family Poaceae (or Gramineae) are known as Si accumulators [28]. Furthermore, plant functional groups strongly affect Si stocks in aboveground biomass, whereby grasses increase and legumes decrease these stocks

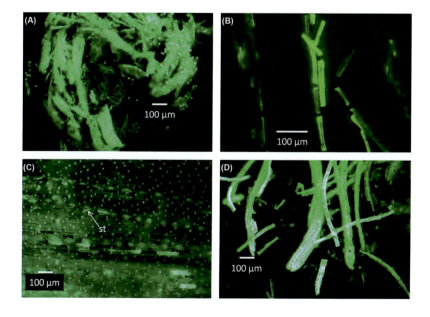

FIGURE 22.2 Micrographs of auto-fluorescent siliceous structures in heated (A) husk, (B) leaf, (C) stem, and (D) root samples of wheat (*Triticum aestivum*). "st" in (C) = stomata in the epidermis of the stem. Micrographs were taken with an inverted epifluorescence microscope. Source: *From Puppe and Kaczorek (2021).*

[9,19,20]. In this context, Si and calcium bioavailability in soils seems to be a key control of shifts in the dominance of grasses (Si accumulators) to the dominance of legumes (calcium accumulators) [33].

In general, the accumulation of Si in plants has been found to increase plants' resistance against several abiotic and biotic stresses. For example, positive effects of Si absorption in relation to water stress, macronutrient imbalance, heavy metal toxicity, herbivory, and bacterial, viral, and fungal diseases have been reported [3,11,34]. In this context, various physiological, morphological, and biochemical mechanisms have been discussed to explain the effect of Si accumulation on plants' increased resistance. However, many of these explanations have been challenged by the circumstance that $Si(OH)_4$ is regarded as biochemically inert contradicting especially biochemical and physiological explanatory approaches. This is why Coskun et al. [35] proposed a unifying model, that is, the "apoplastic obstruction hypothesis" (Box 22.2).

Si uptake and storage in plants have been analyzed for several human-used (herein agricultural) and natural (unused) ecosystems (Table 22.1). While annual Si absorption in aboveground plant biomass in agricultural plant-soil systems is relatively high (i.e., up to several hundred kilograms per hectare per year) [36–39], annual Si accumulation in aboveground plant biomass of natural sites is comparably low (i.e., <50 kg Si ha^{-1} year^{-1} in forests and 60 and 75 kg Si ha^{-1} year^{-1} in *Calamagrostis epigejos* and *Phragmites australis* stands, respectively) [40–44]. Unfortunately, there are almost no data on Si absorption in belowground plant biomass. However, the existing data indicate that Si accumulation in (fine) roots might represent a significant phytogenic Si pool in plant-soil systems (Table 22.1).

These data impressively indicate the potential of agricultural plant-soil systems for Si accumulation. Due to their relatively high biomasses and Si concentrations, crops absorb Si to a large extent. Crops are often harvested without any recycling of plant materials, and thus Si absorbed in crops is removed from the fields without Si compensation in many regions potentially leading to a depletion of bioavailable Si in soils (anthropogenic desilication). This is in contrast to natural ecosystems, where Si is recycled to great amounts. Anthropogenic desilication in turn can have impacts on phytogenic silica pools in soils (Section 22.2.2) and ecosystem functioning and services of agricultural plant-soil systems (see Section 22.4.1 for further details), and thus strategies for prevention of Si depletion of agricultural soils are needed (discussed in Section 22.4.2).

22.2.2 Phytogenic silica in soils—distribution and pool quantities

The distribution of phytoliths in soils is quite variable as they are subject to translocation processes, especially driven by bioturbation and percolation. In natural plant-soil systems, the highest contents of phytogenic Si can be found in the uppermost (organic) soil horizons. Phytolith contents in most soil horizons range between 0.01% and 3%. However, total phytogenic Si contents in soils can be assumed to be even higher, because phytolith analyses usually are restricted to silt-sized particles, that is, particles $>2\,\mu$m. Interestingly, it was found that phytoliths $>5\,\mu$m represent only about 16% of total Si contents of plant materials of *Calamagrostis epigejos* and *Phragmites australis* (Poaceae) [45]. Wilding and Drees [46] showed that about 72% of leaf phytoliths of American beech (*Fagus grandifolia*) are smaller than 5 μm. These findings clearly point to the potential significance of phytogenic Si $<5\,\mu$m for Si cycling in general.

The classification of soil phytoliths is more or less limited to phytoliths $>2\,\mu$m with a well-developed and well-recognizable morphology [27]. Thus there is no definition of soil phytoliths, which are smaller than 2 μm and/or not characterized by a well-recognizable morphology, that is, highly weathered or fragmented phytoliths. Such focus on

---

BOX 22.2

The apoplastic obstruction hypothesis.

This hypothesis assumes that Si deposits in the apoplast lead to increased resistance of plants. Regarding biotic stress these Si deposits are hypothesized to hamper attacks on plants by interfering with (1) the injection of effectors (small proteins secreted by parasites to, for example, suppress plant defense responses) and (2) the formation of haustoria (specialized hyphae formed by fungal parasites to invade plant cells). Regarding abiotic stress, these Si deposits are hypothesized to (1) fortify apoplastic barriers of the plant vasculature precluding the transport and accumulation of toxicants into the shoot and (2) coprecipitate with toxicants in the extracellular matrix. Finally, Si deposits in the cuticle are assumed to prevent water loss, which is particularly important regarding osmotic stress under unfavorable conditions like drought, high salinity, or freezing.

TABLE 22.1 Examples of studies on annual Si absorption in above- and belowground biomass of agricultural and nonagricultural plant-soil systems.

Plant-soil system	Si absorption (kg Si ha^{-1} year^{-1})		References
	aboveground	belowground	
1. Agricultural sites			
Rice	230–470	–	[39]
"	270–500	–	[36]
Sugarcane	379	–	[38]
Wheat	20–113	–	[36]
"	66	–	[37]
2. Nonagricultural sites			
Beech-fir forest	26	–	[41]
Pine forest	8	–	"
Black pine forest	2	–	[40]
Douglas fir forest	31	–	"
European beech forest	23	–	"
Norway spruce forest	44	–	"
Oak forest	19	–	"
Beech forest	35	–	[42]
Deciduous forest (89% beech); soil type: Dystric Cambisol	47	110*	[43]
Deciduous forest (89% beech); soil type: Eutric Cambisol	40	100*	"
Deciduous forest (89% beech); soil type: Rendzic Leptosol	26	68*	"
Initial ecosystem (artificial catchment); *Calamagrostis epigejos* stands	60	–	[44]
Initial ecosystem (artificial catchment); *Phragmites australis* stands	75	–	"

*0–90 cm

well-recognizable phytoliths works well for corresponding taxonomic identifications of plants in archeological or paleoenvironmental studies, for example. However, for a comprehensive understanding of Si cycling in plant-soil systems, all forms of phytogenic silica have to be considered. Consequently Schaller and colleagues [19] proposed a model for the presence of phytogenic silica in soils, termed the "phytogenic Si continuum in soils" (Fig. 22.3).

This model describes a continuum of various forms of phytogenic silica ranging from nanometers to millimeters in size and at different stages of decomposition/dissolution in soils. Phytogenic silica is continuously transformed in soils driven by physicochemical properties of soils (e.g., soil pH, Si bioavailability, and adsorption capacity) and phytoliths (e.g., size, specific surface area, and degree of condensation). For example, the so-called Si double layer represented by relatively small phytogenic Si particles with a potentially lower degree of condensation might dissolve faster compared to other phytoliths. The degree of silica condensation might be influenced by the location of phytolith formation in the cell because cell wall phytoliths seem to be less stable than cell lumen phytoliths [47].

Phytolith properties are highly variable and seem to depend mainly on phytolith morphotypes, that is, phytolith geometry and related surface-area-to-volume ratio, although some studies ascribe differences in phytolith dissolution also to phytolith origin, that is, type of vegetation. For example, grass phytoliths appeared to be less soluble compared to tree phytoliths [42]. In general, the following can be stated as a rule of thumb: the lower the degree of condensation, the higher the specific surface area, and the higher the surface-to-volume ratio, the higher the potential dissolution rate of phytogenic silica particles.

Phytoliths represent a huge pool of relatively soluble silica compared to Si minerals in soils, and thus are considered as one of the main sources of bioavailable Si in terrestrial ecosystems. Phytogenic Si pools have been quantified for numerous plant-soil systems highlighting their relevance for Si bioavailability and cycling in these systems (Table 22.2). While there are some quantitative data on different BSi pools in nonagricultural soils [42,45,48–50], quantitative data on

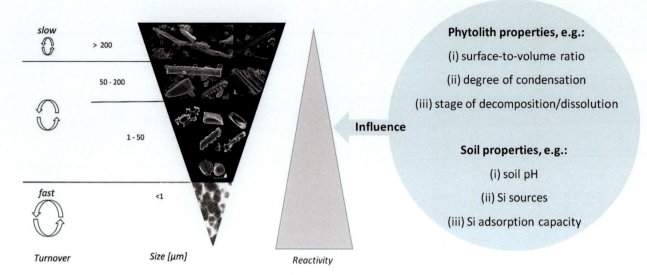

FIGURE 22.3 Conceptual model of the phytogenic Si continuum in soils. Source: *Modified from Schaller J, Puppe D, Kaczorek D, Ellerbrock R, Sommer M. Silicon cycling in soils revisited. Plants 2021;10(2):295. https://doi.org/10.3390/plants10020295.*

TABLE 22.2 Selection of data on various BSi pools and alkaline extractable Si in soils of different plant-soil systems. Values are given in kg Si ha^{-1} unless stated otherwise.

Plant-soil system	Alkaline extractable Si (kg Si ha^{-1})	BSi in soils (kg Si ha^{-1})				References
		Phytogenic	Protozoic	Protophytic	Zoogenic	
1. Agricultural sites						
Rice	–	800 (upper 25 cm)	–	X	–	[52]
"	3.3–4.6 g Si kg^{-1} (upper 20 cm)	4.1–6.9 g Si kg^{-1} (upper 20 cm)	–	X	–	[53]
Wheat	6,500 (upper 50 cm)	7,260 (upper 50 cm)	–	–	–	[36]
"	–	1.1–3.4 g Si kg^{-1} (upper 25 cm)	–	X	X	[37]
Arable land	4.7 g Si kg^{-1} (upper 30 cm)	1,400 (upper 70 cm)	–	–	–	[51]
"	5–25 g Si kg^{-1} (upper 5 cm)	–	3–50 µg Si kg^{-1} (upper 5 cm)	–	–	[54]
2. Nonagricultural sites						
Beech forest	40,000 (upper 100 cm)	660 (upper 20 cm)	1.9 (upper 5 cm)	–	–	[42]
Nature reserve	–	–	50–80 µg Si kg^{-1} (upper 2.5 cm)	0–35 µg Si kg^{-1} (upper 2.5 cm)	–	[48]
Initial ecosystem (artificial catchment)	–	–	0–0.06 (upper 5 cm)	0.025–0.3 (upper 5 cm)	0–0.2 (upper 5 cm)	[49]
"	1,960–5,520 (upper 5 cm)	0–0.5 (upper 5 cm)	0–0.4 (upper 5 cm)	0–1.6 (upper 5 cm)	0–0.5 (upper 5 cm)	[45]
Floodplain (heavy metal contaminated)	–	–	600–700 µg Si kg^{-1} (upper 5 cm)	1,700 µg Si kg^{-1} (upper 5 cm)	–	[50]
Floodplain (uncontaminated)	–	–	900–1,900 µg Si kg^{-1} (upper 5 cm)	4,900–8,400 µg Si kg^{-1} (upper 5 cm)	–	"

X = observed, but not quantified

BSi pools in agricultural soils are mainly restricted to phytogenic Si [36,51]. Sponge spicules and diatom frustules in agricultural soils were observed under the microscope, but there are no quantitative data on corresponding zoogenic and protophytic Si pools in these soils, respectively [37,52,53]. Siliceous testate ameba shells were also observed [37], but quantitative data on corresponding protozoic Si pools are rare [54].

22.3 Further organisms and corresponding BSi pools

Aside from plants, numerous soil-inhabiting organisms have been found to contribute to Si cycling in plant-soil systems. However, while research has been focused on plants and phytogenic silica, comparably little is known about these organisms and their role in the Si cycle. In the following, we will focus on protists (Section 22.3.1) and sponges, fungi, and bacteria (Section 22.3.2) and their potential for Si cycling in plant-soil systems.

22.3.1 Unicellular organisms in soils—the role of protists in terrestrial Si cycling

Regarding protozoic BSi several studies have been conducted just recently (reviewed by one of us [55]). Protozoic silica in soils is mainly formed by testate amoebae (Box 22.3; Fig. 22.4).

The silica platelets of testate amoebae are formed in so-called silica deposition vesicles in the cell cytoplasm and deposited on the cell surface by exocytosis, where they are finally bound together by organic cement [56].

At the beginning of the 21st century, the potential of protozoic silica for Si cycling was hypothesized [7,57]. Shortly thereafter Aoki and colleagues [58] were the first, who quantified BSi in the shells of different testate ameba taxa in the order Euglyphida. Based on these results they further quantified protozoic silica pools in pine-oak forest soil in Japan and calculated annual biosilicification rates of living testate amoebae. In doing so, Aoki and coauthors [58] showed annual biosilicification of idiosomic testate amebae to be comparable to silica released by trees via litter fall, and thus testate amebae to potentially be as important for global Si cycling as trees. Although this potential was recognized by some authors (e.g., Ref. [9]), it took some more years until the quantification of protozoic silica pools and annual biosilicification of testate amebae and implications for Si cycling became the focus of attention of several researchers (e.g., Refs. [42,48,59]). These studies clearly showed that biosilicification by testate amoebae has to be considered in analyses of Si cycling in plant-soil systems because Si fixation in testate ameba shells is comparable to or even can exceed the amounts of Si absorbed by trees year by year [55,60,61]. Recent studies showed the susceptibility of testate amoebae and corresponding protozoic Si pools to land-use changes [54,62]. However, we urgently need detailed data on protozoic silica pools and their contribution to total BSi pools in agricultural plant-soil systems to clarify their significance for Si cycling in these systems (Table 22.2).

Moreover, we still do not know how big protozoic Si pools represented by single idiosomes (the building blocks of siliceous shells) in soils are (Fig. 22.5). Total protozoic Si pools (shells plus single idiosomes) might be comparable to the ones of phytoliths in some soils [55].

BOX 22.3

Testate amebae.

Testate amebae form a polyphyletic group of worldwide occurring, unicellular eukaryotes (protists) with a shell (also termed test) ranging between about 5 and 300 μm. Testate amoebae can be assigned to two supergroups, that is, (1) the Amorphea including the order Arcellinida and (2) TSAR including the order Euglyphida [91]. The order Arcellinida includes testate amoebae with lobose pseudopodia and shells made by secretion (autogenous shells), agglutination of foreign materials collected in the environment (xenogenous shells), or a combination of secretion and agglutination. The order Euglyphida includes testate amoebae with filose pseudopodia and almost all extant species in this order are characterized by siliceous shells made up of self-synthesized silica platelets, the so-called idiosomes. Research on protozoic silica has been focused on species in the order Euglyphida, although a few taxa (i.e., *Lesquereusia*, *Netzelia*, and *Quadrulella*) with autogenous siliceous shells can also be found in the order Arcellinida.

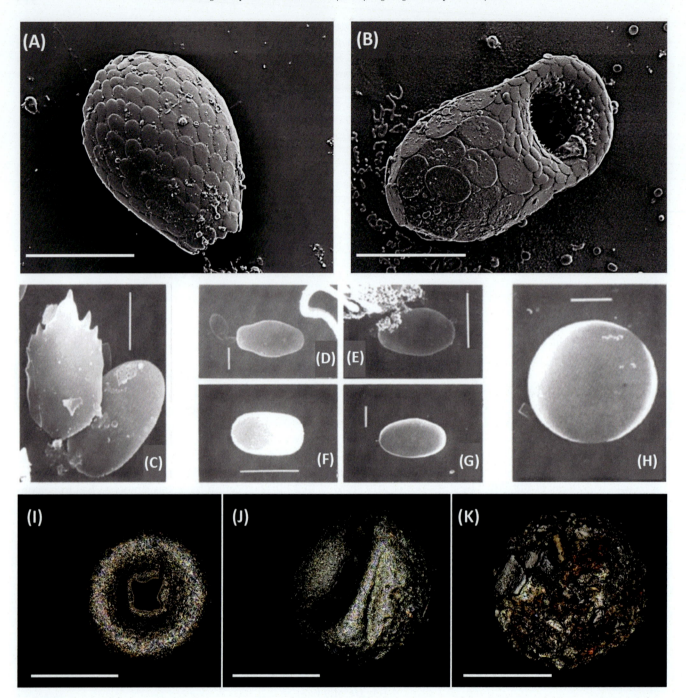

FIGURE 22.4 Micrographs of euglyphid testate ameba shells (A and B, scanning electron microscopy, scale bars = 20 μm), single idiosomes (C-H, scanning electron microscopy, scale bars = 3 μm), and arcellinid testate ameba shells (I-K, confocal laser scanning microscopy, scale bars = 50 μm). (A) *Euglypha rotunda*-like testate ameba, (B) *Puytoracia bonneti*, (C) an apertural platelet of *Euglypha* sp. on top of an unknown platelet, (D) most likely a platelet of *Assulina* sp., (E) platelet of *Assulina muscorum*, (F) platelet of *Corythion* sp. or *Trinema* sp., (G) platelet of *Euglypha strigosa*, (H) a large body platelet of *Trinema* sp., (I) *Trigonopyxis arcula*, apertural view, (J) *T. arcula*, lateral view, and (K) *T. arcula*, dorsal view. *Source: Modified from Puppe D. Review on protozoic silica and its role in silicon cycling. Geoderma 2020;365:114224. https://doi.org/10.1016/j.geoderma.2020.114224.*

Aside from testate amoebae terrestrial diatoms (Fig. 22.5) have been found to play a role in Si cycles of some plant-soil systems [45,48,49]. There are also some hints that bioweathering might play a role in diatoms (see Ref. [6,63]), but there are no studies on bioweathering by other Si accumulating protists like testate amoebae available so far. Regarding agricultural plant-soil systems, Desplanques and coauthors [52] indicated the potential of diatoms as a

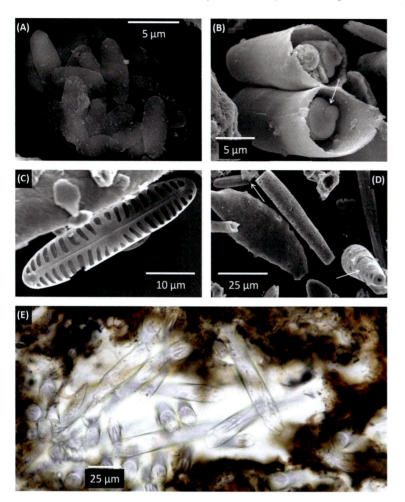

FIGURE 22.5 Micrographs of single idiosomes (platelets), diatom frustules, and sponge spicules in forest [(A)–(C)] and agricultural [(D) and (E)] soils. (A) Assemblage of differently sized, oval siliceous testate ameba platelets, (B) round and oval idiosomes (arrow) within a sleeve-like siliceous structure, (C) pennate diatom frustule, (D) pennate diatom frustule (top left, arrow) and sponge spicule fragment (downright, arrow), and (E) sponge spicules in a soil thin section. Micrographs were taken with a scanning electron [(A)–(D)] and a light (E) microscope. Source: *From Kaczorek and Puppe (2021).*

source of bioavailable Si in a rice field in Camargue, France. These authors calculated Si inputs via irrigation water of about 100 kg Si ha^{-1} year^{-1}, whereby diatoms were identified as main source of Si. Taking into account that about 270 kg Si ha^{-1} were exported via rice harvesting (approximately equaling Si uptake by rice) at this site per year, the reported contribution of diatoms becomes even more impressive. However, there are still no quantitative data on protophytic Si pools in agricultural plant-soil systems to the best of our knowledge (Table 22.2).

22.3.2 Sponges, fungi, and bacteria—the underexplored players in terrestrial Si cycling

While there are few studies on zoogenic Si pools (sponge spicules) and their role in Si cycling in plant-soil systems [45,49] (Table 22.2), there are no comparable studies on fungal and bacterial Si pools in terrestrial ecosystems. At least we know that some bacteria (e.g., *Proteus mirabilis*) and fungi are able to accumulate Si within their cells [6]. Furthermore, these organisms can enhance the dissolution of amorphous and crystalline silica by the release of acidic metabolites (bioweathering). In general, we urgently need more research on BSi formed by sponges, fungi, and bacteria, if we want to understand the importance of biota for Si cycling in agricultural plant-soil systems in detail. In fact, sponge spicules are quite common in some agricultural soils (Fig. 22.5), but we still have no idea about their contribution to total BSi pools in these soils, and thus their relevance for Si cycling in agricultural plant-soil systems.

22.4 Implications for ecosystem functioning and services of agricultural plant-soil systems

In the previous subsections, we discussed the influence of plants, protists, and other organisms on Si bioavailability and Si cycling in plant-soil systems. Now we will focus on human impacts on Si accumulating plants and consequences for Si cycling, especially in agricultural plant-soil systems. In this context, we will discuss how humans influence Si cycling (Section 22.4.1) and how we can prevent anthropogenic desilication (Section 22.4.2).

22.4.1 Anthropogenic desilication—how humans influence Si cycling

Si is accumulated to a large extent in several major biome types, for example, forests, steppes, and cultivated lands. However, humans directly affect the distribution and size of these biomes and thus influence corresponding Si cycling through intensified land use, that is, forestry and agriculture (changes of soil properties and vegetation) [12,13]. In addition, increased greenhouse gas emissions and consequential changes in climate conditions might have severe impacts on Si cycling [20]. Clymans et al. [64] estimated that the total amorphous (biogenic plus minerogenic) Si pool in temperate soils decreased by about 10% within the last 5,000 years due to human land use. Amorphous silica in turn has been found to increase the water-holding capacity of soils, influence nutrient supply (e.g., phosphorus mobility), and act as the main source for bioavailable Si [19,65,66].

Concentrations of amorphous Si are considerably lower in agricultural soils compared to nonagricultural soils, for example, forest or steppe soils. This is because Si exports through harvested crops generally lead to Si loss in agricultural plant-soil systems (= anthropogenic desilication) [13]. However, some agricultural practices might also increase Si availability in soils, for example, humans set fires [67,68], the application of Si-rich fertilizers [19], or liming (pH effect; [69]). On a global scale, about 35% of BSi in plant biomass is accumulated in field crops and this proportion is going to increase with increased agricultural production within the next decades [16]. Si uptakes, which can be assumed to approximately equal Si outputs by harvesting, of cereal crops are quite high and reach up to several 100 kg ha^{-1} in a year (Table 22.1). However, it should be kept in mind that phytogenic Si inputs are not only driven by aboveground plant materials but also by the roots of plants (Table 22.1).

In contrast to natural ecosystems (see Table 22.1), where large amounts of Si are recycled year by year (e.g., Ref. [42]), the annual Si exports in agricultural plant-soil systems are mostly not compensated (Fig. 22.6).

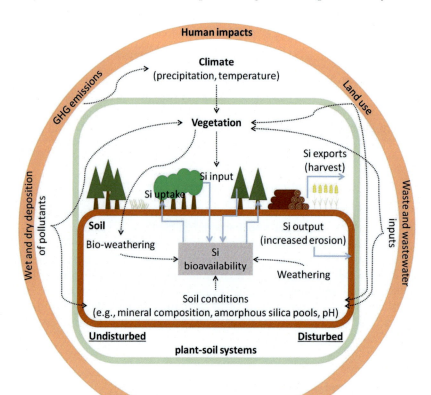

FIGURE 22.6 Schematic overview of Si cycling in undisturbed (unused and natural) and disturbed (used) plant-soil systems. Si bioavailability in soils, and thus Si cycling, is strongly influenced by human impacts, that is, greenhouse gas (GHG) emissions, pollution via wet and dry deposition, waste and wastewater inputs, and land use (agriculture and forestry). Source: *Modified from Katz O, Puppe D, Kaczorek D, Prakash NB, Schaller J. Silicon in the soil—plant continuum: intricate feedback mechanisms within ecosystems. Plants 2021;10.*

However, targeted manipulation of Si cycling (e.g., Si fertilization, straw recycling) might be a promising strategy to both (1) prevent desilication of agricultural plant-soil systems and consequently improve crop resistance against abiotic and biotic stress and (2) enhance C sequestration (Box 22.4) in agricultural biogeosystems to mitigate climate change [70–72].

Guntzer and coauthors [73] analyzed archived soil and plant samples of the long-term "Broadbalk winter wheat experiment" at Rothamsted Research in the United Kingdom. They found that the long-term removal of wheat straw considerably decreased amorphous silica pools in soils. However, they did not observe a distinct relationship between the decrease of amorphous silica and a corresponding decrease in Si concentrations of crop straw. In fact, Guntzer et al. [73] found such a relationship only for the samples taken before the year 1944. After this year Si concentrations in straw tended to increase. From their results, Guntzer et al. [73] concluded an increased soil pH due to periodic liming to increase amorphous silica (i.e., phytoliths) dissolution, and thus to represent the main driver of increased Si uptake by the cultivated wheat plants. This is underpinned by a recent study of Caubet et al. [74], who ascribed an increase of calcium chloride extractable Si in cultivated soils (perennial and annual crops) in France to liming. However, it has to be kept in mind that Si availability in (agricultural) soils is determined by a complex interaction of factors, and thus liming effects on Si availability follow no general rule, that is, there are studies showing negative and other studies reporting positive correlations between pH and Si availability [69].

22.4.2 Anthropogenic desilication—strategies for prevention

A long-term field experiment, established in 1963 in NE Germany, revealed that about 43%–60% of Si exports can be saved by crop straw recycling (herein straw incorporation) [37]. Puppe and colleagues (2021) found crop straw incorporation to become more effective the longer straw incorporation is applied indicated by an increase (or replenishment) of plant-available Si in soils with time, which was also reflected by increasing phytolith contents in these soils. In fact, plant-available Si increased from about 5 mg kg^{-1} (a value that is comparable to other agricultural sites in the temperate zone; [51]) to about 10 mg kg^{-1} (comparable to undisturbed ecosystems like forests under temperate conditions, e.g., 10–40 mg kg^{-1} [75]; 7–40 mg kg^{-1} [60]; and 4–80 mg kg^{-1} [51]) within 42 years of straw incorporation [37]. Thus straw incorporation in combination with soil [76] and foliar [3] Si fertilization might be the most promising strategy to (1) restore natural Si recycling processes in agricultural ecosystems to the highest possible extent and (2) produce resilient crops in modern, sustainable agriculture.

Furthermore, combined Si-phosphorus fertilizers have been found to increase concentrations of plant-available Si in soils leading to higher biomasses and phytolith contents of rice plants [70–72]. However, due to the fact that Si-phosphorus interactions in the plant-soil system are driven by complex biogeochemical processes that are still not fully understood (e.g., Ref. [65]), further studies are needed to enlighten this aspect. In this context, it is of great interest to which degree Si uptake of cultured plants is determined by their phylogenetic position and environmental factors like temperature or Si availability (e.g., Refs. [28,77,78]).

Aside from straw incorporation, open straw burning in the field and the use of biochar have been proposed as suitable ways for Si supply of agricultural plant-soil systems [79]. Every way of straw recycling (incorporation, open burning, or pyrolysis) has its advantages and disadvantages and should be chosen according to site-specific

---

BOX 22.4

Si-induced C sequestration.

C sequestration in agricultural systems might be enhanced by the regulation of (1) weathering (e.g., silicate rock powder amendment), (2) organic C stabilization (e.g., silicon and biochar fertilization), and (3) phytolith-occluded C (e.g., partial straw retention after harvest) [72]. C sequestration in the form of phytolith-occluded C, for example, has been stated to amount to 16–44 Tg CO_2 per year in global agricultural ecosystems, with the big three (maize, rice, and wheat) as main contributors. However, it should be noted that the potential of phytoliths in C sequestration is still under controversial discussion (see the review of Ref. [47] and references therein).

(soil properties, cultured crop, climate) factors. For example, open burning of straw in the field is still a common practice in many countries worldwide, although it has been widely criticized due to health (smog) and climate change (emissions of CO_2) concerns (e.g., Ref. [80]). However, the ban on open straw burning has been discussed controversially (e.g., Ref. [81]).

22.5 Concluding remarks

For a better understanding of biological impacts on Si cycling in agricultural plant-soil systems more empirical data are urgently needed. A combination of field and greenhouse experiments will help us (1) to understand in detail the underlying processes of Si bioavailability, uptake, and functioning in crops and (2) to derive protocols for agricultural practices of Si supply (Si fertilization and crop straw incorporation) considering environmental (e.g., specific soil conditions and climate) and plant-related (e.g., specific Si requirements) conditions. Crop straw incorporation might be a promising strategy to prevent anthropogenic desilication and to replenish plant-available Si stocks of agricultural plant-soil systems. Si supply by crop straw incorporation has the potential to act as a key management practice in sustainable, low fertilization agriculture in the future. Consequently, Si supply through crop straw incorporation might be a key factor to achieve a drastic reduction of greenhouse gas emissions as a response to the escalating climate crisis. In this context, the ultimate goal should be the restoration of natural Si recycling processes in agricultural ecosystems to the extent deemed possible. Moreover, our current knowledge of the role of biota, biosilicification, and BSi pools in Si cycles of agricultural plant-soil systems is still very limited. The vast majority of BSi studies have been focused on phytogenic silica, that is, phytoliths. However, we have no idea about global annual Si cycling rates of, for example, protists (testate amoebae, diatoms) in terrestrial ecosystems. Their significance for Si cycling in terrestrial ecosystems might be comparable to the role of protists for Si cycling in the oceans (marine diatoms). Recent studies indicate that protistic Si pools are strongly affected by land use, but we still do not know which effects protistic Si pool changes have on the ecosystem scale (e.g., impacts on Si availability). Furthermore, we still do not know if unicellular organisms are able to increase weathering rates in soils by bioweathering. Actually, there are some hints that bioweathering might play a role in protists, underpinning their potential significance for Si cycling in terrestrial ecosystems. What we need now to push the field forward is detailed field and laboratory research on protistic BSi in agricultural plant-soil systems.

22.6 Future directions

To put the field forward, the following key questions should be addressed in future studies:

1. *How big are the different BSi pools in agricultural soils?*

   In our current chapter, we emphasized the significance BSi for Si cycling in agricultural plant-soil systems. For a profound understanding of Si cycling in these systems, we urgently need a comprehensive overview of all occurring BSi pools. Thus, quantitative data on (1) soil phytoliths $<2\,\mu m$, (2) Si absorption in belowground biomass of crops, and (3) protistic (testate ameba shells and diatom frustules), zoogenic (sponge spicules), fungal, and bacterial Si pools are needed. Only with this data, it will be possible to evaluate the role of the various BSi pools in agricultural plant-soil systems and to better understand how Si availability in soils is related to these pools. This knowledge in turn will help us to evaluate the Si status (Si availability, Si sources, physic-chemical BSi properties) of a specific plant-soil system and to assess the need for Si fertilization.

2. *How can we reliably test Si availability in agricultural soils?*

   Unfortunately, there is no standard extraction method for the determination of Si availability in soils yet, because these procedures have been developed for specific plants in specific climates, that is, mainly sugarcane and rice in (sub)tropical zones. This is why different studies often show inconsistent results. Crusciol et al. [82] showed that correlations between plant-available Si in soils and Si concentrations in sugarcane were not only depending on soil texture, but also on the used extractant (i.e., $CaCl_2$, deionized water, KCl, Na-acetate buffer at pH 4.0, and acetic acid). Thus we urgently need standardized tests for quantification of plant-available Si in agricultural soils. Due to the fact that several tests are used by researchers worldwide [19,83], it is difficult to determine reliable lower limits for plant-available Si in different soils, and thus for a need for Si supply. In addition, extractions alone might be not suitable for an evaluation of the Si status of a given soil (which Si sources are present? How much Si is plant available?), because some studies showed that extractions do not necessarily reflect, for example, phytolith pool quantities (which are often assumed to represent the main source of plant-available Si) in soils [51,84]. Thus Si extraction methods should be accompanied by analyses of physicochemical properties of (1) biogenic silica as a potential key source of plant-available Si (e.g., Ref. [5]) and (2) soils as their adsorption capacities or retention capabilities directly control Si availability (e.g., Ref. [85]).

3. *Is there an interdependency between Si availability in soils and Si concentrations in plants?*

Miles et al. [86] analyzed 28 sites located throughout the sugarcane-growing areas of South Africa and found a close correlation between plant-available Si in soils and Si contents in sugarcane leaves. Regarding rice production, Korndörfer et al. [87] analyzed 28 field experiments in the Everglades Agriculture Area representing a wide range of available Si in soils. They found plant-available Si in these soils to be correlated with Si contents in rice straw. In contrast to these studies, which considered several study sites with relatively large gradients in plant-available Si, Klotzbücher and coauthors [88] found no relationship between plant-available Si in soils (herein concentration of dissolved Si in soil solution) and Si contents in rice straw in one (i.e., the drier one) of two analyzed cropping seasons in a field experiment in Southern Vietnam. However, they found such a correlation in the second cropping season, that is, the wetter one. From their results, Klotzbücher et al. [88] speculated climatic differences to be responsible for their observation and concluded field experiments to be inconsistent with results from laboratory studies regarding relationships between plant-available Si in soils and Si uptake by plants (e.g., Ref. [89]). This is underpinned by a study of Keeping [90], who found that the uptake of Si by sugarcane in a shade house pot experiment did neither reflect the concentration of plant-available Si in soils nor the Si content of used Si sources (calcium silicate slag, fused magnesium (thermo) phosphate, volcanic rock dust, magnesium silicate, and granular potassium silicate). This impressively indicates that we need more data on the relationship between Si availability in soils and Si concentrations in plants. Only if we know how much bioavailable Si of a specific soil is taken up by a specific crop (uptake efficacy), the need for additional Si supply can be determined. Unfortunately, this is hampered by the fact that there are no standard protocols for both the determination of bioavailable Si in soils (which extractant?) and the Si concentration of plants (e.g., use of plant shoots or leaves? Testing at which growth stage?). Such protocols might be derived from results of *meta-* and big-data analyses of data obtained from interdisciplinary research, practitioner-scientist cooperations, and standardized experiments.

Acknowledgments

DP was funded by the Deutsche Forschungsgemeinschaft (DFG) under grant PU 626/2-1. Many thanks to Patricia Legaspi (Elsa User Support) for technical advice.

References

[1] Epstein E. Silicon. Annu Rev Plant Biol 1999;50:641−64. Available from: https://doi.org/10.1146/annurev.arplant.50.1.641.

[2] Ma JF, Yamaji N. Silicon uptake and accumulation in higher plants. Trends Plant Sci 2006;11(8):392−7. Available from: https://doi.org/10.1016/j.tplants.2006.06.007.

[3] Puppe D, Sommer M. Experiments, uptake mechanisms, and functioning of silicon foliar fertilization—a review focusing on maize, rice, and wheat. Adv Agron 2018;152:1−49. Available from: https://doi.org/10.1016/bs.agron.2018.07.003.

[4] Fraysse F, Pokrovsky OS, Schott J, Meunier JD. Surface chemistry and reactivity of plant phytoliths in aqueous solutions. Chem Geol 2009;258(3−4):197−206. Available from: https://doi.org/10.1016/j.chemgeo.2008.10.003.

[5] Puppe D, Leue M. Physicochemical surface properties of different biogenic silicon structures: results from spectroscopic and microscopic analyses of protistic and phytogenic silica. Geoderma 2018;330:212−20. Available from: https://doi.org/10.1016/j.geoderma.2018.06.001.

[6] Ehrlich H, Demadis KD, Pokrovsky OS, Koutsoukos PG. Modern views on desilicification: biosilica and abiotic silica dissolution in natural and artificial environments. Chem Rev 2010;110(8):4656−89. Available from: https://doi.org/10.1021/cr900334y.

[7] Sommer M, Kaczorek D, Kuzyakov Y, Breuer J. Silicon pools and fluxes in soils and landscapes—a review. J Plant Nutr Soil Sci 2006;169(3):310−29. Available from: https://doi.org/10.1002/jpln.200521981.

[8] Street-Perrott FA, Barker PA. Biogenic silica: a neglected component of the coupled global continental biogeochemical cycles of carbon and silicon. Earth Surf Process Landf 2008;33(9):1436−57. Available from: https://doi.org/10.1002/esp.1712.

[9] Struyf E, Conley DJ. Emerging understanding of the ecosystem silica filter. Biogeochemistry 2012;107(1−3):9−18. Available from: https://doi.org/10.1007/s10533-011-9590-2.

[10] Tréguer PJ, De La Rocha CL. The world ocean silica cycle. Annu Rev Mar Sci 2013;5:477−501. Available from: https://doi.org/10.1146/annurev-marine-121211-172346.

[11] Katz O, Puppe D, Kaczorek D, Prakash NB, Schaller J. Silicon in the soil−plant continuum: intricate feedback mechanisms within ecosystems. Plants 2021;10:652.

[12] Struyf E, Smis A, Van Damme S, Garnier J, Govers G, Van Wesemael B, et al. Historical land use change has lowered terrestrial silica mobilization. Nat Commun 2010;1(8):129. Available from: https://doi.org/10.1038/ncomms1128.

[13] Vandevenne F, Struyf E, Clymans W, Meire P. Agricultural silica harvest: have humans created a new loop in the global silica cycle? Front Ecol Environ 2012;10(5):243–8. Available from: https://doi.org/10.1890/110046.

[14] Meunier JD, Guntzer F, Kirman S, Keller C. Terrestrial plant-Si and environmental changes. Mineralogical Mag 2008;72(1):263–7. Available from: https://doi.org/10.1180/minmag.2008.072.1.263.

[15] Tubana BS, Babu T, Datnoff LE. A review of silicon in soils and plants and its role in us agriculture: history and future perspectives. Soil Sci 2016;181(9–10):393–411. Available from: https://doi.org/10.1097/SS.0000000000000179.

[16] Carey JC, Fulweiler RW. Human appropriation of biogenic silicon—the increasing role of agriculture. Funct Ecol 2016;30(8):1331–9. Available from: https://doi.org/10.1111/1365-2435.12544.

[17] Hinsinger P, Fernandes Barros ON, Benedetti MF, Noack Y, Callot G. Plant-induced weathering of a basaltic rock: experimental evidence. Geochimica et Cosmochimica Acta 2001;65(1):137–52. Available from: https://doi.org/10.1016/S0016-7037(00)00524-X.

[18] Kelly EF, Chadwick OA, Hilinski TE. The effect of plants on mineral weathering Biogeochemistry, 42. Netherlands: Kluwer Academic Publishers; 1998. p. 21–53Issues 1–2. Available from: https://doi.org/10.1007/978-94-017-2691-7_2.

[19] Schaller J, Puppe D, Kaczorek D, Ellerbrock R, Sommer M. Silicon cycling in soils revisited. Plants 2021;10(2):295. Available from: https://doi.org/10.3390/plants10020295.

[20] Struyf E, Smis A, van Damme S, Meire P, Conley DJ. The global biogeochemical silicon cycle. Silicon 2009;1(4):207–13. Available from: https://doi.org/10.1007/s12633-010-9035-x.

[21] Kameník J, Mizera J, Řanda Z. Chemical composition of plant silica phytoliths. Environ Chem Lett 2013;11(2):189–95. Available from: https://doi.org/10.1007/s10311-012-0396-9.

[22] Wu Y, Yang Y, Wang H, Wang C. The effects of chemical composition and distribution on the preservation of phytolith morphology. Appl Phys A 2014;114(2):503–7. Available from: https://doi.org/10.1007/s00339-013-7616-4.

[23] Hodson MJ. The development of phytoliths in plants and its influence on their chemistry and isotopic composition. Implications for palaeoecology and archaeology. J Archaeol Sci 2016;68:62–9. Available from: https://doi.org/10.1016/j.jas.2015.09.002.

[24] Sangster AG, Hodson MJ, Tubb HJ. Chapter 5: Silicon deposition in higher plants. Silicon in agriculture. Amsterdam: Elsevier BV; 2001. p. 85–113. Available from: https://doi.org/10.1016/s0928-3420(01)80009-4.

[25] Piperno DR. Phytolith analysis: an archaeological and geological perspective. San Diego, CA: Academic Press; 1988.

[26] Watteau F, Villemin G. Ultrastructural study of the biogeochemical cycle of silicon in the soil and litter of a temperate forest. Eur J Soil Sci 2001;52(3):385–96. Available from: https://doi.org/10.1046/j.1365-2389.2001.00391.x.

[27] Neumann K, Strömberg CAE, Ball T, Albert RM, Vrydaghs L, Cummings LS. International code for phytolith nomenclature (ICPN) 2.0. Ann Botany 2019;124(2):189–99. Available from: https://doi.org/10.1093/aob/mcz064.

[28] Hodson MJ, White PJ, Mead A, Broadley MR. Phylogenetic variation in the silicon composition of plants. Ann Bot 2005;96(6):1027–46. Available from: https://doi.org/10.1093/aob/mci255.

[29] Ma JF, Yamaji N. A cooperative system of silicon transport in plants. Trends Plant Sci 2015;20(7):435–42. Available from: https://doi.org/10.1016/j.tplants.2015.04.007.

[30] Exley C. A possible mechanism of biological silicification in plants. Front Plant Sci 2015;6:853. Available from: https://doi.org/10.3389/fpls.2015.00853.

[31] Exley C, Guerriero G, Lopez X. How is silicic acid transported in plants? Silicon 2020;12(11):2641–5. Available from: https://doi.org/10.1007/s12633-019-00360-w.

[32] Ma JF, Takahashi E. Soil, fertilizer, and plant silicon research in Japan. Amsterdam; Boston: Elsevier; 2002.

[33] Schaller J, Hodson MJ, Struyf E. Is relative Si/Ca availability crucial to the performance of grassland ecosystems? Ecosphere 2017;8:e01726.

[34] Hwang BC, Metcalfe DB. Reviews and syntheses: impacts of plant-silica-herbivore interactions on terrestrial biogeochemical cycling. Biogeosciences 2021;18(4):1259–68. Available from: https://doi.org/10.5194/bg-18-1259-2021.

[35] Coskun D, Deshmukh R, Sonah H, Menzies JG, Reynolds O, Ma JF, et al. The controversies of silicon's role in plant biology. New Phytol 2019;221(1):67–85. Available from: https://doi.org/10.1111/nph.15343.

[36] Keller C, Guntzer F, Barboni D, Labreuche J, Meunier JD. Impact of agriculture on the Si biogeochemical cycle: input from phytolith studies. C R Geosci 2012;344(11–12):739–46. Available from: https://doi.org/10.1016/j.crte.2012.10.004.

[37] Puppe D, Kaczorek D, Schaller J, Barkusky D, Sommer M. Crop straw recycling prevents anthropogenic desilication of agricultural soil–plant systems in the temperate zone—results from a long-term field experiment in NE Germany. Geoderma 2021;403:115187. Available from: https://doi.org/10.1016/j.geoderma.2021.115187.

[38] Savant NK, Korndörfer GH, Datnoff LE, Snyder GH. Silicon nutrition and sugarcane production: a review. J Plant Nutr 1999;22(12):1853–903. Available from: https://doi.org/10.1080/01904169909365761.

[39] Savant NK, Snyder GH, Datnoff LE. Silicon management and sustainable rice production. Adv Agron 1996;58(C):151–99. Available from: https://doi.org/10.1016/S0065-2113(08)60255-2.

[40] Cornelis JT, Ranger J, Iserentant A, Delvaux B. Tree species impact the terrestrial cycle of silicon through various uptakes. Biogeochemistry 2010;97(2):231–45. Available from: https://doi.org/10.1007/s10533-009-9369-x.

[41] Bartoli F. The biogeochemical cycle of silicon in two temperate forest ecosystems. Environmental biogeochemistry. Ecol Bull (Stockholm) 1983;35:469–76.

[42] Sommer M, Jochheim H, Höhn A, Breuer J, Zagorski Z, Busse J, et al. Si cycling in a forest biogeosystem—the importance of transient state biogenic Si pools. Biogeosciences 2013;10(7):4991–5007. Available from: https://doi.org/10.5194/bg-10-4991-2013.

[43] Turpault MP, Calvaruso C, Kirchen G, Redon PO, Cochet C. Contribution of fine tree roots to the silicon cycle in a temperate forest ecosystem developed on three soil types. Biogeosciences 2018;15(7):2231–49. Available from: https://doi.org/10.5194/bg-15-2231-2018.

[44] Wehrhan M, Puppe D, Kaczorek D, Sommer M. Spatial patterns of aboveground phytogenic Si stocks in a grass-dominated catchment—results from UAS based high resolution remote sensing. Biogeosciences 2021;18:5163–83.

[45] Puppe D, Höhn A, Kaczorek D, Wanner M, Wehrhan M, Sommer M. How big is the influence of biogenic silicon pools on short-term changes in water-soluble silicon in soils? Implications from a study of a 10-year-old soil-plant system. Biogeosciences 2017;14 (22):5239−52. Available from: https://doi.org/10.5194/bg-14-5239-2017.

[46] Wilding LP, Drees LR. Biogenic opal in Ohio soils. Soil Sci Soc Am J 1971;35:1004−10. Available from: https://doi.org/10.2136/sssaj1971.03615995003500060041x.

[47] Hodson MJ. The relative importance of cell wall and lumen phytoliths in carbon sequestration in soil: a hypothesis. Front Earth Sci 2019;7:167. Available from: https://doi.org/10.3389/feart.2019.00167.

[48] Creevy AL, Fisher J, Puppe D, Wilkinson DM. Protist diversity on a nature reserve in NW England—with particular reference to their role in soil biogenic silicon pools. Pedobiologia 2016;59(1−2):51−9. Available from: https://doi.org/10.1016/j.pedobi.2016.02.001.

[49] Puppe D, Höhn A, Kaczorek D, Wanner M, Sommer M. As time goes by—spatiotemporal changes of biogenic Si pools in initial soils of an artificial catchment in NE Germany. Appl Soil Ecol 2016;105:9−16. Available from: https://doi.org/10.1016/j.apsoil.2016.01.020.

[50] Wanner M, Birkhofer K, Fischer T, Shimizu M, Shimano S, Puppe D. Soil testate amoebae and diatoms as bioindicators of an old heavy metal contaminated floodplain in Japan. Microb Ecol 2020;79(1):123−33. Available from: https://doi.org/10.1007/s00248-019-01383-x.

[51] Kaczorek D, Puppe D, Busse J, Sommer M. Effects of phytolith distribution and characteristics on extractable silicon fractions in soils under different vegetation—an exploratory study on loess. Geoderma 2019;356:113917. Available from: https://doi.org/10.1016/j.geoderma.2019.113917.

[52] Desplanques V, Cary L, Mouret JC, Trolard F, Bourrié G, Grauby O, et al. Silicon transfers in a rice field in Camargue (France). J Geochem Explor 2006;88(1−3):190−3. Available from: https://doi.org/10.1016/j.gexplo.2005.08.036.

[53] Yang X, Song Z, Qin Z, Wu L, Yin L, Van Zwieten L, et al. Phytolith-rich straw application and groundwater table management over 36 years affect the soil-plant silicon cycle of a paddy field. Plant Soil 2020;454(1−2):343−58. Available from: https://doi.org/10.1007/s11104-020-04656-4.

[54] Qin Y, Puppe D, Payne R, Li L, Li J, Zhang Z, et al. Land-use change effects on protozoic silicon pools in the Dajiuhu National Wetland Park, China. Geoderma 2020;368:114305. Available from: https://doi.org/10.1016/j.geoderma.2020.114305.

[55] Puppe Daniel. Review on protozoic silica and its role in silicon cycling. Geoderma 2020;365:114224. Available from: https://doi.org/10.1016/j.geoderma.2020.114224.

[56] Meisterfeld R. Testate amoebae with filopodia. In: Lee JJ, Leedale GE, Bradbury P, editors. An illustrated guide to the protozoa. 2nd ed. Lawrence, KS: Allen Press; 2002. p. 1054−84.

[57] Clarke J. The occurrence and significance of biogenic opal in the regolith. Earth-Sci Rev 2003;60(3−4):175−94. Available from: https://doi.org/10.1016/S0012-8252(02)00092-2.

[58] Aoki Y, Hoshino M, Matsubara T. Silica and testate amoebae in a soil under pine-oak forest. Geoderma 2007;142(1−2):29−35. Available from: https://doi.org/10.1016/j.geoderma.2007.07.009.

[59] Puppe D, Kaczorek D, Wanner M, Sommer M. Dynamics and drivers of the protozoic Si pool along a 10-year chronosequence of initial ecosystem states. Ecol Eng 2014;70:477−82. Available from: https://doi.org/10.1016/j.ecoleng.2014.06.011.

[60] Puppe D, Ehrmann O, Kaczorek D, Wanner M, Sommer M. The protozoic Si pool in temperate forest ecosystems—quantification, abiotic controls and interactions with earthworms. Geoderma 2015;243−244:196−204. Available from: https://doi.org/10.1016/j.geoderma.2014.12.018.

[61] Puppe D, Wanner M, Sommer M. Data on euglyphid testate amoeba densities, corresponding protozoic silicon pools, and selected soil parameters of initial and forested biogeosystems. Data Brief 2018;21:1697−703. Available from: https://doi.org/10.1016/j.dib.2018.10.164.

[62] Qin Y, Puppe D, Zhang L, Sun R, Li P, Xie S. How does Sphagnum growing affect testate amoeba communities and corresponding protozoic Si pools? Results from field analyses in SW China. Microb Ecol 2021;82:459−69. Available from: https://doi.org/10.1007/s00248-020-01668-6.

[63] Brehm U, Gorbushina A, Mottershead D. The role of microorganisms and biofilms in the breakdown and dissolution of quartz and glass. Palaeogeog Palaeoclimatol Palaeoecol 2005;219(1−2):117−29. Available from: https://doi.org/10.1016/j.palaeo.2004.10.017.

[64] Clymans W, Struyf E, Govers G, Vandevenne F, Conley DJ. Anthropogenic impact on amorphous silica pools in temperate soils. Biogeosciences 2011;8(8):2281−93. Available from: https://doi.org/10.5194/bg-8-2281-2011.

[65] Schaller J, Faucherre S, Joss H, Obst M, Goeckede M, Planer-Friedrich B, et al. Silicon increases the phosphorus availability of Arctic soils. Sci Rep 2019;9(1):449. Available from: https://doi.org/10.1038/s41598-018-37104-6.

[66] Schaller J, Frei S, Rohn L, Gilfedder BS. Amorphous silica controls water storage capacity and phosphorus mobility in soils. Front Environ Sci 2020;8:94. Available from: https://doi.org/10.3389/fenvs.2020.00094.

[67] Nguyen ATQ, Nguyen MN. Straw phytolith for less hazardous open burning of paddy straw. Sci Rep 2019;9(1):20043. Available from: https://doi.org/10.1038/s41598-019-56735-x.

[68] Schaller J, Puppe D. Heat improves silicon availability in mineral soils. Geoderma 2021;386:114909. Available from: https://doi.org/10.1016/j.geoderma.2020.114909.

[69] Haynes RJ. What effect does liming have on silicon availability in agricultural soils? Geoderma 2019;337:375−83. Available from: https://doi.org/10.1016/j.geoderma.2018.09.026.

[70] Berhane M, Xu M, Liang Z, Shi J, Wei G, Tian X. Effects of long-term straw return on soil organic carbon storage and sequestration rate in North China upland crops: a meta-analysis. Glob Change Biol 2020;26(4):2686−701. Available from: https://doi.org/10.1111/gcb.15018.

[71] Li H, Dai M, Dai S, Dong X. Current status and environment impact of direct straw return in China's cropland—a review. Ecotoxicol Environ Saf 2018;159:293−300. Available from: https://doi.org/10.1016/j.ecoenv.2018.05.014.

[72] Song Z, Müller K, Wang H. Biogeochemical silicon cycle and carbon sequestration in agricultural ecosystems. Earth-Sci Rev 2014;139:268−78. Available from: https://doi.org/10.1016/j.earscirev.2014.09.009.

[73] Guntzer F, Keller C, Poulton PR, McGrath SP, Meunier JD. Long-term removal of wheat straw decreases soil amorphous silica at Broadbalk, Rothamsted. Plant Soil 2012;352(1−2):173−84. Available from: https://doi.org/10.1007/s11104-011-0987-4.

[74] Caubet M, Cornu S, Saby NPA, Meunier JD. Agriculture increases the bioavailability of silicon, a beneficial element for crop, in temperate soils. Sci Rep 2020;10(1):19999. Available from: https://doi.org/10.1038/s41598-020-77059-1.

[75] Cornelis JT, Titeux H, Ranger J, Delvaux B. Identification and distribution of the readily soluble silicon pool in a temperate forest soil below three distinct tree species. Plant Soil 2011;342(1−2):369−78. Available from: https://doi.org/10.1007/s11104-010-0702-x.

[76] Haynes RJ. Significance and role of Si in crop production. Advances in agronomy, 146. Cambridge, MA: Academic Press Inc.; 2017. p. 83−166. Available from: https://doi.org/10.1016/bs.agron.2017.06.001.

[77] Cooke J, Leishman MR. Tradeoffs between foliar silicon and carbon-based defences: evidence from vegetation communities of contrasting soil types. Oikos 2012;121(12):2052−60. Available from: https://doi.org/10.1111/j.1600-0706.2012.20057.x.

[78] Prychid CJ, Rudall PJ, Gregory M. Systematics and biology of silica bodies in monocotyledons. Bot Rev 2003;69(4):377−440. Available from: https://doi.org/10.1663/0006-8101(2004)069[0377:SABOSB]2.0.CO;2.

[79] Li Z, Guo F, Cornelis JT, Song Z, Wang X, Delvaux B. Combined silicon-phosphorus fertilization affects the biomass and phytolith stock of rice plants. Front Plant Sci 2020;11:67. Available from: https://doi.org/10.3389/fpls.2020.00067.

[80] Gustafsson O, Kruså M, Zencak Z, Sheesley RJ, Granat L, Engström E, et al. Brown clouds over South Asia: biomass or fossil fuel combustion? Science 2009;323(5913):495−8. Available from: https://doi.org/10.1126/science.1164857.

[81] Nguyen MN. Worldwide bans of rice straw brning could increase human arsenic exposure. Environmental Science and Technology 2020;. Available from: https://doi.org/10.1021/acs.est.0c00866.

[82] Crusciol CAC, De Arruda DP, Fernandes AM, Antonangelo JA, Alleoni LRF, Nascimento CACD, et al. Methods and extractants to evaluate silicon availability for sugarcane. Sci Rep 2018;8(1):916. Available from: https://doi.org/10.1038/s41598-018-19240-1.

[83] Sauer D, Saccone L, Conley DJ, Herrmann L, Sommer M. Review of methodologies for extracting plant-available and amorphous Si from soils and aquatic sediments. Biogeochemistry 2006;80(1):89−108. Available from: https://doi.org/10.1007/s10533-005-5879-3.

[84] Li Zimin, Unzué-Belmonte D, Cornelis J-T, Linden CV, Struyf E, Ronsse F, et al. Effects of phytolithic rice-straw biochar, soil buffering capacity and pH on silicon bioavailability. Plant Soil 2019;438:187−203. Available from: https://doi.org/10.1007/s11104-019-04013-0.

[85] Haynes RJ. A contemporary overview of silicon availability in agricultural soils. J Plant Nutr Soil Sci 2014;177(6):831−44. Available from: https://doi.org/10.1002/jpln.201400202.

[86] Miles N, Manson AD, Rhodes R, van Antwerpen R, Weigel A. Extractable silicon in soils of the South African sugar industry and relationships with crop uptake. Commun Soil Sci Plant Anal 2014;45(22):2949−58. Available from: https://doi.org/10.1080/00103624.2014.956881.

[87] Korndörfer GH, Snyder GH, Ulloa M, Powell G, Datnoff LE. Calibration of soil and plant silicon analysis for rice production. J Plant Nutr 2001;24(7):1071−84. Available from: https://doi.org/10.1081/PLN-100103804.

[88] Klotzbücher A, Klotzbücher T, Jahn R, Xuan LD, Cuong LQ, Van Chien H, et al. Effects of Si fertilization on Si in soil solution, Si uptake by rice, and resistance of rice to biotic stresses in Southern Vietnam. Paddy Water Environ 2018;16(2):243−52. Available from: https://doi.org/10.1007/s10333-017-0610-2.

[89] Gocke M, Liang W, Sommer M, Kuzyakov Y. Silicon uptake by wheat: Effects of Si pools and pH. J Plant Nutr Soil Sci 2013;176(4):551−60. Available from: https://doi.org/10.1002/jpln.201200098.

[90] Keeping MG. Uptake of silicon by sugarcane from applied sources may not reflect plant-available soil silicon and total silicon content of sources. Front Plant Sci 2017;8:760. Available from: https://doi.org/10.3389/fpls.2017.00760.

[91] Burki F, Roger AJ, Brown MW, Simpson AGB. The new tree of Eukaryotes. Trends Ecol Evolution 2020;35(1):43−55. Available from: https://doi.org/10.1016/j.tree.2019.08.008.

23

Nanosilica-mediated plant growth and environmental stress tolerance in plants: mechanisms of action

Jonas Pereira de Souza Júnior[1], Renato de Mello Prado[1], Cid Naudi Silva Campos[2], Gelza Carliane Marques Teixeira[1] and Patrícia Messias Ferreira[1]

[1]School of Agricultural and Veterinary Sciences, São Paulo State University, Jaboticabal, Brazil
[2]Federal University of Mato Grosso do Sul, Chapadão do Sul, Brazil

23.1 Introduction

The growing global population has increased the demand for food and the need for greater agricultural production worldwide. Research is one of the main drivers of technology in the field, making it possible to increase crop yield and improve the efficient use of fertilizers to provide an adequate amount of nutrients and decrease production costs.

In this respect, recent strategies using nanotechnology in agriculture, particularly in growth promoters, pesticides, additives, and nanofertilizers, have awakened the interest of researchers [1]. Nanoparticles (NPs) exhibit greater sorption capacity, larger surface areas, and an intelligent delivery system to sites, allowing molecules to breach cell wall barriers and penetrate plant tissue [1].

These nanoparticulate materials have direct applications in plant nutrition and may provide devices and mechanisms to synchronize and release nutrients according to plant needs, hindering their premature conversion into chemical/gas forms that cannot be absorbed. A number of published studies with nanoparticulate fertilizers demonstrate the beneficial and harmful effects on agronomic variables [2–7].

The increase in biotic and abiotic stress caused by climate change, deficient soils, or those with excessive nutrients or potentially toxic heavy metals has heightened interest in the use of silicon (Si) in agriculture [8]. Si is a beneficial element that can mitigate different stresses in crops and has prompted the use of NPs that may increase these benefits. The nanoparticulate forms of Si consist of synthesized amorphous silica with diameters between 10 and 100 nm [9] and a specific surface area of $50-500 \text{ m}^2 \text{ g}^{-1}$. They can be produced from several sources, such as tetraethyl ortosilicate ($Si(OC_2H_5)_4$), inorganic salts such as sodium silicate (Na_2SiO_3), or using rice husk as an organic source [9]. Nanoparticles are characterized by their differentiated structure as they have characteristics on a nanometer scale, which promotes a large surface area, that is, a greater ratio between surface area and mass, thus promoting an increase in the contact area with any external material [10]. Due to this reduction in the average particle size of nanomaterials, there is an increase in the surface area per volume, which can be exemplified as follows: taking as an example the surface area of a cube, whose dimensions correspond to 1 cm on the side is 6 cm^2 for a volume of 1 cm^3, when reducing the edge of the cube to 1 mm, there will be 1000 new cubes with a surface area corresponding to 60 cm^2. This implies that as the particle size decreases, the total

Silicon and Nano-silicon in Environmental Stress Management and Crop Quality Improvement.
DOI: https://doi.org/10.1016/B978-0-323-91225-9.00023-6

number of atoms in the particle increases, and, consequently, there is a significant increase in surface area relative to the volume of material [11].

Si nanoparticles may contribute to greater plant absorption capacity, thereby avoiding losses due to the high polymerization rate in solution, which is still common in field conditions. This is because Si in solution starts to polymerize at a concentration of 3 mmol L^{-1} [12], forming dimers, trimers, and polymer chains [13] that are not absorbed by plants.

In this chapter, we will discuss the beneficial effect of Si nanoparticles and the mechanisms of action involved in the morphological, biochemical, or physiological changes that attenuate the harmful effects of environmental stress. In addition, the limitations and future perspectives for Si nanoparticle use in agriculture will be discussed.

23.2 Nanosilica stability in solution and efficiency in providing Si to crops

The use of Si nanoparticles (nano-Si) requires knowledge of the factors that affect their solubility and stability. Nanosilica solubility increases linearly with the temperature of the solution, and rises with pH up to 7.5, declining at higher values. Its stability is negatively affected by the presence of ions such as Mg^{2+}, Al^{3+}, Fe^{3+}, Fe^{2+}, and Ca^{2+} among others, due to their multiple integration possibilities, forming nucleation points that may change the scale of the particles [14].

Nanosilica can be absorbed by plants via a passive process, with different concentrations, or actively in special groups of aquaporins, with energy expenditure and genetic control [13,14]. Nanoparticle absorption by roots or leaves depends on their shape and mainly their size [15,16].

Root absorption may occur through specific carriers via the symplastic pathway [17] or via the apoplastic pathway, through pores with diameters between 5 and 20 nm, located between the cell wall and root epidermis [18].

Leaf absorption occurs primarily through the stomatal pores [17]. Nanoparticle transport from one cell to neighboring cells depends on molecule size, since it occurs via plasmodesmata, in canals with diameters between 20 and 50 nm [15,19].

Thus the polymerization of Si in solution may reduce absorption and transport in plants. Nano-Si sources consist of Si oxide colloidal nanoparticles with dimensions between 1 and 10 nm [17,18] that can be absorbed by plants. However, silanol groups are formed during Si polymerization and condensation. Next, the supersaturated solution is converted into its polymer form, consisting of spherical colloidal particles, and large clusters of these particles form long chains in the network structure [20], which may hinder plant absorption (Fig. 23.1).

Polymerization is a process that depends on the pH and concentration of Si in solution [21] and can be indicated by visible changes in the solution from translucid to opaque or cloudy (Fig. 23.2).

The efficiency of innovative silicate sources for foliar applications in raising leaf Si content in plants is related to the stability of the element in solution [22]. Among the recently studied sources, nano-SiO_2 stands out for its high stability, remaining stable in the spray solution applied without visible changes for up to 180 min after

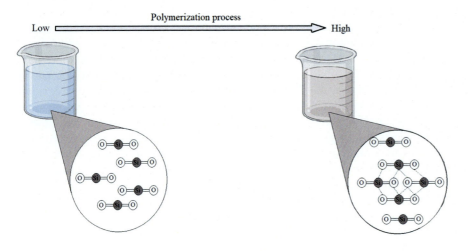

Polymerization process
Low ——————————————————▶ High

FIGURE 23.1 Polymerization of Si oxide in aqueous solution.

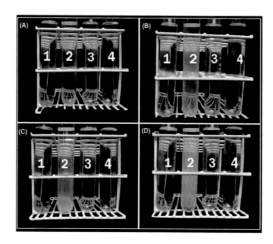

FIGURE 23.2 Visual assessment of silicon solution turbidity at a concentration of $0.8\ g\ L^{-1}$ (pH 6.5 ± 0.2) in different Si sources: (1) sodium silicate and potassium stabilized with sorbitol; (2) monosilicic acid stabilized with PEG 400; (3) bindizil nanosilica; and (4) potassium silicate without stabilizer; determined immediately after preparation of the silicate solution (A); 6 h (B); 12 h (C), and 24 h (D) after preparation. Source: *From Souza Junior J.P., et al. Effect of different foliar silicon sources on cotton plants. J Soil Sci Plant Nutr 2021;21:95–103. https://doi.org/ 10.1007/s42729-020-00345-4. [22].*

TABLE 23.1 Crops that exhibited an increase in Si content in plant organs after application of the element in its nanoparticle form using different application modes.

Crop	Application mode	References
Avena sativa	Root	Asgari et al. [24]
Beta vulgaris	Foliar and seed treatment	Khan and Siddiqui [25]
Citrullus lanatus	Foliar	Kang et al. [26]
Fragaria vesca	Foliar	Avestan et al. [27]
Glycine max	Foliar	Felisberto et al. [23]
Gossypium hirsutum	Root	Le et al. [28]
Lupinus angustifolium	Root	Sun et al. [29]
Oryza sativa	Foliar	Felisberto et al. [23]
Oryza sativa	Root	Abdel-Haliem et al. [7]
Oryza sativa	Root	Félix Alvarez et al. [3]
Polianthes tuberosa	Foliar and Root	Karimian et al. [30]
Saccharum officinarum	Foliar	Santos et al. [31]
Trigonella foenumgraceum	Root	Nazaralian et al. [32]
Triticum aestivum	Root	Sun et al. [29]
Triticum aestivum	Seed treatment	Hussain et al. [1]
Zea mays	Foliar	Aqaei et al. [33]
Zea mays	Root	Sousa et al. [34]

solution preparation, as observed by Felisberto et al. [23] in a study with $42\ mmol\ L^{-1}$ of nano-Si; or for up to 360 min, as reported by Souza Junior et al. [22] in a study with a concentration of $22.4\ mmol\ L^{-1}$. In addition, Souza Junior et al. [22] found no changes in the turbidity index, an indicator of the onset of polymerization, in the first 360 min after solution preparation.

The low polymerization rate of nano-SiO_2 can be confirmed by the efficiency of this source in raising leaf Si content in different crops (Table 23.1).

23.3 Effects of nanosilica on plants grown under environmental stress

23.3.1 Morphological changes

Nanosilica penetrates the cell wall of roots and enters the endotherm and extracellular spaces, reaching the vascular tissue via the apoplastic and symplastic pathways, and is then transported to the shoots [22,25−28]. After absorption, Si nanostructures are found mainly deposited on specialized cells called phytoliths, which store Si (Fig. 23.3) [35].

The increase in phytolith production and consequently, silica nanoparticles inside cells, causes important morphological changes in plants, thereby influencing cell resistance and the integrity and maintenance of plant vessels. The distribution and foliar deposition patterns of Si nanoparticles are similar to that of soluble Si sources (Fig. 23.4) [3].

This nano-Si absorption and deposition on plants increase lignin production and deposition on the cell wall [32], an important and complex phenolic compound essential to the vascular cell wall structure [36]. This occurs because Si raises the production and activity of phenylalanine ammonia lyase, an important enzyme in lignin biosynthesis [27,31]. In addition, classic studies demonstrate that Si is an important component in pectins, calluses, and tannins, which also provide greater cell resistance to plants [32−34].

In this respect, applying nano-Si is promising in reducing pest attacks [37] and fungal diseases [38] due to increased cell resistance. The digestive apparatus of masticating insects is compromised, leading to a high mortality rate, primarily in the adult phase [6,35,36,39,40], and fungal hyphae have difficulty penetrating plant tissues [38].

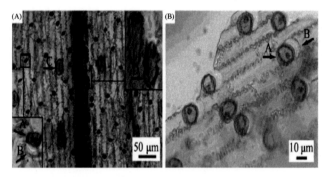

FIGURE 23.3 Phytolith morphology in wheat straw: (A) photomicrographs of combustion samples; (B) acid treatment. Note: (A) silica cells; (B) cork cells; (C) stomata; and (D) oblong phytoliths. Source: *Chen H., Wang F., Zhang C., Shi Y., Jin G., Yuan S. Preparation of nano-silica materials: the concept from wheat straw. J Non Cryst Solids 2010;356(50−51):2781−2785. https://doi.org/10.1016/j.jnoncrysol.2010.09.051. [35].*

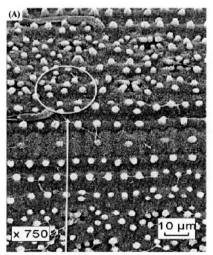

Regular and abundant amorphous silica deposits

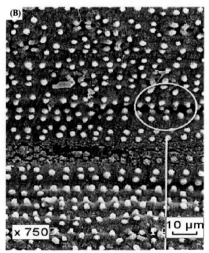

Regular and abundant amorphous silica deposits

FIGURE 23.4 Scanning electron micrographs of the flag leaf of *Oryza sativa* showing the cell surface, the shape of silica cells, and the deposition pattern of nanosilica (A) and soluble Si source (B). Source: *Adapted from Félix Alvarez RDC, Prado RDM, Gelisberto G, Fernandes Deus AC, Lima de Oliveira RL. Effects of soluble silicate and nanosilica application on rice nutrition in an oxisol. Pedosphere 2018;28 (4):597−606 [3].*

Cell wall enrichment, caused by nano-Si, may also hinder the feeding phase of sucking insects [41], lengthening their pubescence period [42]. As a result, insect development is compromised, with lower potential and longer duration of the F1 generation cycle and larval phase [43].

It is important to underscore that applying nanoparticle products, such as nano-Si, has not harmed the development and reproduction of predatory and pollinating insects. Knowledge of the impacts of nanomaterials on the growth, development, parasitism, or predatory efficiency and emergence capacity, as well as the protection of beneficial insects, is essential to protecting parasitoids, predators, and pollinators from nanotoxicity, while reducing insect attacks on crops [44].

In other situations of environmental stress, such as water stress [33], or calcium [45], magnesium [45], or boron deficiency [46], cellular integrity may be compromised by deformation of the bundle sheaths of the xylem and a reduction of water and nutrient transport by plant vessels.

Once absorbed, in addition to being involved in the structure of the leaf cell wall, nano-Si also increases cell wall lignification in the xylem [24], increasing plant vessel diameter in the roots and leaves (Fig. 23.5), as observed in *Avena sativa* L [24].

This beneficial effect of nano-Si increases the absorption and transport of water and nutrients, as observed in *Zea mays* grown under water stress [33].

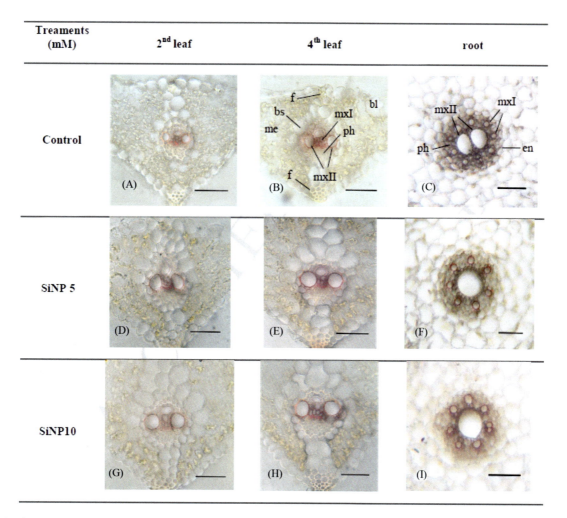

FIGURE 23.5 Cross-section of the leaves (second and fourth leaf pairs) and root of *Avena sativa* L. Grown in the absence (control) and presence of nanosilica at a concentration of 5 nM (SiNP5) and 10 mM (SiNP10). Note: bs = bundle sheath, en = endodermis, f = fibers, mxI = metaxylem I, mxII = metaxylem II, ph = phloem, and me = mesophyll. Scale bars: [(A), (B), (D), (E), (G), and (H)] 100 μM and [(C), (F), and (I)] 50 μM. Source: *Adapted from Asgari F., Majd A., Jonoubi P., Najafi F. Effects of silicon nanoparticles on molecular, chemical, structural and ultrastructural characteristics of oat (Avena sativa L.). Plant Physiol Biochem 2018;127:152−160. https://doi.org/10.1016/j.plaphy.2018.03.021. [24].*

The morphological changes in vascular bundles, increasing water, and nutrient absorption, may act as a mitigating mechanism of action of Si [47], given the reports that the beneficial element applied in its soluble form improves nutrient absorption in different plants grown under nutrient deficiency [41,42,48].

23.3.2 Biochemical changes

Nano-Si particles are apparently better able to penetrate leaf tissues and the increased Si concentration inside the cell may inhibit the movement of other elements in plant organs [49]. This is because high Si concentration promotes the precipitation of a number of elements, forming silicate complexes in metabolically less active tissues, such as the cell wall of the endoderm, pericycle, xylem, and phloem (Fig. 23.6) [50], also observed by other authors [45–47].

The supply of nano-Si may decrease the toxic effects of some elements, by immobilizing them in the roots and stems of plants, thereby reducing the amount in plants and the partitioning of new organs such as leaves, shoots, flowers, and fruits [51–53].

Other important biochemical changes caused by Si are that after absorption, Si can partially substitute carbon (C) in organic compounds [54] and modify C:N:P stoichiometry in plant tissue [50,55–57]. Replacing structural carbon with phytoliths (organic silica compounds) in plant structures requires less metabolic energy for their formation and incorporation [58], meaning plants can use the excess energy to synthesize other nonstructural organic compounds [54].

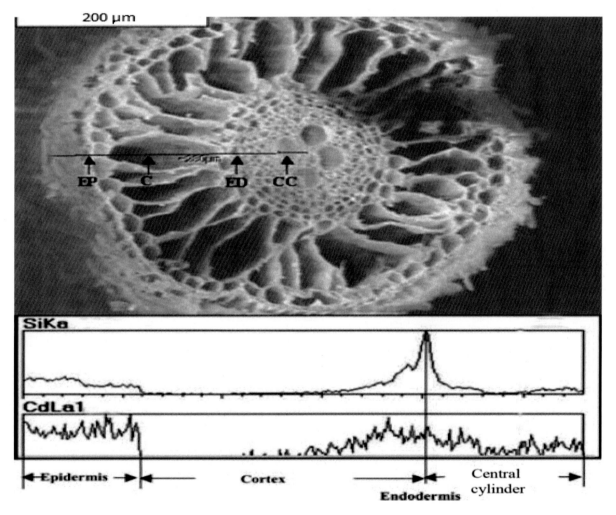

FIGURE 23.6 Cross-section of the root tip of *Oryza sativa* frozen in liquid N$_2$, lyophilized and examined under a scanning electron microscope (top of the photograph) and a profile of the energy-dispersive X-ray line scan of Si and Cd distribution along the epidermis (EP), cortex (C), endodermis (DE), and central cylinder (CC) (bottom of the photograph). *Source: From Shi X., Zhang C., Wang H., Zhang F. Effect of Si on the distribution of Cd in rice seedlings. Plant Soil 2005;272(1−2):53−60. https://doi.org/10.1007/s11104-004-3920-2. [50].*

Although this effect of Si has been widely discussed in the literature, studies using Si nanoparticles are incipient, and different results may be related to the Si sources used [59]. In this respect, Félix Alvarez et al. [3] found no changes in stoichiometry with the supply of nano-Si via soil in rice plants, while Neu et al. [60] observed an important modification in the C:Si, C:N, and C:P ratios in *Triticum aestivum* L. plants that received silicon dioxide in aerosol via fertigation. However, new studies are needed to better elucidate the effects of Si nanoparticles on plant stoichiometry.

23.3.3 Physiological changes

Once absorbed, Si is transported to the leaves along with water, depending on foliar transpiration. When it reaches the leaves, Si is concentrated due to water loss, thereby favoring polymerization, forming amorphous or biogenic silica consisting of 90% of the Si absorbed and maybe in Si-cellulose structures present in the cell wall [61]. However, the 10% of Si not incorporated in the cell wall causes important alterations in plant physiology, influencing the antioxidant defense system, oxidative stress, photosynthetic and plant-water relations, thereby mitigating biotic and abiotic stress.

23.3.3.1 Oxidative stress and antioxidant systems

Under natural conditions, plants continuously produce several reactive oxygen species (ROS) during photosynthesis and respiration in cell organelles such as mitochondria, chloroplasts, and peroxisomes [62]. In photosynthesis, during the reduction of O_2 to H_2O, energy may be transferred to O_2, forming singlet oxygen ($^1O_2$), which exhibits high reactivity compared to O_2. One, two, or three electrons can be transferred to O_2 and form the ROS superoxide anion ($O_2^{\bullet}$), hydrogen peroxide (H_2O_2), and hydroxyl radical ($^{\bullet}OH$), respectively [63]. In respiratory processes, oxygen is gradually reduced to H_2O or via tetravalent mechanisms [alternative oxidase (AOX)], leading to the formation of $O_2^{\bullet}$, which is easily converted to H_2O_2 through the action of superoxide dismutase containing manganese (MnSOD) [64].

Under normal growing conditions, plants maintain homeostasis between ROS production and elimination using two different detoxification mechanisms involving the enzymatic and nonenzymatic antioxidant system [54,65,66]. In the former system, reducing ROS to H_2O occurs via the action of superoxide dismutase (SOD), catalase (CAT), ascorbate peroxidase (APX), and glutathione peroxidase (GPX) (Fig. 23.7) [63].

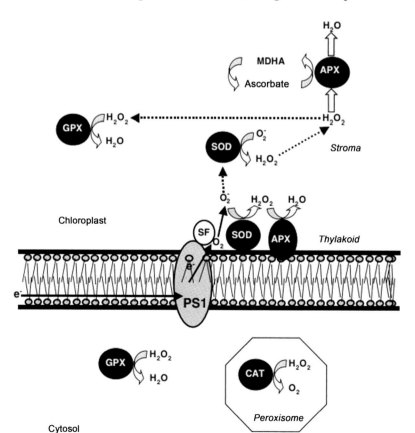

FIGURE 23.7 Reactive oxygen species (ROS) elimination pathway. *SF*, Stomatal factor; *PSI*, photosystem I; *GPX*, glutathione peroxidase; *CAT*, catalase; *SOD*, superoxide dismutase; *APX*, ascorbate peroxidase; *MDHA*, monodehydroascorbate. Source: *Adapted from Gratão P.L., Polle A., Lea P.J., Azevedo R.A. Making the life of heavy metal-stressed plants a little easier. Funct Plant Biol 2005;32(6):481–494. https://doi.org/10.1071/FP05016. [63].*

Under environmental stress, ROS is produced and accumulates [67], damaging the cell before it is eliminated due to lipid peroxidation, considered the greatest degenerative process in plants [7]. In this scenario, the protection Si provides to plants may be associated with its biochemical role in cells, activating a complex antioxidant mechanism that improves the tolerance of plants under stress conditions [2]. In this respect, the function of Si has been amply explored, indicating its beneficial effect in increasing the content of enzymatic and nonenzymatic antioxidants [2,7,68], in addition to reducing lipid peroxidation and malondialdehyde (MDA) content [69].

Studies indicate that nano-Si has an osmoprotective function, namely, increasing the activity of enzymes important for the antioxidant defense mechanism (Table 23.2).

The action of Si nanoparticles in the antioxidant defense mechanism has been widely discussed in the literature. However, studies show that nano-Si also has an important function in regulating the biosynthesis of nonenzymatic antioxidant compounds such as ascorbate [76], proline [2,7,63], glycine-betaine [2], and glutathione [76].

23.3.3.2 Photosynthetic pigments and efficiency

Environmental stresses that cause oxidative stress, such as nutritional disorders [41,64,67,78], potentially toxic heavy metal contamination [61,62,68−70], salinity [23,63], and water stress [2,71,73], are directly related to chlorophyll degradation, due to chloroplast membrane peroxidation and inhibition of photosynthetic pigment biosynthesis [79].

In this sense, nano-Si is efficient in reducing the damage caused by oxidative stress in plants (Section 23.3.3.1), raising chlorophyll production and reducing chloroplast membrane peroxidation. In addition, studies report that morphological changes that increase cell resistance (Section 23.3.1) also contribute to improving leaf architecture, raising absorption, and increasing the use of sunlight, with a consequent rise in pigment production [80].

TABLE 23.2   Enzymes whose activity increases after Si application in nanoparticle form in different crops with or without stress.

Crop	Stress	Enzymes	Author
Beta vulgaris	Pectobacterium betavasculorum and Rhizoctonia solani	SOD, CAT, PPO, PAL	Khan and Siddiqui [25]
Beta vulgaris	Water stress	CAT, SOD, GPX	Namjoyan et al. [2]
Cucumis sativus	Salinity	CAT, SOD	Gengmao et al. [4]
Cucurbita pego	Salinity	CAT POD, SOD, GR, APX	Siddiqui et al. [70]
Glycine max	Salinity	CAT, SOD, APX,	Farhangi-Abriz and Torabian [5]
Hordeum vulgare	Water stress	CAD, POD, APX	Gorbanpour et al. [71]
Oryza sativa	Salinity	SOD	Abdel-Halien et al. [7]
Oryza sativa	Cadmium	CAT, SOD, POD	Wang et al. [72]
Pisum sativum	Chromium	SOD, CAT	Tripathi et al. [73]
Solanum lycopersicum	Clavibacter michiganensis	CAT, SOD, GPX, APX, PAL	Cumplido-Nájera et al. [74]
Triticum aestivum	Cadmium	CAT, SOD, POD	Hussain et al. [1]
Triticum aestivum	UV-B radiation	SOD, APX, CAT, GPX	Tripathi et al. [75]
Zea mays	Aluminum	SOD, APX, POD, CAT, GPX, GR, GST, GSH, DHAR, MDHAR, ASC	Sousa et al. [34]
Zea mays	Arsenic	SOD, APX, GR, DHAR	Tripathi et al. [76]
Zea mays	Fusarium oxysporum and Aspergillus niger	PPO, PAL, POD	Suriyaphabha et al. [77]

CAT, Catalase; SOD, superoxide dismutase; GPX, glutathione peroxidase; APX, ascorbate peroxidase; POD, peroxidase, GST, glutathione-S-transferase; GSH, glutathione; GR, glutathione reductase; DHAR, dehydroascorbate reductase; MDHAR, monodehydroascorbate reductase; ASC, ascorbate; PPO, polyphenol oxidase; PAL, phenylalanine ammonia lyase.

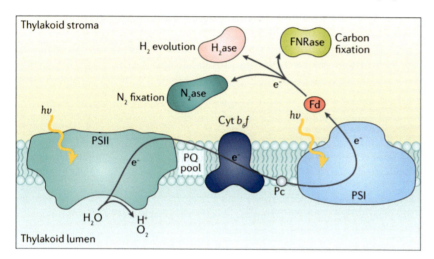

FIGURE 23.8 Photosynthesis in plants grown under normal conditions. *PSII*, Photosystem II; *PQ pool*, plastoquinone pool; *Cytb$_6$f*, cytochrome b$_6$f; *PSI*, photosystem I; *fd*, ferredoxin; *FNRase*, ferredoxin NaDP$^+$ reductase; *H$_2$ase*: hydrogenase, *N$_2$ase*: nitrogenase. Source: *From Zhang J.Z., Reisner E. Advancing photosystem II photoelectrochemistry for semi-artificial photosynthesis. Nat Rev Chem 2019;4(1):6−21. https://doi.org/10.1038/s41570-019-0149-4.*

Another important function of nano-Si is related to the maintenance of photosystem I and II integrity. This is important because at the biochemical level, the energy captured for photosynthesis may be lost and the photosynthesis system compromised in cases of environmental stress. Under normal plant development, photosynthesis starts in the transmembrane complex, denominated photosystem II (PS II), which is embedded in the thylakoid membrane. The light energy collected by PS II is used to extract 4e$^-$ from 2 H$_2$O to produce O$_2$ and 4H$^+$. The accumulation of H$^+$ creates a concentration gradient that drives ATP production via an ATPase bound to the transmembrane. Electrons are transported between PS II and the cytochrome b$_6$f complex, via a plastoquinone pool. Plastocyanin receives electrons from the cytochrome b$_6$f complex one at a time and transports them to PS I. In light irradiation, charge separation occurs in the PS I reaction center and the photoexcited electrons reduce ferredoxin, which can then donate electrons to the NADP$^+$ reductase. The NaDPH produced, along with ATP, leads the Calvin−Benson cycle to CO$_2$ fixation. Ferredoxin can pass electrons to other pathways under certain conditions, such as those involving nitrogenase, for N$_2$ fixation or hydrogenase, and the overall evolution of H$_2$ (Fig. 23.8) [76,77,81].

In plants grown under environmental stress, however, electrons may be lost during transport between PS I and II, which results in a loss of heat energy and fluorescence when transferred to PS I [82].

Initial fluorescence indicates the fluorescence of chlorophyll a [83−85] and the light collection system of PSI [86] and can be used to obtain information on the efficiency of electron transport to this photosystem [84,85]. This variable is inversely proportional to the greater photoreduction of the quinone receptor, which indicates the optimal function of PS II and directly affects the electron flow between PS I and II [82]. Thus the higher the initial fluorescence, the greater the loss of energy in PS II, resulting in less quantum efficiency. Studies have demonstrated that nano-Si increases the quantum efficiency of PS II, thereby reducing energy losses from initial and maximum fluorescence and raising the quantic efficiency of PS II in different crops such as *Fragrariavesca* [81] and *Gossypium hirsutum* [22]. This occurs because nano-Si raises the degree of opening in the PSII reaction centers, reducing energy dissipation through the antenna complex, improving real photochemical efficiency and the rate of electron transfer between the photosystems in addition to increasing the liquid photosynthesis capacity of plants [87].

23.4 Limitations and future perspective

The main limitation in the use of nanoparticles is related to factors that may interfere in the stability of these sources at the moment of preparation, primarily high concentrations, use or not of stabilizers, and the presence of particles in a solution that reach a micrometric scale, or after application, such as particle clusters on the plant surface (leaves or seeds) or in the soil that interfere in the effect and may indicate mistaken conclusions about the ecotoxicity of these nanoparticles, thereby inhibiting the potential beneficial effects.

As such, research should focus on increasing the stability of the nanometric forms of Si using chemical or organic polymerization inhibitors for periods long enough to provide effective absorption of the element.

It is important to establish critical Si levels in the soil and leaves in different crops to help decision making and increase the successful application of Si nanoparticles. In addition, new studies are needed to indicate Si concentration in liquid formulations for use in fertigation, leaf spraying and seed treatment, the doses to apply in the soil of a total area (surface or incorporated) or in the solid formulation applied in the seeding furrow associated with high crop production.

References

[1] Hussain A, Rizwan M, Ali Q, Ali S. Seed priming with silicon nanoparticles improved the biomass and yield while reduced the oxidative stress and cadmium concentration in wheat grains. Environ Sci Pollut Res 2019;26(8):7579—88. Available from: https://doi.org/10.1007/s11356-019-04210-5.

[2] Namjoyan S, Sorooshzadeh A, Rajabi A, Aghaalikhani M. Nano-silicon protects sugar beet plants against water deficit stress by improving the antioxidant systems and compatible solutes. Acta Physiol Plant 2020;42(10):1—16. Available from: https://doi.org/10.1007/S11738-020-03137-6.

[3] Félix Alvarez RDC, Prado RDM, Gelisberto G, Fernandes Deus AC, Lima de Oliveira RL. Effects of soluble silicate and nanosilica application on rice nutrition in an oxisol. Pedosphere 2018;28(4):597—606. Available from: https://doi.org/10.1016/S1002-0160(18)60035-9.

[4] Gengmao Z, Shihui L, Xing S, Yizhou W, Zipan C. The role of silicon in physiology of the medicinal plant (*Lonicera japonica* L.) under salt stress. Sci Rep 2015;5(1):1—11. Available from: https://doi.org/10.1038/srep12696.

[5] Farhangi-Abriz S, Torabian S. Nano-silicon alters antioxidant activities of soybean seedlings under salt toxicity. Protoplasma 2018;255 (3):953—62. Available from: https://doi.org/10.1007/S00709-017-1202-0.

[6] Debnath N, Das S, Patra P, Mitra S, Goswami A. Toxicological evaluation of entomotoxic silica nanoparticle. Toxicol Environ Chem 2012;94(5):944—51. Available from: https://doi.org/10.1080/02772248.2012.682462.

[7] Abdel-Haliem MEF, Hegazy HS, Hassan NS, Naguib DM. Effect of silica ions and nano silica on rice plants under salinity stress. Ecol Eng 2017;99:282—9. Available from: https://doi.org/10.1016/J.ECOLENG.2016.11.060.

[8] de R, Prado M. Mineral nutrition of tropical plants. Cham: Springer Nature Switzerland AG; 2021.

[9] Laane H-M. The effects of foliar sprays with different silicon compounds. Plants 2018;7(2):45. Available from: https://doi.org/10.3390/plants7020045.

[10] Roduner E. Size matters: why nanomaterials are different. Chem Soc Rev 2006;35(7):583—92. Available from: https://doi.org/10.1039/B502142C.

[11] Uskoković V. Nanotechnologies: what we do not know. Technol Soc 2007;29(1):43—61. Available from: https://doi.org/10.1016/J.TECHSOC.2006.10.005.

[12] Birchall JD. The essentiality of silicon in biology. Chem Soc Rev 1995;24(5):351—7. Available from: https://doi.org/10.1039/CS9952400351.

[13] McKeague JA, Cline MG. Silica in soils. Adv Agron 1963;15:339—96.

[14] Weng PF. Silica scale inhibition and colloidal silica dispersion for reverse osmosis systems. Desalination 1995;103(1—2):59—67. Available from: https://doi.org/10.1016/0011-9164(95)00087-9.

[15] Ma JF, Yamaji N. Silicon uptake and accumulation in higher plants. Trends Plant Sci 2006;11(8):392—7. Available from: https://doi.org/10.1016/j.tplants.2006.06.007.

[16] Schaller J, Brackhage C, Paasch S, Brunner E, Bäucker E, Dudel EG. Silica uptake from nanoparticles and silica condensation state in different tissues of *Phragmites australis*. Sci Total Environ 2013;442:6—9. Available from: https://doi.org/10.1016/j.scitotenv.2012.10.016.

[17] Tripathi DK, et al. An overview on manufactured nanoparticles in plants: Uptake, translocation, accumulation and phytotoxicity. Plant Physiol Biochem 2017;110:2—12. Available from: https://doi.org/10.1016/j.plaphy.2016.07.030.

[18] Deng YQ, White JC, Xing BS. Interactions between engineered nanomaterials and agricultural crops: Implications for food safety. J Zhejiang Univ Sci A 2014;15(8):552—72. Available from: https://doi.org/10.1631/jzus.A1400165.

[19] Pacheco I, Buzea C. Nanoparticle uptake by plants: Beneficial or detrimental? Phytotoxicity of nanoparticles. Cham: Springer International Publishing; 2018. p. 1—60.

[20] Zhuravlev LT. The surface chemistry of amorphous silica. Zhuravlev model. Colloids Surf, A Physicochem Eng Asp 2000;173:1—38 [Online]. Available: http://www.elsevier.nl/locate/colsurfa [Accessed November 01, 2020].

[21] Haynes RJ. What effect does liming have on silicon availability in agricultural soils? Geoderma 2019;337:375—83. Available from: https://doi.org/10.1016/j.geoderma.2018.09.026.

[22] Souza Junior JP, et al. Effect of different foliar silicon sources on cotton plants. J Soil Sci Plant Nutr 2021;21:95—103. Available from: https://doi.org/10.1007/s42729-020-00345-4.

[23] Felisberto G, de Mello Prado R, de Oliveira RLL, de Carvalho Felisberto PA. Are nanosilica, potassium silicate and new soluble sources of silicon effective for silicon foliar application to soybean and rice plants? Silicon 2021;13:3217—28. Available from: https://doi.org/10.1007/s12633-020-00668-y.

[24] Asgari F, Majd A, Jonoubi P, Najafi F. Effects of silicon nanoparticles on molecular, chemical, structural and ultrastructural characteristics of oat (*Avena sativa* L.). Plant Physiol Biochem 2018;127:152—60. Available from: https://doi.org/10.1016/j.plaphy.2018.03.021.

[25] Khan MR, Siddiqui ZA. Use of silicon dioxide nanoparticles for the management of *Meloidogyne incognita*, *Pectobacterium betavasculorum* and *Rhizoctonia solani* disease complex of beetroot (*Beta vulgaris* L.). Sci Hortic 2020;265:109211. Available from: https://doi.org/10.1016/J.SCIENTA.2020.109211.

[26] Kang H, et al. Silica nanoparticle dissolution rate controls the suppression of fusarium wilt of watermelon (*Citrullus lanatus*). Environ Sci Technol 2021;55:13513—22. Available from: https://doi.org/10.1021/ACS.EST.0C07126.

[27] Avestan S, Ghasemnezhad M, Esfahani M, Barker AV. Effects of nanosilicon dioxide on leaf anatomy, chlorophyll fluorescence, and mineral element composition of strawberry under salinity stress. J Plant Nutr 2021;44:3005−19. Available from: https://doi.org/10.1080/01904167.2021.1936036.

[28] Le VN, Rui Y, Gui X, Li X, Liu S, Han Y. Uptake, transport, distribution and bio-effects of SiO_2 nanoparticles in Bt-transgenic cotton. J Nanobiotechnol 2014;12(1):1−15. Available from: https://doi.org/10.1186/s12951-014-0050-8.

[29] Sun D, Hussain HI, Yi Z, Rookes JE, Kong L, Cahill DM. Mesoporous silica nanoparticles enhance seedling growth and photosynthesis in wheat and lupin. Chemosphere 2016;152:81−91. Available from: https://doi.org/10.1016/J.CHEMOSPHERE.2016.02.096.

[30] Karimian N, Nazari F, Samadi S. Morphological and biochemical properties, leaf nutrient content, and vase life of tuberose (Polianthes tuberosa L.) affected by root or foliar applications of silicon (Si) and silicon nanoparticles (SiNPs). J Plant Growth Regul 2020;2020:1−15. Available from: https://doi.org/10.1007/S00344-020-10272-4.

[31] Santos LCN, Teixeira GCM, de Mello Prado R, Rocha AMS, dos Santos Pinto RC. Response of pre-sprouted sugarcane seedlings to foliar spraying of potassium silicate, sodium and potassium silicate, nanosilica and monosilicic acid. Sugar Tech 2020;22:773−81. Available from: https://doi.org/10.1007/s12355-020-00833-y.

[32] Nazaralian S, et al. Comparison of silicon nanoparticles and silicate treatments in fenugreek. Plant Physiol Biochem 2017;115:25−33. Available from: https://doi.org/10.1016/j.plaphy.2017.03.009.

[33] Aqaei P, Weisany W, Diyanat M, Razmi J, Struik PC. Response of maize (Zea mays L.) to potassium nano-silica application under drought stress. J Plant Nutr 2020;43(9):1205−16. Available from: https://doi.org/10.1080/01904167.2020.1727508.

[34] de Sousa A, et al. Silicon dioxide nanoparticles ameliorate the phytotoxic hazards of aluminum in maize grown on acidic soil. Sci Total Environ 2019;693:133636. Available from: https://doi.org/10.1016/J.SCITOTENV.2019.133636.

[35] Chen H, Wang F, Zhang C, Shi Y, Jin G, Yuan S. Preparation of nano-silica materials: the concept from wheat straw. J Non Cryst Solids 2010;356(50−51):2781−5. Available from: https://doi.org/10.1016/j.jnoncrysol.2010.09.051.

[36] Boerjan W, Ralph J, Baucher M. Lignin biosynthesis. Annu Rev Plant Biol 2003;54:519−46. Available from: https://doi.org/10.1146/annurev.arplant.54.031902.134938.

[37] Alhousari F, Greger M. Silicon and mechanisms of plant resistance to insect pests. Plants 2018;7(2):33. Available from: https://doi.org/10.3390/plants7020033.

[38] Nguyen NT, Nguyen DH, Pham DD, Dang VP, Nguyen QH, Hoang DQ. New oligochitosan-nanosilica hybrid materials: preparation and application on chili plants for resistance to anthracnose disease and growth enhancement. Polym J 2017;49(12):861−9. Available from: https://doi.org/10.1038/pj.2017.58.

[39] Fleck AT, Nye T, Repenning C, Stahl F, Zahn M, Schenk MK. Silicon enhances suberization and lignification in roots of rice (Oryza sativa). J Exp Bot 2011;62(6):2001−11. Available from: https://doi.org/10.1093/jxb/erq392.

[40] Boylston EK, Hebert JJ, Hensarling TP, Bradow JM, Thibodeaux DP. Role of silicon in developing cotton fibers. J Plant Nutr 1990;13(1):131−48. Available from: https://doi.org/10.1080/01904169009364063.

[41] Massey FP, Hartley SE. Physical defences wear you down: progressive and irreversible impacts of silica on insect hervibores. J Anim Ecol 2009;78(1):281−91.

[42] Reynolds OL, Padula MP, Zeng R, Gurr GM. Silicon: potential to promote direct and indirect effects on plant defense against arthropod pests in agriculture. Front Plant Sci 2016;7:744. Available from: https://doi.org/10.3389/fpls.2016.00744.

[43] Ribeiro EB, et al. Biological aspects of the two-spotted spider mite on strawberry plants under silicon application. Hortic Bras 2021;39(1):5−10. Available from: https://doi.org/10.1590/s0102-0536-20210101.

[44] Kannan M, Elango K, Tamilnayagan T, Preetha S, Kasivelu G. Impact of nanomaterials on beneficial insects in agricultural ecosystems. In: Thangadurai D, Sangeetha J, Prasad R, editors. Nanotechnology for food, agriculture and environment. Singapore: Springer; 2020. p. 379−93.

[45] Martinez HEP, et al. Leaf and stem anatomy of cherry tomato under calcium and magnesium deficiencies. Braz Arch Biol Technol 2020;63:2020. Available from: https://doi.org/10.1590/1678-4324-2020180670.

[46] Li M, et al. Effect of boron deficiency on anatomical structure and chemical composition of petioles and photosynthesis of leaves in cotton (Gossypium hirsutum L.). Sci Rep 2017;7(1):1−9. Available from: https://doi.org/10.1038/s41598-017-04655-z.

[47] Ali N, Réthoré E, Yvin J-C, Hosseini SA. The regulatory role of silicon in mitigating plant nutritional stresses. Plants 2020;9(12):1779. Available from: https://doi.org/10.3390/PLANTS9121779.

[48] El-Naggar ME, et al. Soil application of nano silica on maize yield and its insecticidal activity against some stored insects after the postharvest. Nanomaterials 2020;10(4):739. Available from: https://doi.org/10.3390/nano10040739.

[49] Chen R, Zhang C, Zhao Y, Huang Y, Liu Z. Foliar application with nano-silicon reduced cadmium accumulation in grains by inhibiting cadmium translocation in rice plants. Environ Sci Pollut Res 2018;25(3):2361−8. Available from: https://doi.org/10.1007/s11356-017-0681-z.

[50] Shi X, Zhang C, Wang H, Zhang F. Effect of Si on the distribution of Cd in rice seedlings. Plant Soil 2005;272(1−2):53−60. Available from: https://doi.org/10.1007/s11104-004-3920-2.

[51] Buchelt AC, et al. Silicon contribution via nutrient solution in forage plants to mitigate nitrogen, potassium, calcium, magnesium, and sulfur deficiency. J Soil Sci Plant Nutr 2020;20:1532−48. Available from: https://doi.org/10.1007/s42729-020-00245-7.

[52] Santos Sarah MM, de Mello Prado R, Teixeira GCM, de Souza Júnior JP, de Medeiros RLS, Barreto RF. Silicon supplied via roots or leaves relieves potassium deficiency in maize plants. Silicon 2021;1−10. Available from: https://doi.org/10.1007/s12633-020-00908-1.

[53] Teixeira GCM, de Mello Prado R, Oliveira KS, D'Amico-Damião V, da Silveira Sousa Junior G. Silicon increases leaf chlorophyll content and iron nutritional efficiency and reduces iron deficiency in sorghum plants. J Soil Sci Plant Nutr 2020;20:1311−20. Available from: https://doi.org/10.1007/s42729-020-00214-0.

[54] Klotzbücher T, Klotzbücher A, Kaiser K, Vetterlein D, Jahn R, Mikutta R. Variable silicon accumulation in plants affects terrestrial carbon cycling by controlling lignin synthesis. Glob Chang Biol 2018;24(1):e183−9. Available from: https://doi.org/10.1111/gcb.13845.

[55] Cunha KPV, Nascimento CWA. Silicon effects on metal tolerance and structural changes in maize (Zea mays L.) grown on a cadmium and zinc enriched soil. Water Air Soil Pollut 2009;197(1−4):323−30. Available from: https://doi.org/10.1007/s11270-008-9814-9.

[56] Chen D, et al. Effects of boron, silicon and their interactions on cadmium accumulation and toxicity in rice plants. J Hazard Mater 2019;367:447–55. Available from: https://doi.org/10.1016/j.jhazmat.2018.12.111.

[57] Emamverdian A, Ding Y, Xie Y, Sangari S. Silicon mechanisms to ameliorate heavy metal stress in plants. BioMed Res Int 2018;2018. Available from: https://doi.org/10.1155/2018/8492898.

[58] Schaller J, Schoelynck J, Struyf E, Meire P. Silicon affects nutrient content and ratios of wetland plants. Silicon 2016;8(4):479–85. Available from: https://doi.org/10.1007/s12633-015-9302-y.

[59] Eneji AE, et al. Growth and nutrient use in four grasses under drought stress as mediated by silicon fertilizers. J Plant Nutr 2008;31(2):355–65. Available from: http://doi.org/10.1080/01904160801894913.

[60] Neu S, Schaller J, Dudel EG. Silicon availability modifies nutrient use efficiency and content, C:N:P stoichiometry, and productivity of winter wheat (Triticum aestivum L.). Sci Rep 2017;7(1):1–8. Available from: https://doi.org/10.1038/srep40829.

[61] Yoshida S, Ohnishi Y, Kitagishi K. Histochemistry of silicon in rice plant. Soil Sci Plant Nutr 1962;8(2):1–5. Available from: https://doi.org/10.1080/00380768.1962.10430982.

[62] Kim Y-H, Khan AL, Waqas M, Lee I-J. Silicon regulates antioxidant activities of crop plants under abiotic-induced oxidative stress: a review. Front Plant Sci 2017;8:510. Available from: https://doi.org/10.3389/fpls.2017.00510.

[63] Gratão PL, Polle A, Lea PJ, Azevedo RA. Making the life of heavy metal-stressed plants a little easier. Funct Plant Biol 2005;32(6):481–94. Available from: https://doi.org/10.1071/FP05016.

[64] Navrot N, Rouhier N, Gelhaye E, Jacquot JP. Reactive oxygen species generation and antioxidant systems in plant mitochondria,". Physiol Plant 2007;129(1):185–95. Available from: https://doi.org/10.1111/j.1399-3054.2006.00777.x.

[65] Frazão JJ, de R, Prado M, de Souza Júnior JP, Rossatto DR. Silicon changes C:N:P stoichiometry of sugarcane and its consequences for photosynthesis, biomass partitioning and plant growth. Sci Rep 2020;10(1):12492. Available from: https://doi.org/10.1038/s41598-020-69310-6.

[66] Lata-Tenesaca LF, de Mello Prado R, de Cássia Piccolo M, da Silva DL, da Silva JLF. Silicon modifies C:N:P stoichiometry, and increases nutrient use efficiency and productivity of quinoa. Sci Rep 2021;11(1):9893. Available from: https://doi.org/10.1038/s41598-021-89416-9.

[67] Xie X, He Z, Chen N, Tang Z, Wang Q, Cai Y. The roles of environmental factors in regulation of oxidative stress in plant. BioMed Res Inl, 2019. 2019. Available from: http://doi.org/10.1155/2019/9732325.

[68] Shi Q, Bao Z, Zhu Z, He Y, Qian Q, Yu J. Silicon-mediated alleviation of Mn toxicity in Cucumis sativus in relation to activities of superoxide dismutase and ascorbate peroxidase. Phytochemistry 2005;66(13):1551–9. Available from: https://doi.org/10.1016/J.PHYTOCHEM.2005.05.006.

[69] Liang Y, Chen Q, Liu Q, Zhang W, Ding R. Exogenous silicon (Si) increases antioxidant enzyme activity and reduces lipid peroxidation in roots of salt-stressed barley (Hordeum vulgare L.). J Plant Physiol 2003;160(10):1157–64. Available from: https://doi.org/10.1078/0176-1617-01065.

[70] Siddiqui MH, Al-Whaibi MH, Faisal M, Al Sahli AA. Nano-silicon dioxide mitigates the adverse effects of salt stress on Cucurbita pepo L. Environ Toxicol Chem 2014;33(11):2429–37. Available from: https://doi.org/10.1002/etc.2697.

[71] Ghorbanpour M, Mohammadi H, Kariman K. Nanosilicon-based recovery of barley (Hordeum vulgare) plants subjected to drought stress. Environ Sci Nano 2020;7(2):443–61. Available from: https://doi.org/10.1039/C9EN00973F.

[72] Wang S, Wang F, Gao S. Foliar application with nano-silicon alleviates Cd toxicity in rice seedlings. Environ Sci Pollut Res 2014;22(4):2837–45. Available from: https://doi.org/10.1007/S11356-014-3525-0.

[73] Tripathi DK, Singh VP, Prasad SM, Chauhan DK, Dubey NK. Silicon nanoparticles (SiNp) alleviate chromium (VI) phytotoxicity in Pisum sativum (L.) seedlings. Plant Physiol Biochem 2015;96:189–98. Available from: https://doi.org/10.1016/J.PLAPHY.2015.07.026.

[74] Cumplido-Nájera CF, González-Morales S, Ortega-Ortíz H, Cadenas-Pliego G, Benavides-Mendoza A, Juárez-Maldonado A. The application of copper nanoparticles and potassium silicate stimulate the tolerance to Clavibacter michiganensis in tomato plants. Sci Hortic 2019;245:82–9. Available from: https://doi.org/10.1016/J.SCIENTA.2018.10.007.

[75] Tripathi DK, Singh S, Singh VP, Prasad SM, Dubey NK, Chauhan DK. Silicon nanoparticles more effectively alleviated UV-B stress than silicon in wheat (Triticum aestivum) seedlings. Plant Physiol Biochem 2017;110:70–81. Available from: https://doi.org/10.1016/J.PLAPHY.2016.06.026.

[76] Tripathi DK, Singh S, Singh VP, Prasad SM, Chauhan DK, Dubey NK. Silicon nanoparticles more efficiently alleviate arsenate toxicity than silicon in maize cultiver and hybrid differing in arsenate tolerance. Front Environ Sci 2016;4:46. Available from: https://doi.org/10.3389/FENVS.2016.00046.

[77] Suriyaprabha R, Karunakaran G, Kavitha K, Yuvakkumar R, Rajendran V, Kannan N. Application of silica nanoparticles in maize to enhance fungal resistance. IET Nanobiotechnol 2014;8(3):133–7. Available from: https://doi.org/10.1049/iet-nbt.2013.0004.

[78] Soares C, Carvalho MEA, Azevedo RA, Fidalgo F. Plants facing oxidative challenges—a little help from the antioxidant networks. Environ Exp Bot 2019;161:4–25. Available from: https://doi.org/10.1016/j.envexpbot.2018.12.009.

[79] Rao AR, Dayananda C, Sarada R, Shamala TR, Ravishankar GA. Effect of salinity on growth of green alga Botryococcus braunii and its constituents. Bioresour Technol 2007;98(3):560–4. Available from: https://doi.org/10.1016/J.BIORTECH.2006.02.007.

[80] Raven JA. The transport and function of silicon in plants. Biol Rev 1983;58(2):179–207. Available from: https://doi.org/10.1111/j.1469-185x.1983.tb00385.x.

[81] Avestan S, Ghasemnezhad M, Esfahani M, Byrt CS. Application of nano-silicon dioxide improves salt stress tolerance in strawberry plants. Agronomy 2019;9(5):246. Available from: https://doi.org/10.3390/AGRONOMY9050246.

[82] Gonçalves JFC, Santos UM. Utilization of the chlorophyll a fluorescence technique as a tool for selecting tolerant species to environments of high irradiance. Braz J Plant Physiol 2005;17(3):307–13. Available from: https://doi.org/10.1590/s1677-04202005000300005.

[83] Oliveira KR, et al. Exogenous silicon and salicylic acid applications improve tolerance to boron toxicity in field pea cultivars by intensifying antioxidant defence systems. Ecotoxicol Environ Saf 2020;201:110778. Available from: https://doi.org/10.1016/j.ecoenv.2020.110778.

[84] da Silva JLF, de R, Prado M, de JP, Junior S, Tenesaca LFL, et al. Feasibility of silicon addition to boron foliar spraying in cauliflowers. J Soil Sci Plant Nutr 2021;2021:1–8. Available from: https://doi.org/10.1007/S42729-021-00536-7.

[85] Farooq MA, Ali S, Hameed A, Ishaque W, Mahmood K, Iqbal Z. Alleviation of cadmium toxicity by silicon is related to elevated photosynthesis, antioxidant enzymes; suppressed cadmium uptake and oxidative stress in cotton. Ecotoxicol Environ Saf 2013;96:242–9. Available from: https://doi.org/10.1016/j.ecoenv.2013.07.006.

[86] Netto AT, Campostrini E, De Oliveira JG, Bressan-Smith RE. Photosynthetic pigments, nitrogen, chlorophyll a fluorescence and SPAD-502 readings in coffee leaves. Sci Hortic 2005;104(2):199–209. Available from: https://doi.org/10.1016/j.scienta.2004.08.013.

[87] Elsheery NI, Sunoj VSJ, Wen Y, Zhu JJ, Muralidharan G, Cao KF. Foliar application of nanoparticles mitigates the chilling effect on photosynthesis and photoprotection in sugarcane. Plant Physiol Biochem 2020;149:50−60. Available from: https://doi.org/10.1016/j.plaphy.2020.01.035.

Further reading

Ali MH, Tayeb EH, Kordy AM, Ghitheeth HH. Comparative insecticidal activity of nano and coarse silica on the Chinese beetle *Callosobruchus chinensis* (L)(Coleoptera: Bruchidae). Alex Sci Exch J 2017;38:655−60.

Ball P. Natural strategies for the molecular engineer. Nanotechnology 2002;13(5):R15−28. Available from: https://doi.org/10.1088/0957-4484/13/5/201.

Bharwana SA, Ali S, Farooq MA, Abbas F, Bioremed Biodeg J. Alleviation of lead toxicity by silicon is related to elevated photosynthesis, antioxidant enzymes suppressed lead uptake and oxidative stress in cotton. J Bioremed Biodeg 2013;4:4. Available from: https://doi.org/10.4172/2155-6199.1000187.

de Souza Junior JP, de R, Prado M, dos MM, Sarah S, Felisberto G. Silicon mitigates boron deficiency and toxicity in cotton cultivated in nutrient solution. J Plant Nutr Soil Sci 2019;182(5):805−14. Available from: https://doi.org/10.1002/jpln.201800398.

Happe T, Hemschemeier A, Winkler M, Kaminski A. Hydrogenases in green algae: do they save the algae's life and solve our energy problems? Trends Plant Sci 2002;7(6):246−50. Available from: https://doi.org/10.1016/S1360-1385(02)02274-4.

Kaur R, et al. Heavy metal stress in rice: uptake, transport, signaling and tolerance mechanisms. Physiol Plant 2021;173(1):430−48. Available from: https://doi.org/10.1111/PPL.13491.

Khaliq A, et al. Silicon alleviates nickel toxicity in cotton seedlings through enhancing growth, photosynthesis, and suppressing Ni uptake and oxidative stress. Arch Agron Soil Sci 2016;62(5):633−47. Available from: https://doi.org/10.1080/03650340.2015.1073263.

Meena V, Dotaniya ML, Saha JK, Patra AK. Silicon potential to mitigate plant heavy metals stress for sustainable agriculture: a review. Silicon 2021;2021:1−16. Available from: https://doi.org/10.1007/S12633-021-01170-9.

Nair R, Poulose AC, Nagaoka Y, Yoshida Y, Maekawa T, Kumar DS. Uptake of FITC labeled silica nanoparticles and quantum dots by rice seedlings: effects on seed germination and their potential as biolabels for plants. J Fluoresc 2011;21(6):2057−68. Available from: https://doi.org/10.1007/s10895-011-0904-5.

Nelson N, Yocum CF. Structure and function of photosystems I and II. Annu Rev Plant Biol 2006;57:521−65. Available from: https://doi.org/10.1146/ANNUREV.ARPLANT.57.032905.105350.

Oliveira Filho ASB, et al. Silicon attenuates the effects of water deficit in sugarcane by modifying physiological aspects and C:N:P stoichiometry and its use efficiency. Agric Water Manag 2021;255:107006. Available from: https://doi.org/10.1016/J.AGWAT.2021.107006.

Roco MC. Broader societal issues of nanotechnology. J Nanopart Res 2003;5(3−4):181−9. Available from: https://doi.org/10.1023/A:1025548512438.

Rouhani M, Samih MA, Kalantari S. Insecticidal effect of silica and silver nanoparticles on the cowpea seed beetle, *Callosobruchus maculatus* F. (Col.: Bruchidae). J Entomol Res 2012;4(4):297−305.

Schwarz K. A bound form of silicon in glycosaminoglycans and polyuronides. Proc Natl Acad Sci U S A 1973;70(5):1608−12. Available from: https://doi.org/10.1073/pnas.70.5.1608.

Smith AM, Duan H, Mohs AM, Nie S. Bioconjugated quantum dots for in vivo molecular and cellular imaging. Adv Drug Deliv Rev, 60. 2008. p. 1226−40. Available from: http://doi.org/10.1016/j.addr.2008.03.015.

Stirbet A, Govindjee. Chlorophyll a fluorescence induction: a personal perspective of the thermal phase, the J-I-P rise. Photosynth Res 2012;113(1−3):15−61. Available from: https://doi.org/10.1007/s11120-012-9754-5.

Stirbet A, Govindjee. On the relation between the Kautsky effect (chlorophyll a fluorescence induction) and photosystem II: basics and applications of the OJIP fluorescence transient. J Photochem Photobiol B Biol 2011;104(1−2):236−57. Available from: https://doi.org/10.1016/j.jphotobiol.2010.12.010.

Sun D, et al. Uptake and cellular distribution, in four plant species, of fluorescently labeled mesoporous silica nanoparticles. Plant Cell Rep 2014;33(8):1389−402. Available from: https://doi.org/10.1007/s00299-014-1624-5.

Teixeira GCM, de R, Prado M, Rocha AMS, de M, Piccolo C. Root- and foliar-applied silicon modifies C:N:P ratio and increases the nutritional efficiency of pre-sprouted sugarcane seedlings under water deficit. PLoS One 2020;15(10):e0240847. Available from: https://doi.org/10.1371/journal.pone.0240847 Oct.

Wang Y, Zhang B, Jiang D, Chen G. Silicon improves photosynthetic performance by optimizing thylakoid membrane protein components in rice under drought stress. Environ Exp Bot 2019;158:117−24. Available from: https://doi.org/10.1016/J.ENVEXPBOT.2018.11.022.

Waterkeyn L. Cytochemical localization and function of the 3-linked glucan callose in the developing cotton fibre cell wall. Protoplasma 1981;106(1−2):49−67. Available from: https://doi.org/10.1007/BF02115961.

Zahran NF, Sayed RM. Protective effect of nanosilica on irradiated dates against saw toothed grain beetle, *Oryzaephilus surinamensis* (Coleoptera: Silvanidae) adults. J Stored Prod Res 2021;92:101799. Available from: https://doi.org/10.1016/j.jspr.2021.101799.

Zhang JZ, Reisner E. Advancing photosystem II photoelectrochemistry for semi-artificial photosynthesis. Nat Rev Chem 2019;4(1):6−21. Available from: https://doi.org/10.1038/s41570-019-0149-4.

24

Manipulation of silicon metabolism in plants for stress tolerance

Zahoor Ahmad[1], Asim Abbasi[2], Syeda Refat Sultana[3],
Ejaz Ahmad Waraich[4], Arkadiusz Artyszak[5], Adeel Ahmad[6],
Muhammad Ammir Iqbal[7] and Celaleddin Barutçular[8]

[1]Department of Botany, University of Central Punjab, Punjab Group of College, Bahawalpur, Pakistan [2]Department of Zoology, University of Central Punjab, Punjab Group of College, Bahawalpur, Pakistan [3]Department of Filed Crops, Faculty of Agriculture, Cukurova University Adana, Adana, Turkey [4]Department of Agronomy, University of Agriculture Faisalabad, Faisalabad, Pakistan [5]Department of Agronomy, Warsaw University of Life Sciences—SGGW, Warsaw, Poland [6]Institue of Soil and Environmental Science, University of Agriculture Faisalabad, Faisalabad, Pakistan [7]Department of Agronomy, University of Poonch Rawalakot Azad Kashmir, Rawalakot, Pakistan [8]Department of Filed Crops, Faculty of Agriculture, Cukurova University Adana, Adana, Turkey

24.1 Background

Environmental stresses that affect plant growth and development are a major source of frustration across the world. Both biotic and abiotic stresses result in significant losses in agricultural production through reduced crop productivity. Plant scientists always had difficulty in maintaining optimal crop growth and production particularly under stressful conditions [1]. However, continuous exposure to these stresses enabled plants to evolve certain defensive mechanisms which help plants to cope with these external stressful conditions. Moreover, timely and balanced plant nutrition is also crucial for proper plant growth and stress tolerance mechanisms. Several studies have established the significance of micronutrients in promoting plant tolerance against a score of ecological stressors [2,3]. In this regard, silicon (Si) has gained much popularity due to its ability to enhance plant tolerance against biotic and abiotic stresses [4]. Despite its high abundance in earth crust (0.1−0.6 mM), Si is still not readily available to crop plants. Silicon is taken up by plants as H_4SiO_4 (silicic acid) the lone bioavailable silicon form present in soil solution [5]. Moreover, the concentration of silicon is gradually diminishing in soils that have been extremely weathered or intensively cultivated [6]. To improve Si-uptake by crop plants and its derived benefits, understanding the molecular basis of Si-uptake and tissue transport is essential.

24.2 Impact of stresses on plant growth

Plants work in an oscillating environment where external ecological anomalies are compensated by internal changes. The growth and productivity of crop plants are usually hampered by long-term exposure to different ecological stresses. Abiotic stresses are usually characterized as the ill effects imposed on any living entity by nonliving components of the environment [7]. The nonliving component usually changes the environment outside its usual fluctuation range to influence the demographic output of an organism or its individual physiology.

The global agricultural expansion imposed greater environmental concerns including drought stress, salinity, nutritional imbalance including mineral toxicity and deficiency, and high temperatures [8]. Plants usually respond to water-scarce conditions by altering certain cellular reactions which lead to a sequence of physiological, morphological, molecular, and biochemical changes in plants impacting their normal growth and development [9]. The growth and development of plants are greatly influenced by drought, salinity, temperature, and heavy metal toxicity [8]. Drought has an impact on three main stages of crop plants, that is, growth, preanthesis, and terminal phase [10]. Leaf width, leaf abscission, smaller leaf region, and reduced water loss through transpiration are all physiological reactions that plants adapt to cope with environmental stresses [11].

Drought stress affects water transport from the xylem in higher plants to adjacent elongating cells, which inhibits cell elongation [12,13]. Drought stress is also responsible for lower leaf area and shorter plant sowing to reduced cell division and poor mitosis [8]. Drought stress may also enhance the concentration of toxic ions in plant cells ultimately leading to cellular damage, reduced growth, and development [14]. Water is an essential component for plant growth and plays an important role in the transport of nutrients from the soil to aboveground plant tissues, hence a shortage of water might affect plant sustainability [15]. Moreover, excessive salt concentration might cause plants to suffer water stress. This might be due to the lower osmotic potential of salts as compared to water making it difficult for roots to absorb ample water from the soil. Similarly, high temperatures may also lead to drought stress through increased water loss by evaporation [16,17].

24.3 Metabolic changes under stress

Drought, temperature, and salinity are the major environmental factors that induced different morphological, agronomical, physiological, biochemical, and metabolic alterations in plants by disrupting certain cellular mechanisms that eventually affect transportation and storage of various plant assimilates along with its normal growth and development [18]. Drought and salinity significantly influence the functioning of plant membrane structures, which are involved in regulation of different ion transport and exchange resulting from external temperature stress [19]. The plant usually responds to ecological stresses in complex ways which not only depends on the duration and the intensity of imposed stresses but also depends largely on certain other factors including plant species, cultivars, and their morphology. Metabolic studies are helpful not only in determining plant adaptability to abiotic stress such as drought, salinity, extreme high/low temperature, but also useful for detecting the production of certain defensive compounds which mediate the adaption and tolerance to various stresses [20]. The earliest reaction of plants against water scarcity is the adjustment in photosynthesis and osmoregulation [21] followed by the accumulation of carbohydrates, amino acids, NH_4 compounds, and some polyamines which maintain the turgor pressure by decreasing the osmotic potential and stabilization of protein enzymes and cell membranes [20,22]. Osmolytes are also helpful in controlling ROS levels, managing stress, contributing to repairing processes, and also supporting plant growth by producing different energy compounds [23,24].

Salinity hinders plant growth as a result of cell damage due to osmotic stress on plants. Sodium accumulation affects the absorption of K^+ and the formation of protein [25]. Plants usually respond to salinity and drought stress by initiating certain biochemical pathways which ultimately lead to similar adjustments of plant metabolites. For example, in case of salinity, amassing of osmolytes functions as an adaptive mechanism that enables relocation of water status, cell turgor pressure, membrane stabilization, and control ROS accumulation [26]. Similarly, plants also respond to salinity stress by increasing the concentrations of some carbohydrates. The excessive accumulation of ROS particularly under saline conditions damages certain plant membranes and macromolecules [22]. Metabolomics studies revealed that phenolic compounds act as a powerful antioxidant to cope with the excessive accumulation of ROS, and the phenylpropanoid biosynthetic pathway is stimulated under salinity stress [22].

24.4 Agronomic approaches for abiotic stress management

Abiotic stresses resulting due to current climate change and global warming significantly affected crop growth and development all around the globe [27]. There are several approaches to mitigate drought and salinity stress which are primarily highlighted in the current scenario of global warming and climate change [28]. In this heading, different agronomic approaches will be discussed which can increase the yield and production of different crops under various abiotic stresses.

24.4.1 Planting time

Planting time has a significant effect on crop production as yield gradually declines in the case of late sowing of crops [29]. Each locality has a specific recommended sowing time for different field crops depending on the ecological conditions of that particular area [30]. Different agrometeorological parameters like rainfall, temperature, and photoperiod are the major determinants of crop yield [31]. Optimum sowing time also ensures better seed germination and establishment under adverse soil conditions. A significant reduction in crop yield due to late sowing has been reported in a number of studies [32,33]. Moreover, Liu et al. [34] reported that the grain-filling phase of crops requires a specific temperature and radiation level for the development of grain. Late harvesting of rice usually delays wheat sowing in some parts of the globe exposing wheat plants to heat stress at the grain development stage [35]. However, many researchers have reported that early sowing might result in higher crop production [36]. The prime cause of better crop yield is the gradual transfer of well-accumulated photosynthates from leaves to those parts of plants which mainly contribute to yield enhancement [37,38]. Different sowing dates associated with environmental changes can significantly affect crop yield and its components [39].

24.4.2 Irrigation management

To alleviate the osmotic and heat stress, adequate water supply at regular intervals is an important approach for better crop production [40,41]. Wheat is a water-stress-sensitive crop and requires ample water supply at vegetative, reproductive, and grain formation stages. Seedling establishment, tillering, booting, and grain formation are critical growth stages for moisture deficit [42,43]. A well-developed root system plays an important part in adequate moisture and nutrient uptake [44]. Root growth and development significantly increase nutrient uptake at critical growth stages of crops particularly under moisture stress [42].

Improved irrigation practices and mulching considerably improve moisture and nutrient uptake [45]. To avoid the ill effects of water scarcity, the latest irrigation techniques play a major role in improving the photosynthetic ability of crop plants and their yield potential [32,46]. Drip irrigation is the best solution to avoid moisture stress particularly in rain-fed areas and significant improvement of water use efficiency (WUE) of crop plants [47,48]. Moreover, under drought stress, use of the advanced water application techniques like magnetic water application via artificial magnetization process also proved their worth by improving plant growth and yield. Similarly, remote sensing is also helpful in the assessment of drought stress. Similarly, judicious water application using sprinkler irrigation also improved the growth and yield of crop plants under hot climatic regions where water availability is minimal [49–51]. Moreover, utilization of the current distant detecting instruments like satellite symbolism, microwave far-off sensors for evaluation of dry season pressure is also being utilized with greater ease and success. Furthermore, the use of different crop simulation models like agricultural production systems simulator is also helpful for irrigation scheduling under rain-fed and irrigated areas [52].

24.5 Nutrition role in stress tolerance

All over the world, about 60% of the cultivated soils have problems associated with limiting plant growth mainly due to deficiencies and toxicities of mineral nutrients. Marginal lands receive sunlight which is helpful in electron transport and CO_2 fixation for photosynthesis. In the case of abundant light absorption in chloroplast molecular O_2 is activated which reacts with oxygen species (ROS). The excess level of ROS leads to photooxidative and chlorophyll damage, consequently, leading to cell death. Environmental stressors enhance the damage in chloroplast when the conversion of light energy during photosynthesis is limited. Nutrient availability is essential for electron transport chain maintenance during stress conditions. Hence, the exogenous application of mineral nutrients like nitrogen (N), potassium (K), magnesium (Mg), calcium (Ca), zinc (Zn), and boron (B) is essential under abiotic stress conditions to minimize their adverse effects. Drought and salinity stimulate oxidase of NADPH-dependent molecules which stimulates the production of ROS. Potassium and zinc interfere with oxidizing enzyme of NADPH which results in additional protection against ROS under salinity and drought stress. Balanced nutrition plays a major role in mitigating the detrimental effects of stresses on the growth and yield of crop plants [53].

Under abiotic stresses, the survival of the crop plant is based on the aptitude of the plant to develop an adaptive mechanism to avoid or tolerate stress. It is evident from different studies that the ability to adapt to adverse conditions mainly depends on the plant's mineral nutritional status. In this part, the nutritional status of the plants in regard to nitrogen (N), potassium (K), magnesium (Mg), calcium (Ca), zinc (Zn), and boron (B) will be discussed particularly under stresses.

24.5.1 Nitrogen

Nitrogen is an integral part of chlorophyll and also has a pivotal role in the absorption of radiance energy and metabolism of carbon involved in the process of photosynthesis [54,55]. Nitrogen deficiency leads to photooxidative damage in leaves because of the abundant amount of nonutilized energy. For example, the low level of N is responsible for high lipid peroxidation under high light intensity in rice [54,55]. The form in which N is applied to crop plants also affects their tolerance against photodamage. The light-induced conversion of violaxanthin to zeaxanthin, as a means to dissipate surplus light energy was, found to be lower in bean leaves supplied with ammonium than in those supplied with nitrate [56]. Similarly, Zhu et al. [57]. also reported that bean plants amended with nitrate showed higher tolerance to photodamage than ammonium-grown ones. Under high light intensity plants treated with ammonium had higher contents of antioxidative enzymes and lipid peroxidation levels.

24.5.2 Potassium, magnesium, and zinc

Similar to N deficiency, crop plants are also vulnerable to photo-oxidative damage when subjected to K, Mg, and Zn deficiencies. Limiting the supply of these nutrients causes chlorosis, necrosis, and stunted plant growth when exposed to elevated light intensity [58,59]. Similarly, in the antioxidant system, manganese also acts as an important cofactor. Commencement of Mn-CAT and Mn-SOD takes place in the presence of manganese and it also acts as a scavenger of ROS. Accumulation of biomass, RGR, NAR, and chlorophyll contents, photosynthetic efficiency, and reduced lipid peroxidation may improve by Mn application under salt-stressed conditions [60–62]. Even under high saline conditions, Mn-treated plants improved ionic and osmotic regulation, recovered from chlorosis, and had high activity of antioxidant enzymes [62]. Moreover, a gradual decline in photosynthetic C metabolism and utilization of fixed carbon was recorded under Mg and K deficiencies. The deficiencies of these mineral nutrients result in a massive accumulation of carbohydrates in source leaves, with subsequent inhibition of photosynthetic C reduction [63]. Deficiencies of these nutrients lead to photoactivation of molecular oxygen and the occurrence of photo-oxidative damage due to nonutilized light energy and photoelectrons. That is why leaves of Mg- and K-deficient plants are highly vulnerable to high light intensity [64,65]. These observations strongly suggest that under K and Mg deficiency, photo-oxidative damage to chloroplasts is more profound. However, in contrast, P deficiency had no significant effect on sucrose transport and photosynthates accumulation in leaves. Chlorosis is common in plants deficient in K and Mg, but not in plants deficient in P [65].

24.5.3 Calcium

Many processes of the cell take place in the presence of calcium which is recognized as a secondary messenger in the plant cell [66–69]. Osmotic adjustment under drought stress occurred because of Ca^{2+} which increased soluble sugar and proline content in plants [70]. It activates several cell processes of target proteins which mediate plant responses to drought and salinity [71]. Metabolism of crops is primarily dependent on calcium-dependent protein kinase (CDPK) and is involved in trigging all kinases (MAPK and CIPKs) in drought tolerance [72,73].

Limiting Ca use interrupts many physiological processes, that is, photosynthesis capacity, reduced carboxylation efficiency, opening and closing behavior of stomata, and crop yield [74,75]. Moreover, cation channels in the plasma membrane are activated by auxin, which facilitates Ca^{+2}-K^+ exchange [76].

24.5.4 Salinity

From different studies, it is concluded that some reactive O_2 species are helpful in the mediation of the salt-induced cell damage to crop plants. Application of NaCl at low levels, in several plant species, causes stimulation of certain antioxidative enzymes which suggests a role of salt stress in ROS formation [77]. In rice plants, cell damage induced by salt stress can be mitigated by overexpression of superoxide dismutase (SOD) in chloroplasts [78].

The activity of NADPH oxidase is inhibited by zinc ions. The role of Zn is helpful in the structural integrity of the plasma membrane and it also regulates the uptake of Na and other toxic ions [79]. Zinc not only protects the plant from damaging attacks of ROS but also prevents salt-stressed plants from taking toxic ions from the plasma membrane. Moreover, in Zn-treated plants, total carbohydrates, photosynthetic pigments, and phenolic compounds were higher under stress conditions. By triggering the activities of SOD, CAT, and APX, Zn counteracts the adverse effects of salinity [80]. Similarly, K is also an important nutrient element like Zn that guards plant cells against cell damage induced by salt. In different studies, it is reported that limited K content is affiliated with the production of O^{2-} and leads to an increase in the NADPH oxidase activity in beans [81,82]. Moreover, boron in trace amounts is also essential for crops however, in high concentration is phytotoxic in rice crops under salinity. Under saline conditions, high concentration hindered crop growth and straw production [83].

24.5.5 Drought

Under water-deficit conditions or high light intensity at low temperature, generation of ROS, and photooxidative damage is common [77]. For the maintenance of photosynthetic electron transport, mineral nutrients are the basic necessity. When the plants suffer nutrient deficiencies, photo-oxidative damage can be drastic particularly under stressed environments. Potassium plays an important role in maintaining water relations of plants and also regulates stomatal opening particularly under water-scarce conditions [63]. Potassium deficiency hindered photosynthetic activity in wheat crops under water-deficit condition, but the effect was marginal when the K supply was adequate [84]. Moreover, in breeding genotypes, the presence of K in plant tissues is a useful trait for high tolerance to drought stress. It not only helps plants to cope with water scarcity but also enhances root longevity [84].

24.6 Impact of silicon nutrition under stresses

Despite its high abundance in nature and soil, the beneficial and ameliorative effects of Si, particularly under stressful conditions, have been studied intensively only during the recent decades [85,86]. Silicon accumulation in plant tissues provides mechanical strength and support to cell walls by augmenting lignification, suberization, and silicification processes [87]. Moreover, silicon binding cell wall hemicellulose is related to enhancing structural stability which influences plants to cope with water-scarce conditions [88,89]. Silicon also alters the properties of plant cells by forming a hardened layer beneath the leaf epidermis which prevents water loss through stomatal conductance and cuticular transpiration [90,91] and also augments UV tolerance [92] by alleviating UVB imposed membrane ruptures [93]. Moreover, Si in the form of silicic acid polymerizes in the plant apoplast leading to the formation of amorphous silica barriers which inhibits penetration of certain toxicants such as Al, Cd, and Na into symplast and also prevents plants from pathogenic infection [94,95]. Furthermore, higher salt tolerance in Si amended plants is mainly attributed to the synthesis of suberin and lignin which forms fences against apoplastic Na^+ transport in roots [96,97].

Arthropod herbivores are another class of biotic stressors against which Si provided considerable protection not only by enhancing the surface roughness of plant tissue but also by reducing their nutrient composition and digestibility for insects [98,99]. Moreover, Si also provides mechanical protection to the plant thus lowering insect feeding, probing time, and oviposition preference [100,101]. Furthermore, Si amendments also trigger various molecular reactions in plants leading to the production of certain compounds such as phytoalexins, momilactones, and phenolics which plant utilizes as a prime source of defense against biotic stressors [102]. Apart from these, Si also activates the production of important defensive enzymes such as polyphenol oxidase, peroxidase, lipoxygenase, and phenylalanine ammonia-lyase, which aid plants to avoid biotic stressors via nonpreference and antibiosis mechanisms [103].

An increased attraction of predators and parasitoids was also reported in Si-treated pest-infested plants as compared to untreated plants. The augmented production of herbivore-induced plant volatiles in Si-treated pest-infested plants is believed to be the prime cause of this biological phenomenon [104]. Moreover, Si also slows down the growth and biological cycle of attacking herbivores making them more vulnerable to other management tactics such as biological and chemical control [105].

24.7 Role of silicon in plant metabolism

The potential of silicon (Si) for alleviating ill effects of biotic and abiotic stress in plants is well known [100], however, Si is not readily available to plants as silicic acid due to a number of reasons, hence exogenous application of silicate fertilizers is needed in different crop production systems. Crop plants absorb Si in the form of monosilicic acid, the only available form of silicon in the earth's crust. However, certain plants have developed specialized transporter genes (Lsi1, Lsi2 and Lsi6) for Si absorption and regulation throughout their body [106]. The beneficial effect of Si regarding enhanced uptake of macro- and micronutrients, better plant growth, and yield has been proved by a number of studies [107–111]. Plant species usually vary in their response to Si application however, in general, Si boosts instant water-use efficiency and photosynthesis of plants, as well as alters nutrient disparities [112,113]. To combat oxidative stress, plants have a fully operational defense system comprising of different antioxidant compounds and enzymes [114]. Metal toxicity usually hinders the performance of certain antioxidant enzymes, however, application of silicon sources, particularly under stressful conditions, signifies the production of peroxidases, superoxide dismutase, ascorbate peroxidase, and catalase, which assist plants to cope with metal toxicity [115,116]. Moreover, Si applications also enhanced the photosynthesis and chlorophyll fluorescence yields of maize tomato, wheat, and soybean plants, however, the effect was more pronounced under stressful conditions [117]. Moreover, Si also protects plants from certain biotic and abiotic stresses and provides mechanical strength against lodging, thus increasing plant growth and biomass production of a number of important monocot and dicot plants [86,118]. Moreover, the indirect effects of silicon were mostly related to soil pH adjustment and assimilation of micro- and macronutrients enclosed in silicate fertilizers [119].

Furthermore, the positive impact of silicon in increasing leaf area, stem length, stomatal number and size, stomatal conductance, chlorophyll fluorescence, and pigment concentration under stressful conditions have been reported in a number of studies [120–122]. Moreover, the foliar application of Si through plant leaves also indicated its beneficial effects on plant physiology and plant growth which is mainly attributed to the indirect effect of Si on plant metabolism and photosynthesis [123–126].

24.8 Conclusions and remarks

The currently available research shows that Si can be recognized as a regular fertilizer especially in high-accumulating cereal crops. The beneficial advantages of Si in many plant species have also been demonstrated by recent findings. To enhance applicability and widen the range of plant species that can explore advantages derived from Si, a greater knowledge of underlying molecular processes is necessary. Presently, various theories and processes have been presented to explain how Si provides enhanced protection to plants in stressful conditions. However, none of the proposed theories explains precisely how Si plays a crucial role. Although these concerns are not resolved, the current expertise allows us to investigate the Si-derived advantages by either complementing or creating new cultivars with improved Si absorption. The enhanced advantages of Si-derived plants will assist to build in the future a more sustainable cultivation system.

References

[1] Zargar SM, Mahajan R, Bhat JA, Nazir M, Deshmukh R. Role of silicon in plant stress tolerance: opportunities to achieve a sustainable cropping system. 3 Biotech 2019;9:73.
[2] Vanderschuren H, Boycheva S, Li KT, Szydlowski N, Gruissem W, Fitzpatrick TB. Strategies for vitamin B6 biofortification of plants: a dual role as a micronutrient and a stress protectant. Front Plant Sci 2013;4:143.
[3] Bradacova K, Weber NF, Morad-Talab N, Asim M, Imran M, Weinmann M, et al. Micronutrients (Zn/Mn), seaweed extracts, and plant growth-promoting bacteria as cold-stress protectants in maize. Chem Biol Technol Agric 2016;3:19.
[4] Ma JF. Role of silicon in enhancing the resistance of plants to biotic and abiotic stresses. Soil Sci Plant Nutr 2004;50:11–18.
[5] Epstein E. Silicon: its manifold roles in plants. Ann Appl Biol 2009;155:155–60.
[6] Meena V, Dotaniya M, Coumar V, Rajendiran S, Kundu S, Rao AS. A case for silicon fertilization to improve crop yields in tropical soils. Proc Natl Acad Sci India Sect B Biol Sci 2014;84:505–18.
[7] Verslues PE, Zhu JK. Before and beyond ABA: upstream sensing and internal signals that determine ABA accumulation and response under abiotic stress. Biochem Soc Trans 2005;33:375–9.
[8] Fahad S, Bajwa AA, Nazir U, Anjum SA, Farooq A, Zohaib A. Crop production under drought and heat stress: plant responses and management options. Front Plant Sci 2017;29:8.
[9] Bajguz A, Hayat S. Effects of brassinosteroids on the plant responses to environmental stresses. Plant Physiol Biochem 2009;47:1–8.
[10] Shavrukov Y, Kurishbayev A, Jatayev S, Shvidchenko V, Zotova L, Koekemoer F. Early flowering as a drought escape mechanism in plants: how can it aid wheat production? Front Plant Sci 2017;17:8.
[11] Fghire R, Anaya F, Ali OI, Benlhabib O, Ragab R, Wahbi S. Physiological and photosynthetic response of quinoa to drought stress. Chil J Agric Res 2015;75:174–83.

[12] Jaleel CA, Manivannan P, Wahid A, Farooq M, Al-Juburi HJ, Somasundaram R, et al. Drought stress in plants: a review on morphological characteristics and pigments composition. Int J Agric Biol 2009;11:100−5.

[13] Anjum SA, Xie XY, Wang LC, Saleem MF, Man C, Lei W. Morphological, physiological and biochemical responses of plants to drought stress. Afr J Agric Res 2011;6:2026−32.

[14] Ojuederie OB, Olanrewaju OS, Babalola OO. Plant growth-promoting rhizobacterial mitigation of drought stress in crop plants: implications for sustainable agriculture. Agron 2019;9:712.

[15] Ashkavand P, Zarafshar M, Tabari M, Mirzaie J, Nikpour A, Kazembordbar S. Application of SIO₂ nanoparticles AS pretreatment alleviates the impact of drought on the physiological performance of *Prunus mahaleb* (Rosaceae). Bol Soc Argent Bot 2018;53:207.

[16] Clarke JM, Richards RA, Condon AG. Effect of drought stress on residual transpiration and its relationship with water use of wheat. Can J Plant Sci 1991;71:695−702.

[17] Blum A. Effective use of water (EUW) and not water-use efficiency (WUE) is the target of crop yield improvement under drought stress. Field Crop Res 2009;112:119−23.

[18] Di Ferdinando M, Brunetti C, Agati G, Tattini M. Multiple functions of polyphenols in plants inhabiting unfavorable Mediterranean areas. Environ Exp Bot 2014;103:107−16.

[19] Valitova J, Renkova A, Mukhitova F, Dmitrieva S, Beckett RP, Minibayeva FV. Membrane sterols and genes of sterol biosynthesis are involved in the response of *Triticum aestivum* seedlings to cold stress. Plant Physiol Biochem 2019;142:452−9.

[20] Jorge TF, António C. Plant metabolomics in a changing world: metabolite responses to abiotic stress combinations. In: Andjelkovic V, editor. Plant, abiotic stress and responses to climate change. New York: Intech Open Science; 2018. p. 111−32.

[21] Slama I, Abdelly C, Bouchereau A, Flower T, Savouré A. Diversity, distribution and roles of osmoprotective compounds accumulated in halophytes under abiotic stress. Ann Bot 2015;115:433−47.

[22] Sharma A, Shahzad B, Rehman A, Bhardwaj R, Landi M, Zheng B. Response of phenylpropanoid pathway and the role of polyphenols in plants under abiotic stress. Mol 2019;24:2452.

[23] Silva S, Santos C, Serodio J, Silva AMS, Dias MC. Physiological performance of drought-stressed olive plants when exposed to a combined heat-UV-B shock and after stress relief. Funct Plant Biol 2018;45:1233−40.

[24] Fàbregas N, Fernie AR. The metabolic response to drought. J Exp Bot 2019;70:1077−85.

[25] Deinlein U, Stephan AB, Horie T, Luo W, Xu G, Schroeder JI. Plant salt-tolerance mechanisms. Trends Plant Sci 2014;19:371−9.

[26] Kao CH. Mechanisms of salt tolerance in rice plants: compatible solutes and aquaporins. Crop Env Bioinform 2015;12:73−82.

[27] Tubiello FN, Soussana JF, Howden SM. Crop and pasture response to climate change. Proc Natl Acad Sci U S A 2007;104:19686−90.

[28] Velde B, Barré P. Soils, plants and clay minerals: mineral and biologic interactions. Berlin/Heidelberg, Germany: Springer; 2010. p. 261.

[29] Pal R, Mahajan G, Sardana V, Chauhan BS. Impact of sowing date on yield, dry matter and nitrogen accumulation, and nitrogen translocation in dry-seeded rice in North-West India. Field Crop Res 2017;206:138−48.

[30] Sarkar S, Gaydon DS, Brahmachari K, Nanda MK, Ghosh A, Mainuddin M. Modelling yield and seasonal soil salinity dynamics in rice-grasspea cropping system for the coastal saline zone of West Bengal. India Proc 2019;36:146.

[31] Chen C, Huang J, Zhu L, Shah F, Nie L, Cui K, et al. Varietal difference in the response of rice chalkiness to temperature during ripening phase across different sowing dates. Field Crop Res 2013;15:85−91.

[32] Panhwar QA, Ali A, Naher UA, Memon MY. Fertilizer management strategies for enhancing nutrient use efficiency and sustainable wheat production. In: Sarath, Chandran MR, Unni ST, editors. Food science, technology and nutrition, organic farming. Amsterdam: Woodhead Publishing; 2019.

[33] Shah F, Coulter JA, Ye C, Wu W. Yield penalty due to delayed sowing of winter wheat and the mitigatory role of increased seeding rate. Eur J Agron 2020;119:126120.

[34] Liu S, Lu Y, Yang C, Liu C, Ma L, Dang Z. Effects of modified biochar on rhizosphere microecology of rice (*Oryza sativa* L.) grown in As-contaminated soil. Environ Sci Pollut Res 2017;24:23815−24.

[35] Dubey R, Pathak H, Chakrabarti B, Singh S, Gupta DK, Harit R. Impact of terminal heat stress on wheat yield in India and options for adaptation. Agric Syst 2020;181:102826.

[36] Jat RK, Singh P, Jat ML, Dia M, Sidhu HS, Jat SL, et al. Heat stress and yield stability of wheat genotypes under different sowing dates across agro-ecosystems in India. Field Crop Res 2018;218:33−50.

[37] Singh BS, Humphreys E, Yadav S, Gaydon DS. Options for increasing the productivity of the rice−wheat system of northwest India while reducing groundwater depletion. Part 1. Rice variety duration, sowing date and inclusion mungbean. Field Crop Res 2015;173:68−80.

[38] Gou F, van Ittersum MK, van derWerf W. Simulating potential growth in a relay-strip intercropping system: model description, calibration and testing. Field Crop Res 2017;200:122−42.

[39] Sarkar S, Ghosh A, Brahmachari K, Ray K, Nanda MK, Sarkar D. Weather relation of rice-grass pea crop sequence in Indian Sundarbans. J Agrometeorol 2020;22:148−57.

[40] Rana L, Banerjee H, Ray K, Sarkar S. System of wheat intensification (SWI)—a new approach for increasing wheat yield in small holder farming system. J Appl Nat Sci 2017;9:1453−64.

[41] Tariq A, Pan K, Olatunji OA, Graciano C, Li Z, Sun F, et al. Phosphorous application improves drought tolerance of *Phoebe zhennan*. Front Plant Sci 2017;8:8.

[42] Panda RK, Behera SK, Kashyap PS. Effective management of irrigation water for wheat under stressed conditions. Agric Water Manag 2003;63:37−56.

[43] Ahmed I, Ullah A, Rahman MH, Ahmad B, Wajid SA, Ahmad A, et al. Climate change impacts and adaptation strategies for agronomic crops. Climate change and agriculture. London: Intech open; 2019. p. 1−14.

[44] Ali S, Xu Y, Ma X, Ahmad I, Jia Q, Zhang J, et al. Deficit irrigation strategies to improve winter wheat productivity and regulating root growth under different planting patterns. Agric Water Manage 2019;219:1−11.

[45] Doussan C, Pierret A, Garrigues E, Pagès L. Water uptake by plant roots: II-Modelling of water transfer in the soil root-system with explicit account of flow within the root system—comparison with experiments. Plant and soil. New York, NY: Springer; 2016. p. 99−117.

[46] Liu Y, Zhang X, Xi L, Liao Y, Han J. Ridge-furrow planting promotes wheat grain yield and water productivity in the irrigated sub-humid region of China. Agric Water Manage 2020;231:105935.

[47] Jha SK, Ramatshaba TS, Wang G, Liang Y, Liu H, Duan A, et al. Response of growth, yield and water use efficiency of winter wheat to different irrigation methods and scheduling in North China Plain. Agric Water Manage 2019;217:292–302.

[48] Mondal S, Dutta S, Crespo-Herrera L, Huerta-Espino J, Braun HJ, Singh RP. Fifty years of semi-dwarf spring wheat breeding at CIMMYT: grain yield progress in optimum, drought and heat stress environments. Field Crop Res 2020;250:107757.

[49] Javadipour Z, Balouchi H, Dehnavi MM, Yadavi A. Roles of methyl jasmonate in improving growth and yield of two varieties of bread wheat (Triticum aestivum) under different irrigation regimes. Agric Water Manage 2019;222:336–45.

[50] Selim DAFH, Nassar RMA, Boghdady MS, Bonfill M. Physiological and anatomical studies of two wheat cultivars irrigated with magnetic water under drought stress conditions. Plant Physiol Biochem 2019;135:480–8.

[51] Olivera-Guerra L, Merlin O, Er-Raki S. Irrigation retrieval from Landsat optical/thermal data integrated into a crop water balance model: a case study over winter wheat fields in a semi-arid region. Remote Sens Environ 2020;239:111627.

[52] Gaydon DS, Singh B, Wang E, Poulton P, Ahmad B, Ahmed F, et al. Evaluation of the APSIM model in cropping systems of Asia. Field Crop Res 2017;204:52–75.

[53] Haneklaus SH, Bloem E, Schnug E. Hungry plants—a short treatise on how to feed crops under stress. Agriculture 2018;8:43.

[54] Kato MC, Hikosaka K, Hirotsu N, Makino A, Hirose T. The excess light energy that is neither utilized in photosynthesis nor dissipated by photoprotective mechanisms determines the rate of photoinactivation in photosystem II. Plant Cell Physiol 2003;44:318–25.

[55] Huang ZA, Jiang DA, Yang Y, Sun JW, Jin SH. Effects of nitrogen deficiency on gas exchange, chlorophyll fluorescence, and antioxidant enzymes in leaves of rice plants. Photosynthetica 2004;42:357–64.

[56] Bendixen R, Gerendás J, Schinner K, Sattelmacher B, Hansen UP. Difference in zeaxanthin formation in nitrate-and ammonium-grown Phaseolus vulgaris. Physiol Plant 2001;111:255–61.

[57] Zhu Z, Gerendas J, Bendixen R, Schinner K, Tabrizi H, Sattelmacher B, et al. Different tolerance to light stress in NO_3^- and NH_4^+ grown Phaseolus vulgaris L. Plant Biol 2000;2:558–70.

[58] Cakmak I, Atli M, Kaya R, Evliya H, Marschner H. Association of high light and zinc deficiency in cold-induced leaf chlorosis in grapefruit and mandarin trees. J Plant Physiol 1995;146:355–60.

[59] Polle A. Mehler reaction: friend or foe in photosynthesis? Bot Acta 1996;109:84–9.

[60] Pandya DH, Mer RK, Prajith PK, Pandey AN. Effect of salt stress and manganese supply on growth of barley seedlings. J Plant Nutr 2004;27:1361–79.

[61] Sebastian A, Prasad MN. Iron and manganese-assisted cadmium tolerance in Oryza sativa L.: lowering of rhizotoxicity next to functional photosynthesis. Planta 2015;241:1519–28.

[62] Rahman A, Hossain MS, Mahmud JA, Nahar K, Hasanuzzaman M. Fujita manganese-induced salt stress tolerance in rice seedlings: regulation of ion homeostasis, antioxidant defense and glyoxalase systems. Physiol Mol Biol Plant 2016;22:291–306.

[63] Mengel K, Kirkby EA. Principles of plant nutrition. 5th ed. Dordrecht: Kluwer Academic Publishers; 2001. p. 848.

[64] Marschner H, Cakmak I. High light intensity enhances chlorosis and necrosis in leaves of zinc-, potassium- and magnesium-deficient bean (Phaseolus vulgaris) plants. J Plant Physiol 1989;134:308–15.

[65] Cakmak I. Activity of ascorbate-dependent H_2O_2-scavenging enzymes and leaf chlorosis are enhanced in magnesium and potassium deficient leaves, but not in phosphorus-deficient leaves. J Exp Bot 1994;45:1259–66.

[66] Felle HH. pH: signal and messenger in plant cells. Plant Biol 2001;3:577–91.

[67] Choi WG, Hilleary R, Swanson SJ, Kim SH, Gilroy S. Rapid, long-distance electrical and calcium signaling in plants. Annu Rev Plant Biol 2016;67:287–307.

[68] Gilroy S, Białasek M, Suzuki N, Górecka M, Devireddy AR, Karpiński S. ROS, calcium, and electric signals: key mediators of rapid systemic signaling in plants. Plant Physiol 2016;171:1606–15.

[69] Khushboo BK, Singh P, Raina M, Sharma V, Kumar D. Exogenous application of calcium chloride in wheat genotypes alleviates negative effect of drought stress by modulating antioxidant machinery and enhanced osmolyte accumulation. Vitro Cell Dev Biol Plant 2018;54:495–507.

[70] Jaleel CA, Manivannan P, Sankar B, Kishorekumar A, Panneerselvam R. Calcium chloride effects on salinity-induced oxidative stress, proline metabolism and indole alkaloid accumulation in Catharanthus roseus. Comptes Rendus Biol 2007;330:674–83.

[71] Silva EC, Nogueira RJMC, Silva MA, Alburquerque MB. Drought stress and plant nutrition. Plant Stress 2011;5:32–41.

[72] Ciesla A, Mituła F, Misztal L, Fedorowicz-Strońska O, Janicka S, Tajdel-Zielińska M. A role for barley calcium-dependent protein kinase CPK2a in the response to drought. Front Plant Sci 2016;7:1550.

[73] Kumar S., Sachdeva S., Bhat K.V., Vats S. Plants responses to drought stress: physiological, biochemical and molecular basis in biotic and abiotic stress tolerance in Plants. S. Vats, editor (Singapore: Springer Nature Singapore Pte. Ltd.). 2018.

[74] Eticha D, Kwast A, de Souza CT, Horowitz N, Stützel H. Calcium nutrition of orange and its impact on growth, nutrient uptake and leaf cell wall. Citrus Res Technol 2017;38:62–70.

[75] Song Q, Liu Y, Pang J, Yong JWH, Chen Y, Bai C. Supplementary calcium restores peanut (Arachis hypogaea) growth and photosynthetic capacity under low nocturnal temperature. Front Plant Sci 2019;10:1637.

[76] Vanneste S, Friml J. Calcium: the missing link in auxin action. Plants 2013;2:650–75.

[77] Wang YJ, Wisniewski M, Melian R, Cui MG, Webb R, Fuchigami L. Overexpression of cytosolic ascorbate peroxidase in tomato confers tolerance to chilling and salt stress. J Am Soc Hortic Sci 2005;130:167–73.

[78] Tanaka Y, Hibino T, Hayashi Y, Tanaka A, Kishitani S, Takabe T, et al. Salt tolerance of transgenic rice overexpressing yeast mitochondrial Mn-SOD in chloroplasts. Plant Sci 1999;148:131–8.

[79] Cakmak I, Marschner H. Enhanced superoxide radical production in roots of zinc-deficient plants. J Exp Bot 1988;39:1449–60.

[80] Jan AU, Hadi F, Midrarullah, Nawaz MA, Rahman K. Potassium and zinc increase tolerance to salt stress in wheat (Triticum aestivum L.). Plant Physiol Biochem 2017;116:139–49.

[81] Shin R, Schachtman DP. Hydrogen peroxide mediates plant root cell response to nutrient deprivation. Proc Natl Acad Sci U S A 2004;101:8827–32.

[82] Cakmak I. The role of potassium in alleviating detrimental effects of abiotic stresses in plants. J Plant Nutri Soil Sci 2005;168:521–30.

[83] Mehmood EUH, Kausar R, Akram M, Shahzad SM. Is boron required to improve rice growth and yield in saline environment? Pak J Bot 2009;41:1339–50.

[84] Egilla JN, Davies FT, Drew MC. Effect of potassium on drought resistance of *Hibiscusrosa sinensis* cv. Leprechaun: plant growth, leaf macro- and micronutrient content and root longevity. Plant Soil 2001;229:213–24.

[85] Currie HA, Perry CC. Silica in plants: biological, biochemical and chemical studies. Ann Bot 2007;100:1383–9.

[86] Liang Y, Nikolic M, Belanger R, Gong H, Song A. Effect of silicon on crop growth, yield and quality. Silicon in agriculture. Dordrecht: Springer; 2015. p. 209–23.

[87] Guerriero G, Hausman JF, Legay S. Silicon and the plant extracellular matrix. Front Plant Sci 2016;7:463.

[88] He C, Ma J, Wang L. A hemicellulose-bound form of silicon with potential to improve the mechanical properties and regeneration of the cell wall of rice. New Phytol 2015;206:1051–62.

[89] Ma J, Cai H, He C, Zhang W, Wang L. A hemicellulose-bound form of silicon inhibits cadmium ion uptake in rice (*Oryza sativa*) cells. New Phytol 2015;206:1063–74.

[90] Gong HJ, Chen KM, Chen GC, Wang SM, Zhang CL. Effects of silicon on growth of wheat under drought. J Plant Nutr 2003;26:1055–63.

[91] Zhu Y, Gong H. Beneficial effects of silicon on salt and drought tolerance in plants. Agron Sustain Dev 2014;34:455–72.

[92] Goto M, Ehara H, Karita S, Takabe K, Ogawa N, Yamada Y. Protective effect of silicon on phenolic biosynthesis and ultraviolet spectral stress in rice crop. Plant Sci 2003;164:349–56.

[93] Shen X, Zhou Y, Duan L, Li Z, Eneji AE, Li J. Silicon effects on photosynthesis and antioxidant parameters of soybean seedlings under drought and ultraviolet-B radiation. J Plant Physiol 2010;167:1248–52.

[94] Fauteux F, Remus-Borel W, Menzies JG, Belanger RR. Silicon and plant disease resistance against pathogenic fungi. FEMS Microbiol Lett 2005;249:1–6.

[95] Saqib M, Zoerb C, Schubert S. Silicon-mediated improvement in the salt resistance of wheat (*Triticum aestivum*) results from increased sodium exclusion and resistance to oxidative stress. Funct Plant Biol 2008;35:633–9.

[96] Fleck AT, Nye T, Repenning C, Stahl F, Zahn M, Schenk MK. Silicon enhances suberization and lignification in roots of rice (*Oryza sativa*). J Exp Bot 2011;62:2001–11.

[97] Krishnamurthy P, Ranathunge K, Nayak S, Schreiber L, Mathew MK. Root apoplastic barriers block Na^+ transport to shoots in rice (*Oryza sativa* L.). J Exp Bot 2011;62:4215–28.

[98] Massey FP, Hartley SE. Physical defenses wear you down: progressive and irreversible impacts of silica on insect herbivores. J Anim Ecol 2009;78:281–91.

[99] Hartley SE, Fitt RN, McLarnon EL, Wade RN. Defending the leaf surface: intra-and inter-specific differences in silicon deposition in grasses in response to damage and silicon supply. Front Plant Sci 2015;6:35.

[100] Zhang C, Wang L, Zhang W, Zhan F. Do lignification and silicification of the cell wall precede silicon deposition in the silica cell of the rice (*Oryza sativa* L.) leaf epidermis? Plant Soil 2013;372:137–49.

[101] Schurt DA, Cruz MFA, Nascimento KJT, Filippi MCC, Rodrigues FA. Silicon potentiates the activities of defense enzymes in the leaf sheaths of rice plants infected by *Rhizoctonia solani*. Trop Plant Pathol 2014;39:457–63.

[102] Remus-Borel W, Menzies JG, Bélanger RR. Silicon induces antifungal compounds in powdery mildew-infected wheat. Physiol Mol Plant Pathol 2005;66:108–15.

[103] Cai K, Gao D, Luo S, Zeng R, Yang J, Zhu X. Physiological and cytological mechanisms of silicon-induced resistance in rice against blast disease. Physiol Plant 2008;134:324–33.

[104] Reynolds OL, Padula MP, Zeng R, Gurr GM. Silicon: potential to promote direct and indirect effects on plant defense against arthropod pests in agriculture. Front Plant Sci 2016;7:744.

[105] James DG. Field evaluation of herbivore-induced plant volatiles as attractants for beneficial insects: methyl salicylate and the green lacewing, *Chrysopa nigricornis*. J Chem Ecol 2003;29:1601–9.

[106] Ma JF, Yamaji N. Silicon uptake and accumulation in higher plants. Trends Plant Sci 2006;11:392–7.

[107] Tripathi DK, Singh VP, Ahmad P, Chauhan DK, Prasad SM. Silicon in plants: advances and future prospects. London: CRC Press; 2016.

[108] Chen D, Cao B, Wang S, Liu P, Deng X, Yin L, et al. Silicon moderated the K deficiency by improving the plant-water status in sorghum. Sci Rep 2016;6:22882.

[109] Lavinsky AO, Detmann KC, Reis JV, Ávila RT, Sanglard ML, Pereira LF, et al. Silicon improves rice grain yield and photosynthesis specifically when supplied during the reproductive growth stage. J Plant Physiol 2016;206:125–32.

[110] Carneiro JMT, Chacón-Madrid K, Galazzi RM, Campos BK, Arruda SCC, Azevedo RA, et al. Evaluation of silicon influence on the mitigation of cadmium stress in the development of Arabidopsis thaliana through total metal content, proteomic and enzymatic approaches. J Trace Elem Med Biol 2017;44:50–8.

[111] Viciedo DO, Prado RM, Toledo RL, Nascimento dos Santos LC, Calero Hurtado AC, Nedd LLT, et al. Silicon supplementation alleviates ammonium toxicity in sugar beet (*Beta vulgaris* L.). J Soil Sci Plant Nutr 2019;19:413–19.

[112] Feng J, Shi Q, Wang X, Wei M, Yang F, Xu H. Silicon supplementation ameliorated the inhibition of photosynthesis and nitrate metabolism by cadmium (Cd) toxicity in *Cucumis sativus* L. Sci Hortic 2010;123:521–30.

[113] Farooq MA, Ali S, Hameed A, Ishaque W, Mahmood K, Iqbal Z. Alleviation of cadmium toxicity by silicon is related to elevated photosynthesis, antioxidant enzymes; suppressed cadmium uptake and oxidative stress in cotton. Ecotoxicol Environ Saf 2013;96:242–9.

[114] Hussain S, Pang T, Iqbal N, Shafiq I, Skalicky M, Brestic M. Acclimation strategy and plasticity of different soybean genotypes in intercropping. Funct Plant Biol 2020;47:592–610.

[115] Song A, Li Z, Zhang J, Xue G, Fan F, Liang Y. Silicon-enhanced resistance to cadmium toxicity in *Brassica chinensis* L. is attributed to Si-suppressed cadmium uptake and transport and Si-enhanced antioxidant defense capacity. J Hazard Mat 2009;172:74–83.

[116] Wang S, Liu P, Chen D, Yin L, Li H, Deng X. Silicon enhanced salt tolerance by improving the root water uptake and decreasing the ion toxicity in cucumber. Front Plant Sci 2015;6:759.

[117] Al-aghabary K, Zhu Z, Shi Q. Influence of silicon supply on chlorophyll content, chlorophyll fluorescence, and antioxidative enzyme activities in tomato plants under salt stress. J Plant Nutr 2005;27:2101–15.

[118] Mehrabanjoubani P, Abdolzadeh A, Sadeghipour HR, Aghdasi M. Impacts of silicon nutrition on growth and nutrient status of rice plants grown under varying zinc regimes. Theor Exp Plant Physiol 2015;27:19–29.

[119] Tripathi SC, Sayre KD, Kaul JN, Narang RS. Growth and morphology of spring wheat (*Triticum aestivum* L.) culms and their association with lodging: effects of genotypes, N levels and ethephon. Field Crop Res 2003;84:271–90.

[120] Xiaoqin Y, Jianzhou C, Kunzheng C, Long L, Jiandong S, Wenyue G. Silicon improves the tolerance of wheat seedlings to ultraviolet-B stress. Biol Trace Elem Res 2011;143:507–17.

[121] Mateos-Naranjo E, Andrades-Moreno L, Davy AJ. Silicon alleviates deleterious effects of high salinity on the halophytic grass *Spartina densiflora*. Plant Physiol Biochem 2013;63:115–21.

[122] Shen X, Li Z, Duan L, Eneji AE, Li J. Silicon mitigates ultraviolet-b radiation stress on soybean by enhancing chlorophyll and photosynthesis and reducing transpiration. J Plant Nutr 2014;37:837–49.

[123] Ma JF, Tamai K, Yamaji N, Mitani N, Konishi S, Katsuhara M, et al. A silicon transporter in rice. Nature 2006;440:688–91.

[124] Abdalla MM. Beneficial effects of diatomite on growth, the biochemical contents and polymorphic DNA in *Lupinus albus* plants grown under water stress. Int J Agric Biol 2011;2:207–20.

[125] Flores RA, Arruda EM, Damin V, Junior JPS, Maranhao DDC, Correia MAR, et al. Physiological quality and dry mass production of sorghum bicolor following silicon (Si) foliar application. Aust J Crop Sci 2018;12:631–8.

[126] Camargo MS, Bezerra BKL, Holanda LA, Oliveira AL, Vitti AC, Silva MA. Silicon fertilization improves physiological responses in sugarcane cultivars grown under water deficit. J Soil Sci Plant Nutr 2019;17:99–111.

25

Directions for future research to use silicon and silicon nanoparticles to increase crops tolerance to stresses and improve their quality

Hassan Etesami[1], Fatemeh Noori[2] and Byoung Ryong Jeong[3]

[1]Soil Science Department, College of Agriculture and Natural Resources, University of Tehran, Karaj, Iran
[2]Department of Biotechnology and Plant Breeding, Sari Agricultural Sciences and Natural Resources University, Sari, Iran [3]Department of Horticulture, Division of Applied Life Science (BK21 Four), Graduate School, Gyeongsang National University, Jinju, Republic of Korea

25.1 Introduction

Environmental stresses, including extreme temperatures, salinity, excessive light, UV-B, insufficient mineral nutrition, drought, metal toxicity, and pathogens, threaten global crop production and food security [1–10]. Both biotic and abiotic stresses significantly decrease the yield and productivity of crop production. Abiotic stresses account for more than 50% of the global crop losses [11,12]. Abiotic stresses negatively impact numerous physiological processes of plants, including ion uptake, water uptake, mineral nutrition, photosynthesis, respiration, translocation, transpiration, seed germination, and stomatal behavior [7,13–16]. Plants generate various reactive oxygen species (ROS) in cells in response to these stresses, including singlet oxygen (O_2), superoxide (O_2^-), hydroxyl radicals (OH), and hydrogen peroxide (H_2O_2) [17,18]. This in turn inflicts serious oxidative damages to the lipids, proteins, and DNA of plant cell components [6,8,19].

A constant, great challenge for researchers in plant and agricultural sciences is maintaining a stable and healthy growth of crops under different stresses. Plants have also developed different mechanisms to survive various stresses; healthy plants are able to survive better or sustain themselves in stressed conditions [20]. A plant's nutrition is crucial in maintaining healthy growth and enhancing stress resistance. Several studies in the literature report how nutrients help plants build tolerance against different stresses [21,22].

Silicon is a metalloid, which has intermediate chemical and physical properties of nonmetals and metals and is the second most abundant element in the earth's crust [23]. Silicon application is regarded as an ecofriendly approach for improving crop stress tolerance and crop quality [10,24,25]. The essentiality of silicon for plants is dubious, as most plants do not require the element for survival, but silicon provides benefits to plants and helps them adapt to different environmental stresses [10,24,26,27]. Silicon is not essential for the normal vegetative growth or yield of plants, but as is the case for nitrogen, phosphorous, and potassium, soluble silicon is a major component of the plant defense system, as the systemic acquired resistance of plants treated with silicon stems from the site of deposition in varying shapes and sizes [28]. It is noteworthy that plants of the *Equisetaceae* family cannot survive without silicon, and therefore silicon is an essential element for them [23]. Si was previously thought to be not essential for plant growth [29]; however, numerous studies conducted in recent decades observed that silicon provides benefits to several crops [30]. The benefits of silicon application to plants are more obvious under stress, and silicon is widely regarded as a quasiessential element [30]. The International Plant Nutrition Institute has recently declared silicon as a nutritive element for plants (http://www.ipni.net/). The

349

Association of American Plant Food Control Officials has also officially announced that silicon is a "beneficial substance" for plants (http://www.aapfco.org/). Contrary to previous understandings, silicon plays a substantial role in helping plants combat biotic and abiotic stresses. The latest available literature provides evidence that silicon could be regarded as a regular fertilizer for high-accumulating species, such as most monocots and cereals. Recent studies have also highlighted how silicon provides benefits to some dicot species [20,31]. Silicon helps plants grow, produce better yields, increase crop quality, improve photosynthesis and nitrogen fixation, and resist different aforementioned biotic and abiotic stresses [20,24,32–41]. Silicon helps enhance plant growth, promote biomass accumulation, maintain structural rigidity, manage the essential nutrients, increase the photosynthetic efficiency, balance the ion homeostasis, resist lodging, activate the antioxidant system, elicit stress-resistance-related secondary metabolites, and regulate genes to resultantly provide a resistance mechanism to different stresses [6,8,26,27,42–44].

Most silicon in the soil is present as silicon dioxide, (SiO_2), and hence plants cannot uptake silicon despite its abundance in the earth's crust. Furthermore, the H_4SiO_4 concentration and the soil Si quantity are not correlated [31,45]. Silicon is absorbed from the soil in amounts that are even higher than those of the essential macronutrients [24]. The accumulation and transport of silicon occur in the upward direction, as identified in several crops. Different silicon uptake mechanisms in plants account for the observed variations in plants' silicon levels, which range from 0.1% to 10% [37]. Plants take up dissolved silicon in the form of monosilicic acid at pH levels lower than pH 9 [26]. Different silicon transporter genes (LSi1, LSi2, LSi6, etc.), which have been identified in rice, barley, and maize roots, were observed to aid silicon transportation in plants with a high capacity for metalloid accumulation [46–49]. Plants are classified into accumulators (e.g., Equisetales, Cyperales, and Poales), excluders (e.g., tomato), and intermediate (e.g., Urtica dioica) types [26,50] with respect to silicon, depending on the amount of biogenic silica found in their tissues.

Nanotechnology is an emerging field of technology that provides extraordinary applications. In agriculture, nanotechnology may help improve global food production, food quality, as well as monitor plant growth, protect plants, reduce waste, and detect diseases [51,52]. Nanoparticles (NPs) display markedly different properties in comparison to their bulk counterparts because of their small size, high surface area-to-weight ratio, and different shapes [53]. The high surface-to-volume ratio of NPs increases their reactivity, as well as their biochemical activities [54]. Because of their small size, NPs may facilitate effective absorption by crossing cell walls and plasma membranes [55]. Therefore NPs may be used to help supply elements to plants. Silicon nanoparticles (SiNPs) also exhibit different chemical and physical properties than bulk silicon [56]. SiNPs have a great potential in agriculture and may work better than bulk silicon in alleviating different biotic and abiotic stresses because of their unique properties [6,57–59].

Numerous studies have demonstrated that biotic and abiotic stresses and their detrimental effects may potentially be controlled by nanosilica [60]. Recent research has displayed that SiNPs may directly interact with plants in various ways to impact their morphology and physiology [61–64]. SiNPs may also deliver nucleotides, proteins, and other various chemicals in plants. SiNPs application to agriculture may potentially aid in developing improved crops with relatively higher productivity and lead to increased worldwide food security [65]. SiNPs have easy access to plant cells and impact plant metabolism through diverse interactions to help improve their development and growth, and resulting in trigger plant potential to better combat different stresses [66]. Plants may cope better with damages stemming from abiotic stresses or climate change with the help of SiNPs and their unique properties [67]. SiNPs were observed to be able to combat damages from UV-B exposure [6], heavy metal toxicity [58], dehydration [68], salinity [59], etc. SiNPs may also be used as herbicides (nanoherbicides), pesticides (nanopesticides), and fertilizers (nanofertilizers) [66,69]. The application of organic fertilizers with nanosilicon dioxide was observed to enhance plant productivity [70]. Mesoporous silica NPs with pore sizes between 2 and 10 nm effectively delivered fertilizers based on boron, nitrogen, and urea [69,71]. Silicon is commonly used in contemporary farming systems, especially in organic farming systems, to inhibit the residual pesticide problem in food products [20]. As discussed before, SiNPs may help sustainable agriculture and improve crop production, and potentially be used as the sole fertilizer for certain crops, and as a facilitator for herbicides and other fertilizers for other plants [66].

Among the various NPs that are employed to keep different plant diseases under control [72], nanosilica has received a lot of attention because how silicon nanopesticides help plants maintain their health is already known [66,73]. Because it is known how silicon and its NPs help protect plants against biotic stresses [26,35,61–63,66,74–91], a broad spectrum of research is currently in progress and is promoted to establish a prophylactic, environmentally friendly agriculture with the application of Si/SiNPs. There is a potential for Si/SiNPs application to provide a thorough resistance of a wide range of plants to various diseases and pasts and be integrated to be used in diseases and pest management programs.

The current knowledge on silicon is that it is nonpolluting, noncorrosive, and even in excess does not pose any damage to plants. Silicon is a major component of the soil and provides great benefits to plants, especially in mitigating abiotic stresses. Silicon is used as a sustainable fertilizer to help crops deal with various environmental stresses. Agricultural silicon application increases the resistance of crops to environmental adversities and improves their adaptability to different stresses and, in turn, may help enhance food security. Silicon fertilizers carry with them economic and ecological benefits, and can be effectively used in silicon-deficient areas as silicon fertilizer can provide economic as well as ecological benefits [92].

So far, many research aspects for silicon have been investigated. For example, silicon application is well-documented in the literature to boost the growth and yield of a wide spectrum of crops under different biotic and abiotic stresses (Table 25.1). Furthermore, omic technologies (the basis for transcriptomic, proteomic, and

TABLE 25.1 A list of some studies done on improved the growth of plants by silicon application under biotic and abiotic stresses.

Crop	Type of environmental stress	Action mechanism(s)	References
Maize	Drought	Decreased transpiration rate and conductance from leaf surfaces	Gao et al. [93,94]
Wheat	Drought	Increased transpiration and stomatal conductance	Gong et al. [95]
Sorghum	Drought	Increased transpiration and stomatal conductance	Hattori et al. [96] and Liu et al. [97]
Maize	Drought	Decreased transpiration rate and membrane permeability	Amin et al. [98]
Wheat	Drought	Decreased transpiration rate and membrane permeability	Maghsoudi et al. [99]
Rice	Drought	Decreased transpiration rate and membrane permeability	Agarie et al. [100]
Tomato	Drought	Decreases transpiration rate and membrane permeability	Silva et al. [101]
Melon	Drought	Decreased transpiration rate and membrane permeability	Neocleous [102]
Oil palm	Drought	Decreased transpiration rate and membrane permeability	Putra and Purwanto [103]
Wheat	Drought	Enhanced the antioxidant enzyme	Ahmad and Haddad [104]
Sunflower	Drought	Enhanced the antioxidant enzyme	Gunes et al. [105]
Tomato	Drought	Enhanced the antioxidant enzyme	Shi et al. [106]
Chickpea	Drought	Enhanced the antioxidant enzymes	Gunes et al. [107]
Wheat	Drought	Increased leaf P contents	Gong and Chen [108]
Maize	Drought	Increased leaf K and Ca content and increased proline content	Kaya et al. [109]
Sunflower	Drought	Increased macro- (P, K, Ca, and Mg) and micronutrients (Fe, Cu, and Mn) uptake	Gunes et al. [105]
Rice	Drought	Increased K, Ca, Mg, and Fe contents	Chen et al. [110]
Rice	Drought	Increased osmotic potential	Ming et al. [111]
Wheat	Drought	Increased leaf water content and water potential	Gong et al. [112]
Pepper	Drought	Increased proline content	Pereira et al. [113]
Rice	Salinity	Increased concentrations of tissue ABA and enhanced the activities of antioxidant enzymes	Kim et al. [114]
Borage	Salinity	Increased activity of SOD	Torabi et al. [44]
Faba bean	Salinity	Increased K^+ concentrations and decreased Na^+ uptake	Shahzad et al. [115]
Rice	Salinity	Increased K^+ concentrations and decreased Na^+ uptake	Shi et al. [116]
Wheat	Salinity	Accumulated Na^+ in roots	Tuna et al. [117]
Egyptian clover	Salinity	Increased the uptake and translocation of mineral elements	Abdalla [118]

(Continued)

TABLE 25.1   (*Continued*)

Crop	Type of environmental stress	Action mechanism(s)	References
Halophytic	Salinity	Increased the uptake and translocation of mineral elements	Mateos-Naranjo et al. [119]
Tobacco	Salinity	Adjusted the levels of solutes and phytohormones	Hajiboland and Cheraghvareh [120]
Soybean	Salinity	Adjusted the levels of solutes and phytohormones	Lee et al. [121]
Maize	Salinity	Adjusted the levels of solutes and phytohormones	Moussa [122]
Okra	Salinity	Increased photosynthetic rate	Abbas et al. [123]
Maize	Salinity	Increased photosynthetic rate	Xie et al. [124]
Alfalfa	Salinity	Enhanced the activities of antioxidant enzymes	Wang et al. [125]
Wheat	Salinity	Enhanced the activities of antioxidant enzymes	Ali et al. [126]
Maize	Al	Co-deposition of Si and metals	Wang et al. [127]
Rice	Zn	Co-deposition of Si and metals	Gu et al. [128]
Barley	Cr	Enhanced gas-exchange characteristics	Ali et al. [129]
Wheat	Cr, Cu, and Cd	Increased uptake and translocation of micronutrients and macronutrients	Rizwan et al. [130] Tripathi et al. [131] and Keller et al. [132]
Rice	Mn	Promoted Mn oxidizing power of roots	Okuda and Takahashi [133]
Salvia splendens	High temperature	Increased activities of SOD, APX, and GPX	Soundararajan et al. [134]
Sugarcane	Chilling	Regulated the activity of antioxidant defense system	Liang et al. [135]
Maize	P deficiency	Increased the amount of bioavailable P and soil pH	Owino-Gerroh and Gascho [136]
Perennial ryegrass	Pathogen-infected	Increased the activities of enzymes	Cai et al. [137]
Miniature roses	Powdery mildew	Accumulation of several phenolics and flavonoids	Shetty et al. [138]
Rice	Blast fungus	Promoted photosynthesis	Liu et al. [139]
Arabidopsis	Powdery mildew	Stimulated biosynthesis of SA, JA, and ET	Fauteux et al. [140]
Rice	Insect pests	Increased silica deposition in the plant	Reynolds et al. [80]
Rice	Rice leaf folder	Reduced food quality of the insects	Han et al. [141]
Wheat	Greenbug aphid	Increased the activity of peroxidase and polyphenol oxidase	Gomes et al. [142]
Tomato	*Ralstonia solanacearum*	Upregulated expression of the jasmonic acid/ethylene marker genes (*JERF3*, *TSRF1*, and *ACCO*)	Wydra et al. [143]

APX, Ascorbate peroxidase; ET, ethylene; GPX, glutathione peroxidase; JA, jasmonate acid; SA, salicylic acid; SOD, superoxide dismutase.

metabolomic studies) also help the extraordinary role that silicon plays in plants. Also, various plant hormones and their signaling in response to silicon establish the phytohormone crosstalk [144]. Hence the focus of this chapter is to provide systematic future research directions regarding silicon and nanosilicon in crop stress tolerance and crop quality improvements.

25.2 Future directions of silicon/nanosilicon application in agriculture

Agricultural use of silicon seems to be a viable, sustainable future strategy for alleviating stresses for plants. Furthermore, silicon application may improve the health of plants and be a major component of sustainable low-input agriculture of various crops [10,24]. Several avenues of future research approaches are suggested here, based on the relevant understandings generated up to now (the currently published results).

25.2.1 Silicon and biotic stress

Silicon's overlooked role in boosting plant protection against pests, pathogens, and other biotic stresses has been more understood recently, but the exact genomic, metabolomic, and proteomic mechanism(s) with which silicon helps the host defense mechanisms of plants and how it modulates plant physiology still warrants further research [145].

To help enhance the tolerance of important crop plants to biotic stresses, the determinants and regulatory mechanisms with which silicon improves the stress tolerance in plants need to be identified and characterized. Many such determinants have yet to be researched in detail.

A considerable number of studies now exist in the literature that supports how silicon acts as a physical defense mechanism in plants. Furthermore, a growing number of studies document how silicon enhances the biochemical defense in plants. However, there is little in the literature that notes the role silicon plays in tritrophic interactions. The relative importance of the biochemical and physical defense of plants, and how (if) this differs with respect to different herbivores should be the subject of future research. An insight into how silicon's effectiveness varies for different feeding guilds or taxons may be obtained through a *meta*-analysis of the available literature on silicon and would be valuable. Future research should also examine the different ways in which silicon interacts with plant defense pathways. There is a wealth of literature that examines how silicon and pathogens interact; future studies should look into whether silicon helps plants defend themselves against insects in a similar manner [146].

Little research exists on how silicon application to plants affects the below-ground defense of plants against predators. There is a lack of understanding of how silicon application affects root toughness and chemical defenses through the root. Studies should also test how foliar silicon deposits from foliar silicon applications affect natural foraging while keeping in mind that changing the surface with denser or more robust trichomes may negatively impact natural enemy foraging. In general terms, researchers should investigate how silicon application affects plant defenses in field conditions and move the research focus from greenhouse and laboratory settings, as is the case for research on mammals in natural ecological systems. This will help take into account the natural predators of herbivores into silicon application and how it affects plant defenses. Currently, there are no published studies that investigate how silicon application affects the production of herbivore-induced plant volatiles (HIPVs), but such research is known to be in progress. If research indicates that there is strong evidence of silicon promoting HIPV production, greater attention should be given to future research on silicon's impact to plant defenses to the third trophic level. The role silicon plays in the production of defense-related chemicals and HIPVs, as well as the associated energy costs in plants, may be determined from a system-wide analysis or an omic technological viewpoint. This may enable manipulating plants to maximize the impact of natural enemies and minimize herbivory. Modern metabolomic, transcriptomic, proteomic, and transgenic mutant approaches will be powerful tools with which to investigate the underlying mechanisms of silicon's involvement in plant defense systems. Large restrictions have recently been put on the use of toxic pesticides due to the damages it causes to human health and the environment, and sustainable pest management is of growing interest; extensive research should be conducted to investigate how silicon application may be used for pest management [146].

25.2.2 Silicon and salinity and drought stress

A comprehensive literature survey shows that silicon application could benefit the development of different plant species under drought and salinity stress. A lot of work has been done with respect to the application of silicon through hydroponics and through soil amendment mostly in pot culture to study its effect on salinity and water deficit alleviation. Different mechanisms may be attributed to the increased growth and biomass accumulation of silicon-treated plants under drought and salinity stress. Maintaining nutrient homeostasis, enhancing antioxidant enzyme activities to reduce oxidative stresses, adjusting the osmotic potential for a higher leaf water content, modifying gas-exchange attributes and phytohormones, and upregulating gene expressions are commonly known mechanisms with which silicon helps alleviate stresses of drought and high salinity. In addition to the aforementioned mechanisms, silicon may reduce Na^+ uptake, translocation, and free Na^+ in plants, and increase the K^+ uptake and translocation to increase the growth and biomass accumulation of plants under salt stress. This demonstrates the various common mechanisms with which drought and salinity stresses are dealt with by plants with the help of silicon. Many aspects of physiological changes caused by drought and salt stress in plants are shared, including reduced photosynthesis, nutrient uptake, water absorption, etc. [9,10,24]. This overlap explains the similar mechanisms that silicon works with to alleviate stresses from drought and salinity in plants. A plant's stress tolerance levels vary among different species, and the positive effects of exogenous silicon depend on these stress tolerance levels. Differing silicon uptake capabilities for different species may be the reason for this. The silicon concentration and application method, cultivation method, plant species, genotype, stress intensity, and duration are different factors that affect the effects of silicon on a plant's stress resistance [147,148]. Silicon application may be a pathway to improve the growth, development, and yield of crops undergoing drought and salt stresses. Focused research is necessary to settle the debate on the main mechanisms that form the basis of silicon's role in mitigating drought and salinity stresses in plants. Lately, a new wave of foliar silicon application has emerged, and the mechanisms underlying it should be critically investigated. Also, the effects should be confirmed in the field conditions. Current approaches such as transcriptomics may be applied to study the complete plant responses in terms of silicon applications and to validate its use in drought and salinity-stressed agriculture [149].

A need exists to further investigate and validate how silicon helps alleviate drought and salt stresses in plants on a large scale, for a wider spectrum of plant species, and under different environmental conditions. Existing studies studied short-term silicon applications for isolated drought or salinity stress, but future studies should keep in mind that plants are simultaneously exposed to multiple stresses in field conditions. Thus the long-term role of silicon on plant responses to both drought and salt stress needs to be researched. Studies should be conducted in detail to determine the molecular mechanisms with which drought and salt tolerance are mitigated in plants. Future research should also be able to provide recommendations for silicon application based on the soil type, plant species, and environmental conditions. The expressions of plant genes following silicon application under drought and salinity stress, and how plants uptake the silicon, should be investigated as they may provide important information. This may enable scientists to genetically modify economically important crops like wheat to lend highly profitable drought- and salinity-tolerant cultivars in the near future [150].

A more detailed, focused research should be directed to investigate how silicon helps mitigate drought and salt stresses in plants at the metabolomic, proteomic, and transcriptomic levels with the development of advanced omics technologies. The salt overly sensitive (SOS) signaling pathway is known to be vital to a plant's tolerance of salt stress. But little is known of how exogenous silicon interacts with the SOS pathway and other salinity stress sensors in plants. Most of the studies to date have focused on the short-term roles that silicon plays on its own in alleviating salinity stress. Because plants are exposed to multiple stresses at any given time in nature, it would be ideal to establish long-term stress resistance for plants, so that plants can predict and react to the changing global climate, where one stress can lead to other forms of stress. This necessitates in-depth research of how silicon interacts with plants in the long term in response to multiple stresses. Future research should analyze how silicon affects osmotic stresses induced by salinity, and demonstrate how it adjusts a plant's osmotic potential. Genetically engineering conduits for different compatible solutes like glycine betaine, proline, and sorbitol may be employed to produce salinity-resistant plants. Molecular and transcriptional changes at the plant level following silicon application, including the metabolomic and proteomic changes in plant organs have also yet to be investigated. The genetic and molecular mechanisms that form the basis of silicon-induced salt stress alleviation have yet to be well-understood. For real-world effectiveness, future research should focus on silicon's effects when applied in the field conditions rather than in isolated greenhouse or laboratory conditions [147].

Study on how silicon alleviates salinity stresses in plants has extended to the metabolomic, proteomic, and transcriptomic levels with the advances in omics technology. However, academia has mainly focused on how silicon mitigates biotic stresses in plants, and only a few preliminary reports are available on how silicon helps plants tolerate abiotic stresses like salinity. How silicon directly affects the expression of hepatoma transmembrane kinase and SOS genes was first experimentally demonstrated by a recent study. The detailed mechanisms with which silicon regulates the SOS signaling pathways and interacts with other salinity stress sensors still remain unknown. Silicon's regulatory mechanisms for salinity-induced osmotic stresses need to be analyzed by future research. Silicon transporters are members of the aquaporin family's NOD26-like intrinsic proteins (NIPs) subfamily. Studies exploring cloning and functions of aquaporins and silicon transporters in different plant species may build the foundation that aids the understanding of how water metabolism is regulated. Furthermore, the silicon permeability within the NIPs in higher plants needs to be conveyed by classifying plants as silicon accumulators or silicon excluders according to their respective structural features.

Studies have shown that silicon can promote the development of suberized structures in the root exodermis and endodermis, root growth, and regulate the ion distribution to different root parts under stressed conditions. Further research is still necessary to explore the molecular mechanisms involved in silicon's regulation of ion uptake/distribution in the root, and the root structure, including cell wall components, carbohydrate (starch, soluble sugars, etc.) storage may be important for osmoregulation and energy storage, as well as for signaling for plants under salinity stress. This highlights the need to research how silicon application to stressed plants affects carbohydrate mechanisms. Future research should study foliar-stabilized silicic acid, a biostimulant, as studies have demonstrated its effectiveness in protecting plants from biotic and abiotic stresses. In conclusion, the basic knowledge on how silicon mitigates plants' salinity stresses at the molecular level should be expanded, and a theoretical foundation needs to be established to determine how to practically apply silicon in agriculture [151].

25.2.3 Silicon and UV-B irradiation stress

UV-B irradiation is serious abiotic stress that has a far-reaching impact on plant biochemistry and physiology. Research shows how silicon helps plants respond to UV-B since increased UV-B levels are known to severely decrease plant growth and yield [152]. However, there is little in the literature regarding the physiological role silicon application plays for plants irradiated by UV-B, and if/how silicon can alleviate UV-B damages to plants, which needs to be studied in the future.

25.2.4 Silicon and its biochemical, physiological, and molecular aspects

Studies regarding the biochemical, molecular, and physiological bases on which silicon helps plants mitigate various biotic and abiotic stresses are still in their infancy. The interrelationship of plant responses to different individual stresses is also not very well understood. Future proteomic/transcription level studies are necessary to understand how silicon helps plants grow resistance to stresses by interacting with defense priming. Although it is established that there is a close interrelationship between the different induced defense responses to biotic and abiotic stresses, mainly via interactions during signal transduction, it is yet unclear how silicon can affect such a complex system of interactions [42].

It remains obscure whether silicon interacts with intercellular processes like gene expression, or purely at the physicochemical level, such as being deposited on the cell walls to alter ion and water fluxes. There is little evidence to date that supports silicon plays a biochemical role in plants; a survey of the available literature points to the contradictory outcomes by different studies, and generally, not many genes or proteins are affected by silicon application. Changes in protein and transcript levels are measured after long treatment periods (days and weeks), which presents a problem in the methodology because this increases the likelihood that secondary effects are observed, from factors like modified ion and water fluxes. Ca, K, or Na silicates are commonly used as silicon sources, which may cause increased cation concentrations and severely modified plant nutrition, and proper control experiments should be performed to negate these effects. For credibility, studies conducted over a short time (minutes to hours) should present reproducible, hard evidence that silicon directly affects gene transcription. The problem could also be addressed by using more controllable systems like cell cultures. Putative biochemical roles that silicon plays in plants could be obtained with the extension of gathering convincing evidence of the specific genes and proteins involved in mutational studies [148]. Results of a study done by Thorne et al. [148] demonstrate that yield gains exceeding 10% can be expected to incur a positive cost-benefit. Strict parallel to agronomy may not be easily drawn, biomass gains of 10% or more have commonly been reported in laboratory settings. Because the species and the cultivar impact effectiveness of silicon, the critical soil silicon levels for maximal yield gains need to be studied for different species and cultivars. Further research, with a focus on the long-term benefits of silicon application in field conditions, is needed to identify geographical and climatological conditions where silicon fertilization can benefit agriculture.

Because silicon has several benefits, it has been added to supplement nutrition in many contemporary agricultural applications, such as soilless cultivation systems. In-depth insight into how silicon helps mitigate stresses at the molecular level needs to be gained. The recent developments made in the field of omics technology could help us better understand how silicon affects the biology of various plants under stress. For the benefit of agricultural and horticultural crop production, research may be conducted on innovative mechanisms to incorporate silicon into the culture or growth media to enhance yield and stress resistance [153]. Furthermore, because most studies on silicon application to date were carried out in hydroponic systems or pots, studies in field and ecosystem scales are now necessary.

The efficiency of carbon restoration by increased plant biomass from silicon application for different plant species and at different stress intensities, the relationship between the terrestrial ecosystem resilience and the biogeochemical silicon cycle, and the coupling relations between silicon and essential elements for plants, are several challenging areas that require further study, especially in fragmented landscapes. An evaluation model that predicts how recovery following silicon application contributes to carbon accumulation in the form of plant biomass in different ecosystems should be developed [154].

Studies of how biogenic silica and its NPs affect proteomics would be of interest and would add to the knowledge of how silicon affects stressed and unstressed plants. Another area of research interest that has yet to be explored is silicon recycling and how it affects genetic engineering and proteomics [152].

A more comprehensive understanding of the underlying molecular mechanisms responsible for benefits to plants following silicon application is critical to increasing the applicability for a wider range of plant species. Different mechanisms and models are currently proposed to attempt to explain how silicon helps plants better tolerate stresses, but the exact biochemical reactions or pathways where silicon plays a key role are not yet identified by any of them. Advanced technologies utilized for research may help discover these pathways and reactions [20].

Critical physiological and molecular studies are needed to elucidate the unclear mechanisms behind the tolerances and responses of plants to different abiotic stresses. Exploring the most effective and easiest ways to overcome stressors is one of the emerging tasks for plant scientists. The published literature indicates that silicon plays a vital role in numerous plant physiological processes. Silicon has a prominent role in protecting plants from damages induced by abiotic stresses. How silicon affects plants under abiotic stresses is well-documented, but the underlying physiological mechanisms by which silicon could protect plants from stressful conditions are elusive. Exogenous silicon application has recently received great attention for its potential in plant protection, and many studies have discovered how it helps tolerate oxidative stresses. However, the exact dose and methods of application are still under study [155].

Silicon application to monocots and dicots under adverse soil and climate conditions can improve their growth and production. Signal transduction of phytohormones has been suggested as the reason for increased plant tolerance to stresses following silicon application. Studies with different crops and vegetables reveal that such responses vary. Crosstalk of abscisic acid (ABA), jasmonate (JA), and salicylic acid (SA) are important for regulating stress defenses. Further research at the molecular level is necessary to fully uncover the mechanisms behind stress tolerances, especially after silicon application. Higher silicon concentrations led to greater tolerance to salinity stresses by reducing JA synthesis and regulating genes responsible for ABA biosynthesis. However, SA displayed irregular responses to silicon applications, warranting further biochemical and molecular level research. Different external stimuli lead to the ever-present hormonal crosstalk variations. These dynamic interactions hamper studies on how silicon affects stressed plants. However, with the help of advanced proteomic and transcriptomic techniques, a better understanding of silicon-plant-stress interactions and the answers to the aforementioned questions can be obtained [156].

Silicon application to normal fertigation regimes consistently resulted in the yield [157–159]. It is thought that silicon sequestration modes differ between monocots and dicots. Monocots store increased silicon concentrations at various parts of the plant without transporters and still display stronger direct and indirect tolerance to biotic and abiotic stresses [145,158]. There are numerous studies in the literature that report how silicon application acts as a biostimulant, but much is left yet to explore in terms of silicon's various action modes in plants [145,158–162]. The potential complexity of silicon's action modes in plants is due to the presence of an intricately intertwined set of various interactions in biological systems. One example is the triggering of the phytochemical pathways that produce HIPVs to prevent herbivory; another is prophylactic actions that interact act with the JA defense pathways to increase plant protection against stresses [146,163,164]. Production of HIPVs is a well-documented response, and the extremely low concentrations of silicon may still induce different responses of the host plant and play a vital role [158,161]. Coskun et al. [159] argued that the roles silicon plays in plants reported in the literature thus far have led to confusion in the scientific community regarding a thorough understanding of the biological roles that silicon plays for plants. The authors proposed a so-called "apoplastic obstruction hypothesis," which is a working model made in an attempt to unify the different observations made in the literature regarding how silicon benefits the growth, development, and yield of plants. The model claims that silicon plays a fundamental role for plants in dealing with biotic and abiotic stresses as an extracellular prophylactic agent, whose cascading effects are significant factors affecting plant form and function [28,160].

Silicon plays multiple roles regulating plant genes involved in different physiological functions, such as transcription, photosynthesis, polyamine biosynthesis, and secondary metabolism. To improve plant growth under various stresses, the silicon-triggered modulations at the molecular level need to be investigated in depth. A *meta*-analysis of how silicon mitigates abiotic stresses in plants revealed that most of the research was for a 1-species, 1-stress models; further research should be conducted to compare how silicon affects stress mitigation for different species and different stresses. Researchers and farmers alike should take caution in potentially applying silicon in agriculture. Farmers should take care to apply an appropriate concentration of silicon for a particular crop, as well as to choose an appropriate mode of application and apt time, to increase the productivity of different crops [92].

The current understanding of the role silicon plays in plant biology is dominated by the different silicon-induced effects in plants to alleviate stresses. Studies have demonstrated how silicon affects gene expression and plant metabolism to a degree, but the mechanisms with which silicon affects the development and growth of plants remain obscure. Therefore it is suggested that approaches to research the role silicon plays in plants take a paradigm shift. Regarding Si to be "nonessential" to the nutrition of plants has been challenged [165–167]. Even for nonsilicon-accumulators, it has been extensively observed that silicon notably affects numerous plant species under various environmental conditions [168], which indicates that silicon concentration may not necessarily decide the magnitude and effectiveness of silicon for plants [169]. Silicon acts in a multitude of ways to alleviate stresses in plants: increasing phytolith deposition, altering the expression of defense-related enzymes and metabolites, assimilating CO_2, increasing antioxidant enzyme activities, and changing the transpiration rate. There is growing evidence that suggests silicon plays a critical role in the growth, development, and primary metabolism of plants [35,170,171], which necessitates further investigation for a stronger basis. There are emerging avenues for future research regarding silicon and its diversity of benefits on plants. How silicon nutrition influences gene regulation in plants, and how this differs according to the species, is only partially understood currently. Silicon could be involved in important mechanisms of a plant's primary metabolism for some species that accumulate high levels of silicon [172]. Currently, studies involving rice and other nonaccumulators of silicon that respond to silicon treatments, such as *Arabidopsis thaliana* [140,171] and tomato (*Solanum lycopersicum*) [168,173] provide a basis on which future studies of such mechanisms involving silicon application could be built. The nonaccumulators of silicon that respond to silicon treatments should be incrementally studied across multiple experiments, in conjunction with high accumulators of silicon barley, rice, and wheat, to evaluate how silicon impacts plants in response to no stress, single stress, or multiple stresses, both biotic and abiotic. These studies should particularly investigate the primary plant metabolic processes that may potentially interact with silicon, such as oxidation metabolism. As previously discussed, silicon may interact with primary photosynthetic processes like photorespiration; but this needs to be further clarified as C3, C4, and CAM (crassulacean acid metabolism) type plants have different enzymatic and physiological carbon fixation processes and therefore are likely to react differently in response to silicon application. The different formulations via which silicon is delivered to the plant may also significantly influence how it affects plant metabolism. How silicon is delivered to plants (as calcium silicates, as sodium silicates, etc.) and how the different formulations affect plants should be investigated in-depth, with a focus on how to optimally translocate and accumulate silicon in plants, as the defense and growth of plants may critically be affected by where and in what concentration silicon is delivered. It is also of similar importance to explore how silicon affects the plant community structure, and how it ecologically impacts mutualists and inter-plant interactions, including the allelopathic effects of invasive species.

Silicon's far-reaching effects on multiple fields that span agronomy, ecology, entomology, evolutionary biology, plant biochemistry, and plant pathology, may be uncovered through incremental studies that are cross-disciplinary and appropriately designed. How a plant's silicon uptake capacity impacts its breeding, and agriculture and agronomy is an area of research interest [174], as is producing transgenic species with an enhanced silicon uptake capacity, by perhaps introducing silicon-transporter genes *Lsi1* and *Lsi2*. Targeted profiling of significant markers that are vital to primary metabolic processes like oxidation and photosynthesis should be used purposefully in conjunction with untargeted omics that utilize multivariate analyses, and initial comparisons should be made with specific stresses (e.g., herbivores, biotrophic pathogens, etc.) or with specific trait groups (e.g., C3 and C4 plants, silicon accumulators and silicon nonaccumulators, etc.). Such a comparative, thorough set of studies will not only facilitate the understanding of how silicon affects plant growth and development, and how the different effects interact but also shed light on the future research directions to fully uncover the fundamental roles silicon plays in helping plants grow and develop. Accumulated silicon in plants may interfere with effector proteins of (hemi)biotrophic pathogens or herbivores which suppress plant effector-triggered immunity (ETI); this may be another way in which silicon enhances the resistance of plants to certain biotic stresses [175]. Considering how this is closely associated with defense phytohormone signaling and ROS production, this may lead to the understanding of silicon to help plants alleviate certain abiotic stresses. Targeted transcriptomics may be used as the first step, investigating how genes associated with ETI are regulated in plants under stresses of (hemi)biotrophic pathogens or herbivores are in environments with high and low silicon levels. Studies from the last decade indicate that silicon may have an unknown, but important, role in the primary metabolism of plants, specifically related to ROS generation and oxidation metabolism that reaches farther than biotic stress alleviation. That silicon does indeed impact the primary metabolism of plants may be confirmed and new insights may be obtained through untargeted omics that selectively compare stress types. The unifying mechanisms and pathways involved with which silicon affects plant metabolism will be identified with the help of subsequent *meta*-analyses. Many researchers in the field of plant sciences still remain unaware or indifferent of the numerous ways that silicon may help plant growth and development, despite past efforts to raise awareness on silicon's importance [165,167], as well as certain studies that identified the element's role in regulating genes for enhanced plant development and growth. To further raise awareness and research interest in the importance of silicon for improved agriculture, Frew et al. [160] suggest that collaborating researchers should put together thoughtful, focused multidisciplinary experiments that lead to a better understanding of how silicon benefits the growth and development of higher plants, as well as their resistance to different stresses.

25.2.5 Silicon and its foliar application

It has generally been recognized that silicon fertilization improves plant health. In this respect, the number of studies that demonstrate the importance of foliar silicon application is increasing. Little is currently known on how foliar silicon application interchanges with bioavailable soil silicon content and mineral nutrition. Because harvesting removes plant materials, the soil silicon content decreases over time. Further research into a purposeful foliar silicon application and providing concrete guidelines for the bioavailable soil silicon content depending on the type of silicon fertilizers can potentially stabilize crop yield. To this extent, appropriate concentrations, volumes, and timing for foliar application of silicon fertilizers are of great interest and should be further investigated. A clearer understanding of how silicon is taken up through the leaves, and how foliar applications of silicon are transported in the plant afterward is of great importance. It is worth noting that silicon-accumulating plants take up silicon at levels comparable to their nitrogen and potassium uptake. Because foliar fertilizers are applied in very small amounts, it is very unlikely that foliar silicon fertilizers will affect the shoot silicon concentration in shoots (i.e., the total plant silicon status). However, some authors have noted that foliar silicon fertilization led to increased leaf silicon concentrations, which suggests that selective or supplementary (in addition to soil silicon fertilizers) foliar silicon fertilization is a promising strategy in modern sustainable agriculture. This potentially combines physical barrier formation, enhanced biochemical and molecular defense mechanisms, and antifungal/osmotic effects. Unless the mechanisms underlying foliar silicon fertilization are fully understood with further research, the practice will remain arcane [176].

A greater amount of empirical data is needed to better understand foliar silicon application. Experiments in the field and in the greenhouse will help (1) compile detailed protocols for foliar silicon fertilization with the environmental (e.g., plant-available silicon, climate, etc.) and plant-specific (e.g., leaf morphology, plant silicon status, etc.) conditions in mind and (2) analyze the detailed mechanisms underlying the foliar uptake and functioning of silicon fertilizers. To better document and scientifically evaluate foliar silicon fertilization in agricultural practices, scientists and farming practitioners should cooperate more intensely. The cooperation between scientists and farming practitioners will help further understanding of different aspects of silicon application to benefit plants. These include (1) factors that potentially influence silicon application, for example, species-specific leaf morphology, the silicon source, environmental factors, etc., (2) appropriate foliar application methods, e.g., fertilizer quantity, time of application, etc., (3) effects of silicon fertilizers, and (4) silicon uptake efficiency. Gaining knowledge of the mechanisms involved in foliar silicon fertilization is promising with the application of interdisciplinary research. A scientific review [176] demonstrates that the pathways of foliar silicon absorption and the functioning of silicon fertilizers are complex processes that require detailed interdisciplinary research. These disciplines may include, but are not limited to, genetics, biochemistry, agricultural economics, and ecology. Modern tools show promise in gathering further information; isotope analysis may shed light on the pathways of foliar silicon uptake, and real-time polymerase chain reaction may provide insight into the activation of defense-related genes. The basis for analyzing the results of multiple studies, like *meta*-analysis, big data, are provided by the data from standardized experiments, scientist-practitioner cooperations, and interdisciplinary research. For this avenue, advanced analytical tools, (scientific) networks, and noncommercial data repositories should be utilized [176].

Supplementing conventional fertilizers with silicon fertilizers has been proven effective for improving the yield and quality of crops. However, large-scale silicon fertilization for major field crops is not yet conventional. Field trials are necessary to determine the appropriate, optimal silicon fertilization rate for different crops in different soils. Research and development of more cost-effective silicon fertilizers should also be conducted.

Furthermore, how silicon fertilizers may cause ecological issues in natural and agricultural ecosystems need to be explored attention [177].

25.2.6 Silicon and vegetables

Studies have examined how silicon affects plant health in open field conditions to an extent, in greenhouse/glasshouse settings, and in hydroponic systems [26]. There is little research conducted on how silicon application can potentially benefit greenhouse crops. Silicon's role for vegetables (dicots) is much less investigated in comparison to that of model plants such as rice and *Arabidopsis*. Therefore many recent reviews on silicon's effects on plants report little on the matter regarding vegetable crops [160,178,179]. To collate relevant information on how silicon application affects vegetables, the role silicon application plays in stress management, agronomy, and quality of vegetables.

25.2.7 Silicon and its uptake, transportation, distribution, and accumulation in plant

Silicon uptake, transportation, and accumulation in plants are crucial to silicon-mediated stress mitigation in plants. Although there is extensive literature on the silicon uptake and transport in plants, the mechanisms behind these processes occur is still a matter of research because the nature of uptake and transport and the concentration of silicon in plant tissues vary greatly between species (i.e., 0.1%–10% of the shoot dry weight) [180]. These differences are mainly due to the differential characteristics and capacities of silicon uptake and transport among various species.

Understanding the mechanisms behind how silicon accumulates and deposits in plant tissues is shallow. Silicon transport in plants is an active process that is facilitated by specific silicon transporters such as Lsi1, Lsi2, and Lsi6, which are found in various plant parts [181]. Silicon transporter types responsible for silicon absorption have been differentiated in several plant species, the channel protein or transporter for silicon loading has yet to be identified. Furthermore, most of the silicon transporters identified thus far are from monocots. Logically, further work should be conducted to clarify how silicon is absorbed and transported (including xylem loading of silicon) for different species and cultivars, with the silicon accumulation capacities taken into account (e.g., monocots vs dicots) [41].

All plants rooted in the soil contain silicon in their tissues, but the molecular mechanisms underlying silicon accumulation are still poorly understood. There are indications for various transporters involved in silicon accumulation in different tissues, but only a few have been identified [180].

A transporter's function may be improved by cloning a transporter with another's species transporter; if molecular techniques of cloning transporters of two different species could be harnessed, it may be effectively used in the future for improving transporter functions. Improved transport functions may increase plants' silicon uptake, leading to enhanced stress tolerances and yield. Furthermore, incapable transporters for silicon uptake may be changed to capable transporters with the process of cloning. Silicon accumulation is cell-specific, so there exists a need for future research to further investigate how different silicon transporters may facilitate silicon transport in other plant parts [181].

A rough model has been established regarding the absorption, transport, distribution, and storage of silicon has been established. However, little is known of the molecular mechanisms responsible for regulating the transcription of silicon transporter genes, the structural basis of transmembrane silicon transport, polar localization mechanisms, and silicon transporter activity regulators. Silicon transporters for different plant species need to be further identified, as currently, only a few species have known silicon transporters. Further research regarding these factors would establish the basis for genetic engineering geared toward improving silicon accumulation in plants and consequently confer a wide range of benefits to plants [177].

The detailed relationship between how silicon application affects root anatomy (e.g., lignification, suberization Casparian band development, etc.), and the ensuing enhanced stress tolerances in plants should be determined in future studies. Most of the silicon-related research at the subcellular level has focused on the cell wall. Studies regarding how silicon is distributed in the cell nucleus and other organelles are also of interest, as they may further help in the understanding of how silicon biologically helps plants develop stronger tolerances against environmental stresses [41].

Genetic modifications that enhance the silicon uptake capacity of roots and storage capacity in different plant parts can enhance plants' tolerance to various environmental stresses. Such modifications will reduce costs for using silicon in the crop field.

The present knowledge enables research using supplementary silicon nutrition or new cultivars with improved silicon uptake capacities to explore how silicon benefits plants. Genetically modifying the roots of dicots would help develop dicots with improved silicon uptake and accumulation capacities. A sustainable crop agriculture system may be developed with improvements in silicon-derived benefits for plants [20]. There is a need for further study on mechanisms with which silicon is regulated in plants, during normal and under stress conditions. Also, genetically engineering plants for altered silicon uptake and accumulation should also explore the metabolic pathways for different plant attributes [144].

An intricate network of feedback mechanisms in ecosystems is formed as many plants take up silicon present in soils, store it within their tissues where silicon may affect the physiology, and then reincorporate it back into the soil. A true soil-plant continuum is supported by the bidirectional effects of plant litter on soils, and soil properties on silicon uptake by plants. The effects of silicon in plants and soils on ecosystems and global processes should be taken into consideration in the aforementioned concept of a soil-plant continuum. Silicon in the soil-plant continuum is an important factor of the various functions of an ecosystem and a key driver of some ecosystem services due to the intricate feedback mechanisms in ecosystems, and the effects are far-reaching, where herbivores and the atmosphere are also affected in addition to the plants and the soils involved [182]. Katz et al. emphasize the need for detailed, interdisciplinary research with a focus on (1) developing a standard protocol for determining bioavailable silicon content in soils, (2) understanding the modes with which silicon acts in plants, that is, the pathways that increase plant resistance to stresses with silicon accumulation, and (3) obtaining profound insights on how silicon influences the functioning and structure of ecosystems.

25.2.8 Silicon nanoparticles

The possibilities of formulating SiNPs for use in enhancing plant stress tolerance can be explored using applications related to nanotechnology [183,184]. However, research has yet to be conducted to explore how SiNPs can potentially alleviate abiotic stresses and the associated mechanisms [6]. Not many nanoscale products have been commercialized for agriculture despite the potential advantages associated with the use of NPs. Underutilized pest-crop host systems, insufficient field trials, etc. may explain the scarcity of commercial NP-based agricultural products. Nanotechnological applications have progressed rapidly for other industries. For advancing applications of nanotechnology to agriculture, fundamental questions regarding the use of nanotechnology in agriculture must be addressed to provide the rationale for facilitating research and development of commercial NP applications for agriculture such as SiNPs [185]. Research into biogenic silica and its NPs on proteomics may further contribute to understanding the mechanisms with which silicon affects stressed and unstressed crops. Another interesting avenue for research that has yet to be explored is silicon recycling and how it affects proteomics and genetic engineering [152]. There are studies in the literature that report how negatively SiNPs affect plants [186,187]. Slomberg and Schoenfisch [186] found that adding SiNPs changed the pH of the growth media and was toxic to plants. The authors observed that SiNPs did not cause any changes in *Arabidopsis* besides changing the pH. Le et al. [187] lacked information on the SiNP properties and therefore did not investigate the pH maintenance or the zeta potential with respect to SiNP applications. This raises questions about the phytotoxicity of SiNPs, which should be addressed by future studies. A heavy focus should be put future research on evaluating the positive and phytotoxic effects of silica NPs for different crops. Meticulously developing silica-NP-based fertilizers (nanoherbicides, nanoinsectisides, nanopesticides, etc.) and increasing their use in agriculture are another important target of our forthcoming research. A novel area of research opportunities has been opened for agronomists to sustainably enhance agricultural production and help boost global food security with the advent of silica-based nanofertilizers [60].

The underlying molecular mechanisms of how nanosilica regulate genes to affect plant growth, development, and stress responses should be researched. At the same time, crop improvement strategies should be adopted in lieu of increasing nanosilica uptake by plants [60]. Simultaneously, instead of increasing plant uptake of nanosilica, other strategies should be employed to improve crop yield and quality. In addition, the metabolically active roles that SiNPs play in plants under biotic and abiotic stresses, and particularly how SiNPs mediate plant nutrition, should be the subject of future studies. Applied research to determine the optimal Si/SiNP levels, application duration, and methods for plants under biotic and abiotic stresses is also needed [188].

25.2.9 Interaction between silicon and plant growth-promoting microorganisms

Silicon and plant growth-promoting microorganisms (PGPMs) have nearly identical capacities to mitigate the biotic and abiotic stresses in plants [7,10,24,189–191]. Hence PGPMs and silicon may be used together for expected synergistic benefits in stress alleviation [189–191].

The use of silicon, phosphate-solubilizing bacteria, silica-solubilizing bacteria, and arbuscular mycorrhizal fungi should be studied separately and together for their effectiveness in alleviating various biotic and abiotic stresses for plants. The mechanisms behind how silicon regulates the interactions between plants and microbes have yet to be identified for higher plants. How silicon affects plant biochemistry and microbial gene expression also needs to be further studied [24,191]. Whether a combined use of PGPMs and silicon is effective for improving the tolerance of crops to various stresses should also be investigated. Little is done to compare how silicon and PGPM interact, and further research in this area would lend the understanding of the plant mechanisms governing biotic and abiotic stress responses. Research regarding silicon's role in plant biology should emphasize making full use of silicon (along with PGPM) as an agent to confer enhanced tolerance of abiotic stresses, and therefore aiding environmental remediation in a more extended perspective. Most PGPMs are heterotrophic and exposed to environmental stresses. Also, PGPMs compete for limited resources in the soil. Therefore the benefits of inoculating with PGPMs should be investigated for poor soils. No research has been conducted to date to identify how silicon and PGPMs actively affect the metabolism of plants under biotic and abiotic stresses, especially on how they affect plant nutrition. The mechanisms with which silicon and PGPMs help plants mitigate abiotic stresses are poorly understood at the molecular and genetic levels and require further research. The research interest for PGPMs-and silicon-based plant stress tolerance mechanisms has been foreseen.

25.3 Concluding remarks

A comprehensive review of research in the literature revealed that silicon could alleviate the various biotic and abiotic stresses, stimulate growth, and improve the quality of plants. In general, various aspects of silicon have been addressed to some extent. But compared to silicon, nanosilica has received less attention. In addition, the studies mentioned thus far were for a short-term, specific stress source but in nature, plants are exposed to multiple stresses at any given time. Therefore more research should be conducted to examine how silicon affects plant resistance to a set of various concurrent environmental stresses. Most studies conducted to date are on how silicon affects silicon-accumulation plants. The effects that silicon elicits in nonsilicon-accumulators should also be researched, to see if similar mechanisms in silicon accumulators can be effective in nonaccumulators. Generally, a better understanding of how to alleviate biotic and abiotic stresses in plants with silicon, and a better prediction of how plants respond to silicon applications may be facilitated by the development of more detailed knowledge of how plants and silicon fundamentally interact. We anticipate that in the near future, the interest in the study of silicon/nanosilica, its combined use with microorganisms, and the mechanisms with which it mediates stress tolerance in plants would grow. To summarize, long-term, large-scale, well-designed field trials are required to evaluate the feasibility of Si application to plants under biotic and abiotic stresses, including the economic feasibility of different Si sources.

Acknowledgments

The authors wish to thank the University of Tehran for providing the necessary facilities for this study.

References

[1] Wang L, Chen W, Zhou W, Huang G. Teleconnected influence of tropical Northwest Pacific sea surface temperature on interannual variability of autumn precipitation in Southwest China. Clim Dyn 2015;45(9):2527–39.
[2] Ranjan R. Adapting to catastrophic water scarcity in agriculture through social networking and inter-generational occupational transitioning. J Nat Resour Policy Res 2015;7(1):71–92.
[3] Wang L, Chen W, Zhou W. Assessment of future drought in Southwest China based on CMIP5 multimodel projections. Adv Atmos Sci 2014;31(5):1035–50.
[4] Zhang Z, Chen Y, Wang P, Zhang S, Tao F, Liu X. Spatial and temporal changes of agro-meteorological disasters affecting maize production in China since 1990. Nat Hazards 2014;71(3):2087–100.
[5] Gernot B, Alireza A, Kaul H. Management of crop water under drought: a review. Agron Sustain Dev 2015;35:401–42.
[6] Tripathi DK, Singh S, Singh VP, Prasad SM, Dubey NK, Chauhan DK. Silicon nanoparticles more effectively alleviated UV-B stress than silicon in wheat (Triticum aestivum) seedlings. Plant Physiol Biochem 2017;110:70–81. Available from: https://doi.org/10.1016/j.plaphy.2016.06.026.
[7] Etesami H, Maheshwari DK. Use of plant growth promoting rhizobacteria (PGPRs) with multiple plant growth promoting traits in stress agriculture: action mechanisms and future prospects. Ecotoxicol Environ Saf 2018;156:225–46. Available from: https://doi.org/10.1016/j.ecoenv.2018.03.013.

[8] Kim Y-H, Khan AL, Waqas M, Lee I-J. Silicon regulates antioxidant activities of crop plants under abiotic-induced oxidative stress: a review. Front Plant Sci 2017;8:510.

[9] Rizwan M, Ali S, Ibrahim M, Farid M, Adrees M, Bharwana SA, et al. Mechanisms of silicon-mediated alleviation of drought and salt stress in plants: a review. Environ Sci Pollut Res 2015;22(20):15416–31.

[10] Etesami H, Jeong BR, Rizwan M. The use of silicon in stressed agriculture management: action mechanisms and future prospects. Metalloids in plants: advances and future prospects. Wiley; 2020. p. 381–431.

[11] Wang W, Vinocur B, Altman A. Plant responses to drought, salinity and extreme temperatures: towards genetic engineering for stress tolerance. Planta 2003;218(1):1–14. Available from: https://doi.org/10.1007/s00425-003-1105-5.

[12] Allahmoradi P, Ghobadi M, Taherabadi S, Taherabadi S, editors. Physiological aspects of mungbean (Vigna radiata L.) in response to drought stress. In: International conference on food engineering and biotechnology—ICFEB 2011, Bangkok, Thailand; 2011.

[13] Saud S, Li X, Chen Y, Zhang L, Fahad S, Hussain S, et al. Silicon application increases drought tolerance of Kentucky bluegrass by improving plant water relations and morphophysiological functions. Sci World J 2014;2014.

[14] Hayat R, Ali S, Amara U, Khalid R, Ahmed I. Soil beneficial bacteria and their role in plant growth promotion: a review. Ann Microbiol 2010;60(4):579–98. Available from: https://doi.org/10.1007/s13213-010-0117-1.

[15] Singh S, Parihar P, Singh R, Singh VP, Prasad SM. Heavy metal tolerance in plants: role of transcriptomics, proteomics, metabolomics, and ionomics. Front Plant Sci 2016;6:1143.

[16] Etesami H. Chapter 15—Plant–microbe interactions in plants and stress tolerance. In: Tripathi DK, Pratap Singh V, Chauhan DK, Sharma S, Prasad SM, Dubey NK, et al., editors. Plant life under changing environment. London: Academic Press; 2020. p. 355–96.

[17] Sharma P, Jha AB, Dubey RS, Pessarakli M. Reactive oxygen species, oxidative damage, and antioxidative defense mechanism in plants under stressful conditions. J Bot 2012;2012:217037. Available from: https://doi.org/10.1155/2012/217037.

[18] Das K, Roychoudhury A. Reactive oxygen species (ROS) and response of antioxidants as ROS-scavengers during environmental stress in plants. Front Environ Sci 2014;2:53.

[19] Lobo V, Patil A, Phatak A, Chandra N. Free radicals, antioxidants and functional foods: impact on human health. Pharmacognosy Rev 2010;4(8):118.

[20] Zargar SM, Mahajan R, Bhat JA, Nazir M, Deshmukh R. Role of silicon in plant stress tolerance: opportunities to achieve a sustainable cropping system. 3 Biotech 2019;9(3):73.

[21] Vanderschuren H, Boycheva S, Li K-T, Szydlowski N, Gruissem W, Fitzpatrick TB. Strategies for vitamin B6 biofortification of plants: a dual role as a micronutrient and a stress protectant. Front Plant Sci 2013;4:143.

[22] Bradáčová K, Weber NF, Morad-Talab N, Asim M, Imran M, Weinmann M, et al. Micronutrients (Zn/Mn), seaweed extracts, and plant growth-promoting bacteria as cold-stress protectants in maize. Chem Biol Technol Agric 2016;3(1):1–10.

[23] Epstein E. The anomaly of silicon in plant biology. Proc Natl Acad Sci U S A 1994;91(1):11–17.

[24] Etesami H, Jeong BR. Silicon (Si): review and future prospects on the action mechanisms in alleviating biotic and abiotic stresses in plants. Ecotoxicol Environ Saf 2018;147:881–96.

[25] Etesami H, Jeong BR. Importance of silicon in fruit nutrition: agronomic and physiological implications. Fruit crops. Cambridge, MA: Elsevier; 2020. p. 255–77.

[26] Luyckx M, Hausman J-F, Lutts S, Guerriero G. Silicon and plants: current knowledge and technological perspectives. Front Plant Sci 2017;8:411.

[27] Ma JF. Role of silicon in enhancing the resistance of plants to biotic and abiotic stresses. Soil Sci Plant Nutr 2004;50(1):11–18.

[28] Murali-Baskaran RK, Senthil-Nathan S, Hunter WB. Anti-herbivore activity of soluble silicon for crop protection in agriculture: a review. Environ Sci Pollut Res 2021;28(3):2626–37.

[29] Arnon DI, Stout PR. The essentiality of certain elements in minute quantity for plants with special reference to copper. Plant Physiol 1939;14(2):371.

[30] Liang Y, Nikolic M, Bélanger R, Gong H, Song A. Silicon in agriculture, 10. Dordrecht: Springer; 2015. p. 978–94.

[31] Meena VD, Dotaniya ML, Coumar V, Rajendiran S, Kundu S, Rao AS. A case for silicon fertilization to improve crop yields in tropical soils. Proc Natl Acad Sci India Sect B Biol Sci 2014;84(3):505–18.

[32] Korndörfer GH, Snyder GH, Ulloa M, Powell G, Datnoff LE. Calibration of soil and plant silicon analysis for rice production. J Plant Nutr 2001;24(7):1071–84.

[33] Richmond KE, Sussman M. Got silicon? The non-essential beneficial plant nutrient. Curr Opplant Biol 2003;6(3):268–72.

[34] Etesami H, Jeong BR. Chapter 19—Importance of silicon in fruit nutrition: agronomic and physiological implications. In: Srivastava A.K., Hu C., editors. Fruit crops. Amsterdam, CA: Elsevier; 2020. p. 255–277.

[35] Van Bockhaven J, De Vleesschauwer D, Höfte M. Towards establishing broad-spectrum disease resistance in plants: silicon leads the way. J Exp Bot 2013;64(5):1281–93.

[36] Ma JF, Yamaji N. Functions and transport of silicon in plants. Cell Mol Life Sci 2008;65(19):3049–57.

[37] Liang Y, Sun W, Zhu Y-G, Christie P. Mechanisms of silicon-mediated alleviation of abiotic stresses in higher plants: a review. Environ Pollut 2007;147(2):422–8.

[38] Bélanger R, Benhamou N, Menzies J. Cytological evidence of an active role of silicon in wheat resistance to powdery mildew (Blumeria graminis f. sp. tritici). Phytopathology. 2003;93(4):402–12.

[39] Côté-Beaulieu C, Chain F, Menzies J, Kinrade S, Bélanger R. Absorption of aqueous inorganic and organic silicon compounds by wheat and their effect on growth and powdery mildew control. Environ Exp Bot 2009;65(2–3):155–61.

[40] Hernandez-Apaolaza L. Can silicon partially alleviate micronutrient deficiency in plants? A review. Planta 2014;240(3):447–58.

[41] Zhu Y, Gong H. Beneficial effects of silicon on salt and drought tolerance in plants. Agron Sustain Dev 2014;34(2):455–72. Available from: https://doi.org/10.1007/s13593-013-0194-1.

[42] Haynes RJ. Chapter three—significance and role of Si in crop production. In: Sparks DL, editor. Advances in agronomy, 146. San Diego: Academic Press;; 2017. p. 83–166.

[43] Kim Y-H, Khan A, Waqas M, Shahzad R, Lee I-J. Silicon-mediated mitigation of wounding stress acts by up-regulating the rice antioxidant system. Cereal Res Commun 2016;44(1):111–21.

[44] Torabi F, Majd A, Enteshari S. The effect of silicon on alleviation of salt stress in borage (*Borago officinalis* L.). Soil Sci Plant Nutr 2015;61(5):788–98.

[45] Carpinteri A, Manuello A. Reply to "comments on 'geomechanical and geochemical evidence of piezonuclear fission reactions in the earth's crust' by A. Carpinteri and A. Manuello" by U. Bardi and G. Comoretto. Strain 2013;49(6):548–51.

[46] Ma JF, Tamai K, Yamaji N, Mitani N, Konishi S, Katsuhara M, et al. A silicon transporter in rice. Nature 2006;440(7084):688–91.

[47] Mitani N, Chiba Y, Yamaji N, Ma JF. Identification and characterization of maize and barley Lsi2-like silicon efflux transporters reveals a distinct silicon uptake system from that in rice. Plant Cell 2009;21(7):2133–42.

[48] Cooke J, Leishman MR. Is plant ecology more siliceous than we realise? Trends Plant Sci 2011;16(2):61–8.

[49] Yamaji N, Chiba Y, Mitani-Ueno N, Ma JF. Functional characterization of a silicon transporter gene implicated in silicon distribution in barley. Plant Physiol 2012;160(3):1491–7.

[50] Mitani N, Ma JF. Uptake system of silicon in different plant species. J Exp Bot 2005;56(414):1255–61.

[51] Sekhon BS. Nanotechnology in agri-food production: an overview. Nanotechnol Sci Appl 2014;7:31–53. Available from: https://doi.org/10.2147/NSA.S39406 PubMed PMID: 24966671.

[52] Siddiqui H, Ahmed KBM, Sami F, Hayat S. Silicon nanoparticles and plants: current knowledge and future perspectives. Sustain Agric Rev 2020;41:129–42.

[53] Roduner E. Size matters: why nanomaterials are different. Chem Soc Rev 2006;35(7):583–92. Available from: https://doi.org/10.1039/B502142C.

[54] Dubchak S, Ogar A, Mietelski JW, Turnau K. Influence of silver and titanium nanoparticles on arbuscular mycorrhiza colonization and accumulation of radiocaesium in *Helianthus annuus*. Span J Agric Res 2010;1:103–8.

[55] Monica RC, Cremonini R. Nanoparticles and higher plants. Caryologia. 2009;62(2):161–5.

[56] O'Farrell N, Houlton A, Horrocks BR. Silicon nanoparticles: applications in cell biology and medicine. Int J Nanomed 2006;1(4):451–72. Available from: https://doi.org/10.2147/nano.2006.1.4.451 Epub 2007/08/28. PubMed PMID: 17722279; PubMed Central PMCID: PMCPMC2676646.

[57] Tripathi DK, Singh VP, Prasad SM, Chauhan DK, Dubey NK. Silicon nanoparticles (SiNp) alleviate chromium (VI) phytotoxicity in *Pisum sativum* (L.) seedlings. Plant Physiol Biochem 2015;96:189–98. Available from: https://doi.org/10.1016/j.plaphy.2015.07.026.

[58] Cui J, Liu T, Li F, Yi J, Liu C, Yu H. Silica nanoparticles alleviate cadmium toxicity in rice cells: mechanisms and size effects. Environ Pollut 2017;228:363–9. Available from: https://doi.org/10.1016/j.envpol.2017.05.014 Epub 2017/05/30. PubMed PMID: 28551566.

[59] Abdel-Haliem MEF, Hegazy HS, Hassan NS, Naguib DM. Effect of silica ions and nano silica on rice plants under salinity stress. Ecol Eng 2017;99:282–9. Available from: https://doi.org/10.1016/j.ecoleng.2016.11.060.

[60] Mathur P, Roy S. Nanosilica facilitates silica uptake, growth and stress tolerance in plants. Plant Physiol Biochem 2020;157:114–27.

[61] Bao-shan L, Shao-qi D, Chun-hui L, Li-jun F, Shu-chun Q, Min Y. Effect of TMS (nanostructured silicon dioxide) on growth of *Changbai larch* seedlings. J For Res 2004;15(2):138–40. Available from: https://doi.org/10.1007/BF02856749.

[62] Suriyaprabha R, Karunakaran G, Yuvakkumar R, Rajendran V, Kannan N. Foliar application of silica nanoparticles on the phytochemical responses of maize (*Zea mays* L.) and its toxicological behavior. Synth React Inorg Metal-Org Nano-Met Chem 2014;44(8):1128–31. Available from: https://doi.org/10.1080/15533174.2013.799197.

[63] Siddiqui MH, Al-Whaibi MH. Role of nano-SiO$_2$ in germination of tomato (*Lycopersicum esculentum* seeds Mill.). Saudi J Biol Sci 2014;21 (1):13–17. Available from: https://doi.org/10.1016/j.sjbs.2013.04.005.

[64] Strout G, Russell SD, Pulsifer DP, Erten S, Lakhtakia A, Lee DW. Silica nanoparticles aid in structural leaf coloration in the Malaysian tropical rainforest understorey herb *Mapania caudata*. Ann Bot 2013;112(6):1141–8. Available from: https://doi.org/10.1093/aob/mct172 Epub 2013/08/21. PubMed PMID: 23960046; PubMed Central PMCID: PMCPMC3783236.

[65] Parisi C, Vigani M, Rodríguez-Cerezo E. Agricultural nanotechnologies: what are the current possibilities? Nano Today 2015;10(2):124–7. Available from: https://doi.org/10.1016/j.nantod.2014.09.009.

[66] Rastogi A, Tripathi DK, Yadav S, Chauhan DK, Živčák M, Ghorbanpour M, et al. Application of silicon nanoparticles in agriculture. 3 Biotech 2019;9(3):1–11.

[67] Tripathi DK, Singh VP, Kumar D, Chauhan DK. Impact of exogenous silicon addition on chromium uptake, growth, mineral elements, oxidative stress, antioxidant capacity, and leaf and root structures in rice seedlings exposed to hexavalent chromium. Acta Physiol Plant 2012;34(1):279–89.

[68] Jullok N, Van Hooghten R, Luis P, Volodin A, Van Haesendonck C, Vermant J, et al. Effect of silica nanoparticles in mixed matrix membranes for pervaporation dehydration of acetic acid aqueous solution: plant-inspired dewatering systems. J Clean Prod 2016;112:4879–89. Available from: https://doi.org/10.1016/j.jclepro.2015.09.019.

[69] Wanyika H, Gatebe E, Kioni P, Tang Z, Gao Y. Mesoporous silica nanoparticles carrier for urea: potential applications in agrochemical delivery systems. J Nanosci Nanotechnol 2012;12(3):2221–8.

[70] Janmohammadi M, Amanzadeh T, Sabaghnia N, Ion V. Effect of nano-silicon foliar application on safflower growth under organic and inorganic fertilizer regimes. Botanica 2016;22(1):53–64.

[71] Torney F, Trewyn BG, Lin VSY, Wang K. Mesoporous silica nanoparticles deliver DNA and chemicals into plants. Nat Nanotechnol 2007;2(5):295–300.

[72] Elmer W, Ma C, White J. Nanoparticles for plant disease management. Curr Opin Environ Sci Health 2018;6:66–70.

[73] Datnoff LE, Rodrigues FA, Seebold KW. Silicon and plant disease. Mineral nutrition and plant disease. St. Paul, MI: APS Press; 2007. p. 233–46.

[74] Rouhani M, Samih MA, Kalantari S. Insecticidal effect of silica and silver nanoparticles on the cowpea seed beetle, *Callosobruchus maculatus* F. (Col.: Bruchidae). Entomol Res 2013;4(4):297–305.

[75] Ye M, Song Y, Long J, Wang R, Baerson SR, Pan Z, et al. Priming of jasmonate-mediated antiherbivore defense responses in rice by silicon. Proc Natl Acad Sci U S A 2013;110(38):E3631–9.

[76] Datnoff LE, Deren CW, Snyder GH. Silicon fertilization for disease management of rice in Florida. Crop Prot 1997;16(6):525–31.

[77] Miyake Y, Takahashi E. Effect of silicon on the growth of solution-cultured cucumber plant. Soil Sci Plant Nutr 1983;29(1):71–83.

[78] Fawe A, Abou-Zaid M, Menzies J, Bélanger R. Silicon-mediated accumulation of flavonoid phytoalexins in cucumber. Phytopathology 1998;88(5):396–401.

[79] Rodrigues FÁ, McNally DJ, Datnoff LE, Jones JB, Labbé C, Benhamou N, et al. Silicon enhances the accumulation of diterpenoid phytoalexins in rice: a potential mechanism for blast resistance. Phytopathology 2004;94(2):177–83.

[80] Reynolds OL, Keeping MG, Meyer JH. Silicon-augmented resistance of plants to herbivorous insects: a review. Ann Appl Biol 2009;155(2):171–86.

[81] Barman K, Sharma S, Siddiqui MW. Emerging postharvest treatment of fruits and vegetables. Boca Raton, FL: CRC Press; 2018.

[82] James A, Zikankuba V. Postharvest management of fruits and vegetable: A potential for reducing poverty, hidden hunger and malnutrition in sub-Sahara Africa. Cogent Food Agric 2017;3(1):1312052.

[83] Heath MC, Stumpf MA. Ultrastructural observations of penetration sites of the cowpea rust fungus in untreated and silicon-depleted French bean cells. Physiol Mol Plant Pathol 1986;29(1):27–39.

[84] Chérif M, Asselin A, Bélanger RR. Defense responses induced by soluble silicon in cucumber roots infected by Pythium spp. Phytopathology 1994;84(3):236–42.

[85] Dann EK, Muir S. Peas grown in media with elevated plant-available silicon levels have higher activities of chitinase and β-1, 3-glucanase, are less susceptible to a fungal leaf spot pathogen and accumulate more foliar silicon. Austral Plant Pathol 2002;31(1): 9–13.

[86] Law C, Exley C. New insight into silica deposition in horsetail (*Equisetum arvense*). BMC Plant Biol 2011;11(1):112. Available from: https://doi.org/10.1186/1471-2229-11-112.

[87] Datnoff LE, Rodrigues FA. History of silicon and plant disease. Silicon and plant disease. New York: Springer; 2015. p. 1–5.

[88] Li ZZ, Chen JF, Liu F, Liu AQ, Wang Q, Sun HY, et al. Study of UV-shielding properties of novel porous hollow silica nanoparticle carriers for avermectin. Pest Manag Sci 2007;63(3):241–6.

[89] Chen J, Wang W, Xu Y, Zhang X. Slow-release formulation of a new biological pesticide, pyoluteorin, with mesoporous silica. J Agric Food Chem 2011;59(1):307–11. Available from: https://doi.org/10.1021/jf103640t.

[90] Ulrichs C, Mewis I, Goswami A. Crop diversification aiming nutritional security in West Bengal: biotechnology of stinging capsules in nature's water-blooms. Ann Tech Issue State Agri Technologists Serv Assoc 2005;1–18.

[91] Rai M, Ingle A. Role of nanotechnology in agriculture with special reference to management of insect pests. Appl Microbiol Biotechnol 2012;94(2):287–93.

[92] Malhotra C, Kapoor RT. Silicon: a sustainable tool in abiotic stress tolerance in plants. Plant abiotic stress tolerance. Cham: Springer; 2019. p. 333–56.

[93] Gao X, Zou C, Wang L, Zhang F. Silicon improves water use efficiency in maize plants. J Plant Nutr 2005;27(8):1457–70.

[94] Gao X, Zou C, Wang L, Zhang F. Silicon decreases transpiration rate and conductance from stomata of maize plants. J Plant Nutr 2006;29(9):1637–47.

[95] Gong H, Zhu X, Chen K, Wang S, Zhang C. Silicon alleviates oxidative damage of wheat plants in pots under drought. Plant Sci 2005;169(2):313–21.

[96] Hattori T, Inanaga S, Tanimoto E, Lux A, Luxová M, Sugimoto Y. Silicon-induced changes in viscoelastic properties of sorghum root cell walls. Plant Cell Physiol 2003;44(7):743–9. Available from: https://doi.org/10.1093/pcp/pcg090 Epub 2003/07/26. PubMed PMID: 12881502.

[97] Liu P, Yin L, Deng X, Wang S, Tanaka K, Zhang S. Aquaporin-mediated increase in root hydraulic conductance is involved in silicon-induced improved root water uptake under osmotic stress in *Sorghum bicolor* L. J Exp Bot 2014;65(17):4747–56.

[98] Amin M, Ahmad R, Basra S, Murtaza G. Silicon induced improvement in morpho-physiological traits of maize (*Zea mays* L.) under water deficit. Pak J Agric Sci 2014;51(1).

[99] Maghsoudi K, Emam Y, Ashraf M. Foliar application of silicon at different growth stages alters growth and yield of selected wheat cultivars. J Plant Nutr 2016;39(8):1194–203.

[100] Agarie S, Hanaoka N, Ueno O, Miyazaki A, Kubota F, Agata W, et al. Effects of silicon on tolerance to water deficit and heat stress in rice plants (*Oryza sativa* L.), monitored by electrolyte leakage. Plant Prod Sci 1998;1(2):96–103.

[101] Silva ON, Lobato AKS, Ávila FW, Costa RCL, Neto CFO, Santos Filho BG, et al. Silicon-induced increase in chlorophyll is modulated by the leaf water potential in two water-deficient tomato cultivars. Plant Soil Environ 2012;58(11):481–6.

[102] Neocleous D. Grafting and silicon improve photosynthesis and nitrate absorption in melon (*Cucumis melo* L.) plants. J Agric Sci Technol 2015;17(7):1815–24.

[103] Putra ETS, Purwanto BH. Physiological responses of oil palm seedlings to the drought stress using boron and silicon applications. J Agron 2015;14(2):49–61.

[104] Ahmad ST, Haddad R. Study of silicon effects on antioxidant enzyme activities and osmotic adjustment of wheat under drought stress. Czech J Genet Plant Breed 2011;47(1):17–27.

[105] Gunes A, Pilbeam DJ, Inal A, Coban S. Influence of silicon on sunflower cultivars under drought stress, I: growth, antioxidant mechanisms, and lipid peroxidation. Commun Soil Sci Plant Anal 2008;39(13–14):1885–903.

[106] Shi Y, Zhang Y, Yao H, Wu J, Sun H, Gong H. Silicon improves seed germination and alleviates oxidative stress of bud seedlings in tomato under water deficit stress. Plant Physiol Biochem 2014;78:27–36.

[107] Gunes A, Inal A, Bagci EG, Pilbeam DJ. Silicon-mediated changes of some physiological and enzymatic parameters symptomatic for oxidative stress in spinach and tomato grown in sodic-B toxic soil. Plant Soil 2007;290(1–2):103–14.

[108] Gong H, Chen K. The regulatory role of silicon on water relations, photosynthetic gas exchange, and carboxylation activities of wheat leaves in field drought conditions. Acta Physiol Plant 2012;34(4):1589–94.

[109] Kaya C, Tuna L, Higgs D. Effect of silicon on plant growth and mineral nutrition of maize grown under water-stress conditions. J Plant Nutr 2006;29(8):1469–80. Available from: https://doi.org/10.1080/01904160600837238.

[110] Chen W, Yao X, Cai K, Chen J. Silicon alleviates drought stress of rice plants by improving plant water status, photosynthesis and mineral nutrient absorption. Biol Trace Elem Res 2011;142(1):67–76.

[111] Ming DF, Pei ZF, Naeem MS, Gong HJ, Zhou WJ. Silicon alleviates PEG-induced water-deficit stress in upland rice seedlings by enhancing osmotic adjustment. J Agron Crop Sci 2012;198(1):14–26.

[112] Gong Hj, Chen km, Chen Gc, Wang Sm, Zhang Cl. Effects of silicon on growth of wheat under drought. J Plant Nutr 2003;26(5):1055–63.

[113] Pereira TS, da Silva Lobato AK, Tan DKY, da Costa DV, Uchoa EB, do Nascimento Ferreira R, et al. Positive interference of silicon on water relations, nitrogen metabolism, and osmotic adjustment in two pepper ('Capsicum annum') cultivars under water deficit. Aust J Crop Sci 2013;7(8):1064.

[114] Kim Y-H, Khan AL, Kim D-H, Lee S-Y, Kim K-M, Waqas M, et al. Silicon mitigates heavy metal stress by regulating P-type heavy metal ATPases, Oryza sativa low silicon genes, and endogenous phytohormones. BMC Plant Biol 2014;14(1):1–13.

[115] Shahzad M, Zörb C, Geilfus CM, Mühling KH. Apoplastic Na$^+$ in Vicia faba leaves rises after short-term salt stress and is remedied by silicon. J Agron Crop Sci 2013;199(3):161–70.

[116] Shi Y, Wang Y, Flowers TJ, Gong H. Silicon decreases chloride transport in rice (Oryza sativa L.) in saline conditions. J Plant Physiol 2013;170(9):847–53. Available from: https://doi.org/10.1016/j.jplph.2013.01.018.

[117] Tuna AL, Kaya C, Higgs D, Murillo-Amador B, Aydemir S, Girgin AR. Silicon improves salinity tolerance in wheat plants. Environ Exp Bot 2008;62(1):10–16.

[118] Abdalla MM. Impact of diatomite nutrition on two Trifolium alexandrinum cultivars differing in salinity tolerance. Int J Plant Physiol Biochem 2011;3(13):233–46.

[119] Mateos-Naranjo E, Andrades-Moreno L, Davy AJ. Silicon alleviates deleterious effects of high salinity on the halophytic grass Spartina densiflora. Plant Physiol Biochem 2013;63:115–21.

[120] Hajiboland R, Cheraghvareh L. Influence of Si supplementation on growth and some physiological and biochemical parameters in salt-stressed tobacco (Nicotiana rustica L.) plants. J Sci Islam Repub Iran 2014;25(3):205–17.

[121] Lee SK, Sohn EY, Hamayun M, Yoon JY, Lee IJ. Effect of silicon on growth and salinity stress of soybean plant grown under hydroponic system. Agrofor Syst 2010;80(3):333–40.

[122] Moussa HR. Influence of exogenous application of silicon on physiological response of salt-stressed maize (Zea mays L.). Int J Agric Biol 2006;8(3):293–7.

[123] Abbas T, Balal RM, Shahid MA, Pervez MA, Ayyub CM, Aqueel MA, et al. Silicon-induced alleviation of NaCl toxicity in okra (Abelmoschus esculentus) is associated with enhanced photosynthesis, osmoprotectants and antioxidant metabolism. Acta Physiol Plant 2015;37(2):6.

[124] Xie Z, Song R, Shao H, Song F, Xu H, Lu Y. Silicon improves maize photosynthesis in saline-alkaline soils. Sci World J 2015;2015.

[125] Wang X, Wei Z, Liu D, Zhao G. Effects of NaCl and silicon on activities of antioxidative enzymes in roots, shoots and leaves of alfalfa. Afr J Biotechnol 2011;10(4):545–9.

[126] Ali A, Basra SMA, Hussain S, Iqbal J, Bukhsh MAAHA, Sarwar M. Salt stress alleviation in field crops through nutritional supplementation of silicon. Pak J Nutr 2012;11(8):637.

[127] Wang Y, Stass A, Horst WJ. Apoplastic binding of aluminum is involved in silicon-induced amelioration of aluminum toxicity in maize. Plant Physiol 2004;136(3):3762–70. Available from: https://doi.org/10.1104/pp.104.045005 Epub 10/22. PubMed PMID: 15502015.

[128] Gu H-H, Qiu H, Tian T, Zhan S-S, Chaney RL, Wang S-Z, et al. Mitigation effects of silicon rich amendments on heavy metal accumulation in rice (Oryza sativa L.) planted on multi-metal contaminated acidic soil. Chemosphere. 2011;83(9):1234–40.

[129] Ali S, Farooq MA, Yasmeen T, Hussain S, Arif MS, Abbas F, et al. The influence of silicon on barley growth, photosynthesis and ultra-structure under chromium stress. Ecotoxicol Environ Saf 2013;89:66–72.

[130] Rizwan M, Meunier J-D, Miche H, Keller C. Effect of silicon on reducing cadmium toxicity in durum wheat (Triticum turgidum L. cv. Claudio W.) grown in a soil with aged contamination. J Hazard Mater 2012;209:326–34.

[131] Tripathi DK, Singh VP, Prasad SM, Chauhan DK, Kishore Dubey N, Rai AK. Silicon-mediated alleviation of Cr(VI) toxicity in wheat seedlings as evidenced by chlorophyll florescence, laser induced breakdown spectroscopy and anatomical changes. Ecotoxicol Environ Saf 2015;113:133–44. Available from: https://doi.org/10.1016/j.ecoenv.2014.09.029.

[132] Keller C, Rizwan M, Davidian JC, Pokrovsky OS, Bovet N, Chaurand P, et al. Effect of silicon on wheat seedlings (Triticum turgidum L.) grown in hydroponics and exposed to 0 to 30 μM Cu. Planta 2015;241(4):847–60.

[133] Okuda A. Effect of silicon supply on the injuries due to excessive amounts of Fe, Mn, Cu, As, Al, Co of barley and rice plant. Jpn J Soil Sci Plant Nutr 1962;33:1–8.

[134] Soundararajan P, Sivanesan I, Jana S, Jeong BR. Influence of silicon supplementation on the growth and tolerance to high temperature in Salvia splendens. Hortic Environ Biotechnol 2014;55(4):271–9. Available from: https://doi.org/10.1007/s13580-014-0023-8.

[135] Liang Y, Zhu J, Li Z, Chu G, Ding Y, Zhang J, et al. Role of silicon in enhancing resistance to freezing stress in two contrasting winter wheat cultivars. Environ Exp Bot 2008;64(3):286–94. Available from: https://doi.org/10.1016/j.envexpbot.2008.06.005.

[136] Owino-Gerroh C, Gascho GJ. Effect of silicon on low pH soil phosphorus sorption and on uptake and growth of maize. Commun Soil Sci Plant Anal 2005;35(15–16):2369–78. Available from: https://doi.org/10.1081/LCSS-200030686.

[137] Cai K, Gao D, Luo S, Zeng R, Yang J, Zhu X. Physiological and cytological mechanisms of silicon-induced resistance in rice against blast disease. Physiol Plant 2008;134(2):324–33.

[138] Shetty R, Fretté X, Jensen B, Shetty NP, Jensen JD, Jørgensen HJ, et al. Silicon-induced changes in antifungal phenolic acids, flavonoids, and key phenylpropanoid pathway genes during the interaction between miniature roses and the biotrophic pathogen Podosphaera pannosa. Plant Physiol 2011;157(4):2194–205. Available from: https://doi.org/10.1104/pp.111.185215 Epub 2011/10/25. PubMed PMID: 22021421; PubMed Central PMCID: PMCPMC3327176.

[139] Liu M, Cai K, Chen Y, Luo S, Zhang Z, Lin W. Proteomic analysis of silicon-mediated resistance to Magnaporthe oryzae in rice (Oryza sativa L.). Eur J Plant Pathol 2014;139(3):579–92. Available from: https://doi.org/10.1007/s10658-014-0414-9.

[140] Fauteux F, Chain F, Belzile F, Menzies JG, Bélanger RR. The protective role of silicon in the Arabidopsis–powdery mildew pathosystem. Proc Natl Acad Sci U S A 2006;103(46):17554–9.

[141] Han Y, Lei W, Wen L, Hou M. Silicon-mediated resistance in a susceptible rice variety to the rice leaf folder, Cnaphalocrocis medinalis Guenée (Lepidoptera: Pyralidae). PLoS One 2015;10(4):e0120557.

[142] Gomes FB, Moraes JCd, Santos CDd, Goussain MM. Resistance induction in wheat plants by silicon and aphids. Sci Agricola 2005;62 (6):547–51.

[143] Wydra K, Beri H. Structural changes of homogalacturonan, rhamnogalacturonan I and arabinogalactan protein in xylem cell walls of tomato genotypes in reaction to *Ralstonia solanacearum*. Physiol Mol Plant Pathol 2006;68(1–3):41–50.

[144] Souri Z, Khanna K, Karimi N, Ahmad P. Silicon and plants: current knowledge and future prospects. J Plant Growth Regul 2021;40 (3):906–25.

[145] Debona D, Rodrigues FA, Datnoff LE. Silicon's role in abiotic and biotic plant stresses. Annu Rev Phytopathol 2017;55:85–107.

[146] Reynolds OL, Padula MP, Zeng R, Gurr GM. Silicon: potential to promote direct and indirect effects on plant defense against arthropod pests in agriculture. Front Plant Sci 2016;7:744.

[147] Khan A, Khan AL, Muneer S, Kim Y-H, Al-Rawahi A, Al-Harrasi A. Silicon and salinity: crosstalk in crop-mediated stress tolerance mechanisms. Front Plant Sci 2019;10:1429.

[148] Thorne SJ, Hartley SE, Maathuis FJM. Is silicon a panacea for alleviating drought and salt stress in crops? Front Plant Sci 2020;11:1221.

[149] Sapre SS, Vakharia DN. Role of silicon under water deficit stress in wheat: (biochemical perspective): a review. Agric Rev 2016;37 (2):109–16.

[150] Tayyab M, Islam W, Zhang H. Promising role of silicon to enhance drought resistance in wheat. Commun Soil Sci Plant Anal 2018;49 (22):2932–41.

[151] Zhu Y-X, Gong H-J, Yin J-L. Role of silicon in mediating salt tolerance in plants: a review. Plants 2019;8(6):147.

[152] Tripathi DK, Singh VP, Gangwar S, Prasad SM, Maurya JN, Chauhan DK. Role of silicon in enrichment of plant nutrients and protection from biotic and abiotic stresses. Improvement of crops in the era of climatic changes. New York: Springer; 2014. p. 39–56.

[153] Liu B, Soundararajan P, Manivannan A. Mechanisms of silicon-mediated amelioration of salt stress in plants. Plants. 2019;8(9):307.

[154] Li Z, Song Z, Yan Z, Hao Q, Song A, Liu L, et al. Silicon enhancement of estimated plant biomass carbon accumulation under abiotic and biotic stresses. A *meta*-analysis. Agron Sustain Dev 2018;38(3):1–19.

[155] Hasanuzzaman M, Nahar K, Fujita M. Silicon and selenium: two vital trace elements that confer abiotic stress tolerance to plants. Emerging technologies and management of crop stress tolerance. Elsevier; 2014. p. 377–422.

[156] Kim Y-H, Khan AL, Lee I-J. Silicon: a duo synergy for regulating crop growth and hormonal signaling under abiotic stress conditions. Crit Rev Biotechnol 2016;36(6):1099–109.

[157] Nazaralian S, Majd A, Irian S, Najafi F, Ghahremaninejad F, Landberg T, et al. Comparison of silicon nanoparticles and silicate treatments in fenugreek. Plant Physiol Biochem 2017;115:25–33.

[158] Laane H-M. The effects of foliar sprays with different silicon compounds. Plants. 2018;7(2):45.

[159] Coskun D, Deshmukh R, Sonah H, Menzies JG, Reynolds O, Ma JF, et al. The controversies of silicon's role in plant biology. New Phytol 2019;221(1):67–85.

[160] Frew A, Weston LA, Reynolds OL, Gurr GM. The role of silicon in plant biology: a paradigm shift in research approach. Ann Bot 2018;121(7):1265–73.

[161] Laane H-M. The effects of the application of foliar sprays with stabilized silicic acid: an overview of the results from 2003–2014. Silicon. 2017;9(6):803–7.

[162] Hall CR, Waterman JM, Vandegeer RK, Hartley SE, Johnson SN. The role of silicon in antiherbivore phytohormonal signalling. Front Plant Sci 2019;10:1132.

[163] Leroy N, de Tombeur F, Walgraffe Y, Cornélis J-T, Verheggen FJ. Silicon and plant natural defenses against insect pests: Impact on plant volatile organic compounds and cascade effects on multitrophic interactions. Plants. 2019;8(11):444.

[164] Ma JF, Yamaji N. Silicon uptake and accumulation in higher plants. Trends Plant Sci 2006;11(8):392–7.

[165] Epstein E. Silicon. Annu Rev Plant Biol 1999;50(1):641–64.

[166] Takahashi E, Ma JF, Miyake Y. The possibility of silicon as an essential element for higher plants. Comments Agric Food Chem 1990;2 (2):99–102.

[167] Cooke J, Leishman MR. Silicon concentration and leaf longevity: is silicon a player in the leaf dry mass spectrum? Funct Ecol 2011;25 (6):1181–8.

[168] Li H, Zhu Y, Hu Y, Han W, Gong H. Beneficial effects of silicon in alleviating salinity stress of tomato seedlings grown under sand culture. Acta Physiol Plant 2015;37(4):71.

[169] Katz O. Beyond grasses: the potential benefits of studying silicon accumulation in non-grass species. Front Plant Sci 2014;5:376.

[170] Detmann KC, Araújo WL, Martins SCV, Sanglard LMVP, Reis JV, Detmann E, et al. Silicon nutrition increases grain yield, which, in turn, exerts a feed-forward stimulation of photosynthetic rates via enhanced mesophyll conductance and alters primary metabolism in rice. New Phytol 2012;196(3):752–62.

[171] Markovich O, Steiner E, Kouřil Š, Tarkowski P, Aharoni A, Elbaum R. Silicon promotes cytokinin biosynthesis and delays senescence in Arabidopsis and Sorghum. Plant Cell Environ 2017;40(7):1189–96.

[172] Van Bockhaven J, Steppe K, Bauweraerts I, Kikuchi S, Asano T, Höfte M, et al. Primary metabolism plays a central role in moulding silicon-inducible brown spot resistance in rice. Mol Plant Pathol 2015;16(8):811–24.

[173] Ghareeb H, Bozsó Z, Ott PG, Repenning C, Stahl F, Wydra K. Transcriptome of silicon-induced resistance against *Ralstonia solanacearum* in the silicon non-accumulator tomato implicates priming effect. Physiol Mol Plant Pathol 2011;75(3):83–9.

[174] Simpson KJ, Wade RN, Rees M, Osborne CP, Hartley SE. Still armed after domestication? Impacts of domestication and agronomic selection on silicon defences in cereals. Funct Ecol 2017;31(11):2108–17.

[175] Vivancos J, Labbé C, Menzies JG, Bélanger RR. Silicon-mediated resistance of Arabidopsis against powdery mildew involves mechanisms other than the salicylic acid (SA)-dependent defence pathway. Mol Plant Pathol 2015;16(6):572–82.

[176] Puppe D, Sommer M. Experiments, uptake mechanisms, and functioning of silicon foliar fertilization—a review focusing on maize, rice, and wheat. Adv Agron 2018;152:1–49.

[177] Yan G-c, Nikolic M, Ye M-j, Xiao Z-x, Liang Y-c. Silicon acquisition and accumulation in plant and its significance for agriculture. J Integr Agric 2018;17(10):2138–50.

[178] Sivanesan I, Park SW. The role of silicon in plant tissue culture. Front Plant Sci 2014;5:571.

[179] Hall AD, Morison CGT. On the function of silica in the nutrition of cereals.—Part I. Proc R Soc Lond Ser B Biol Sci 1906;77(520):455–77.

[180] Ma JF, Yamaji N. A cooperative system of silicon transport in plants. Trends Plant Sci 2015;20(7):435–42.

[181] Gaur S, Kumar J, Kumar D, Chauhan DK, Prasad SM, Srivastava PK. Fascinating impact of silicon and silicon transporters in plants: a review. Ecotoxicol Environ Saf 2020;202:110885.

[182] Katz O, Puppe D, Kaczorek D, Prakash NB, Schaller J. Silicon in the soil–plant continuum: intricate feedback mechanisms within ecosystems. Plants. 2021;10(4):652.

[183] Das A, Das B. Nanotechnology a potential tool to mitigate abiotic stress in crop plants. Abiotic and biotic stress in plants. IntechOpen; 2019.

[184] Kumar A, Gupta K, Dixit S, Mishra K, Srivastava S. A review on positive and negative impacts of nanotechnology in agriculture. Int J Environ Sci Technol 2019;16(4):2175–84.

[185] Worrall EA, Hamid A, Mody KT, Mitter N, Pappu HR. Nanotechnology for plant disease management. Agronomy 2018;8(12):285.

[186] Slomberg DL, Schoenfisch MH. Silica nanoparticle phytotoxicity to Arabidopsis thaliana. Environ Sci Technol 2012;46(18):10247–54. Available from: https://doi.org/10.1021/es300949f Epub 2012/08/15. PubMed PMID: 22889047.

[187] Le VN, Rui Y, Gui X, Li X, Liu S, Han Y. Uptake, transport, distribution and bio-effects of SiO_2 nanoparticles in Bt-transgenic cotton. J Nanobiotechnol 2014;12(1):50. Available from: https://doi.org/10.1186/s12951-014-0050-8.

[188] Chanchal Malhotra C, Kapoor R, Ganjewala D. Alleviation of abiotic and biotic stresses in plants by silicon supplementation. Scientia 2016;13(2):59–73.

[189] Etesami H. Can interaction between silicon and plant growth promoting rhizobacteria benefit in alleviating abiotic and biotic stresses in crop plants? Agric Ecosyst Environ 2018;253:98–112.

[190] Etesami H, Adl SM. Can interaction between silicon and non–rhizobial bacteria benefit in improving nodulation and nitrogen fixation in salinity–stressed legumes? A review. Rhizosphere 2020;15:100229.

[191] Etesami H, Jeong BR, Glick BR. Contribution of arbuscular mycorrhizal fungi, phosphate–solubilizing bacteria, and silicon to P uptake by plant. Front Plant Sci 2021;12(1355):699618. Available from: https://doi.org/10.3389/fpls.2021.699618.

Index

Note: Page numbers followed by "*f*" and "*t*" refer to figures and tables, respectively.

Printed in the United States
by Baker & Taylor Publisher Services